Student Solutions Manual

to accompany

Chemistry

Ninth Edition

Raymond Chang
Williams College

Prepared by
Brandon Cruickshank
Northern Arizona University

Boston Burr Ridge, IL Dubuque, IA Madison, WI New York San Francisco St. Louis
Bangkok Bogotá Caracas Kuala Lumpur Lisbon London Madrid Mexico City
Milan Montreal New Delhi Santiago Seoul Singapore Sydney Taipei Toronto

The McGraw·Hill Companies

Student Solutions Manual to accompany
CHEMISTRY, NINTH EDITION
RAYMOND CHANG

Published by McGraw-Hill Higher Education, an imprint of The McGraw-Hill Companies, Inc., 1221 Avenue of the Americas, New York, NY 10020.

Recycled/acid free paper
This book is printed on recycled, acid-free paper containing 10% postconsumer waste.

2 3 4 5 6 7 8 9 0 QPD/QPD 0 9 8 7

ISBN-13: 978-0-07-298061-5
ISBN-10: 0-07-298061-3

www.mhhe.com

FOR THE STUDENT

The *Student Solutions Manual* is intended for use with the ninth edition of Raymond Chang's *Chemistry*. For this manual to be of maximum assistance, you should incorporate it into your overall plan for studying general chemistry. The *Student Solutions Manual* contains material to help you develop problem-solving skills, material to practice your problem-solving skills, and solutions to the even-numbered problems in the text. Each chapter in the manual corresponds to one in the text. Most chapters of the manual contain the following main features:

Problem-Solving Strategies

You will encounter several different types of problems in each chapter. To make your problem-solving task easier, we will break down each chapter into the most common problem types. We will present step-by-step methods for solving each problem type.

Example Problems

Incorporated into each problem type are example problems with detailed step-by-step tutorial solutions. These example problems and tutorial solutions are intended to reinforce the problem-solving strategy presented with each problem type. The solutions often contain further explanation of important concepts.

Practice Exercises

Throughout all chapters are numerous practice exercises to allow you to test your problem-solving skills and your knowledge of the material. Answers to the practice exercises are found at the end of the problem-solving strategy section of each chapter.

Text Solutions

Solutions to all the even-numbered text problems are presented following the problem-solving strategies for each chapter. Usually one text problem of each problem type is solved in the step-by-step tutorial manner presented in the problem-solving strategy section. The even-numbered problems with tutorial solutions refer back to the problem type presented in the strategy section. This will allow you to refer back to the appropriate material if you are having difficulty solving the problem.

Similar Problems

Following each problem type is a list of similar even-numbered problems in the text. Problems in bold-face type are solved in the step-by-step tutorial manner presented in the problem-solving strategy section. Similar problems are listed to allow you to practice more problems that follow a comparable strategy.

I hope that the *Student Solutions Manual* helps you succeed in general chemistry. I have attempted to present detailed problem-solving strategies to help you become a better problem-solver. Improved problem-solving will not only help you succeed in general chemistry, but also in many other courses throughout your college career. Probably one of the best ways to learn general chemistry is to work through, and sometimes struggle through, the problems assigned by your instructor. It always looks easy when your instructor solves a problem in class, but you really do not learn the material until you prove to yourself that you can solve the problem.

When solving problems, always try to have a flexible problem-solving approach. Please do not think that all problems that you encounter in general chemistry will fall into one of the problem types presented in each chapter. However, take the knowledge that you gain from the tutorial solutions and learn to apply that knowledge to different and perhaps more difficult problems. Also, try to work a few problems *every* day. (OK, you can take Friday off!) You cannot afford to fall behind in a general chemistry course.

If you have any comments or suggestions regarding the *Student Solutions Manual*, I would like to hear from you. My e-mail address is Brandon.Cruickshank@nau.edu or you can write me at the following address:

Dr. Brandon Cruickshank
Northern Arizona University
Department of Chemistry
P.O. Box 5698
Flagstaff, AZ 86011-5698

Good luck in all your college endeavors! And try to enjoy chemistry. You might be pleasantly surprised.

ACKNOWLEDGMENTS

My most sincere thanks go to Dr. Raymond Chang who provided solutions for all the new problems in this edition and reviewed the entire manuscript. His insights, helpful suggestions, and corrections greatly improved this manual.

My deepest thanks go to Ms Shirley Oberbroeckling, Managing Developmental Editor at McGraw-Hill, for directing and organizing this project.

I also wish to thank Mr. Kent Peterson, Publisher, and Mr. Thomas Timp, Sponsoring Editor, for developing and managing an excellent "Chemistry Team" at McGraw-Hill.

Finally, I express my love and thanks to my wife, Liz, and our sons, Baden and Camden, for their support and encouragement during work on this project.

Brandon J. Cruickshank

CONTENTS

CHAPTER 1
CHEMISTRY: THE STUDY OF CHANGE

PROBLEM-SOLVING STRATEGIES AND TUTORIAL SOLUTIONS

TYPES OF PROBLEMS

Problem Type 1: Density Calculations.

Problem Type 2: Temperature Conversions.
(a) °C → °F
(b) °F → °C

Problem Type 3: Scientific Notation.
(a) Expressing a number in scientific notation.
(b) Addition and subtraction.
(c) Multiplication and division.

Problem Type 4: Significant Figures.
(a) Addition and subtraction.
(b) Multiplication and division.

Problem Type 5: The Dimensional Analysis Method of Solving Problems.

PROBLEM TYPE 1: DENSITY CALCULATIONS

Density is the mass of an object divided by its volume.

$$\text{density} = \frac{\text{mass}}{\text{volume}}$$

$$d = \frac{m}{V}$$

Densities of solids and liquids are typically expressed in units of grams per cubic centimeter (g/cm^3) or equivalently grams per milliliter (g/mL). Because gases are much less dense than solids and liquids, typical units are grams per liter (g/L).

EXAMPLE 1.1

A lead brick with dimensions of 5.08 cm by 10.2 cm by 20.3 cm has a mass of 11,950 g. What is the density of lead in g/cm^3?

Strategy: You are given the mass of the lead brick in the problem. You need to calculate the volume of the lead brick to solve for the density. The volume of a rectangular object is equal to the length × width × height.

$$\text{density} = \frac{\text{mass}}{\text{volume}}$$

Solution:

$$\text{Volume} = \text{length} \times \text{width} \times \text{height}$$

$$\text{Volume} = 5.08\text{ cm} \times 10.2\text{ cm} \times 20.3\text{ cm} = \mathbf{1052\ cm^3}$$

Calculate the density by substituting the mass and the volume into the equation.

$$d = \frac{m}{V} = \frac{11{,}950 \text{ g}}{1052 \text{ cm}^3} = \mathbf{11.4 \text{ g/cm}^3}$$

PRACTICE EXERCISE

1. Platinum has a density of 21.4 g/cm^3. What is the mass of a small piece of platinum that has a volume of 7.50 cm^3?

Text Problem: 1.22

PROBLEM TYPE 2: TEMPERATURE CONVERSIONS

To convert between the Fahrenheit scale and the Celsius scale, you must account for two differences between the two scales.

(1) The Fahrenheit scale defines the normal freezing point of water to be exactly 32°F, whereas the Celsius scale defines it to be exactly 0°C.

(2) A Fahrenheit degree is 5/9 the size of a Celsius degree.

A. Converting degrees Fahrenheit to degrees Celsius

The equation needed to complete a conversion from degrees Fahrenheit to degrees Celsius is:

$$\mathbf{?\ °C = (°F - 32°F) \times \frac{5°C}{9°F}}$$

32°F is subtracted to compensate for the normal freezing point of water being 32°F, compared to 0° on the Celsius scale. We multiply by (5/9) because a Fahrenheit degree is 5/9 the size of a Celsius degree.

EXAMPLE 1.2
Convert 20°F to degrees Celsius.

$$?\ °C = (°F - 32°F) \times \frac{5°C}{9°F}$$

$$?\ °C = (20°F - 32°F) \times \frac{5°C}{9°F} = \mathbf{-6.7°C}$$

B. Converting degrees Celsius to degrees Fahrenheit

The equation needed to complete a conversion from degrees Celsius to degrees Fahrenheit is:

$$\mathbf{?\ °F = \left(°C \times \frac{9°F}{5°C}\right) + 32°F}$$

°C is multiplied by (9/5) because a Celsius degree is 9/5 the size of a Fahrenheit degree. 32°F is then added to compensate for the normal freezing point of water being 32°F, compared to 0° on the Celsius scale.

EXAMPLE 1.3
Normal human body temperature on the Celsius scale is 37.0°C. Convert this to the Fahrenheit scale.

$$?\ °F = \left(°C \times \frac{9°F}{5°C}\right) + 32°F$$

$$? \, °\mathbf{F} = \left(37.0°\text{C} \times \frac{9°\text{F}}{5°\text{C}}\right) + 32°\text{F} = \mathbf{98.6°F}$$

PRACTICE EXERCISE

2. Convert –40°F to degrees Celsius.

Text Problems: 1.24, 1.26

PROBLEM TYPE 3: SCIENTIFIC NOTATION

Scientific notation is typically used when working with small or large numbers. All numbers can be expressed in the form

$$N \times 10^n$$

where N is a number between 1 and 10 and n is an exponent that can be a positive or negative integer, or zero.

A. Expressing a number in scientific notation

Strategy: Writing scientific notation as $N \times 10^n$, we determine n by counting the number of places that the decimal point must be moved to give N, a number between 1 and 10.

If the decimal point is moved to the left, n is a positive integer, the number you are working with is larger than 10. If the decimal point is moved to the right, n is a negative integer. The number you are working with is smaller than 1.

EXAMPLE 1.4
Express 0.000105 in scientific notation.

Solution: The decimal point must be moved four places to the right to give N, a number between 1 and 10. In this case,

$$N = \mathbf{1.05}$$

Since 0.000105 is a number less than one, n is a negative integer. In this case, $n = -4$ (The decimal point was moved four places to the right to give $N = 1.05$).

Combining the above two steps:

$$0.000105 = \mathbf{1.05 \times 10^{-4}}$$

Tip: The notation 1.05×10^{-4} means the following: Take 1.05 and multiply by 10^{-4} (0.0001).

$$1.05 \times 0.0001 = 0.000105$$

EXAMPLE 1.5
Express 4224 in scientific notation.

Solution: The decimal point must be moved three places to the left to give N, a number between 1 and 10. In this case,

$$N = \mathbf{4.224}$$

Since 4,224 is a number greater than one, n is a positive integer. In this case, $n = 3$ (the decimal point was moved three places to the left to give $N = 4.224$).

Combining the above two steps:

$$4224 = \mathbf{4.224 \times 10^3}$$

Tip: The notation 4.224×10^3 means the following: Take 4.224 and multiply by 10^3 (1000).

$$4.224 \times 1000 = 4{,}224$$

PRACTICE EXERCISE

3. Express the following numbers in scientific notation:

(a) 45,781 **(b)** 0.0000430

Text Problem: 1.30

B. Addition and subtraction using scientific notation

Strategy: Let's express scientific notation as $N \times 10^n$. When adding or subtracting numbers using scientific notation, we must write each quantity with the same exponent, n. We can then add or subtract the N parts of the numbers, keeping the exponent, n, the same.

EXAMPLE 1.6

Express the answer to the following calculation in scientific notation. $(2.43 \times 10^1) + (5.955 \times 10^2) = ?$

Solution: Write each quantity with the same exponent, n. Let's write 2.43×10^1 in such a way that $n = 2$.

Tip: We are *increasing* 10^n by a factor of 10, so we must *decrease* N by a factor of 10. We move the decimal point one place to the left.

$$2.43 \times 10^1 = \mathbf{0.243 \times 10^2}$$

(n was increased by 1. Move the decimal point one place to the left.)

Add or subtract, as required, the N parts of the numbers, keeping the exponent, n, the same. In this example, the process is addition.

$$\begin{array}{r} 0.243 \times 10^2 \\ +\ 5.955 \times 10^2 \\ \hline \mathbf{6.198 \times 10^2} \end{array}$$

C. Multiplication and division using scientific notation

Strategy: Let's express scientific notation as $N \times 10^n$. Multiply or divide the N parts of the numbers in the usual way. To come up with the correct exponent n, when multiplying, *add* the exponents, when dividing, *subtract* the exponents.

EXAMPLE 1.7

Divide 4.2×10^{-7} **by** 5.0×10^{-5}.

Solution: Divide the N parts of the numbers in the usual way.

$$4.2 \div 5.0 = 0.84$$

When dividing the 10^n parts, *subtract* the exponents.

$$0.84 \times 10^{-7-(-5)} = 0.84 \times 10^{-7+5} = \mathbf{0.84 \times 10^{-2}}$$

The usual practice is to express N as a number between 1 and 10. Therefore, it is more appropriate to move the decimal point of the above number one place to the right, decreasing the exponent by 1.

$$0.84 \times 10^{-2} = \mathbf{8.4 \times 10^{-3}}$$

Tip: In the answer, we moved the decimal point to the right, *increasing* N by a factor of 10. Therefore, we must *decrease* 10^n by a factor of 10. The exponent, n, is changed from –2 to –3.

EXAMPLE 1.8

Multiply 2.2×10^{-3} by 1.4×10^6.

Solution: Multiply the N parts of the numbers in the usual way.

$$2.2 \times 1.4 = 3.1$$

When multiplying the 10^n parts, *add* the exponents.

$$3.1 \times 10^{-3+6} = \mathbf{3.1 \times 10^3}$$

PRACTICE EXERCISE

4. Express the answer to the following calculations in scientific notation. Try these without using a calculator.

(a) $2.20 \times 10^3 - 4.54 \times 10^2 =$

(b) $4.78 \times 10^5 \div 6.332 \times 10^{-7} =$

Text Problem: 1.32

PROBLEM TYPE 4: SIGNIFICANT FIGURES

See Section 1.8 of the text for guidelines for using significant figures.

A. Addition and subtraction

Strategy: The number of significant figures to the right of the decimal point in the answer is determined by the lowest number of digits to the right of the decimal point in any of the original numbers.

EXAMPLE 1.9

Carry out the following operations and express the answer to the correct number of significant figures.

102.226 + 2.51 + 736.0 =

Solution:

$$\begin{array}{r} 102.226 \\ 2.51 \\ +\ \underline{736.0} \\ 840.7\mathbf{36} \end{array} \quad \leftarrow \text{fewest digits to the right of the decimal point}$$

The 3 and 6 are nonsignificant digits, since 736.0 only has one digit to the right of the decimal point. The answer should only have one digit to the right of the decimal point.

The correct answer rounded off to the correct number of significant figures is **840.7**

Tip: To round off a number at a certain point, simply drop the digits that follow if the first of them is less than 5. If the first digit following the point of rounding off is equal to or greater than 5, add 1 to the preceding digit.

B. Multiplying and dividing

Strategy: The number of significant figures in the answer is determined by the original number having the smallest number of significant figures.

EXAMPLE 1.10

Carry out the following operations and express the answer to the correct number of significant figures.

$$\mathbf{12 \times 2143.1 \div 3.11 = ?}$$

Solution:

$$12 \times 2143.1 \div 3.11 = 8269.2 = 8.\mathbf{2692} \times 10^3$$

The 6, 9, and 2 (bolded) are nonsignificant digits because the original number 12 only has two significant figures. Therefore, the answer has only two significant figures.

The correct answer rounded off to the correct number of significant figures is $\mathbf{8.3 \times 10^3}$

PRACTICE EXERCISE

5. Carry out the following operations and express the answer to the correct number of significant figures.

(a) $90.25 - 83 + 1.0015 =$

(b) $55.6 \times 3.482 \div 505.34 =$

Text Problem: 1.36

PROBLEM TYPE 5: THE DIMENSIONAL ANALYSIS METHOD OF SOLVING PROBLEMS

In order to convert from one unit to another, you need to be proficient at applying dimensional analysis. See Section 1.9 of the text. Conversion factors can seem daunting, but if you keep track of the units, making sure that the appropriate units cancel, your effort will be rewarded.

Step 1: Map out a strategy to proceed from initial units to final units based on available conversion factors.

Step 2: Use the following method as many times as is necessary to ensure that you obtain the desired unit.

$$\text{Given unit} \times \left(\frac{\text{desired unit}}{\text{given unit}}\right) = \text{desired unit}$$

EXAMPLE 1.11

How long will it take to fly from Denver to New York, a distance of 1631 miles, at a speed of 815 km/hr?

Strategy: One conversion factor is given in the problem, 815 km/hr. This conversion factor can be used to convert from distance (in km) to time (in hr). If you can convert the distance of 1631 miles to km, then you can use the conversion factor (815 km/hr) to convert to time in hours. Another conversion factor that you can look up is

$$1 \text{ mi} = 1.61 \text{ km}$$

You should come up with the following strategy.

$$\text{miles} \rightarrow \text{km} \rightarrow \text{hours}$$

Solution: Carry out the necessary conversions, making sure that units cancel.

$$\textbf{? hours} = 1631 \text{ mi (given)} \times \frac{1.61 \text{ km (desired)}}{1 \text{ mi (given)}} \times \frac{1 \text{ h (desired)}}{815 \text{ km (given)}} = \mathbf{3.22\ h}$$

Tip: In the first conversion factor (km/mi), km is the desired unit. When moving on to the next conversion factor (h/km), km is now given, and the desired unit is h.

EXAMPLE 1.12

The *Voyager II* mission to the outer planets of our solar system transmitted by radio signals many spectacular photographs of Neptune. Radio waves, like light waves, travel at a speed of 3.00×10^8 m/s. If Neptune was 2.75 billion miles from Earth during these transmissions, how many hours were required for radio signals to travel from Neptune to Earth?

Strategy: One conversion factor is given in the problem, 3.00×10^8 m/s. This conversion factor will allow you to convert from distance (in m) to time (in seconds). If you can convert the distance of 2.75 billion miles to meters, then the speed of light (3.00×10^8 m/s) can be used to convert to time in seconds. Other conversion factors that you can look up are:

1 billion = 1×10^9 60 s = 1 min
1 mi = 1.61 km 60 min = 1 h
1 km = 1000 m

You should come up with the following strategy.

miles → km → meters → seconds → min → hours

Solution: Carry out the necessary conversions, making sure that units cancel.

$$?\,\mathbf{h} = (2.75 \times 10^9\ \text{mi}) \times \frac{1.61\ \text{km}}{1\ \text{mi}} \times \frac{1000\ \text{m}}{1\ \text{km}} \times \frac{1\ \text{s}}{3.00 \times 10^8\ \text{m}} \times \frac{1\ \text{min}}{60\ \text{s}} \times \frac{1\ \text{h}}{60\ \text{min}} = \mathbf{4.10\ h}$$

PRACTICE EXERCISES

6. On a certain day, the concentration of carbon monoxide, CO, in the air over Denver reached 1.8×10^{-5} g/L. Convert this concentration to mg/m^3.

7. Copper (Cu) is a trace element that is essential for nutrition. Newborn infants require 80 μg of Cu per kilogram of body mass per day. The Cu content of a popular baby formula is 0.48 μg of Cu per milliliter. How many milliliters should a 7.0 lb baby consume per day to obtain the minimum daily Cu requirement?

Text Problems: 1.38, 1.40, 1.42, 1.44, 1.46, 1.48, 1.50

ANSWERS TO PRACTICE EXERCISES

1. 161 g Pt

2. −40°C

3. **(a)** 4.5781×10^4
(b) 4.30×10^{-5}

4. **(a)** 1.75×10^3
(b) 7.55×10^{11}

5. **(a)** 8
(b) 0.383

6. 18 mg/m^3

7. 530 mL/day

SOLUTIONS TO SELECTED TEXT PROBLEMS

1.4 **(a)** hypothesis **(b)** law **(c)** theory

1.12 **(a)** Physical change. The helium isn't changed in any way by leaking out of the balloon.

(b) Chemical change in the battery.

(c) Physical change. The orange juice concentrate can be regenerated by evaporation of the water.

(d) Chemical change. Photosynthesis changes water, carbon dioxide, etc., into complex organic matter.

(e) Physical change. The salt can be recovered unchanged by evaporation.

1.14 **(a)** K **(b)** Sn **(c)** Cr **(d)** B **(e)** Ba
(f) Pu **(g)** S **(h)** Ar **(i)** Hg

1.16 **(a)** homogeneous mixture **(b)** element **(c)** compound
(d) homogeneous mixture **(e)** heterogeneous mixture **(f)** homogeneous mixture
(g) heterogeneous mixture

1.22 Density Calculation, Problem Type 1.

Strategy: We are given the density and volume of a liquid and asked to calculate the mass of the liquid. Rearrange the density equation, Equation (1.1) of the text, to solve for mass.

$$\text{density} = \frac{\text{mass}}{\text{volume}}$$

Solution:

$$\textbf{mass} = \text{density} \times \text{volume}$$

$$\textbf{mass of ethanol} = \frac{0.798\ \text{g}}{1\ \cancel{\text{mL}}} \times 17.4\ \cancel{\text{mL}} = \mathbf{13.9\ g}$$

1.24 Temperature Conversion, Problem Type 2.

Strategy: Find the appropriate equations for converting between Fahrenheit and Celsius and between Celsius and Fahrenheit given in Section 1.7 of the text. Substitute the temperature values given in the problem into the appropriate equation.

(a) Conversion from Fahrenheit to Celsius.

$$?\ ^\circ\text{C} = (^\circ\text{F} - 32^\circ\text{F}) \times \frac{5^\circ\text{C}}{9^\circ\text{F}}$$

$$\mathbf{?\ ^\circ C} = (105^\circ\text{F} - 32^\circ\text{F}) \times \frac{5^\circ\text{C}}{9^\circ\text{F}} = \mathbf{41^\circ C}$$

(b) Conversion from Celsius to Fahrenheit.

$$?\ ^\circ\text{F} = \left(^\circ\text{C} \times \frac{9^\circ\text{F}}{5^\circ\text{C}}\right) + 32^\circ\text{F}$$

$$\textbf{? °F} = \left(-11.5\text{°}\cancel{\text{C}} \times \frac{9\text{°F}}{5\text{°}\cancel{\text{C}}}\right) + 32\text{°F} = \mathbf{11.3°F}$$

(c) Conversion from Celsius to Fahrenheit.

$$?\ \text{°F} = \left(\text{°C} \times \frac{9\text{°F}}{5\text{°C}}\right) + 32\text{°F}$$

$$\textbf{? °F} = \left(6.3 \times 10^3\ \text{°}\cancel{\text{C}} \times \frac{9\text{°F}}{5\text{°}\cancel{\text{C}}}\right) + 32\text{°F} = \mathbf{1.1 \times 10^4\ °F}$$

(d) Conversion from Fahrenheit to Celsius.

$$?\ \text{°C} = (\text{°F} - 32\text{°F}) \times \frac{5\text{°C}}{9\text{°F}}$$

$$\textbf{? °C} = (451\text{°F} - 32\text{°F}) \times \frac{5\text{°C}}{9\text{°F}} = \mathbf{233°C}$$

1.26 **(a)** $\text{K} = (\text{°C} + 273\text{°C})\dfrac{1\ \text{K}}{1\text{°C}}$

$\textbf{°C} = \text{K} - 273 = 77\ \text{K} - 273 = \mathbf{-196°C}$

(b) $\textbf{°C} = 4.2\ \text{K} - 273 = \mathbf{-269°C}$

(c) $\textbf{°C} = 601\ \text{K} - 273 = \mathbf{328°C}$

1.30 **(a)** 10^{-2} indicates that the decimal point must be moved two places to the left.

$$1.52 \times 10^{-2} = \mathbf{0.0152}$$

(b) 10^{-8} indicates that the decimal point must be moved 8 places to the left.

$$7.78 \times 10^{-8} = \mathbf{0.0000000778}$$

1.32 Scientific Notation, Problem Types 3B and 3C

(a) Addition using scientific notation.

Strategy: Let's express scientific notation as $N \times 10^n$. When adding numbers using scientific notation, we must write each quantity with the same exponent, n. We can then add the N parts of the numbers, keeping the exponent, n, the same.

Solution: Write each quantity with the same exponent, n.

Let's write 0.0095 in such a way that $n = -3$. We have decreased 10^n by 10^3, so we must increase N by 10^3. Move the decimal point 3 places to the right.

$$0.0095 = 9.5 \times 10^{-3}$$

Add the N parts of the numbers, keeping the exponent, n, the same.

$$\begin{array}{r} 9.5 \times 10^{-3} \\ +\ \ 8.5 \times 10^{-3} \\ \hline \mathbf{18.0 \times 10^{-3}} \end{array}$$

The usual practice is to express N as a number between 1 and 10. Since we must *decrease* N by a factor of 10 to express N between 1 and 10 (1.8), we must *increase* 10^n by a factor of 10. The exponent, n, is increased by 1 from −3 to −2.

$$18.0 \times 10^{-3} = \mathbf{1.8 \times 10^{-2}}$$

(b) Division using scientific notation.

Strategy: Let's express scientific notation as $N \times 10^n$. When dividing numbers using scientific notation, divide the N parts of the numbers in the usual way. To come up with the correct exponent, n, we *subtract* the exponents.

Solution: Make sure that all numbers are expressed in scientific notation.

$$653 = 6.53 \times 10^2$$

Divide the N parts of the numbers in the usual way.

$$6.53 \div 5.75 = 1.14$$

Subtract the exponents, n.

$$1.14 \times 10^{+2-(-8)} = 1.14 \times 10^{+2+8} = \mathbf{1.14 \times 10^{10}}$$

(c) Subtraction using scientific notation.

Strategy: Let's express scientific notation as $N \times 10^n$. When subtracting numbers using scientific notation, we must write each quantity with the same exponent, n. We can then subtract the N parts of the numbers, keeping the exponent, n, the same.

Solution: Write each quantity with the same exponent, n.

Let's write 850,000 in such a way that $n = 5$. This means to move the decimal point five places to the left.

$$850{,}000 = 8.5 \times 10^5$$

Subtract the N parts of the numbers, keeping the exponent, n, the same.

$$\begin{array}{r} 8.5 \times 10^5 \\ -\ 9.0 \times 10^5 \\ \hline \mathbf{-0.5 \times 10^5} \end{array}$$

The usual practice is to express N as a number between 1 and 10. Since we must *increase* N by a factor of 10 to express N between 1 and 10 (5), we must *decrease* 10^n by a factor of 10. The exponent, n, is decreased by 1 from 5 to 4.

$$-0.5 \times 10^5 = \mathbf{-5 \times 10^4}$$

(d) Multiplication using scientific notation.

Strategy: Let's express scientific notation as $N \times 10^n$. When multiplying numbers using scientific notation, multiply the N parts of the numbers in the usual way. To come up with the correct exponent, n, we *add* the exponents.

Solution: Multiply the N parts of the numbers in the usual way.

$$3.6 \times 3.6 = 13$$

Add the exponents, *n*.

$$13 \times 10^{-4+(+6)} = \mathbf{13 \times 10^{2}}$$

The usual practice is to express *N* as a number between 1 and 10. Since we must *decrease* *N* by a factor of 10 to express *N* between 1 and 10 (1.3), we must *increase* 10^n by a factor of 10. The exponent, *n*, is increased by 1 from 2 to 3.

$$\mathbf{13 \times 10^{2} = 1.3 \times 10^{3}}$$

1.34 **(a)** one **(b)** three **(c)** three **(d)** four
(e) two or three **(f)** one **(g)** one or two

1.36 Significant Figures, Problem Types 4B and 4C

(a) Division

Strategy: The number of significant figures in the answer is determined by the original number having the smallest number of significant figures.

Solution:

$$\frac{7.310\text{ km}}{5.70\text{ km}} = 1.28\mathbf{3}$$

The 3 (bolded) is a nonsignificant digit because the original number 5.70 only has three significant digits. Therefore, the answer has only three significant digits.

The correct answer rounded off to the correct number of significant figures is:

1.28 (Why are there no units?)

(b) Subtraction

Strategy: The number of significant figures to the right of the decimal point in the answer is determined by the lowest number of digits to the right of the decimal point in any of the original numbers.

Solution: Writing both numbers in decimal notation, we have

$$\begin{array}{r} 0.00326\text{ mg} \\ -\ 0.0000788\text{ mg} \\ \hline 0.00318\mathbf{12}\text{ mg} \end{array}$$

The bolded numbers are nonsignificant digits because the number 0.00326 has five digits to the right of the decimal point. Therefore, we carry five digits to the right of the decimal point in our answer.

The correct answer rounded off to the correct number of significant figures is:

$$\mathbf{0.00318\text{ mg} = 3.18 \times 10^{-3}\text{ mg}}$$

(c) Addition

Strategy: The number of significant figures to the right of the decimal point in the answer is determined by the lowest number of digits to the right of the decimal point in any of the original numbers.

Solution: Writing both numbers with exponents = +7, we have

$$(0.402 \times 10^{7}\text{ dm}) + (7.74 \times 10^{7}\text{ dm}) = \mathbf{8.14 \times 10^{7}\text{ dm}}$$

Since 7.74×10^7 has only two digits to the right of the decimal point, two digits are carried to the right of the decimal point in the final answer.

1.38 Factor-Label Method, Problem Type 5.

(a)

Strategy: The problem may be stated as

$$? \text{ mg} = 242 \text{ lb}$$

A relationship between pounds and grams is given on the end sheet of your text (1 lb = 453.6 g). This relationship will allow conversion from pounds to grams. A metric conversion is then needed to convert grams to milligrams (1 mg = 1×10^{-3} g). Arrange the appropriate conversion factors so that pounds and grams cancel, and the unit milligrams is obtained in your answer.

Solution: The sequence of conversions is

$$\text{lb} \rightarrow \text{grams} \rightarrow \text{mg}$$

Using the following conversion factors,

$$\frac{453.6 \text{ g}}{1 \text{ lb}} \qquad \frac{1 \text{ mg}}{1 \times 10^{-3} \text{ g}}$$

we obtain the answer in one step:

$$\mathbf{? \text{ mg}} = 242 \text{ lb} \times \frac{453.6 \text{ g}}{1 \text{ lb}} \times \frac{1 \text{ mg}}{1 \times 10^{-3} \text{ g}} = \mathbf{1.10 \times 10^8 \text{ mg}}$$

Check: Does your answer seem reasonable? Should 242 lb be equivalent to 110 million mg? How many mg are in 1 lb? There are 453,600 mg in 1 lb.

(b)

Strategy: The problem may be stated as

$$? \text{ m}^3 = 68.3 \text{ cm}^3$$

Recall that 1 cm = 1×10^{-2} m. We need to set up a conversion factor to convert from cm^3 to m^3.

Solution: We need the following conversion factor so that centimeters cancel and we end up with meters.

$$\frac{1 \times 10^{-2} \text{ m}}{1 \text{ cm}}$$

Since this conversion factor deals with length and we want volume, it must therefore be cubed to give

$$\frac{1 \times 10^{-2} \text{ m}}{1 \text{ cm}} \times \frac{1 \times 10^{-2} \text{ m}}{1 \text{ cm}} \times \frac{1 \times 10^{-2} \text{ m}}{1 \text{ cm}} = \left(\frac{1 \times 10^{-2} \text{ m}}{1 \text{ cm}}\right)^3$$

We can write

$$\mathbf{? \text{ m}^3} = 68.3 \text{ cm}^3 \times \left(\frac{1 \times 10^{-2} \text{ m}}{1 \text{ cm}}\right)^3 = \mathbf{6.83 \times 10^{-5} \text{ m}^3}$$

Check: We know that $1\ \text{cm}^3 = 1 \times 10^{-6}\ \text{m}^3$. We started with $6.83 \times 10^1\ \text{cm}^3$. Multiplying this quantity by 1×10^{-6} gives 6.83×10^{-5}.

(c)
Strategy: The problem may be stated as

$$? \text{ L} = 7.2 \text{ m}^3$$

In Chapter 1 of the text, a conversion is given between liters and cm^3 ($1\ \text{L} = 1000\ \text{cm}^3$). If we can convert m^3 to cm^3, we can then convert to liters. Recall that $1\ \text{cm} = 1 \times 10^{-2}\ \text{m}$. We need to set up two conversion factors to convert from m^3 to L. Arrange the appropriate conversion factors so that m^3 and cm^3 cancel, and the unit liters is obtained in your answer.

Solution: The sequence of conversions is

$$\text{m}^3 \rightarrow \text{cm}^3 \rightarrow \text{L}$$

Using the following conversion factors,

$$\left(\frac{1\ \text{cm}}{1 \times 10^{-2}\ \text{m}}\right)^3 \qquad \frac{1\ \text{L}}{1000\ \text{cm}^3}$$

the answer is obtained in one step:

$$\textbf{? L} = 7.2\ \text{m}^3 \times \left(\frac{1\ \text{cm}}{1 \times 10^{-2}\ \text{m}}\right)^3 \times \frac{1\ \text{L}}{1000\ \text{cm}^3} = \mathbf{7.2 \times 10^3\ L}$$

Check: From the above conversion factors you can show that $1\ \text{m}^3 = 1 \times 10^3\ \text{L}$. Therefore, $7\ \text{m}^3$ would equal $7 \times 10^3\ \text{L}$, which is close to the answer.

(d)
Strategy: The problem may be stated as

$$? \text{ lb} = 28.3\ \mu\text{g}$$

A relationship between pounds and grams is given on the end sheet of your text (1 lb = 453.6 g). This relationship will allow conversion from grams to pounds. If we can convert from μg to grams, we can then convert from grams to pounds. Recall that $1\ \mu\text{g} = 1 \times 10^{-6}\ \text{g}$. Arrange the appropriate conversion factors so that μg and grams cancel, and the unit pounds is obtained in your answer.

Solution: The sequence of conversions is

$$\mu\text{g} \rightarrow \text{g} \rightarrow \text{lb}$$

Using the following conversion factors,

$$\frac{1 \times 10^{-6}\ \text{g}}{1\ \mu\text{g}} \qquad \frac{1\ \text{lb}}{453.6\ \text{g}}$$

we can write

$$\textbf{? lb} = 28.3\ \mu\text{g} \times \frac{1 \times 10^{-6}\ \text{g}}{1\ \mu\text{g}} \times \frac{1\ \text{lb}}{453.6\ \text{g}} = \mathbf{6.24 \times 10^{-8}\ lb}$$

Check: Does the answer seem reasonable? What number does the prefix μ represent? Should 28.3 μg be a very small mass?

1.40 Factor-Label Method, Problem Type 5.

Strategy: The problem may be stated as

$$? \text{ s} = 365.24 \text{ days}$$

You should know conversion factors that will allow you to convert between days and hours, between hours and minutes, and between minutes and seconds. Make sure to arrange the conversion factors so that days, hours, and minutes cancel, leaving units of seconds for the answer.

Solution: The sequence of conversions is

$$\text{days} \rightarrow \text{hours} \rightarrow \text{minutes} \rightarrow \text{seconds}$$

Using the following conversion factors,

$$\frac{24 \text{ h}}{1 \text{ day}} \qquad \frac{60 \text{ min}}{1 \text{ h}} \qquad \frac{60 \text{ s}}{1 \text{ min}}$$

we can write

$$? \text{ s} = 365.24 \text{ day} \times \frac{24 \text{ h}}{1 \text{ day}} \times \frac{60 \text{ min}}{1 \text{ h}} \times \frac{60 \text{ s}}{1 \text{ min}} = \mathbf{3.1557 \times 10^7 \text{ s}}$$

Check: Does your answer seem reasonable? Should there be a very large number of seconds in 1 year?

1.42 (a) $$? \text{ in/s} = \frac{1 \text{ mi}}{13 \text{ min}} \times \frac{5280 \text{ ft}}{1 \text{ mi}} \times \frac{12 \text{ in}}{1 \text{ ft}} \times \frac{1 \text{ min}}{60 \text{ s}} = \mathbf{81 \text{ in/s}}$$

(b) $$? \text{ m/min} = \frac{1 \text{ mi}}{13 \text{ min}} \times \frac{1609 \text{ m}}{1 \text{ mi}} = \mathbf{1.2 \times 10^2 \text{ m/min}}$$

(c) $$? \text{ km/h} = \frac{1 \text{ mi}}{13 \text{ min}} \times \frac{1609 \text{ m}}{1 \text{ mi}} \times \frac{1 \text{ km}}{1000 \text{ m}} \times \frac{60 \text{ min}}{1 \text{ h}} = \mathbf{7.4 \text{ km/h}}$$

1.44 $$? \text{ km/h} = \frac{55 \text{ mi}}{1 \text{ h}} \times \frac{1.609 \text{ km}}{1 \text{ mi}} = \mathbf{88 \text{ km/h}}$$

1.46 $$0.62 \text{ ppm Pb} = \frac{0.62 \text{ g Pb}}{1 \times 10^6 \text{ g blood}}$$

$$6.0 \times 10^3 \text{ g of blood} \times \frac{0.62 \text{ g Pb}}{1 \times 10^6 \text{ g blood}} = \mathbf{3.7 \times 10^{-3} \text{ g Pb}}$$

1.48 (a) $$? \text{ m} = 185 \text{ nm} \times \frac{1 \times 10^{-9} \text{ m}}{1 \text{ nm}} = \mathbf{1.85 \times 10^{-7} \text{ m}}$$

(b) $$? \text{ s} = (4.5 \times 10^9 \text{ yr}) \times \frac{365 \text{ day}}{1 \text{ yr}} \times \frac{24 \text{ h}}{1 \text{ day}} \times \frac{3600 \text{ s}}{1 \text{ h}} = \mathbf{1.4 \times 10^{17} \text{ s}}$$

(c) $? \, \mathbf{m^3} = 71.2 \, \text{cm}^3 \times \left(\frac{0.01 \text{ m}}{1 \text{ cm}}\right)^3 = \mathbf{7.12 \times 10^{-5} \, m^3}$

(d) $? \, \mathbf{L} = 88.6 \, \text{m}^3 \times \left(\frac{1 \text{ cm}}{1 \times 10^{-2} \text{ m}}\right)^3 \times \frac{1 \text{ L}}{1000 \text{ cm}^3} = \mathbf{8.86 \times 10^4 \, L}$

1.50 $\textbf{density} = \frac{0.625 \text{ g}}{1 \text{ L}} \times \frac{1 \text{ L}}{1000 \text{ mL}} \times \frac{1 \text{ mL}}{1 \text{ cm}^3} = \mathbf{6.25 \times 10^{-4} \, g/cm^3}$

1.52 See Section 1.6 of your text for a discussion of these terms.

(a) Chemical property. Iron has changed its composition and identity by chemically combining with oxygen and water.

(b) Chemical property. The water reacts with chemicals in the air (such as sulfur dioxide) to produce acids, thus changing the composition and identity of the water.

(c) Physical property. The color of the hemoglobin can be observed and measured without changing its composition or identity.

(d) Physical property. The evaporation of water does not change its chemical properties. Evaporation is a change in matter from the liquid state to the gaseous state.

(e) Chemical property. The carbon dioxide is chemically converted into other molecules.

1.54 Volume of rectangular bar = length × width × height

$$\textbf{density} = \frac{m}{V} = \frac{52.7064 \text{ g}}{(8.53 \text{ cm})(2.4 \text{ cm})(1.0 \text{ cm})} = \mathbf{2.6 \, g/cm^3}$$

1.56 You are asked to solve for the inner diameter of the tube. If you can calculate the volume that the mercury occupies, you can calculate the radius of the cylinder, $V_{\text{cylinder}} = \pi r^2 h$ (r is the inner radius of the cylinder, and h is the height of the cylinder). The cylinder diameter is $2r$.

$$\text{volume of Hg filling cylinder} = \frac{\text{mass of Hg}}{\text{density of Hg}}$$

$$\text{volume of Hg filling cylinder} = \frac{105.5 \text{ g}}{13.6 \text{ g/cm}^3} = 7.76 \text{ cm}^3$$

Next, solve for the radius of the cylinder.

$$\text{Volume of cylinder} = \pi r^2 h$$

$$r = \sqrt{\frac{\text{volume}}{\pi \times h}}$$

$$r = \sqrt{\frac{7.76 \text{ cm}^3}{\pi \times 12.7 \text{ cm}}} = 0.441 \text{ cm}$$

The cylinder diameter equals $2r$.

$$\textbf{Cylinder diameter} = 2r = 2(0.441 \text{ cm}) = \mathbf{0.882 \text{ cm}}$$

1.58 $$\frac{343\ \text{m}}{1\ \text{s}} \times \frac{1\ \text{mi}}{1609\ \text{m}} \times \frac{3600\ \text{s}}{1\ \text{h}} = \mathbf{767\ mph}$$

1.60 In order to work this problem, you need to understand the physical principles involved in the experiment in Problem 1.59. The volume of the water displaced must equal the volume of the piece of silver. If the silver did not sink, would you have been able to determine the volume of the piece of silver?

The liquid must be *less dense* than the ice in order for the ice to sink. The temperature of the experiment must be maintained at or below *0°C* to prevent the ice from melting.

1.62 $$\text{Volume} = \frac{\text{mass}}{\text{density}}$$

$$\textbf{Volume occupied by Li} = \frac{1.20 \times 10^3\ \text{g}}{0.53\ \text{g/cm}^3} = \mathbf{2.3 \times 10^3\ cm^3}$$

1.64 To work this problem, we need to convert from cubic feet to L. Some tables will have a conversion factor of 28.3 L = 1 ft^3, but we can also calculate it using the dimensional analysis method described in Section 1.9 of the text.

First, converting from cubic feet to liters:

$$(5.0 \times 10^7\ \text{ft}^3) \times \left(\frac{12\ \text{in}}{1\ \text{ft}}\right)^3 \times \left(\frac{2.54\ \text{cm}}{1\ \text{in}}\right)^3 \times \frac{1\ \text{mL}}{1\ \text{cm}^3} \times \frac{1 \times 10^{-3}\ \text{L}}{1\ \text{mL}} = 1.4 \times 10^9\ \text{L}$$

The mass of vanillin (in g) is:

$$\frac{2.0 \times 10^{-11}\ \text{g vanillin}}{1\ \text{L}} \times (1.4 \times 10^9\ \text{L}) = 2.8 \times 10^{-2}\ \text{g vanillin}$$

The cost is:

$$(2.8 \times 10^{-2}\ \text{g vanillin}) \times \frac{\$112}{50\ \text{g vanillin}} = \mathbf{\$0.063} = \mathbf{6.3¢}$$

1.66 There are 78.3 + 117.3 = 195.6 Celsius degrees between 0°S and 100°S. We can write this as a unit factor.

$$\left(\frac{195.6^\circ\text{C}}{100^\circ\text{S}}\right)$$

Set up the equation like a Celsius to Fahrenheit conversion. We need to subtract 117.3°C, because the zero point on the new scale is 117.3°C lower than the zero point on the Celsius scale.

$$?\ ^\circ\text{C} = \left(\frac{195.6^\circ\text{C}}{100^\circ\text{S}}\right)(?\ ^\circ\text{S}) - 117.3^\circ\text{C}$$

Solving for ? °S gives: $$?\ ^\circ\text{S} = (?\ ^\circ\text{C} + 117.3^\circ\text{C})\left(\frac{100^\circ\text{S}}{195.6^\circ\text{C}}\right)$$

For 25°C we have: $$\mathbf{?\ ^\circ S} = (25^\circ\text{C} + 117.3^\circ\text{C})\left(\frac{100^\circ\text{S}}{195.6^\circ\text{C}}\right) = \mathbf{73^\circ S}$$

1.68 **(a)** $$\frac{6000\text{ mL of inhaled air}}{1\text{ min}} \times \frac{0.001\text{ L}}{1\text{ mL}} \times \frac{60\text{ min}}{1\text{ h}} \times \frac{24\text{ h}}{1\text{ day}} = \mathbf{8.6 \times 10^3\ L\ of\ air/day}$$

(b) $$\frac{8.6 \times 10^3\text{ L of air}}{1\text{ day}} \times \frac{2.1 \times 10^{-6}\text{ L CO}}{1\text{ L of air}} = \mathbf{0.018\ L\ CO/day}$$

1.70 First, calculate the volume of 1 kg of seawater from the density and the mass. We chose 1 kg of seawater, because the problem gives the amount of Mg in every kg of seawater. The density of seawater is given in Problem 1.69.

$$\text{volume} = \frac{\text{mass}}{\text{density}}$$

$$\text{volume of 1 kg of seawater} = \frac{1000\text{ g}}{1.03\text{ g/mL}} = 971\text{ mL} = 0.971\text{ L}$$

In other words, there are 1.3 g of Mg in every 0.971 L of seawater.

Next, let's convert tons of Mg to grams of Mg.

$$(8.0 \times 10^4\text{ tons Mg}) \times \frac{2000\text{ lb}}{1\text{ ton}} \times \frac{453.6\text{ g}}{1\text{ lb}} = 7.3 \times 10^{10}\text{ g Mg}$$

Volume of seawater needed to extract 8.0 × 10^4 ton Mg =

$$(7.3 \times 10^{10}\text{ g Mg}) \times \frac{0.971\text{ L seawater}}{1.3\text{ g Mg}} = \mathbf{5.5 \times 10^{10}\ L\ of\ seawater}$$

1.72 Volume = surface area × depth

Recall that 1 L = 1 dm^3. Let's convert the surface area to units of dm^2 and the depth to units of dm.

$$\text{surface area} = (1.8 \times 10^8\text{ km}^2) \times \left(\frac{1000\text{ m}}{1\text{ km}}\right)^2 \times \left(\frac{1\text{ dm}}{0.1\text{ m}}\right)^2 = 1.8 \times 10^{16}\text{ dm}^2$$

$$\text{depth} = (3.9 \times 10^3\text{ m}) \times \frac{1\text{ dm}}{0.1\text{ m}} = 3.9 \times 10^4\text{ dm}$$

$$\textbf{Volume} = \text{surface area} \times \text{depth} = (1.8 \times 10^{16}\text{ dm}^2)(3.9 \times 10^4\text{ dm}) = 7.0 \times 10^{20}\text{ dm}^3 = \mathbf{7.0 \times 10^{20}\ L}$$

1.74 $$\text{Volume of sphere} = \frac{4}{3}\pi r^3$$

$$\text{Volume} = \frac{4}{3}\pi\left(\frac{15\text{ cm}}{2}\right)^3 = 1.8 \times 10^3\text{ cm}^3$$

$$\text{mass} = \text{volume} \times \text{density} = (1.8 \times 10^3\text{ cm}^3) \times \frac{22.57\text{ g Os}}{1\text{ cm}^3} \times \frac{1\text{ kg}}{1000\text{ g}} = \mathbf{41\ kg\ Os}$$

$$41\text{ kg Os} \times \frac{2.205\text{ lb}}{1\text{ kg}} = \mathbf{9.0 \times 10^1\ lb\ Os}$$

1.76 $62\text{ kg} = 6.2 \times 10^4\text{ g}$

O: $(6.2 \times 10^4\text{ g})(0.65) = \mathbf{4.0 \times 10^4\text{ g O}}$

C: $(6.2 \times 10^4\text{ g})(0.18) = \mathbf{1.1 \times 10^4\text{ g C}}$

H: $(6.2 \times 10^4\text{ g})(0.10) = \mathbf{6.2 \times 10^3\text{ g H}}$

N: $(6.2 \times 10^4\text{ g})(0.03) = \mathbf{2 \times 10^3\text{ g N}}$

Ca: $(6.2 \times 10^4\text{ g})(0.016) = \mathbf{9.9 \times 10^2\text{ g Ca}}$

P: $(6.2 \times 10^4\text{ g})(0.012) = \mathbf{7.4 \times 10^2\text{ g P}}$

1.78 $?\ ^\circ\text{C} = (7.3 \times 10^2 - 273)\text{ K} = \mathbf{4.6 \times 10^2\ ^\circ C}$

$$?\ ^\circ\text{F} = \left((4.6 \times 10^2\ \not{^\circ\text{C}}) \times \frac{9^\circ\text{F}}{5\not{^\circ\text{C}}}\right) + 32^\circ\text{F} = \mathbf{8.6 \times 10^2\ ^\circ F}$$

1.80 $$(8.0 \times 10^4\text{ tons Au}) \times \frac{2000\text{ lb Au}}{1\text{ ton Au}} \times \frac{16\text{ oz Au}}{1\text{ lb Au}} \times \frac{\$350}{1\text{ oz Au}} = \mathbf{\$9.0 \times 10^{11}}\text{ or } \textbf{900 billion dollars}$$

1.82 $$?\text{ Fe atoms} = 4.9\text{ g Fe} \times \frac{1.1 \times 10^{22}\text{ Fe atoms}}{1.0\text{ g Fe}} = \mathbf{5.4 \times 10^{22}\text{ Fe atoms}}$$

1.84 10 cm = 0.1 m. We need to find the number of times the 0.1 m wire must be cut in half until the piece left is 1.3×10^{-10} m long. Let n be the number of times we can cut the Cu wire in half. We can write:

$$\left(\frac{1}{2}\right)^n \times 0.1\text{ m} = 1.3 \times 10^{-10}\text{ m}$$

$$\left(\frac{1}{2}\right)^n = 1.3 \times 10^{-9}\text{ m}$$

Taking the log of both sides of the equation:

$$n\log\left(\frac{1}{2}\right) = \log(1.3 \times 10^{-9})$$

$$\boldsymbol{n} = \textbf{30 times}$$

1.86 Volume = area × thickness.

From the density, we can calculate the volume of the Al foil.

$$\text{Volume} = \frac{\text{mass}}{\text{density}} = \frac{3.636\text{ g}}{2.699\text{ g/cm}^3} = 1.347\text{ cm}^3$$

Convert the unit of area from ft^2 to cm^2.

$$1.000\text{ ft}^2 \times \left(\frac{12\text{ in}}{1\text{ ft}}\right)^2 \times \left(\frac{2.54\text{ cm}}{1\text{ in}}\right)^2 = 929.0\text{ cm}^2$$

$$\textbf{thickness} = \frac{\text{volume}}{\text{area}} = \frac{1.347\text{ cm}^3}{929.0\text{ cm}^2} = 1.450 \times 10^{-3}\text{ cm} = \mathbf{1.450 \times 10^{-2}\text{ mm}}$$

1.88 First, let's calculate the mass (in g) of water in the pool. We perform this conversion because we know there is 1 g of chlorine needed per million grams of water.

$$(2.0 \times 10^4 \text{ gallons } H_2O) \times \frac{3.79 \text{ L}}{1 \text{ gallon}} \times \frac{1 \text{ mL}}{0.001 \text{ L}} \times \frac{1 \text{ g}}{1 \text{ mL}} = 7.6 \times 10^7 \text{ g } H_2O$$

Next, let's calculate the mass of chlorine that needs to be added to the pool.

$$(7.6 \times 10^7 \text{ g } H_2O) \times \frac{1 \text{ g chlorine}}{1 \times 10^6 \text{ g } H_2O} = 76 \text{ g chlorine}$$

The chlorine solution is only 6 percent chlorine by mass. We can now calculate the volume of chlorine solution that must be added to the pool.

$$76 \text{ g chlorine} \times \frac{100\% \text{ soln}}{6\% \text{ chlorine}} \times \frac{1 \text{ mL soln}}{1 \text{ g soln}} = \mathbf{1.3 \times 10^3 \text{ mL of chlorine solution}}$$

1.90 We assume that the thickness of the oil layer is equivalent to the length of one oil molecule. We can calculate the thickness of the oil layer from the volume and surface area.

$$40 \text{ m}^2 \times \left(\frac{1 \text{ cm}}{0.01 \text{ m}}\right)^2 = 4.0 \times 10^5 \text{ cm}^2$$

$$0.10 \text{ mL} = 0.10 \text{ cm}^3$$

$$\text{Volume} = \text{surface area} \times \text{thickness}$$

$$\text{thickness} = \frac{\text{volume}}{\text{surface area}} = \frac{0.10 \text{ cm}^3}{4.0 \times 10^5 \text{ cm}^2} = 2.5 \times 10^{-7} \text{ cm}$$

Converting to nm:

$$(2.5 \times 10^{-7} \text{ cm}) \times \frac{0.01 \text{ m}}{1 \text{ cm}} \times \frac{1 \text{ nm}}{1 \times 10^{-9} \text{ m}} = \mathbf{2.5 \text{ nm}}$$

1.92 **(a)**

$$\frac{\$1.30}{15.0 \text{ ft}^3} \times \left(\frac{1 \text{ ft}}{12 \text{ in}}\right)^3 \times \left(\frac{1 \text{ in}}{2.54 \text{ cm}}\right)^3 \times \frac{1 \text{ cm}^3}{1 \text{ mL}} \times \frac{1 \text{ mL}}{0.001 \text{ L}} = \mathbf{\$3.06 \times 10^{-3}/L}$$

(b)

$$2.1 \text{ L water} \times \frac{0.304 \text{ ft}^3 \text{ gas}}{1 \text{ L water}} \times \frac{\$1.30}{15.0 \text{ ft}^3} = \mathbf{\$0.055 = 5.5¢}$$

1.94 This problem is similar in concept to a limiting reagent problem. We need sets of coins with 3 quarters, 1 nickel, and 2 dimes. First, we need to find the total number of each type of coin.

$$\text{Number of quarters} = (33.871 \times 10^3 \text{ g}) \times \frac{1 \text{ quarter}}{5.645 \text{ g}} = 6000 \text{ quarters}$$

$$\text{Number of nickels} = (10.432 \times 10^3 \text{ g}) \times \frac{1 \text{ nickel}}{4.967 \text{ g}} = 2100 \text{ nickels}$$

$$\text{Number of dimes} = (7.990 \times 10^3 \text{ g}) \times \frac{1 \text{ dime}}{2.316 \text{ g}} = 3450 \text{ dimes}$$

Next, we need to find which coin limits the number of sets that can be assembled. For each set of coins, we need 2 dimes for every 1 nickel.

$$2100 \text{ nickels} \times \frac{2 \text{ dimes}}{1 \text{ nickel}} = 4200 \text{ dimes}$$

We do not have enough dimes.

For each set of coins, we need 2 dimes for every 3 quarters.

$$6000 \text{ quarters} \times \frac{2 \text{ dimes}}{3 \text{ quarters}} = 4000 \text{ dimes}$$

Again, we do not have enough dimes, and therefore the number of dimes is our "limiting reagent".

If we need 2 dimes per set, the number of sets that can be assembled is:

$$3450 \text{ dimes} \times \frac{1 \text{ set}}{2 \text{ dimes}} = \mathbf{1725 \text{ sets}}$$

The mass of each set is:

$$\left(3 \text{ quarters} \times \frac{5.645 \text{ g}}{1 \text{ quarter}}\right) + \left(1 \text{ nickel} \times \frac{4.967 \text{ g}}{1 \text{ nickel}}\right) + \left(2 \text{ dimes} \times \frac{2.316 \text{ g}}{1 \text{ dime}}\right) = 26.53 \text{ g/set}$$

Finally, the total mass of 1725 sets of coins is:

$$1725 \text{ sets} \times \frac{26.53 \text{ g}}{1 \text{ set}} = \mathbf{4.576 \times 10^4 \text{ g}}$$

1.96 We want to calculate the mass of the cylinder, which can be calculated from its volume and density. The volume of a cylinder is $\pi r^2 l$. The density of the alloy can be calculated using the mass percentages of each element and the given densities of each element.

The volume of the cylinder is:

$$V = \pi r^2 l$$

$$V = \pi(6.44 \text{ cm})^2(44.37 \text{ cm})$$

$$V = 5.78 \times 10^3 \text{ cm}^3$$

The density of the cylinder is:

$$\text{density} = (0.7942)(8.94 \text{ g/cm}^3) + (0.2058)(7.31 \text{ g/cm}^3) = 8.605 \text{ g/cm}^3$$

Now, we can calculate the mass of the cylinder.

$$\text{mass} = \text{density} \times \text{volume}$$

$$\textbf{mass} = (8.605 \text{ g/cm}^3)(5.78 \times 10^3 \text{ cm}^3) = \mathbf{4.97 \times 10^4 \text{ g}}$$

The assumption made in the calculation is that the alloy must be homogeneous in composition.

1.98 The density of the mixed solution should be based on the percentage of each liquid and its density. Because the solid object is suspended in the mixed solution, it should have the same density as this solution. The density of the mixed solution is:

$$(0.4137)(2.0514\ \text{g/mL}) + (0.5863)(2.6678\ \text{g/mL}) = 2.413\ \text{g/mL}$$

As discussed, the density of the object should have the same density as the mixed solution (**2.413 g/mL**).

Yes, this procedure can be used in general to determine the densities of solids. This procedure is called the flotation method. It is based on the assumptions that the liquids are totally miscible and that the volumes of the liquids are additive.

1.100 When the carbon dioxide gas is released, the mass of the solution will decrease. If we know the starting mass of the solution and the mass of solution after the reaction is complete (given in the problem), we can calculate the mass of carbon dioxide produced. Then, using the density of carbon dioxide, we can calculate the volume of carbon dioxide released.

$$\text{Mass of hydrochloric acid} = 40.00\ \text{mL} \times \frac{1.140\ \text{g}}{1\ \text{mL}} = 45.60\ \text{g}$$

$$\text{Mass of solution before reaction} = 45.60\ \text{g} + 1.328\ \text{g} = 46.93\ \text{g}$$

We can now calculate the mass of carbon dioxide by difference.

$$\text{Mass of CO}_2\ \text{released} = 46.93\ \text{g} - 46.699\ \text{g} = 0.23\ \text{g}$$

Finally, we use the density of carbon dioxide to convert to liters of CO_2 released.

$$\text{Volume of CO}_2\ \text{released} = 0.23\ \cancel{\text{g}} \times \frac{1\ \text{L}}{1.81\ \cancel{\text{g}}} = \mathbf{0.13\ L}$$

CHAPTER 2
ATOMS, MOLECULES, AND IONS

PROBLEM-SOLVING STRATEGIES AND TUTORIAL SOLUTIONS

TYPES OF PROBLEMS

Problem Type 1: Atomic number, Mass number, and Isotopes.

Problem Type 2: Empirical and Molecular Formulas.

Problem Type 3: Naming Compounds.
- **(a)** Ionic compounds.
- **(b)** Molecular compounds.
- **(c)** Acids.
- **(d)** Bases.

Problem Type 4: Formulas of Ionic Compounds.

PROBLEM TYPE 1: ATOMIC NUMBER, MASS NUMBER, AND ISOTOPES

The **atomic number (*Z*)** is the number of protons in the nucleus of each atom of an element.

EXAMPLE 2.1
What is the atomic number of an oxygen atom?

Solution: The atomic number is listed above each element in the periodic table. For oxygen, the atomic number is **8**, meaning that an oxygen atom has eight protons in the nucleus.

The **mass number (*A*)** is the total number of neutrons and protons present in the nucleus of an atom of an element.

$$\textbf{mass number} = \text{number of protons} + \text{number of neutrons}$$
$$\textbf{mass number} = \text{atomic number} + \text{number of neutrons}$$

EXAMPLE 2.2
A particular oxygen atom has nine neutrons in the nucleus. What is the mass number of this atom?

Strategy: Looking at a periodic table, you should find that every oxygen atom has an atomic number of 8. The number of neutrons is given, so we can solve for the mass number of this atom.

Solution:

$$\text{mass number} = \text{atomic number} + \text{number of neutrons}$$

$$\textbf{mass number} = 8 + 9 = \mathbf{17}$$

Isotopes are atoms that have the same atomic number, but different mass numbers. For example, there are three isotopes of oxygen found in nature, oxygen-16, oxygen-17, and oxygen-18. The accepted way to denote the atomic number and mass number of an element X is as follows:

$$^{A}_{Z}\mathrm{X}$$

where, A = mass number
Z = atomic number

EXAMPLE 2.3
The three isotopes of oxygen found in nature are oxygen-16, -17, and -18. Write their isotopic symbols.

Strategy: The atomic number of oxygen is 8, so all isotopes of oxygen contain eight protons. The mass numbers are 16, 17, and 18, respectively.

Solution: $^{16}_{8}\mathrm{O}$ $^{17}_{8}\mathrm{O}$ $^{18}_{8}\mathrm{O}$

The number of **electrons** in an *atom* is equal to the number of protons.

number of electrons (atom) = number of protons = atomic number

The number of **electrons** in an *ion* is equal to the number of protons minus the charge on the ion.

number of electrons (ion) = number of protons – charge on the ion

EXAMPLE 2.4
What is the total number of fundamental particles (protons, neutrons, and electrons) in (a) an atom of $^{56}_{26}\mathrm{Fe}$ and (b) an $^{56}_{26}\mathrm{Fe}^{3+}$ ion?

Strategy: Both $^{56}_{26}\mathrm{Fe}$ and $^{56}_{26}\mathrm{Fe}^{3+}$ have the same atomic number and mass number, but the number of electrons will be different because one species is neutral and the other has a +3 charge.

Solution: For both (a) and (b):

number of protons = atomic number = **26**

and

number of neutrons = mass number – atomic number = 56 – 26 = **30**

However, the number of electrons for the above species differ.

(a) $^{56}_{26}\mathrm{Fe}$ is a neutral atom. Therefore,

number of electrons = number of protons = **26**

(b) $^{56}_{26}\mathrm{Fe}^{3+}$ is an ion with a +3 charge. Therefore,

number of electrons = number of protons – charge = 26 – (+3) = **23**

PRACTICE EXERCISE

1. How many protons, neutrons, and electrons are contained in each of the following atoms or ions?

(a) $^{19}\mathrm{F}$ **(b)** $^{79}\mathrm{Se}^{2-}$ **(c)** $^{40}\mathrm{Ca}$ **(d)** $^{48}\mathrm{Ti}^{4+}$

Text Problems: 2.14, 2.16, 2.18, 2.36

PROBLEM TYPE 2: EMPIRICAL AND MOLECULAR FORMULAS

A *molecular* formula shows the exact number of atoms of each element in the smallest unit of a substance. An *empirical formula* tells us which elements are present and the simplest whole-number ratio of their atoms. Empirical formulas are therefore the simplest chemical formulas; they are always written so that the subscripts in the molecular formulas are converted to the smallest possible whole numbers.

EXAMPLE 2.5

What is the empirical formula of each of the following compounds: (a) H_2O_2, (b) $C_6H_8O_6$, (c) $MgCl_2$, and (d) C_6H_6?

Strategy: An *empirical formula* tells us which elements are present and the *simplest* whole-number ratio of their atoms. Can you divide the subscripts in the formula by some factor to end up with smaller whole-number subscripts?

Solution:

(a) The simplest whole number ratio of the atoms in H_2O_2 is **HO**.

(b) The simplest whole number ratio of the atoms in $C_6H_8O_6$ is $\mathbf{C_3H_4O_3}$.

(c) The molecular formula as written contains the simplest whole number ratio of the atoms present. In this case, the molecular formula and the empirical formula are the same.

(d) The simplest whole number ratio of the atoms in C_6H_6 is **CH**.

PRACTICE EXERCISE

2. What is the empirical formula of each of the following compounds?

(a) C_2H_6O **(b)** $C_6H_{12}O_6$ **(c)** CH_3COOH **(d)** $C_{12}H_{22}O_{11}$

Text Problems: 2.46, 2.48

PROBLEM TYPE 3: NAMING COMPOUNDS

A. Naming ionic compounds

(1) **Metal cation has only one charge**. When naming ionic compounds, our reference for the names of cations and anions is Table 2.3 of the text. You should memorize the metal cations that have only one charge when they form ionic compounds. These include the alkali metals (Group 1A), which always have a +1 charge in ionic compounds, the alkaline earth metals (Group 2A), which always have a +2 charge in ionic compounds, and Al^{3+}, Ag^+, Cd^{2+}, and Zn^{2+}.

Since the metal cation has only one possible charge, we do not need to specify this charge in the compound. Therefore, the name of this type of ionic compound can be written simply by first naming the metal cation as it appears on the periodic table, followed by the nonmetallic anion. Anions from elements are named by changing the suffix to "-ide".

EXAMPLE 2.6

Name the following ionic compounds: (a) AlF_3, (b) Na_3N, (c) $Ba(NO_3)_2$.

Strategy: The metal cation in each of the compounds given has only one charge when forming ionic compounds. We do not specify this charge in naming the compound.

Solution:

(a) aluminum fluor*ide*
(b) sodium nitr*ide*
(c) barium nitrate

> **Tip:** There are a number of ions that contain more than one atom. These ions are called **polyatomic ions**. Nitrate, NO_3^-, is an example. Ask your instructor which polyatomic ions you should know.

(2) **Metals that form more than one type of cation**. Transition metals typically can form more than one type of cation when forming ionic compounds. If a metal can form cations of different charges, we need to use the Stock system. In the Stock system, Roman numerals are used to specify the charge of the cation. The Roman numeral (I) is used for one positive charge, (II) for two positive charges, (III) for three positive charges, and so on.

EXAMPLE 2.7

Name the following compounds: (a) FeO, (b) Fe_2O_3, (c) $HgSO_4$.

Strategy: The metals in the compounds above can form cations of different charges. We use the Stock system to name these compounds.

Solution:

(a) In this compound, the iron cation has a +2 charge, since oxide has a −2 charge (see Table 2.2 of the text). Therefore, the compound is named **iron(II) oxide**.

(b) In this compound, the iron cation has a +3 charge. Therefore, the compound is named **iron(III) oxide**.

(c) This compound contains the polyatomic ion sulfate, which has a −2 charge (SO_4^{2-}). Thus, the charge on mercury (Hg) is +2. The compound is named **mercury(II) sulfate**.

B. Naming molecular compounds

Unlike ionic compounds, molecular compounds contain discrete molecular units. They are usually composed of nonmetallic elements. There are two types of molecular compounds to consider.

(1) **Only one compound of the two elements exists.** If this is the case, you simply name the first element in the formula as it appears on the periodic table, followed by naming the second element with an "-ide" suffix.

EXAMPLE 2.8

Name the following compounds: (a) HF, (b) SiC.

(a) hydrogen fluoride
(b) silicon carbide

(2) **More than one compound composed of the two elements exists.** It is quite common for one pair of elements to form several different compounds. Therefore, we must be able to differentiate between the compounds. Greek prefixes are used to denote the number of atoms of each element present. See Table 2.4 in the text for the Greek prefixes used.

EXAMPLE 2.9

Name the following compounds: (a) CO, (b) CO_2, (c) N_2O_5, (d) SF_6.

(a) carbon *mono*xide
(b) carbon *di*oxide
(c) *di*nitrogen *pent*oxide
(d) sulfur *hexa*fluoride

> **Tip:** The prefix "mono-" may be omitted when naming the first element in a molecular compound (see carbon monoxide and carbon dioxide above). The absence of a prefix for the first element usually means that only one atom of that element is present in the molecule.

C. Naming acids

An acid is a substance that yields hydrogen ions (H^+) when dissolved in water. There are two types of acids to consider.

(1) Acids that do not contain oxygen. This type of acid contains one or more hydrogen atoms as well as an anionic group. To name these acids, you add the prefix "hydro-" to the anion name, change the "-ide" suffix of the anion to "-ic", and then add the word "acid" at the end.

EXAMPLE 2.10

Name the following binary acids: (a) HF(*aq*), (b) HCN(*aq*).

(a) *hydro*fluor*ic* acid

(b) CN^- is a polyatomic ion called cyanide. In the acid, the "-ide" suffix is changed to "-ic". The correct name is ***hydro*****cyan*****ic*** **acid**.

> **Tip:** The (*aq*) above means that the substance is dissolved in water. HF dissolved in water is an acid and is named hydrofluoric acid. However, HF in its pure state is a molecular compound and is named hydrogen fluoride.

(2) Oxoacids. This type of acid contains hydrogen, oxygen, and another element. To name oxoacids, you must look carefully at the anion name. If the suffix of the anion is "-ate", change the suffix to "-ic" and add the word "acid" at the end. If the suffix of the anion is "-ite", change the suffix to "-ous" and add the word "acid" at the end.

EXAMPLE 2.11

Name the following oxoacids: (a) HNO_2(*aq*), (b) $HClO_3$(*aq*).

(a) The NO_2^- polyatomic ion is called nitr*ite*. Simply change the suffix to "-ous" and add the word "acid". The correct name is **nitr*ous* acid**.

(b) The ClO_3^- polyatomic ion is called chlor*ate*. Simply change the suffix to "-ic" and add the work "acid". The correct name is **chlor*ic* acid**.

> **Tip:** If one O atom is added to the "-ic" acid, the acid is called "per...ic" acid. For example, $HClO_4$ is named perchloric acid. Compare this acid to chloric acid above. If one O atom is removed from the "-ous" acid, the acid is called "hypo...ous" acid. For example, HClO is named hypochlorous acid. Compare this to chlorous acid, $HClO_2$.

D. Naming Bases

A base is a substance that produces the hydroxide ion (OH^-) when dissolved in water. At this point, for naming purposes, we will only consider bases that contain the hydroxide ion. To name this type of base, simply name the metal cation first as it appears on the periodic table, then add "hydroxide".

EXAMPLE 2.12

Name the following bases: (a) KOH, (b) $Sr(OH)_2$.

(a) potassium hydroxide

(b) strontium hydroxide

PRACTICE EXERCISE

3. Name the following compounds:

(a) $MgCl_2$ **(b)** $CuCl_2$ **(c)** HNO_3 **(d)** P_2O_5 **(e)** $Ca(OH)_2$

Text Problems: 2.58, 2.60

PROBLEM TYPE 4: FORMULAS OF IONIC COMPOUNDS

The formulas of ionic compounds are usually the same as their empirical formulas because ionic compounds do not consist of discrete molecular units. See Section 2.6 of the text if you need further information.

Ionic compounds are electrically neutral. In order for ionic compounds to be electrically neutral, the sum of the charges on the cation and anion in each formula unit must add up to zero. There are two possibilities to consider.

(1) If the charges on the cation and anion are numerically equal, no subscripts are necessary in the formula.
(2) If the charges on the cation and anion are numerically different, the subscript of the cation is numerically equal to the charge on the anion, and the subscript of the anion is numerically equal to the charge on the cation.

You should memorize the metal cations that have only one charge when they form ionic compounds. These include the alkali metals (Group 1A), which always have a +1 charge in ionic compounds, the alkaline earth metals (Group 2A), which always have a +2 charge in ionic compounds, and Al^{3+}, Ag^{+}, Cd^{2+}, and Zn^{2+}. All other metals can have more than one possible positive charge when forming ionic compounds. This positive charge will be specified using a Roman numeral in the name of the compound.

You should also memorize the charges of polyatomic ions (see Table 2.3 of the text) and the common charges of monatomic anions based on their positions in the periodic table (see Table 2.2 of the text).

EXAMPLE 2.13
Write the formula for the ionic compound, magnesium oxide.

Strategy: The magnesium cation is an alkaline earth metal cation which always has a +2 charge in an ionic compound, and the oxide anion is −2 in an ionic compound (see Table 2.2 of the text).

Solution: Mg^{2+} and the oxide anion, O^{2-}, combine to form the ionic compound magnesium oxide. The sum of the charges is +2 + (−2) = 0, so no subscripts are necessary. The formula is **MgO**.

EXAMPLE 2.14
Write the formula for the ionic compound, iron(II) chloride.

Strategy: An iron cation can either have a +2 or +3 charge in an ionic compound. The Roman numeral, II, specifies that in this compound it is the +2 cation, Fe^{2+}. The chloride anion has a −1 charge in an ionic compound (see Table 2.2 of the text).

Solution: Fe^{2+} and the chloride anion, Cl^{-}, combine to form the ionic compound iron(II) chloride. The charges on the cation and anion are numerically different, so make the subscript of the cation (Fe^{2+}) numerically equal to the charge of the anion (subscript = 1). Also, make the subscript of the anion (Cl^{-}) numerically equal to the charge of the cation (subscript = 2). The formula is $\mathbf{FeCl_2}$.

Tip: Check to make sure that the compound is electrically neutral by multiplying the charge of each ion by its subscript and then adding them together. The sum should equal zero.

$$(+2)(1) + (-1)(2) = 0$$

PRACTICE EXERCISE

4. Write the correct formulas for the following ionic compounds:

(a) Sodium oxide **(b)** Copper(II) nitrate **(c)** Aluminum oxide

Text Problems: 2.60 a, b, f, g, i, j

ANSWERS TO PRACTICE EXERCISES

1. **(a)** 9p, 10n, 9e
(b) 34p, 45n, 36e
(c) 20p, 20n, 20e
(d) 22p, 26n, 18e

2. **(a)** C_2H_6O
(b) CH_2O
(c) CH_2O
(d) $C_{12}H_{22}O_{11}$

3. **(a)** magnesium chloride
(b) copper(II) chloride
(c) nitric acid
(d) diphosphorus pentoxide
(e) calcium hydroxide

4. **(a)** Na_2O
(b) $Cu(NO_3)_2$
(c) Al_2O_3

SOLUTIONS TO SELECTED TEXT PROBLEMS

2.8 Note that you are given information to set up the unit factor relating meters and miles.

$$r_{\textbf{atom}} = 10^4 r_{\text{nucleus}} = 10^4 \times 2.0\ \cancel{\text{cm}} \times \frac{1\ \cancel{\text{m}}}{100\ \cancel{\text{cm}}} \times \frac{1\ \text{mi}}{1609\ \cancel{\text{m}}} = \mathbf{0.12\ mi}$$

2.14 Problem Type 1, Atomic number, Mass number, and Isotopes.

Strategy: The 239 in Pu-239 is the mass number. The **mass number (*A*)** is the total number of neutrons and protons present in the nucleus of an atom of an element. You can look up the atomic number (number of protons) on the periodic table.

Solution:

mass number = number of protons + number of neutrons

number of neutrons = mass number – number of protons = 239 – 94 = **145**

2.16

Isotope	$^{15}_{7}\text{N}$	$^{33}_{16}\text{S}$	$^{63}_{29}\text{Cu}$	$^{84}_{38}\text{Sr}$	$^{130}_{56}\text{Ba}$	$^{186}_{74}\text{W}$	$^{202}_{80}\text{Hg}$
No. Protons	7	16	29	38	56	74	80
No. Neutrons	8	17	34	46	74	112	122
No. Electrons	7	16	29	38	56	74	80

2.18 The accepted way to denote the atomic number and mass number of an element X is as follows:

$$^{A}_{Z}\mathbf{X}$$

where,

A = mass number
Z = atomic number

(a) $^{186}_{74}\text{W}$ **(b)** $^{201}_{80}\text{Hg}$

2.24 **(a)** Metallic character increases as you progress down a group of the periodic table. For example, moving down Group 4A, the nonmetal carbon is at the top and the metal lead is at the bottom of the group.

(b) Metallic character decreases from the left side of the table (where the metals are located) to the right side of the table (where the nonmetals are located).

2.26 F and Cl are Group 7A elements; they should have similar chemical properties. Na and K are both Group 1A elements; they should have similar chemical properties. P and N are both Group 5A elements; they should have similar chemical properties.

2.32 **(a)** This is a diatomic molecule that is a compound.
(b) This is a polyatomic molecule that is a compound.
(c) This is a polyatomic molecule that is the elemental form of the substance. It is not a compound.

2.34 There are more than two correct answers for each part of the problem.

(a) H_2 and F_2 **(b)** HCl and CO **(c)** S_8 and P_4
(d) H_2O and $C_{12}H_{22}O_{11}$ (sucrose)

2.36 The **atomic number (*Z*)** is the number of protons in the nucleus of each atom of an element. You can find this on a periodic table. The number of **electrons** in an *ion* is equal to the number of protons minus the charge on the ion.

number of electrons (ion) = number of protons – charge on the ion

Ion	K^+	Mg^{2+}	Fe^{3+}	Br^-	Mn^{2+}	C^{4-}	Cu^{2+}
No. protons	19	12	26	35	25	6	29
No. electrons	18	10	23	36	23	10	27

2.44
(a) The copper ion has a +1 charge and bromide has a –1 charge. The correct formula is CuBr.
(b) The manganese ion has a +3 charge and oxide has a –2 charge. The correct formula is Mn_2O_3.
(c) We have the Hg_2^{2+} ion and iodide (I^-). The correct formula is Hg_2I_2.
(d) Magnesium ion has a +2 charge and phosphate has a –3 charge. The correct formula is $Mg_3(PO_4)_2$.

2.46 Problem Type 2, Empirical and Molecular Formulas.

Strategy: An *empirical formula* tells us which elements are present and the *simplest* whole-number ratio of their atoms. Can you divide the subscripts in the formula by some factor to end up with smaller whole-number subscripts?

Solution:

(a) Dividing both subscripts by 2, the simplest whole number ratio of the atoms in Al_2Br_6 is $\mathbf{AlBr_3}$.
(b) Dividing all subscripts by 2, the simplest whole number ratio of the atoms in $Na_2S_2O_4$ is $\mathbf{NaSO_2}$.
(c) The molecular formula as written, $\mathbf{N_2O_5}$, contains the simplest whole number ratio of the atoms present. In this case, the molecular formula and the empirical formula are the same.
(d) The molecular formula as written, $\mathbf{K_2Cr_2O_7}$, contains the simplest whole number ratio of the atoms present. In this case, the molecular formula and the empirical formula are the same.

2.48 The molecular formula of ethanol is $\mathbf{C_2H_6O}$.

2.50 Compounds of metals with nonmetals are usually ionic. Nonmetal-nonmetal compounds are usually molecular.

Ionic: NaBr, BaF_2, CsCl.
Molecular: CH_4, CCl_4, ICl, NF_3

2.58 Problem Type 3, Naming Compounds.

Strategy: When naming ionic compounds, our reference for the names of cations and anions is Table 2.3 of the text. Keep in mind that if a metal can form cations of different charges, we need to use the Stock system. In the Stock system, Roman numerals are used to specify the charge of the cation. The metals that have only one charge in ionic compounds are the alkali metals (+1), the alkaline earth metals (+2), Ag^+, Zn^{2+}, Cd^{2+}, and Al^{3+}.

When naming acids, binary acids are named differently than oxoacids. For binary acids, the name is based on the nonmetal. For oxoacids, the name is based on the polyatomic anion. For more detail, see Section 2.7 of the text.

Solution:

(a) This is an ionic compound in which the metal cation (K^+) has only one charge. The correct name is **potassium hypochlorite**. Hypochlorite is a polyatomic ion with one less O atom than the chlorite ion, ClO_2^-

(b) **silver carbonate**

(c) This is an oxoacid that contains the nitrite ion, NO_2^-. The "-ite" suffix is changed to "-ous". The correct name is **nitrous acid**.

(d) **potassium permanganate** **(e)** **cesium chlorate** **(f)** **potassium ammonium sulfate**

(g) This is an ionic compound in which the metal can form more than one cation. Use a Roman numeral to specify the charge of the Fe ion. Since the oxide ion has a –2 charge, the Fe ion has a +2 charge. The correct name is **iron(II) oxide**.

(h) **iron(III) oxide**

(i) This is an ionic compound in which the metal can form more than one cation. Use a Roman numeral to specify the charge of the Ti ion. Since each of the four chloride ions has a –1 charge (total of –4), the Ti ion has a +4 charge. The correct name is **titanium(IV) chloride**.

(j) **sodium hydride** **(k)** **lithium nitride** **(l)** **sodium oxide**

(m) This is an ionic compound in which the metal cation (Na^+) has only one charge. The O_2^{2-} ion is called the peroxide ion. Each oxygen has a –1 charge. You can determine that each oxygen only has a –1 charge, because each of the two Na ions has a +1 charge. Compare this to sodium oxide in part (l). The correct name is **sodium peroxide**.

2.60 Problem Types 3 and 4.

Strategy: When writing formulas of molecular compounds, the prefixes specify the number of each type of atom in the compound.

When writing formulas of ionic compounds, the subscript of the cation is numerically equal to the charge of the anion, and the subscript of the anion is numerically equal to the charge on the cation. If the charges of the cation and anion are numerically equal, then no subscripts are necessary. Charges of common cations and anions are listed in Table 2.3 of the text. Keep in mind that Roman numerals specify the charge of the cation, *not* the number of metal atoms. Remember that a Roman numeral is not needed for some metal cations, because the charge is known. These metals are the alkali metals (+1), the alkaline earth metals (+2), Ag^+, Zn^{2+}, Cd^{2+}, and Al^{3+}.

When writing formulas of oxoacids, you must know the names and formulas of polyatomic anions (see Table 2.3 of the text).

Solution:

(a) The Roman numeral I tells you that the Cu cation has a +1 charge. Cyanide has a –1 charge. Since, the charges are numerically equal, no subscripts are necessary in the formula. The correct formula is **CuCN**.

(b) Strontium is an alkaline earth metal. It only forms a +2 cation. The polyatomic ion chlorite, ClO_2^-, has a –1 charge. Since the charges on the cation and anion are numerically different, the subscript of the cation is numerically equal to the charge on the anion, and the subscript of the anion is numerically equal to the charge on the cation. The correct formula is $\mathbf{Sr(ClO_2)_2}$.

(c) Perbromic tells you that the anion of this oxoacid is perbromate, BrO_4^-. The correct formula is $\mathbf{HBrO_4}$***(aq)***. Remember that (*aq*) means that the substance is dissolved in water.

(d) Hydroiodic tells you that the anion of this binary acid is iodide, I^-. The correct formula is **HI(*aq*)**.

(e) Na is an alkali metal. It only forms a +1 cation. The polyatomic ion ammonium, NH_4^+, has a +1 charge and the polyatomic ion phosphate, PO_4^{3-}, has a −3 charge. To balance the charge, you need 2 Na^+ cations. The correct formula is **$Na_2(NH_4)PO_4$**.

(f) The Roman numeral II tells you that the Pb cation has a +2 charge. The polyatomic ion carbonate, CO_3^{2-}, has a −2 charge. Since, the charges are numerically equal, no subscripts are necessary in the formula. The correct formula is **$PbCO_3$**.

(g) The Roman numeral II tells you that the Sn cation has a +2 charge. Fluoride has a −1 charge. Since the charges on the cation and anion are numerically different, the subscript of the cation is numerically equal to the charge on the anion, and the subscript of the anion is numerically equal to the charge on the cation. The correct formula is **SnF_2**.

(h) This is a molecular compound. The Greek prefixes tell you the number of each type of atom in the molecule. The correct formula is **P_4S_{10}**.

(i) The Roman numeral II tells you that the Hg cation has a +2 charge. Oxide has a −2 charge. Since, the charges are numerically equal, no subscripts are necessary in the formula. The correct formula is **HgO**.

(j) The Roman numeral I tells you that the Hg cation has a +1 charge. However, this cation exists as Hg_2^{2+}. Iodide has a −1 charge. You need two iodide ion to balance the +2 charge of Hg_2^{2+}. The correct formula is **Hg_2I_2**.

(k) This is a molecular compound. The Greek prefixes tell you the number of each type of atom in the molecule. The correct formula is **SeF_6**.

2.62 Changing the electrical charge of an atom usually has a major effect on its chemical properties. The two electrically neutral carbon isotopes should have nearly identical chemical properties.

2.64 Atomic number = 127 − 74 = 53. This anion has 53 protons, so it is an iodide ion. Since there is one more electron than protons, the ion has a −1 charge. The correct symbol is **I^-**.

2.66 NaCl is an ionic compound; it doesn't form molecules.

2.68 The species and their identification are as follows:

(a)	SO_2	molecule and compound	**(g)**	O_3	element and molecule
(b)	S_8	element and molecule	**(h)**	CH_4	molecule and compound
(c)	Cs	element	**(i)**	KBr	compound
(d)	N_2O_5	molecule and compound	**(j)**	S	element
(e)	O	element	**(k)**	P_4	element and molecule
(f)	O_2	element and molecule	**(l)**	LiF	compound

2.70 **(a)** Ammonium is NH_4^+, not NH_3^+. The formula should be **$(NH_4)_2CO_3$**.

(b) Calcium has a +2 charge and hydroxide has a −1 charge. The formula should be **$Ca(OH)_2$**.

(c) Sulfide is S^{2-}, not SO_3^{2-}. The correct formula is **CdS**.

(d) Dichromate is $Cr_2O_7^{2-}$, not $Cr_2O_4^{2-}$. The correct formula is **$ZnCr_2O_7$**.

2.72 **(a)** Ionic compounds are typically formed between metallic and nonmetallic elements.

(b) In general the transition metals, the actinides and lanthanides have variable charges.

2.74 The symbol ^{23}Na provides more information than $_{11}Na$. The mass number plus the chemical symbol identifies a specific isotope of Na (sodium) while combining the atomic number with the chemical symbol tells you nothing new. Can other isotopes of sodium have different atomic numbers?

2.76 Mercury (Hg) and bromine (Br_2)

2.78 H_2, N_2, O_2, F_2, Cl_2, He, Ne, Ar, Kr, Xe, Rn

2.80 They do not have a strong tendency to form compounds. Helium, neon, and argon are chemically inert.

2.82 All isotopes of radium are radioactive. It is a radioactive decay product of uranium-238. Radium itself does *not* occur naturally on Earth.

2.84 Argentina is named after silver (argentum, Ag).

2.86 **(a)** NaH, sodium hydride **(b)** B_2O_3, diboron trioxide **(c)** Na_2S, sodium sulfide
(d) AlF_3, aluminum fluoride **(e)** OF_2, oxygen difluoride **(f)** $SrCl_2$, strontium chloride

2.88 All of these are molecular compounds. We use prefixes to express the number of each atom in the molecule. The names are nitrogen trifluoride (NF_3), phosphorus pentabromide (PBr_5), and sulfur dichloride (SCl_2).

2.90

Cation	Anion	Formula	Name
$\mathbf{Mg^{2+}}$	$\mathbf{HCO_3^-}$	$\mathbf{Mg(HCO_3)_2}$	Magnesium bicarbonate
$\mathbf{Sr^{2+}}$	$\mathbf{Cl^-}$	$SrCl_2$	**Strontium chloride**
Fe^{3+}	NO_2^-	$\mathbf{Fe(NO_2)_3}$	**Iron(III) nitrite**
$\mathbf{Mn^{2+}}$	$\mathbf{ClO_3^-}$	$\mathbf{Mn(ClO_3)_2}$	Manganese(II) chlorate
$\mathbf{Sn^{4+}}$	$\mathbf{Br^-}$	$SnBr_4$	**Tin(IV) bromide**
Co^{2+}	PO_4^{3-}	$\mathbf{Co_3(PO_4)_2}$	**Cobalt(II) phosphate**
Hg_2^{2+}	I^-	$\mathbf{Hg_2I_2}$	**Mercury(I) iodide**
$\mathbf{Cu^+}$	$\mathbf{CO_3^{2-}}$	Cu_2CO_3	**Copper(I) carbonate**
$\mathbf{Li^+}$	$\mathbf{N^{3-}}$	$\mathbf{Li_3N}$	Lithium nitride
Al^{3+}	S^{2-}	$\mathbf{Al_2S_3}$	**Aluminum sulfide**

2.92 The change in energy is equal to the energy released. We call this ΔE. Similarly, Δm is the change in mass. Because $m = \frac{E}{c^2}$, we have

$$\Delta m = \frac{\Delta E}{c^2} = \frac{(1.715 \times 10^3 \text{ kJ}) \times \frac{1000 \text{ J}}{1 \text{ kJ}}}{(3.00 \times 10^8 \text{ m/s})^2} = 1.91 \times 10^{-11} \text{ kg} = \mathbf{1.91 \times 10^{-8} \text{ g}}$$

Note that we need to convert kJ to J so that we end up with units of kg for the mass. $\left(1 \text{ J} = \frac{1 \text{ kg} \cdot \text{m}^2}{\text{s}^2}\right)$

We can add together the masses of hydrogen and oxygen to calculate the mass of water that should be formed.

$$12.096 \text{ g} + 96.000 = 108.096 \text{ g}$$

The predicted change (loss) in mass is only 1.91×10^{-8} g which is too small a quantity to measure. Therefore, for all practical purposes, the law of conservation of mass is assumed to hold for ordinary chemical processes.

2.94 **(a)** Rutherford's experiment is described in detail in Section 2.2 of the text. From the average magnitude of scattering, Rutherford estimated the number of protons (based on electrostatic interactions) in the nucleus.

(b) Assuming that the nucleus is spherical, the volume of the nucleus is:

$$V = \frac{4}{3}\pi r^3 = \frac{4}{3}\pi(3.04 \times 10^{-13} \text{ cm})^3 = 1.18 \times 10^{-37} \text{ cm}^3$$

The density of the nucleus can now be calculated.

$$d = \frac{m}{V} = \frac{3.82 \times 10^{-23} \text{ g}}{1.18 \times 10^{-37} \text{ cm}^3} = \mathbf{3.24 \times 10^{14} \text{ g/cm}^3}$$

To calculate the density of the space occupied by the electrons, we need both the mass of 11 electrons, and the volume occupied by these electrons.

The mass of 11 electrons is:

$$11 \text{ electrons} \times \frac{9.1095 \times 10^{-28} \text{ g}}{1 \text{ electron}} = 1.0020 \times 10^{-26} \text{ g}$$

The volume occupied by the electrons will be the difference between the volume of the atom and the volume of the nucleus. The volume of the nucleus was calculated above. The volume of the atom is calculated as follows:

$$186 \text{ pm} \times \frac{1 \times 10^{-12} \text{ m}}{1 \text{ pm}} \times \frac{1 \text{ cm}}{1 \times 10^{-2} \text{ m}} = 1.86 \times 10^{-8} \text{ cm}$$

$$V_{\text{atom}} = \frac{4}{3}\pi r^3 = \frac{4}{3}\pi(1.86 \times 10^{-8} \text{ cm})^3 = 2.70 \times 10^{-23} \text{ cm}^3$$

$$V_{\text{electrons}} = V_{\text{atom}} - V_{\text{nucleus}} = (2.70 \times 10^{-23} \text{ cm}^3) - (1.18 \times 10^{-37} \text{ cm}^3) = 2.70 \times 10^{-23} \text{ cm}^3$$

As you can see, the volume occupied by the nucleus is insignificant compared to the space occupied by the electrons.

The density of the space occupied by the electrons can now be calculated.

$$d = \frac{m}{V} = \frac{1.0020 \times 10^{-26} \text{ g}}{2.70 \times 10^{-23} \text{ cm}^3} = \mathbf{3.71 \times 10^{-4} \text{ g/cm}^3}$$

The above results do support Rutherford's model. Comparing the space occupied by the electrons to the volume of the nucleus, it is clear that most of the atom is empty space. Rutherford also proposed that the nucleus was a *dense* central core with most of the mass of the atom concentrated in it. Comparing the density of the nucleus with the density of the space occupied by the electrons also supports Rutherford's model.

2.96 **(a)**

Ethane	Acetylene
2.65 g C	4.56 g C
0.665 g H	0.383 g H

Let's compare the ratio of the hydrogen masses in the two compounds. To do this, we need to start with the same mass of carbon. If we were to start with 4.56 g of C in ethane, how much hydrogen would combine with 4.56 g of carbon?

$$0.665 \text{ g H} \times \frac{4.56 \text{ g C}}{2.65 \text{ g C}} = 1.14 \text{ g H}$$

We can calculate the ratio of H in the two compounds.

$$\frac{1.14 \text{ g}}{0.383 \text{ g}} \approx 3$$

This is consistent with the Law of Multiple Proportions which states that if two elements combine to form more than one compound, the masses of one element that combine with a fixed mass of the other element are in ratios of small whole numbers. In this case, the ratio of the masses of hydrogen in the two compounds is 3:1.

(b) For a given amount of carbon, there is 3 times the amount of hydrogen in ethane compared to acetylene. Reasonable formulas would be:

Ethane	Acetylene
CH_3	CH
C_2H_6	C_2H_2

2.98 The mass number is the sum of the number of protons and neutrons in the nucleus.

$$\text{Mass number} = \text{number of protons} + \text{number of neutrons}$$

Let the atomic number (number of protons) equal A. The number of neutrons will be $1.2A$. Plug into the above equation and solve for A.

$$55 = A + 1.2A$$
$$A = 25$$

The element with atomic number 25 is **manganese, Mn.**

2.100 The acids, from left to right, are chloric acid, nitrous acid, hydrocyanic acid, and sulfuric acid.

CHAPTER 3
MASS RELATIONSHIPS IN CHEMICAL REACTIONS

PROBLEM-SOLVING STRATEGIES

TYPES OF PROBLEMS

Problem Type 1: Calculating Average Atomic Mass.

Problem Type 2: Calculations Involving Molar Mass of an Element and Avogadro's Number.
- **(a)** Converting between moles of atoms and mass of atoms.
- **(b)** Calculating the mass of a single atom.
- **(c)** Converting mass in grams to number of atoms.

Problem Type 3: Calculations Involving Molecular Mass.
- **(a)** Calculating molecular mass.
- **(b)** Calculating the number of moles in a given amount of a compound.
- **(c)** Calculating the number of atoms in a given amount of a compound.

Problem Type 4: Calculations Involving Percent Composition
- **(a)** Calculating percent composition of a compound.
- **(b)** Determining empirical formula from percent composition.
- **(c)** Calculating mass from percent composition.

Problem Type 5: Experimental Determination of Empirical Formulas.

Problem Type 6: Determining the Molecular Formula of a Compound.

Problem Type 7: Calculating the Amounts of Reactants and Products.

Problem Type 8: Limiting Reagent Calculations.

Problem Type 9: Calculating the Percent Yield of a Reaction.

PROBLEM TYPE 1: CALCULATING AVERAGE ATOMIC MASS

The atomic mass you look up on a periodic table is an average atomic mass. The reason for this is that most naturally occurring elements have more than one isotope. The average atomic mass can be calculated as follows:

Step 1: Convert the percentage of each isotope to fractions. For example, an isotope that is 69.09 percent abundant becomes 69.09/100 = 0.6909.

Step 2: Multiply the mass of each isotope by its abundance and add them together.

average atomic mass = (fraction of isotope A)(mass of isotope A) + (fraction of isotope B)(mass of isotope B) + . . . + (fraction of isotope Z)(mass of isotope Z).

EXAMPLE 3.1

The element lithium has two isotopes that occur in nature: $^{6}_{3}Li$ with 7.5 percent abundance and $^{7}_{3}Li$ with 92.5 percent abundance. The atomic mass of $^{6}_{3}Li$ is 6.01513 amu and that of $^{7}_{3}Li$ is 7.01601 amu. Calculate the average atomic mass of lithium.

Strategy: Each isotope contributes to the average atomic mass based on its relative abundance. Multiplying the mass of an isotope by its fractional abundance (not percent) will give the contribution to the average atomic mass of that particular isotope.

Solution: Convert the percentage of each isotope to fractions.

$$^{6}_{3}\textbf{Li}: \ 7.5/100 = 0.075$$

$$^{7}_{3}\textbf{Li}: \ 92.5/100 = 0.925$$

Multiply the mass of each isotope by its abundance and add them together.

$$\textbf{average atomic mass} = (0.075)(6.01513) + (0.925)(7.01601) = \textbf{6.94 amu}$$

PRACTICE EXERCISE

1. The element boron (B) consists of two stable isotopes with atomic masses of 10.0129 amu and 11.0093 amu. The average atomic mass of B is 10.81 amu. Which isotope is more abundant?

Text Problem: 3.6

PROBLEM TYPE 2: CALCULATIONS INVOLVING MOLAR MASS OF AN ELEMENT AND AVOGADRO'S NUMBER

A. Converting between moles of atoms and mass of atoms

In order to convert from one unit to another, you need to be proficient at the dimensional analysis method. See Section 1.9 of your text and Problem Type 5, Chapter 1. Unit conversions can seem daunting, but if you keep track of the units, making sure that the appropriate units cancel, your effort will be rewarded.

Step 1: Map out a strategy to proceed from initial units to final units based on available conversion factors.

Step 2: Use the following method to ensure that you obtain the desired unit.

$$\text{Given unit} \times \left(\frac{\text{desired unit}}{\text{given unit}}\right) = \text{desired unit}$$

To convert between moles and mass, you need to use the molar mass of the element as a conversion factor.

$$\text{mol} \times \frac{\text{g}}{\text{mol}} = \text{g}$$

Also, going in the opposite direction

$$\text{g} \times \frac{\text{mol}}{\text{g}} = \text{mol}$$

Tip: Whether you are converting from g → mol or from mol → g, you will need to use the molar mass as the conversion factor. The molar mass of an element can be found directly on the periodic table.

EXAMPLE 3.2
How many grams are there in 0.130 mole of Cu?

Strategy: We are given moles of copper and asked to solve for grams of copper. What conversion factor do we need to convert between moles and grams? Arrange the appropriate conversion factor so moles cancel, and the unit grams is obtained for the answer.

Solution: The conversion factor needed to covert between moles and grams is the molar mass. In the periodic table (see inside front cover of the text), we see that the molar mass of Cu is 63.55 g. This can be expressed as

$$1 \text{ mol Cu} = 63.55 \text{ g Cu}$$

From this equality, we can write two conversion factors.

$$\frac{1 \text{ mol Cu}}{63.55 \text{ g Cu}} \quad \text{and} \quad \frac{63.55 \text{ g Cu}}{1 \text{ mol Cu}}$$

The conversion factor on the right is the correct one. Moles will cancel, leaving the unit grams for the answer.

We write

$$\textbf{? g Cu} = 0.130 \text{ mol Cu} \times \frac{63.55 \text{ g Cu}}{1 \text{ mol Cu}} = \textbf{8.26 g Cu}$$

Check: Does a mass of 8.26 g for 0.130 mole of Cu seem reasonable? What is the mass of 1 mole of Cu?

PRACTICE EXERCISE

2. How many moles of Cu are in 125 g of Cu?

Text Problem: 3.16

B. Calculating the mass of a single atom

To calculate the mass of a single atom, you can use Avogadro's number. The conversion factor is

$$\frac{1 \text{ mol}}{6.022 \times 10^{23} \text{ atoms}}$$

EXAMPLE 3.3
Copper is a minor component of pennies minted since 1981, and it is also used in electrical cables. Calculate the mass (in grams) of a single Cu atom.

Strategy: We can look up the molar mass of copper (Cu) on the periodic table (63.55 g/mol). We want to find the mass of a single atom of copper (unit of g/atom). Therefore, we need to convert from the unit mole in the denominator to the unit atom in the denominator. What conversion factor is needed to convert between moles and atoms? Arrange the appropriate conversion factor so mole in the denominator cancels, and the unit atom is obtained in the denominator.

Solution: The conversion factor needed is Avogadro's number. We have

$$1 \text{ mol} = 6.022 \times 10^{23} \text{ particles (atoms)}$$

From this equality, we can write two conversion factors.

$$\frac{1 \text{ mol Cu}}{6.022 \times 10^{23} \text{ Cu atoms}} \quad \text{and} \quad \frac{6.022 \times 10^{23} \text{ Cu atoms}}{1 \text{ mol Cu}}$$

The conversion factor on the left is the correct one. Moles will cancel, leaving the unit atoms in the denominator of the answer.

We write

$$? \textbf{ g/Cu atom} = \frac{63.55 \text{ g Cu}}{1 \text{ mol Cu}} \times \frac{1 \text{ mol Cu}}{6.022 \times 10^{23} \text{ Cu atoms}} = \mathbf{1.055 \times 10^{-22}} \textbf{ g/Cu atom}$$

Check: Should the mass of a single atom of Cu be a very small mass?

PRACTICE EXERCISE

3. Titanium (Ti) is a transition metal with a very high strength-to-weight ratio. For this reason, titanium is used in the construction of aircraft. What is the mass (in grams) of one Ti atom?

Text Problem: 3.18

C. Converting mass in grams to number of atoms

To complete the following conversion, you need to use both molar mass and Avogadro's number as conversion factors.

EXAMPLE 3.4

Zinc is the main component of pennies minted after 1981. How many zinc atoms are present in 20.0 g of Zn?

Strategy: The question asks for atoms of Zu. We cannot convert directly from grams to atoms of zinc. What unit do we need to convert grams of Zn to in order to convert to atoms? What does Avogadro's number represent?

Solution: To calculate the number of Zn atoms, we first must convert grams of Zn to moles of Zn. We use the molar mass of zinc as a conversion factor. Once moles of Zn are obtained, we can use Avogadro's number to convert from moles of zinc to atoms of zinc.

$$1 \text{ mol Zn} = 65.39 \text{ g Zn}$$

The conversion factor needed is

$$\frac{1 \text{ mol Zn}}{65.39 \text{ g Zn}}$$

Avogadro's number is the key to the second conversion. We have

$$1 \text{ mol} = 6.022 \times 10^{23} \text{ particles (atoms)}$$

From this equality, we can write two conversion factors.

$$\frac{1 \text{ mol Zn}}{6.022 \times 10^{23} \text{ Zn atoms}} \quad \text{and} \quad \frac{6.022 \times 10^{23} \text{ Zn atoms}}{1 \text{ mol Zn}}$$

The conversion factor on the right is the one we need because it has number of Zn atoms in the numerator, which is the unit we want for the answer.

Let's complete the two conversions in one step.

$$\text{grams of Zn} \rightarrow \text{moles of Zn} \rightarrow \text{number of Zn atoms}$$

$$\textbf{? atoms of Zn} = 20.0\ \text{g Zn} \times \frac{1\ \text{mol Zn}}{65.39\ \text{g Zn}} \times \frac{6.022 \times 10^{23}\ \text{Zn atoms}}{1\ \text{mol Zn}} = \mathbf{1.84 \times 10^{23}\ Zn\ atoms}$$

Check: Should 20.0 g of Zn contain fewer than Avogadro's number of atoms? What mass of Zn would contain Avogadro's number of atoms?

PRACTICE EXERCISE

4. What is the mass (in grams) of 9.09×10^{23} atoms of Zn?

Text Problem: 3.20

PROBLEM TYPE 3: CALCULATIONS INVOLVING MOLECULAR MASS

A. Calculating molecular mass

The molecular mass is simply the sum of the atomic masses (in amu) of all the atoms in the molecule.

EXAMPLE 3.5

Calculate the molecular mass of carbon tetrachloride (CCl_4).

Strategy: How do atomic masses of different elements combine to give the molecular mass of a compound?

Solution: To calculate the molecular mass of a compound, we need to sum all the atomic masses of the elements in the molecule. For each element, we multiply its atomic mass by the number of atoms of that element in one molecule of the compound. We find atomic masses for the elements in the periodic table (inside front cover of the text).

$$\text{molecular mass } CCl_4 = (\text{mass of C}) + 4(\text{mass of Cl})$$

$$\textbf{molecular mass } \mathbf{CCl_4} = (12.01\ \text{amu}) + 4(35.45\ \text{amu}) = \mathbf{153.8\ amu}$$

PRACTICE EXERCISE

5. Bananas owe their characteristic smell and flavor to the ester, isopentyl acetate [$CH_3COOCH_2CH_2CH(CH_3)_2$]. Calculate the molecular mass of isopentyl acetate.

Text Problem: 3.24

B. Calculating the number of moles in a given amount of a compound

To complete this conversion, the only conversion factor needed is the molar mass in units of g/mol. Remember, the molar mass of a compound (in grams) is numerically equal to its molecular mass (in atomic mass units). For example, the molar mass of CCl_4 is 153.8 g/mol, compared to its molecular mass of 153.8 amu.

EXAMPLE 3.6

How many moles of ethane (C_2H_6) are present in 50.3 g of ethane?

Strategy: First, calculate the molar mass of ethane. Then, arrange the molar mass as a conversion factor to convert from grams of ethane to moles of ethane.

Solution:

$$\text{molar mass of } C_2H_6 = 2(12.01\ \text{g}) + 6(1.008\ \text{g}) = 30.07\ \text{g}$$

Hence, the conversion factor is

$$\frac{1 \text{ mol } C_2H_6}{30.07 \text{ g } C_2H_6}$$

Using this conversion factor, convert from grams to moles.

$$\textbf{? mol of } \mathbf{C_2H_6} = 50.3 \text{ g } C_2H_6 \times \frac{1 \text{ mol } C_2H_6}{30.07 \text{ g } C_2H_6} = \mathbf{1.67 \text{ mol}}$$

PRACTICE EXERCISE

6. What is the mass (in grams) of 0.436 moles of ethane (C_2H_6)?

Text Problem: 3.26

C. Calculating the number of atoms in a given amount of a compound

Again, this is a unit conversion problem. This calculation is more difficult than the conversions above, because you must convert from *grams of compound* to *moles of compound* to *moles of a particular atom* to *number of atoms*. Sound tough? Let's try an example.

EXAMPLE 3.7

How many carbon atoms are present in 50.3 g of ethane (C_2H_6)?

Strategy: We started this problem in Example 3.6 when we calculated the moles of ethane in 50.3 g ethane. To continue, we need two additional conversion factors. One should represent the mole ratio between moles of C atoms and moles of ethane molecules. The other conversion factor needed is Avogadro's number.

Solution: The two conversion factors needed are:

$$\frac{2 \text{ mol C}}{1 \text{ mol } C_2H_6} \qquad \frac{6.022 \times 10^{23} \text{ C atoms}}{1 \text{ mol C}}$$

You should come up with the following strategy.

grams of C_2H_6 → moles of C_2H_6 → moles of C → atoms of C

$$\textbf{? C atoms} = 50.3 \text{ g } C_2H_6 \times \frac{1 \text{ mol } C_2H_6}{30.07 \text{ g } C_2H_6} \times \frac{2 \text{ mol C}}{1 \text{ mol } C_2H_6} \times \frac{6.022 \times 10^{23} \text{ C atoms}}{1 \text{ mol C}}$$

$$= \mathbf{2.01 \times 10^{24} \text{ C atoms}}$$

Check: Does the answer seem reasonable? We have 50.3 g ethane. How many atoms of C would 30.07 g of ethane contain?

PRACTICE EXERCISE

7. Glucose, the sugar used by the cells of our bodies for energy, has the molecular formula, $C_6H_{12}O_6$. How many atoms of *carbon* are present in a 3.50 g sample of glucose?

Text Problem: 3.28

PROBLEM TYPE 4: CALCULATIONS INVOLVING PERCENT COMPOSITION

A. Calculating percent composition of a compound

The *percent composition by mass* is the percent by mass of each element the compound contains. Percent composition is obtained by dividing the mass of each element in 1 mole of the compound by the molar mass of the compound, then multiplying by 100 percent.

$$\textbf{percent by mass of each element} = \frac{\text{mass of element in 1 mol of compound}}{\text{molar mass of compound}} \times 100\%$$

EXAMPLE 3.8

Calculate the percent composition by mass of all the elements in sodium bicarbonate, $NaHCO_3$.

Strategy: First, calculate the molar mass of sodium bicarbonate. Then, calculate the percent by mass of each element.

Solution:

$$\text{molar mass sodium bicarbonate} = 22.99\text{ g} + 1.008\text{ g} + 12.01\text{ g} + 3(16.00\text{ g}) = 84.01\text{ g}$$

$$\%\text{Na} = \frac{22.99\text{ g}}{84.01\text{ g}} \times 100\% = \mathbf{27.37\%}$$

$$\%\text{H} = \frac{1.008\text{ g}}{84.01\text{ g}} \times 100\% = \mathbf{1.200\%}$$

$$\%\text{C} = \frac{12.01\text{ g}}{84.01\text{ g}} \times 100\% = \mathbf{14.30\%}$$

$$\%\text{O} = \frac{3(16.00\text{ g})}{84.01\text{ g}} \times 100\% = \mathbf{57.14\%}$$

Tip: You can check your work by making sure that the mass percents of all the elements added together equals 100%. Checking above, 27.37% + 1.200% + 14.30% + 57.14% = 100.01% ≈ 100%.

PRACTICE EXERCISE

8. Cinnamic alcohol is used mainly in perfumes, particularly for soaps and cosmetics. Its molecular formula is $C_9H_{10}O$. Calculate the percent composition by mass of *hydrogen* in cinnamic alcohol.

Text Problems: 3.40, 3.42

B. Determining empirical formula from percent composition

The procedure used above to calculate the percent composition of a compound can be reversed. Given the percent composition by mass of a compound, you can determine the empirical formula of the compound.

EXAMPLE 3.9

Dieldrin, like DDT, is an insecticide that contains only C, H, Cl, and O. It is composed of 37.84 percent C, 2.12 percent H, 55.84 percent Cl, and 4.20 percent O. Determine its empirical formula.

Strategy: In a chemical formula, the subscripts represent the ratio of the number of moles of each element that combine to form the compound. Therefore, we need to convert from mass percent to moles in order to determine the empirical formula. If we assume an exactly 100 g sample of the compound, do we know the mass of each element in the compound? How do we then convert from grams to moles?

Solution: If we have 100 g of the compound, then each percentage can be converted directly to grams. In this sample, there will be 37.84 g of C, 2.12 g of H, 55.84 g Cl, and 4.20 g of O. Because the subscripts in the formula represent a mole ratio, we need to convert the grams of each element to moles. The conversion factor needed is the molar mass of each element. Let *n* represent the number of moles of each element so that

$$n_C = 37.84\ \text{g C} \times \frac{1\ \text{mol C}}{12.01\ \text{g C}} = \mathbf{3.151\ mol\ C}$$

$$n_H = 2.12\ \text{g H} \times \frac{1\ \text{mol H}}{1.008\ \text{g H}} = \mathbf{2.10\ mol\ H}$$

$$n_{Cl} = 55.84\ \text{g Cl} \times \frac{1\ \text{mol Cl}}{35.45\ \text{g Cl}} = \mathbf{1.575\ mol\ Cl}$$

$$n_O = 4.20\ \text{g O} \times \frac{1\ \text{mol O}}{16.00\ \text{g O}} = \mathbf{0.263\ mol\ O}$$

Thus, we arrive at the formula $C_{3.151}H_{2.10}Cl_{1.575}O_{0.263}$, which gives the identity and the ratios of atoms present. However, chemical formulas are written with whole numbers.

Try to convert to whole numbers by dividing all the subscripts by the smallest subscript.

$$\textbf{C}: \frac{3.151}{0.263} = 12.0 \qquad \textbf{H}: \frac{2.10}{0.263} = 7.98 \approx 8 \qquad \textbf{Cl}: \frac{1.575}{0.263} = 5.99 \approx 6 \qquad \textbf{O}: \frac{0.263}{0.263} = 1$$

This gives us the empirical for dieldrin, $\mathbf{C_{12}H_8Cl_6O}$.

Check: Are the subscripts in $C_{12}H_8Cl_6O$ reduced to the smallest whole numbers?

> **Tip:** It's not always this easy. Dividing by the smallest subscript often does not give all whole numbers. If this is the case, you must multiply all the subscripts by some *integer* to come up with whole number subscripts. Try the practice exercise below.

PRACTICE EXERCISE

9. The substance responsible for the green color on the yolk of a boiled egg is composed of 53.58 percent Fe and 46.42 percent S. Determine its empirical formula.

Text Problems: 3.44, **3.50**, 3.54

C. Calculating mass from percent composition

Step 1: Convert the mass percentage to a fraction. For example, if the mass percent of an element in a compound were 54.73 percent, you would convert this to 54.73/100 = 0.5473.

Step 2: Multiply the fraction by the total mass of the compound. This gives the mass of the particular element in the compound.

EXAMPLE 3.10

Calculate the mass of carbon in exactly 10 g of glucose ($C_6H_{12}O_6$).

Strategy: Glucose is composed of C, H, and O. The mass due to C is based on its percentage by mass in the compound. How do we calculate mass percent of an element?

Solution: First, we must find the mass % of carbon in $C_6H_{12}O_6$. Then, we convert this percentage to a fraction and multiply by the mass of the compound (10 g), to find the mass of carbon in 10 g of $C_6H_{12}O_6$.

The percent by mass of carbon in glucose, is calculated as follows:

$$\text{mass \% C} = \frac{\text{mass of C in 1 mol of glucose}}{\text{molar mass of glucose}} \times 100\%$$

$$\textbf{mass \% C} = \frac{6(12.01\text{ g})}{180.16\text{ g}} \times 100\% = \textbf{40.00\% C}$$

Converting this percentage to a fraction, we obtain 40.00/100 = **0.4000**

Next, multiply the fraction by the total mass of the compound.

$$\textbf{? g C in 10 g glucose} = (0.4000)(10\text{ g}) = \textbf{4.000 g C}$$

Check: Note that the mass percent of C is 40 percent. 40% of 10 g is 4 g.

PRACTICE EXERCISE

10. Calculate the mass of hydrogen in exactly 10 grams of glucose ($C_6H_{12}O_6$).

Text Problems: 3.46, **3.48**

PROBLEM TYPE 5: EXPERIMENTAL DETERMINATION OF EMPIRICAL FORMULAS

See Section 3.6 of your text for a description of the experimental setup. To solve this type of problem, you must recognize that all of the carbon in the sample is converted to CO_2 and all the hydrogen in the sample is converted to H_2O. Then, you can calculate the mass of C in CO_2 and the mass of H in H_2O. Finally, you can calculate the mass of oxygen by difference, if necessary.

EXAMPLE 3.11

When a 0.761-g sample of a compound containing only carbon and hydrogen is burned in an apparatus with CO_2 and H_2O absorbers, 2.23 g CO_2 and 1.37 g H_2O are collected. Determine the empirical formula of the compound.

Strategy: Calculate the moles of C in 2.23 g CO_2, and the moles of H in 1.37 g H_2O. In this problem, we do not need to convert to grams of C and H, because there are no other elements in the compound. To calculate the moles of each component, you need the molar masses and the correct mole ratio.

You should come up with the following strategy.

$$\text{g CO}_2 \rightarrow \text{mol CO}_2 \rightarrow \text{mol C}$$

Next, determine the smallest whole number ratio in which the elements combine.

Solution:

$$\text{? mol C} = 2.23\text{ g CO}_2 \times \frac{1\text{ mol CO}_2}{44.01\text{ g CO}_2} \times \frac{1\text{ mol C}}{1\text{ mol CO}_2} = 0.507\text{ mol C}$$

Similarly,

$$\text{? mol H} = 1.37\text{ g H}_2\text{O} \times \frac{1\text{ mol H}_2\text{O}}{18.02\text{ g H}_2\text{O}} \times \frac{2\text{ mol H}}{1\text{ mol H}_2\text{O}} = 0.152\text{ mol H}$$

Thus, we arrive at the formula $C_{0.0507}H_{0.152}$, which gives the identity and the ratios of atoms present. However, chemical formulas are written with whole numbers.

Try to convert to whole numbers by dividing all the subscripts by the smallest subscript.

$$\mathbf{C}: \frac{0.0507}{0.0507} = 1.00 \qquad \mathbf{H}: \frac{0.152}{0.0507} = 3.00$$

This gives the empirical formula, $\mathbf{CH_3}$.

PRACTICE EXERCISE

11. Diethyl ether, commonly known as "ether", was used as an anesthetic for many years. Diethyl ether contains C, H, and O. When a 1.45 g sample of ether is burned in an apparatus such as that shown in Figure 3.6 of the text, 2.77 g of CO_2 and 1.70 g of H_2O are collected. Determine the empirical formula of diethyl ether.

Text Problem: 3.136

PROBLEM TYPE 6: DETERMINING THE MOLECULAR FORMULA OF A COMPOUND

To determine the molecular formula of a compound, we must know both the *approximate* molar mass and the empirical formula of the compound. The molecular formula will either be equal to the empirical formula or be some integral multiple of it. Thus, the molar mass divided by the empirical mass will be an integer greater than or equal to one.

$$\frac{\text{molar mass}}{\text{empirical molar mass}} \geq 1 \text{ (integer values)}$$

EXAMPLE 3.12

A mass spectrum obtained on the compound in Example 3.11, shows its molecular mass to be about 31 g/mol. What is its molecular formula?

Strategy: First, determine the empirical formula. Then compare the molar mass to the empirical molar mass to determine the molecular formula.

Solution: The empirical formula was determined in the previous example to be $\mathbf{CH_3}$.

Next, calculate the empirical molar mass.

$$\text{empirical molar mass} = 12.01 \text{ g} + 3(1.008 \text{ g}) = 15.03 \text{ g/mol}$$

Determine the number of (CH_3) units present in the molecular formula. This number is found by taking the ratio

$$\frac{\text{molar mass}}{\text{empirical molar mass}} = \frac{31 \text{ g}}{15.03 \text{ g}} = 2.1 \approx 2$$

Thus, there are two CH_3 units in each molecule of the compound, so the molecular formula is $(CH_3)_2$, or $\mathbf{C_2H_6}$.

PRACTICE EXERCISE

12. In Example 3.9, the empirical formula of dieldrin was determined to be $C_{12}H_8Cl_6O$. If the molar mass of dieldrin is 381 ± 10 g/mol, what is the molecular formula of dieldrin?

Text Problems: 3.52, 3.54

PROBLEM TYPE 7: CALCULATING THE AMOUNTS OF REACTANTS AND PRODUCTS

These types of problems are dimensional analysis problems. You must always remember to start this type of problem with a balanced chemical equation. The typical approach is given below. See Section 3.8 of your text for a step-by-step method.

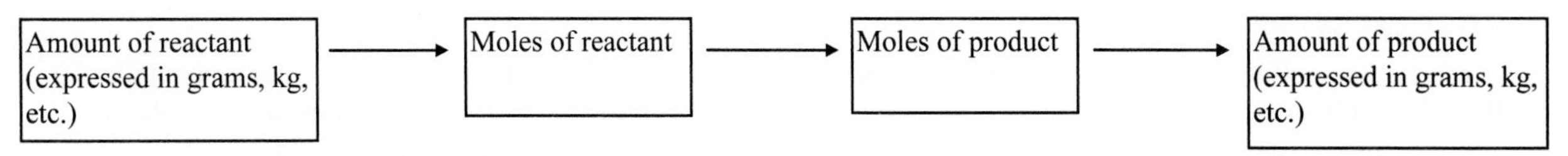

> **Tip:** Always try to be flexible when solving problems. Most problems of this type will follow an approach similar to the one above, but you may have to modify it sometimes.

EXAMPLE 3.13

Sulfur dioxide can be removed from stack gases by reaction with quicklime (CaO):

$$\mathbf{SO_2(\mathit{g}) + CaO(\mathit{s}) \longrightarrow CaSO_3(\mathit{s})}$$

If 975 kg of SO_2 are to be removed from stack gases by the above reaction, how many kilograms of CaO are required?

Strategy: We compare SO_2 and CaO based on the *mole ratio* in the balanced equation. Before we can determine moles of CaO required, we need to convert to moles of SO_2. What conversion factor is needed to convert from grams of SO_2 to moles of SO_2? Once moles of CaO are obtained, another conversion factor is needed to convert from moles of CaO to grams of CaO.

Solution: The molar mass of SO_2 will allow us to convert from grams of SO_2 to moles of SO_2. The molar mass of SO_2 = 32.07 g + 2(16.00 g) = 64.07 g. The balanced equation is given, so the mole ratio between SO_2 and CaO is known, that is, 1 mole SO_2 ≏ 1 mole CaO. Finally, the molar mass of CaO will convert moles of CaO to grams of CaO. This sequence of conversions is summarized as follows:

$$\text{kg } SO_2 \rightarrow \text{g } SO_2 \rightarrow \text{moles } SO_2 \rightarrow \text{moles CaO} \rightarrow \text{g CaO} \rightarrow \text{kg CaO}$$

$$\textbf{? kg CaO} = 975 \text{ kg } SO_2 \times \frac{1000 \text{ g } SO_2}{1 \text{ kg } SO_2} \times \frac{1 \text{ mol } SO_2}{64.07 \text{ g } SO_2} \times \frac{1 \text{ mol CaO}}{1 \text{ mol } SO_2} \times \frac{56.08 \text{ g CaO}}{1 \text{ mol CaO}} \times \frac{1 \text{ kg CaO}}{1000 \text{ g CaO}}$$

$$= \textbf{853 kg CaO}$$

> **Tip:** Notice that the approach followed was a slight modification of the flow diagram given above. We went from mass of one reactant, to moles of that reactant, to moles of a second reactant, and finally to mass of second reactant.

PRACTICE EXERCISE

13. Carbon dioxide in the air of a spacecraft can be removed by its reaction with a lithium hydroxide solution.

$$CO_2(g) + 2LiOH(aq) \longrightarrow Li_2CO_3(aq) + H_2O(l)$$

On average, a person will exhale about 1 kg of CO_2/day. How many kilograms of LiOH are required to react with 1.0 kg of CO_2?

Text Problems: 3.66, 3.68, 3.70, **3.72**, 3.74, 3.76, 3.78

PROBLEM TYPE 8: LIMITING REAGENT CALCULATIONS

When a chemist carries out a reaction, the reactants are usually not present in exact **stoichiometric amounts**, that is, in the proportions indicated by the balanced equation. The reactant used up first in a reaction is called the **limiting reagent**. When this reactant is used up, no more product can be formed.

Typically, the only difference between this type of problem and Problem Type 7, Calculating the Amounts of Reactants and Products, is that you must first determine which reactant is the limiting reagent.

EXAMPLE 3.14

Phosphine (PH_3) burns in oxygen (O_2) to produce phosphorus pentoxide and water.

$$2PH_3(g) + 4O_2(g) \longrightarrow P_2O_5(s) + 3H_2O(l)$$

How many grams of P_2O_5 will be produced when 17.0 g of phosphine are reacted with 16.0 g of O_2?

Strategy: Note that this reaction gives the amounts of both reactants, so it is likely to be a limiting reagent problem. The reactant that produces fewer moles of product is the limiting reagent because it limits the amount of product that can be produced. How do we convert from the amount of reactant to amount of product? Perform this calculation for each reactant, and then compare the moles of product, P_2O_5, formed by the given amounts of PH_3 and O_2 to determine which reactant is the limiting reagent.

Solution: We carry out two separate calculations. First, starting with 17.0 g PH_3, we calculate the number of moles of P_2O_5 that could be produced if all the PH_3 reacted. We complete the following conversions.

$$\text{grams of } PH_3 \rightarrow \text{moles of } PH_3 \rightarrow \text{moles of } P_2O_5$$

Combining these two conversions into one calculation, we write

$$?\ \text{mol } P_2O_5 = 17.0\ \text{g } PH_3 \times \frac{1\ \text{mol } PH_3}{33.99\ \text{g } PH_3} \times \frac{1\ \text{mol } P_2O_5}{2\ \text{mol } PH_3} = 0.250\ \text{mol } P_2O_5$$

Second, starting with 16.0 g of O_2, we complete similar conversions.

$$\text{grams of } O_2 \rightarrow \text{moles of } O_2 \rightarrow \text{moles of } P_2O_5$$

Combining these two conversions into one calculation, we write

$$?\ \text{mol } P_2O_5 = 16.0\ \text{g } O_2 \times \frac{1\ \text{mol } O_2}{32.0\ \text{g } O_2} \times \frac{1\ \text{mol } P_2O_5}{4\ \text{mol } O_2} = 0.125\ \text{mol } P_2O_5$$

The initial amount of O_2 limits the amount of product that can be formed; therefore, it is the limiting reagent.

The problem asks for grams of P_2O_5 produced. We already know the moles of P_2O_5 produced, 0.125 mole. Use the molar mass of P_2O_5 as a conversion factor to convert to grams.

$$\mathbf{?\ g\ P_2O_5} = 0.125\ \text{mol } P_2O_5 \times \frac{141.94\ \text{g } P_2O_5}{1\ \text{mol } P_2O_5} = \mathbf{17.7\ g\ P_2O_5}$$

Check: Does your answer seem reasonable? 0.125 mole of product is formed. What is the mass of 1 mole of P_2O_5?

PRACTICE EXERCISE

14. Iron can be produced by reacting iron ore with carbon. The iron produced can then be used to make steel. The reaction is

$$2Fe_2O_3(s) + 3C(s) \xrightarrow{\text{heat}} 4Fe(l) + 3CO_2(g)$$

(a) How many grams of Fe can be produced from a mixture of 200.0 g of Fe_2O_3 and 300.0 g C?
(b) How many grams of excess reagent will remain after the reaction ceases?

Text Problems: 3.84, 3.86

PROBLEM TYPE 9: CALCULATING THE PERCENT YIELD OF A REACTION

The **theoretical yield** is the amount of product that would result if all the limiting reagent reacted. This is the maximum obtainable yield predicted by the balanced equation. However, the amount of product obtained is almost always less than the theoretical yield. The **actual yield** is the quantity of product that actually results from a reaction.

To determine the efficiency of a reaction, chemists often calculate the **percent yield**, which describes the proportion of the actual yield to the theoretical yield. The percent yield is calculated as follows:

$$\%\text{ yield} = \frac{\text{actual yield}}{\text{theoretical yield}} \times 100\%$$

EXAMPLE 3.15

In Example 3.14, the theoretical yield of P_2O_5 was determined to be 17.7 g. If only 12.6 g of P_2O_5 are actually obtained, what is the percent yield of the reaction?

Solution:

$$\%\text{ yield} = \frac{\text{actual yield}}{\text{theoretical yield}} \times 100\%$$

$$\textbf{\% yield} = \frac{12.6\text{ g}}{17.7\text{ g}} \times 100\% = \textbf{71.2\%}$$

PRACTICE EXERCISE

15. Refer back to Practice Exercise 14 to answer this question. If the actual yield of Fe is 110 g, what is the percent yield of Fe?

Text Problems: 3.90, 3.92, 3.94

ANSWERS TO PRACTICE EXERCISES

1. ^{11}B

2. 1.97 moles Cu

3. 7.951×10^{-23} g/Ti atom

4. 98.7 g Zn

5. 130.18 amu

6. 13.1 g ethane

7. 7.02×10^{22} C atoms

8. 7.513 percent H by mass

9. Fe_2S_3

10. 0.67 g H

11. C_2H_6O

12. $C_{12}H_8Cl_6O$

13. 1.1 kg LiOH

14. **(a)** 139.9 g Fe
(b) 277 g C

15. 78.6 percent yield

SOLUTIONS TO SELECTED TEXT PROBLEMS

3.6 This is a variation of Problem Type 1, Calculating Average Atomic Mass.

Strategy: Each isotope contributes to the average atomic mass based on its relative abundance. Multiplying the mass of an isotope by its fractional abundance (not percent) will give the contribution to the average atomic mass of that particular isotope.

It would seem that there are two unknowns in this problem, the fractional abundance of ^{6}Li and the fractional abundance of ^{7}Li. However, these two quantities are not independent of each other; they are related by the fact that they must sum to 1. Start by letting x be the fractional abundance of ^{6}Li. Since the sum of the two abundance's must be 1, we can write

$$\text{Abundance }^7\text{Li} = (1 - x)$$

Solution:

$$\textbf{Average atomic mass of Li} = 6.941 \text{ amu} = x(6.0151 \text{ amu}) + (1 - x)(7.0160 \text{ amu})$$
$$6.941 = -1.0009x + 7.0160$$
$$1.0009x = 0.075$$
$$\boldsymbol{x = 0.075}$$

$x = 0.075$ corresponds to a natural abundance of ^{6}Li of **7.5 percent**. The natural abundance of ^{7}Li is $(1 - x) = 0.925$ or **92.5 percent**.

3.8 The unit factor required is $\left(\dfrac{6.022 \times 10^{23} \text{ amu}}{1 \text{ g}}\right)$

$$\textbf{? amu} = 8.4 \text{ g} \times \frac{6.022 \times 10^{23} \text{ amu}}{1 \text{ g}} = \mathbf{5.1 \times 10^{24} \text{ amu}}$$

3.12 The thickness of the book in miles would be:

$$\frac{0.0036 \text{ in}}{1 \text{ page}} \times \frac{1 \text{ ft}}{12 \text{ in}} \times \frac{1 \text{ mi}}{5280 \text{ ft}} \times (6.022 \times 10^{23} \text{ pages}) = 3.4 \times 10^{16} \text{ mi}$$

The distance, in miles, traveled by light in one year is:

$$1.00 \text{ yr} \times \frac{365 \text{ day}}{1 \text{ yr}} \times \frac{24 \text{ h}}{1 \text{ day}} \times \frac{3600 \text{ s}}{1 \text{ h}} \times \frac{3.00 \times 10^{8} \text{ m}}{1 \text{ s}} \times \frac{1 \text{ mi}}{1609 \text{ m}} = 5.88 \times 10^{12} \text{ mi}$$

The thickness of the book in light-years is:

$$(3.4 \times 10^{16} \text{ mi}) \times \frac{1 \text{ light-yr}}{5.88 \times 10^{12} \text{ mi}} = \mathbf{5.8 \times 10^{3} \text{ light-yr}}$$

It will take light 5.8×10^3 years to travel from the first page to the last one!

3.14

$$(6.00 \times 10^{9} \text{ Co atoms}) \times \frac{1 \text{ mol Co}}{6.022 \times 10^{23} \text{ Co atoms}} = \mathbf{9.96 \times 10^{-15} \text{ mol Co}}$$

3.16 Converting between moles of atoms and mass of atoms, Problem Type 2A.

Strategy: We are given moles of gold and asked to solve for grams of gold. What conversion factor do we need to convert between moles and grams? Arrange the appropriate conversion factor so moles cancel, and the unit grams is obtained for the answer.

Solution: The conversion factor needed to covert between moles and grams is the molar mass. In the periodic table (see inside front cover of the text), we see that the molar mass of Au is 197.0 g. This can be expressed as

$$1 \text{ mol Au} = 197.0 \text{ g Au}$$

From this equality, we can write two conversion factors.

$$\frac{1 \text{ mol Au}}{197.0 \text{ g Au}} \quad \text{and} \quad \frac{197.0 \text{ g Au}}{1 \text{ mol Au}}$$

The conversion factor on the right is the correct one. Moles will cancel, leaving the unit grams for the answer.

We write

$$\textbf{? g Au} = 15.3 \cancel{\text{mol}} \text{ Au} \times \frac{197.0 \text{ g Au}}{1 \cancel{\text{mol}} \text{ Au}} = \mathbf{3.01 \times 10^3 \text{ g Au}}$$

Check: Does a mass of 3010 g for 15.3 moles of Au seem reasonable? What is the mass of 1 mole of Au?

3.18 Calculating the mass of a single atom, Problem Type 2B.

(a)
Strategy: We can look up the molar mass of arsenic (As) on the periodic table (74.92 g/mol). We want to find the mass of a single atom of arsenic (unit of g/atom). Therefore, we need to convert from the unit mole in the denominator to the unit atom in the denominator. What conversion factor is needed to convert between moles and atoms? Arrange the appropriate conversion factor so mole in the denominator cancels, and the unit atom is obtained in the denominator.

Solution: The conversion factor needed is Avogadro's number. We have

$$1 \text{ mol} = 6.022 \times 10^{23} \text{ particles (atoms)}$$

From this equality, we can write two conversion factors.

$$\frac{1 \text{ mol As}}{6.022 \times 10^{23} \text{ As atoms}} \quad \text{and} \quad \frac{6.022 \times 10^{23} \text{ As atoms}}{1 \text{ mol As}}$$

The conversion factor on the left is the correct one. Moles will cancel, leaving the unit atoms in the denominator of the answer.

We write

$$\textbf{? g/As atom} = \frac{74.92 \text{ g As}}{1 \cancel{\text{mol}} \text{ As}} \times \frac{1 \cancel{\text{mol}} \text{ As}}{6.022 \times 10^{23} \text{ As atoms}} = \mathbf{1.244 \times 10^{-22} \text{ g/As atom}}$$

(b) Follow same method as part (a).

$$\textbf{? g/Ni atom} = \frac{58.69 \text{ g Ni}}{1 \cancel{\text{mol}} \text{ Ni}} \times \frac{1 \cancel{\text{mol}} \text{ Ni}}{6.022 \times 10^{23} \text{ Ni atoms}} = \mathbf{9.746 \times 10^{-23} \text{ g/Ni atom}}$$

Check: Should the mass of a single atom of As or Ni be a very small mass?

3.20 Converting mass in grams to number of atoms, Problem Type 2a.

Strategy: The question asks for atoms of Cu. We cannot convert directly from grams to atoms of copper. What unit do we need to convert grams of Cu to in order to convert to atoms? What does Avogadro's number represent?

Solution: To calculate the number of Cu atoms, we first must convert grams of Cu to moles of Cu. We use the molar mass of copper as a conversion factor. Once moles of Cu are obtained, we can use Avogadro's number to convert from moles of copper to atoms of copper.

$$1 \text{ mol Cu} = 63.55 \text{ g Cu}$$

The conversion factor needed is

$$\frac{1 \text{ mol Cu}}{63.55 \text{ g Cu}}$$

Avogadro's number is the key to the second conversion. We have

$$1 \text{ mol} = 6.022 \times 10^{23} \text{ particles (atoms)}$$

From this equality, we can write two conversion factors.

$$\frac{1 \text{ mol Cu}}{6.022 \times 10^{23} \text{ Cu atoms}} \quad \text{and} \quad \frac{6.022 \times 10^{23} \text{ Cu atoms}}{1 \text{ mol Cu}}$$

The conversion factor on the right is the one we need because it has number of Cu atoms in the numerator, which is the unit we want for the answer.

Let's complete the two conversions in one step.

grams of Cu → moles of Cu → number of Cu atoms

$$\textbf{? atoms of Cu} = 3.14 \text{ g Cu} \times \frac{1 \text{ mol Cu}}{63.55 \text{ g Cu}} \times \frac{6.022 \times 10^{23} \text{ Cu atoms}}{1 \text{ mol Cu}} = \mathbf{2.98 \times 10^{22} \textbf{ Cu atoms}}$$

Check: Should 3.14 g of Cu contain fewer than Avogadro's number of atoms? What mass of Cu would contain Avogadro's number of atoms?

3.22
$$2 \text{ Pb atoms} \times \frac{1 \text{ mol Pb}}{6.022 \times 10^{23} \text{ Pb atoms}} \times \frac{207.2 \text{ g Pb}}{1 \text{ mol Pb}} = 6.881 \times 10^{-22} \text{ g Pb}$$

$$(5.1 \times 10^{-23} \text{ mol He}) \times \frac{4.003 \text{ g He}}{1 \text{ mol He}} = 2.0 \times 10^{-22} \text{ g He}$$

2 atoms of lead have a greater mass than 5.1×10^{-23} mol of helium.

3.24 Calculating molar mass, modification of Problem Type 3A.

Strategy: How do molar masses of different elements combine to give the molar mass of a compound?

Solution: To calculate the molar mass of a compound, we need to sum all the molar masses of the elements in the molecule. For each element, we multiply its molar mass by the number of moles of that element in one mole of the compound. We find molar masses for the elements in the periodic table (inside front cover of the text).

(a) **molar mass Li_2CO_3** = 2(6.941 g) + 12.01 g + 3(16.00 g) = **73.89 g**

(b) **molar mass CS_2** = 12.01 g + 2(32.07 g) = **76.15 g**

(c) **molar mass $CHCl_3$** = 12.01 g + 1.008 g + 3(35.45 g) = **119.37 g**

(d) **molar mass $C_6H_8O_6$** = 6(12.01 g) + 8(1.008 g) + 6(16.00 g) = **176.12 g**

(e) **molar mass KNO_3** = 39.10 g + 14.01 g + 3(16.00 g) = **101.11 g**

(f) **molar mass Mg_3N_2** = 3(24.31 g) + 2(14.01 g) = **100.95 g**

3.26 Calculating the number of molecules in a given amount of compound, similar to Problem Type 3B.

Strategy: We are given grams of ethane and asked to solve for molecules of ethane. We cannot convert directly from grams ethane to molecules of ethane. What unit do we need to obtain first before we can convert to molecules? How should Avogadro's number be used here?

Solution: To calculate number of ethane molecules, we first must convert grams of ethane to moles of ethane. We use the molar mass of ethane as a conversion factor. Once moles of ethane are obtained, we can use Avogadro's number to convert from moles of ethane to molecules of ethane.

$$\text{molar mass of } C_2H_6 = 2(12.01 \text{ g}) + 6(1.008 \text{ g}) = 30.068 \text{ g}$$

The conversion factor needed is

$$\frac{1 \text{ mol } C_2H_6}{30.068 \text{ g } C_2H_6}$$

Avogadro's number is the key to the second conversion. We have

$$1 \text{ mol} = 6.022 \times 10^{23} \text{ particles (molecules)}$$

From this equality, we can write the conversion factor:

$$\frac{6.022 \times 10^{23} \text{ ethane molecules}}{1 \text{ mol ethane}}$$

Let's complete the two conversions in one step.

$$\text{grams of ethane} \rightarrow \text{moles of ethane} \rightarrow \text{number of ethane molecules}$$

$$\textbf{? molecules of } \mathbf{C_2H_6} = 0.334 \text{ g } C_2H_6 \times \frac{1 \text{ mol } C_2H_6}{30.07 \text{ g } C_2H_6} \times \frac{6.022 \times 10^{23} \text{ } C_2H_6 \text{ molecules}}{1 \text{ mol } C_2H_6}$$

$$= \mathbf{6.69 \times 10^{21} \text{ } C_2H_6 \text{ molecules}}$$

Check: Should 0.334 g of ethane contain fewer than Avogadro's number of molecules? What mass of ethane would contain Avogadro's number of molecules?

3.28 Calculating the number of atoms in a given amount of a compound, Problem Type 3C.

Strategy: We are asked to solve for the number of N, C, O, and H atoms in 1.68×10^4 g of urea. We cannot convert directly from grams urea to atoms. What unit do we need to obtain first before we can convert to atoms? How should Avogadro's number be used here? How many atoms of N, C, O, or H are in 1 molecule of urea?

Solution: Let's first calculate the number of N atoms in 1.68×10^4 g of urea. First, we must convert grams of urea to number of molecules of urea. This calculation is similar to Problem 3.26. The molecular formula of urea shows there are two N atoms in one urea molecule, which will allow us to convert to atoms of N. We need to perform three conversions:

grams of urea → moles of urea → molecules of urea → atoms of N

The conversion factors needed for each step are: 1) the molar mass of urea, 2) Avogadro's number, and 3) the number of N atoms in 1 molecule of urea.

We complete the three conversions in one calculation.

$$\textbf{? atoms of N} = (1.68 \times 10^4 \text{ g urea}) \times \frac{1 \text{ mol urea}}{60.06 \text{ g urea}} \times \frac{6.022 \times 10^{23} \text{ urea molecules}}{1 \text{ mol urea}} \times \frac{2 \text{ N atoms}}{1 \text{ molecule urea}}$$

$$= \mathbf{3.37 \times 10^{26} \text{ N atoms}}$$

The above method utilizes the ratio of molecules (urea) to atoms (nitrogen). We can also solve the problem by reading the formula as the ratio of moles of urea to moles of nitrogen by using the following conversions:

grams of urea → moles of urea → moles of N → atoms of N

Try it.

Check: Does the answer seem reasonable? We have 1.68×10^4 g urea. How many atoms of N would 60.06 g of urea contain?

We could calculate the number of atoms of the remaining elements in the same manner, or we can use the atom ratios from the molecular formula. The carbon atom to nitrogen atom ratio in a urea molecule is 1:2, the oxygen atom to nitrogen atom ratio is 1:2, and the hydrogen atom to nitrogen atom ration is 4:2.

$$\textbf{? atoms of C} = (3.37 \times 10^{26} \text{ N atoms}) \times \frac{1 \text{ C atom}}{2 \text{ N atoms}} = \mathbf{1.69 \times 10^{26} \text{ C atoms}}$$

$$\textbf{? atoms of O} = (3.37 \times 10^{26} \text{ N atoms}) \times \frac{1 \text{ O atom}}{2 \text{ N atoms}} = \mathbf{1.69 \times 10^{26} \text{ O atoms}}$$

$$\textbf{? atoms of H} = (3.37 \times 10^{26} \text{ N atoms}) \times \frac{4 \text{ H atoms}}{2 \text{ N atoms}} = \mathbf{6.74 \times 10^{26} \text{ H atoms}}$$

3.30 Mass of water $= 2.56 \text{ mL} \times \frac{1.00 \text{ g}}{1.00 \text{ mL}} = 2.56 \text{ g}$

Molar mass of H_2O = (16.00 g) + 2(1.008 g) = 18.02 g/mol

$$\textbf{? H}_2\textbf{O molecules} = 2.56 \text{ g } H_2O \times \frac{1 \text{ mol } H_2O}{18.02 \text{ g } H_2O} \times \frac{6.022 \times 10^{23} \text{ molecules } H_2O}{1 \text{ mol } H_2O}$$

$$= \mathbf{8.56 \times 10^{22} \text{ molecules}}$$

3.34 Since there are two hydrogen isotopes, they can be paired in three ways: $^{1}H-^{1}H$, $^{1}H-^{2}H$, and $^{2}H-^{2}H$. There will then be three choices for each sulfur isotope. We can make a table showing all the possibilities (masses in amu):

	^{32}S	^{33}S	^{34}S	^{36}S
$^{1}H_2$	34	35	36	38
$^{1}H^{2}H$	35	36	37	39
$^{2}H_2$	36	37	38	40

There will be **seven peaks** of the following mass numbers: 34, 35, 36, 37, 38, 39, and 40.

Very accurate (and expensive!) mass spectrometers can detect the mass difference between two ^{1}H and one ^{2}H. How many peaks would be detected in such a "high resolution" mass spectrum?

3.40 Calculating percent composition of a compound, Problem Type 4A.

Strategy: Recall the procedure for calculating a percentage. Assume that we have 1 mole of $CHCl_3$. The percent by mass of each element (C, H, and Cl) is given by the mass of that element in 1 mole of $CHCl_3$ divided by the molar mass of $CHCl_3$, then multiplied by 100 to convert from a fractional number to a percentage.

Solution: The molar mass of $CHCl_3$ = 12.01 g/mol + 1.008 g/mol + 3(35.45 g/mol) = 119.4 g/mol. The percent by mass of each of the elements in $CHCl_3$ is calculated as follows:

$$\%C = \frac{12.01 \text{ g/mol}}{119.4 \text{ g/mol}} \times 100\% = \mathbf{10.06\%}$$

$$\%H = \frac{1.008 \text{ g/mol}}{119.4 \text{ g/mol}} \times 100\% = \mathbf{0.8442\%}$$

$$\%Cl = \frac{3(35.45) \text{ g/mol}}{119.4 \text{ g/mol}} \times 100\% = \mathbf{89.07\%}$$

Check: Do the percentages add to 100%? The sum of the percentages is (10.06% + 0.8442% + 89.07%) = 99.97%. The small discrepancy from 100% is due to the way we rounded off.

3.42

	Compound	Molar mass (g)	N% by mass
(a)	$(NH_2)_2CO$	60.06	$\frac{2(14.01 \text{ g})}{60.06 \text{ g}} \times 100\% = 46.65\%$
(b)	NH_4NO_3	80.05	$\frac{2(14.01 \text{ g})}{80.05 \text{ g}} \times 100\% = 35.00\%$
(c)	$HNC(NH_2)_2$	59.08	$\frac{3(14.01 \text{ g})}{59.08 \text{ g}} \times 100\% = 71.14\%$
(d)	NH_3	17.03	$\frac{14.01 \text{ g}}{17.03 \text{ g}} \times 100\% = 82.27\%$

Ammonia, NH_3, is the richest source of nitrogen on a mass percentage basis.

3.44 METHOD 1:

Step 1: Assume you have exactly 100 g of substance. 100 g is a convenient amount, because all the percentages sum to 100%. The percentage of oxygen is found by difference:

$$100\% - (19.8\% + 2.50\% + 11.6\%) = 66.1\%$$

In 100 g of PAN there will be 19.8 g C, 2.50 g H, 11.6 g N, and 66.1 g O.

Step 2: Calculate the number of moles of each element in the compound. Remember, an *empirical formula* tells us which elements are present and the simplest whole-number ratio of their atoms. This ratio is also a mole ratio. Use the molar masses of these elements as conversion factors to convert to moles.

$$n_C = 19.8\ \text{g C} \times \frac{1\ \text{mol C}}{12.01\ \text{g C}} = 1.65\ \text{mol C}$$

$$n_H = 2.50\ \text{g H} \times \frac{1\ \text{mol H}}{1.008\ \text{g H}} = 2.48\ \text{mol H}$$

$$n_N = 11.6\ \text{g N} \times \frac{1\ \text{mol N}}{14.01\ \text{g N}} = 0.828\ \text{mol N}$$

$$n_O = 66.1\ \text{g O} \times \frac{1\ \text{mol O}}{16.00\ \text{g O}} = 4.13\ \text{mol O}$$

Step 3: Try to convert to whole numbers by dividing all the subscripts by the smallest subscript. The formula is $C_{1.65}H_{2.48}N_{0.828}O_{4.13}$. Dividing the subscripts by 0.828 gives the empirical formula, $\mathbf{C_2H_3NO_5}$.

To determine the molecular formula, remember that the molar mass/empirical mass will be an integer greater than or equal to one.

$$\frac{\text{molar mass}}{\text{empirical molar mass}} \geq 1\ \text{(integer values)}$$

In this case,

$$\frac{\text{molar mass}}{\text{empirical molar mass}} = \frac{120\ \text{g}}{121.05\ \text{g}} \approx 1$$

Hence, the molecular formula and the empirical formula are the same, $\mathbf{C_2H_3NO_5}$.

METHOD 2:

Step 1: Multiply the mass % (converted to a decimal) of each element by the molar mass to convert to grams of each element. Then, use the molar mass to convert to moles of each element.

$$n_C = (0.198) \times (120\ \text{g}) \times \frac{1\ \text{mol C}}{12.01\ \text{g C}} = 1.98\ \text{mol C} \approx \mathbf{2\ mol\ C}$$

$$n_H = (0.0250) \times (120\ \text{g}) \times \frac{1\ \text{mol H}}{1.008\ \text{g H}} = 2.98\ \text{mol H} \approx \mathbf{3\ mol\ H}$$

$$n_N = (0.116) \times (120\ \text{g}) \times \frac{1\ \text{mol N}}{14.01\ \text{g N}} = 0.994\ \text{mol N} \approx \mathbf{1\ mol\ N}$$

$$n_O = (0.661) \times (120\ \text{g}) \times \frac{1\ \text{mol O}}{16.00\ \text{g O}} = 4.96\ \text{mol O} \approx \mathbf{5\ mol\ O}$$

Step 2: Since we used the molar mass to calculate the moles of each element present in the compound, this method directly gives the molecular formula. The formula is $\mathbf{C_2H_3NO_5}$.

Step 3: Try to reduce the molecular formula to a simpler whole number ratio to determine the empirical formula. The formula is already in its simplest whole number ratio. The molecular and empirical formulas are the same. The empirical formula is **$C_2H_3NO_5$**.

3.46 Using unit factors we convert:

g of Hg → mol Hg → mol S → g S

$$\textbf{? g S} = 246\ \cancel{g}\ Hg \times \frac{1\ \cancel{mol}\ Hg}{200.6\ \cancel{g}\ Hg} \times \frac{1\ \cancel{mol}\ S}{1\ \cancel{mol}\ Hg} \times \frac{32.07\ g\ S}{1\ \cancel{mol}\ S} = \textbf{39.3 g S}$$

3.48 Calculating mass from percent composition, Problem Type 4C.

Strategy: Tin(II) fluoride is composed of Sn and F. The mass due to F is based on its percentage by mass in the compound. How do we calculate mass percent of an element?

Solution: First, we must find the mass % of fluorine in SnF_2. Then, we convert this percentage to a fraction and multiply by the mass of the compound (24.6 g), to find the mass of fluorine in 24.6 g of SnF_2.

The percent by mass of fluorine in tin(II) fluoride, is calculated as follows:

$$\text{mass \% F} = \frac{\text{mass of F in 1 mol SnF}_2}{\text{molar mass of SnF}_2} \times 100\%$$

$$= \frac{2(19.00\ g)}{156.7\ g} \times 100\% = 24.25\%\ F$$

Converting this percentage to a fraction, we obtain 24.25/100 = 0.2425.

Next, multiply the fraction by the total mass of the compound.

$$\textbf{? g F in 24.6 g SnF}_2 = (0.2425)(24.6\ g) = \textbf{5.97 g F}$$

Check: As a ball-park estimate, note that the mass percent of F is roughly 25 percent, so that a quarter of the mass should be F. One quarter of approximately 24 g is 6 g, which is close to the answer.

Note: This problem could have been worked in a manner similar to Problem 3.46. You could complete the following conversions:

g of SnF_2 → mol of SnF_2 → mol of F → g of F

3.50 Determining empirical formula from percent composition, Problem Type 4C.

(a)

Strategy: In a chemical formula, the subscripts represent the ratio of the number of moles of each element that combine to form the compound. Therefore, we need to convert from mass percent to moles in order to determine the empirical formula. If we assume an exactly 100 g sample of the compound, do we know the mass of each element in the compound? How do we then convert from grams to moles?

Solution: If we have 100 g of the compound, then each percentage can be converted directly to grams. In this sample, there will be 40.1 g of C, 6.6 g of H, and 53.3 g of O. Because the subscripts in the formula represent a mole ratio, we need to convert the grams of each element to moles. The conversion factor needed is the molar mass of each element. Let *n* represent the number of moles of each element so that

$$n_C = 40.1\ \text{g C} \times \frac{1 \text{ mol C}}{12.01\ \text{g C}} = 3.34 \text{ mol C}$$

$$n_H = 6.6\ \text{g H} \times \frac{1 \text{ mol H}}{1.008\ \text{g H}} = 6.55 \text{ mol H}$$

$$n_O = 53.3\ \text{g O} \times \frac{1 \text{ mol O}}{16.00\ \text{g O}} = 3.33 \text{ mol O}$$

Thus, we arrive at the formula $C_{3.34}H_{6.5}O_{3.33}$, which gives the identity and the mole ratios of atoms present. However, chemical formulas are written with whole numbers. Try to convert to whole numbers by dividing all the subscripts by the smallest subscript (3.331).

$$\textbf{C:}\ \frac{3.34}{3.33} \approx 1 \qquad \textbf{H:}\ \frac{6.55}{3.33} \approx 2 \qquad \textbf{O:}\ \frac{3.33}{3.33} = 1$$

This gives the empirical formula, $\mathbf{CH_2O}$.

Check: Are the subscripts in CH_2O reduced to the smallest whole numbers?

(b) Following the same procedure as part (a), we find:

$$n_C = 18.4\ \text{g C} \times \frac{1 \text{ mol C}}{12.01\ \text{g C}} = 1.53 \text{ mol C}$$

$$n_N = 21.5\ \text{g N} \times \frac{1 \text{ mol N}}{14.01\ \text{g N}} = 1.53 \text{ mol N}$$

$$n_K = 60.1\ \text{g K} \times \frac{1 \text{ mol K}}{39.10\ \text{g K}} = 1.54 \text{ mol K}$$

Dividing by the smallest number of moles (1.53 mol) gives the empirical formula, **KCN**.

3.52 The empirical molar mass of CH is approximately 13.02 g. Let's compare this to the molar mass to determine the molecular formula.

Recall that the molar mass divided by the empirical mass will be an integer greater than or equal to one.

$$\frac{\text{molar mass}}{\text{empirical molar mass}} \geq 1 \text{ (integer values)}$$

In this case,

$$\frac{\text{molar mass}}{\text{empirical molar mass}} = \frac{78 \text{ g}}{13.02 \text{ g}} \approx 6$$

Thus, there are six CH units in each molecule of the compound, so the molecular formula is $(CH)_6$, or $\mathbf{C_6H_6}$.

3.54 **METHOD 1:**

Step 1: Assume you have exactly 100 g of substance. 100 g is a convenient amount, because all the percentages sum to 100%. In 100 g of MSG there will be 35.51 g C, 4.77 g H, 37.85 g O, 8.29 g N, and 13.60 g Na.

Step 2: Calculate the number of moles of each element in the compound. Remember, an *empirical formula* tells us which elements are present and the simplest whole-number ratio of their atoms. This ratio is also a mole ratio. Let n_C, n_H, n_O, n_N, and n_{Na} be the number of moles of elements present. Use the molar masses of these elements as conversion factors to convert to moles.

$$n_C = 35.51 \not{g}\,C \times \frac{1 \text{ mol C}}{12.01 \not{g}\,C} = 2.957 \text{ mol C}$$

$$n_H = 4.77 \not{g}\,H \times \frac{1 \text{ mol H}}{1.008 \not{g}\,H} = 4.73 \text{ mol H}$$

$$n_O = 37.85 \not{g}\,O \times \frac{1 \text{ mol O}}{16.00 \not{g}\,O} = 2.366 \text{ mol O}$$

$$n_N = 8.29 \not{g}\,N \times \frac{1 \text{ mol N}}{14.01 \not{g}\,N} = 0.592 \text{ mol N}$$

$$n_{Na} = 13.60 \not{g}\,Na \times \frac{1 \text{ mol Na}}{22.99 \not{g}\,Na} = 0.5916 \text{ mol Na}$$

Thus, we arrive at the formula $C_{2.957}H_{4.73}O_{2.366}N_{0.592}Na_{0.5916}$, which gives the identity and the ratios of atoms present. However, chemical formulas are written with whole numbers.

Step 3: Try to convert to whole numbers by dividing all the subscripts by the smallest subscript.

$$\textbf{C}: \frac{2.957}{0.59156} = 4.998 \approx 5 \qquad \textbf{H}: \frac{4.73}{0.59156} = 8.00 \qquad \textbf{O}: \frac{2.366}{0.59156} = 3.999 \approx 4$$

$$\textbf{N}: \frac{0.592}{0.59156} = 1.00 \qquad \textbf{Na}: \frac{0.5916}{0.5916} = 1$$

This gives us the empirical formula for MSG, $C_5H_8O_4NNa$.

To determine the molecular formula, remember that the molar mass/empirical mass will be an integer greater than or equal to one.

$$\frac{\text{molar mass}}{\text{empirical molar mass}} \geq 1 \text{ (integer values)}$$

In this case,

$$\frac{\text{molar mass}}{\text{empirical molar mass}} = \frac{169 \text{ g}}{169.11 \text{ g}} \approx 1$$

Hence, the molecular formula and the empirical formula are the same, $\mathbf{C_5H_8O_4NNa}$. It should come as no surprise that the empirical and molecular formulas are the same since MSG stands for *monosodium*glutamate.

METHOD 2:

Step 1: Multiply the mass % (converted to a decimal) of each element by the molar mass to convert to grams of each element. Then, use the molar mass to convert to moles of each element.

$$n_C = (0.3551) \times (169 \not{g}) \times \frac{1 \text{ mol C}}{12.01 \not{g}\,C} = 5.00 \text{ mol C}$$

$$n_H = (0.0477) \times (169 \not{g}) \times \frac{1 \text{ mol H}}{1.008 \not{g}\,H} = 8.00 \text{ mol H}$$

$$n_O = (0.3785) \times (169 \not{g}) \times \frac{1 \text{ mol O}}{16.00 \not{g}\,O} = 4.00 \text{ mol O}$$

$$n_N = (0.0829) \times (169\ \text{g}) \times \frac{1 \text{ mol N}}{14.01 \text{ g N}} = 1.00 \text{ mol N}$$

$$n_{Na} = (0.1360) \times (169\ \text{g}) \times \frac{1 \text{ mol Na}}{22.99 \text{ g Na}} = 1.00 \text{ mol Na}$$

Step 2: Since we used the molar mass to calculate the moles of each element present in the compound, this method directly gives the molecular formula. The formula is $\mathbf{C_5H_8O_4NNa}$.

3.60 The balanced equations are as follows:

(a) $2N_2O_5 \rightarrow 2N_2O_4 + O_2$

(b) $2KNO_3 \rightarrow 2KNO_2 + O_2$

(c) $NH_4NO_3 \rightarrow N_2O + 2H_2O$

(d) $NH_4NO_2 \rightarrow N_2 + 2H_2O$

(e) $2NaHCO_3 \rightarrow Na_2CO_3 + H_2O + CO_2$

(f) $P_4O_{10} + 6H_2O \rightarrow 4H_3PO_4$

(g) $2HCl + CaCO_3 \rightarrow CaCl_2 + H_2O + CO_2$

(h) $2Al + 3H_2SO_4 \rightarrow Al_2(SO_4)_3 + 3H_2$

(i) $CO_2 + 2KOH \rightarrow K_2CO_3 + H_2O$

(j) $CH_4 + 2O_2 \rightarrow CO_2 + 2H_2O$

(k) $Be_2C + 4H_2O \rightarrow 2Be(OH)_2 + CH_4$

(l) $3Cu + 8HNO_3 \rightarrow 3Cu(NO_3)_2 + 2NO + 4H_2O$

(m) $S + 6HNO_3 \rightarrow H_2SO_4 + 6NO_2 + 2H_2O$

(n) $2NH_3 + 3CuO \rightarrow 3Cu + N_2 + 3H_2O$

3.64 On the reactants side there are 6 A atoms and 4 B atoms. On the products side, there are 4 C atoms and 2 D atoms. Writing an equation,

$$6A + 4B \rightarrow 4C + 2D$$

Chemical equations are typically written with the smallest set of whole number coefficients. Dividing the equation by two gives,

$$\mathbf{3A + 2B \rightarrow 2C + D}$$

The correct answer is choice **(d)**.

3.66 Calculating the Amounts of Reactants and Products, Problem Type 7.

$$Si(s) + 2Cl_2(g) \longrightarrow SiCl_4(l)$$

Strategy: Looking at the balanced equation, how do we compare the amounts of Cl_2 and $SiCl_4$? We can compare them based on the mole ratio from the balanced equation.

Solution: Because the balanced equation is given in the problem, the mole ratio between Cl_2 and $SiCl_4$ is known: 2 moles $Cl_2 \simeq$ 1 mole $SiCl_4$. From this relationship, we have two conversion factors.

$$\frac{2 \text{ mol } Cl_2}{1 \text{ mol } SiCl_4} \quad \text{and} \quad \frac{1 \text{ mol } SiCl_4}{2 \text{ mol } Cl_2}$$

Which conversion factor is needed to convert from moles of $SiCl_4$ to moles of Cl_2? The conversion factor on the left is the correct one. Moles of $SiCl_4$ will cancel, leaving units of "mol Cl_2" for the answer. We calculate moles of Cl_2 reacted as follows:

$$\textbf{? mol Cl}_2\textbf{ reacted} = 0.507 \text{ mol SiCl}_4 \times \frac{2 \text{ mol Cl}_2}{1 \text{ mol SiCl}_4} = \textbf{1.01 mol Cl}_2$$

Check: Does the answer seem reasonable? Should the moles of Cl_2 reacted be *double* the moles of $SiCl_4$ produced?

3.68 Starting with the 5.0 moles of C_4H_{10}, we can use the mole ratio from the balanced equation to calculate the moles of CO_2 formed.

$2C_4H_{10}(g) + 13O_2(g) \rightarrow 8CO_2(g) + 10H_2O(l)$

$$? \text{ mol CO}_2 = 5.0 \text{ mol C}_4\text{H}_{10} \times \frac{8 \text{ mol CO}_2}{2 \text{ mol C}_4\text{H}_{10}} = 20 \text{ mol CO}_2 = \mathbf{2.0 \times 10^1 \textbf{ mol CO}_2}$$

3.70 **(a)** $2NaHCO_3 \longrightarrow Na_2CO_3 + H_2O + CO_2$

(b) Molar mass $NaHCO_3$ = 22.99 g + 1.008 g + 12.01 g + 3(16.00 g) = 84.01 g
Molar mass CO_2 = 12.01 g + 2(16.00 g) = 44.01 g

The balanced equation shows one mole of CO_2 formed from two moles of $NaHCO_3$.

$$\textbf{mass NaHCO}_3 = 20.5 \text{ g CO}_2 \times \frac{1 \text{ mol CO}_2}{44.01 \text{ g CO}_2} \times \frac{2 \text{ mol NaHCO}_3}{1 \text{ mol CO}_2} \times \frac{84.01 \text{ g NaHCO}_3}{1 \text{ mol NaHCO}_3}$$

$$= \textbf{78.3 g NaHCO}_3$$

3.72 Calculating the Amounts of Reactants and Products, Problem Type 7.

$C_6H_{12}O_6 \longrightarrow 2C_2H_5OH + 2CO_2$
glucose ethanol

Strategy: We compare glucose and ethanol based on the *mole ratio* in the balanced equation. Before we can determine moles of ethanol produced, we need to convert to moles of glucose. What conversion factor is needed to convert from grams of glucose to moles of glucose? Once moles of ethanol are obtained, another conversion factor is needed to convert from moles of ethanol to grams of ethanol.

Solution: The molar mass of glucose will allow us to convert from grams of glucose to moles of glucose. The molar mass of glucose = 6(12.01 g) + 12(1.008 g) + 6(16.00 g) = 180.16 g. The balanced equation is given, so the mole ratio between glucose and ethanol is known; that is 1 mole glucose $\simeq$ 2 moles ethanol. Finally, the molar mass of ethanol will convert moles of ethanol to grams of ethanol. This sequence of three conversions is summarized as follows:

grams of glucose → moles of glucose → moles of ethanol → grams of ethanol

$$\textbf{? g C}_2\textbf{H}_5\textbf{OH} = 500.4 \text{ g C}_6\text{H}_{12}\text{O}_6 \times \frac{1 \text{ mol C}_6\text{H}_{12}\text{O}_6}{180.16 \text{ g C}_6\text{H}_{12}\text{O}_6} \times \frac{2 \text{ mol C}_2\text{H}_5\text{OH}}{1 \text{ mol C}_6\text{H}_{12}\text{O}_6} \times \frac{46.07 \text{ g C}_2\text{H}_5\text{OH}}{1 \text{ mol C}_2\text{H}_5\text{OH}}$$

$$= \textbf{255.9 g C}_2\textbf{H}_5\textbf{OH}$$

Check: Does the answer seem reasonable? Should the mass of ethanol produced be approximately half the mass of glucose reacted? Twice as many moles of ethanol are produced compared to the moles of glucose reacted, but the molar mass of ethanol is about one-fourth that of glucose.

The liters of ethanol can be calculated from the density and the mass of ethanol.

$$\text{volume} = \frac{\text{mass}}{\text{density}}$$

$$\textbf{Volume of ethanol obtained} = \frac{255.9\ \cancel{g}}{0.789\ \cancel{g}/\text{mL}} = 324\ \text{mL} = \textbf{0.324 L}$$

3.74 The balanced equation shows that eight moles of KCN are needed to combine with four moles of Au.

$$\textbf{? mol KCN} = 29.0\ \cancel{g}\ \text{Au} \times \frac{1\ \cancel{mol}\ \text{Au}}{197.0\ \cancel{g}\ \text{Au}} \times \frac{8\ \text{mol KCN}}{4\ \cancel{mol}\ \text{Au}} = \textbf{0.294 mol KCN}$$

3.76 **(a)** $NH_4NO_3(s) \longrightarrow N_2O(g) + 2H_2O(g)$

(b) Starting with moles of NH_4NO_3, we can use the mole ratio from the balanced equation to find moles of N_2O. Once we have moles of N_2O, we can use the molar mass of N_2O to convert to grams of N_2O. Combining the two conversions into one calculation, we have:

$$\text{mol } NH_4NO_3 \rightarrow \text{mol } N_2O \rightarrow \text{g } N_2O$$

$$\textbf{? g } \mathbf{N_2O} = 0.46\ \cancel{mol}\ NH_4NO_3 \times \frac{1\ \cancel{mol}\ N_2O}{1\ \cancel{mol}\ NH_4NO_3} \times \frac{44.02\ \text{g}\ N_2O}{1\ \cancel{mol}\ N_2O} = \mathbf{2.0 \times 10^1\ g\ N_2O}$$

3.78 The balanced equation for the decomposition is :

$$2KClO_3(s) \longrightarrow 2KCl(s) + 3O_2(g)$$

$$\textbf{? g } \mathbf{O_2} = 46.0\ \cancel{g}\ KClO_3 \times \frac{1\ \cancel{mol}\ KClO_3}{122.55\ \cancel{g}\ KClO_3} \times \frac{3\ \cancel{mol}\ O_2}{2\ \cancel{mol}\ KClO_3} \times \frac{32.00\ \text{g}\ O_2}{1\ \cancel{mol}\ O_2} = \mathbf{18.0\ g\ O_2}$$

3.82 $N_2 + 3H_2 \rightarrow 2NH_3$

(a) The number of N_2 molecules shown in the diagram is 3. The balanced equation shows 3 moles $H_2 \simeq 1$ mole N_2. Therefore, we need 9 molecules of H_2 to react completely with 3 molecules of N_2. There are 10 molecules of H_2 present in the diagram. H_2 is in excess.

N_2 is the limiting reagent.

(b) 9 molecules of H_2 will react with 3 molecules of N_2, leaving 1 molecule of H_2 in excess. The mole ratio between N_2 and NH_3 is 1:2. When 3 molecules of N_2 react, 6 molecules of NH_3 will be produced.

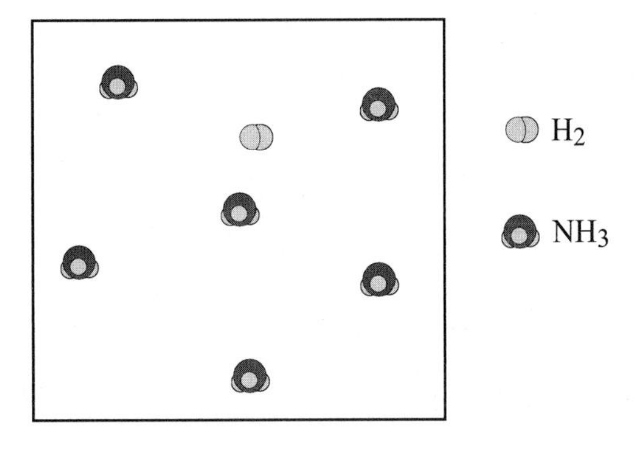

3.84 Limiting Reagent Calculation, Problem Type 8.

Strategy: Note that this reaction gives the amounts of both reactants, so it is likely to be a limiting reagent problem. The reactant that produces fewer moles of product is the limiting reagent because it limits the amount of product that can be produced. How do we convert from the amount of reactant to amount of product? Perform this calculation for each reactant, then compare the moles of product, NO_2, formed by the given amounts of O_3 and NO to determine which reactant is the limiting reagent.

Solution: We carry out two separate calculations. First, starting with 0.740 g O_3, we calculate the number of moles of NO_2 that could be produced if all the O_3 reacted. We complete the following conversions.

$$\text{grams of } O_3 \rightarrow \text{moles of } O_3 \rightarrow \text{moles of } NO_2$$

Combining these two conversions into one calculation, we write

$$? \text{ mol } NO_2 = 0.740 \text{ g } O_3 \times \frac{1 \text{ mol } O_3}{48.00 \text{ g } O_3} \times \frac{1 \text{ mol } NO_2}{1 \text{ mol } O_3} = 0.0154 \text{ mol } NO_2$$

Second, starting with 0.670 g of NO, we complete similar conversions.

$$\text{grams of NO} \rightarrow \text{moles of NO} \rightarrow \text{moles of } NO_2$$

Combining these two conversions into one calculation, we write

$$? \text{ mol } NO_2 = 0.670 \text{ g NO} \times \frac{1 \text{ mol NO}}{30.01 \text{ g NO}} \times \frac{1 \text{ mol } NO_2}{1 \text{ mol NO}} = 0.0223 \text{ mol } NO_2$$

The initial amount of $\mathbf{O_3}$ limits the amount of product that can be formed; therefore, it is the **limiting reagent**.

The problem asks for grams of NO_2 produced. We already know the moles of NO_2 produced, 0.0154 mole. Use the molar mass of NO_2 as a conversion factor to convert to grams (Molar mass NO_2 = 46.01 g).

$$\mathbf{? \text{ g } NO_2} = 0.0154 \text{ mol } NO_2 \times \frac{46.01 \text{ g } NO_2}{1 \text{ mol } NO_2} = \mathbf{0.709 \text{ g } NO_2}$$

Check: Does your answer seem reasonable? 0.0154 mole of product is formed. What is the mass of 1 mole of NO_2?

Strategy: Working backwards, we can determine the amount of NO that reacted to produce 0.0154 mole of NO_2. The amount of NO left over is the difference between the initial amount and the amount reacted.

Solution: Starting with 0.0154 mole of NO_2, we can determine the moles of NO that reacted using the mole ratio from the balanced equation. We can calculate the initial moles of NO starting with 0.670 g and using molar mass of NO as a conversion factor.

$$\text{mol NO reacted} = 0.0154 \text{ mol } NO_2 \times \frac{1 \text{ mol NO}}{1 \text{ mol } NO_2} = 0.0154 \text{ mol NO}$$

$$\text{mol NO initial} = 0.670 \text{ g NO} \times \frac{1 \text{ mol NO}}{30.01 \text{ g NO}} = 0.0223 \text{ mol NO}$$

$$\text{mol NO remaining} = \text{mol NO initial} - \text{mol NO reacted.}$$

$$\textbf{mol NO remaining} = 0.0223 \text{ mol NO} - 0.0154 \text{ mol NO} = \mathbf{0.0069 \text{ mol NO}}$$

3.86 This is a limiting reagent problem. Let's calculate the moles of Cl_2 produced assuming complete reaction for each reactant.

$$0.86 \text{ mol } MnO_2 \times \frac{1 \text{ mol } Cl_2}{1 \text{ mol } MnO_2} = 0.86 \text{ mol } Cl_2$$

$$48.2 \text{ g HCl} \times \frac{1 \text{ mol HCl}}{36.458 \text{ g HCl}} \times \frac{1 \text{ mol } Cl_2}{4 \text{ mol HCl}} = 0.330 \text{ mol } Cl_2$$

HCl is the limiting reagent; it limits the amount of product produced. It will be used up first. The amount of product produced is 0.330 mole Cl_2. Let's convert this to grams.

$$\textbf{? g Cl}_2 = 0.330 \text{ mol } Cl_2 \times \frac{70.90 \text{ g } Cl_2}{1 \text{ mol } Cl_2} = \textbf{23.4 g Cl}_2$$

3.90 **(a)** Start with a balanced chemical equation. It's given in the problem. We use NG as an abbreviation for nitroglycerin. The molar mass of NG = 227.1 g/mol.

$$4C_3H_5N_3O_9 \longrightarrow 6N_2 + 12CO_2 + 10H_2O + O_2$$

Map out the following strategy to solve this problem.

$$\text{g NG} \rightarrow \text{mol NG} \rightarrow \text{mol } O_2 \rightarrow \text{g } O_2$$

Calculate the grams of O_2 using the strategy above.

$$\textbf{? g O}_2 = 2.00 \times 10^2 \text{ g NG} \times \frac{1 \text{ mol NG}}{227.1 \text{ g NG}} \times \frac{1 \text{ mol } O_2}{4 \text{ mol NG}} \times \frac{32.00 \text{ g } O_2}{1 \text{ mol } O_2} = \textbf{7.05 g O}_2$$

(b) The theoretical yield was calculated in part (a), and the actual yield is given in the problem (6.55 g). The percent yield is:

$$\% \text{ yield} = \frac{\text{actual yield}}{\text{theoretical yield}} \times 100\%$$

$$\textbf{\% yield} = \frac{6.55 \text{ g } O_2}{7.05 \text{ g } O_2} \times 100\% = \textbf{92.9\%}$$

3.92 The actual yield of ethylene is 481 g. Let's calculate the yield of ethylene if the reaction is 100 percent efficient. We can calculate this from the definition of percent yield. We can then calculate the mass of hexane that must be reacted.

$$\% \text{ yield} = \frac{\text{actual yield}}{\text{theoretical yield}} \times 100\%$$

$$42.5\% \text{ yield} = \frac{481 \text{ g } C_2H_4}{\text{theoretical yield}} \times 100\%$$

$$\text{theoretical yield } C_2H_4 = 1.13 \times 10^3 \text{ g } C_2H_4$$

The mass of hexane that must be reacted is:

$$(1.13 \times 10^3 \text{ g } C_2H_4) \times \frac{1 \text{ mol } C_2H_4}{28.05 \text{ g } C_2H_4} \times \frac{1 \text{ mol } C_6H_{14}}{1 \text{ mol } C_2H_4} \times \frac{86.17 \text{ g } C_6H_{14}}{1 \text{ mol } C_6H_{14}} = \mathbf{3.47 \times 10^3 \text{ g } C_6H_{14}}$$

3.94 This is a limiting reagent problem. Let's calculate the moles of S_2Cl_2 produced assuming complete reaction for each reactant.

$S_8(l) + 4Cl_2(g) \rightarrow 4S_2Cl_2(l)$

$$4.06 \text{ g } S_8 \times \frac{1 \text{ mol } S_8}{256.6 \text{ g } S_8} \times \frac{4 \text{ mol } S_2Cl_2}{1 \text{ mol } S_8} = 0.0633 \text{ mol } S_2Cl_2$$

$$6.24 \text{ g } Cl_2 \times \frac{1 \text{ mol } Cl_2}{70.90 \text{ g } Cl_2} \times \frac{4 \text{ mol } S_2Cl_2}{4 \text{ mol } Cl_2} = 0.0880 \text{ mol } S_2Cl_2$$

S_8 is the limiting reagent; it limits the amount of product produced. The amount of product produced is 0.0633 mole S_2Cl_2. Let's convert this to grams.

$$? \text{ g } S_2Cl_2 = 0.0633 \text{ mol } S_2Cl_2 \times \frac{135.04 \text{ g } S_2Cl_2}{1 \text{ mol } S_2Cl_2} = \mathbf{8.55 \text{ g } S_2Cl_2}$$

This is the theoretical yield of S_2Cl_2. The actual yield is given in the problem (6.55 g). The percent yield is:

$$\% \text{ yield} = \frac{\text{actual yield}}{\text{theoretical yield}} \times 100\% = \frac{6.55 \text{ g}}{8.55 \text{ g}} \times 100\% = \mathbf{76.6\%}$$

3.96 $2H_2(g) + O_2(g) \rightarrow 2H_2O(g)$

We start with 8 molecules of H_2 and 3 molecules of O_2. The balanced equation shows 2 moles $H_2 \simeq$ 1 mole O_2. If 3 molecules of O_2 react, 6 molecules of H_2 will react, leaving 2 molecules of H_2 in excess. The balanced equation also shows 1 mole $O_2 \simeq$ 2 moles H_2O. If 3 molecules of O_2 react, 6 molecules of H_2O will be produced.

After complete reaction, there will be **2 molecules of H_2** and **6 molecules of H_2O**. The correct diagram is choice **(b)**.

3.98 We assume that all the Cl in the compound ends up as HCl and all the O ends up as H_2O. Therefore, we need to find the number of moles of Cl in HCl and the number of moles of O in H_2O.

$$\text{mol Cl} = 0.233 \text{ g HCl} \times \frac{1 \text{ mol HCl}}{36.46 \text{ g HCl}} \times \frac{1 \text{ mol Cl}}{1 \text{ mol HCl}} = 0.00639 \text{ mol Cl}$$

$$\text{mol O} = 0.403 \text{ g } H_2O \times \frac{1 \text{ mol } H_2O}{18.02 \text{ g } H_2O} \times \frac{1 \text{ mol O}}{1 \text{ mol } H_2O} = 0.0224 \text{ mol O}$$

Dividing by the smallest number of moles (0.00639 mole) gives the formula, $ClO_{3.5}$. Multiplying both subscripts by two gives the empirical formula, $\mathbf{Cl_2O_7}$.

3.100 The symbol "O" refers to moles of oxygen atoms, not oxygen molecule (O_2). Look at the molecular formulas given in parts (a) and (b). What do they tell you about the relative amounts of carbon and oxygen?

(a) $0.212 \text{ mol C} \times \frac{1 \text{ mol O}}{1 \text{ mol C}} = \mathbf{0.212 \text{ mol O}}$

(b) $0.212 \text{ mol C} \times \dfrac{2 \text{ mol O}}{1 \text{ mol C}} = \mathbf{0.424 \text{ mol O}}$

3.102 This is a calculation involving percent composition. Remember,

$$\text{percent by mass of each element} = \frac{\text{mass of element in 1 mol of compound}}{\text{molar mass of compound}} \times 100\%$$

The molar masses are: Al, 26.98 g/mol; $Al_2(SO_4)_3$, 342.17 g/mol; H_2O, 18.016 g/mol. Thus, using x as the number of H_2O molecules,

$$\text{mass \% Al} = \left(\frac{2(\text{molar mass of Al})}{\text{molar mass of } Al_2(SO_4)_3 + x(\text{molar mass of } H_2O)}\right) \times 100\%$$

$$8.20\% = \left(\frac{2(26.98 \text{ g})}{342.2 \text{ g} + x(18.02 \text{ g})}\right) \times 100\%$$

$$\mathbf{x = 17.53}$$

Rounding off to a whole number of water molecules, $\mathbf{x = 18}$. Therefore, the formula is $\mathbf{Al_2(SO_4)_3 \cdot 18\ H_2O}$.

3.104 The number of carbon atoms in a 24-carat diamond is:

$$24 \text{ carat} \times \frac{200 \text{ mg C}}{1 \text{ carat}} \times \frac{0.001 \text{ g C}}{1 \text{ mg C}} \times \frac{1 \text{ mol C}}{12.01 \text{ g C}} \times \frac{6.022 \times 10^{23} \text{ atoms C}}{1 \text{ mol C}} = \mathbf{2.4 \times 10^{23} \text{ atoms C}}$$

3.106 The mass of oxygen in MO is 39.46 g − 31.70 g = 7.76 g O. Therefore, for every 31.70 g of M, there is 7.76 g of O in the compound MO. The molecular formula shows a mole ratio of 1 mole M : 1 mole O. First, calculate moles of M that react with 7.76 g O.

$$\text{mol M} = 7.76 \text{ g O} \times \frac{1 \text{ mol O}}{16.00 \text{ g O}} \times \frac{1 \text{ mol M}}{1 \text{ mol O}} = 0.485 \text{ mol M}$$

$$\text{molar mass M} = \frac{31.70 \text{ g M}}{0.485 \text{ mol M}} = 65.4 \text{ g/mol}$$

Thus, the atomic mass of M is **65.4 amu**. The metal is most likely **Zn**.

3.108 The wording of the problem suggests that the actual yield is less than the theoretical yield. The percent yield will be equal to the percent purity of the iron(III) oxide. We find the theoretical yield :

$$(2.62 \times 10^3 \text{ kg } Fe_2O_3) \times \frac{1000 \text{ g } Fe_2O_3}{1 \text{ kg } Fe_2O_3} \times \frac{1 \text{ mol } Fe_2O_3}{159.7 \text{ g } Fe_2O_3} \times \frac{2 \text{ mol Fe}}{1 \text{ mol } Fe_2O_3} \times \frac{55.85 \text{ g Fe}}{1 \text{ mol Fe}} \times \frac{1 \text{ kg Fe}}{1000 \text{ g Fe}}$$

$$= 1.83 \times 10^3 \text{ kg Fe}$$

$$\text{percent yield} = \frac{\text{actual yield}}{\text{theoretical yield}} \times 100\%$$

$$\textbf{percent yield} = \frac{1.64 \times 10^3 \text{ kg Fe}}{1.83 \times 10^3 \text{ kg Fe}} \times 100\% = \mathbf{89.6\%} = \textbf{purity of } \mathbf{Fe_2O_3}$$

3.110 The carbohydrate contains 40 percent carbon; therefore, the remaining 60 percent is hydrogen and oxygen. The problem states that the hydrogen to oxygen ratio is 2:1. We can write this 2:1 ratio as H_2O.

Assume 100 g of compound.

$$40.0\ \text{g C} \times \frac{1\ \text{mol C}}{12.01\ \text{g C}} = 3.33\ \text{mol C}$$

$$60.0\ \text{g H}_2\text{O} \times \frac{1\ \text{mol H}_2\text{O}}{18.016\ \text{g H}_2\text{O}} = 3.33\ \text{mol H}_2\text{O}$$

Dividing by 3.33 gives **CH_2O** for the empirical formula.

To find the molecular formula, divide the molar mass by the empirical mass.

$$\frac{\text{molar mass}}{\text{empirical mass}} = \frac{178\ \text{g}}{30.03\ \text{g}} \approx 6$$

Thus, there are six CH_2O units in each molecule of the compound, so the molecular formula is $(CH_2O)_6$, or **$C_6H_{12}O_6$**.

3.112 If we assume 100 g of compound, the masses of Cl and X are 67.2 g and 32.8 g, respectively. We can calculate the moles of Cl.

$$67.2\ \text{g Cl} \times \frac{1\ \text{mol Cl}}{35.45\ \text{g Cl}} = 1.90\ \text{mol Cl}$$

Then, using the mole ratio from the chemical formula (XCl_3), we can calculate the moles of X contained in 32.8 g.

$$1.90\ \text{mol Cl} \times \frac{1\ \text{mol X}}{3\ \text{mol Cl}} = 0.633\ \text{mol X}$$

0.633 mole of X has a mass of 32.8 g. Calculating the molar mass of X:

$$\frac{32.8\ \text{g X}}{0.633\ \text{mol X}} = \mathbf{51.8\ g/mol}$$

The element is most likely **chromium** (molar mass = 52.00 g/mol).

3.114 A 100 g sample of myoglobin contains 0.34 g of iron (0.34% Fe). The number of moles of Fe is:

$$0.34\ \text{g Fe} \times \frac{1\ \text{mol Fe}}{55.85\ \text{g Fe}} = 6.1 \times 10^{-3}\ \text{mol Fe}$$

Since there is one Fe atom in a molecule of myoglobin, the moles of myoglobin also equal 6.1×10^{-3} mole. The molar mass of myoglobin can be calculated.

$$\textbf{molar mass myoglobin} = \frac{100\ \text{g myoglobin}}{6.1 \times 10^{-3}\ \text{mol myoglobin}} = \mathbf{1.6 \times 10^4\ g/mol}$$

3.116 If we assume 100 g of the mixture, then there are 29.96 g of Na in the mixture (29.96% Na by mass). This amount of Na is equal to the mass of Na in NaBr plus the mass of Na in Na_2SO_4.

$$29.96\ \text{g Na} = \text{mass of Na in NaBr} + \text{mass of Na in Na}_2\text{SO}_4$$

To calculate the mass of Na in each compound, grams of compound need to be converted to grams of Na using the mass percentage of Na in the compound. If x equals the mass of NaBr, then the mass of Na_2SO_4 is $100 - x$. Recall that we assumed 100 g of the mixture. We set up the following expression and solve for x.

$$29.96 \text{ g Na} = \text{mass of Na in NaBr} + \text{mass of Na in } Na_2SO_4$$

$$29.96 \text{ g Na} = \left[x \text{ g NaBr} \times \frac{22.99 \text{ g Na}}{102.89 \text{ g NaBr}}\right] + \left[(100 - x) \text{ g } Na_2SO_4 \times \frac{(2)(22.99 \text{ g Na})}{142.05 \text{ g } Na_2SO_4}\right]$$

$$29.96 = 0.2234x + 32.37 - 0.3237x$$

$$0.1003x = 2.41$$

$x = 24.0$ g, which equals the mass of NaBr.

The mass of Na_2SO_4 is $100 - x$ which equals 76.0 g.

Because we assumed 100 g of compound, the mass % of NaBr in the mixture is **24.0%** and the mass % of Na_2SO_4 is **76.0%**.

3.118 The mass percent of an element in a compound can be calculated as follows:

$$\text{percent by mass of each element} = \frac{\text{mass of element in 1 mol of compound}}{\text{molar mass of compound}} \times 100\%$$

The molar mass of $Ca_3(PO_4)_2$ = 310.18 g/mol

$$\textbf{\% Ca} = \frac{(3)(40.08 \text{ g})}{310.2 \text{ g}} \times 100\% = \textbf{38.76\% Ca}$$

$$\textbf{\% P} = \frac{(2)(30.97 \text{ g})}{310.2 \text{ g}} \times 100\% = \textbf{19.97\% P}$$

$$\textbf{\% O} = \frac{(8)(16.00 \text{ g})}{310.2 \text{ g}} \times 100\% = \textbf{41.26\% O}$$

3.120 **Yes**. The number of hydrogen atoms in one gram of hydrogen molecules is the same as the number in one gram of hydrogen atoms. There is no difference in mass, only in the way that the particles are arranged. Would the mass of 100 dimes be the same if they were stuck together in pairs instead of separated?

3.122 Since we assume that water exists as either H_2O or D_2O, the natural abundances are 99.985 percent and 0.015 percent, respectively. If we convert to molecules of water (both H_2O or D_2O), we can calculate the molecules that are D_2O from the natural abundance (0.015%).

The necessary conversions are:

mL water $\rightarrow$ g water $\rightarrow$ mol water $\rightarrow$ molecules water $\rightarrow$ molecules D_2O

$$400 \text{ mL water} \times \frac{1 \text{ g water}}{1 \text{ mL water}} \times \frac{1 \text{ mol water}}{18.02 \text{ g water}} \times \frac{6.022 \times 10^{23} \text{ molecules}}{1 \text{ mol water}} \times \frac{0.015\% \text{ molecules } D_2O}{100\% \text{ molecules water}}$$

$$= \mathbf{2.01 \times 10^{21} \textbf{ molecules } D_2O}$$

3.124 First, we can calculate the moles of oxygen.

$$2.445\ \text{g C} \times \frac{1\ \text{mol C}}{12.01\ \text{g C}} \times \frac{1\ \text{mol O}}{1\ \text{mol C}} = 0.2036\ \text{mol O}$$

Next, we can calculate the molar mass of oxygen.

$$\text{molar mass O} = \frac{3.257\ \text{g O}}{0.2036\ \text{mol O}} = 16.00\ \text{g/mol}$$

If 1 mole of oxygen atoms has a mass of 16.00 g, then 1 atom of oxygen has an **atomic mass of 16.00 amu**.

3.126 **(a)** The mass of chlorine is **5.0 g**.

(b) From the percent by mass of Cl, we can calculate the mass of chlorine in 60.0 g of $NaClO_3$.

$$\text{mass \% Cl} = \frac{35.45\ \text{g Cl}}{106.44\ \text{g compound}} \times 100\% = 33.31\%\ \text{Cl}$$

$$\textbf{mass Cl} = 60.0\ \text{g} \times 0.3331 = \textbf{20.0 g Cl}$$

(c) 0.10 mol of KCl contains 0.10 mol of Cl.

$$0.10\ \text{mol Cl} \times \frac{35.45\ \text{g Cl}}{1\ \text{mol Cl}} = \textbf{3.5 g Cl}$$

(d) From the percent by mass of Cl, we can calculate the mass of chlorine in 30.0 g of $MgCl_2$.

$$\text{mass \% Cl} = \frac{(2)(35.45\ \text{g Cl})}{95.21\ \text{g compound}} \times 100\% = 74.47\%\ \text{Cl1}$$

$$\textbf{mass Cl} = 30.0\ \text{g} \times 0.7447 = \textbf{22.3 g Cl}$$

(e) The mass of Cl can be calculated from the molar mass of Cl_2.

$$0.50\ \text{mol Cl}_2 \times \frac{70.90\ \text{g Cl}}{1\ \text{mol Cl}_2} = \textbf{35.45 g Cl}$$

Thus, **(e) 0.50 mol Cl_2** contains the greatest mass of chlorine.

3.128 Both compounds contain only Pt and Cl. The percent by mass of Pt can be calculated by subtracting the percent Cl from 100 percent.

Compound A: Assume 100 g of compound.

$$26.7\ \text{g Cl} \times \frac{1\ \text{mol Cl}}{35.45\ \text{g Cl}} = 0.753\ \text{mol Cl}$$

$$73.3\ \text{g Pt} \times \frac{1\ \text{mol Pt}}{195.1\ \text{g Pt}} = 0.376\ \text{mol Pt}$$

Dividing by the smallest number of moles (0.376 mole) gives the empirical formula, **$PtCl_2$**.

Compound B: Assume 100 g of compound.

$$42.1\ \text{g Cl} \times \frac{1\ \text{mol Cl}}{35.45\ \text{g Cl}} = 1.19\ \text{mol Cl}$$

$$57.9\ \text{g Pt} \times \frac{1\ \text{mol Pt}}{195.1\ \text{g Pt}} = 0.297\ \text{mol Pt}$$

Dividing by the smallest number of moles (0.297 mole) gives the empirical formula, $\mathbf{PtCl_4}$.

3.130 Both compounds contain only Mn and O. When the first compound is heated, oxygen gas is evolved. Let's calculate the empirical formulas for the two compounds, then we can write a balanced equation.

(a) Compound X: Assume 100 g of compound.

$$63.3\ \text{g Mn} \times \frac{1\ \text{mol Mn}}{54.94\ \text{g Mn}} = 1.15\ \text{mol Mn}$$

$$36.7\ \text{g O} \times \frac{1\ \text{mol O}}{16.00\ \text{g O}} = 2.29\ \text{mol O}$$

Dividing by the smallest number of moles (1.15 moles) gives the empirical formula, $\mathbf{MnO_2}$.

Compound Y: Assume 100 g of compound.

$$72.0\ \text{g Mn} \times \frac{1\ \text{mol Mn}}{54.94\ \text{g Mn}} = 1.31\ \text{mol Mn}$$

$$28.0\ \text{g O} \times \frac{1\ \text{mol O}}{16.00\ \text{g O}} = 1.75\ \text{mol O}$$

Dividing by the smallest number of moles gives $MnO_{1.33}$. Recall that an empirical formula must have whole number coefficients. Multiplying by a factor of 3 gives the empirical formula $\mathbf{Mn_3O_4}$.

(b) The unbalanced equation is: $MnO_2 \longrightarrow Mn_3O_4 + O_2$

Balancing by inspection gives: $\mathbf{3MnO_2 \longrightarrow Mn_3O_4 + O_2}$

3.132 SO_2 is converted to H_2SO_4 by reaction with water. The mole ratio between SO_2 and H_2SO_4 is 1:1.

This is a unit conversion problem. You should come up with the following strategy to solve the problem.

tons SO_2 → ton-mol SO_2 → ton-mol H_2SO_4 → tons H_2SO_4

$$\textbf{? tons } \mathbf{H_2SO_4} = (4.0 \times 10^5\ \text{tons SO}_2) \times \frac{1\ \text{ton-mol SO}_2}{64.07\ \text{tons SO}_2} \times \frac{1\ \text{ton-mol H}_2\text{SO}_4}{1\ \text{ton-mol SO}_2} \times \frac{98.09\ \text{tons H}_2\text{SO}_4}{1\ \text{ton-mol H}_2\text{SO}_4}$$

$$= \mathbf{6.1 \times 10^5\ tons\ H_2SO_4}$$

Tip: You probably won't come across a ton-mol that often in chemistry. However, it was convenient to use in this problem. We normally use a g-mol. 1 g-mol SO_2 has a mass of 64.07 g. In a similar manner, 1 ton-mol of SO_2 has a mass of 64.07 tons.

3.134 We assume that the increase in mass results from the element nitrogen. The mass of nitrogen is:

$$0.378\ \text{g} - 0.273\ \text{g} = 0.105\ \text{g N}$$

The empirical formula can now be calculated. Convert to moles of each element.

$$0.273\ \text{g Mg} \times \frac{1\ \text{mol Mg}}{24.31\ \text{g Mg}} = 0.0112\ \text{mol Mg}$$

$$0.105\ \text{g N} \times \frac{1\ \text{mol N}}{14.01\ \text{g N}} = 0.00749\ \text{mol N}$$

Dividing by the smallest number of moles gives $Mg_{1.5}N$. Recall that an empirical formula must have whole number coefficients. Multiplying by a factor of 2 gives the empirical formula $\mathbf{Mg_3N_2}$. The name of this compound is **magnesium nitride**.

3.136 *Step 1:* Calculate the mass of C in 55.90 g CO_2, and the mass of H in 28.61 g H_2O. This is a dimensional analysis problem. To calculate the mass of each component, you need the molar masses and the correct mole ratio.

You should come up with the following strategy:

$$\text{g } CO_2 \rightarrow \text{mol } CO_2 \rightarrow \text{mol C} \rightarrow \text{g C}$$

Step 2:

$$?\ \text{g C} = 55.90\ \text{g } CO_2 \times \frac{1\ \text{mol } CO_2}{44.01\ \text{g } CO_2} \times \frac{1\ \text{mol C}}{1\ \text{mol } CO_2} \times \frac{12.01\ \text{g C}}{1\ \text{mol C}} = 15.25\ \text{g C}$$

Similarly,

$$?\ \text{g H} = 28.61\ \text{g } H_2O \times \frac{1\ \text{mol } H_2O}{18.02\ \text{g } H_2O} \times \frac{2\ \text{mol H}}{1\ \text{mol } H_2O} \times \frac{1.008\ \text{g H}}{1\ \text{mol H}} = 3.201\ \text{g H}$$

Since the compound contains C, H, and Pb, we can calculate the mass of Pb by difference.

$$51.36\ \text{g} = \text{mass C} + \text{mass H} + \text{mass Pb}$$

$$51.36\ \text{g} = 15.25\ \text{g} + 3.201\ \text{g} + \text{mass Pb}$$

$$\text{mass Pb} = 32.91\ \text{g Pb}$$

Step 3: Calculate the number of moles of each element present in the sample. Use molar mass as a conversion factor.

$$?\ \text{mol C} = 15.25\ \text{g C} \times \frac{1\ \text{mol C}}{12.01\ \text{g C}} = 1.270\ \text{mol C}$$

Similarly,

$$?\ \text{mol H} = 3.201\ \text{g H} \times \frac{1\ \text{mol H}}{1.008\ \text{g H}} = 3.176\ \text{mol H}$$

$$?\ \text{mol Pb} = 32.91\ \text{g Pb} \times \frac{1\ \text{mol Pb}}{207.2\ \text{g Pb}} = 0.1588\ \text{mol Pb}$$

Thus, we arrive at the formula $Pb_{0.1588}C_{1.270}H_{3.176}$, which gives the identity and the ratios of atoms present. However, chemical formulas are written with whole numbers.

Step 4: Try to convert to whole numbers by dividing all the subscripts by the smallest subscript.

$$\text{Pb: } \frac{0.1588}{0.1588} = 1.00 \qquad \text{C: } \frac{1.270}{0.1588} \approx 8 \qquad \text{H: } \frac{3.176}{0.1588} \approx 20$$

This gives the empirical formula, $\mathbf{PbC_8H_{20}}$.

3.138 **(a)** The following strategy can be used to convert from the volume of the Mg cube to the number of Mg atoms.

$$\text{cm}^3 \rightarrow \text{grams} \rightarrow \text{moles} \rightarrow \text{atoms}$$

$$1.0\ \text{cm}^3 \times \frac{1.74\ \text{g Mg}}{1\ \text{cm}^3} \times \frac{1\ \text{mol Mg}}{24.31\ \text{g Mg}} \times \frac{6.022 \times 10^{23}\ \text{Mg atoms}}{1\ \text{mol Mg}} = \mathbf{4.3 \times 10^{22}\ Mg\ atoms}$$

(b) Since 74 percent of the available space is taken up by Mg atoms, 4.3×10^{22} atoms occupy the following volume:

$$0.74 \times 1.0\ \text{cm}^3 = 0.74\ \text{cm}^3$$

We are trying to calculate the radius of a single Mg atom, so we need the volume occupied by a single Mg atom.

$$\text{volume Mg atom} = \frac{0.74\ \text{cm}^3}{4.3 \times 10^{22}\ \text{Mg atoms}} = 1.7 \times 10^{-23}\ \text{cm}^3/\text{Mg atom}$$

The volume of a sphere is $\frac{4}{3}\pi r^3$. Solving for the radius:

$$V = 1.7 \times 10^{-23}\ \text{cm}^3 = \frac{4}{3}\pi r^3$$

$$r^3 = 4.1 \times 10^{-24}\ \text{cm}^3$$

$$r = 1.6 \times 10^{-8}\ \text{cm}$$

Converting to picometers:

$$\textbf{radius Mg atom} = (1.6 \times 10^{-8}\ \text{cm}) \times \frac{0.01\ \text{m}}{1\ \text{cm}} \times \frac{1\ \text{pm}}{1 \times 10^{-12}\ \text{m}} = \mathbf{1.6 \times 10^2\ pm}$$

3.140 The molar mass of air can be calculated by multiplying the mass of each component by its abundance and adding them together. Recall that nitrogen gas and oxygen gas are diatomic.

$$\textbf{molar mass air} = (0.7808)(28.02\ \text{g/mol}) + (0.2095)(32.00\ \text{g/mol}) + (0.0097)(39.95\ \text{g/mol}) = \mathbf{28.97\ g/mol}$$

3.142 The surface area of the water can be calculated assuming that the dish is circular.

$$\text{surface area of water} = \pi r^2 = \pi(10\ \text{cm})^2 = 3.1 \times 10^2\ \text{cm}^2$$

The cross-sectional area of one stearic acid molecule in cm^2 is:

$$0.21\ \text{nm}^2 \times \left(\frac{1 \times 10^{-9}\ \text{m}}{1\ \text{nm}}\right)^2 \times \left(\frac{1\ \text{cm}}{0.01\ \text{m}}\right)^2 = 2.1 \times 10^{-15}\ \text{cm}^2/\text{molecule}$$

Assuming that there is no empty space between molecules, we can calculate the number of stearic acid molecules that will fit in an area of 3.1×10^2 cm^2.

$$(3.1 \times 10^2 \text{ cm}^2) \times \frac{1 \text{ molecule}}{2.1 \times 10^{-15} \text{ cm}^2} = 1.5 \times 10^{17} \text{ molecules}$$

Next, we can calculate the moles of stearic acid in the 1.4×10^{-4} g sample. Then, we can calculate Avogadro's number (the number of molecules per mole).

$$1.4 \times 10^{-4} \text{ g stearic acid} \times \frac{1 \text{ mol stearic acid}}{284.5 \text{ g stearic acid}} = 4.9 \times 10^{-7} \text{ mol stearic acid}$$

$$\textbf{Avogadro's number } (N_A) = \frac{1.5 \times 10^{17} \text{ molecules}}{4.9 \times 10^{-7} \text{ mol}} = \mathbf{3.1 \times 10^{23} \textbf{ molecules/mol}}$$

3.144 **(a)** The balanced chemical equation is:

$$C_3H_8(g) + 3H_2O(g) \longrightarrow 3CO(g) + 7H_2(g)$$

(b) You should come up with the following strategy to solve this problem. In this problem, we use kg-mol to save a couple of steps.

$$\text{kg } C_3H_8 \rightarrow \text{mol } C_3H_8 \rightarrow \text{mol } H_2 \rightarrow \text{kg } H_2$$

$$\textbf{? kg } \mathbf{H_2} = (2.84 \times 10^3 \text{ kg } C_3H_8) \times \frac{1 \text{ kg-mol } C_3H_8}{44.09 \text{ kg } C_3H_8} \times \frac{7 \text{ kg-mol } H_2}{1 \text{ kg-mol } C_3H_8} \times \frac{2.016 \text{ kg } H_2}{1 \text{ kg-mol } H_2}$$

$$= \mathbf{9.09 \times 10^2 \textbf{ kg } H_2}$$

3.146 **(a)** 16 amu, CH_4 17 amu, NH_3 18 amu, H_2O 64 amu, SO_2

(b) The formula C_3H_8 can also be written as $CH_3CH_2CH_3$. A CH_3 fragment could break off from this molecule giving a peak at 15 amu. No fragment of CO_2 can have a mass of 15 amu. Therefore, the substance responsible for the mass spectrum is most likely C_3H_8.

(c) First, let's calculate the masses of CO_2 and C_3H_8.

$$\text{molecular mass } CO_2 = 12.00000 \text{ amu} + 2(15.99491 \text{ amu}) = 43.98982 \text{ amu}$$

$$\text{molecular mass } C_3H_8 = 3(12.00000 \text{ amu}) + 8(1.00797 \text{ amu}) = 44.06376 \text{ amu}$$

These masses differ by only 0.07394 amu. The measurements must be precise to **±0.030 amu.**

$$43.98982 + 0.030 \text{ amu} = 44.02 \text{ amu}$$

$$44.06376 - 0.030 \text{ amu} = 44.03 \text{ amu}$$

3.148 When magnesium burns in air, magnesium oxide (MgO) and magnesium nitride (Mg_3N_2) are produced. Magnesium nitride reacts with water to produce ammonia gas.

$$Mg_3N_2(s) + 6H_2O(l) \rightarrow 3Mg(OH)_2(s) + 2NH_3(g)$$

From the amount of ammonia produced, we can calculate the mass of Mg_3N_2 produced. The mass of Mg in that amount of Mg_3N_2 can be determined, and then the mass of Mg in MgO can be determined by difference. Finally, the mass of MgO can be calculated.

$$2.813\ \text{g NH}_3 \times \frac{1\ \text{mol NH}_3}{17.03\ \text{g NH}_3} \times \frac{1\ \text{mol Mg}_3\text{N}_2}{2\ \text{mol NH}_3} \times \frac{100.95\ \text{g Mg}_3\text{N}_2}{1\ \text{mol Mg}_3\text{N}_2} = \mathbf{8.337\ g\ Mg_3N_2}$$

The mass of Mg in 8.337 g Mg_3N_2 can be determined from the mass percentage of Mg in Mg_3N_2.

$$\frac{(3)(24.31\ \text{g Mg})}{100.95\ \text{g Mg}_3\text{N}_2} \times 8.337\ \text{g Mg}_3\text{N}_2 = 6.023\ \text{g Mg}$$

The mass of Mg in the product MgO is obtained by difference: 21.496 g Mg − 6.023 g Mg = 15.473 g Mg

The mass of MgO produced can now be determined from this mass of Mg and the mass percentage of Mg in MgO.

$$\frac{40.31\ \text{g MgO}}{24.31\ \text{g Mg}} \times 15.473\ \text{g Mg} = \mathbf{25.66\ g\ MgO}$$

3.150 The decomposition of $KClO_3$ produces oxygen gas (O_2) which reacts with Fe to produce Fe_2O_3.

$$4\text{Fe} + 3\text{O}_2 \rightarrow 2\text{Fe}_2\text{O}_3$$

When the 15.0 g of Fe is heated in the presence of O_2 gas, any increase in mass is due to oxygen. The mass of oxygen reacted is:

$$17.9\ \text{g} - 15.0\ \text{g} = 2.9\ \text{g O}_2$$

From this mass of O_2, we can now calculate the mass of Fe_2O_3 produced and the mass of $KClO_3$ decomposed.

$$2.9\ \text{g O}_2 \times \frac{1\ \text{mol O}_2}{32.00\ \text{g O}_2} \times \frac{2\ \text{mol Fe}_2\text{O}_3}{3\ \text{mol O}_2} \times \frac{159.7\ \text{g Fe}_2\text{O}_3}{1\ \text{mol Fe}_2\text{O}_3} = \mathbf{9.6\ g\ Fe_2O_3}$$

The balanced equation for the decomposition of $KClO_3$ is: $2KClO_3 \rightarrow 2KCl + 3O_2$. The mass of $KClO_3$ decomposed is:

$$2.9\ \text{g O}_2 \times \frac{1\ \text{mol O}_2}{32.00\ \text{g O}_2} \times \frac{2\ \text{mol KClO}_3}{3\ \text{mol O}_2} \times \frac{122.55\ \text{g KClO}_3}{1\ \text{mol KClO}_3} = \mathbf{7.4\ g\ KClO_3}$$

CHAPTER 4
REACTIONS IN AQUEOUS SOLUTIONS

PROBLEM-SOLVING STRATEGIES AND TUTORIAL SOLUTIONS

TYPES OF PROBLEMS

Problem Type 1: Applying Solubility Rules.

Problem Type 2: Writing Molecular, Ionic, and Net Ionic Equations.

Problem Type 3: Acid-Base Reactions.
(a) Identifying Brønsted acids and bases.
(b) Writing acid/base reactions.

Problem Type 4: Oxidation-Reduction Reactions.
(a) Assigning oxidation numbers.
(b) Writing oxidation/reduction half-reactions.
(c) Using an activity series.

Problem Type 5: Concentration of Solutions.

Problem Type 6: Dilution of Solutions.

Problem Type 7: Gravimetric Analysis.

Problem Type 8: Acid-Base Titrations.

Problem Type 9: Redox Titrations.

PROBLEM TYPE 1: APPLYING SOLUBILITY RULES

Ionic compounds are classified as "soluble", "slightly soluble", or "insoluble". Table 4.2 of your text provides solubility rules that will help you determine how a given compound behaves in aqueous solution.

EXAMPLE 4.1

According to the solubility rules, which of the following compounds are soluble in water?

(a) $MgCO_3$ (b) $AgNO_3$ (c) $MgCl_2$ (d) $Ca_3(PO_4)_2$ (e) KOH

Strategy: Although it is not necessary to memorize the solubilities of compounds, you should keep in mind the following useful rules: all ionic compounds containing alkali metal cations, the ammonium ion, and the nitrate, bicarbonate, and chlorate ions are soluble. For other compounds, refer to Table 4.2 of the text.

Solution:

(a) $MgCO_3$ is *insoluble* (Most ionic compounds containing carbonate ions are *insoluble*).
(b) $AgNO_3$ is *soluble* (All ionic compounds containing nitrate ions are *soluble*).
(c) $MgCl_2$ is *soluble* (Most ionic compounds containing chloride ions are *soluble*).
(d) $Ca_3(PO_4)_2$ is *insoluble* (Most ionic compounds containing phosphate ions are *insoluble*).
(e) KOH is *soluble* (All ionic compounds containing alkali metal ions are *soluble*).

PRACTICE EXERCISE

1. Predict whether the following ionic compounds are soluble or insoluble in water.

(a) $NaNO_3$ **(b)** AgCl **(c)** $Ba(OH)_2$ **(d)** $CaCO_3$

Text Problems: 4.18, **4.20**, 4.24

PROBLEM TYPE 2: WRITING MOLECULAR, IONIC, AND NET IONIC EQUATIONS

In a *molecular equation*, the formulas are written as though all species existed as molecules or whole units. However, a molecular equation does not accurately describe what actually happens at the microscopic level. To better describe the reaction in solution, the equation should show the dissociation of dissolved ionic compounds into ions. An *ionic equation* shows dissolved ionic compounds in terms of their free ions. A *net ionic equation* shows only the species that actually take part in the reaction.

EXAMPLE 4.2

Write balanced molecular, ionic, and net ionic equations for the reaction that occurs when a $BaCl_2$ solution is mixed with a Na_2SO_4 solution.

Strategy: Recall that an *ionic equation* shows dissolved ionic compounds in terms of their free ions. A *net ionic equation* shows only the species that actually take part in the reaction. What happens when ionic compounds dissolve in water? What ions are formed from the dissociation of $BaCl_2$ and Na_2SO_4? What happens when the cations encounter the anions in solution?

Solution: In solution, $BaCl_2$ dissociates into Ba^{2+} and Cl^- ions and Na_2SO_4 dissociates into Na^+ and SO_4^{2-} ions. According to Table 4.2 of the text, barium ions (Ba^{2+}) and sulfate ions (SO_4^{2-}) will form an insoluble compound, barium sulfate ($BaSO_4$), while the other product, NaCl, is soluble and remains in solution. This is a precipitation reaction. The balanced molecular equation is:

$$BaCl_2(aq) + Na_2SO_4(aq) \longrightarrow BaSO_4(s) + 2NaCl(aq)$$

The *ionic equation* should show dissolved ionic compounds in terms of their free ions.

$$Ba^{2+}(aq) + 2Cl^-(aq) + 2Na^+(aq) + SO_4^{2-}(aq) \longrightarrow BaSO_4(s) + 2Na^+(aq) + 2Cl^-(aq)$$

As you write out the ionic equation above, you should notice that some ions (Na^+ and Cl^-) are not involved in the overall reaction. These ions are called *spectator ions*. Since the spectator ions appear on both sides of the equation and are unchanged in the chemical reaction, they can be canceled from both sides of the equation. A *net ionic equation* shows only the species that actually take part in the reaction.

Cancel the spectator ions to write the *net ionic equation*.

$$Ba^{2+}(aq) + SO_4^{2-}(aq) \longrightarrow BaSO_4(s)$$

Check: Note that because we balanced the molecular equation first, the net ionic equation is balanced as to the number of atoms on each side, and the number of positive and negative charges on the left-hand side of the equation is the same.

Tip: To help pick out the spectator ions, think about spectators at a sporting event. The spectators are at the stadium, watching the action, but they do *not* participate in the game.

PRACTICE EXERCISE

2. Write the balanced molecular, ionic, and net ionic equations for the following reaction:

$$CaCl_2(aq) + Na_2CO_3(aq) \longrightarrow$$

Text Problem: 4.22

PROBLEM TYPE 3: ACID-BASE REACTIONS

A. Identifying Brønsted acids and bases

A **Brønsted acid** is a proton donor, and a **Brønsted base** is a proton acceptor. To identify a Brønsted acid, you should look for a substance that contains hydrogen. The formula of inorganic acids will begin with H. For example, consider HCl (hydrochloric acid), HNO_2 (nitrous acid), and H_3PO_4 (phosphoric acid). Carboxylic acids contain the carboxyl group, –COOH. The hydrogen from the carboxyl group can be donated. Examples of carboxylic acids are CH_3COOH (acetic acid) and HCOOH (formic acid).

To identify a Brønsted base, you should look for soluble hydroxide salts. The hydroxide ion (OH^-) will accept a proton to form H_2O. Also look for weak bases, which are amines. Ammonia (NH_3) is an example. Finally, look for anions from acids. These negatives ions can accept a proton (H^+). Some examples are $H_2PO_4^{2-}$, NO_2^-, and HCO_3^-.

EXAMPLE 4.3

Identify each of the following species as a Brønsted acid, base, or both: (a) HNO_3, (b) $Ba(OH)_2$, (c) SO_4^{2-}, (d) $CH_3CH_2CH_2COOH$, (e) HPO_4^{2-}.

Strategy: What are the characteristics of a Brønsted acid? Does it contain at least an H atom? With the exception of ammonia, most Brønsted bases that you will encounter at this stage are anions or soluble hydroxide salts.

Solution:

(a) Brønsted acid. The formula of this compound starts with H; this indicates that it is probably an acid.

(b) Brønsted base. This is a soluble hydroxide salt.

(c) Brønsted base. This negative ion can accept a proton; therefore, it is a base.

(d) Brønsted acid. This is a carboxylic acid. It contains a carboxyl group, –COOH.

(e) Both a Brønsted acid and base. This ion has a proton (H^+) that it can donate. It also has a negative charge and therefore can accept a proton.

PRACTICE EXERCISE

3. Identify each of the following species as a Brønsted acid, base, or both.

(a) CH_3CH_2COOH **(b)** HF **(c)** KOH **(d)** HCO_3^-

Text Problem: 4.32

B. Writing Acid-Base Reactions

An acid-base reaction is called a **neutralization reaction**. The typical products of an acid-base reaction are a salt and water.

$$\text{acid} + \text{base} \longrightarrow \text{salt} + \text{water}$$

Let's consider a generic acid, HA, reacted with a generic base, MOH.

$$HA + MOH \longrightarrow MA(salt) + H_2O$$

The H^+ from the acid combines with OH^- from the base to produce water. The anion from the acid, A^-, combines with the metal cation from the base, M^+, to form the salt, MA.

EXAMPLE 4.4

Complete and balance the following equations and write the corresponding ionic and net ionic equations:

(a) $HBr(aq) + Ba(OH)_2(aq) \longrightarrow$

(b) $HCOOH(aq) + NaOH(aq) \longrightarrow$

Strategy: Recall that strong acids and strong bases are strong electrolytes. They are completely ionized in solution. An *ionic equation* will show strong acids and strong bases in terms of their free ions. Weak acids and weak bases are weak electrolytes. They only ionize to a small extent in solution. Weak acids and weak bases are shown as molecules in ionic and net ionic equations. A *net ionic equation* shows only the species that actually take part in the reaction.

(a)

Solution: HBr is a strong acid. It completely ionizes to H^+ and Br^- ions. $Ba(OH)_2$ is a strong base. It completely ionizes to Ba^{2+} and OH^- ions. Since HBr is an acid, it donates an H^+ to the base, OH^-, producing water. The other product is the salt, $BaBr_2$, which is soluble and remains in solution. The balanced molecular equation is:

$$2HBr(aq) + Ba(OH)_2(aq) \longrightarrow BaBr_2(aq) + 2H_2O(l)$$

The ionic and net ionic equations are:

$$2H^+(aq) + 2Br^-(aq) + Ba^{2+}(aq) + 2OH^-(aq) \longrightarrow Ba^{2+}(aq) + 2Br^-(aq) + 2H_2O(l)$$

$$2H^+(aq) + 2OH^-(aq) \longrightarrow 2H_2O(l) \quad \text{or} \quad H^+(aq) + OH^-(aq) \longrightarrow H_2O(l)$$

(b)

Solution: HCOOH is a weak acid. It will be shown as a molecule in the ionic equation. NaOH is a strong base. It completely ionizes to Na^+ and OH^- ions. Since HCOOH is an acid, it donates an H^+ to the base, OH^-, producing water. The other product is the salt, HCOONa, which is soluble and remains in solution. The balanced molecular equation is:

$$HCOOH(aq) + NaOH(aq) \longrightarrow HCOONa(aq) + H_2O(l)$$

The ionic and net ionic equations are:

$$HCOOH(aq) + Na^+(aq) + OH^-(aq) \longrightarrow Na^+(aq) + HCOO^-(aq) + H_2O(l)$$

$$HCOOH(aq) + OH^-(aq) \longrightarrow HCOO^-(aq) + H_2O(l)$$

PRACTICE EXERCISE

4. Write the balanced molecular, ionic, and net ionic equations for the following acid-base reaction:

$$HNO_2(aq) + Ba(OH)_2(aq) \longrightarrow$$

Text Problem: 4.34

PROBLEM TYPE 4: OXIDATION-REDUCTION REACTIONS

A. Assigning oxidation numbers

Oxidation numbers are assigned to reactants and products in oxidation-reduction (redox) reactions to keep track of electrons. An oxidation number refers to the number of charges an atom would have in a molecule (or an ionic compound) if electrons were transferred completely.

Rules for assigning oxidation numbers are in Section 4.4 of your text. These rules will be used in the following example.

To assign oxidation numbers you should refer to the following *two* steps:

Step 1: Use the rules in Section 4.4 to assign oxidation numbers to as many atoms as possible.

Step 2: Often times, one atom does not follow any rules outlined in Section 4.4. To assign an oxidation number to this atom, follow rule 6 of the text. In a neutral molecule, the sum of the oxidation numbers of all the atoms must be zero. In a polyatomic ion, the sum of the oxidation numbers of all the elements in the ion must be equal to the net charge of the ion.

EXAMPLE 4.5

Assign oxidation numbers to all the atoms in the following compounds and ion:

(a) Na_2SO_4, (b) CuCl, (c) SO_3^{2-}

Strategy: In general, we follow the rules listed in Section 4.4 of the text for assigning oxidation numbers. Remember that all alkali metals have an oxidation number of +1 in ionic compounds, and in most cases hydrogen has an oxidation number of +1 and oxygen has an oxidation number of −2 in their compounds.

Solution:

(a) Na always has an oxidation number of +1 (Rule 2). The oxidation number of oxygen in most compounds is −2 (Rule 3).

You can now assign an oxidation number to S based on Na having a +1 oxidation number and O having a −2 oxidation number. This is a neutral ionic compound, so the sum of the oxidation numbers of all the atoms must be zero.

$$2(\text{oxi. no. Na}) + (\text{oxi. no. S}) + 4(\text{oxi. no. O}) = 0$$

$$2(+1) + (\text{oxi. no. S}) + 4(-2) = 0$$

$$\mathbf{(oxi.\ no.\ S)} = 8 - 2 = \mathbf{+6}$$

(b) An oxidation number of −1 can be assigned to Cl (Rule 5).

You can now assign an oxidation number to Cu. This is a neutral ionic compound.

$$(\text{oxi. no. Cu}) + (\text{oxi. no. Cl}) = 0$$
$$(\text{oxi. no. Cu}) + (-1) = 0$$

$$\mathbf{(oxi.\ no.\ Cu)} = \mathbf{+1}$$

(c) An oxidation number of −2 can be assigned to oxygen (Rule 3).

You can now assign an oxidation number to S. SO_3^{2-} is a polyatomic ion. The sum of the oxidation numbers of all elements in the ion must be equal to the net charge of the ion, in this case −2.

$$(\text{oxi. no. S}) + 3(\text{oxi. no. O}) = -2$$
$$(\text{oxi. no. S}) + 3(-2) = -2$$

$$\mathbf{(oxi.\ no.\ S)} = -2 + 6 = \mathbf{+4}$$

PRACTICE EXERCISE

5. Assign oxidation numbers to the underlined atoms in the following molecules or ions:

   (a) $\underline{C}O_3^{2-}$ **(b)** $\underline{Cu}Cl_2$ **(c)** $\underline{Ti}O_2$ **(d)** $\underline{N}O_3^{-}$

Text Problems: 4.46, 4.48, 4.50

B. Writing oxidation-reduction half-reactions

Strategy: In order to break a redox reaction down into an oxidation half-reaction and a reduction half-reaction, you must first assign oxidation numbers to all the atoms in the reaction. In this way, you can determine which element is oxidized (loses electrons) and which element is reduced (gains electrons).

EXAMPLE 4.6

For the following redox reaction, break down the reaction into its half-reactions.

$$2Al + Fe_2O_3 \longrightarrow Al_2O_3 + 2Fe$$

Solution: Reactants, the oxidation number of Al is 0 (Rule 1), and the oxidation number of O in a compound is −2 (Rule 3). Solve for the oxidation number of Fe in Fe_2O_3. This is a neutral ionic compound.

$$2(\text{oxi. no. Fe}) + 3(\text{oxi. no. O}) = 0$$
$$2(\text{oxi. no. Fe}) + 3(-2) = 0$$
$$2(\text{oxi. no. Fe}) = +6$$
$$\mathbf{(oxi.\ no.\ Fe) = +3}$$

Products, the oxidation number of Fe is 0 (Rule 1), and the oxidation number of O in a compound is −2 (Rule 3). Solve for the oxidation number of Al in Al_2O_3. This is a neutral ionic compound.

$$2(\text{oxi. no. Al}) + 3(\text{oxi. no. O}) = 0$$
$$2(\text{oxi. no. Al}) + 3(-2) = 0$$
$$2(\text{oxi. no. Al}) = +6$$
$$\mathbf{(oxi.\ no.\ Al) = +3}$$

$$Al \longrightarrow Al^{3+} + 3e^- \quad \text{(oxidation half-reaction, } 3e^- \text{ lost)}$$

$$Fe^{3+} + 3e^- \longrightarrow Fe \quad \text{(reduction half-reaction, } 3e^- \text{ gained)}$$

Tip: When a species is oxidized, the oxidation number will *increase*. In this example, the oxidation number of Al *increased* from 0 to +3. When a species is reduced, the oxidation number will *decrease*. In this example, the oxidation number of Fe *decreased* from +3 to 0.

PRACTICE EXERCISE

6. The nickel-cadmium (nicad) battery, a popular rechargeable "dry cell" used in battery-operated tools, uses the following redox reaction to generate electricity:

$$Cd(s) + NiO_2(s) + 2H_2O(l) \longrightarrow Cd(OH)_2(s) + Ni(OH)_2(s)$$

   Assign oxidation numbers to all the atoms and ions, identify the substances that are oxidized and reduced, and write oxidation and reduction half-reactions.

Text Problem: 4.44

C. Using an activity series

An activity series is used to predict whether a metal or hydrogen displacement reaction will occur (see Figure 4.16 of the text). An activity series can be described as a convenient summary of the results of many possible displacement reactions.

(1) Hydrogen displacement. Any metal above hydrogen in the activity series will displace it from water or from an acid. Metals below hydrogen will *not* react with either water or an acid.

(2) Metal displacement. Any metal will react with a compound containing any metal ion listed below it.

EXAMPLE 4.7

Predict the outcome of the reactions represented by the following equations by using the activity series, and balance the equations.

Strategy: *Hydrogen displacement*: Any metal above hydrogen in the activity series will displace it from water or from an acid. Metals below hydrogen will *not* react with either water or an acid.

Solution:

(a) $Mg(s) + HCl(aq) \longrightarrow$

Since Mg is above hydrogen in the activity series, it will displace hydrogen from the acid.

$$Mg(s) + 2HCl(aq) \longrightarrow MgCl_2(aq) + H_2(g)$$

(b) $Au(s) + H_2O(l) \longrightarrow$

Since Au (gold) is below hydrogen in the activity series, it will *not* react with water. You probably already knew this.

$$Au(s) + H_2O(l) \longrightarrow \text{No reaction}$$

Strategy: *Metal displacement*: Any metal will react with a compound containing any metal ion listed below it.

Solution:

(c) $Cu(s) + NiCl_2(aq) \longrightarrow$

In this case Ni^{2+} is listed *above* Cu in the activity series. No reaction will occur.

$$Cu(s) + NiCl_2(aq) \longrightarrow \text{No reaction}$$

PRACTICE EXERCISE

7. Predict the outcome of the following reactions using the activity series. If a reaction occurs, balance the equation.

(a) $Al(s) + HCl(aq) \longrightarrow$
(b) $Au(s) + KCl(aq) \longrightarrow$

Text Problems: 4.52, 4.54

PROBLEM TYPE 5: CONCENTRATION OF SOLUTIONS

Solutions are characterized by their concentration, that is, the amount of solute dissolved in a given quantity of solvent. One of the most common units of concentration in chemistry is **molarity (*M*)**. Molarity is the number of moles of solute in 1 liter of solution:

$$M = \text{molarity} = \frac{\text{moles of solute}}{\text{liters of solution}}$$

Sometimes, it is useful to rearrange the above equation to the following form:

$$\text{moles solute} = (\text{molarity}) \times (\text{liters of solution})$$

EXAMPLE 4.8

What is the molarity of a solution made by dissolving 32.1 g of KNO_3 in enough water to make 500 mL of solution?

Strategy: Since the definition of molarity is moles solute per liters of solution, we need to convert grams of solute to moles of solute and convert mL of solution to L of solution.

Solution:

$\mathcal{M}(KNO_3) = 101.1$ g/mol

$$?\ \text{moles solute} = 32.1\ \text{g}\ KNO_3 \times \frac{1\ \text{mol}\ KNO_3}{101.1\ \text{g}\ KNO_3} = 0.318\ \text{mol}\ KNO_3$$

$$?\ \text{liters of solution} = 500\ \text{mL solution} \times \frac{1\ \text{L}}{1000\ \text{mL}} = 0.500\ \text{L solution}$$

Substitute the above values into the molarity equation.

$$M = \frac{\text{moles of solute}}{\text{liters of solution}}$$

$$\mathbf{M} = \frac{0.318\ \text{mol}\ KNO_3}{0.500\ \text{L solution}} = \mathbf{0.636\ M}$$

This is normally written 0.636 M KNO_3.

PRACTICE EXERCISE

8. An aqueous nutrient solution is prepared by adding 50.23 g of KNO_3 (molar mass = 101.1 g/mol) to enough water to fill a 40.0 L container. What is the molarity of the KNO_3 solution?

EXAMPLE 4.9

How many moles of solute are in 2.50×10^2 mL of 0.100 M KCl?

Strategy: Since the problem asks for moles of solute, you must solve the equation algebraically for moles of solute.

$$\text{moles solute} = (\text{molarity}) \times (\text{liters of solution})$$

Substitute the molarity and liters of solution into the above equation to solve for moles solute.

Solution:

$2.50 \times 10^2\ \text{mL} = 0.250\ \text{L}$

$$\mathbf{?\ moles\ KCl\ solute} = \frac{0.100\ \text{moles solute}}{1\ \text{L solution}} \times 0.250\ \text{L solution} = \mathbf{0.0250\ mol}$$

PRACTICE EXERCISE

9. You need to prepare 1.00 L of a 0.500 M NaCl solution. What mass of NaCl (in g) must you weigh out to prepare this solution?

Text Problems: 4.60, 4.62, 4.64, 4.66

PROBLEM TYPE 6: DILUTION OF SOLUTIONS

Dilution refers to the procedure for preparing a less-concentrated solution from a more-concentrated one. The key to solving a dilution problem is to realize that

moles of solute *before* dilution = moles of solute *after* dilution

In Problem Type 5 above, we discussed how to calculate moles of solute from the molarity and the volume of solution.

moles solute = (molarity) × (volume of solution (in L))

Thus,

moles of solute *before* dilution (initial) = moles of solute *after* dilution (final)

$$\boldsymbol{M_{\text{initial}}V_{\text{initial}} = M_{\text{final}}V_{\text{final}}}$$

EXAMPLE 4.10

What volume of a concentrated (12.0 *M*) hydrochloric acid stock solution is needed to prepare 8.00 × 10^2 mL of 0.120 *M* HCl?

Strategy: Recognize that the problem asks for the initial volume of stock solution needed to prepare the dilute solution. Solve the above equation algebraically for V_{initial}, then substitute in the appropriate values from the problem.

Solution: We prepare for the calculation by tabulating our data.

$M_i = 12.0\ M$ $\qquad M_f = 0.120\ M$

$V_i = ?$ $\qquad V_f = 8.00 \times 10^2\ \text{mL}$

$$V_{\text{initial}} = \frac{M_{\text{final}}V_{\text{final}}}{M_{\text{initial}}}$$

$$\boldsymbol{V_{\text{initial}}} = \frac{(0.120\ \cancel{M})(8.00 \times 10^2\ \text{mL})}{12.0\ \cancel{M}} = \mathbf{8.00\ mL}$$

Tip: The units of V_{initial} and V_{final} can be milliliters or liters for a dilution problem as long as they are the same. Be consistent. Also, make sure to check whether your results seem reasonable. Be sure that $M_{\text{initial}} > M_{\text{final}}$ and $V_{\text{final}} > V_{\text{initial}}$.

PRACTICE EXERCISE

10. A 20.0 mL sample of 0.127 *M* $Ca(NO_3)_2$ is diluted to 5.00 L. What is the molarity of the resulting solution?

Text Problems: 4.70, 4.72, 4.74

PROBLEM TYPE 7: GRAVIMETRIC ANALYSIS

Gravimetric analysis is an analytical technique based on the measurement of mass. The type of gravimetric analysis discussed in your text involves the formation, isolation, and mass determination of a precipitate. This procedure is applicable only to reactions that go to completion, or have nearly a 100 percent yield. Thus, the precipitate must be insoluble rather than slightly soluble. See Section 4.6 of your text for further discussion of this technique.

The typical problem involves determining the mass percent of an element in one of the reactants. The element (ion) of interest is completely precipitated from solution. Use the following approach to solve this type of problem.

Step 1: From the measured mass of precipitate, calculate the mass of the element of interest in the precipitate. This is the same amount of the element that was present in the original sample.

Step 2: Calculate the mass percent of the element of interest in the original sample. See Problem Type 4A, Chapter 3.

> **Tip:** Try to be flexible when solving problems. Some gravimetric analysis problems may ask you to calculate the *molar concentration* of the component of interest in the sample, rather than mass percent. For this type of problem, you must modify your approach by converting grams of the component of interest to moles, then dividing by the volume of solution in liters.

EXAMPLE 4.11

A 0.7469 g sample of an ionic compound containing Pb ions is dissolved in water and treated with excess Na_2SO_4. If the mass of $PbSO_4$ that precipitates is 0.6839 g, what is the percent by mass of Pb in the original sample?

Strategy: We want to calculate the mass % of Pb in the original compound. Let's start with the definition of mass %.

$$\text{mass \% Pb} = \frac{\text{mass Pb}}{\text{mass of sample}} \times 100\%$$

want to calculate (mass % Pb); need to find (mass Pb); given (mass of sample)

The mass of the sample is given in the problem (0.7469 g). Therefore we need to find the mass of Pb in the original sample. We assume the precipitation is quantitative, that is, that all of the lead in the sample has been precipitated as lead sulfate. From the mass of $PbSO_4$ produced, we can calculate the mass of Pb. There is 1 mole of Pb in 1 mole of $PbSO_4$.

Solution: First, we calculate the mass of Pb in 0.6839 g of the $PbSO_4$ precipitate. The molar mass of $PbSO_4$ is 303.27 g/mol.

$$\textbf{? mass of Pb} = 0.6839\ \cancel{\text{g}}\ \text{PbSO}_4 \times \frac{1\ \cancel{\text{mol}}\ \text{PbSO}_4}{303.27\ \cancel{\text{g}}\ \text{PbSO}_4} \times \frac{1\ \cancel{\text{mol}}\ \text{Pb}}{1\ \cancel{\text{mol}}\ \text{PbSO}_4} \times \frac{207.2\ \text{g Pb}}{1\ \cancel{\text{mol}}\ \text{Pb}} = \textbf{0.4673 g Pb}$$

Next, we calculate the mass percent of Pb in the unknown compound.

$$\textbf{\%Pb by mass} = \frac{0.4673\ \text{g}}{0.7469\ \text{g}} \times 100\% = \textbf{62.57\%}$$

PRACTICE EXERCISE

11. The concentration of Pb^{2+} ions in tap water could be determined by adding excess sodium sulfate solution to water. Excess sodium sulfate solution is added to 0.250 L of tap water. Write the net ionic equation and calculate the molar concentration of Pb^{2+} in the water sample if 0.01685 g of solid $PbSO_4$ is formed.

Text Problems: 4.78, 4.80

PROBLEM TYPE 8: ACID-BASE TITRATIONS

You must try to convince yourself that a titration problem follows the same thought process as the stoichiometry problems discussed in Chapter 3. The difference is that for acid and base solutions, you will typically be given the molarity rather than grams of substance.

Remember, you cannot directly compare grams of one substance to grams of another. Similarly, you typically cannot compare molarity or volume of one substance to that of another. Therefore, you must convert to *moles* of one substance, and then apply the correct mole ratio from the balanced chemical equation to convert to *moles* of the other substance.

A typical approach to a stoichiometry problem is outlined below.

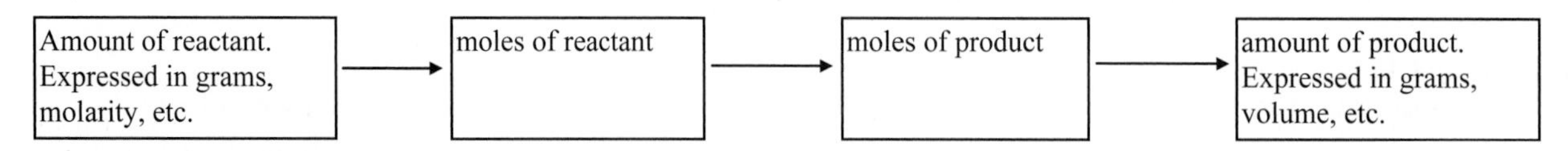

EXAMPLE 4.12

What volume of 0.900 *M* HCl is required to completely neutralize 50.0 mL of a 0.500 *M* $Ba(OH)_2$ solution?

Strategy: We know the molarity of the HCl solution, and we want to calculate the volume of the HCl solution.

given → M of HCl; need to find → mol HCl; want to calculate → L of HCl soln

$$M \text{ of HCl} = \frac{\text{mol HCl}}{\text{L of HCl soln}}$$

If we can determine the moles of HCl, we can then use the definition of molarity to calculate the volume of HCl needed. From the volume and molarity of $Ba(OH)_2$, we can calculate moles of $Ba(OH)_2$. Then, using the mole ratio from the balanced equation, we can calculate moles of HCl.

Solution: In order to have the correct mole ratio to solve the problem, you must start with a balanced chemical equation.

$$2HCl(aq) + Ba(OH)_2(aq) \longrightarrow BaCl_2(aq) + 2H_2O(l)$$

From the molarity and volume of the $Ba(OH)_2$ solution, you can calculate moles of $Ba(OH)_2$. Then, using the mole ratio from the balanced equation above, you can calculate moles of HCl.

50.0 mL = 0.0500 L

$$?\text{ mol HCl} = 0.0500\ \cancel{\text{L}} \times \frac{0.500\ \cancel{\text{mol}}\ Ba(OH)_2}{1\ \cancel{\text{L}}\text{ of solution}} \times \frac{2\text{ mol HCl}}{1\ \cancel{\text{mol}}\ Ba(OH)_2} = 0.0500\text{ mol HCl}$$

Thus, 0.0500 mol of HCl are required to neutralize 50.0 mL of 0.500 *M* $Ba(OH)_2$.

Solve the molarity equation algebraically for liters of solution. Then, substitute in the moles of HCl and molarity of HCl to solve for volume of HCl.

$$\text{liters of solution} = \frac{\text{moles of solute}}{M}$$

$$\textbf{volume of HCl} = \frac{0.0500\ \cancel{\text{mol}}\text{ HCl}}{0.900\ \cancel{\text{mol}}/\text{L}} = \mathbf{0.0556\ L = 55.6\ mL}$$

PRACTICE EXERCISE

12. The distinctive odor of vinegar is due to acetic acid, CH_3COOH. Acetic acid reacts with sodium hydroxide in the following fashion:

$$CH_3COOH(aq) + NaOH(aq) \longrightarrow H_2O(l) + CH_3COONa(aq)$$

If 2.50 mL of vinegar requires 34.9 mL of 0.0960 *M* NaOH to reach the equivalence point in a titration, how many grams of acetic acid are in the 2.50 mL sample?

Text Problems: 4.86, **4.88**

PROBLEM TYPE 9: REDOX TITRATIONS

Redox titration problems are solved in a similar manner to acid-base titration problems. A redox reaction is an *electron* transfer reaction; whereas an acid-base reaction is typically a *proton* transfer reaction. In a redox titration, an oxidizing agent is titrated against a reducing agent.

EXAMPLE 4.13

A 20.32 mL volume of 0.2002 M $KMnO_4$ solution is needed to oxidize 10.00 mL of an oxalic acid ($H_2C_2O_4$) solution. What is the concentration of the oxalic acid solution? The net ionic equation is:

$$2MnO_4^- + 5C_2O_4^{2-} + 16H^+ \longrightarrow 2Mn^{2+} + 10CO_2 + 8H_2O$$

Strategy: We want to calculate the molarity of the oxalic acid solution. From the molarity and volume of $KMnO_4$, we can calculate moles of $KMnO_4$. Then, using the mole ratio from the balanced equation, we can calculate moles of $H_2C_2O_4$. From the moles and volume of oxalic acid, we can calculate the molarity?

Solution: The balanced equation is given in the problem. From the molarity and volume of the $KMnO_4$ solution, you can calculate moles of MnO_4^-. Then, using the mole ratio from the balanced equation above, you can calculate moles of $C_2O_4^{2-}$.

20.32 mL = 0.02032 L

$$0.02032\ \cancel{L}\ \text{soln} \times \frac{0.2002\ \cancel{\text{mol}}\ MnO_4^{2-}}{1\ \cancel{L}\ \text{soln}} \times \frac{5\ \text{mol}\ C_2O_4^{2-}}{2\ \cancel{\text{mol}}\ MnO_4^{2-}} = \mathbf{0.01017\ mol\ C_2O_4^{2-}}$$

The mole ratio between $C_2O_4^{2-}$ and oxalic acid, $H_2C_2O_4$, is 1:1; therefore, the number of moles of oxalic acid is 0.01017 mole.

We can now calculate the molarity of the oxalic acid solution from the moles of oxalic acid and the volume of the solution.

$$M = \frac{\text{moles of solute}}{\text{liters of solution}} = \frac{0.01017\ \text{mol oxalic acid}}{10.00 \times 10^{-3}\ \text{L soln}} = \mathbf{1.017\ \mathit{M}\ oxalic\ acid}$$

PRACTICE EXERCISE

13. Fe metal reacts with hydrochloric acid to produce Fe^{2+} ions and hydrogen gas. It takes 55.6 mL of 1.15 M HCl to completely react with a piece of Fe. What is the mass of the Fe? The balanced equation is:

$$Fe(s) + 2H^+(aq) + 2Cl^-(aq) \longrightarrow Fe^{2+}(aq) + H_2(g) + 2Cl^-(aq)$$

Text Problems: 4.92, 4.94, 4.96, 4.98

ANSWERS TO PRACTICE EXERCISES

1. **(a)** soluble **(b)** insoluble **(c)** soluble **(d)** insoluble

2. $CaCl_2(aq) + Na_2CO_3(aq) \longrightarrow 2NaCl(aq) + CaCO_3(s)$

$Ca^{2+}(aq) + 2Cl^-(aq) + 2Na^+(aq) + CO_3^{2-}(aq) \longrightarrow 2Na^+(aq) + 2Cl^-(aq) + CaCO_3(s)$

$Ca^{2+}(aq) + CO_3^{2-}(aq) \longrightarrow CaCO_3(s)$

3. **(a)** acid **(b)** acid **(c)** base **(d)** both

4. $2HNO_2(aq) + Ba(OH)_2(aq) \longrightarrow Ba(NO_2)_2(aq) + 2H_2O(l)$

$2HNO_2(aq) + Ba^{2+}(aq) + 2OH^-(aq) \longrightarrow Ba^{2+}(aq) + 2NO_2^-(aq) + 2H_2O(l)$

$2HNO_2(aq) + 2OH^-(aq) \longrightarrow 2NO_2^-(aq) + 2H_2O(l)$

5. **(a)** +4 **(b)** +2 **(c)** +4 **(d)** +5

6. The oxidation numbers of the atoms and ions are:

$$Cd^0(s) + Ni^{4+}O_2^{2-}(s) + 2H_2^+O^{2-}(l) \longrightarrow Cd^{2+}(OH)_2^-(s) + Ni^{2+}(OH)_2^-(s)$$

$Cd^0(s)$ is oxidized, and Ni^{4+} is reduced. The oxidation and reduction half-reactions are:

$$Cd \longrightarrow Cd^{2+} + 2e^-$$
$$Ni^{4+} + 2e^- \longrightarrow Ni^{2+}$$

7. **(a)** $2Al(s) + 6HCl(aq) \longrightarrow 2AlCl_3(aq) + 3H_2(g)$ **(b)** No reaction

8. 0.0124 *M* **9.** 29.22 g NaCl **10.** 5.08×10^{-4} *M*

11. $Pb^{2+}(aq) + SO_4^{2-}(aq) \longrightarrow PbSO_4(s)$

$[Pb^{2+}] = 2.222 \times 10^{-4}\ M$

12. 0.201 g acetic acid **13.** 1.79 g Fe

SOLUTIONS TO SELECTED TEXT PROBLEMS

4.8 When NaCl dissolves in water it dissociates into Na^+ and Cl^- ions. When the ions are hydrated, the water molecules will be oriented so that the negative end of the water dipole interacts with the positive sodium ion, and the positive end of the water dipole interacts with the negative chloride ion. The negative end of the water dipole is near the oxygen atom, and the positive end of the water dipole is near the hydrogen atoms. The diagram that best represents the hydration of NaCl when dissolved in water is choice **(c)**.

4.10 Ionic compounds, strong acids, and strong bases (metal hydroxides) are strong electrolytes (completely broken up into ions of the compound). Weak acids and weak bases are weak electrolytes. Molecular substances other than acids or bases are nonelectrolytes.

(a) strong electrolyte (ionic)

(b) nonelectrolyte

(c) weak electrolyte (weak base)

(d) strong electrolyte (strong base)

4.12 **(a)** Solid NaCl does not conduct. The ions are locked in a rigid lattice structure.

(b) Molten NaCl conducts. The ions can move around in the liquid state.

(c) Aqueous NaCl conducts. NaCl dissociates completely to $Na^+(aq)$ and $Cl^-(aq)$ in water.

4.14 Since HCl dissolved in water conducts electricity, then HCl(*aq*) must actually exists as $H^+(aq)$ cations and $Cl^-(aq)$ anions. Since HCl dissolved in benzene solvent does not conduct electricity, then we must assume that the HCl molecules in benzene solvent do not ionize, but rather exist as un-ionized molecules.

4.18 Refer to Table 4.2 of the text to solve this problem. $Mg(OH)_2$ is insoluble in water. It will precipitate from solution. KCl is soluble in water and will remain as K^+ and Cl^- ions in solution. Diagram **(b)** best represents the mixture.

4.20 Applying solubility rules, Problem Type 1.

Strategy: Although it is not necessary to memorize the solubilities of compounds, you should keep in mind the following useful rules: all ionic compounds containing alkali metal cations, the ammonium ion, and the nitrate, bicarbonate, and chlorate ions are soluble. For other compounds, refer to Table 4.2 of the text.

Solution:

(a) $CaCO_3$ is **insoluble**. Most carbonate compounds are insoluble.

(b) $ZnSO_4$ is **soluble**. Most sulfate compounds are soluble.

(c) $Hg(NO_3)_2$ is **soluble**. All nitrate compounds are soluble.

(d) $HgSO_4$ is **insoluble**. Most sulfate compounds are soluble, but those containing Ag^+, Ca^{2+}, Ba^{2+}, Hg^{2+}, and Pb^{2+} are insoluble.

(e) NH_4ClO_4 is **soluble**. All ammonium compounds are soluble.

4.22 Writing Molecular, Ionic, and Net Ionic Equations, Problem Type 2.

(a)

Strategy: Recall that an *ionic equation* shows dissolved ionic compounds in terms of their free ions. A *net ionic equation* shows only the species that actually take part in the reaction. What happens when ionic compounds dissolve in water? What ions are formed from the dissociation of Na_2S and $ZnCl_2$? What happens when the cations encounter the anions in solution?

Solution: In solution, Na_2S dissociates into Na^+ and S^{2-} ions and $ZnCl_2$ dissociates into Zn^{2+} and Cl^- ions. According to Table 4.2 of the text, zinc ions (Zn^{2+}) and sulfide ions (S^{2-}) will form an insoluble compound, zinc sulfide (ZnS), while the other product, NaCl, is soluble and remains in solution. This is a precipitation reaction. The balanced molecular equation is:

$$Na_2S(aq) + ZnCl_2(aq) \longrightarrow ZnS(s) + 2NaCl(aq)$$

The ionic and net ionic equations are:

Ionic: $2Na^+(aq) + S^{2-}(aq) + Zn^{2+}(aq) + 2Cl^-(aq) \longrightarrow ZnS(s) + 2Na^+(aq) + 2Cl^-(aq)$

Net ionic: $Zn^{2+}(aq) + S^{2-}(aq) \longrightarrow ZnS(s)$

Check: Note that because we balanced the molecular equation first, the net ionic equation is balanced as to the number of atoms on each side, and the number of positive and negative charges on the left-hand side of the equation is the same.

(b)
Strategy: What happens when ionic compounds dissolve in water? What ions are formed from the dissociation of K_3PO_4 and $Sr(NO_3)_2$? What happens when the cations encounter the anions in solution?

Solution: In solution, K_3PO_4 dissociates into K^+ and PO_4^{3-} ions and $Sr(NO_3)_2$ dissociates into Sr^{2+} and NO_3^- ions. According to Table 4.2 of the text, strontium ions (Sr^{2+}) and phosphate ions (PO_4^{3-}) will form an insoluble compound, strontium phosphate [$Sr_3(PO_4)_2$], while the other product, KNO_3, is soluble and remains in solution. This is a precipitation reaction. The balanced molecular equation is:

$$2K_3PO_4(aq) + 3Sr(NO_3)_2(aq) \longrightarrow Sr_3(PO_4)_2(s) + 6KNO_3(aq)$$

The ionic and net ionic equations are:

Ionic: $6K^+(aq) + 2PO_4^{3-}(aq) + 3Sr^{2+}(aq) + 6NO_3^-(aq) \longrightarrow Sr_3(PO_4)_2(s) + 6K^+(aq) + 6NO_3^-(aq)$

Net ionic: $3Sr^{2+}(aq) + 2PO_4^{3-}(aq) \longrightarrow Sr_3(PO_4)_2(s)$

Check: Note that because we balanced the molecular equation first, the net ionic equation is balanced as to the number of atoms on each side, and the number of positive and negative charges on the left-hand side of the equation is the same.

(c)
Strategy: What happens when ionic compounds dissolve in water? What ions are formed from the dissociation of $Mg(NO_3)_2$ and NaOH? What happens when the cations encounter the anions in solution?

Solution: In solution, $Mg(NO_3)_2$ dissociates into Mg^{2+} and NO_3^- ions and NaOH dissociates into Na^+ and OH^- ions. According to Table 4.2 of the text, magnesium ions (Mg^{2+}) and hydroxide ions (OH^-) will form an insoluble compound, magnesium hydroxide [$Mg(OH)_2$], while the other product, $NaNO_3$, is soluble and remains in solution. This is a precipitation reaction. The balanced molecular equation is:

$$Mg(NO_3)_2(aq) + 2NaOH(aq) \longrightarrow Mg(OH)_2(s) + 2NaNO_3(aq)$$

The ionic and net ionic equations are:

Ionic: $Mg^{2+}(aq) + 2NO_3^-(aq) + 2Na^+(aq) + 2OH^-(aq) \longrightarrow Mg(OH)_2(s) + 2Na^+(aq) + 2NO_3^-(aq)$

Net ionic: $Mg^{2+}(aq) + 2OH^-(aq) \longrightarrow Mg(OH)_2(s)$

Check: Note that because we balanced the molecular equation first, the net ionic equation is balanced as to the number of atoms on each side, and the number of positive and negative charges on the left-hand side of the equation is the same.

4.24 **(a)** Add chloride ions. KCl is soluble, but AgCl is not.

(b) Add hydroxide ions. $Ba(OH)_2$ is soluble, but $Pb(OH)_2$ is insoluble.

(c) Add carbonate ions. $(NH_4)_2CO_3$ is soluble, but $CaCO_3$ is insoluble.

(d) Add sulfate ions. $CuSO_4$ is soluble, but $BaSO_4$ is insoluble.

4.32 Identifying Brønsted acids and bases, Problem Type 3A.

Strategy: What are the characteristics of a Brønsted acid? Does it contain at least an H atom? With the exception of ammonia, most Brønsted bases that you will encounter at this stage are anions.

Solution:

(a) PO_4^{3-} in water can accept a proton to become HPO_4^{2-}, and is thus a **Brønsted base**.

(b) ClO_2^- in water can accept a proton to become $HClO_2$, and is thus a **Brønsted base**.

(c) NH_4^+ dissolved in water can donate a proton H^+, thus behaving as a **Brønsted acid**.

(d) HCO_3^- can either accept a proton to become H_2CO_3, thus behaving as a **Brønsted base**. Or, HCO_3^- can donate a proton to yield H^+ and CO_3^{2-}, thus behaving as a **Brønsted acid**.

Comment: The HCO_3^- species is said to be *amphoteric* because it possesses both acidic and basic properties.

4.34 Writing acid-base reactions, Problem Type 3B.

Strategy: Recall that strong acids and strong bases are strong electrolytes. They are completely ionized in solution. An *ionic equation* will show strong acids and strong bases in terms of their free ions. Weak acids and weak bases are weak electrolytes. They only ionize to a small extent in solution. Weak acids and weak bases are shown as molecules in ionic and net ionic equations. A *net ionic equation* shows only the species that actually take part in the reaction.

(a)

Solution: CH_3COOH is a weak acid. It will be shown as a molecule in the ionic equation. KOH is a strong base. It completely ionizes to K^+ and OH^- ions. Since CH_3COOH is an acid, it donates an H^+ to the base, OH^-, producing water. The other product is the salt, CH_3COOK, which is soluble and remains in solution. The balanced molecular equation is:

$$CH_3COOH(aq) + KOH(aq) \longrightarrow CH_3COOK(aq) + H_2O(l)$$

The ionic and net ionic equations are:

Ionic: $CH_3COOH(aq) + K^+(aq) + OH^-(aq) \longrightarrow CH_3COO^-(aq) + K^+(aq) + H_2O(l)$

Net ionic: $CH_3COOH(aq) + OH^-(aq) \longrightarrow CH_3COO^-(aq) + H_2O(l)$

(b)

Solution: H_2CO_3 is a weak acid. It will be shown as a molecule in the ionic equation. NaOH is a strong base. It completely ionizes to Na^+ and OH^- ions. Since H_2CO_3 is an acid, it donates an H^+ to the base, OH^-, producing water. The other product is the salt, Na_2CO_3, which is soluble and remains in solution. The balanced molecular equation is:

$$H_2CO_3(aq) + 2NaOH(aq) \longrightarrow Na_2CO_3(aq) + 2H_2O(l)$$

The ionic and net ionic equations are:

Ionic: $H_2CO_3(aq) + 2Na^+(aq) + 2OH^-(aq) \longrightarrow 2Na^+(aq) + CO_3^{2-}(aq) + 2H_2O(l)$

Net ionic: $H_2CO_3(aq) + 2OH^-(aq) \longrightarrow CO_3^{2-}(aq) + 2H_2O(l)$

(c)

Solution: HNO_3 is a strong acid. It completely ionizes to H^+ and NO_3^- ions. $Ba(OH)_2$ is a strong base. It completely ionizes to Ba^{2+} and OH^- ions. Since HNO_3 is an acid, it donates an H^+ to the base, OH^-, producing water. The other product is the salt, $Ba(NO_3)_2$, which is soluble and remains in solution. The balanced molecular equation is:

$$2HNO_3(aq) + Ba(OH)_2(aq) \longrightarrow Ba(NO_3)_2(aq) + 2H_2O(l)$$

The ionic and net ionic equations are:

Ionic: $2H^+(aq) + 2NO_3^-(aq) + Ba^{2+}(aq) + 2OH^-(aq) \longrightarrow Ba^{2+}(aq) + 2NO_3^-(aq) + 2H_2O(l)$

Net ionic: $2H^+(aq) + 2OH^-(aq) \longrightarrow 2H_2O(l)$ or $H^+(aq) + OH^-(aq) \longrightarrow H_2O(l)$

4.44 Writing oxidation/reduction half-reactions, Problem Type 4B.

Strategy: In order to break a redox reaction down into an oxidation half-reaction and a reduction half-reaction, you should first assign oxidation numbers to all the atoms in the reaction. In this way, you can determine which element is oxidized (loses electrons) and which element is reduced (gains electrons).

Solution: In each part, the reducing agent is the reactant in the first half-reaction and the oxidizing agent is the reactant in the second half-reaction. The coefficients in each half-reaction have been reduced to smallest whole numbers.

(a) The product is an ionic compound whose ions are Fe^{3+} and O^{2-}.

$$Fe \longrightarrow Fe^{3+} + 3e^-$$

$$O_2 + 4e^- \longrightarrow 2O^{2-}$$

O_2 is the oxidizing agent; Fe is the reducing agent.

(b) Na^+ does not change in this reaction. It is a "spectator ion."

$$2Br^- \longrightarrow Br_2 + 2e^-$$

$$Cl_2 + 2e^- \longrightarrow 2Cl^-$$

Cl_2 is the oxidizing agent; Br^- is the reducing agent.

(c) Assume SiF_4 is made up of Si^{4+} and F^-.

$$Si \longrightarrow Si^{4+} + 4e^-$$

$$F_2 + 2e^- \longrightarrow 2F^-$$

F_2 is the oxidizing agent; Si is the reducing agent.

(d) Assume HCl is made up of H^+ and Cl^-.

$$H_2 \longrightarrow 2H^+ + 2e^-$$
$$Cl_2 + 2e^- \longrightarrow 2Cl^-$$

Cl_2 is the oxidizing agent; H_2 is the reducing agent.

4.46 Assigning oxidation numbers, Problem Type 4A.

Strategy: In general, we follow the rules listed in Section 4.4 of the text for assigning oxidation numbers. Remember that all alkali metals have an oxidation number of +1 in ionic compounds, and in most cases hydrogen has an oxidation number of +1 and oxygen has an oxidation number of −2 in their compounds.

Solution: All the compounds listed are neutral compounds, so the oxidation numbers must sum to zero (Rule 6, Section 4.4 of the text).

Let the oxidation number of P = x.

(a) $x + 1 + (3)(-2) = 0,\ \mathbf{x = +5}$

(b) $x + (3)(+1) + (2)(-2) = 0,\ \mathbf{x = +1}$

(c) $x + (3)(+1) + (3)(-2) = 0,\ \mathbf{x = +3}$

(d) $x + (3)(+1) + (4)(-2) = 0,\ \mathbf{x = +5}$

(e) $2x + (4)(+1) + (7)(-2) = 0,\ 2x = 10,\ \mathbf{x = +5}$

(f) $3x + (5)(+1) + (10)(-2) = 0,\ 3x = 15,\ \mathbf{x = +5}$

The molecules in part (a), (e), and (f) can be made by strongly heating the compound in part (d). Are these oxidation-reduction reactions?

Check: In each case, does the sum of the oxidation numbers of all the atoms equal the net charge on the species, in this case zero?

4.48 All are free elements, so all have an oxidation number of **zero**.

4.50 **(a)** N: −3 **(b)** O: −1/2 **(c)** C: −1 **(d)** C: +4

(e) C: +3 **(f)** O: −2 **(g)** B: +3 **(h)** W: +6

4.52 Using an activity series, Problem Type 4C.

Strategy: *Hydrogen displacement*: Any metal above hydrogen in the activity series will displace it from water or from an acid. Metals below hydrogen will *not* react with either water or an acid.

Solution: Only **(b)** Li and **(d)** Ca are above hydrogen in the activity series, so they are the only metals in this problem that will react with water.

4.54 **(a)** $Cu(s) + HCl(aq) \rightarrow$ no reaction, since $Cu(s)$ is less reactive than the hydrogen from acids.

(b) $I_2(s) + NaBr(aq) \rightarrow$ no reaction, since $I_2(s)$ is less reactive than $Br_2(l)$.

(c) $Mg(s) + CuSO_4(aq) \rightarrow MgSO_4(aq) + Cu(s)$, since $Mg(s)$ is more reactive than Cu(s).

Net ionic equation: $Mg(s) + Cu^{2+}(aq) \rightarrow Mg^{2+}(aq) + Cu(s)$

(d) $Cl_2(g) + 2KBr(aq) \rightarrow Br_2(l) + 2KCl(aq)$, since $Cl_2(g)$ is more reactive than $Br_2(l)$

Net ionic equation: $Cl_2(g) + 2Br^-(aq) \rightarrow 2Cl^-(aq) + Br_2(l)$

4.56 **(a)** Combination reaction
(b) Decomposition reaction
(c) Displacement reaction
(d) Disproportionation reaction

4.60 Concentration of Solutions, Problem Type 5.

Strategy: How many moles of $NaNO_3$ does 250 mL of a 0.707 *M* solution contain? How would you convert moles to grams?

Solution: From the molarity (0.707 *M*), we can calculate the moles of $NaNO_3$ needed to prepare 250 mL of solution.

$$\text{Moles NaNO}_3 = \frac{0.707 \text{ mol NaNO}_3}{1000 \text{ mL soln}} \times 250 \text{ mL soln} = 0.177 \text{ mol}$$

Next, we use the molar mass of $NaNO_3$ as a conversion factor to convert from moles to grams.

$\mathcal{M}(NaNO_3) = 85.00$ g/mol.

$$0.177 \text{ mol NaNO}_3 \times \frac{85.00 \text{ g NaNO}_3}{1 \text{ mol NaNO}_3} = 15.0 \text{ g NaNO}_3$$

To make the solution, **dissolve 15.0 g of $NaNO_3$ in enough water to make 250 mL of solution**.

Check: As a ball-park estimate, the mass should be given by [molarity (mol/L) × volume (L) = moles × molar mass (g/mol) = grams]. Let's round the molarity to 1 *M* and the molar mass to 80 g, because we are simply making an estimate. This gives: [1 mol/L × (1/4)L × 80 g = 20 g]. This is close to our answer of 15.0 g.

4.62 Since the problem asks for grams of solute (KOH), you should be thinking that you can calculate moles of solute from the molarity and volume of solution. Then, you can convert moles of solute to grams of solute.

$$\text{? moles KOH solute} = \frac{5.50 \text{ moles solute}}{1000 \text{ mL solution}} \times 35.0 \text{ mL solution} = 0.193 \text{ mol KOH}$$

The molar mass of KOH is 56.11 g/mol. Use this conversion factor to calculate grams of KOH.

$$\textbf{? grams KOH} = 0.193 \text{ mol KOH} \times \frac{56.108 \text{ g KOH}}{1 \text{ mol KOH}} = \mathbf{10.8\ g\ KOH}$$

4.64 **(a)** $$\text{? mol CH}_3\text{OH} = 6.57 \text{ g CH}_3\text{OH} \times \frac{1 \text{ mol CH}_3\text{OH}}{32.042 \text{ g CH}_3\text{OH}} = 0.205 \text{ mol CH}_3\text{OH}$$

$$M = \frac{0.205 \text{ mol CH}_3\text{OH}}{0.150 \text{ L}} = \mathbf{1.37}\ \boldsymbol{M}$$

(b) $$\text{? mol CaCl}_2 = 10.4 \text{ g CaCl}_2 \times \frac{1 \text{ mol CaCl}_2}{110.98 \text{ g CaCl}_2} = 0.0937 \text{ mol CaCl}_2$$

$$M = \frac{0.0937 \text{ mol CaCl}_2}{0.220 \text{ L}} = \mathbf{0.426}\ \boldsymbol{M}$$

(c) $? \text{ mol } C_{10}H_8 = 7.82 \text{ g } C_{10}H_8 \times \frac{1 \text{ mol } C_{10}H_8}{128.16 \text{ g } C_{10}H_8} = 0.0610 \text{ mol } C_{10}H_8$

$$\boldsymbol{M} = \frac{0.0610 \text{ mol } C_{10}H_8}{0.0852 \text{ L}} = \mathbf{0.716}\ \boldsymbol{M}$$

4.66 A 250 mL sample of 0.100 M solution contains 0.0250 mol of solute (mol = $M \times$ L). The computation in each case is the same:

(a) $0.0250 \text{ mol CsI} \times \frac{259.8 \text{ g CsI}}{1 \text{ mol CsI}} = \mathbf{6.50 \text{ g CsI}}$

(b) $0.0250 \text{ mol } H_2SO_4 \times \frac{98.086 \text{ g } H_2SO_4}{1 \text{ mol } H_2SO_4} = \mathbf{2.45 \text{ g } H_2SO_4}$

(c) $0.0250 \text{ mol } Na_2CO_3 \times \frac{105.99 \text{ g } Na_2CO_3}{1 \text{ mol } Na_2CO_3} = \mathbf{2.65 \text{ g } Na_2CO_3}$

(d) $0.0250 \text{ mol } K_2Cr_2O_7 \times \frac{294.2 \text{ g } K_2Cr_2O_7}{1 \text{ mol } K_2Cr_2O_7} = \mathbf{7.36 \text{ g } K_2Cr_2O_7}$

(e) $0.0250 \text{ mol } KMnO_4 \times \frac{158.04 \text{ g } KMnO_4}{1 \text{ mol } KMnO_4} = \mathbf{3.95 \text{ g } KMnO_4}$

4.70 Dilution of Solutions, Problem Type 6.

Strategy: Because the volume of the final solution is greater than the original solution, this is a dilution process. Keep in mind that in a dilution, the concentration of the solution decreases, but the number of moles of the solute remains the same.

Solution: We prepare for the calculation by tabulating our data.

$M_i = 0.866\ M \qquad M_f = ?$

$V_i = 25.0 \text{ mL} \qquad V_f = 500 \text{ mL}$

We substitute the data into Equation (4.3) of the text.

$$M_iV_i = M_fV_f$$

$$(0.866\ M)(25.0 \text{ mL}) = M_f(500 \text{ mL})$$

$$\boldsymbol{M_f} = \frac{(0.866\ M)(25.0 \text{ mL})}{500 \text{ mL}} = \mathbf{0.0433}\ \boldsymbol{M}$$

4.72 You need to calculate the final volume of the dilute solution. Then, you can subtract 505 mL from this volume to calculate the amount of water that should be added.

$$V_{\text{final}} = \frac{M_{\text{initial}}V_{\text{initial}}}{M_{\text{final}}} = \frac{(0.125\ M)(505 \text{ mL})}{(0.100\ M)} = 631 \text{ mL}$$

$$(631 - 505) \text{ mL} = \mathbf{126 \text{ mL of water}}$$

4.74 Moles of calcium nitrate in the first solution:

$$\frac{0.568 \text{ mol}}{1000 \text{ mL soln}} \times 46.2 \text{ mL soln} = 0.0262 \text{ mol } Ca(NO_3)_2$$

Moles of calcium nitrate in the second solution:

$$\frac{1.396 \text{ mol}}{1000 \text{ mL soln}} \times 80.5 \text{ mL soln} = 0.112 \text{ mol } Ca(NO_3)_2$$

The volume of the combined solutions = 46.2 mL + 80.5 mL = 126.7 mL. The concentration of the final solution is:

$$M = \frac{(0.0262 + 0.112) \text{ mol}}{0.1267 \text{ L}} = \mathbf{1.09\ M}$$

4.78 Gravimetric Analysis, Problem Type 7.

Strategy: We want to calculate the mass % of Ba in the original compound. Let's start with the definition of mass %.

want to calculate (→ mass % Ba) need to find (→ mass Ba)

$$\text{mass \% Ba} = \frac{\text{mass Ba}}{\text{mass of sample}} \times 100\%$$

given (→ mass of sample)

The mass of the sample is given in the problem (0.6760 g). Therefore we need to find the mass of Ba in the original sample. We assume the precipitation is quantitative, that is, that all of the barium in the sample has been precipitated as barium sulfate. From the mass of $BaSO_4$ produced, we can calculate the mass of Ba. There is 1 mole of Ba in 1 mole of $BaSO_4$.

Solution: First, we calculate the mass of Ba in 0.4105 g of the $BaSO_4$ precipitate. The molar mass of $BaSO_4$ is 233.4 g/mol.

$$? \text{ mass of Ba} = 0.4105 \text{ g } BaSO_4 \times \frac{1 \text{ mol } BaSO_4}{233.37 \text{ g } BaSO_4} \times \frac{1 \text{ mol Ba}}{1 \text{ mol } BaSO_4} \times \frac{137.3 \text{ g Ba}}{1 \text{ mol Ba}}$$

$$= 0.2415 \text{ g Ba}$$

Next, we calculate the mass percent of Ba in the unknown compound.

$$\textbf{\%Ba by mass} = \frac{0.2415 \text{ g}}{0.6760 \text{ g}} \times 100\% = \mathbf{35.72\%}$$

4.80 The net ionic equation is: $Cu^{2+}(aq) + S^{2-}(aq) \longrightarrow CuS(s)$

The answer sought is the molar concentration of Cu^{2+}, that is, moles of Cu^{2+} ions per liter of solution. The factor-label method is used to convert, in order:

g of CuS → moles CuS → moles Cu^{2+} → moles Cu^{2+} per liter soln

$$[\mathbf{Cu^{2+}}] = 0.0177 \text{ g CuS} \times \frac{1 \text{ mol CuS}}{95.62 \text{ g CuS}} \times \frac{1 \text{ mol } Cu^{2+}}{1 \text{ mol CuS}} \times \frac{1}{0.800 \text{ L}} = \mathbf{2.31 \times 10^{-4}\ M}$$

4.86 The reaction between HCl and NaOH is:

$$HCl(aq) + NaOH(aq) \rightarrow H_2O(l) + NaCl(aq)$$

We know the volume of the NaOH solution, and we want to calculate the molarity of the NaOH solution.

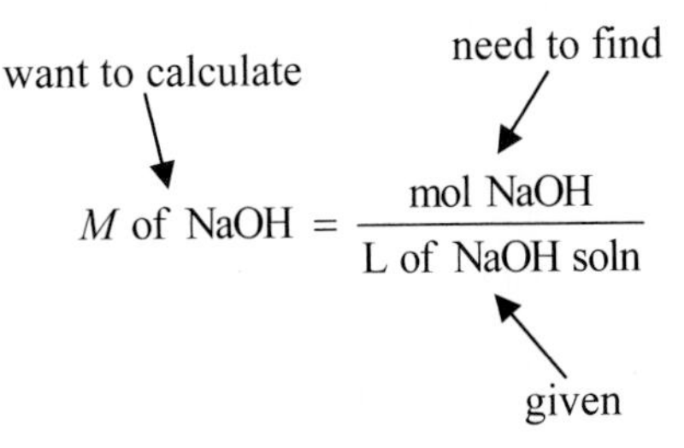

If we can determine the moles of NaOH in the solution, we can then calculate the molarity of the solution. From the volume and molarity of HCl, we can calculate moles of HCl. Then, using the mole ratio from the balanced equation, we can calculate moles of NaOH.

$$? \text{ mol NaOH} = 17.4 \text{ mL HCl} \times \frac{0.312 \text{ mol HCl}}{1000 \text{ mL soln}} \times \frac{1 \text{ mol NaOH}}{1 \text{ mol HCl}} = 5.43 \times 10^{-3} \text{ mol NaOH}$$

From the moles and volume of NaOH, we calculate the molarity of the NaOH solution.

$$\boldsymbol{M} \textbf{ of NaOH} = \frac{\text{mol NaOH}}{\text{L of NaOH soln}} = \frac{5.43 \times 10^{-3} \text{ mol NaOH}}{25.0 \times 10^{-3} \text{ L soln}} = \mathbf{0.217}\ \boldsymbol{M}$$

4.88 Acid-Base Titrations, Problem Type 8.

Strategy: We know the molarity of the HCl solution, and we want to calculate the volume of the HCl solution.

given → M of HCl; need to find → mol HCl; want to calculate → L of HCl soln

$$M \text{ of HCl} = \frac{\text{mol HCl}}{\text{L of HCl soln}}$$

If we can determine the moles of HCl, we can then use the definition of molarity to calculate the volume of HCl needed. From the volume and molarity of NaOH or $Ba(OH)_2$, we can calculate moles of NaOH or $Ba(OH)_2$. Then, using the mole ratio from the balanced equation, we can calculate moles of HCl.

Solution:

(a) In order to have the correct mole ratio to solve the problem, you must start with a balanced chemical equation.

$$HCl(aq) + NaOH(aq) \longrightarrow NaCl(aq) + H_2O(l)$$

$$? \text{ mol HCl} = 10.0 \text{ mL} \times \frac{0.300 \text{ mol NaOH}}{1000 \text{ mL of solution}} \times \frac{1 \text{ mol HCl}}{1 \text{ mol NaOH}} = 3.00 \times 10^{-3} \text{ mol HCl}$$

From the molarity and moles of HCl, we calculate volume of HCl required to neutralize the NaOH.

$$\text{liters of solution} = \frac{\text{moles of solute}}{M}$$

$$\textbf{volume of HCl} = \frac{3.00 \times 10^{-3} \text{ mol HCl}}{0.500 \text{ mol/L}} = \mathbf{6.00 \times 10^{-3}\ L = 6.00\ mL}$$

(b) This problem is similar to part (a). The difference is that the mole ratio between acid and base is 2:1.

$$2HCl(aq) + Ba(OH)_2(aq) \longrightarrow BaCl_2(aq) + 2H_2O(l)$$

$$? \text{ mol HCl} = 10.0 \text{ mL} \times \frac{0.200 \text{ mol } Ba(OH)_2}{1000 \text{ mL of solution}} \times \frac{2 \text{ mol HCl}}{1 \text{ mol } Ba(OH)_2} = 4.00 \times 10^{-3} \text{ mol HCl}$$

$$\textbf{volume of HCl} = \frac{4.00 \times 10^{-3} \text{ mol HCl}}{0.500 \text{ mol/L}} = \mathbf{8.00 \times 10^{-3}\ L = 8.00\ mL}$$

4.92 Redox Titrations, Problem Type 9.

Strategy: We want to calculate the grams of SO_2 in the sample of air. From the molarity and volume of $KMnO_4$, we can calculate moles of $KMnO_4$. Then, using the mole ratio from the balanced equation, we can calculate moles of SO_2. How do we convert from moles of SO_2 to grams of SO_2?

Solution: The balanced equation is given in the problem.

$$5SO_2 + 2MnO_4^- + 2H_2O \longrightarrow 5SO_4^{2-} + 2Mn^{2+} + 4H^+$$

The moles of $KMnO_4$ required for the titration are:

$$\frac{0.00800 \text{ mol } KMnO_4}{1000 \text{ mL soln}} \times 7.37 \text{ mL} = 5.90 \times 10^{-5} \text{ mol } KMnO_4$$

We use the mole ratio from the balanced equation and the molar mass of SO_2 as conversion factors to convert to grams of SO_2.

$$(5.90 \times 10^{-5} \text{ mol } KMnO_4) \times \frac{5 \text{ mol } SO_2}{2 \text{ mol } KMnO_4} \times \frac{64.07 \text{ g } SO_2}{1 \text{ mol } SO_2} = \mathbf{9.45 \times 10^{-3}\ g\ SO_2}$$

4.94 The balanced equation is given in the problem.

$$2MnO_4^- + 5H_2O_2 + 6H^+ \longrightarrow 5O_2 + 2Mn^{2+} + 8H_2O$$

First, calculate the moles of potassium permanganate in 36.44 mL of solution.

$$\frac{0.01652 \text{ mol } KMnO_4}{1000 \text{ mL soln}} \times 36.44 \text{ mL} = 6.020 \times 10^{-4} \text{ mol } KMnO_4$$

Next, calculate the moles of hydrogen peroxide using the mole ratio from the balanced equation.

$$(6.020 \times 10^{-4} \text{ mol } KMnO_4) \times \frac{5 \text{ mol } H_2O_2}{2 \text{ mol } KMnO_4} = 1.505 \times 10^{-3} \text{ mol } H_2O_2$$

Finally, calculate the molarity of the H_2O_2 solution. The volume of the solution is 0.02500 L.

$$\textbf{Molarity of H}_2\textbf{O}_2 = \frac{1.505 \times 10^{-3}\text{ mol H}_2\text{O}_2}{0.02500\text{ L}} = \mathbf{0.06020}\ \boldsymbol{M}$$

4.96 From the reaction of oxalic acid with NaOH, the moles of oxalic acid in 15.0 mL of solution can be determined. Then, using this number of moles and other information given, the volume of the $KMnO_4$ solution needed to react with a second sample of oxalic acid can be calculated.

First, calculate the moles of oxalic acid in the solution. $H_2C_2O_4(aq) + 2NaOH(aq) \rightarrow Na_2C_2O_4(aq) + 2H_2O(l)$

$$0.0252\text{ L} \times \frac{0.149\text{ mol NaOH}}{1\text{ L soln}} \times \frac{1\text{ mol H}_2\text{C}_2\text{O}_4}{2\text{ mol NaOH}} = 1.88 \times 10^{-3}\text{ mol H}_2\text{C}_2\text{O}_4$$

Because we are reacting a second sample of equal volume (15.0 mL), the moles of oxalic acid will also be 1.88×10^{-3} mole in this second sample. The balanced equation for the reaction between oxalic acid and $KMnO_4$ is:

$$2MnO_4^- + 16H^+ + 5C_2O_4^{2-} \rightarrow 2Mn^{2+} + 10CO_2 + 8H_2O$$

Let's calculate the moles of $KMnO_4$ first, and then we will determine the volume of $KMnO_4$ needed to react with the 15.0 mL sample of oxalic acid.

$$(1.88 \times 10^{-3}\text{ mol H}_2\text{C}_2\text{O}_4) \times \frac{2\text{ mol KMnO}_4}{5\text{ mol H}_2\text{C}_2\text{O}_4} = 7.52 \times 10^{-4}\text{ mol KMnO}_4$$

Using Equation (4.2) of the text:

$$M = \frac{n}{V}$$

$$V_{\text{KMnO}_4} = \frac{n}{M} = \frac{7.52 \times 10^{-4}\text{ mol}}{0.122\text{ mol/L}} = \mathbf{0.00616\ L} = \mathbf{6.16\ mL}$$

4.98 The balanced equation is:

$$2MnO_4^- + 16H^+ + 5C_2O_4^{2-} \longrightarrow 2Mn^{2+} + 10CO_2 + 8H_2O$$

$$\text{mol MnO}_4^- = \frac{9.56 \times 10^{-4}\text{ mol MnO}_4^-}{1000\text{ mL of soln}} \times 24.2\text{ mL} = 2.31 \times 10^{-5}\text{ mol MnO}_4^-$$

Using the mole ratio from the balanced equation, we can calculate the mass of Ca^{2+} in the 10.0 mL sample of blood.

$$(2.31 \times 10^{-5}\text{ mol MnO}_4^-) \times \frac{5\text{ mol C}_2\text{O}_4^{2-}}{2\text{ mol MnO}_4^-} \times \frac{1\text{ mol Ca}^{2+}}{1\text{ mol C}_2\text{O}_4^{2-}} \times \frac{40.08\text{ g Ca}^{2+}}{1\text{ mol Ca}^{2+}} = 2.31 \times 10^{-3}\text{ g Ca}^{2+}$$

Converting to mg/mL:

$$\frac{2.31 \times 10^{-3}\text{ g Ca}^{2+}}{10.0\text{ mL of blood}} \times \frac{1\text{ mg}}{0.001\text{ g}} = \mathbf{0.231\ mg\ Ca^{2+}/mL\ of\ blood}$$

4.100 First, the gases could be tested to see if they supported combustion. O_2 would support combustion, CO_2 would not. Second, if CO_2 is bubbled through a solution of calcium hydroxide [$Ca(OH)_2$], a white precipitate of $CaCO_3$ forms. No reaction occurs when O_2 is bubbled through a calcium hydroxide solution.

4.102 Starting with a balanced chemical equation:

$$Mg(s) + 2HCl(aq) \longrightarrow MgCl_2(aq) + H_2(g)$$

From the mass of Mg, you can calculate moles of Mg. Then, using the mole ratio from the balanced equation above, you can calculate moles of HCl reacted.

$$4.47\ \text{g Mg} \times \frac{1\ \text{mol Mg}}{24.31\ \text{g Mg}} \times \frac{2\ \text{mol HCl}}{1\ \text{mol Mg}} = 0.368\ \text{mol HCl reacted}$$

Next we can calculate the number of moles of HCl in the original solution.

$$\frac{2.00\ \text{mol HCl}}{1000\ \text{mL soln}} \times (5.00 \times 10^2\ \text{mL}) = 1.00\ \text{mol HCl}$$

$$\text{Moles HCl remaining} = 1.00\ \text{mol} - 0.368\ \text{mol} = 0.632\ \text{mol HCl}$$

$$\textbf{conc. of HCl after reaction} = \frac{\text{mol HCl}}{\text{L soln}} = \frac{0.632\ \text{mol HCl}}{0.500\ \text{L}} = 1.26\ \text{mol/L} = \mathbf{1.26}\ \boldsymbol{M}$$

4.104 The balanced equation is:

$$2HCl(aq) + Na_2CO_3(s) \longrightarrow CO_2(g) + H_2O(l) + 2NaCl(aq)$$

The mole ratio from the balanced equation is 2 moles HCl : 1 mole Na_2CO_3. The moles of HCl needed to react with 0.256 g of Na_2CO_3 are:

$$0.256\ \text{g Na}_2\text{CO}_3 \times \frac{1\ \text{mol Na}_2\text{CO}_3}{105.99\ \text{g Na}_2\text{CO}_3} \times \frac{2\ \text{mol HCl}}{1\ \text{mol Na}_2\text{CO}_3} = 4.83 \times 10^{-3}\ \text{mol HCl}$$

$$\textbf{Molarity HCl} = \frac{\text{moles HCl}}{\text{L soln}} = \frac{4.83 \times 10^{-3}\ \text{mol HCl}}{0.0283\ \text{L soln}} = 0.171\ \text{mol/L} = \mathbf{0.171}\ \boldsymbol{M}$$

4.106 Starting with a balanced chemical equation:

$$CH_3COOH(aq) + NaOH(aq) \longrightarrow CH_3COONa(aq) + H_2O(l)$$

From the molarity and volume of the NaOH solution, you can calculate moles of NaOH. Then, using the mole ratio from the balanced equation above, you can calculate moles of CH_3COOH.

$$5.75\ \text{mL solution} \times \frac{1.00\ \text{mol NaOH}}{1000\ \text{mL of solution}} \times \frac{1\ \text{mol CH}_3\text{COOH}}{1\ \text{mol NaOH}} = 5.75 \times 10^{-3}\ \text{mol CH}_3\text{COOH}$$

$$\textbf{Molarity CH}_3\textbf{COOH} = \frac{5.75 \times 10^{-3}\ \text{mol CH}_3\text{COOH}}{0.0500\ \text{L}} = \mathbf{0.115}\ \boldsymbol{M}$$

4.108 The balanced equation is:

$$Zn(s) + 2AgNO_3(aq) \longrightarrow Zn(NO_3)_2(aq) + 2Ag(s)$$

Let x = mass of Ag produced. We can find the mass of Zn reacted in terms of the amount of Ag produced.

$$x \text{ g Ag} \times \frac{1 \text{ mol Ag}}{107.9 \text{ g Ag}} \times \frac{1 \text{ mol Zn}}{2 \text{ mol Ag}} \times \frac{65.39 \text{ g Zn}}{1 \text{ mol Zn}} = 0.303x \text{ g Zn reacted}$$

The mass of Zn remaining will be:

$$2.50 \text{ g} - \text{amount of Zn reacted} = 2.50 \text{ g Zn} - 0.303x \text{ g Zn}$$

The final mass of the strip, 3.37 g, equals the mass of Ag produced + the mass of Zn remaining.

$$3.37 \text{ g} = x \text{ g Ag} + (2.50 \text{ g Zn} - 0.303\, x \text{ g Zn})$$

$$\mathbf{x = 1.25 \text{ g} = \text{mass of Ag produced}}$$

$$\textbf{mass of Zn remaining} = 3.37 \text{ g} - 1.25 \text{ g} = \mathbf{2.12 \text{ g Zn}}$$

or

$$\textbf{mass of Zn remaining} = 2.50 \text{ g Zn} - 0.303x \text{ g Zn} = 2.50 \text{ g} - (0.303)(1.25 \text{ g}) = \mathbf{2.12 \text{ g Zn}}$$

4.110 The balanced equation is: $HNO_3(aq) + NaOH(aq) \longrightarrow NaNO_3(aq) + H_2O(l)$

$$\text{mol } HNO_3 = \frac{0.211 \text{ mol } HNO_3}{1000 \text{ mL soln}} \times 10.7 \text{ mL soln} = 2.26 \times 10^{-3} \text{ mol } HNO_3$$

$$\text{mol NaOH} = \frac{0.258 \text{ mol NaOH}}{1000 \text{ mL soln}} \times 16.3 \text{ mL soln} = 4.21 \times 10^{-3} \text{ mol NaOH}$$

Since the mole ratio from the balanced equation is 1 mole NaOH : 1 mole HNO_3, then 2.26×10^{-3} mol HNO_3 will react with 2.26×10^{-3} mol NaOH.

$$\text{mol NaOH remaining} = (4.21 \times 10^{-3} \text{ mol}) - (2.26 \times 10^{-3} \text{ mol}) = 1.95 \times 10^{-3} \text{ mol NaOH}$$

$$10.7 \text{ mL} + 16.3 \text{ mL} = 27.0 \text{ mL} = 0.0270 \text{ L}$$

$$\textbf{molarity NaOH} = \frac{1.95 \times 10^{-3} \text{ mol NaOH}}{0.0270 \text{ L}} = \mathbf{0.0722\ \mathit{M}}$$

4.112 The balanced equations for the two reactions are:

$$X(s) + H_2SO_4(aq) \longrightarrow XSO_4(aq) + H_2(g)$$

$$H_2SO_4(aq) + 2NaOH(aq) \longrightarrow Na_2SO_4(aq) + 2H_2O(l)$$

First, let's find the number of moles of excess acid from the reaction with NaOH.

$$0.0334 \text{ L} \times \frac{0.500 \text{ mol NaOH}}{1 \text{ L soln}} \times \frac{1 \text{ mol } H_2SO_4}{2 \text{ mol NaOH}} = 8.35 \times 10^{-3} \text{ mol } H_2SO_4$$

The original number of moles of acid was:

$$0.100\ \cancel{L} \times \frac{0.500\ \text{mol } H_2SO_4}{1\ \cancel{L}\ \text{soln}} = 0.0500\ \text{mol } H_2SO_4$$

The amount of sulfuric acid that reacted with the metal, X, is

$$(0.0500\ \text{mol } H_2SO_4) - (8.35 \times 10^{-3}\ \text{mol } H_2SO_4) = 0.0417\ \text{mol } H_2SO_4.$$

Since the mole ratio from the balanced equation is 1 mole X : 1 mole H_2SO_4, then the amount of X that reacted is 0.0417 mol X.

$$\textbf{molar mass X} = \frac{1.00\ \text{g X}}{0.0417\ \text{mol X}} = \mathbf{24.0\ g/mol}$$

The element is **magnesium**.

4.114 First, calculate the number of moles of glucose present.

$$\frac{0.513\ \text{mol glucose}}{1000\ \cancel{mL}\ \text{soln}} \times 60.0\ \cancel{mL} = 0.0308\ \text{mol glucose}$$

$$\frac{2.33\ \text{mol glucose}}{1000\ \cancel{mL}\ \text{soln}} \times 120.0\ \cancel{mL} = 0.280\ \text{mol glucose}$$

Add the moles of glucose, then divide by the total volume of the combined solutions to calculate the molarity.

60.0 mL + 120.0 mL = 180.0 mL = 0.180 L

$$\textbf{Molarity of final solution} = \frac{(0.0308 + 0.280)\ \text{mol glucose}}{0.180\ \text{L}} = 1.73\ \text{mol/L} = \mathbf{1.73}\ \boldsymbol{M}$$

4.116 Iron(II) compounds can be oxidized to iron(III) compounds. The sample could be tested with a small amount of a strongly colored oxidizing agent like a $KMnO_4$ solution, which is a deep purple color. A loss of color would imply the presence of an oxidizable substance like an iron(II) salt.

4.118 Since both of the original solutions were strong electrolytes, you would expect a mixture of the two solutions to also be a strong electrolyte. However, since the light dims, the mixture must contain fewer ions than the original solution. Indeed, H^+ from the sulfuric acid reacts with the OH^- from the barium hydroxide to form water. The barium cations react with the sulfate anions to form insoluble barium sulfate.

$$2H^+(aq) + SO_4^{2-}(aq) + Ba^{2+}(aq) + 2OH^-(aq) \longrightarrow 2H_2O(l) + BaSO_4(s)$$

Thus, the reaction depletes the solution of ions and the conductivity decreases.

4.120 You could test the conductivity of the solutions. Sugar is a nonelectrolyte and an aqueous sugar solution will not conduct electricity; whereas, NaCl is a strong electrolyte when dissolved in water. Silver nitrate could be added to the solutions to see if silver chloride precipitated. In this particular case, the solutions could also be tasted.

4.122 In a redox reaction, the oxidizing agent gains one or more electrons. In doing so, the oxidation number of the element gaining the electrons must become more negative. In the case of chlorine, the −1 oxidation number is already the most negative state possible. The chloride ion *cannot* accept any more electrons; therefore, hydrochloric acid is *not* an oxidizing agent.

4.124 The reaction is too violent. This could cause the hydrogen gas produced to ignite, and an explosion could result.

4.126 The solid sodium bicarbonate would be the better choice. The hydrogen carbonate ion, HCO_3^-, behaves as a Brønsted base to accept a proton from the acid.

$$HCO_3^-(aq) + H^+(aq) \longrightarrow H_2CO_3(aq) \longrightarrow H_2O(l) + CO_2(g)$$

The heat generated during the reaction of hydrogen carbonate with the acid causes the carbonic acid, H_2CO_3, that was formed to decompose to water and carbon dioxide.

The reaction of the spilled sulfuric acid with sodium hydroxide would produce sodium sulfate, Na_2SO_4, and water. There is a possibility that the Na_2SO_4 could precipitate. Also, the sulfate ion, SO_4^{2-} is a weak base; therefore, the "neutralized" solution would actually be *basic*.

$$H_2SO_4(aq) + 2NaOH(aq) \longrightarrow Na_2SO_4(aq) + 2H_2O(l)$$

Also, NaOH is a caustic substance and therefore is not safe to use in this manner.

4.128 **(a)** Table salt, NaCl, is very soluble in water and is a strong electrolyte. Addition of $AgNO_3$ will precipitate AgCl.

(b) Table sugar or sucrose, $C_{12}H_{22}O_{11}$, is soluble in water and is a nonelectrolyte.

(c) Aqueous acetic acid, CH_3COOH, the primary ingredient of vinegar, is a weak electrolyte. It exhibits all of the properties of acids (Section 4.3).

(d) Baking soda, $NaHCO_3$, is a water-soluble strong electrolyte. It reacts with acid to release CO_2 gas. Addition of $Ca(OH)_2$ results in the precipitation of $CaCO_3$.

(e) Washing soda, $Na_2CO_3 \cdot 10H_2O$, is a water-soluble strong electrolyte. It reacts with acids to release CO_2 gas. Addition of a soluble alkaline-earth salt will precipitate the alkaline-earth carbonate. Aqueous washing soda is also slightly basic (Section 4.3).

(f) Boric acid, H_3BO_3, is weak electrolyte and a weak acid.

(g) Epsom salt, $MgSO_4 \cdot 7H_2O$, is a water-soluble strong electrolyte. Addition of $Ba(NO_3)_2$ results in the precipitation of $BaSO_4$. Addition of hydroxide precipitates $Mg(OH)_2$.

(h) Sodium hydroxide, NaOH, is a strong electrolyte and a strong base. Addition of $Ca(NO_3)_2$ results in the precipitation of $Ca(OH)_2$.

(i) Ammonia, NH_3, is a sharp-odored gas that when dissolved in water is a weak electrolyte and a weak base. NH_3 in the gas phase reacts with HCl gas to produce solid NH_4Cl.

(j) Milk of magnesia, $Mg(OH)_2$, is an insoluble, strong base that reacts with acids. The resulting magnesium salt may be soluble or insoluble.

(k) $CaCO_3$ is an insoluble salt that reacts with acid to release CO_2 gas. $CaCO_3$ is discussed in the Chemistry in Action essays entitled, "An Undesirable Precipitation Reaction" and "Metal from the Sea" in Chapter 4.

With the exception of NH_3 and vinegar, all the compounds in this problem are white solids.

4.130 We carry an additional significant figure throughout this calculation to minimize rounding errors. The balanced equation for the reaction is:

$$XCl(aq) + AgNO_3(aq) \longrightarrow AgCl(s) + XNO_3(aq) \qquad \text{where X = Na, or K}$$

From the amount of AgCl produced, we can calculate the moles of XCl reacted (X = Na, or K).

$$1.913\ \text{g AgCl} \times \frac{1\ \text{mol AgCl}}{143.35\ \text{g AgCl}} \times \frac{1\ \text{mol XCl}}{1\ \text{mol AgCl}} = 0.013345\ \text{mol XCl}$$

Let x = number of moles NaCl. Then, the number of moles of KCl = 0.013345 mol − x. The sum of the NaCl and KCl masses must equal the mass of the mixture, 0.8870 g. We can write:

$$\text{mass NaCl} + \text{mass KCl} = 0.8870\ \text{g}$$

$$\left[x\ \text{mol NaCl} \times \frac{58.44\ \text{g NaCl}}{1\ \text{mol NaCl}}\right] + \left[(0.013345 - x)\ \text{mol KCl} \times \frac{74.55\ \text{g KCl}}{1\ \text{mol KCl}}\right] = 0.8870\ \text{g}$$

$$x = 6.6958 \times 10^{-3} = \text{moles NaCl}$$

$$\text{mol KCl} = 0.013345 - x = 0.013345\ \text{mol} - (6.6958 \times 10^{-3}\ \text{mol}) = 6.6492 \times 10^{-3}\ \text{mol KCl}$$

Converting moles to grams:

$$\text{mass NaCl} = (6.6958 \times 10^{-3}\ \text{mol NaCl}) \times \frac{58.44\ \text{g NaCl}}{1\ \text{mol NaCl}} = 0.3913\ \text{g NaCl}$$

$$\text{mass KCl} = (6.6492 \times 10^{-3}\ \text{mol KCl}) \times \frac{74.55\ \text{g KCl}}{1\ \text{mol KCl}} = 0.4957\ \text{g KCl}$$

The percentages by mass for each compound are:

$$\textbf{\% NaCl} = \frac{0.3913\ \text{g}}{0.8870\ \text{g}} \times 100\% = \textbf{44.11\% NaCl}$$

$$\textbf{\% KCl} = \frac{0.4957\ \text{g}}{0.8870\ \text{g}} \times 100\% = \textbf{55.89\% KCl}$$

4.132 The number of moles of oxalic acid in 5.00×10^2 mL is:

$$\frac{0.100\ \text{mol}\ H_2C_2O_4}{1000\ \text{mL soln}} \times (5.00 \times 10^2\ \text{mL}) = 0.0500\ \text{mol}\ H_2C_2O_4$$

The balanced equation shows a mole ratio of 1 mol Fe_2O_3 : 6 mol $H_2C_2O_4$. The mass of rust that can be removed is:

$$0.0500\ \text{mol}\ H_2C_2O_4 \times \frac{1\ \text{mol}\ Fe_2O_3}{6\ \text{mol}\ H_2C_2O_4} \times \frac{159.7\ \text{g}\ Fe_2O_3}{1\ \text{mol}\ Fe_2O_3} = \mathbf{1.33\ g\ Fe_2O_3}$$

4.134 The precipitation reaction is: $Ag^+(aq) + Br^-(aq) \longrightarrow AgBr(s)$

In this problem, the relative amounts of NaBr and $CaBr_2$ are not known. However, the total amount of Br^- in the mixture can be determined from the amount of AgBr produced. Let's find the number of moles of Br^-. We carry an additional significant figure throughout this calculation to minimize rounding errors.

$$1.6930 \text{ g AgBr} \times \frac{1 \text{ mol AgBr}}{187.8 \text{ g AgBr}} \times \frac{1 \text{ mol Br}^-}{1 \text{ mol AgBr}} = 9.0149 \times 10^{-3} \text{ mol Br}^-$$

The amount of Br^- comes from both NaBr and $CaBr_2$. Let x = number of moles NaBr. Then, the number of moles of $CaBr_2 = \dfrac{9.0149 \times 10^{-3} \text{ mol} - x}{2}$. The moles of $CaBr_2$ are divided by 2, because 1 mol of $CaBr_2$ produces 2 moles of Br^-. The sum of the NaBr and $CaBr_2$ masses must equal the mass of the mixture, 0.9157 g. We can write:

$$\text{mass NaBr} + \text{mass CaBr}_2 = 0.9157 \text{ g}$$

$$\left[x \text{ mol NaBr} \times \frac{102.89 \text{ g NaBr}}{1 \text{ mol NaBr}}\right] + \left[\left(\frac{9.0149 \times 10^{-3} - x}{2}\right) \text{mol CaBr}_2 \times \frac{199.88 \text{ g CaBr}_2}{1 \text{ mol CaBr}_2}\right] = 0.9157 \text{ g}$$

$$2.95x = 0.014751$$

$$x = 5.0003 \times 10^{-3} = \text{moles NaBr}$$

Converting moles to grams:

$$\text{mass NaBr} = (5.0003 \times 10^{-3} \text{ mol NaBr}) \times \frac{102.89 \text{ g NaBr}}{1 \text{ mol NaBr}} = 0.51448 \text{ g NaBr}$$

The percentage by mass of NaBr in the mixture is:

$$\textbf{\% NaBr} = \frac{0.51448 \text{ g}}{0.9157 \text{ g}} \times 100\% = \mathbf{56.18\% \ NaBr}$$

4.136 There are two moles of Cl^- per one mole of $CaCl_2$.

(a) $$25.3 \text{ g CaCl}_2 \times \frac{1 \text{ mol CaCl}_2}{110.98 \text{ g CaCl}_2} \times \frac{2 \text{ mol Cl}^-}{1 \text{ mol CaCl}_2} = 0.456 \text{ mol Cl}^-$$

$$\textbf{Molarity Cl}^- = \frac{0.456 \text{ mol Cl}^-}{0.325 \text{ L soln}} = 1.40 \text{ mol/L} = \mathbf{1.40}\ \boldsymbol{M}$$

(b) We need to convert from mol/L to grams in 0.100 L.

$$\frac{1.40 \text{ mol Cl}^-}{1 \text{ L soln}} \times \frac{35.45 \text{ g Cl}}{1 \text{ mol Cl}^-} \times 0.100 \text{ L soln} = \mathbf{4.96 \ g \ Cl^-}$$

4.138 **(a)** $NH_4^+(aq) + OH^-(aq) \longrightarrow NH_3(aq) + H_2O(l)$

(b) From the amount of NaOH needed to neutralize the 0.2041 g sample, we can find the amount of the 0.2041 g sample that is NH_4NO_3.

First, calculate the moles of NaOH.

$$\frac{0.1023 \text{ mol NaOH}}{1000 \text{ mL of soln}} \times 24.42 \text{ mL soln} = 2.498 \times 10^{-3} \text{ mol NaOH}$$

Using the mole ratio from the balanced equation, we can calculate the amount of NH_4NO_3 that reacted.

$$(2.498 \times 10^{-3} \text{ mol NaOH}) \times \frac{1 \text{ mol } NH_4NO_3}{1 \text{ mol NaOH}} \times \frac{80.05 \text{ g } NH_4NO_3}{1 \text{ mol } NH_4NO_3} = 0.2000 \text{ g } NH_4NO_3$$

The purity of the NH_4NO_3 sample is:

$$\textbf{\% purity} = \frac{0.2000 \text{ g}}{0.2041 \text{ g}} \times 100\% = \textbf{97.99\%}$$

4.140 Using the rules for assigning oxidation numbers given in Section 4.4, H is +1, F is −1, so the oxidation number of O must be **zero**.

4.142 The balanced equation is:

$$3CH_3CH_2OH + 2K_2Cr_2O_7 + 8H_2SO_4 \longrightarrow 3CH_3COOH + 2Cr_2(SO_4)_3 + 2K_2SO_4 + 11H_2O$$

From the amount of $K_2Cr_2O_7$ required to react with the blood sample, we can calculate the mass of ethanol (CH_3CH_2OH) in the 10.0 g sample of blood.

First, calculate the moles of $K_2Cr_2O_7$ reacted.

$$\frac{0.07654 \text{ mol } K_2Cr_2O_7}{1000 \text{ mL soln}} \times 4.23 \text{ mL} = 3.24 \times 10^{-4} \text{ mol } K_2Cr_2O_7$$

Next, using the mole ratio from the balanced equation, we can calculate the mass of ethanol that reacted.

$$3.24 \times 10^{-4} \text{ mol } K_2Cr_2O_7 \times \frac{3 \text{ mol ethanol}}{2 \text{ mol } K_2Cr_2O_7} \times \frac{46.07 \text{ g ethanol}}{1 \text{ mol ethanol}} = 0.0224 \text{ g ethanol}$$

The percent ethanol by mass is:

$$\textbf{\% by mass ethanol} = \frac{0.0224 \text{ g}}{10.0 \text{ g}} \times 100\% = \textbf{0.224\%}$$

This is well above the legal limit of 0.1 percent by mass ethanol in the blood. The individual should be prosecuted for drunk driving.

4.144 **(a)** $Zn(s) + H_2SO_4(aq) \longrightarrow ZnSO_4(aq) + H_2(g)$

(b) $2KClO_3(s) \longrightarrow 2KCl(s) + 3O_2(g)$

(c) $Na_2CO_3(s) + 2HCl(aq) \longrightarrow 2NaCl(aq) + CO_2(g) + H_2O(l)$

(d) $NH_4NO_2(s) \xrightarrow{\text{heat}} N_2(g) + 2H_2O(g)$

4.146 NH_4Cl exists as NH_4^+ and Cl^-. To form NH_3 and HCl, a proton (H^+) is transferred from NH_4^+ to Cl^-. Therefore, this is a Brønsted acid-base reaction.

4.148 We carry an additional significant figure throughout this calculation to minimize rounding errors.

(a)
First Solution:

$$0.8214 \text{ g KMnO}_4 \times \frac{1 \text{ mol KMnO}_4}{158.04 \text{ g KMnO}_4} = 5.1974 \times 10^{-3} \text{ mol KMnO}_4$$

$$M = \frac{\text{mol solute}}{\text{L of soln}} = \frac{5.1974 \times 10^{-3} \text{ mol KMnO}_4}{0.5000 \text{ L}} = 1.0395 \times 10^{-2}\ M$$

Second Solution:

$$M_1V_1 = M_2V_2$$
$$(1.0395 \times 10^{-2}\ M)(2.000 \text{ mL}) = M_2(1000 \text{ mL})$$
$$M_2 = 2.079 \times 10^{-5}\ M$$

Third Solution:

$$M_1V_1 = M_2V_2$$
$$(2.079 \times 10^{-5}\ M)(10.00 \text{ mL}) = M_2(250.0 \text{ mL})$$
$$\mathbf{M_2 = 8.316 \times 10^{-7}\ M}$$

(b) From the molarity and volume of the final solution, we can calculate the moles of $KMnO_4$. Then, the mass can be calculated from the moles of $KMnO_4$.

$$\frac{8.316 \times 10^{-7} \text{ mol KMnO}_4}{1000 \text{ mL of soln}} \times 250 \text{ mL} = 2.079 \times 10^{-7} \text{ mol KMnO}_4$$

$$2.079 \times 10^{-7} \text{ mol KMnO}_4 \times \frac{158.04 \text{ g KMnO}_4}{1 \text{ mol KMnO}_4} = \mathbf{3.286 \times 10^{-5}\ g\ KMnO_4}$$

This mass is too small to directly weigh accurately.

4.150 The first titration oxidizes Fe^{2+} to Fe^{3+}. This titration gives the amount of Fe^{2+} in solution. Zn metal is added to reduce all Fe^{3+} back to Fe^{2+}. The second titration oxidizes all the Fe^{2+} back to Fe^{3+}. We can find the amount of Fe^{3+} in the original solution by difference.

Titration #1: The mole ratio between Fe^{2+} and MnO_4^- is 5:1.

$$23.0 \text{ mL soln} \times \frac{0.0200 \text{ mol MnO}_4^-}{1000 \text{ mL soln}} \times \frac{5 \text{ mol Fe}^{2+}}{1 \text{ mol MnO}_4^-} = 2.30 \times 10^{-3} \text{ mol Fe}^{2+}$$

$$[\mathbf{Fe^{2+}}] = \frac{\text{mol solute}}{\text{L of soln}} = \frac{2.30 \times 10^{-3}\ \text{mol Fe}^{2+}}{25.0 \times 10^{-3}\ \text{L soln}} = \mathbf{0.0920\ \mathit{M}}$$

Titration #2: The mole ratio between Fe^{2+} and MnO_4^- is 5:1.

$$40.0\ \text{mL soln} \times \frac{0.0200\ \text{mol MnO}_4^-}{1000\ \text{mL soln}} \times \frac{5\ \text{mol Fe}^{2+}}{1\ \text{mol MnO}_4^-} = 4.00 \times 10^{-3}\ \text{mol Fe}^{2+}$$

In this second titration, there are more moles of Fe^{2+} in solution. This is due to Fe^{3+} in the original solution being reduced by Zn to Fe^{2+}. The number of moles of Fe^{3+} in solution is:

$$(4.00 \times 10^{-3}\ \text{mol}) - (2.30 \times 10^{-3}\ \text{mol}) = 1.70 \times 10^{-3}\ \text{mol Fe}^{3+}$$

$$[\mathbf{Fe^{3+}}] = \frac{\text{mol solute}}{\text{L of soln}} = \frac{1.70 \times 10^{-3}\ \text{mol Fe}^{3+}}{25.0 \times 10^{-3}\ \text{L soln}} = \mathbf{0.0680\ \mathit{M}}$$

4.152 **(a)** The precipitation reaction is: $Mg^{2+}(aq) + 2OH^-(aq) \longrightarrow Mg(OH)_2(s)$

The acid-base reaction is: $Mg(OH)_2(s) + 2HCl(aq) \longrightarrow MgCl_2(aq) + 2H_2O(l)$

The redox reactions are:

$$Mg^{2+} + 2e^- \longrightarrow Mg$$
$$2Cl^- \longrightarrow Cl_2 + 2e^-$$
$$\overline{MgCl_2 \longrightarrow Mg + Cl_2}$$

(b) NaOH is much more expensive than CaO.

(c) Dolomite has the advantage of being an additional source of magnesium that can also be recovered.

4.154 Let's set up a table showing each reaction, the volume of solution added, and the species responsible for any electrical conductance of the solution. Note that if a substance completely dissociates into +1 ions and –1 ions in solution, its conductance unit will be twice its molarity. Similarly, if a substance completely dissociates into +2 ions and –2 ions in solution, its conductance unit will be four times its molarity.

(1) $CH_3COOH(aq) + KOH(aq) \rightarrow CH_3COOK(aq) + H_2O(l)$

Volume (added)	Conductance unit	
0 L, KOH	$[CH_3COOH] = 1.0\ M$, (negligible ions, weak acid)	0 unit
1 L, KOH	$[CH_3COOK] = \frac{1.0\ \text{mol}}{2.0\ \text{L}} = 0.50\ M$, (CH_3COO^-, K^+)	1 unit
2 L, KOH	$[CH_3COOK] = \frac{1.0\ \text{mol}}{3.0\ \text{L}} = \frac{1}{3}M$, $[KOH] = \frac{1.0\ \text{mol}}{3.0\ \text{L}} = \frac{1}{3}M$, (K^+, OH^-)	1.3 units

(2) $NaOH(aq) + HCl(aq) \rightarrow NaCl(aq) + H_2O(l)$

Volume (added)	Conductance unit	
0 L, NaOH	$[HCl] = 1.0\ M, (H^+, Cl^-)$	2 units
1 L, NaOH	$[NaCl] = \dfrac{1.0\text{ mol}}{2.0\text{ L}} = 0.50\ M, (Na^+, Cl^-)$	1 unit
2 L, NaOH	$[NaCl] = \dfrac{1.0\text{ mol}}{3.0\text{ L}} = \dfrac{1}{3}M,\ [NaOH] = \dfrac{1.0\text{ mol}}{3.0\text{ L}} = \dfrac{1}{3}M, (Na^+, OH^-)$	1.3 units

(3) $BaCl_2(aq) + K_2SO_4(aq) \rightarrow BaSO_4(s) + 2KCl(aq)$

Volume (added)	Conductance unit	
0 L, $BaCl_2$	$[K_2SO_4] = 1.0\ M, (2K^+, SO_4^{2-})$	4 units
1 L, $BaCl_2$	$[KCl] = \dfrac{2.0\text{ mol}}{2.0\text{ L}} = 1.0\ M, (K^+, Cl^-)$	2 units
2 L, $BaCl_2$	$[KCl] = \dfrac{2.0\text{ mol}}{3.0\text{ L}} = \dfrac{2}{3}M,\ [BaCl_2] = \dfrac{1.0\text{ mol}}{3.0\text{ L}} = \dfrac{1}{3}M, (Ba^{2+}, 2Cl^-)$	2.7 units

(4) $NaCl(aq) + AgNO_3(aq) \rightarrow AgCl(s) + NaNO_3(aq)$

Volume (added)	Conductance unit	
0 L, NaCl	$[AgNO_3] = 1.0\ M, (Ag^+, NO_3^-)$	2 units
1 L, NaCl	$[NaNO_3] = \dfrac{1.0\text{ mol}}{2.0\text{ L}} = 0.50\ M, (Na^+, NO_3^-)$	1 unit
2 L, NaCl	$[NaNO_3] = \dfrac{1.0\text{ mol}}{3.0\text{ L}} = \dfrac{1}{3}M,\ [NaCl] = \dfrac{1.0\text{ mol}}{3.0\text{ L}} = \dfrac{1}{3}M, (Na^+, Cl^-)$	1.3 units

(5) $CH_3COOH(aq) + NH_3(aq) \rightarrow CH_3COONH_4$

Volume (added)	Conductance unit	
0 L, CH_3COOH	$[NH_3] = 1.0\ M$, (negligible ions, weak base)	0 unit
1 L, CH_3COOH	$[CH_3COONH_4] = \dfrac{1.0\text{ mol}}{2.0\text{ L}} = 0.50\ M, (CH_3COO^-, NH_4^+)$	1 unit
2 L, CH_3COOH	$[CH_3COONH_4] = \dfrac{1.0\text{ mol}}{3.0\text{ L}} = \dfrac{1}{3}\ M$	0.67 unit

Matching this data to the diagrams shown, we find:

Diagram **(a)**: Reactions **(2)** and **(4)**
Diagram **(b)**: Reaction **(5)**
Diagram **(c)**: Reaction **(3)**
Diagram **(d)**: Reaction **(1)**

CHAPTER 5
GASES

PROBLEM-SOLVING STRATEGIES AND TUTORIAL SOLUTIONS

TYPES OF PROBLEMS

Problem Type 1: Pressure Conversions (Also see Problem Type 5, Chapter 1).

Problem Type 2: Calculations Using the Ideal Gas Law.
- **(a)** Given three of the four variable quantities in the equation (P, V, n, and T).
- **(b)** Only one or two variable quantities have fixed values.
- **(c)** Calculations using density or molar mass (Also see Density calculations, Problem Type 1, Chapter 1).
- **(d)** Stoichiometry involving gases (Also see Stoichiometry problems, Problem Type 7, Chapter 3).

Problem Type 3: Dalton's Law of Partial Pressures.
- **(a)** Introduction
- **(b)** Collecting a gas over water.

Problem Type 4: Root-Mean-Square (RMS) Speed.

Problem Type 5: Deviations from Ideal Behavior.

PROBLEM TYPE 1: PRESSURE CONVERSIONS

(Also see Problem Type 5, Chapter 1)

In order to convert from one unit of pressure to another, you need to be proficient at the dimensional analysis method. See Section 1.9 of your text. Unit conversions can seem daunting, but if you keep track of the units, making sure that the appropriate units cancel, your effort will be rewarded.

EXAMPLE 5.1
Convert 555 mmHg to kPa.

Strategy: Map out a strategy to proceed from initial units to final units based on available conversion factors. Looking in Section 5.2 of your text, you should find the following conversions.

$$1\text{ atm} = 760\text{ mmHg}$$
$$1\text{ atm} = 1.01325 \times 10^2\text{ kPa}$$

You should come up with the following strategy:

$$\text{mm Hg} \rightarrow \text{atm} \rightarrow \text{kPa}$$

Solution: Use the following method to ensure that you obtain the desired unit.

$$\text{Given unit} \times \left(\frac{\text{desired unit}}{\text{given unit}}\right) = \text{desired unit}$$

$$\textbf{? kPa} = 555\text{ mmHg (given)} \times \left(\frac{1\text{ atm (desired)}}{760\text{ mmHg (given)}}\right) \times \left(\frac{1.01325 \times 10^2\text{ kPa (desired)}}{1\text{ atm (given)}}\right) = \textbf{74.0 kPa}$$

Note: In the first conversion factor (atm/mmHg), atm is the desired unit. When moving on to the next conversion factor (kPa/atm), atm is now given, and the desired unit is kPa.

PRACTICE EXERCISE

1. Convert 5.32 atm to mmHg.

Text Problem: 5.14

PROBLEM TYPE 2: CALCULATIONS USING THE IDEAL GAS LAW

You will encounter five different types of problems in this chapter that use the Ideal Gas Law. Each type will be addressed individually. Remember that R (the gas constant) has the units L·atm/mol·K. Therefore, in problems that contain R, you *must* use the following units: volume in liters (L), pressure in atmospheres (atm), and temperature in Kelvin (K).

A. Given three of the four variable quantities in the equation (*P*, *V*, *n*, and *T*)

If you know the values of three of the variable quantities, you can solve $PV = nRT$ algebraically for the fourth one. Then, you can calculate its value by substituting in the three known quantities.

EXAMPLE 5.2

Carbon monoxide gas, CO, stored in a 2.00 L container at 25.0°C exerts a pressure of 15.5 atm. How many moles of CO(g) are in the container?

Strategy: This problem gives the volume, temperature, and pressure of CO gas. Is the gas undergoing a change in any of its properties? What equation should we use to solve for moles of CO? What temperature unit should be used?

Solution: Because no changes in gas properties occur, we can use the ideal gas equation to calculate moles. Rearranging Equation (5.8) of the text, we write:

$$n = \frac{PV}{RT}$$

Check that the three known quantities (P, V, and T) have the appropriate units. P and V have the correct units, but T must be converted to units of K.

$$T(\text{K}) = {}^\circ\text{C} + 273^\circ$$
$$T(\text{K}) = 25.0^\circ + 273^\circ = 298\text{ K}$$

Calculate the value of n by substituting the three known quantities into the equation.

$$\boldsymbol{n} = \frac{PV}{RT} = \frac{(15.5\ \cancel{\text{atm}})(2.00\ \cancel{\text{L}})}{298\ \cancel{\text{K}}} \times \frac{\text{mol}\cdot\cancel{\text{K}}}{0.0821\ \cancel{\text{L}}\cdot\cancel{\text{atm}}} = \textbf{1.27 mol CO}$$

PRACTICE EXERCISE

2. Carbon dioxide gas, CO_2, and water vapor are the two species primarily responsible for the greenhouse effect. Combustion of fossil fuels adds 22 billion tons of carbon dioxide to the atmosphere each year. How many liters of CO_2 is this at 25°C and 1.0 atm pressure?

Text Problems: 5.32, 5.40, 5.42

B. Only one or two variable quantities have fixed values.

Let's look at the case where n and P are constant (Charles' law). We start with Equation (5.9) of the text.

$$\frac{P_1V_1}{n_1T_1} = \frac{P_2V_2}{n_2T_2} \qquad \text{(5.9, text)}$$

Because $n_1 = n_2$ and $P_1 = P_2$,

$$\frac{V_1}{T_1} = \frac{V_2}{T_2}$$

This is Charles' law.

Solve the equation algebraically for the missing quantity. Then, you can calculate its value by substituting in the known quantities.

EXAMPLE 5.3

Given 10.0 L of neon gas at 5.0°C and 630 mmHg, calculate the new volume at 400°C and 2.5 atm.

Strategy: The amount of gas remains constant, but the pressure, temperature, and volume change. What equation would you use to solve for the final volume?

Solution: We start with Equation (5.9) of the text.

$$\frac{P_1V_1}{n_1T_1} = \frac{P_2V_2}{n_2T_2}$$

Because $n_1 = n_2$,

$$\frac{P_1V_1}{T_1} = \frac{P_2V_2}{T_2}$$

Solve the equation algebraically for V_2, and then substitute in the known quantities to solve for V_2.

$$P_2 = 2.5\ \text{atm} \times \frac{760\ \text{mmHg}}{1\ \text{atm}} = 1.9 \times 10^3\ \text{mmHg}$$

$$V_2 = \frac{P_1V_1T_2}{P_2T_1}$$

$$V_2 = \frac{(630\ \text{mmHg})(10.0\ \text{L})(673\ \text{K})}{(1.9 \times 10^3\ \text{mmHg})(278\ \text{K})} = \mathbf{8.0\ L}$$

Tip: Since R is not in this equation, the units of pressure and volume do *not* have to be atm and liters, respectively. However, temperature must *always* be in units of Kelvin for gas law calculations.

PRACTICE EXERCISE

3. After flying in an airplane, a passenger finds that his sticky hair-gel has opened in his bag making a complete mess. This happens because the cabin is pressurized to about 8000 feet above sea level. During flight, the volume of air inside the bottle increases due to the decreased pressure. Sometimes, this can cause the flip-top cap to open. If the volume of air inside the bottle is 125 mL at sea level (P = 760 mmHg), what volume does the air in the bottle occupy in the airplane (P = 595 mmHg)? Assume that the temperature is kept constant.

Text Problems: 5.20, **5.22**, **5.24**, 5.34, 5.36, 5.38

C. Calculations involving density (*d*) or molar mass ($\mathcal{M}$)

(Also see Problem Type 1, Chapter 1)

Step 1: Algebraically solve the ideal gas equation for either density $\left(\frac{m}{V}\right)$ or molar mass ($\mathcal{M}$).

$$PV = nRT$$

$$\frac{n}{V} = \frac{P}{RT} \quad \text{and} \quad n = \frac{m}{\mathcal{M}}$$

Substituting for *n*,

$$\frac{m}{\mathcal{M}V} = \frac{P}{RT}$$

and,

$$\boldsymbol{d = \frac{m}{V} = \frac{P\mathcal{M}}{RT}} \qquad \text{(5.11, text)}$$

Furthermore,

$$\boldsymbol{\mathcal{M} = \frac{dRT}{P}}$$

Step 2: Calculate the value of density or molar mass by substituting in the known quantities.

EXAMPLE 5.4

What is the density of methane gas, CH_4, at STP.

Strategy: This problem is simplified if you recall that 1 mole of an ideal gas occupies a volume of 22.4 L at STP. If we can calculate the mass of 1 mole of methane, we can solve for its density.

Solution: Let's assume that we have 1 mole of methane gas. At STP, 1 mole of methane will occupy a volume of 22.4 L. First, we calculate the molar mass of methane.

$$\mathcal{M}(CH_4) = 12.01 \text{ g} + 4(1.008 \text{ g}) = 16.04 \text{ g}$$

Knowing the mass of 1 mole of methane and the volume it occupies, we can calculate the density.

$$d = \frac{\text{mass}}{\text{volume}}$$

$$d = \frac{16.04 \text{ g}}{22.4 \text{ L}} = \mathbf{0.716\ g/L}$$

This problem could also be solved using Equation (5.11) of the text derived above.

$$\boldsymbol{d = \frac{P\mathcal{M}}{RT}}$$

Calculate the density by substituting in the known quantities.

STP: $P = 1.00$ atm
$T = 273$ K

$\mathcal{M} = 16.04$ g/mol

$$d = \frac{(1.00 \text{ atm})\left(16.04 \frac{\text{g}}{\text{mol}}\right)}{273 \text{ K}} \times \frac{\text{mol} \cdot \text{K}}{0.0821 \text{ L} \cdot \text{atm}} = \mathbf{0.716\ g/L}$$

PRACTICE EXERCISE

4. An element that exists as a diatomic gas at room temperature has a density of 1.553 g/L at 25°C and 1 atm pressure. Identify the unknown gas.

Text Problems: 5.44, 5.48, 5.50

D. Stoichiometry Involving Gases

(Also see Problem Type 7, Chapter 3)

A typical gas stoichiometry problem involves the following approach:

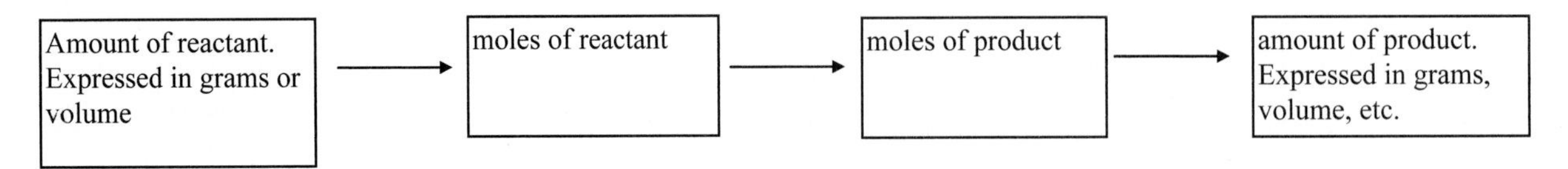

EXAMPLE 5.5

Oxygen gas was discovered by decomposing mercury (II) oxide.

$$2HgO(s) \longrightarrow 2Hg(l) + O_2(g)$$

What volume of oxygen gas would be produced by the reaction of 35.2 g of the oxide if the gas is collected at STP?

Strategy: From the moles of HgO reacted, we can calculate the moles of O_2 produced. From the balanced equation, we see that 2 mol HgO $\simeq$ 1 mol O_2. Once moles of O_2 are determined, we can use the ideal gas equation to calculate the volume of O_2.

Solution: Since stoichiometry uses mole ratios, make sure that you start with a balanced equation. In this problem, you are given the balanced equation. First let's calculate moles of O_2 produced.

$$\text{grams HgO} \rightarrow \text{moles HgO} \rightarrow \text{moles } O_2$$

$$? \text{ mol } O_2 = 35.2 \text{ g HgO} \times \frac{1 \text{ mol HgO}}{216.6 \text{ g HgO}} \times \frac{1 \text{ mol } O_2}{2 \text{ mol HgO}} = 0.0813 \text{ mol } O_2$$

Solve the ideal gas equation algebraically for V_{O_2}. Then, calculate the volume by substituting the known quantities into the equation.

STP: $P = 1.00$ atm
$T = 273$ K

$$V_{O_2} = \frac{n_{O_2}RT}{P}$$

$$V_{O_2} = \frac{(0.0813 \text{ mol})(273 \text{ K})}{1.00 \text{ atm}} \times \frac{0.0821 \text{ L}\cdot\text{atm}}{\text{mol}\cdot\text{K}} = \mathbf{1.82 \text{ L}}$$

EXAMPLE 5.6

Calculate the volume of methane, CH_4, at STP required to completely consume 3.50 L of oxygen at STP.

Strategy: You could follow the strategy used in the previous example to solve this problem. You could find the moles of oxygen in 3.50 L, apply the correct mole ratio to find moles of methane, and then use the ideal gas equation to find the volume of methane. However, there is a short cut if you remember Avogadro's Law. Avogadro's Law states that the volume of a gas is directly proportional to the number of moles of gas at constant temperature and pressure.

Solution: Write the balanced chemical equation.

$$CH_4(g) + 2O_2(g) \longrightarrow CO_2(g) + 2H_2O(l)$$

The stoichiometric ratio, $\frac{1 \text{ mol } CH_4}{2 \text{ mol } O_2}$, can be written in terms of volume, $\frac{1 \text{ L } CH_4}{2 \text{ L } O_2}$. Calculate the volume of methane using the above conversion factor.

$$\textbf{? L CH}_4 = 3.50 \cancel{\text{L}}\ O_2 \times \frac{1 \text{ L } CH_4}{2 \cancel{\text{L}}\ O_2} = \textbf{1.75 L CH}_4$$

PRACTICE EXERCISE

5. The decomposition of sodium azide (NaN_3) to sodium and nitrogen gas was one of the first reactions used to inflate air-bag systems in automobiles.

$$2NaN_3(s) \longrightarrow 2Na(l) + 3N_2(g)$$

What mass of NaN_3 in grams would be needed to inflate a 100 L air bag with nitrogen at 25.0°C and 755 mmHg?

Text Problems: **5.52**, 5.54, **5.56**, 5.58, 5.60

PROBLEM TYPE 3: DALTON'S LAW OF PARTIAL PRESSURES

A. Introduction

In order to understand how to approach problems with partial pressures, it is important to know (1) the derivation of Dalton's law of partial pressures and (2) the relationship between partial pressure and the total pressure.

(1) Dalton's law for two gases, **A** and **B**, in a container of volume, V.

$$P_{\text{Total}} = \frac{n_{\text{Total}}RT}{V}$$

and, $n_{\text{Total}} = n_A + n_B$

$$P_{\text{Total}} = \frac{(n_A + n_B)RT}{V}$$

$$P_{\text{Total}} = \frac{n_A RT}{V} + \frac{n_B RT}{V}$$

$$\boldsymbol{P_{\text{Total}} = P_A + P_B}$$

In general, $\boldsymbol{P_{\text{Total}} = P_1 + P_2 + P_3 + \ldots + P_n}$, where $P_1, P_2, P_3, \ldots$ are the partial pressures of components 1, 2, 3, . .

(2) The relationship between partial pressure and P_{Total} for two gases, **A** and **B**, in a container of volume, V.

$$\frac{P_A}{P_T} = \frac{\frac{n_A RT}{V}}{\frac{(n_A + n_B)RT}{V}}$$

$$\frac{P_A}{P_T} = \frac{n_A}{(n_A + n_B)} \quad \text{and} \quad \frac{n_A}{(n_A + n_B)} = X_A$$

where X_A = mole fraction of component A.

$$\frac{P_A}{P_T} = X_A$$

$$\mathbf{P_A = X_A P_T}$$

Similarly,

$$\mathbf{P_B = X_B P_T}$$

In general, $\mathbf{P_i = X_i P_T}$, where X_i is the mole fraction of substance i.

EXAMPLE 5.7

A 2.00 L container at 22.0°C contains a mixture of 1.00 g $H_2(g)$ and 1.00 g He(g).

(a) What are the partial pressures of H_2 and He? What is the total pressure?

Strategy: This is a mixture of two gases that obeys Dalton's law of partial pressures.

$$P_T = P_{H_2(g)} + P_{He(g)}$$

The partial pressure of each component can be calculated from the amount of each gas and the given conditions. The total pressure is the sum of the two partial pressures calculated.

Solution: First, calculate the moles of H_2 and He.

$$n_{H_2} = 1.00\,\text{g} \times \frac{1\text{ mol } H_2}{2.016\text{ g } H_2} = 0.496\text{ mol } H_2$$

$$n_{He} = 1.00\,\text{g} \times \frac{1\text{ mol He}}{4.003\text{ g He}} = 0.250\text{ mol He}$$

Solve the ideal gas equation algebraically for P, and then substitute in the known quantities to solve for P_{H_2} or P_{He}.

$$T(\text{K}) = 22.0° + 273° = 295\text{ K}$$

$$P_{H_2} = \frac{n_{H_2}RT}{V} \qquad P_{He} = \frac{n_{He}RT}{V}$$

$$P_{H_2} = \frac{(0.496\text{ mol})(295\text{ K})}{2.00\text{ L}} \times \frac{0.0821\text{ L}\cdot\text{atm}}{\text{mol}\cdot\text{K}} \qquad P_{He} = \frac{(0.250\text{ mol})(295\text{ K})}{2.00\text{ L}} \times \frac{0.0821\text{ L}\cdot\text{atm}}{\text{mol}\cdot\text{K}}$$

$$\mathbf{P_{H_2} = 6.01\text{ atm}} \qquad \mathbf{P_{He} = 3.03\text{ atm}}$$

$$P_{Total} = P_{H_2} + P_{He}$$

$$\mathbf{P_{Total}} = 6.01\text{ atm} + 3.03\text{ atm} = \mathbf{9.04\text{ atm}}$$

(b) What are the mole fractions of H_2 and He?

Strategy: Remember that $P_i = X_i P_T$.

Solution:

$$P_{H_2} = X_{H_2}P_T \qquad P_{He} = X_{He}P_T$$

$$X_{H_2} = \frac{P_{H_2}}{P_T} \qquad X_{He} = \frac{P_{He}}{P_T}$$

$$\mathbf{X_{H_2}} = \frac{6.01 \text{ atm}}{9.04 \text{ atm}} = \mathbf{0.665} \qquad \mathbf{X_{He}} = \frac{3.03 \text{ atm}}{9.04 \text{ atm}} = \mathbf{0.335}$$

PRACTICE EXERCISE

6. About two-thirds of the carbon monoxide (CO) emissions in the United States come from automobiles. CO is extremely toxic to humans because it binds 210 times more strongly to hemoglobin than does O_2. Hemoglobin is the iron-containing protein responsible for oxygen transport in blood. In a typical urban environment, the CO concentration is 10 parts per million (ppm). Assuming an atmospheric pressure of 750 torr, calculate the partial pressure of CO.

> **Hint:** 1 ppm of a gas refers to one part by volume in 1 million volume units of the whole (volume fraction).

Text Problems: 5.64, 5.66, 5.68, 5.70

B. Collecting a gas over water

In this type of problem, you must account for the fact that

$$P_{TOTAL} = P_{SAMPLE} + P_{H_2O(g)}$$

The pressure of the gas sample of interest is:

$$P_{SAMPLE} = P_{TOTAL} - P_{H_2O(g)}$$

EXAMPLE 5.8

When heated, calcium carbonate decomposes forming solid calcium oxide and carbon dioxide gas. A sample of calcium carbonate is completely decomposed by heating and 50.5 mL of gas is collected over water at 755.0 torr and 25.0°C. How much did the sample of calcium carbonate weigh? The vapor pressure of water at 25°C is 23.76 torr.

Strategy: If we calculate the pressure of CO_2 gas collected, we can then calculate moles of CO_2 using the ideal gas equation. Then, using the mole ratio from the balanced equation, we can convert to moles calcium carbonate. Finally, we convert to grams of calcium carbonate.

Solution: We need to start with a balanced equation to have the correct mole ratio between CO_2 and $CaCO_3$.

$$CaCO_3(s) \longrightarrow CaO(s) + CO_2(g)$$

Since CO_2 is collected over water, water vapor is collected in addition to the CO_2 gas.

$$P_T = P_{CO_2} + P_{H_2O(g)}$$

$$P_{CO_2} = P_T - P_{H_2O(g)} = (755.0 - 23.76)\text{ torr} = 731.2 \text{ torr}$$

Solve the ideal gas equation algebraically for moles of CO_2. Then, substitute in the known values.

$$P_{CO_2} = 731.2 \text{ torr} \times \frac{1 \text{ atm}}{760 \text{ torr}} = 0.9621 \text{ atm}$$

$$T(\text{K}) = 25.0° + 273° = 298 \text{ K}$$

$$V = 50.5 \text{ mL} \times \frac{1 \text{ L}}{1000 \text{ mL}} = 0.0505 \text{ L}$$

$$n_{CO_2} = \frac{P_{CO_2}V}{RT}$$

$$n_{CO_2} = \frac{(0.9621 \text{ atm})(0.0505 \text{ L})}{298 \text{ K}} \times \frac{\text{mol} \cdot \text{K}}{0.0821 \text{ L} \cdot \text{atm}} = 1.99 \times 10^{-3} \text{ mol } CO_2$$

Calculate the mass of $CaCO_3$ using the mole ratio from the balanced equation and the molar mass of $CaCO_3$.

$$\textbf{? g CaCO}_3 = (1.99 \times 10^{-3} \text{ mol } CO_2) \times \frac{1 \text{ mol } CaCO_3}{1 \text{ mol } CO_2} \times \frac{100.09 \text{ g } CaCO_3}{1 \text{ mol } CaCO_3} = \mathbf{0.199 \text{ g } CaCO_3}$$

PRACTICE EXERCISE

7. Fermentation is one of the oldest studied biochemical processes. The fermentation process is a catabolic pathway, which involves the use of yeast to convert sugars into ethanol and carbon dioxide. In brewing beer, malt sugar is fermented to form ethanol (CH_3CH_2OH) and carbon dioxide (CO_2) according to the following overall reaction:

$$C_6H_{12}O_6(aq) \longrightarrow 2CH_3CH_2OH(aq) + 2CO_2(g)$$

A micro-brewery is making a batch of beer, and they wish to determine the mass percentage of ethanol in their brew by collecting the $CO_2(g)$ evolved. They start the process by mixing malt sugar, hops, and yeast in water. The mixture has a total mass of 800 lbs. During fermentation, 10,000 L of CO_2 is collected over water at 25.0°C and 750 mmHg. What is the mass percentage of ethanol in the beer? (The vapor pressure of water at 25°C is 23.76 mm Hg.)

Text Problems: 5.66, **5.68**

PROBLEM TYPE 4: ROOT-MEAN-SQUARE (RMS) SPEED

This type of problem typically involves substituting known values into the following equation:

$$u_{rms} = \sqrt{\frac{3RT}{\mathcal{M}}}$$

You must remember to use R in units of $\frac{\text{J}}{\text{mol} \cdot \text{K}}$ and $\mathcal{M}$ in units of kg/mol. 1 joule $= 1 \frac{\text{kg} \cdot \text{m}^2}{\text{s}^2}$. If these units are used, the units of u_{rms} are m/s.

$$u_{rms}\text{ (units)} = \sqrt{\frac{\left(\frac{\text{J}}{\text{mol} \cdot \text{K}}\right)(\text{K})}{\left(\frac{\text{kg}}{\text{mol}}\right)}}$$

$$u_{rms}\text{ (units)} = \sqrt{\frac{\left(\frac{\text{kg}\cdot\text{m}^2}{\text{s}^2} \times \frac{1}{\text{mol}\cdot\text{K}}\right)(\text{K})}{\left(\frac{\text{kg}}{\text{mol}}\right)}}$$

$$u_{rms}\text{ (units)} = \sqrt{\frac{\text{m}^2}{\text{s}^2}}$$

$$\boldsymbol{u_{rms}\text{ (units)} = \frac{\text{m}}{\text{s}}}$$

EXAMPLE 5.9

Which has a higher root mean square velocity, $H_2(g)$ at 150 K or He(g) at 650K?

Strategy: To calculate the root-mean-square speed, we use Equation (5.16) of the text. What units should we use for R and $\mathcal{M}$ so the u_{rms} will be expressed in units of m/s.

Solution: To calculate u_{rms}, the units of R should be 8.314 J/mol·K, and because 1 J = 1 kg·m^2/s^2, the units of molar mass must be kg/mol. Let's convert the molar masses in g/mol to kg/mol.

$$\mathcal{M}_{H_2} = \frac{2.016 \text{ g } H_2}{1 \text{ mol } H_2} \times \frac{1 \text{ kg}}{1000 \text{ g}} = 2.016 \times 10^{-3} \text{ kg/mol}$$

$$\mathcal{M}_{He} = \frac{4.003 \text{ g He}}{1 \text{ mol He}} \times \frac{1 \text{ kg}}{1000 \text{ g}} = 4.003 \times 10^{-3} \text{ kg/mol}$$

Substitute the appropriate values into the equation to solve for u_{rms}.

$$u_{rms}(H_2) = \sqrt{\frac{(3)\left(8.314 \frac{\text{J}}{\text{mol}\cdot\text{K}}\right)(150 \text{ K})}{\left(2.016 \times 10^{-3} \frac{\text{kg}}{\text{mol}}\right)}}$$

$$u_{rms}(\text{He}) = \sqrt{\frac{(3)\left(8.314 \frac{\text{J}}{\text{mol}\cdot\text{K}}\right)(650 \text{ K})}{\left(4.003 \times 10^{-3} \frac{\text{kg}}{\text{mol}}\right)}}$$

$$\boldsymbol{u_{rms}(H_2) = 1.36 \times 10^3 \text{ m/s}}$$

$$\boldsymbol{u_{rms}(\text{He}) = 2.01 \times 10^3 \text{ m/s}}$$

He(g) has the higher root mean square velocity.

PRACTICE EXERCISE

8. What is the root-mean-square velocity of $F_2(g)$ at 298 K?

Text Problems: 5.78, 5.80

PROBLEM TYPE 5: DEVIATIONS FROM IDEAL BEHAVIOR

Van der Waals' equation is a simple modification of the ideal gas equation. It takes into account intermolecular forces and finite molecular volumes.

Starting with the ideal gas equation:

$$P_{ideal}V_{ideal} = nRT$$

van der Waals made two corrections.

(1) $P_{\text{ideal}} = P_{\text{real}} + \text{correction for attraction between particles}$

$$P_{\text{ideal}} = P_{\text{real}} + \frac{an^2}{V_{\text{real}}^2} \quad \text{(see Section 5.8 of the text for discussion)}$$

(2) $V_{\text{ideal}} = V_{\text{real}} - \text{correction for finite volume of particles}$

$$V_{\text{ideal}} = V_{\text{real}} - nb \quad \text{(see Section 5.8 of the text for discussion)}$$

where,

a and b are constants for a particular gas (See Table 5.4 of the text)
n = moles of gas

Substituting these corrections into the ideal gas equation leads to the **van der Waals equation.**

$$\left(P_{\text{real}} + \frac{an^2}{V_{\text{real}}^2}\right)(V_{\text{real}} - nb) = nRT$$

EXAMPLE 5.10

The molar volume of isopentane (C_5H_{12}) is 1.00 L at 503 K and 30.0 atm.

(a) Does isopentane behave like an ideal gas?

Strategy: In this problem we can determine if the gas deviates from ideal behavior, by comparing the ideal pressure with the actual pressure. We can calculate the ideal gas pressure using the ideal gas equation, and then compare it to the actual pressure given in the problem.

Solution: Solve the ideal gas equation algebraically for P, and then substitute in the known values.

$$P = \frac{nRT}{V} = \frac{(1.00\ \text{mol})(503\ \text{K})}{1.00\ \text{L}} \times \frac{0.0821\ \text{L} \cdot \text{atm}}{\text{mol} \cdot \text{K}}$$

$$P = 41.3\ \text{atm}$$

Compare the ideal pressure to the actual pressure. Calculating the percentage error would be helpful. Percentage of error is the difference between the two values divided by the actual value.

$$\%\ \text{error} = \frac{41.3\ \text{atm} - 30.0\ \text{atm}}{30.0\ \text{atm}} \times 100 = \mathbf{37.7\%}$$

Because of the large percent error, we conclude that under these conditions, C_5H_{12} behaves in a nonideal manner.

(b) Given that $a = 17.0\ \text{L}^2\cdot\text{atm/mol}^2$ and $b = 0.136$ L/mol, calculate the pressure of isopentane as predicted by the van der Waals' equation.

Strategy: Calculate the pressure of 1 mol of gas at 503 K that occupies 1.00 L using van der Waals' equation.

Solution: Calculate the correction terms.

$$\frac{an^2}{V_{\text{obs}}^2} = \frac{\left(17.0\frac{\text{L}^2 \cdot \text{atm}}{\text{mol}^2}\right)(1.00\ \text{mol})^2}{1.00\ \text{L}^2} = 17.0\ \text{atm}$$

$$nb = (1.00 \cancel{\text{mol}})\left(0.136 \frac{\text{L}}{\cancel{\text{mol}}}\right) = 0.136 \text{ L}$$

Substitute the values into the van der Waals' equation and solve for P_{real}.

$$\left(P_{\text{real}} + \frac{an^2}{V_{\text{real}}^2}\right)(V_{\text{real}} - nb) = nRT$$

$$(P_{\text{real}} + 17.0 \text{ atm})(1.00 \text{ L} - 0.136 \text{ L}) = (1.00 \text{ mol})\left(\frac{0.0821 \text{ L} \cdot \text{atm}}{\text{mol} \cdot \text{K}}\right)(503 \text{ K})$$

$$(P_{\text{real}} + 17.0 \text{ atm})(0.864) = 41.3 \text{ atm}$$

$$P_{\text{real}} + 17.0 \text{ atm} = 47.8 \text{ atm}$$

$$\boldsymbol{P_{\text{real}}} = 47.8 \text{ atm} - 17.0 \text{ atm} = \mathbf{30.8 \text{ atm}}$$

The pressure calculated for 1 mole of this "real" gas at 503 K and 1.00 L using van der Waals' equation is much closer to the actual value of 30.0 atm. The percent error is only 2.7 percent.

PRACTICE EXERCISE

9. Using van der Waals' equation, calculate the pressure exerted by 1.00 mol of water vapor that occupies a volume of 5.55 L at 150°C.

Text Problem: 5.90

ANSWERS TO PRACTICE EXERCISES

1. 4.04×10^3 mmHg

2. $V_{CO_2} = 1.1 \times 10^{16}$ L

3. $V_{\text{air}} = 160$ mL

4. The gas is fluorine, F_2.

5. 176 g NaN_3

6. $P_{CO} = 7.5 \times 10^{-3}$ torr

7. mass % ethanol = 5.0 %

8. $u_{\text{rms}} = 442$ m/s

9. $P_{\text{real}} = 6.11$ atm

SOLUTIONS TO SELECTED TEXT PROBLEMS

5.14 Pressure Conversions, Problem Type 1.

Strategy: Because 1 atm = 760 mmHg, the following conversion factor is needed to obtain the pressure in atmospheres.

$$\frac{1\text{ atm}}{760\text{ mmHg}}$$

For the second conversion, 1 atm = 101.325 kPa.

Solution:

$$\textbf{? atm} = 606\text{ mmHg} \times \frac{1\text{ atm}}{760\text{ mmHg}} = \textbf{0.797 atm}$$

$$\textbf{? kPa} = 0.797\text{ atm} \times \frac{101.325\text{ kPa}}{1\text{ atm}} = \textbf{80.8 kPa}$$

5.18 **(1)** Recall that $V \propto \frac{1}{P}$. As the pressure is tripled, the volume will decrease to $\frac{1}{3}$ of its original volume, assuming constant n and T. The correct choice is **(b)**.

(2) Recall that $V \propto T$. As the temperature is doubled, the volume will also double, assuming constant n and P. The correct choice is **(a)**. The depth of color indicates the density of the gas. As the volume increases at constant moles of gas, the density of the gas will decrease. This decrease in gas density is indicated by the lighter shading.

(3) Recall that $V \propto n$. Starting with n moles of gas, adding another n moles of gas ($2n$ total) will double the volume. The correct choice is **(c)**. The density of the gas will remain the same as moles are doubled and volume is doubled.

(4) Recall that $V \propto T$ and $V \propto \frac{1}{P}$. Halving the temperature would decrease the volume to $\frac{1}{2}$ its original volume. However, reducing the pressure to $\frac{1}{4}$ its original value would increase the volume by a factor of 4. Combining the two changes, we have

$$\frac{1}{2} \times 4 = 2$$

The volume will double. The correct choice is **(a)**.

5.20 Temperature and amount of gas do not change in this problem ($T_1 = T_2$ and $n_1 = n_2$). Pressure and volume change; it is a Boyle's law problem.

$$\frac{P_1V_1}{n_1T_1} = \frac{P_2V_2}{n_2T_2}$$

$$P_1V_1 = P_2V_2$$

$V_2 = 0.10\ V_1$

$$P_2 = \frac{P_1V_1}{V_2}$$

$$P_2 = \frac{(5.3 \text{ atm})V_1}{0.10V_1} = \mathbf{53 \text{ atm}}$$

5.22 Only one or two variable quantities have fixed values, Problem Type 2B.

(a)
Strategy: The amount of gas and its temperature remain constant, but both the pressure and the volume change. What equation would you use to solve for the final volume?

Solution: We start with Equation (5.9) of the text.

$$\frac{P_1V_1}{n_1T_1} = \frac{P_2V_2}{n_2T_2}$$

Because $n_1 = n_2$ and $T_1 = T_2$,

$$P_1V_1 = P_2V_2$$

which is Boyle's Law. The given information is tabulated below.

Initial conditions	Final Conditions
$P_1 = 1.2$ atm	$P_2 = 6.6$ atm
$V_1 = 3.8$ L	$V_2 = ?$

The final volume is given by:

$$V_2 = \frac{P_1V_1}{P_2}$$

$$V_2 = \frac{(1.2 \text{ atm})(3.8 \text{ L})}{(6.6 \text{ atm})} = \mathbf{0.69 \text{ L}}$$

Check: When the pressure applied to the sample of air is increased from 1.2 atm to 6.6 atm, the volume occupied by the sample will decrease. Pressure and volume are inversely proportional. The final volume calculated is less than the initial volume, so the answer seems reasonable.

(b)
Strategy: The amount of gas and its temperature remain constant, but both the pressure and the volume change. What equation would you use to solve for the final pressure?

Solution: You should also come up with the equation $P_1V_1 = P_2V_2$ for this problem. The given information is tabulated below.

Initial conditions	Final Conditions
$P_1 = 1.2$ atm	$P_2 = ?$
$V_1 = 3.8$ L	$V_2 = 0.075$ L

The final pressure is given by:

$$P_2 = \frac{P_1V_1}{V_2}$$

$$P_2 = \frac{(1.2 \text{ atm})(3.8 \text{ L})}{(0.075 \text{ L})} = \mathbf{61 \text{ atm}}$$

Check: To decrease the volume of the gas fairly dramatically from 3.8 L to 0.075 L, the pressure must be increased substantially. A final pressure of 61 atm seems reasonable.

5.24 Only one or two variable quantities have fixed values, Problem Type 2B.

Strategy: The amount of gas and its pressure remain constant, but both the temperature and the volume change. What equation would you use to solve for the final temperature? What temperature unit should we use?

Solution: We start with Equation (5.9) of the text.

$$\frac{P_1V_1}{n_1T_1} = \frac{P_2V_2}{n_2T_2}$$

Because $n_1 = n_2$ and $P_1 = P_2$,

$$\frac{V_1}{T_1} = \frac{V_2}{T_2}$$

which is Charles' Law. The given information is tabulated below.

Initial conditions	Final Conditions
$T_1 = (88 + 273)\text{K} = 361\text{ K}$	$T_2 = ?$
$V_1 = 9.6\text{ L}$	$V_2 = 3.4\text{ L}$

The final temperature is given by:

$$T_2 = \frac{T_1V_2}{V_1}$$

$$T_2 = \frac{(361\text{ K})(3.4\text{ L})}{(9.6\text{ L})} = \mathbf{1.3 \times 10^2\ K}$$

5.26 This is a gas stoichiometry problem that requires knowledge of Avogadro's law to solve. Avogadro's law states that the volume of a gas is directly proportional to the number of moles of gas at constant temperature and pressure.

The volume ratio, 1 vol. Cl_2 : 3 vol. F_2 : 2 vol. product, can be written as a mole ratio, 1 mol Cl_2 : 3 mol F_2 : 2 mol product.

Attempt to write a balanced chemical equation. The subscript of F in the product will be three times the Cl subscript, because there are three times as many F atoms reacted as Cl atoms.

$$\mathbf{1}Cl_2(g) + \mathbf{3}F_2(g) \longrightarrow \mathbf{2}Cl_xF_{3x}(g)$$

Balance the equation. The x must equal one so that there are two Cl atoms on each side of the equation. If $x = 1$, the subscript on F is 3.

$$Cl_2(g) + 3F_2(g) \longrightarrow 2ClF_3(g)$$

The formula of the product is **ClF_3**.

5.32 Given three of the four variable quantities in the equation (P, V, n, and T), Problem Type 2A.

Strategy: This problem gives the amount, volume, and temperature of CO gas. Is the gas undergoing a change in any of its properties? What equation should we use to solve for the pressure? What temperature unit should be used?

Solution: Because no changes in gas properties occur, we can use the ideal gas equation to calculate the pressure. Rearranging Equation (5.8) of the text, we write:

$$P = \frac{nRT}{V}$$

$$\boldsymbol{P} = \frac{(6.9\ \cancel{\text{mol}})\left(0.0821\frac{\cancel{\text{L}}\cdot\text{atm}}{\cancel{\text{mol}}\cdot\cancel{\text{K}}}\right)(62+273)\cancel{\text{K}}}{30.4\ \cancel{\text{L}}} = \mathbf{6.2\ atm}$$

5.34 In this problem, the moles of gas and the volume the gas occupies are constant ($V_1 = V_2$ and $n_1 = n_2$). Temperature and pressure change.

$$\frac{P_1V_1}{n_1T_1} = \frac{P_2V_2}{n_2T_2}$$

$$\frac{P_1}{T_1} = \frac{P_2}{T_2}$$

The given information is tabulated below.

Initial conditions	Final Conditions
$T_1 = (25 + 273)\text{K} = 298\ \text{K}$	$T_2 = ?$
$P_1 = 0.800\ \text{atm}$	$P_2 = 2.00\ \text{atm}$

The final temperature is given by:

$$T_2 = \frac{T_1P_2}{P_1}$$

$$\boldsymbol{T_2} = \frac{(298\ \text{K})(2.00\ \cancel{\text{atm}})}{(0.800\ \cancel{\text{atm}})} = \mathbf{745\ K} = \mathbf{472°C}$$

5.36 In this problem, the moles of gas and the volume the gas occupies are constant ($V_1 = V_2$ and $n_1 = n_2$). Temperature and pressure change.

$$\frac{P_1V_1}{n_1T_1} = \frac{P_2V_2}{n_2T_2}$$

$$\frac{P_1}{T_1} = \frac{P_2}{T_2}$$

The given information is tabulated below.

Initial conditions	Final Conditions
$T_1 = 273\text{ K}$	$T_2 = (250 + 273)\text{K} = 523\text{ K}$
$P_1 = 1.0\text{ atm}$	$P_2 = ?$

The final pressure is given by:

$$P_2 = \frac{P_1T_2}{T_1}$$

$$\boldsymbol{P_2} = \frac{(1.0\text{ atm})(523\text{ K})}{273\text{ K}} = \mathbf{1.9\text{ atm}}$$

5.38 In this problem, the moles of gas and the pressure on the gas are constant ($n_1 = n_2$ and $P_1 = P_2$). Temperature and volume are changing.

$$\frac{P_1V_1}{n_1T_1} = \frac{P_2V_2}{n_2T_2}$$

$$\frac{V_1}{T_1} = \frac{V_2}{T_2}$$

The given information is tabulated below.

Initial conditions	Final Conditions
$T_1 = (20.1 + 273)\text{ K} = 293.1\text{ K}$	$T_2 = (36.5 + 273)\text{K} = 309.5\text{ K}$
$V_1 = 0.78\text{ L}$	$V_2 = ?$

The final volume is given by:

$$V_2 = \frac{V_1T_2}{T_1}$$

$$\boldsymbol{V_2} = \frac{(0.78\text{ L})(309.5\text{ K})}{(293.1\text{ K})} = \mathbf{0.82\text{ L}}$$

5.40 In the problem, temperature and pressure are given. If we can determine the moles of CO_2, we can calculate the volume it occupies using the ideal gas equation.

$$?\text{ mol }CO_2 = 88.4\text{ g }CO_2 \times \frac{1\text{ mol }CO_2}{44.01\text{ g }CO_2} = 2.01\text{ mol }CO_2$$

We now substitute into the ideal gas equation to calculate volume of CO_2.

$$\boldsymbol{V_{CO_2}} = \frac{nRT}{P} = \frac{(2.01\text{ mol})\left(0.0821\frac{\text{L}\cdot\text{atm}}{\text{mol}\cdot\text{K}}\right)(273\text{ K})}{(1\text{ atm})} = \mathbf{45.1\text{ L}}$$

Alternatively, we could use the fact that 1 mole of an ideal gas occupies a volume of 22.41 L at STP. After calculating the moles of CO_2, we can use this fact as a conversion factor to convert to volume of CO_2.

$$? \text{ L } CO_2 = 2.01 \text{ mol } CO_2 \times \frac{22.41 \text{ L}}{1 \text{ mol}} = \mathbf{45.0 \text{ L } CO_2}$$

The slight difference in the results of our two calculations is due to rounding the volume occupied by 1 mole of an ideal gas to 22.41 L.

5.42 The molar mass of CO_2 = 44.01 g/mol. Since $PV = nRT$, we write:

$$P = \frac{nRT}{V}$$

$$\boldsymbol{P} = \frac{\left(0.050 \text{ g} \times \frac{1 \text{ mol}}{44.01 \text{ g}}\right)\left(0.0821 \frac{\text{L}\cdot\text{atm}}{\text{mol}\cdot\text{K}}\right)(30 + 273)\text{K}}{4.6 \text{ L}} = \mathbf{6.1 \times 10^{-3} \text{ atm}}$$

5.44 Calculations involving molar mass, Problem Type 2C.

Strategy: We can calculate the molar mass of a gas if we know its density, temperature, and pressure. What temperature and pressure units should we use?

Solution: We need to use Equation (5.12) of the text to calculate the molar mass of the gas.

$$\mathcal{M} = \frac{dRT}{P}$$

Before substituting into the above equation, we need to calculate the density and check that the other known quantities (P and T) have the appropriate units.

$$d = \frac{7.10 \text{ g}}{5.40 \text{ L}} = 1.31 \text{ g/L}$$

$$T = 44° + 273° = 317 \text{ K}$$

$$P = 741 \text{ torr} \times \frac{1 \text{ atm}}{760 \text{ torr}} = 0.975 \text{ atm}$$

Calculate the molar mass by substituting in the known quantities.

$$\mathcal{M} = \frac{\left(1.31 \frac{\text{g}}{\text{L}}\right)\left(0.0821 \frac{\text{L}\cdot\text{atm}}{\text{mol}\cdot\text{K}}\right)(317 \text{ K})}{0.975 \text{ atm}} = \mathbf{35.0 \text{ g/mol}}$$

Alternatively, we can solve for the molar mass by writing:

$$\text{molar mass of compound} = \frac{\text{mass of compound}}{\text{moles of compound}}$$

Mass of compound is given in the problem (7.10 g), so we need to solve for moles of compound in order to calculate the molar mass.

$$n = \frac{PV}{RT}$$

$$n = \frac{(0.975\ \text{atm})(5.40\ \text{L})}{\left(0.0821\frac{\text{L}\cdot\text{atm}}{\text{mol}\cdot\text{K}}\right)(317\ \text{K})} = 0.202\ \text{mol}$$

Now, we can calculate the molar mass of the gas.

$$\textbf{molar mass of compound} = \frac{\text{mass of compound}}{\text{moles of compound}} = \frac{7.10\ \text{g}}{0.202\ \text{mol}} = \textbf{35.1 g/mol}$$

5.46 The number of particles in 1 L of gas at STP is:

$$\text{Number of particles} = 1.0\ \text{L} \times \frac{1\ \text{mol}}{22.414\ \text{L}} \times \frac{6.022\times10^{23}\ \text{particles}}{1\ \text{mol}} = 2.7\times10^{22}\ \text{particles}$$

$$\textbf{Number of N}_2\textbf{ molecules} = \left(\frac{78\%}{100\%}\right)(2.7\times10^{22}\ \text{particles}) = \mathbf{2.1\times10^{22}\ N_2\ molecules}$$

$$\textbf{Number of O}_2\textbf{ molecules} = \left(\frac{21\%}{100\%}\right)(2.7\times10^{22}\ \text{particles}) = \mathbf{5.7\times10^{21}\ O_2\ molecules}$$

$$\textbf{Number of Ar atoms} = \left(\frac{1\%}{100\%}\right)(2.7\times10^{22}\ \text{particles}) = \mathbf{3\times10^{20}\ Ar\ atoms}$$

5.48 The density can be calculated from the ideal gas equation.

$$d = \frac{P\mathcal{M}}{RT}$$

$\mathcal{M} = 1.008\ \text{g/mol} + 79.90\ \text{g/mol} = 80.91\ \text{g/mol}$

$T = 46° + 273° = 319\ \text{K}$

$$P = 733\ \text{mmHg} \times \frac{1\ \text{atm}}{760\ \text{mmHg}} = 0.964\ \text{atm}$$

$$\boldsymbol{d} = \frac{(0.964\ \text{atm})\left(\frac{80.91\ \text{g}}{1\ \text{mol}}\right)}{319\ \text{K}} \times \frac{\text{mol}\cdot\text{K}}{0.0821\ \text{L}\cdot\text{atm}} = \textbf{2.98 g/L}$$

Alternatively, we can solve for the density by writing:

$$\text{density} = \frac{\text{mass}}{\text{volume}}$$

Assuming that we have 1 mole of HBr, the mass is 80.91 g. The volume of the gas can be calculated using the ideal gas equation.

$$V = \frac{nRT}{P}$$

$$V = \frac{(1\ \text{mol})\left(0.0821\frac{\text{L}\cdot\text{atm}}{\text{mol}\cdot\text{K}}\right)(319\ \text{K})}{0.964\ \text{atm}} = 27.2\ \text{L}$$

Now, we can calculate the density of HBr gas.

$$\textbf{density} = \frac{\text{mass}}{\text{volume}} = \frac{80.91\text{ g}}{27.2\text{ L}} = \textbf{2.97 g/L}$$

5.50 This is an extension of an ideal gas law calculation involving molar mass. If you determine the molar mass of the gas, you will be able to determine the molecular formula from the empirical formula (see Determining the Molecular Formula of a Compound, Problem Type 6, Chapter 3).

$$\mathcal{M} = \frac{dRT}{P}$$

Calculate the density, then substitute its value into the equation above.

$$d = \frac{0.100\text{ g}}{22.1\text{ mL}} \times \frac{1000\text{ mL}}{1\text{ L}} = 4.52\text{ g/L}$$

$$T(\text{K}) = 20° + 273° = 293\text{ K}$$

$$\mathcal{M} = \frac{\left(4.52\,\frac{\text{g}}{\text{L}}\right)\left(0.0821\,\frac{\text{L}\cdot\text{atm}}{\text{mol}\cdot\text{K}}\right)(293\text{ K})}{1.02\text{ atm}} = 107\text{ g/mol}$$

Compare the empirical mass to the molar mass.

$$\text{empirical mass} = 32.07\text{ g/mol} + 4(19.00\text{ g/mol}) = 108.07\text{ g/mol}$$

Remember, the molar mass will be a whole number multiple of the empirical mass. In this case, the $\frac{\text{molar mass}}{\text{empirical mass}} \approx 1$. Therefore, the molecular formula is the same as the empirical formula, $\mathbf{SF_4}$.

5.52 Gas stoichiometry, Problem Type 2D.

Strategy: From the moles of CH_4 reacted, we can calculate the moles of CO_2 produced. From the balanced equation, we see that 1 mol $CH_4 \simeq$ 1 mol CO_2. Once moles of CO_2 are determined, we can use the ideal gas equation to calculate the volume of CO_2.

Solution: First let's calculate moles of CO_2 produced.

$$?\text{ mol CO}_2 = 15.0\text{ mol CH}_4 \times \frac{1\text{ mol CO}_2}{1\text{ mol CH}_4} = 15.0\text{ mol CO}_2$$

Now, we can substitute moles, temperature, and pressure into the ideal gas equation to solve for volume of CO_2.

$$V = \frac{nRT}{P}$$

$$V_{\text{CO}_2} = \frac{(15.0\text{ mol})\left(0.0821\,\frac{\text{L}\cdot\text{atm}}{\text{mol}\cdot\text{K}}\right)(23 + 273)\text{K}}{0.985\text{ atm}} = \mathbf{3.70 \times 10^2\ L}$$

5.54 From the amount of glucose reacted (5.97 g), we can calculate the theoretical yield of CO_2. We can then compare the theoretical yield to the actual yield given in the problem (1.44 L) to determine the percent yield.

First, let's determine the moles of CO_2 that can be produced theoretically. Then, we can use the ideal gas equation to determine the volume of CO_2.

$$? \text{ mol CO}_2 = 5.97 \text{ g glucose} \times \frac{1 \text{ mol glucose}}{180.2 \text{ g glucose}} \times \frac{2 \text{ mol CO}_2}{1 \text{ mol glucose}} = 0.0663 \text{ mol CO}_2$$

Now, substitute moles, pressure, and temperature into the ideal gas equation to calculate the volume of CO_2.

$$V = \frac{nRT}{P}$$

$$V_{CO_2} = \frac{(0.0663 \text{ mol})\left(0.0821 \frac{\text{L} \cdot \text{atm}}{\text{mol} \cdot \text{K}}\right)(293 \text{ K})}{0.984 \text{ atm}} = 1.62 \text{ L}$$

This is the theoretical yield of CO_2. The actual yield, which is given in the problem, is 1.44 L. We can now calculate the percent yield.

$$\text{percent yield} = \frac{\text{actual yield}}{\text{theoretical yield}} \times 100\%$$

$$\textbf{percent yield} = \frac{1.44 \text{ L}}{1.62 \text{ L}} \times 100\% = \mathbf{88.9\%}$$

5.56 Gas stoichiometry, Problem Type 2D.

Strategy: We can calculate the moles of M reacted, and the moles of H_2 gas produced. By comparing the number of moles of M reacted to the number of moles H_2 produced, we can determine the mole ratio in the balanced equation.

Solution: First let's calculate the moles of the metal (M) reacted.

$$\text{mol M} = 0.225 \text{ g M} \times \frac{1 \text{ mol M}}{27.0 \text{ g M}} = 8.33 \times 10^{-3} \text{ mol M}$$

Solve the ideal gas equation algebraically for n_{H_2}. Then, calculate the moles of H_2 by substituting the known quantities into the equation.

$$P = 741 \text{ mmHg} \times \frac{1 \text{ atm}}{760 \text{ mmHg}} = 0.975 \text{ atm}$$

$$T = 17° + 273° = 290 \text{ K}$$

$$n_{H_2} = \frac{PV_{H_2}}{RT}$$

$$n_{H_2} = \frac{(0.975 \text{ atm})(0.303 \text{ L})}{\left(0.0821 \frac{\text{L} \cdot \text{atm}}{\text{mol} \cdot \text{K}}\right)(290 \text{ K})} = 1.24 \times 10^{-2} \text{ mol H}_2$$

Compare the number moles of H_2 produced to the number of moles of M reacted.

$$\frac{1.24 \times 10^{-2} \text{ mol } H_2}{8.33 \times 10^{-3} \text{ mol M}} \approx 1.5$$

This means that the mole ratio of H_2 to M is 1.5 : 1.

We can now write the balanced equation since we know the mole ratio between H_2 and M.

The unbalanced equation is:

$$M(s) + HCl(aq) \longrightarrow 1.5H_2(g) + M_xCl_y(aq)$$

We have 3 atoms of H on the products side of the reaction, so a 3 must be placed in front of HCl. The ratio of M to Cl on the reactants side is now 1 : 3. Therefore the formula of the metal chloride must be MCl_3. The balanced equation is:

$$\mathbf{M(s) + 3HCl(aq) \longrightarrow 1.5H_2(g) + MCl_3(aq)}$$

From the formula of the metal chloride, we determine that the charge of the metal is +3. Therefore, the formula of the metal oxide and the metal sulfate are $\mathbf{M_2O_3}$ and $\mathbf{M_2(SO_4)_3}$, respectively.

5.58 From the moles of CO_2 produced, we can calculate the amount of calcium carbonate that must have reacted. We can then determine the percent by mass of $CaCO_3$ in the 3.00 g sample.

The balanced equation is:

$$CaCO_3(s) + 2HCl(aq) \longrightarrow CO_2(g) + CaCl_2(aq) + H_2O(l)$$

The moles of CO_2 produced can be calculated using the ideal gas equation.

$$n_{CO_2} = \frac{PV_{CO_2}}{RT}$$

$$n_{CO_2} = \frac{\left(792 \text{ mmHg} \times \dfrac{1 \text{ atm}}{760 \text{ mmHg}}\right)(0.656 \text{ L})}{\left(0.0821 \dfrac{\text{L} \cdot \text{atm}}{\text{mol} \cdot \text{K}}\right)(20 + 273 \text{ K})} = \mathbf{2.84 \times 10^{-2} \text{ mol } CO_2}$$

The balanced equation shows a 1:1 mole ratio between CO_2 and $CaCO_3$. Therefore, 2.84×10^{-2} mole of $CaCO_3$ must have reacted.

$$? \text{ g } CaCO_3 \text{ reacted} = (2.84 \times 10^{-2} \text{ mol } CaCO_3) \times \frac{100.1 \text{ g } CaCO_3}{1 \text{ mol } CaCO_3} = 2.84 \text{ g } CaCO_3$$

The percent by mass of the $CaCO_3$ sample is:

$$\mathbf{\% \ CaCO_3} = \frac{2.84 \text{ g}}{3.00 \text{ g}} \times 100\% = \mathbf{94.7\%}$$

Assumption: The impurity (or impurities) must not react with HCl to produce CO_2 gas.

5.60 The balanced equation is:

$$C_2H_5OH(l) + 3O_2(g) \longrightarrow 2CO_2(g) + 3H_2O(l)$$

The moles of O_2 needed to react with 227 g ethanol are:

$$227 \text{ g } C_2H_5OH \times \frac{1 \text{ mol } C_2H_5OH}{46.07 \text{ g } C_2H_5OH} \times \frac{3 \text{ mol } O_2}{1 \text{ mol } C_2H_5OH} = 14.8 \text{ mol } O_2$$

14.8 moles of O_2 correspond to a volume of:

$$V_{O_2} = \frac{n_{O_2}RT}{P} = \frac{(14.8 \text{ mol } O_2)\left(0.0821 \frac{\text{L}\cdot\text{atm}}{\text{mol}\cdot\text{K}}\right)(35 + 273 \text{ K})}{\left(790 \text{ mmHg} \times \frac{1 \text{ atm}}{760 \text{ mmHg}}\right)} = 3.60 \times 10^2 \text{ L } O_2$$

Since air is 21.0 percent O_2 by volume, we can write:

$$V_{air} = V_{O_2}\left(\frac{100\% \text{ air}}{21\% \ O_2}\right) = (3.60 \times 10^2 \text{ L } O_2)\left(\frac{100\% \text{ air}}{21\% \ O_2}\right) = \mathbf{1.71 \times 10^3 \text{ L air}}$$

5.64 Dalton's law states that the total pressure of the mixture is the sum of the partial pressures.

(a) $P_{total} = 0.32 \text{ atm} + 0.15 \text{ atm} + 0.42 \text{ atm} = \mathbf{0.89 \text{ atm}}$

(b) We know:

Initial conditions	Final Conditions
$P_1 = (0.15 + 0.42)\text{atm} = 0.57 \text{ atm}$	$P_2 = 1.0 \text{ atm}$
$T_1 = (15 + 273)\text{K} = 288 \text{ K}$	$T_2 = 273 \text{ K}$
$V_1 = 2.5 \text{ L}$	$V_2 = ?$

$$\frac{P_1V_1}{n_1T_1} = \frac{P_2V_2}{n_2T_2}$$

Because $n_1 = n_2$, we can write:

$$V_2 = \frac{P_1V_1T_2}{P_2T_1}$$

$$V_2 = \frac{(0.57 \text{ atm})(2.5 \text{ L})(273 \text{ K})}{(1.0 \text{ atm})(288 \text{ K})} = \mathbf{1.4 \text{ L at STP}}$$

5.66 $P_{Total} = P_1 + P_2 + P_3 + \ldots + P_n$

In this case,

$$P_{Total} = P_{Ne} + P_{He} + P_{H_2O}$$

$$P_{Ne} = P_{Total} - P_{He} - P_{H_2O}$$

$$P_{Ne} = 745 \text{ mm Hg} - 368 \text{ mmHg} - 28.3 \text{ mmHg} = \mathbf{349 \text{ mmHg}}$$

5.68 Collecting a gas over water, Problem Type 2E.

Strategy: To solve for moles of H_2 generated, we must first calculate the partial pressure of H_2 in the mixture. What gas law do we need? How do we convert from moles of H_2 to amount of Zn reacted?

Solution: Dalton's law of partial pressure states that

$$P_{\text{Total}} = P_1 + P_2 + P_3 + \ldots + P_n$$

In this case,

$$P_{\text{Total}} = P_{H_2} + P_{H_2O}$$

$$P_{H_2} = P_{\text{Total}} - P_{H_2O}$$

$$P_{H_2} = 0.980 \text{ atm} - (23.8 \text{ mmHg})\left(\frac{1 \text{ atm}}{760 \text{ mmHg}}\right) = 0.949 \text{ atm}$$

Now that we know the pressure of H_2 gas, we can calculate the moles of H_2. Then, using the mole ratio from the balanced equation, we can calculate moles of Zn.

$$n_{H_2} = \frac{P_{H_2}V}{RT}$$

$$n_{H_2} = \frac{(0.949 \text{ atm})(7.80 \text{ L})}{(25 + 273) \text{ K}} \times \frac{\text{mol} \cdot \text{K}}{0.0821 \text{ L} \cdot \text{atm}} = 0.303 \text{ mol } H_2$$

Using the mole ratio from the balanced equation and the molar mass of zinc, we can now calculate the grams of zinc consumed in the reaction.

$$\textbf{? g Zn} = 0.303 \text{ mol } H_2 \times \frac{1 \text{ mol Zn}}{1 \text{ mol } H_2} \times \frac{65.39 \text{ g Zn}}{1 \text{ mol Zn}} = \mathbf{19.8 \text{ g Zn}}$$

5.70 $P_i = X_i P_T$

We need to determine the mole fractions of each component in order to determine their partial pressures. To calculate mole fraction, write the balanced chemical equation to determine the correct mole ratio.

$$2NH_3(g) \longrightarrow N_2(g) + 3H_2(g)$$

The mole fractions of H_2 and N_2 are:

$$X_{H_2} = \frac{3 \text{ mol}}{3 \text{ mol} + 1 \text{ mol}} = 0.750$$

$$X_{N_2} = \frac{1 \text{ mol}}{3 \text{ mol} + 1 \text{ mol}} = 0.250$$

The partial pressures of H_2 and N_2 are:

$$\boldsymbol{P_{H_2}} = X_{H_2}P_T = (0.750)(866 \text{ mmHg}) = \mathbf{650 \text{ mmHg}}$$

$$\boldsymbol{P_{N_2}} = X_{N_2}P_T = (0.250)(866 \text{ mmHg}) = \mathbf{217 \text{ mmHg}}$$

5.78 Root-Mean-Square (RMS) Speed, Problem Type 4.

Strategy: To calculate the root-mean-square speed, we use Equation (5.16) of the text. What units should we use for R and $\mathcal{M}$ so the u_{rms} will be expressed in units of m/s.

Solution: To calculate u_{rms}, the units of R should be 8.314 J/mol·K, and because 1 J = 1 kg·m^2/s^2, the units of molar mass must be kg/mol.

First, let's calculate the molar masses ($\mathcal{M}$) of N_2, O_2, and O_3. Remember, $\mathcal{M}$ must be in units of kg/mol.

$$\mathcal{M}_{N_2} = 2(14.01\text{ g/mol}) = 28.02\frac{\text{g}}{\text{mol}} \times \frac{1\text{ kg}}{1000\text{ g}} = 0.02802\text{ kg/mol}$$

$$\mathcal{M}_{O_2} = 2(16.00\text{ g/mol}) = 32.00\frac{\text{g}}{\text{mol}} \times \frac{1\text{ kg}}{1000\text{ g}} = 0.03200\text{ kg/mol}$$

$$\mathcal{M}_{O_3} = 3(16.00\text{ g/mol}) = 48.00\frac{\text{g}}{\text{mol}} \times \frac{1\text{ kg}}{1000\text{ g}} = 0.04800\text{ kg/mol}$$

Now, we can substitute into Equation (5.16) of the text.

$$u_{rms} = \sqrt{\frac{3RT}{\mathcal{M}}}$$

$$u_{rms}(N_2) = \sqrt{\frac{(3)\left(8.314\frac{\text{J}}{\text{mol}\cdot\text{K}}\right)(-23+273)\text{K}}{\left(0.02802\frac{\text{kg}}{\text{mol}}\right)}}$$

$$\boldsymbol{u_{rms}(N_2) = 472\text{ m/s}}$$

Similarly,

$$\boldsymbol{u_{rms}(O_2) = 441\text{ m/s}} \qquad \boldsymbol{u_{rms}(O_3) = 360\text{ m/s}}$$

Check: Since the molar masses of the gases increase in the order: $N_2 < O_2 < O_3$, we expect the lightest gas (N_2) to move the fastest on average and the heaviest gas (O_3) to move the slowest on average. This is confirmed in the above calculation.

5.80 **RMS speed** $= \sqrt{\frac{(2.0^2 + 2.2^2 + 2.6^2 + 2.7^2 + 3.3^2 + 3.5^2)(\text{m/s})^2}{6}} = \mathbf{2.8\ m/s}$

Average speed $= \frac{(2.0 + 2.2 + 2.6 + 2.7 + 3.3 + 3.5)\text{m/s}}{6} = \mathbf{2.7\ m/s}$

The root-mean-square value is always greater than the average value, because squaring favors the larger values compared to just taking the average value.

5.82 The separation factor is given by:

$$s = \frac{r_1}{r_2} = \sqrt{\frac{\mathcal{M}_2}{\mathcal{M}_1}}$$

This equation is the same as Graham's Law, Equation (5.17) of the text. For $^{235}UF_6$ and $^{238}UF_6$, we have:

$$s = \sqrt{\frac{238 + (6)(19.00)}{235 + (6)(19.00)}} = \mathbf{1.0043}$$

This is a very small separation factor, which is why many (thousands) stages of effusion are needed to enrich ^{235}U.

5.84 The rate of effusion is the number of molecules passing through a porous barrier in a given time. The molar mass of CH_4 is 16.04 g/mol. Using Equation (5.17) of the text, we find the molar mass of $Ni(CO)_x$.

$$\frac{r_1}{r_2} = \sqrt{\frac{\mathcal{M}_2}{\mathcal{M}_1}}$$

$$\frac{3.3}{1.0} = \sqrt{\frac{\mathcal{M}_{Ni(CO)_x}}{16.04 \text{ g/mol}}}$$

$$10.89 = \frac{\mathcal{M}_{Ni(CO)_x}}{16.04 \text{ g/mol}}$$

$$\mathcal{M}_{Ni(CO)_x} = 174.7 \text{ g/mol}$$

To find the value of x, we first subtract the molar mass of Ni from 174.7 g/mol.

$$174.7 \text{ g} - 58.69 \text{ g} = 116.0 \text{ g}$$

116.0 g is the mass of CO in 1 mole of the compound. The mass of 1 mole of CO is 28.01 g.

$$\frac{116.0 \text{ g}}{28.01 \text{ g}} = 4.141 \approx \mathbf{4}$$

This calculation indicates that there are 4 moles of CO in 1 mole of the compound. The value of $\boldsymbol{x}$ is **4**.

5.90 Deviations from Ideal Behavior, Problem Type 5.

Strategy: In this problem we can determine if the gas deviates from ideal behavior, by comparing the ideal pressure with the actual pressure. We can calculate the ideal gas pressure using the ideal gas equation, and then compare it to the actual pressure given in the problem. What temperature unit should we use in the calculation?

Solution: We convert the temperature to units of Kelvin, then substitute the given quantities into the ideal gas equation.

$$T(\text{K}) = 27°\text{C} + 273° = 300 \text{ K}$$

$$P = \frac{nRT}{V} = \frac{(10.0 \text{ mol})\left(0.0821 \frac{\text{L}\cdot\text{atm}}{\text{mol}\cdot\text{K}}\right)(300 \text{ K})}{1.50 \text{ L}} = 164 \text{ atm}$$

Now, we can compare the ideal pressure to the actual pressure by calculating the percent error.

$$\% \text{ error} = \frac{164 \text{ atm} - 130 \text{ atm}}{130 \text{ atm}} \times 100\% = 26.2\%$$

Based on the large percent error, we conclude that under this condition of high pressure, the gas behaves in a **non-ideal** manner.

5.92 When a and b are zero, the van der Waals equation simply becomes the ideal gas equation. In other words, an ideal gas has zero for the a and b values of the van der Waals equation. It therefore stands to reason that the gas with the smallest values of a and b will behave most like an ideal gas under a specific set of pressure and temperature conditions. Of the choices given in the problem, the gas with the smallest a and b values is **Ne** (see Table 5.4).

5.94 We need to determine the molar mass of the gas. Comparing the molar mass to the empirical mass will allow us to determine the molecular formula.

$$n = \frac{PV}{RT} = \frac{(0.74\ \text{atm})\left(97.2\ \text{mL} \times \dfrac{0.001\ \text{L}}{1\ \text{mL}}\right)}{\left(0.0821\dfrac{\text{L}\cdot\text{atm}}{\text{mol}\cdot\text{K}}\right)(200 + 273)\text{K}} = 1.85 \times 10^{-3}\ \text{mol}$$

$$\text{molar mass} = \frac{0.145\ \text{g}}{1.85 \times 10^{-3}\ \text{mol}} = 78.4\ \text{g/mol}$$

The empirical mass of CH = 13.02 g/mol

Since $\dfrac{78.4\ \text{g/mol}}{13.02\ \text{g/mol}} = 6.02 \approx 6$, the molecular formula is $(CH)_6$ or $\mathbf{C_6H_6}$.

5.96 The reaction is: $HCO_3^-(aq) + H^+(aq) \longrightarrow H_2O(l) + CO_2(g)$

The mass of HCO_3^- reacted is:

$$3.29\ \text{g tablet} \times \frac{32.5\%\ HCO_3^-}{100\%\ \text{tablet}} = 1.07\ \text{g}\ HCO_3^-$$

$$\text{mol } CO_2 \text{ produced} = 1.07\ \text{g}\ HCO_3^- \times \frac{1\ \text{mol}\ HCO_3^-}{61.02\ \text{g}\ HCO_3^-} \times \frac{1\ \text{mol}\ CO_2}{1\ \text{mol}\ HCO_3^-} = 0.0175\ \text{mol}\ CO_2$$

$$V_{CO_2} = \frac{n_{CO_2}RT}{P} = \frac{(0.0175\ \text{mol}\ CO_2)\left(0.0821\dfrac{\text{L}\cdot\text{atm}}{\text{mol}\cdot\text{K}}\right)(37 + 273)\text{K}}{(1.00\ \text{atm})} = 0.445\ \text{L} = \mathbf{445\ mL}$$

5.98 **(a)** The number of moles of $Ni(CO)_4$ formed is:

$$86.4\ \text{g Ni} \times \frac{1\ \text{mol Ni}}{58.69\ \text{g Ni}} \times \frac{1\ \text{mol}\ Ni(CO)_4}{1\ \text{mol Ni}} = 1.47\ \text{mol}\ Ni(CO)_4$$

The pressure of $Ni(CO)_4$ is:

$$P = \frac{nRT}{V} = \frac{(1.47\ \text{mol})\left(0.0821\dfrac{\text{L}\cdot\text{atm}}{\text{mol}\cdot\text{K}}\right)(43 + 273)\text{K}}{4.00\ \text{L}} = \mathbf{9.53\ atm}$$

(b) $Ni(CO)_4$ decomposes to produce more moles of gas (CO), which increases the pressure.

$$Ni(CO)_4(g) \longrightarrow Ni(s) + 4CO(g)$$

5.100 Using the ideal gas equation, we can calculate the moles of gas.

$$n = \frac{PV}{RT} = \frac{(1.1\ \text{atm})\left(5.0 \times 10^2\ \text{mL} \times \frac{0.001\ \text{L}}{1\ \text{mL}}\right)}{\left(0.0821 \frac{\text{L} \cdot \text{atm}}{\text{mol} \cdot \text{K}}\right)(37 + 273)\text{K}} = 0.0216\ \text{mol gas}$$

Next, use Avogadro's number to convert to molecules of gas.

$$0.0216\ \text{mol gas} \times \frac{6.022 \times 10^{23}\ \text{molecules}}{1\ \text{mol gas}} = \mathbf{1.30 \times 10^{22}\ molecules\ of\ gas}$$

The most common gases present in exhaled air are: $\mathbf{CO_2}$, $\mathbf{O_2}$, $\mathbf{N_2}$, and $\mathbf{H_2O}$.

5.102 Mass of the Earth's atmosphere = (surface area of the earth in cm^2) × (mass per 1 cm^2 column)

Mass of a single column of air with a surface area of 1 cm^2 area is:

$$76.0\ \text{cm} \times 13.6\ \text{g/cm}^3 = 1.03 \times 10^3\ \text{g/cm}^2$$

The surface area of the Earth in cm^2 is:

$$4\pi r^2 = 4\pi(6.371 \times 10^8\ \text{cm})^2 = 5.10 \times 10^{18}\ \text{cm}^2$$

$$\textbf{Mass of atmosphere} = (5.10 \times 10^{18}\ \text{cm}^2)(1.03 \times 10^3\ \text{g/cm}^2) = 5.25 \times 10^{21}\ \text{g} = \mathbf{5.25 \times 10^{18}\ kg}$$

5.104 To calculate the molarity of NaOH, we need moles of NaOH and volume of the NaOH solution. The volume is given in the problem; therefore, we need to calculate the moles of NaOH. The moles of NaOH can be calculated from the reaction of NaOH with HCl. The balanced equation is:

$$\text{NaOH}(aq) + \text{HCl}(aq) \longrightarrow \text{H}_2\text{O}(l) + \text{NaCl}(aq)$$

The number of moles of HCl gas is found from the ideal gas equation. $V = 0.189$ L, $T = (25 + 273)\text{K} = 298$ K, and $P = 108\ \text{mmHg} \times \frac{1\ \text{atm}}{760\ \text{mmHg}} = 0.142\ \text{atm}$.

$$n_{\text{HCl}} = \frac{PV_{\text{HCl}}}{RT} = \frac{(0.142\ \text{atm})(0.189\ \text{L})}{\left(0.0821 \frac{\text{L} \cdot \text{atm}}{\text{mol} \cdot \text{K}}\right)(298\ \text{K})} = 1.10 \times 10^{-3}\ \text{mol HCl}$$

The moles of NaOH can be calculated using the mole ratio from the balanced equation.

$$(1.10 \times 10^{-3}\ \text{mol HCl}) \times \frac{1\ \text{mol NaOH}}{1\ \text{mol HCl}} = 1.10 \times 10^{-3}\ \text{mol NaOH}$$

The molarity of the NaOH solution is:

$$M = \frac{\text{mol NaOH}}{\text{L of soln}} = \frac{1.10 \times 10^{-3}\ \text{mol NaOH}}{0.0157\ \text{L soln}} = 0.0701\ \text{mol/L} = \mathbf{0.0701}\ \boldsymbol{M}$$

5.106 To calculate the partial pressures of He and Ne, the total pressure of the mixture is needed. To calculate the total pressure of the mixture, we need the total number of moles of gas in the mixture (mol He + mol Ne).

$$n_{He} = \frac{PV}{RT} = \frac{(0.63\ \text{atm})(1.2\ \text{L})}{\left(0.0821\frac{\text{L}\cdot\text{atm}}{\text{mol}\cdot\text{K}}\right)(16+273)\text{K}} = 0.032\ \text{mol He}$$

$$n_{Ne} = \frac{PV}{RT} = \frac{(2.8\ \text{atm})(3.4\ \text{L})}{\left(0.0821\frac{\text{L}\cdot\text{atm}}{\text{mol}\cdot\text{K}}\right)(16+273)\text{K}} = 0.40\ \text{mol Ne}$$

The total pressure is:

$$P_{Total} = \frac{(n_{He}+n_{Ne})RT}{V_{Total}} = \frac{(0.032+0.40)\text{mol}\left(0.0821\frac{\text{L}\cdot\text{atm}}{\text{mol}\cdot\text{K}}\right)(16+273)\text{K}}{(1.2+3.4)\text{L}} = 2.2\ \text{atm}$$

$P_i = X_iP_T$. The partial pressures of He and Ne are:

$$\boldsymbol{P_{He}} = \frac{0.032\ \text{mol}}{(0.032+0.40)\text{mol}} \times 2.2\ \text{atm} = \mathbf{0.16\ atm}$$

$$\boldsymbol{P_{Ne}} = \frac{0.40\ \text{mol}}{(0.032+0.40)\text{mol}} \times 2.2\ \text{atm} = \mathbf{2.0\ atm}$$

5.108 When the water enters the flask from the dropper, some hydrogen chloride dissolves, creating a partial vacuum. Pressure from the atmosphere forces more water up the vertical tube.

5.110 Use the ideal gas equation to calculate the moles of water produced. We carry an extra significant figure in the first step of the calculation to limit rounding errors.

$$n_{H_2O} = \frac{PV}{RT} = \frac{(24.8\ \text{atm})(2.00\ \text{L})}{\left(0.0821\frac{\text{L}\cdot\text{atm}}{\text{mol}\cdot\text{K}}\right)(120+273)\text{K}} = 1.537\ \text{mol H}_2\text{O}$$

Next, we can determine the mass of H_2O in the 54.2 g sample. Subtracting the mass of H_2O from 54.2 g will give the mass of $MgSO_4$ in the sample.

$$1.537\ \text{mol H}_2\text{O} \times \frac{18.02\ \text{g H}_2\text{O}}{1\ \text{mol H}_2\text{O}} = 27.7\ \text{g H}_2\text{O}$$

$$\text{Mass MgSO}_4 = 54.2\ \text{g sample} - 27.7\ \text{g H}_2\text{O} = 26.5\ \text{g MgSO}_4$$

Finally, we can calculate the moles of $MgSO_4$ in the sample. Comparing moles of $MgSO_4$ to moles of H_2O will allow us to determine the correct mole ratio in the formula.

$$26.5\ \text{g MgSO}_4 \times \frac{1\ \text{mol MgSO}_4}{120.4\ \text{g MgSO}_4} = 0.220\ \text{mol MgSO}_4$$

$$\frac{\text{mol H}_2\text{O}}{\text{mol MgSO}_4} = \frac{1.54\ \text{mol}}{0.220\ \text{mol}} = 7.00$$

Therefore, the mole ratio between H_2O and $MgSO_4$ in the compound is 7 : 1. Thus, the value of $\boldsymbol{x} = \mathbf{7}$, and the formula is $\mathbf{MgSO_4 \cdot 7H_2O}$.

5.112 The circumference of the cylinder is $= 2\pi r = 2\pi\left(\frac{15.0\text{ cm}}{2}\right) = 47.1\text{ cm}$

(a) The speed at which the target is moving equals:

$$\text{speed of target} = \text{circumference} \times \text{revolutions/sec}$$

$$\textbf{speed of target} = \frac{47.1\text{ cm}}{1\text{ revolution}} \times \frac{130\text{ revolutions}}{1\text{ s}} \times \frac{0.01\text{ m}}{1\text{ cm}} = \mathbf{61.2\text{ m/s}}$$

(b) $$2.80\text{ cm} \times \frac{0.01\text{ m}}{1\text{ cm}} \times \frac{1\text{ s}}{61.2\text{ m}} = \mathbf{4.58 \times 10^{-4}\text{ s}}$$

(c) The Bi atoms must travel across the cylinder to hit the target. This distance is the diameter of the cylinder, which is 15.0 cm. The Bi atoms travel this distance in 4.58×10^{-4} s.

$$\frac{15.0\text{ cm}}{4.58 \times 10^{-4}\text{ s}} \times \frac{0.01\text{ m}}{1\text{ cm}} = \mathbf{328\text{ m/s}}$$

$$u_{\text{rms}} = \sqrt{\frac{3RT}{\mathcal{M}}} = \sqrt{\frac{3(8.314\text{ J/K}\cdot\text{mol})(850 + 273)\text{K}}{209.0 \times 10^{-3}\text{ kg/mol}}} = \mathbf{366\text{ m/s}}$$

The magnitudes of the speeds are comparable, but not identical. This is not surprising since 328 m/s is the velocity of a particular Bi atom, and u_{rms} is an average value.

5.114 The moles of O_2 can be calculated from the ideal gas equation. The mass of O_2 can then be calculated using the molar mass as a conversion factor.

$$n_{O_2} = \frac{PV}{RT} = \frac{(132\text{ atm})(120\text{ L})}{\left(0.0821\frac{\text{L}\cdot\text{atm}}{\text{mol}\cdot\text{K}}\right)(22 + 273)\text{K}} = 654\text{ mol } O_2$$

$$?\ \mathbf{g\ O_2} = 654\text{ mol } O_2 \times \frac{32.00\text{ g } O_2}{1\text{ mol } O_2} = \mathbf{2.09 \times 10^4\text{ g } O_2}$$

The volume of O_2 gas under conditions of 1.00 atm pressure and a temperature of 22°C can be calculated using the ideal gas equation. The moles of O_2 = 654 moles.

$$V_{O_2} = \frac{n_{O_2}RT}{P} = \frac{(654\text{ mol})\left(0.0821\frac{\text{L}\cdot\text{atm}}{\text{mol}\cdot\text{K}}\right)(22 + 273)\text{K}}{1.00\text{ atm}} = \mathbf{1.58 \times 10^4\text{ L } O_2}$$

5.116 The fruit ripens more rapidly because the quantity (partial pressure) of ethylene gas inside the bag increases.

5.118 As the pen is used the amount of ink decreases, increasing the volume inside the pen. As the volume increases, the pressure inside the pen decreases. The hole is needed to equalize the pressure as the volume inside the pen increases.

5.120 **(a)** $NH_4NO_3(s) \longrightarrow N_2O(g) + 2H_2O(l)$

(b)

$$R = \frac{PV}{nT} = \frac{\left(718\ \text{mmHg} \times \dfrac{1\ \text{atm}}{760\ \text{mmHg}}\right)(0.340\ \text{L})}{\left(0.580\ \text{g N}_2\text{O} \times \dfrac{1\ \text{mol N}_2\text{O}}{44.02\ \text{g N}_2\text{O}}\right)(24 + 273)\text{K}} = \mathbf{0.0821\frac{L \cdot atm}{mol \cdot K}}$$

5.122 The value of *a* indicates how strongly molecules of a given type of gas attract one anther. $\mathbf{C_6H_6}$ has the greatest intermolecular attractions due to its larger size compared to the other choices. Therefore, it has the largest *a* value.

5.124 The gases inside the mine were a mixture of carbon dioxide, carbon monoxide, methane, and other harmful compounds. The low atmospheric pressure caused the gases to flow out of the mine (the gases in the mine were at a higher pressure), and the man suffocated.

5.126 **(a)** This is a Boyle's law problem.

$$P_{\text{tire}}V_{\text{tire}} = P_{\text{air}}V_{\text{air}}$$

$$(5.0\ \text{atm})(0.98\ \text{L}) = (1.0\ \text{atm})V_{\text{air}}$$

$$\mathbf{V_{air} = 4.90\ L}$$

(b) Pressure in the tire – atmospheric pressure = gauge pressure

Pressure in the tire – 1.0 atm = 5.0 atm

Pressure in the tire = 6.0 atm

(c) Again, this is a Boyle's law problem.

$$P_{\text{pump}}V_{\text{pump}} = P_{\text{gauge}}V_{\text{gauge}}$$

$$(1\ \text{atm})(0.33V_{\text{tire}}) = P_{\text{gauge}}V_{\text{gauge}}$$

$$P_{\text{gauge}} = 0.33\ \text{atm}$$

This is the gauge pressure after one pump stroke. After three strokes, the gauge pressure will be (3 × 0.33 atm), or approximately **1 atm**. This is assuming that the initial gauge pressure was zero.

5.128 **(a)** First, let's convert the concentration of hydrogen from atoms/cm^3 to mol/L. The concentration in mol/L can be substituted into the ideal gas equation to calculate the pressure of hydrogen.

$$\frac{1\ \text{H atom}}{1\ \text{cm}^3} \times \frac{1\ \text{mol H}}{6.022 \times 10^{23}\ \text{H atoms}} \times \frac{1\ \text{cm}^3}{1\ \text{mL}} \times \frac{1\ \text{mL}}{0.001\ \text{L}} = \frac{2 \times 10^{-21}\ \text{mol H}}{\text{L}}$$

The pressure of H is:

$$P = \left(\frac{n}{V}\right)RT = \left(\frac{2 \times 10^{-21}\ \text{mol}}{1\ \text{L}}\right)\left(0.0821\frac{\text{L} \cdot \text{atm}}{\text{mol} \cdot \text{K}}\right)(3\ \text{K}) = \mathbf{5 \times 10^{-22}\ atm}$$

(b) From part (a), we know that 1 L contains 1.66×10^{-21} mole of H atoms. We convert to the volume that contains 1.0 g of H atoms.

$$\frac{1\ \text{L}}{2 \times 10^{-21}\ \text{mol H}} \times \frac{1\ \text{mol H}}{1.008\ \text{g H}} = \mathbf{5 \times 10^{20}\ L/g\ of\ H}$$

Note: This volume is about that of all the water on Earth!

5.130 From Table 5.3, the equilibrium vapor pressure at 30°C is 31.82 mmHg.

Converting 3.9×10^3 Pa to units of mmHg:

$$(3.9 \times 10^3\ \text{Pa}) \times \frac{760\ \text{mmHg}}{1.01325 \times 10^5\ \text{Pa}} = 29\ \text{mmHg}$$

$$\textbf{Relative Humidity} = \frac{\text{partial pressure of water vapor}}{\text{equilibrium vapor pressure}} \times 100\% = \frac{29\ \text{mmHg}}{31.82\ \text{mmHg}} \times 100\% = \mathbf{91\%}$$

5.132 The volume of one alveoli is:

$$V = \frac{4}{3}\pi r^3 = \frac{4}{3}\pi(0.0050\ \text{cm})^3 = (5.2 \times 10^{-7}\ \text{cm}^3) \times \frac{1\ \text{mL}}{1\ \text{cm}^3} \times \frac{0.001\ \text{L}}{1\ \text{mL}} = 5.2 \times 10^{-10}\ \text{L}$$

The number of moles of air in one alveoli can be calculated using the ideal gas equation.

$$n = \frac{PV}{RT} = \frac{(1.0\ \text{atm})(5.2 \times 10^{-10}\ \text{L})}{\left(0.0821\frac{\text{L}\cdot\text{atm}}{\text{mol}\cdot\text{K}}\right)(37 + 273)\text{K}} = 2.0 \times 10^{-11}\ \text{mol of air}$$

Since the air inside the alveoli is 14 percent oxygen, the moles of oxygen in one alveoli equals:

$$(2.0 \times 10^{-11}\ \text{mol of air}) \times \frac{14\%\ \text{oxygen}}{100\%\ \text{air}} = 2.8 \times 10^{-12}\ \text{mol}\ O_2$$

Converting to O_2 molecules:

$$(2.8 \times 10^{-12}\ \text{mol}\ O_2) \times \frac{6.022 \times 10^{23}\ O_2\ \text{molecules}}{1\ \text{mol}\ O_2} = \mathbf{1.7 \times 10^{12}\ O_2\ molecules}$$

5.134 The molar mass of a gas can be calculated using Equation (5.12) of the text.

$$\mathcal{M} = \frac{dRT}{P} = \frac{\left(1.33\ \frac{\text{g}}{\text{L}}\right)\left(0.0821\frac{\text{L}\cdot\text{atm}}{\text{mol}\cdot\text{K}}\right)(150 + 273)\text{K}}{\left(764\ \text{mmHg} \times \frac{1\ \text{atm}}{760\ \text{mmHg}}\right)} = 45.9\ \text{g/mol}$$

Some nitrogen oxides and their molar masses are:

NO 30 g/mol N_2O 44 g/mol NO_2 46 g/mol

The nitrogen oxide is most likely $\mathbf{NO_2}$, although N_2O cannot be completely ruled out.

5.136 When calculating root-mean-square speed, remember that the molar mass must be in units of kg/mol.

$$u_{\text{rms}} = \sqrt{\frac{3RT}{\mathcal{M}}} = \sqrt{\frac{3(8.314\ \text{J/mol}\cdot\text{K})(1.7\times10^{-7}\ \text{K})}{85.47\times10^{-3}\ \text{kg/mol}}} = \mathbf{7.0\times10^{-3}\ m/s}$$

The mass of one Rb atom in kg is:

$$\frac{85.47\ \text{g Rb}}{1\ \text{mol Rb}}\times\frac{1\ \text{mol Rb}}{6.022\times10^{23}\ \text{Rb atoms}}\times\frac{1\ \text{kg}}{1000\ \text{g}} = 1.419\times10^{-25}\ \text{kg/Rb atom}$$

$$\overline{\textbf{KE}} = \frac{1}{2}m\overline{u^2} = \frac{1}{2}(1.419\times10^{-25}\ \text{kg})(7.0\times10^{-3}\ \text{m/s})^2 = \mathbf{3.5\times10^{-30}\ J}$$

5.138 The molar volume is the volume of 1 mole of gas under the specified conditions.

$$V = \frac{nRT}{P} = \frac{(1\ \text{mol})\left(0.0821\frac{\text{L}\cdot\text{atm}}{\text{mol}\cdot\text{K}}\right)(220\ \text{K})}{\left(6.0\ \text{mmHg}\times\frac{1\ \text{atm}}{760\ \text{mmHg}}\right)} = \mathbf{2.3\times10^{3}\ L}$$

5.140 The volume of the bulb can be calculated using the ideal gas equation. Pressure and temperature are given in the problem. Moles of air must be calculated before the volume can be determined.

Mass of air = 91.6843 g − 91.4715 g = 0.2128 g air

Molar mass of air = (0.78 × 28.02 g/mol) + (0.21 × 32.00 g/mol) + (0.01 × 39.95 g/mol) = 29 g/mol

$$\text{moles air} = 0.2128\ \text{g air}\times\frac{1\ \text{mol air}}{29\ \text{g air}} = 7.3\times10^{-3}\ \text{mol air}$$

Now, we can calculate the volume of the bulb.

$$V_{\textbf{bulb}} = \frac{nRT}{P} = \frac{(7.3\times10^{-3}\ \text{mol})\left(0.0821\frac{\text{L}\cdot\text{atm}}{\text{mol}\cdot\text{K}}\right)(23+273)\text{K}}{\left(744\ \text{mmHg}\times\frac{1\ \text{atm}}{760\ \text{mmHg}}\right)} = 0.18\ \text{L} = \mathbf{1.8\times10^{2}\ mL}$$

5.142 In Problem 5.102, the mass of the Earth's atmosphere was determined to be 5.25×10^{18} kg. Assuming that the molar mass of air is 29.0 g/mol, we can calculate the number of molecules in the atmosphere.

(a) $$(5.25\times10^{18}\ \text{kg air})\times\frac{1000\ \text{g}}{1\ \text{kg}}\times\frac{1\ \text{mol air}}{29.0\ \text{g air}}\times\frac{6.022\times10^{23}\ \text{molecules air}}{1\ \text{mol air}} = \mathbf{1.09\times10^{44}\ molecules}$$

(b) First, calculate the moles of air exhaled in every breath. (500 mL = 0.500 L)

$$n = \frac{PV}{RT} = \frac{(1\ \text{atm})(0.500\ \text{L})}{\left(0.0821\frac{\text{L}\cdot\text{atm}}{\text{mol}\cdot\text{K}}\right)(37+273)\text{K}} = 1.96\times10^{-2}\ \text{mol air/breath}$$

Next, convert to molecules of air per breath.

$$1.96 \times 10^{-2} \text{ mol air/breath} \times \frac{6.022 \times 10^{23} \text{ molecules air}}{1 \text{ mol air}} = \mathbf{1.18 \times 10^{22} \text{ molecules/breath}}$$

(c) $$\frac{1.18 \times 10^{22} \text{ molecules}}{1 \text{ breath}} \times \frac{12 \text{ breaths}}{1 \text{ min}} \times \frac{60 \text{ min}}{1 \text{ h}} \times \frac{24 \text{ h}}{1 \text{ day}} \times \frac{365 \text{ days}}{1 \text{ yr}} \times 35 \text{ yr} = \mathbf{2.60 \times 10^{30} \text{ molecules}}$$

(d) Fraction of molecules in the atmosphere exhaled by Mozart is:

$$\frac{2.60 \times 10^{30} \text{ molecules}}{1.09 \times 10^{44} \text{ molecules}} = \mathbf{2.39 \times 10^{-14}}$$

Or,

$$\frac{1}{2.39 \times 10^{-14}} = 4.18 \times 10^{13}$$

Thus, about 1 molecule of air in every 4×10^{13} molecules was exhaled by Mozart.

In a single breath containing 1.18×10^{22} molecules, we would breathe in on average:

$$(1.18 \times 10^{22} \text{ molecules}) \times \frac{1 \text{ Mozart air molecule}}{4 \times 10^{13} \text{ air molecules}} = \mathbf{3 \times 10^{8} \text{ molecules that Mozart exhaled}}$$

(e) We made the following assumptions:

1. Complete mixing of air in the atmosphere.
2. That no molecules escaped to the outer atmosphere.
3. That no molecules were used up during metabolism, nitrogen fixation, and so on.

5.144 The ideal gas law can be used to calculate the moles of water vapor per liter.

$$\frac{n}{V} = \frac{P}{RT} = \frac{1.0 \text{ atm}}{\left(0.0821 \frac{\text{L} \cdot \text{atm}}{\text{mol} \cdot \text{K}}\right)(100 + 273)\text{K}} = 0.033 \frac{\text{mol}}{\text{L}}$$

We eventually want to find the distance between molecules. Therefore, let's convert moles to molecules, and convert liters to a volume unit that will allow us to get to distance (m^3).

$$\left(\frac{0.033 \text{ mol}}{1 \text{ L}}\right)\left(\frac{6.022 \times 10^{23} \text{ molecules}}{1 \text{ mol}}\right)\left(\frac{1000 \text{ L}}{1 \text{ m}^3}\right) = 2.0 \times 10^{25} \frac{\text{molecules}}{\text{m}^3}$$

This is the number of ideal gas molecules in a cube that is 1 meter on each side. Assuming an equal distribution of molecules along the three mutually perpendicular directions defined by the cube, a linear density in one direction may be found:

$$\left(\frac{2.0 \times 10^{25} \text{ molecules}}{1 \text{ m}^3}\right)^{\frac{1}{3}} = 2.7 \times 10^{8} \frac{\text{molecules}}{\text{m}}$$

This is the number of molecules on a line *one* meter in length. The distance between each molecule is given by:

$$\frac{1\text{ m}}{2.70 \times 10^{8}} = 3.7 \times 10^{-9}\text{ m} = \mathbf{3.7\text{ nm}}$$

Assuming a water molecule to be a sphere with a diameter of 0.3 nm, the water molecules are separated by over 12 times their diameter: $\frac{3.7\text{ nm}}{0.3\text{ nm}} \approx 12$ times.

A similar calculation is done for liquid water. Starting with density, we convert to molecules per cubic meter.

$$\frac{0.96\text{ g}}{1\text{ cm}^3} \times \frac{1\text{ mol H}_2\text{O}}{18.02\text{ g H}_2\text{O}} \times \frac{6.022 \times 10^{23}\text{ molecules}}{1\text{ mol H}_2\text{O}} \times \left(\frac{100\text{ cm}}{1\text{ m}}\right)^3 = 3.2 \times 10^{28}\,\frac{\text{molecules}}{\text{m}^3}$$

This is the number of liquid water molecules in *one* cubic meter. From this point, the calculation is the same as that for water vapor, and the space between molecules is found using the same assumptions.

$$\left(\frac{3.2 \times 10^{28}\text{ molecules}}{1\text{ m}^3}\right)^{\frac{1}{3}} = 3.2 \times 10^{9}\,\frac{\text{molecules}}{\text{m}}$$

$$\frac{1\text{ m}}{3.2 \times 10^{9}} = 3.1 \times 10^{-10}\text{ m} = \mathbf{0.31\text{ nm}}$$

Assuming a water molecule to be a sphere with a diameter of 0.3 nm, to one significant figure, the water molecules are touching each other in the liquid phase.

5.146 Since the $R = 8.314$ J/mol·K and $1\text{ J} = 1\,\frac{\text{kg}\cdot\text{m}^2}{\text{s}^2}$, then the mass substituted into the equation must have units of kg and the height must have units of meters.

29 g/mol = 0.029 kg/mol
5.0 km = 5.0×10^3 m

Substituting the given quantities into the equation, we find the atmospheric pressure at 5.0 km to be:

$$P = P_0 e^{-\frac{g\mathcal{M}h}{RT}}$$

$$P = (1.0\text{ atm})e^{-\left(\frac{(9.8\text{ m/s}^2)(0.029\text{ kg/mol})(5.0 \times 10^3\text{ m})}{(8.314\text{ J/mol}\cdot\text{K})(278\text{ K})}\right)}$$

$$\boldsymbol{P = 0.54\text{ atm}}$$

5.148 The reaction between Zn and HCl is: $\text{Zn}(s) + 2\text{HCl}(aq) \rightarrow \text{H}_2(g) + \text{ZnCl}_2(aq)$

From the amount of $\text{H}_2(g)$ produced, we can determine the amount of Zn reacted. Then, using the original mass of the sample, we can calculate the mass % of Zn in the sample.

$$n_{\text{H}_2} = \frac{PV_{\text{H}_2}}{RT}$$

$$n_{H_2} = \frac{\left(728\text{ mmHg} \times \dfrac{1\text{ atm}}{760\text{ mmHg}}\right)(1.26\text{ L})}{\left(0.0821\dfrac{\text{L}\cdot\text{atm}}{\text{mol}\cdot\text{K}}\right)(22 + 273)\text{K}} = 0.0498\text{ mol H}_2$$

Since the mole ratio between H_2 and Zn is 1:1, the amount of Zn reacted is also 0.0498 mole. Converting to grams of Zn, we find:

$$0.0498\text{ mol Zn} \times \frac{65.39\text{ g Zn}}{1\text{ mol Zn}} = 3.26\text{ g Zn}$$

The mass percent of Zn in the 6.11 g sample is:

$$\textbf{mass \% Zn} = \frac{\text{mass Zn}}{\text{mass sample}} \times 100\% = \frac{3.26\text{ g}}{6.11\text{ g}} \times 100\% = \mathbf{53.4\%}$$

5.150 We start with Graham's Law as this problem relates to effusion of gases. Using Graham's Law, we can calculate the effective molar mass of the mixture of CO and CO_2. Once the effective molar mass of the mixture is known, we can determine the mole fraction of each component. Because $n \propto V$ at constant T and P, the volume fraction = mole fraction.

$$\frac{r_{He}}{r_{mix}} = \sqrt{\frac{\mathcal{M}_{mix}}{\mathcal{M}_{He}}}$$

$$\mathcal{M}_{mix} = \left(\frac{r_{He}}{r_{mix}}\right)^2 \mathcal{M}_{He}$$

$$\mathcal{M}_{mix} = \left(\frac{\dfrac{29.7\text{ mL}}{2.00\text{ min}}}{\dfrac{10.0\text{ mL}}{2.00\text{ min}}}\right)^2 (4.003\text{ g/mol}) = 35.31\text{ g/mol}$$

Now that we know the molar mass of the mixture, we can calculate the mole fraction of each component.

$$X_{CO} + X_{CO_2} = 1$$

and

$$X_{CO_2} = 1 - X_{CO}$$

The mole fraction of each component multiplied by its molar mass will give the contribution of that component to the effective molar mass.

$$X_{CO}\mathcal{M}_{CO} + X_{CO_2}\mathcal{M}_{CO_2} = \mathcal{M}_{mix}$$

$$X_{CO}\mathcal{M}_{CO} + (1 - X_{CO})\mathcal{M}_{CO_2} = \mathcal{M}_{mix}$$

$$X_{CO}(28.01\text{ g/mol}) + (1 - X_{CO})(44.01\text{ g/mol}) = 35.31\text{ g/mol}$$

$$28.01X_{CO} + 44.01 - 44.01X_{CO} = 35.31$$

$$16.00X_{CO} = 8.70$$

$$X_{CO} = 0.544$$

At constant P and T, $n \propto V$. Therefore, volume fraction = mole fraction. As a result,

$$\% \text{ of CO by volume} = \mathbf{54.4\%}$$

$$\% \text{ of } CO_2 \text{ by volume} = 1 - \% \text{ of CO by volume} = \mathbf{45.6\%}$$

5.152 The reactions are:

$$CH_4 + 2O_2 \rightarrow CO_2 + 2H_2O$$

$$2C_2H_6 + 7O_2 \rightarrow 4CO_2 + 6H_2O$$

For a given volume and temperature, $n \propto P$. This means that the greater the pressure of reactant, the more moles of reactant, and hence the more product (CO_2) that will be produced. The pressure of CO_2 produced comes from both the combustion of methane and ethane. We set up an equation using the mole ratios from the balanced equation to convert to pressure of CO_2.

$$\left(P_{CH_4} \times \frac{1 \text{ mol } CO_2}{1 \text{ mol } CH_4}\right) + \left(P_{C_2H_6} \times \frac{4 \text{ mol } CO_2}{2 \text{ mol } C_2H_6}\right) = 356 \text{ mmHg } CO_2$$

$$(1) \quad P_{CH_4} + 2P_{C_2H_6} = 356 \text{ mmHg}$$

Also,

$$(2) \quad P_{CH_4} + P_{C_2H_6} = 294 \text{ mmHg}$$

Subtracting equation (2) from equation (1) gives:

$$P_{C_2H_6} = 356 - 294 = 62 \text{ mmHg}$$

$$P_{CH_4} = 294 - 62 = 232 \text{ mmHg}$$

Lastly, because $n \propto P$, we can solve for the mole fraction of each component using partial pressures.

$$X_{CH_4} = \frac{232}{294} = \mathbf{0.789} \qquad X_{C_2H_6} = \frac{62}{294} = \mathbf{0.211}$$

5.154 **(a)** We see from the figure that two hard spheres of radius r cannot approach each other more closely than $2r$ (measured from the centers). Thus, there is a sphere of radius $2r$ surrounding each hard sphere from which other hard spheres are excluded. The excluded volume/pair of molecules is:

$$V_{\text{excluded/pair}} = \frac{4}{3}\pi(2r)^3 = \frac{32}{3}\pi r^3 = \mathbf{8\left(\frac{4}{3}\pi r^3\right)}$$

This is eight times the volume of an individual molecule.

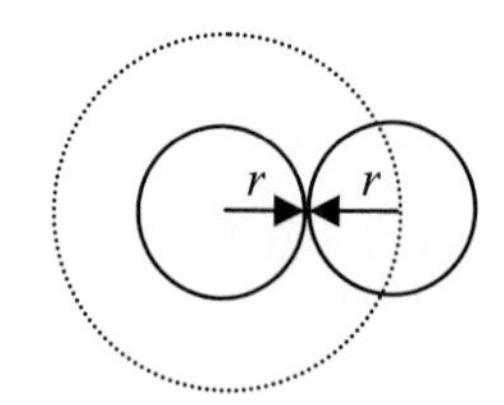

(b) The result in part (a) is for a pair of molecules, so the excluded volume/molecule is:

$$V_{\text{excluded/molecule}} = \frac{1}{2}\left(\frac{32}{3}\pi r^3\right) = \frac{16}{3}\pi r^3$$

To convert from excluded volume per molecule to excluded volume per mole, we need to multiply by Avogadro's number, N_A.

$$V_{\text{excluded/mole}} = \mathbf{\frac{16}{3}N_A\pi r^3}$$

The sum of the volumes of a mole of molecules (treated as hard spheres of radius r) is $\frac{4}{3}N_A\pi r^3$. The excluded volume is **four times** the volume of the spheres themselves.

5.156 From the root-mean-square speed, we can calculate the molar mass of the gaseous oxide.

$$u_{\text{rms}} = \sqrt{\frac{3RT}{\mathcal{M}}}$$

$$\mathcal{M} = \frac{3RT}{(u_{\text{rms}})^2} = \frac{3(8.314\ \text{J/mol}\cdot\text{K})(293\ \text{K})}{(493\ \text{m/s})^2} = 0.0301\ \text{kg/mol} = 30.1\ \text{g/mol}$$

The compound must be a monoxide because 2 moles of oxygen atoms would have a mass of 32.00 g. The molar mass of the other element is:

$$30.1\ \text{g/mol} - 16.00\ \text{g/mol} = 14.01\ \text{g/mol}$$

The compound is nitrogen monoxide, **NO**.

5.158 Pressure and volume are constant. We start with Equation (5.9) of the text.

$$\frac{P_1V_1}{n_1T_1} = \frac{P_2V_2}{n_2T_2}$$

Because $P_1 = P_2$ and $V_1 = V_2$, this equation reduces to:

$$\frac{1}{n_1T_1} = \frac{1}{n_2T_2}$$

or,

$$n_1T_1 = n_2T_2$$

Because $T_1 = 2T_2$, substituting into the above equation gives:

$$2n_1T_2 = n_2T_2$$

or,

$$2n_1 = n_2$$

This equation indicates that the number of moles of gas after reaction is twice the number of moles of gas before reaction. Only reaction **(b)** fits this description.

CHAPTER 6
THERMOCHEMISTRY

PROBLEM-SOLVING STRATEGIES AND TUTORIAL SOLUTIONS

TYPES OF PROBLEMS

Problem Type 1: The First Law of Thermodynamics.
- **(a)** Applying the First Law of Thermodynamics.
- **(b)** Calculating the work done in gas expansion.
- **(c)** Enthalpy and the First Law of Thermodynamics. Calculating the internal energy change of a gaseous reaction.

Problem Type 2: Thermochemical Equations.

Problem Type 3: Calculating Heat Absorbed or Released Using Specific Heat Data.

Problem Type 4: Calorimetry.
- **(a)** Constant-volume calorimetry.
- **(b)** Constant-pressure calorimetry.

Problem Type 5: Standard Enthalpy of Formation and Reaction.
- **(a)** Calculating the standard enthalpy of reaction.
- **(b)** Direct method of calculating the standard enthalpy of formation.
- **(c)** Indirect method of calculating the standard enthalpy of formation, Hess's law.

PROBLEM TYPE 1: THE FIRST LAW OF THERMODYNAMICS

A. Applying the First Law of Thermodynamics

The **First Law of Thermodynamics** states that energy can be converted from one form to another, but cannot be created or destroyed. Another way of stating the first law is that the energy of the universe is constant. The universe is composed of both the system and the surroundings.

$$\Delta E_{sys} + \Delta E_{surr} = 0$$

where,

the subscripts "sys" and "surr" denote system and surroundings, respectively.

However, in chemistry, we are normally interested in the changes associated with the *system* (which may be a flask containing reactants and products), not with its surroundings. Therefore, a more useful form of the first law is

$$\Delta E = q + w \qquad (6.1, \text{text})$$

where,

ΔE is the change in the internal energy of the system
q is the heat exchange between the system and surroundings
w is the work done on (or by) the system

Using the sign convention for thermochemical processes (see Section 6.3 of your text for discussion), q is positive for an endothermic process and negative for an exothermic process. For work, w is positive for work done *on* the system *by* the surroundings and negative for work done *by* the system *on* the surroundings. Try to understand the sign convention in this manner. If a *system* loses heat to the surroundings or does work on the surroundings, we expect its internal energy to decrease since both processes are energy depleting. Conversely, if heat is added to the *system* or if work is done on the *system*, then the internal energy of the system would increase.

EXAMPLE 6.1

A system does 975 kJ of work on its surroundings while at the same time it absorbs 625 kJ of heat. What is the change in energy, ΔE, for the system?

Strategy: The system does work on the surroundings, so what is the sign for w? Heat is absorbed by the gas from the surroundings. Is this an endothermic or exothermic process? What is the sign for q?

Solution: To calculate the energy change of the gas (ΔE), we need Equation (6.1) of the text. To solve this problem, you must make sure to get the sign convention correct. The system does work on the surroundings; this is an energy-depleting process.

$$\mathbf{w = -975\ kJ}$$

The system absorbs 625 kJ of heat. Therefore, the internal energy of the system would increase.

$$\mathbf{q = +625\ kJ}$$

Finally,

$$\Delta E = q + w = 625\ \text{kJ} + (-975\ \text{kJ}) = \mathbf{-350\ kJ}$$

PRACTICE EXERCISE

1. The surroundings do 455 kJ of work on the system while at the same time the system releases 253 kJ of heat. What is the change in energy, ΔE, for the system?

Text Problem: 6.18

B. Calculating the work done in gas expansion

A useful example of mechanical work is the expansion of a gas. Picture a gas-filled cylinder that is fitted with a weightless, frictionless, movable piston, at a certain temperature, pressure, and volume. As the gas expands, it pushes the piston upward against a constant, opposing, external atmospheric pressure, P. The gas (system) is doing work on the surroundings. The work can be calculated as follows:

$$w = -P\Delta V \qquad \text{(6.3, text)}$$

where,

P is the external pressure

ΔV is the change in volume ($V_f - V_i$)

Note: The minus sign in the equation takes care of the sign convention for w. For gas expansion, $\Delta V > 0$, so $-P\Delta V$ is a negative quantity. When a gas expands, it's doing work on the surroundings; the internal energy of the system decreases. For gas compression, $\Delta V < 0$, so $-P\Delta V$ is a positive quantity. When a gas is compressed, the surroundings are doing work on the system, increasing the internal energy.

EXAMPLE 6.2

A gas initially at a pressure of 10.0 atm and occupying a volume of 5.0 L is allowed to expand at constant temperature against a constant external pressure of 4.0 atm. After expansion, the gas occupies a volume of 12.5 L. Calculate the work done by the gas on the surroundings.

Strategy: The work done in gas expansion is equal to the product of the external, opposing pressure and the change in volume [Equation (6.3) of the text].

$$w = -P\Delta V$$

What is the conversion factor between L·atm and J?

Solution: We are given the external pressure in the problem, but we must calculate ΔV.

$$\Delta V = V_f - V_i = 12.5\text{ L} - 5.0\text{ L} = \mathbf{7.5\text{ L}}$$

Substitute P and ΔV into Equation (6.3) of the text and solve for w.

$$w = -P\Delta V = -(4.0\text{ atm})(7.5\text{ L}) = \mathbf{-3.0 \times 10^1\text{ L·atm}}$$

It would be more convenient to express w in units of joules. The following conversion factor can be obtained from Appendix 1 of the text:

$$1\text{ L·atm} = 101.3\text{ J}$$

Thus, we can write:

$$w = (-3.0 \times 10^1\text{ L·atm}) \times \frac{101.3\text{ J}}{1\text{ L·atm}} = \mathbf{-3.0 \times 10^3\text{ J}}$$

Check: Because this is gas expansion (work is done by the system on the surroundings), the work done has a negative sign.

PRACTICE EXERCISE

2. Calculate the work done on the system when 6.0 L of a gas is compressed to 1.0 L by a constant external pressure of 2.0 atm.

Text Problem: 6.20

C. Enthalpy and the First Law of Thermodynamics—Calculating the internal energy change of a gaseous reaction

Let's return to the following form of the first law of thermodynamics.

$$\Delta E_{sys} = q + w$$

Under constant-pressure conditions we can write:

$$\Delta E = q_p + w$$

Recall that the heat evolved or absorbed (q) by a reaction carried out under constant-pressure conditions is equal to the enthalpy change of the system, ΔH.

Thus,

$$\Delta E = \Delta H + w$$

Also, we know that for gas expansion or compression under a constant external pressure, $w = -P\Delta V$. Substituting into the above equation, we have:

$$\Delta E = \Delta H - P\Delta V$$

Also, for an ideal gas at constant pressure,

$$P\Delta V = \Delta(nRT)$$

$$\Delta V = \frac{\Delta(nRT)}{P}$$

Substituting gives:

$$\Delta E = \Delta H - \Delta(nRT)$$

Finally, at constant temperature,

$$\mathbf{\Delta E = \Delta H - RT\Delta n} \qquad (6.10,\ \text{text})$$

where Δn is defined as

$$\Delta n = \text{number of moles of product gases} - \text{number of moles of reactant gases}$$

EXAMPLE 6.3

Calculate the change in internal energy when 1 mole of H_2 and 1/2 mole of O_2 are converted to 1 mole of H_2O at 1 atm and 25°C.

$$\mathbf{H_2(g) + \tfrac{1}{2}O_2(g) \longrightarrow H_2O(l) \qquad \Delta H° = -286\ kJ/mol}$$

Strategy: Calculate the total change in the number of moles of ***gas***. Note that the product is a liquid. Substitute the values of $\Delta H°$ and Δn into Equation (6.10) of the text.

Solution:

$$\Delta n = 0\ \text{mol} - (1\ \text{mol} + 1/2\ \text{mol}) = -1.5\ \text{mol}$$

Substitute the values for $\Delta H°$ and Δn into Equation (6.10).

$T = 25° + 273° = 298\ \text{K}$

$$\Delta E° = \Delta H° - RT\Delta n$$

$$\mathbf{\Delta E°} = -286\ \text{kJ/mol} - \left(8.314\frac{\cancel{\text{J}}}{\text{mol}\cdot\cancel{\text{K}}}\right)(298\ \cancel{\text{K}})(-1.5) \times \frac{1\ \text{kJ}}{1000\ \cancel{\text{J}}} = \mathbf{-282\ kJ/mol}$$

PRACTICE EXERCISE

3. Calculate the change in the internal energy when 1.0 mole of water vaporizes at 1.0 atm and 100°C. Assume that water vapor is an ideal gas and that the volume of liquid water is negligible compared with that of steam at 100°C. [$\Delta H_{vap}(H_2O) = 40.67$ kJ/mol at 100°C].

Text Problem: 6.28

PROBLEM TYPE 2: THERMOCHEMICAL EQUATIONS

Equations showing both the mass and enthalpy relations are called **thermochemical equations**. The following guidelines are helpful in writing and interpreting thermochemical equations:

1. The stoichiometric coefficients always refer to the number of moles of each substance.
2. When an equation is reversed, the roles of reactants and products change. Consequently, the magnitude of ΔH for the equation remains the same, but its sign changes.
3. If both sides of a thermochemical equation are multiplied by a factor n, then ΔH must also be multiplied by the same factor.
4. When writing thermochemical equations, the physical states of all reactants and products must be specified, because they help determine the actual enthalpy changes.

See Section 6.4 of your text for further discussion.

EXAMPLE 6.4
Given the thermochemical equation

$$SO_2(g) + \frac{1}{2}O_2(g) \longrightarrow SO_3(g) \quad \Delta H = -99 \text{ kJ/mol}$$

how much heat is evolved when (a) 1/2 mol of SO_2 reacts and (b) 3 mol of SO_2 reacts?

Strategy: The thermochemical equation shows that for every 1 mole of SO_2 reacted, 99 kJ of heat are given off (note the negative sign). We can write a conversion factor from this information.

$$\frac{-99 \text{ kJ}}{1 \text{ mol } SO_2}$$

Solution:

(a) If 1/2 mole of SO_2 reacts, that means that we are multiplying the equation by 1/2. Therefore, we must multiply ΔH by 1/2.

$$\textbf{heat evolved} = 0.5 \text{ mol } SO_2 \times \frac{99 \text{ kJ}}{1 \text{ mol } SO_2} = \mathbf{5.0 \times 10^1 \text{ kJ}}$$

(b) Following the same argument as in part (a):

$$\textbf{heat evolved} = 3(99 \text{ kJ}) = \mathbf{3.0 \times 10^2 \text{ kJ}}$$

Why isn't there a negative sign in our answer? The sign convention for an exothermic reaction (energy, as heat, is released by the system) is negative ($-\Delta H$). However, in the above example, we state that heat is evolved, so a negative sign is unnecessary.

Tip: Remember that the heat evolved or absorbed (q) by a reaction carried out under constant-pressure conditions is equal to the enthalpy change of the system, ΔH.

PRACTICE EXERCISE

4. Given the thermochemical equation

$$SO_2(g) + \frac{1}{2}O_2(g) \longrightarrow SO_3(g) \quad \Delta H = -99 \text{ kJ/mol}$$

how much heat is evolved when 75 g of SO_2 is combusted?

Text Problem: 6.26

PROBLEM TYPE 3: CALCULATING HEAT ABSORBED OR RELEASED USING SPECIFIC HEAT DATA

If the specific heat (s) and the amount of substance is known, then the change in the sample's temperature (Δt) will tell us the amount of heat (q) that has been absorbed or released in a particular process. The equation for calculating the heat change is given by:

$$q = ms\Delta t \qquad (6.12, \text{text})$$

or

$$q = C\Delta t$$

where m is the mass of the sample and Δt is the temperature change.

$$\Delta t = t_{\text{final}} - t_{\text{initial}}$$

EXAMPLE 6.5

How much heat is absorbed by 80.0 g of iron (Fe) when its temperature is raised from 25°C to 500°C? The specific heat of iron is 0.444 J/g·°C.

Strategy: We know the mass, specific heat, and the change in temperature for iron. We can use Equation (6.12) of the text to solve this problem.

Solution: Substitute the known values into Equation (6.12).

$$q = m_{Fe}s_{Fe}\Delta t$$

$$q = (80.0\ \cancel{g})(0.444\ J/\cancel{g\cdot°C})(500-25)\cancel{°C} = \mathbf{1.69 \times 10^4\ J}$$

PRACTICE EXERCISE

5. A piece of iron initially at a temperature of 25°C absorbs 10.0 kJ of heat. If its mass is 50.0 g, calculate the final temperature of the piece of iron. The specific heat of iron is 0.444 J/g·°C.

Text Problems: 6.34, **6.36**

PROBLEM TYPE 4: CALORIMETRY

A. Constant-volume calorimetry

For a discussion of constant-volume calorimetry, see Section 6.5 of your text. Heat of combustion is usually measured in a constant-volume calorimeter. The heat released during combustion is absorbed by the calorimeter. Because no heat enters or leaves the system throughout the process, we can write:

$$q_{system} = 0 = q_{cal} + q_{rxn}$$

or,

$$q_{rxn} = -q_{cal}$$

The heat absorbed by the calorimeter can be calculated using the heat capacity of the bomb calorimeter and the temperature rise.

$$q_{cal} = C_{cal}\Delta t$$

> **Note:** The negative sign for q_{rxn} indicates that heat was released during the combustion. You should expect this, because all combustion processes are exothermic. Thermal energy is transferred from the system to the surroundings.

EXAMPLE 6.6

0.500 g of ethanol [$CH_3CH_2OH(l)$] was burned in a bomb calorimeter. The temperature of the water rose 1.60°C. The heat capacity of the calorimeter plus water was 9.06 kJ/°C.

(a) Write a balanced equation for the combustion of ethanol.

(b) Calculate the molar heat of combustion of ethanol.

(a) Recall that a **combustion reaction** is typically a vigorous and exothermic reaction that takes place between certain substances and oxygen. If the reactant contains only C, H, and/or O, then the products are CO_2 and H_2O. Therefore, the balanced equation for the combustion of ethanol is:

$$CH_3CH_2OH(l) + 3O_2(g) \longrightarrow 2CO_2(g) + 3H_2O(g)$$

(b)

Strategy: Knowing the heat capacity and the temperature rise, how do we calculate the heat absorbed by the calorimeter? What is the heat generated by the combustion of 0.500 g ethanol? What is the conversion factor between grams and moles of ethanol?

Solution: The heat absorbed by the calorimeter and water is equal to the product of the heat capacity and the temperature change. From Equation (6.16) of the text, assuming no heat is lost to the surroundings, we write

$$q_{cal} = C_{cal}\Delta t$$

$$q_{cal} = (9.06\ kJ/°C)(1.60°C) = 14.5\ kJ$$

Because $q_{sys} = 0 = q_{cal} + q_{rxn}$, $q_{rxn} = -q_{cal}$. The heat change of the reaction is –14.5 kJ. This is the heat released by the combustion of 0.500 g of ethanol; therefore, we can write the conversion factor as

$$\frac{-14.5\ kJ}{0.500\ g\ ethanol}$$

From the molar mass of ethanol and the above conversion factor, the heat of combustion of 1 mole of ethanol can be calculated.

$$\mathcal{M}_{ethanol} = 2(12.01\ g) + 6(1.008\ g) + 16.00\ g = 46.07\ g$$

$$\textbf{Molar heat of combustion} = \frac{-14.5\ kJ}{0.500\ \cancel{g}\ ethanol} \times \frac{46.07\ \cancel{g}\ ethanol}{1\ mol\ ethanol} = \mathbf{-1.34 \times 10^3\ kJ/mol\ ethanol}$$

PRACTICE EXERCISE

6. The combustion of benzoic acid is often used as a standard source of heat for calibrating combustion bomb calorimeters. The heat of combustion of benzoic acid has been accurately determined to be 26.42 kJ/g. When 0.8000 g of benzoic acid was burned in a calorimeter containing water, a temperature rise of 4.08°C was observed. What is the heat capacity of the bomb calorimeter plus water?

Text Problem: 6.102

B. Constant-pressure calorimetry

A simpler device than the constant-volume calorimeter is the constant-pressure calorimeter that is used to determine the heat changes for noncombustion reactions. The reactions usually occur in solution. Because the measurements are carried out under constant atmospheric pressure conditions, the heat change for the process (q_{rxn}) is equal to the enthalpy change (ΔH).

The heat released during reaction is absorbed both by the solution in the calorimeter. We ignore the small heat capacity of the calorimeter in our calculations. Because no heat enters or leaves the system throughout the process, we can write:

$$q_{system} = 0 = q_{soln} + q_{rxn}$$

or,

$$q_{soln} = -q_{rxn}$$

The heat absorbed by the solution can be calculated using the equation

$$q_{soln} = m_{soln}s_{soln}\Delta t$$

EXAMPLE 6.7

The heat of neutralization for the following reaction is –56.2 kJ/mol.

$$\mathbf{NaOH}(aq) + \mathbf{HCl}(aq) \longrightarrow \mathbf{NaCl}(aq) + \mathbf{H_2O}(l)$$

1.00×10^2 mL of 1.50 *M* HCl is mixed with 1.00×10^2 mL of 1.50 *M* NaOH in a constant-pressure calorimeter. The initial temperature of the HCl and NaOH solutions is the same, 23.2°C. Calculate the final temperature of the mixed solution. Assume that the density and specific heat of the mixed solution is the same as for water (1.00 g/mL and 4.184 J/g·°C, respectively).

Strategy: The neutralization reaction is exothermic. 56.2 kJ of heat are released when 1 mole of H^+ reacts with 1 mole of OH^-. Assuming no heat is lost to the surroundings, we can equate the heat lost by the reaction to the heat gained by the combined solution. How do we calculate the heat released during the reaction? Are we reacting 1 mole of H^+ with 1 mole of OH^-? How do we calculate the heat absorbed by the combined solution?

Solution: Assuming no heat is lost to the surroundings, we can write:

$$q_{\text{soln}} + q_{\text{rxn}} = 0$$

or

$$q_{\text{soln}} = -q_{\text{rxn}}$$

First, let's set up how we would calculate the heat gained by the solution,

$$q_{\text{soln}} = m_{\text{soln}} s_{\text{soln}} \Delta t$$

where m and s are the mass and specific heat of the solution and $\Delta t = t_f - t_i$.

We assume that the specific heat of the solution is the same as the specific heat of water, and we assume that the density of the solution is the same as the density of water (1.00 g/mL). Since the density is 1.00 g/mL, the mass of 200 mL of solution (100 mL + 100 mL) is 200 g.

Substituting into the equation above, the heat gained by the solution can be represented as:

$$q_{\text{soln}} = (2.00 \times 10^2 \text{ g})(4.184 \text{ J/g}\cdot°\text{C})(t_f - 23.2°\text{C})$$

Next, let's calculate q_{rxn}, the heat released when 100 mL of 1.50 *M* HCl are mixed with 100 mL of 1.50 *M* NaOH. There is exactly enough NaOH to neutralize all the HCl. Note that 1 mole HCl ≏ 1 mole NaOH. The number of moles of HCl is:

$$(1.00 \times 10^2 \text{ mL}) \times \frac{1.50 \text{ mol HCl}}{1000 \text{ mL}} = 0.150 \text{ mol HCl}$$

The amount of heat released when 1 mole of H^+ is reacted is given in the problem (–56.2 kJ/mol). The amount of heat liberated when 0.150 mole of H^+ is reacted is:

$$q_{\text{rxn}} = 0.150 \text{ mol} \times \frac{-56.2 \times 10^3 \text{ J}}{1 \text{ mol}} = -8.43 \times 10^3 \text{ J}$$

Finally, knowing that the heat lost by the reaction equals the heat gained by the solution, we can solve for the final temperature of the mixed solution.

$$q_{\text{soln}} = -q_{\text{rxn}}$$

$$(2.00 \times 10^2 \text{ g})(4.184 \text{ J/g}\cdot°\text{C})(t_f - 23.2°\text{C}) = -(-8.43 \times 10^3 \text{ J})$$

$$(8.37 \times 10^2)t_f - (1.94 \times 10^4) = 8.43 \times 10^3 \text{ J}$$

$$\mathbf{t_f = 33.2°C}$$

PRACTICE EXERCISE

7. A 10.4 g sample of an unknown metal at 99.0°C was placed in a constant-pressure calorimeter containing 75.0 g of water at 23.5°C. The final temperature of the system was found to be 25.7°C. Calculate the specific heat of the metal, then use Table 6.2 of the text to predict the identity of the metal.

Text Problem: 6.38

PROBLEM TYPE 5: STANDARD ENTHALPY OF FORMATION AND REACTION

The **standard enthalpy of formation (ΔH_f°)** is defined as the heat change that results when one mole of a compound is formed from its elements at a pressure of 1 atm. The standard enthalpy of formation of any element in its most stable form is zero.

A. Calculating the standard enthalpy of reaction

From standard enthalpies of formation, we can calculate the **standard enthalpy of reaction, ΔH_{rxn}°**.

Consider the hypothetical reaction

$$aA + bB \longrightarrow cC + dD$$

where a, b, c, and d are stoichiometric coefficients.

The standard enthalpy of reaction is given by

$$\Delta H_{rxn}^\circ = [c\Delta H_f^\circ(C) + d\Delta H_f^\circ(D)] - [a\Delta H_f^\circ(A) + b\Delta H_f^\circ(B)]$$

Note that in calculations, the stoichiometric coefficients are just numbers without units.

The equation can be written in the general form:

$$\Delta H_{rxn}^\circ = \Sigma n\Delta H_f^\circ(\text{products}) - \Sigma m\Delta H_f^\circ(\text{reactants}) \qquad (6.18, \text{text})$$

where m and n denote the stoichiometric coefficients for the reactants and products, and Σ (sigma) means "the sum of".

EXAMPLE 6.8
A reaction used for rocket engines is

$$N_2H_4(l) + 2H_2O_2(l) \longrightarrow N_2(g) + 4H_2O(l)$$

What is the standard enthalpy of reaction in kilojoules? The standard enthalpies of formation are ΔH_f° [$N_2H_4(l)$] = 95.1 kJ/mol, ΔH_f° [$H_2O_2(l)$] = −187.8 kJ/mol, and ΔH_f° [$H_2O(l)$] = −285.8 kJ/mol.

Strategy: The enthalpy of a reaction is the difference between the sum of the enthalpies of the products and the sum of the enthalpies of the reactants. The enthalpy of each species (reactant or product) is given by the product of the stoichiometric coefficient and the standard enthalpy of formation, ΔH_f°, of the species.

Solution: We use the ΔH_f° values in Appendix 3 and Equation (6.18) of the text.

$$\Delta H_{rxn}^\circ = \sum n\Delta H_f^\circ(\text{products}) - \sum m\Delta H_f^\circ(\text{reactants})$$

$$\Delta H_{rxn}^\circ = \Delta H_f^\circ[N_2(g)] + 4\Delta H_f^\circ[H_2O(l)] - \{\Delta H_f^\circ[N_2H_4(l) + 2\Delta H_f^\circ[H_2O_2(l)]\}$$

Remember, the standard enthalpy of formation of any element in its most stable form is zero. Therefore, $\Delta H^\circ_f\,[N_2(g)] = 0$.

$$\boldsymbol{\Delta H^\circ_{rxn}} = [0 + 4(-285.8\text{ kJ/mol})] - [95.1\text{ kJ/mol} + 2(-187.8\text{ kJ/mol})] = \mathbf{-862.7\ kJ/mol}$$

PRACTICE EXERCISE

8. The combustion of methane, the main component of natural gas, occurs according to the equation

$$CH_4(g) + 2O_2(g) \longrightarrow CO_2(g) + 2H_2O(l) \qquad \Delta H^\circ_{rxn} = -890\text{ kJ/mol}$$

Use standard enthalpies of formation for CO_2 and H_2O to determine the standard enthalpy of formation of methane.

Text Problems: 6.52, 6.54, 6.56, 6.60

B. Direct method of calculating the standard enthalpy of formation

This method of measuring ΔH°_f applies to compounds that can be readily synthesized from their elements. The best way to describe this direct method is to look at an example.

EXAMPLE 6.9

The combustion of sulfur occurs according to the following thermochemical equation:

$$\mathbf{S(rhombic) + O_2(g) \longrightarrow SO_2(g) \qquad \Delta H^\circ_{rxn} = -296\ kJ/mol}$$

What is the enthalpy of formation of SO_2?

Strategy: What is the ΔH°_f value for an element in its standard state?

Solution: Knowing that the standard enthalpy of formation of any element in its most stable form is zero, and using Equation (6.18) of the text, we write:

$$\Delta H^\circ_{rxn} = \Sigma n\Delta H^\circ_f(\text{products}) - \Sigma m\Delta H^\circ_f(\text{reactants})$$

$$\Delta H^\circ_{rxn} = [\Delta H^\circ_f(SO_2)] - [\Delta H^\circ_f(S) + \Delta H^\circ_f(O_2)]$$

$$-296\text{ kJ/mol} = [\Delta H^\circ_f(SO_2) - [0 + 0]$$

$$\boldsymbol{\Delta H^\circ_f(SO_2) = -296\ \textbf{kJ/mol}\ SO_2}$$

Note: You should recognize that this chemical equation as written meets the definition of a *formation* reaction. Thus, ΔH°_{rxn} *is* ΔH°_f of $SO_2(g)$.

PRACTICE EXERCISE

9. Hydrogen iodide (HI) can be produced according to the following equation:

$$H_2(g) + I_2(s) \longrightarrow 2HI(g) \qquad \Delta H^\circ_{rxn} = 51.8\text{ kJ/mol}$$

What is the enthalpy of formation (ΔH°_f) of HI?

Text Problem: 6.50

C. Indirect method of calculating the standard enthalpy of formation, Hess's law

Many compounds cannot be directly synthesized from their elements. In these cases, ΔH_f° can be determined by an indirect approach using **Hess's law**. Hess's law states that when reactants are converted to products, the change in enthalpy is the same whether the reaction takes place in one step or in a series of steps. This means that if we can break down the reaction of interest into a series of reactions for which ΔH_{rxn}° can be measured, we can calculate ΔH_{rxn}° for the overall reaction. Let's look at an example.

EXAMPLE 6.10

From the following heats of combustion with fluorine, calculate the enthalpy of formation of methane, CH_4.

$$\text{(a)}\quad CH_4(g) + 4F_2(g) \longrightarrow CF_4(g) + 4HF(g) \qquad \Delta H_{rxn}^\circ = -1942 \text{ kJ/mol}$$

$$\text{(b)}\quad C(\text{graphite}) + 2F_2(g) \longrightarrow CF_4(g) \qquad \Delta H_{rxn}^\circ = -933 \text{ kJ/mol}$$

$$\text{(c)}\quad H_2(g) + F_2(g) \longrightarrow 2HF(g) \qquad \Delta H_{rxn}^\circ = -542 \text{ kJ/mol}$$

Strategy: Our goal is to calculate the enthalpy change for the formation of CH_4 from its elements C and H_2. This reaction does not occur directly, however, so we must use an indirect route using the information given in the three equations, which we will call equations (a), (b), and (c).

Solution: The enthalpy of formation of methane can be determined from the following equation.

$$C(\text{graphite}) + 2H_2(g) \longrightarrow CH_4(g) \qquad \Delta H_{rxn}^\circ = ?$$

First, we need one mole of C(graphite) as a reactant. Equation (b) has C(graphite) on the reactant side so let's keep that equation as written. Next, we need two moles of H_2 as a reactant. Equation (c) has 1 mole of H_2 as a reactant, so let's multiply this equation by 2.

$$\text{(d)}\quad 2H_2(g) + 2F_2(g) \longrightarrow 4HF(g) \qquad \Delta H_{rxn}^\circ = 2(-542 \text{ kJ/mol}) = -1084 \text{ kJ/mol}$$

Last, we need one mole of CH_4 as a product. Equation (a) has one mole of CH_4 as a reactant, so we need to reverse the equation.

$$\text{(e)}\quad CF_4(g) + 4HF(g) \longrightarrow CH_4(g) + 4F_2(g) \qquad \Delta H_{rxn}^\circ = +1942 \text{ kJ/mol}$$

Note: ΔH_{rxn}° changed sign when reversing the direction of the reaction.

Adding Equations (b), (d), and (e) together, we have:

$$\text{(b)}\quad C(\text{graphite}) + \cancel{2F_2}(g) \longrightarrow \cancel{CF_4}(g) \qquad \Delta H_{rxn}^\circ = -933 \text{ kJ/mol}$$

$$\text{(d)}\quad 2H_2(g) + \cancel{2F_2}(g) \longrightarrow \cancel{4HF}(g) \qquad \Delta H_{rxn}^\circ = -1084 \text{ kJ/mol}$$

$$\text{(e)}\quad \cancel{CF_4}(g) + \cancel{4HF}(g) \longrightarrow CH_4(g) + \cancel{4F_2}(g) \qquad \Delta H_{rxn}^\circ = +1942 \text{ kJ/mol}$$

$$C(\text{graphite}) + 2H_2(g) \longrightarrow CH_4(g) \qquad \Delta H_{rxn}^\circ = -75 \text{ kJ/mol}$$

Since the above equation represents the synthesis of CH_4 from its elements, the ΔH_{rxn}° calculated is the ΔH_f° of methane.

$$\Delta H_{rxn}^\circ = \mathbf{\Delta H_f^\circ(CH_4) = -75 \text{ kJ/mol}}$$

PRACTICE EXERCISE

10. From the following enthalpies of reaction, calculate the enthalpy of combustion of methane (CH_4) with F_2:

$$CH_4(g) + 4F_2(g) \longrightarrow CF_4(g) + 4HF(g) \qquad \Delta H^\circ_{rxn} = ?$$

$$C(\text{graphite}) + 2H_2(g) \longrightarrow CH_4(g) \qquad \Delta H^\circ_{rxn} = -75 \text{ kJ/mol}$$

$$C(\text{graphite}) + 2F_2(g) \longrightarrow CF_4(g) \qquad \Delta H^\circ_{rxn} = -933 \text{ kJ/mol}$$

$$H_2(g) + F_2(g) \longrightarrow 2HF(g) \qquad \Delta H^\circ_{rxn} = -542 \text{ kJ/mol}$$

Text Problems: 6.62, 6.64

ANSWERS TO PRACTICE EXERCISES

1. $\Delta E = 202$ kJ

2. $w = 1.0 \times 10^3$ J

3. $\Delta E = 37.57$ kJ

4. heat evolved $= 1.2 \times 10^2$ kJ

5. 475°C

6. $C_{cal} = 5.18$ kJ/°C

7. $s_{metal} = 0.906$ J/g°C. The metal is probably aluminum.

8. $\Delta H^\circ_f(CH_4) = -75.1$ kJ/mol

9. $\Delta H^\circ_f(HI) = 25.9$ kJ/mol

10. $\Delta H^\circ_{rxn} = -1942$ kJ/mol

SOLUTIONS TO SELECTED TEXT PROBLEMS

6.16 **(a)** Because the external pressure is zero, no work is done in the expansion.

$$w = -P\Delta V = -(0)(89.3 - 26.7)\text{mL}$$

$$\boldsymbol{w} = \mathbf{0}$$

(b) The external, opposing pressure is 1.5 atm, so

$$w = -P\Delta V = -(1.5\text{ atm})(89.3 - 26.7)\text{mL}$$

$$w = -94\ \cancel{\text{mL}}\cdot\text{atm} \times \frac{0.001\text{ L}}{1\ \cancel{\text{mL}}} = -0.094\text{ L}\cdot\text{atm}$$

To convert the answer to joules, we write:

$$\boldsymbol{w} = -0.094\ \cancel{\text{L}\cdot\text{atm}} \times \frac{101.3\text{ J}}{1\ \cancel{\text{L}\cdot\text{atm}}} = \mathbf{-9.5\ J}$$

(c) The external, opposing pressure is 2.8 atm, so

$$w = -P\Delta V = -(2.8\text{ atm})(89.3 - 26.7)\text{mL}$$

$$w = (-1.8 \times 10^2\ \cancel{\text{mL}}\cdot\text{atm}) \times \frac{0.001\text{ L}}{1\ \cancel{\text{mL}}} = -0.18\text{ L}\cdot\text{atm}$$

To convert the answer to joules, we write:

$$\boldsymbol{w} = -0.18\ \cancel{\text{L}\cdot\text{atm}} \times \frac{101.3\text{ J}}{1\ \cancel{\text{L}\cdot\text{atm}}} = \mathbf{-18\ J}$$

6.18 Applying the First Law of Thermodynamics, Problem Type 1A.

Strategy: Compression is work done on the gas, so what is the sign for w? Heat is released by the gas to the surroundings. Is this an endothermic or exothermic process? What is the sign for q?

Solution: To calculate the energy change of the gas (ΔE), we need Equation (6.1) of the text. Work of compression is positive and because heat is given off by the gas, q is negative. Therefore, we have:

$$\boldsymbol{\Delta E} = q + w = -26\text{ kJ} + 74\text{ kJ} = \mathbf{48\ kJ}$$

As a result, the energy of the gas increases by 48 kJ.

6.20 Calculating the work done in gas expansion, Problem Type 1B.

Strategy: The work done in gas expansion is equal to the product of the external, opposing pressure and the change in volume.

$$w = -P\Delta V$$

We assume that the volume of liquid water is zero compared to that of steam. How do we calculate the volume of the steam? What is the conversion factor between L·atm and J?

Solution: First, we need to calculate the volume that the water vapor will occupy (V_f).

Using the ideal gas equation:

$$V_{H_2O} = \frac{n_{H_2O}RT}{P} = \frac{(1\ \text{mol})\left(0.0821\dfrac{\text{L}\cdot\text{atm}}{\text{mol}\cdot\text{K}}\right)(373\ \text{K})}{(1.0\ \text{atm})} = 31\ \text{L}$$

It is given that the volume occupied by liquid water is negligible. Therefore,

$$\Delta V = V_f - V_i = 31\ \text{L} - 0\ \text{L} = 31\ \text{L}$$

Now, we substitute P and ΔV into Equation (6.3) of the text to solve for w.

$$w = -P\Delta V = -(1.0\ \text{atm})(31\ \text{L}) = -31\ \text{L}\cdot\text{atm}$$

The problems asks for the work done in units of joules. The following conversion factor can be obtained from Appendix 2 of the text.

$$1\ \text{L}\cdot\text{atm} = 101.3\ \text{J}$$

Thus, we can write:

$$w = -31\ \text{L}\cdot\text{atm} \times \frac{101.3\ \text{J}}{1\ \text{L}\cdot\text{atm}} = \mathbf{-3.1 \times 10^3\ J}$$

Check: Because this is gas expansion (work is done by the system on the surroundings), the work done has a negative sign.

6.26 Thermochemical Equations, Problem Type 2.

Strategy: The thermochemical equation shows that for every 2 moles of NO_2 produced, 114.6 kJ of heat are given off (note the negative sign). We can write a conversion factor from this information.

$$\frac{-114.6\ \text{kJ}}{2\ \text{mol NO}_2}$$

How many moles of NO_2 are in 1.26×10^4 g of NO_2? What conversion factor is needed to convert between grams and moles?

Solution: We need to first calculate the number of moles of NO_2 in 1.26×10^4 g of the compound. Then, we can convert to the number of kilojoules produced from the exothermic reaction. The sequence of conversions is:

grams of NO_2 $\rightarrow$ moles of NO_2 $\rightarrow$ kilojoules of heat generated

Therefore, the heat given off is:

$$(1.26 \times 10^4\ \text{g NO}_2) \times \frac{1\ \text{mol NO}_2}{46.01\ \text{g NO}_2} \times \frac{-114.6\ \text{kJ}}{2\ \text{mol NO}_2} = \mathbf{-1.57 \times 10^4\ kJ}$$

6.28 We initially have 6 moles of gas (3 moles of chlorine and 3 moles of hydrogen). Since our product is 6 moles of hydrogen chloride, there is no change in the number of moles of gas. Therefore there is no volume change; $\Delta V = 0$.

$$w = -P\Delta V = -(1\ \text{atm})(0\ \text{L}) = 0$$

$$\Delta E^\circ = \Delta H^\circ - P\Delta V$$

$-P\Delta V = 0$, so

$$\mathbf{\Delta E = \Delta H}$$

$$\Delta H = 3\Delta H^\circ_{rxn} = 3(-184.6 \text{ kJ/mol}) = -553.8 \text{ kJ/mol}$$

We need to multiply ΔH°_{rxn} by three, because the question involves the formation of 6 moles of HCl; whereas, the equation as written only produces 2 moles of HCl.

$$\Delta E^\circ = \Delta H^\circ = \mathbf{-553.8 \text{ kJ/mol}}$$

6.34 $q = m_{Cu}s_{Cu}\Delta t = (6.22 \times 10^3 \text{ g})(0.385 \text{ J/g}\cdot{}^\circ\text{C})(324.3^\circ\text{C} - 20.5^\circ\text{C}) = 7.28 \times 10^5 \text{ J} = \mathbf{728 \text{ kJ}}$

6.36 Calculating Heat Absorbed Using Specific Heat Data, Problem Type 3.

Strategy: We know the masses of gold and iron as well as the initial temperatures of each. We can look up the specific heats of gold and iron in Table 6.2 of the text. Assuming no heat is lost to the surroundings, we can equate the heat lost by the iron sheet to the heat gained by the gold sheet. With this information, we can solve for the final temperature of the combined metals.

Solution: Treating the calorimeter as an isolated system (no heat lost to the surroundings), we can write:

$$q_{Au} + q_{Fe} = 0$$

or

$$q_{Au} = -q_{Fe}$$

The heat gained by the gold sheet is given by:

$$q_{Au} = m_{Au}s_{Au}\Delta t = (10.0 \text{ g})(0.129 \text{ J/g}\cdot{}^\circ\text{C})(t_f - 18.0)^\circ\text{C}$$

where m and s are the mass and specific heat, and $\Delta t = t_{final} - t_{initial}$.

The heat lost by the iron sheet is given by:

$$q_{Fe} = m_{Fe}s_{Fe}\Delta t = (20.0 \text{ g})(0.444 \text{ J/g}\cdot{}^\circ\text{C})(t_f - 55.6)^\circ\text{C}$$

Substituting into the equation derived above, we can solve for t_f.

$$q_{Au} = -q_{Fe}$$

$$(10.0 \text{ g})(0.129 \text{ J/g}\cdot{}^\circ\text{C})(t_f - 18.0)^\circ\text{C} = -(20.0 \text{ g})(0.444 \text{ J/g}\cdot{}^\circ\text{C})(t_f - 55.6)^\circ\text{C}$$

$$1.29\ t_f - 23.2 = -8.88\ t_f + 494$$

$$10.2\ t_f = 517$$

$$\mathbf{t_f = 50.7^\circ C}$$

Check: Must the final temperature be between the two starting values?

6.38 Constant-pressure calorimetry, Problem Type 4B.

Strategy: The neutralization reaction is exothermic. 56.2 kJ of heat are released when 1 mole of H^+ reacts with 1 mole of OH^-. Assuming no heat is lost to the surroundings, we can equate the heat lost by the reaction to the heat gained by the combined solution. How do we calculate the heat released during the reaction? Are we reacting 1 mole of H^+ with 1 mole of OH^-? How do we calculate the heat absorbed by the combined solution?

Solution: Assuming no heat is lost to the surroundings, we can write:

$$q_{soln} + q_{rxn} = 0$$

or

$$q_{soln} = -q_{rxn}$$

First, let's set up how we would calculate the heat gained by the solution,

$$q_{soln} = m_{soln}s_{soln}\Delta t$$

where m and s are the mass and specific heat of the solution and $\Delta t = t_f - t_i$.

We assume that the specific heat of the solution is the same as the specific heat of water, and we assume that the density of the solution is the same as the density of water (1.00 g/mL). Since the density is 1.00 g/mL, the mass of 400 mL of solution (200 mL + 200 mL) is 400 g.

Substituting into the equation above, the heat gained by the solution can be represented as:

$$q_{soln} = (4.00 \times 10^2 \text{ g})(4.184 \text{ J/g}\cdot{}^\circ\text{C})(t_f - 20.48^\circ\text{C})$$

Next, let's calculate q_{rxn}, the heat released when 200 mL of 0.862 M HCl are mixed with 200 mL of 0.431 M $Ba(OH)_2$. The equation for the neutralization is:

$$2HCl(aq) + Ba(OH)_2(aq) \longrightarrow 2H_2O(l) + BaCl_2(aq)$$

There is exactly enough $Ba(OH)_2$ to neutralize all the HCl. Note that 2 mole HCl $\simeq$ 1 mole $Ba(OH)_2$, and that the concentration of HCl is double the concentration of $Ba(OH)_2$. The number of moles of HCl is:

$$(2.00 \times 10^2 \text{ mL}) \times \frac{0.862 \text{ mol HCl}}{1000 \text{ mL}} = 0.172 \text{ mol HCl}$$

The amount of heat released when 1 mole of H^+ is reacted is given in the problem (−56.2 kJ/mol). The amount of heat liberated when 0.172 mole of H^+ is reacted is:

$$q_{rxn} = 0.172 \text{ mol} \times \frac{-56.2 \times 10^3 \text{ J}}{1 \text{ mol}} = -9.67 \times 10^3 \text{ J}$$

Finally, knowing that the heat lost by the reaction equals the heat gained by the solution, we can solve for the final temperature of the mixed solution.

$$q_{soln} = -q_{rxn}$$

$$(4.00 \times 10^2 \text{ g})(4.184 \text{ J/g}\cdot{}^\circ\text{C})(t_f - 20.48^\circ\text{C}) = -(-9.67 \times 10^3 \text{ J})$$

$$(1.67 \times 10^3)t_f - (3.43 \times 10^4) = 9.67 \times 10^3 \text{ J}$$

$$\mathbf{t_f = 26.3^\circ C}$$

6.46 The standard enthalpy of formation of any element in its most stable form is zero. Therefore, since $\Delta H_f^\circ(O_2) = 0$, $\mathbf{O_2}$ is the more stable form of the element oxygen at this temperature.

6.48 **(a)** $Br_2(l)$ is the most stable form of bromine at 25°C; therefore, $\Delta H_f^\circ[Br_2(l)] = 0$. Since $Br_2(g)$ is less stable than $Br_2(l)$, $\Delta H_f^\circ[Br_2(g)] > 0$.

(b) $I_2(s)$ is the most stable form of iodine at 25°C; therefore, $\Delta H_f^\circ[I_2(s)] = 0$. Since $I_2(g)$ is less stable than $I_2(s)$, $\Delta H_f^\circ[I_2(g)] > 0$.

6.50 Direct method of calculating the standard enthalpy of formation, Problem Type 5B.

Strategy: What is the reaction for the formation of Ag_2O from its elements? What is the ΔH_f° value for an element in its standard state?

Solution: The balanced equation showing the formation of $Ag_2O(s)$ from its elements is:

$$2Ag(s) + \frac{1}{2}O_2(g) \longrightarrow Ag_2O(s)$$

Knowing that the standard enthalpy of formation of any element in its most stable form is zero, and using Equation (6.18) of the text, we write:

$$\Delta H_{rxn}^\circ = \Sigma n\Delta H_f^\circ(\text{products}) - \Sigma m\Delta H_f^\circ(\text{reactants})$$

$$\Delta H_{rxn}^\circ = [\Delta H_f^\circ(Ag_2O)] - [2\Delta H_f^\circ(Ag) + \frac{1}{2}\Delta H_f^\circ(O_2)]$$

$$\Delta H_{rxn}^\circ = [\Delta H_f^\circ(Ag_2O)] - [0 + 0]$$

$$\boldsymbol{\Delta H_f^\circ(Ag_2O) = \Delta H_{rxn}^\circ}$$

In a similar manner, you should be able to show that $\boldsymbol{\Delta H_f^\circ(CaCl_2) = \Delta H_{rxn}^\circ}$ for the reaction

$$Ca(s) + Cl_2(g) \longrightarrow CaCl_2(s)$$

6.52 Calculating the standard enthalpy of reaction, Problem Type 5A.

Strategy: The enthalpy of a reaction is the difference between the sum of the enthalpies of the products and the sum of the enthalpies of the reactants. The enthalpy of each species (reactant or product) is given by the product of the stoichiometric coefficient and the standard enthalpy of formation, ΔH_f°, of the species.

Solution: We use the ΔH_f° values in Appendix 3 and Equation (6.18) of the text.

$$\Delta H_{rxn}^\circ = \Sigma n\Delta H_f^\circ(\text{products}) - \Sigma m\Delta H_f^\circ(\text{reactants})$$

(a) $HCl(g) \rightarrow H^+(aq) + Cl^-(aq)$

$$\Delta H_{rxn}^\circ = \Delta H_f^\circ(H^+) + \Delta H_f^\circ(Cl^-) - \Delta H_f^\circ(HCl)$$

$$-74.9 \text{ kJ/mol} = 0 + \Delta H_f^\circ(Cl^-) - (1)(-92.3 \text{ kJ/mol})$$

$$\boldsymbol{\Delta H_f^\circ(Cl^-) = -167.2 \text{ kJ/mol}}$$

(b) The neutralization reaction is:

$$H^+(aq) + OH^-(aq) \rightarrow H_2O(l)$$

and,

$$\Delta H^\circ_{rxn} = \Delta H^\circ_f[H_2O(l)] - [\Delta H^\circ_f(H^+) + \Delta H^\circ_f(OH^-)]$$

$\Delta H^\circ_f[H_2O(l)] = -285.8$ kJ/mol (See Appendix 3 of the text.)

$$\mathbf{\Delta H^\circ_{rxn}} = (1)(-285.8 \text{ kJ/mol}) - [(1)(0 \text{ kJ/mol}) + (1)(-229.6 \text{ kJ/mol})] = \mathbf{-56.2 \text{ kJ/mol}}$$

6.54 **(a)** $\Delta H^\circ = [2\Delta H^\circ_f(CO_2) + 2\Delta H^\circ_f(H_2O)] - [\Delta H^\circ_f(C_2H_4) + 3\Delta H^\circ_f(O_2)]]$

$\Delta H^\circ = [(2)(-393.5 \text{ kJ/mol}) + (2)(-285.8 \text{ kJ/mol})] - [(1)(52.3 \text{ kJ/mol}) + (3)(0)]$

$\mathbf{\Delta H^\circ = -1411 \text{ kJ/mol}}$

(b) $\Delta H^\circ = [2\Delta H^\circ_f(H_2O) + 2\Delta H^\circ_f(SO_2)] - [2\Delta H^\circ_f(H_2S) + 3\Delta H^\circ_f(O_2)]$

$\Delta H^\circ = [(2)(-285.8 \text{ kJ/mol}) + (2)(-296.1 \text{ kJ/mol})] - [(2)(-20.15 \text{ kJ/mol}) + (3)(0)]$

$\mathbf{\Delta H^\circ = -1124 \text{ kJ/mol}}$

6.56 $\Delta H^\circ_{rxn} = \sum n\Delta H^\circ_f(\text{products}) - \sum m\Delta H^\circ_f(\text{reactants})$

The reaction is:

$$H_2(g) \longrightarrow H(g) + H(g)$$

and,

$$\Delta H^\circ_{rxn} = [\Delta H^\circ_f(H) + \Delta H^\circ_f(H)] - \Delta H^\circ_f(H_2)$$

$\Delta H^\circ_f(H_2) = 0$

$$\Delta H^\circ_{rxn} = 436.4 \text{ kJ/mol} = 2\Delta H^\circ_f(H) - (1)(0)$$

$$\mathbf{\Delta H^\circ_f(H)} = \frac{436.4 \text{ kJ/mol}}{2} = \mathbf{218.2 \text{ kJ/mol}}$$

6.58 Using the ΔH°_f values in Appendix 3 and Equation (6.18) of the text, we write

$\Delta H^\circ_{rxn} = [5\Delta H^\circ_f(B_2O_3) + 9\Delta H^\circ_f(H_2O)] - [2\Delta H^\circ_f(B_5H_9) + 12\Delta H^\circ_f(O_2)]$

$\Delta H^\circ = [(5)(-1263.6 \text{ kJ/mol}) + (9)(-285.8 \text{ kJ/mol})] - [(2)(73.2 \text{ kJ/mol}) + (12)(0 \text{ kJ/mol})]$

$\mathbf{\Delta H^\circ} = -9036.6$ kJ/mol

Looking at the balanced equation, this is the amount of heat released for every 2 moles of B_5H_9 reacted. We can use the following ratio

$$\frac{-9036.6 \text{ kJ}}{2 \text{ mol } B_5H_9}$$

to convert to kJ/g B_5H_9. The molar mass of B_5H_9 is 63.12 g, so

$$\text{heat released per gram } B_5H_9 = \frac{-9036.6 \text{ kJ}}{2 \text{ mol } B_5H_9} \times \frac{1 \text{ mol } B_5H_9}{63.12 \text{ g } B_5H_9} = \mathbf{-71.58 \text{ kJ/g } B_5H_9}$$

6.60 $\Delta H^\circ_{rxn} = \sum n\Delta H^\circ_f(\text{products}) - \sum m\Delta H^\circ_f(\text{reactants})$

The balanced equation for the reaction is:

$$CaCO_3(s) \longrightarrow CaO(s) + CO_2(g)$$

$$\Delta H^\circ_{rxn} = [\Delta H^\circ_f(CaO) + \Delta H^\circ_f(CO_2)] - \Delta H^\circ_f(CaCO_3)$$

$$\Delta H^\circ_{rxn} = [(1)(-635.6\text{ kJ/mol}) + (1)(-393.5\text{ kJ/mol})] - (1)(-1206.9\text{ kJ/mol}) = 177.8\text{ kJ/mol}$$

The enthalpy change calculated above is the enthalpy change if 1 mole of CO_2 is produced. The problem asks for the enthalpy change if 66.8 g of CO_2 are produced. We need to use the molar mass of CO_2 as a conversion factor.

$$\boldsymbol{\Delta H^\circ} = 66.8\text{ g }CO_2 \times \frac{1\text{ mol }CO_2}{44.01\text{ g }CO_2} \times \frac{177.8\text{ kJ}}{1\text{ mol }CO_2} = \mathbf{2.70 \times 10^2\text{ kJ}}$$

6.62 Indirect method of calculating the standard enthalpy of formation, Hess's law. Problem Type 5C.

Strategy: Our goal is to calculate the enthalpy change for the formation of C_2H_6 from is elements C and H_2. This reaction does not occur directly, however, so we must use an indirect route using the information given in the three equations, which we will call equations (a), (b), and (c).

Solution: Here is the equation for the formation of C_2H_6 from its elements.

$$2C(\text{graphite}) + 3H_2(g) \longrightarrow C_2H_6(g) \qquad \Delta H^\circ_{rxn} = ?$$

Looking at this reaction, we need two moles of graphite as a reactant. So, we multiply Equation (a) by two to obtain:

$$\text{(d)}\quad 2C(\text{graphite}) + 2O_2(g) \longrightarrow 2CO_2(g) \qquad \Delta H^\circ_{rxn} = 2(-393.5\text{ kJ/mol}) = -787.0\text{ kJ/mol}$$

Next, we need three moles of H_2 as a reactant. So, we multiply Equation (b) by three to obtain:

$$\text{(e)}\quad 3H_2(g) + \tfrac{3}{2}O_2(g) \longrightarrow 3H_2O(l) \qquad \Delta H^\circ_{rxn} = 3(-285.8\text{ kJ/mol}) = -857.4\text{ kJ/mol}$$

Last, we need one mole of C_2H_6 as a product. Equation (c) has two moles of C_2H_6 as a reactant, so we need to reverse the equation and divide it by 2.

$$\text{(f)}\quad 2CO_2(g) + 3H_2O(l) \longrightarrow C_2H_6(g) + \tfrac{7}{2}O_2(g) \qquad \Delta H^\circ_{rxn} = \tfrac{1}{2}(3119.6\text{ kJ/mol}) = 1559.8\text{ kJ/mol}$$

Adding Equations (d), (e), and (f) together, we have:

Reaction	ΔH° (kJ/mol)
(d) $2C(\text{graphite}) + 2\cancel{O_2}(g) \longrightarrow 2\cancel{CO_2}(g)$	−787.0
(e) $3H_2(g) + \tfrac{3}{2}\cancel{O_2}(g) \longrightarrow 3\cancel{H_2O}(l)$	−857.4
(f) $2\cancel{CO_2}(g) + 3\cancel{H_2O}(l) \longrightarrow C_2H_6(g) + \tfrac{7}{2}\cancel{O_2}(g)$	1559.8
$2C(\text{graphite}) + 3H_2(g) \longrightarrow C_2H_6(g)$	**$\Delta H^\circ = -84.6$ kJ/mol**

6.64 The second and third equations can be combined to give the first equation.

$$2Al(s) + \frac{3}{2}\cancel{O_2}(g) \longrightarrow Al_2O_3(s) \qquad \Delta H° = -1669.8 \text{ kJ/mol}$$

$$Fe_2O_3(s) \longrightarrow 2Fe(s) + \frac{3}{2}\cancel{O_2}(g) \qquad \Delta H° = 822.2 \text{ kJ/mol}$$

$$2Al(s) + Fe_2O_3(s) \longrightarrow 2Fe(s) + Al_2O_3(s) \qquad \mathbf{\Delta H° = -847.6 \text{ kJ/mol}}$$

6.72 Rearrange the equations as necessary so they can be added to yield the desired equation.

$$2B \longrightarrow \cancel{A} \qquad -\Delta H_1$$

$$\cancel{A} \longrightarrow C \qquad \Delta H_2$$

$$2B \longrightarrow C \qquad \mathbf{\Delta H° = \Delta H_2 - \Delta H_1}$$

6.74 **(a)** $\Delta H°_{rxn} = \sum n\Delta H°_f(\text{products}) - \sum m\Delta H°_f(\text{reactants})$

$$\Delta H°_{rxn} = [4\Delta H°_f(NH_3) + \Delta H°_f(N_2)] - 3\Delta H°_f(N_2H_4)$$

$$\mathbf{\Delta H°_{rxn}} = [(4)(-46.3 \text{ kJ/mol}) + (0)] - (3)(50.42 \text{ kJ/mol}) = \mathbf{-336.5 \text{ kJ/mol}}$$

(b) The balanced equations are:

(1) $N_2H_4(l) + O_2(g) \longrightarrow N_2(g) + 2H_2O(l)$

(2) $4NH_3(g) + 3O_2(g) \longrightarrow 2N_2(g) + 6H_2O(l)$

The standard enthalpy change for equation (1) is:

$$\Delta H°_{rxn} = \Delta H°_f(N_2) + 2\Delta H°_f[H_2O(l)] - \{\Delta H°_f[N_2H_4(l)] + \Delta H°_f(O_2)\}$$

$$\Delta H°_{rxn} = [(1)(0) + (2)(-285.8 \text{ kJ/mol})] - [(1)(50.42 \text{ kJ/mol}) + (1)(0)] = -622.0 \text{ kJ/mol}$$

The standard enthalpy change for equation (2) is:

$$\Delta H°_{rxn} = [2\Delta H°_f(N_2) + 6\Delta H°_f(H_2O)] - [4\Delta H°_f(NH_3) + 3\Delta H°_f(O_2)]$$

$$\Delta H°_{rxn} = [(2)(0) + (6)(-285.8 \text{ kJ/mol})] - [(4)(-46.3 \text{ kJ/mol}) + (3)(0)] = -1529.6 \text{ kJ/mol}$$

We can now calculate the enthalpy change per kilogram of each substance. $\Delta H°_{rxn}$ above is in units of kJ/mol. We need to convert to kJ/kg.

$$N_2H_4(l){:}\ \mathbf{\Delta H°_{rxn}} = \frac{-622.0 \text{ kJ}}{1 \cancel{\text{mol}}\ N_2H_4} \times \frac{1 \cancel{\text{mol}}\ N_2H_4}{32.05 \cancel{\text{g}}\ N_2H_4} \times \frac{1000 \cancel{\text{g}}}{1 \text{ kg}} = \mathbf{-1.941 \times 10^4 \text{ kJ/kg } N_2H_4}$$

$$NH_3(g){:}\ \mathbf{\Delta H°_{rxn}} = \frac{-1529.6 \text{ kJ}}{4 \cancel{\text{mol}}\ NH_3} \times \frac{1 \cancel{\text{mol}}\ NH_3}{17.03 \cancel{\text{g}}\ NH_3} \times \frac{1000 \cancel{\text{g}}}{1 \text{ kg}} = \mathbf{-2.245 \times 10^4 \text{ kJ/kg } NH_3}$$

Since **ammonia, NH_3**, releases more energy per kilogram of substance, it would be a better fuel.

6.76 The reaction is, $2Na(s) + Cl_2(g) \rightarrow 2NaCl(s)$. First, let's calculate $\Delta H°$ for this reaction using ΔH_f° values in Appendix 3.

$$\Delta H_{rxn}^\circ = 2\Delta H_f^\circ(\text{NaCl}) - [2\Delta H_f^\circ(\text{Na}) + \Delta H_f^\circ(\text{Cl}_2)]$$

$$\Delta H_{rxn}^\circ = 2(-411.0 \text{ kJ/mol}) - [2(0) + 0] = -822.0 \text{ kJ/mol}$$

This is the amount of heat released when 1 mole of Cl_2 reacts (see balanced equation). We are not reacting 1 mole of Cl_2, however. From the volume and density of Cl_2, we can calculate grams of Cl_2. Then, using the molar mass of Cl_2 as a conversion factor, we can calculate moles of Cl_2. Combining these two calculations into one step, we find moles of Cl_2 to be:

$$2.00 \text{ L Cl}_2 \times \frac{1.88 \text{ g Cl}_2}{1 \text{ L Cl}_2} \times \frac{1 \text{ mol Cl}_2}{70.90 \text{ g Cl}_2} = 0.0530 \text{ mol Cl}_2$$

Finally, we can use the ΔH_{rxn}° calculated above to find the amount of heat released when 0.0530 mole of Cl_2 reacts.

$$0.0530 \text{ mol Cl}_2 \times \frac{-822.0 \text{ kJ}}{1 \text{ mol Cl}_2} = \mathbf{-43.6 \text{ kJ}}$$

6.78 The initial and final states of this system are identical. Since enthalpy is a state function, its value depends only upon the state of the system. The enthalpy change is **zero**.

6.80 $H(g) + Br(g) \longrightarrow HBr(g) \qquad \Delta H_{rxn}^\circ = ?$

Rearrange the equations as necessary so they can be added to yield the desired equation.

$$H(g) \longrightarrow \tfrac{1}{2}H_2(g) \qquad \Delta H_{rxn}^\circ = \tfrac{1}{2}(-436.4 \text{ kJ/mol}) = -218.2 \text{ kJ/mol}$$

$$Br(g) \longrightarrow \tfrac{1}{2}Br_2(g) \qquad \Delta H_{rxn}^\circ = \tfrac{1}{2}(-192.5 \text{ kJ/mol}) = -96.25 \text{ kJ/mol}$$

$$\tfrac{1}{2}H_2(g) + \tfrac{1}{2}Br_2(g) \longrightarrow HBr(g) \qquad \Delta H_{rxn}^\circ = \tfrac{1}{2}(-72.4 \text{ kJ/mol}) = -36.2 \text{ kJ/mol}$$

$$H(g) + Br(g) \longrightarrow HBr(g) \qquad \mathbf{\Delta H° = -350.7 \text{ kJ/mol}}$$

6.82 $q_{system} = 0 = q_{metal} + q_{water} + q_{calorimeter}$

$q_{metal} + q_{water} + q_{calorimeter} = 0$

$m_{metal}s_{metal}(t_{final} - t_{initial}) + m_{water}s_{water}(t_{final} - t_{initial}) + C_{calorimeter}(t_{final} - t_{initial}) = 0$

All the needed values are given in the problem. All you need to do is plug in the values and solve for s_{metal}.

$(44.0 \text{ g})(s_{metal})(28.4 - 99.0)°\text{C} + (80.0 \text{ g})(4.184 \text{ J/g·°C})(28.4 - 24.0)°\text{C} + (12.4 \text{ J/°C})(28.4 - 24.0)°\text{C} = 0$

$(-3.11 \times 10^3)s_{metal} (\text{g·°C}) = -1.53 \times 10^3 \text{ J}$

$\mathbf{s_{metal} = 0.492 \text{ J/g·°C}}$

6.84 A good starting point would be to calculate the standard enthalpy for both reactions.

Calculate the standard enthalpy for the reaction: $C(s) + \tfrac{1}{2}O_2(g) \longrightarrow CO(g)$

This reaction corresponds to the standard enthalpy of formation of CO, so we use the value of −110.5 kJ/mol (see Appendix 3 of the text).

Calculate the standard enthalpy for the reaction: $C(s) + H_2O(g) \longrightarrow CO(g) + H_2(g)$

$$\Delta H^\circ_{rxn} = [\Delta H^\circ_f(CO) + \Delta H^\circ_f(H_2)] - [\Delta H^\circ_f(C) + \Delta H^\circ_f(H_2O)]$$

$$\Delta H^\circ_{rxn} = [(1)(-110.5\text{ kJ/mol}) + (1)(0)] - [(1)(0) + (1)(-241.8\text{ kJ/mol})] = 131.3\text{ kJ/mol}$$

The first reaction, which is exothermic, can be used to promote the second reaction, which is endothermic. Thus, the two gases are produced alternately.

6.86 First, calculate the energy produced by 1 mole of octane, C_8H_{18}.

$$C_8H_{18}(l) + \tfrac{25}{2}O_2(g) \longrightarrow 8CO_2(g) + 9H_2O(l)$$

$$\Delta H^\circ_{rxn} = 8\Delta H^\circ_f(CO_2) + 9\Delta H^\circ_f[H_2O(l)] - [\Delta H^\circ_f(C_8H_{18}) + \tfrac{25}{2}\Delta H^\circ_f(O_2)]$$

$$\Delta H^\circ_{rxn} = [(8)(-393.5\text{ kJ/mol}) + (9)(-285.8\text{ kJ/mol})] - [(1)(-249.9\text{ kJ/mol}) + (\tfrac{25}{2})(0)]$$
$$= -5470\text{ kJ/mol}$$

The problem asks for the energy produced by the combustion of 1 gallon of octane. ΔH°_{rxn} above has units of kJ/mol octane. We need to convert from kJ/mol octane to kJ/gallon octane. The heat of combustion for 1 gallon of octane is:

$$\Delta H^\circ = \frac{-5470\text{ kJ}}{1\text{ mol octane}} \times \frac{1\text{ mol octane}}{114.2\text{ g octane}} \times \frac{2660\text{ g}}{1\text{ gal}} = -1.274 \times 10^5\text{ kJ/gal}$$

The combustion of hydrogen corresponds to the standard heat of formation of water:

$$H_2(g) + \tfrac{1}{2}O_2(g) \longrightarrow H_2O(l)$$

Thus, ΔH°_{rxn} is the same as ΔH°_f for $H_2O(l)$, which has a value of −285.8 kJ/mol. The number of moles of hydrogen required to produce 1.274×10^5 kJ of heat is:

$$n_{H_2} = (1.274 \times 10^5\text{ kJ}) \times \frac{1\text{ mol }H_2}{285.8\text{ kJ}} = 445.8\text{ mol }H_2$$

Finally, use the ideal gas law to calculate the volume of gas corresponding to 445.8 moles of H_2 at 25°C and 1 atm.

$$V_{H_2} = \frac{n_{H_2}RT}{P} = \frac{(445.8\text{ mol})\left(0.0821\frac{\text{L}\cdot\text{atm}}{\text{mol}\cdot\text{K}}\right)(298\text{ K})}{(1\text{ atm})} = \mathbf{1.09 \times 10^4\text{ L}}$$

That is, the volume of hydrogen that is energy-equivalent to 1 gallon of gasoline is over **10,000 liters** at 1 atm and 25°C!

6.88 The combustion reaction is: $C_2H_6(l) + \tfrac{7}{2}O_2(g) \longrightarrow 2CO_2(g) + 3H_2O(l)$

The heat released during the combustion of 1 mole of ethane is:

$$\Delta H^\circ_{rxn} = [2\Delta H^\circ_f(CO_2) + 3\Delta H^\circ_f(H_2O)] - [\Delta H^\circ_f(C_2H_6) + \tfrac{7}{2}\Delta H^\circ_f(O_2)]$$

$$\Delta H^\circ_{rxn} = [(2)(-393.5 \text{ kJ/mol}) + (3)(-285.8 \text{ kJ/mol})] - [(1)(-84.7 \text{ kJ/mol} + (\tfrac{7}{2})(0)]$$
$$= -1560 \text{ kJ/mol}$$

The heat required to raise the temperature of the water to 98°C is:

$$q = m_{H_2O}s_{H_2O}\Delta t = (855 \text{ g})(4.184 \text{ J/g}\cdot{}^\circ\text{C})(98.0 - 25.0)^\circ\text{C} = 2.61 \times 10^5 \text{ J} = 261 \text{ kJ}$$

The combustion of 1 mole of ethane produces 1560 kJ; the number of moles required to produce 261 kJ is:

$$261 \text{ kJ} \times \frac{1 \text{ mol ethane}}{1560 \text{ kJ}} = 0.167 \text{ mol ethane}$$

The volume of ethane is:

$$V_{\textbf{ethane}} = \frac{nRT}{P} = \frac{(0.167 \text{ mol})\left(0.0821 \frac{\text{L}\cdot\text{atm}}{\text{mol}\cdot\text{K}}\right)(296 \text{ K})}{\left(752 \text{ mmHg} \times \frac{1 \text{ atm}}{760 \text{ mmHg}}\right)} = \mathbf{4.10 \text{ L}}$$

6.90 The heat gained by the liquid nitrogen must be equal to the heat lost by the water.

$$q_{N_2} = -q_{H_2O}$$

If we can calculate the heat lost by the water, we can calculate the heat gained by 60.0 g of the nitrogen.

$$\text{Heat lost by the water} = q_{H_2O} = m_{H_2O}s_{H_2O}\Delta t$$

$$q_{H_2O} = (2.00 \times 10^2 \text{ g})(4.184 \text{ J/g}\cdot{}^\circ\text{C})(41.0 - 55.3)^\circ\text{C} = -1.20 \times 10^4 \text{ J}$$

The heat gained by 60.0 g nitrogen is the opposite sign of the heat lost by the water.

$$q_{N_2} = -q_{H_2O}$$
$$q_{N_2} = 1.20 \times 10^4 \text{ J}$$

The problem asks for the molar heat of vaporization of liquid nitrogen. Above, we calculated the amount of heat necessary to vaporize 60.0 g of liquid nitrogen. We need to convert from J/60.0 g N_2 to J/mol N_2.

$$\Delta H_{\textbf{vap}} = \frac{1.20 \times 10^4 \text{ J}}{60.0 \text{ g } N_2} \times \frac{28.02 \text{ g } N_2}{1 \text{ mol } N_2} = \mathbf{5.60 \times 10^3 \text{ J/mol} = 5.60 \text{ kJ/mol}}$$

6.92 Recall that the standard enthalpy of formation (ΔH°_f) is defined as the heat change that results when 1 mole of a compound is formed from its elements at a pressure of 1 atm. Only in choice **(a)** does $\Delta H^\circ_{rxn} = \Delta H^\circ_f$. In choice (b), C(diamond) is *not* the most stable form of elemental carbon under standard conditions; C(graphite) is the most stable form.

6.94 **(a)** No work is done by a gas expanding in a vacuum, because the pressure exerted on the gas is zero.

(b) $w = -P\Delta V$

$$w = -(0.20 \text{ atm})(0.50 - 0.050)\text{L} = -0.090 \text{ L}\cdot\text{atm}$$

Converting to units of joules:

$$w = -0.090\,\text{L}\cdot\text{atm} \times \frac{101.3\text{ J}}{\text{L}\cdot\text{atm}} = \mathbf{-9.1\text{ J}}$$

(c) The gas will expand until the pressure is the same as the applied pressure of 0.20 atm. We can calculate its final volume using the ideal gas equation.

$$V = \frac{nRT}{P} = \frac{(0.020\text{ mol})\left(0.0821\frac{\text{L}\cdot\text{atm}}{\text{mol}\cdot\text{K}}\right)(273 + 20)\text{K}}{0.20\text{ atm}} = \mathbf{2.4\text{ L}}$$

The amount of work done is:

$$w = -P\Delta V = (0.20\text{ atm})(2.4 - 0.050)\text{L} = -0.47\text{ L}\cdot\text{atm}$$

Converting to units of joules:

$$w = -0.47\,\text{L}\cdot\text{atm} \times \frac{101.3\text{ J}}{\text{L}\cdot\text{atm}} = \mathbf{-48\text{ J}}$$

6.96 **(a)** The more closely packed, the greater the mass of food. Heat capacity depends on both the mass and specific heat.

$$C = ms$$

The heat capacity of the food is greater than the heat capacity of air; hence, the cold in the freezer will be retained longer.

(b) Tea and coffee are mostly water; whereas, soup might contain vegetables and meat. Water has a higher heat capacity than the other ingredients in soup; therefore, coffee and tea retain heat longer than soup.

6.98 $4Fe(s) + 3O_2(g) \rightarrow 2Fe_2O_3(s)$. This equation represents twice the standard enthalpy of formation of Fe_2O_3. From Appendix 3, the standard enthalpy of formation of Fe_2O_3 = −822.2 kJ/mol. So, $\Delta H°$ for the given reaction is:

$$\Delta H^\circ_{\text{rxn}} = (2)(-822.2\text{ kJ/mol}) = -1644\text{ kJ/mol}$$

Looking at the balanced equation, this is the amount of heat released when four moles of Fe react. But, we are reacting 250 g of Fe, not 4 moles. We can convert from grams of Fe to moles of Fe, then use $\Delta H°$ as a conversion factor to convert to kJ.

$$250\text{ g Fe} \times \frac{1\text{ mol Fe}}{55.85\text{ g Fe}} \times \frac{-1644\text{ kJ}}{4\text{ mol Fe}} = \mathbf{-1.84 \times 10^3\text{ kJ}}$$

6.100 The heat required to raise the temperature of 1 liter of water by 1°C is:

$$4.184\frac{\text{J}}{\text{g}\cdot{}^\circ\text{C}} \times \frac{1\text{ g}}{1\text{ mL}} \times \frac{1000\text{ mL}}{1\text{ L}} \times 1^\circ\text{C} = 4184\text{ J/L}$$

Next, convert the volume of the Pacific Ocean to liters.

$$(7.2 \times 10^8\text{ km}^3) \times \left(\frac{1000\text{ m}}{1\text{ km}}\right)^3 \times \left(\frac{100\text{ cm}}{1\text{ m}}\right)^3 \times \frac{1\text{ L}}{1000\text{ cm}^3} = 7.2 \times 10^{20}\text{ L}$$

The amount of heat needed to raise the temperature of 7.2×10^{20} L of water is:

$$(7.2 \times 10^{20}\ \text{L}) \times \frac{4184\ \text{J}}{1\ \text{L}} = 3.0 \times 10^{24}\ \text{J}$$

Finally, we can calculate the number of atomic bombs needed to produce this much heat.

$$(3.0 \times 10^{24}\ \text{J}) \times \frac{1\ \text{atomic bomb}}{1.0 \times 10^{15}\ \text{J}} = \mathbf{3.0 \times 10^{9}\ atomic\ bombs} = \mathbf{3.0\ billion\ atomic\ bombs}$$

6.102 Constant-volume calorimetry, Problem Type 4A.

Strategy: The heat released during the reaction is absorbed by both the water and the calorimeter. How do we calculate the heat absorbed by the water? How do we calculate the heat absorbed by the calorimeter? How much heat is released when 1.9862 g of benzoic acid are reacted? The problem gives the amount of heat that is released when 1 mole of benzoic acid is reacted (−3226.7 kJ/mol).

Solution: The heat of the reaction (combustion) is absorbed by both the water and the calorimeter.

$$q_{rxn} = -(q_{water} + q_{cal})$$

If we can calculate both q_{water} and q_{rxn}, then we can calculate q_{cal}. First, let's calculate the heat absorbed by the water.

$$q_{water} = m_{water}s_{water}\Delta t$$

$$q_{water} = (2000\ \text{g})(4.184\ \text{J/g}\cdot{}^\circ\text{C})(25.67 - 21.84)^\circ\text{C} = 3.20 \times 10^{4}\ \text{J} = 32.0\ \text{kJ}$$

Next, let's calculate the heat released (q_{rxn}) when 1.9862 g of benzoic acid are burned. ΔH_{rxn} is given in units of kJ/mol. Let's convert to q_{rxn} in kJ.

$$q_{rxn} = 1.9862\ \text{g benzoic acid} \times \frac{1\ \text{mol benzoic acid}}{122.1\ \text{g benzoic acid}} \times \frac{-3226.7\ \text{kJ}}{1\ \text{mol benzoic acid}} = -52.49\ \text{kJ}$$

And,

$$q_{cal} = -q_{rxn} - q_{water}$$

$$q_{cal} = 52.49\ \text{kJ} - 32.0\ \text{kJ} = 20.5\ \text{kJ}$$

To calculate the heat capacity of the bomb calorimeter, we can use the following equation:

$$q_{cal} = C_{cal}\Delta t$$

$$\mathbf{C_{cal}} = \frac{q_{cal}}{\Delta t} = \frac{20.5\ \text{kJ}}{(25.67 - 21.84)^\circ\text{C}} = \mathbf{5.35\ kJ/^\circ C}$$

6.104 First, let's calculate the standard enthalpy of reaction.

$$\Delta H^\circ_{rxn} = 2\Delta H^\circ_f(CaSO_4) - [2\Delta H^\circ_f(CaO) + 2\Delta H^\circ_f(SO_2) + \Delta H^\circ_f(O_2)]$$

$$= (2)(-1432.7\ \text{kJ/mol}) - [(2)(-635.6\ \text{kJ/mol}) + (2)(-296.1\ \text{kJ/mol}) + 0]$$

$$= -1002\ \text{kJ/mol}$$

This is the enthalpy change for every 2 moles of SO_2 that are removed. The problem asks to calculate the enthalpy change for this process if 6.6×10^{5} g of SO_2 are removed.

$$(6.6 \times 10^5 \text{ g SO}_2) \times \frac{1 \text{ mol SO}_2}{64.07 \text{ g SO}_2} \times \frac{-1002 \text{ kJ}}{2 \text{ mol SO}_2} = \mathbf{-5.2 \times 10^6 \text{ kJ}}$$

6.106 First, we need to calculate the volume of the balloon.

$$V = \frac{4}{3}\pi r^3 = \frac{4}{3}\pi(8 \text{ m})^3 = (2.1 \times 10^3 \text{ m}^3) \times \frac{1000 \text{ L}}{1 \text{ m}^3} = 2.1 \times 10^6 \text{ L}$$

(a) We can calculate the mass of He in the balloon using the ideal gas equation.

$$n_{\text{He}} = \frac{PV}{RT} = \frac{\left(98.7 \text{ kPa} \times \frac{1 \text{ atm}}{1.01325 \times 10^2 \text{ kPa}}\right)(2.1 \times 10^6 \text{ L})}{\left(0.0821 \frac{\text{L} \cdot \text{atm}}{\text{mol} \cdot \text{K}}\right)(273 + 18)\text{K}} = 8.6 \times 10^4 \text{ mol He}$$

$$\textbf{mass He} = (8.6 \times 10^4 \text{ mol He}) \times \frac{4.003 \text{ g He}}{1 \text{ mol He}} = \mathbf{3.4 \times 10^5 \text{ g He}}$$

(b) Work done $= -P\Delta V$

$$= -\left(98.7 \text{ kPa} \times \frac{1 \text{ atm}}{1.01325 \times 10^2 \text{ kPa}}\right)(2.1 \times 10^6 \text{ L})$$

$$= (-2.0 \times 10^6 \text{ L} \cdot \text{atm}) \times \frac{101.3 \text{ J}}{1 \text{ L} \cdot \text{atm}}$$

$$\textbf{Work done} = \mathbf{-2.0 \times 10^8 \text{ J}}$$

6.108 **(a)** The heat needed to raise the temperature of the water from 3°C to 37°C can be calculated using the equation:

$$q = ms\Delta t$$

First, we need to calculate the mass of the water.

$$4 \text{ glasses of water} \times \frac{2.5 \times 10^2 \text{ mL}}{1 \text{ glass}} \times \frac{1 \text{ g water}}{1 \text{ mL water}} = 1.0 \times 10^3 \text{ g water}$$

The heat needed to raise the temperature of 1.0×10^3 g of water is:

$$\mathbf{q} = ms\Delta t = (1.0 \times 10^3 \text{ g})(4.184 \text{ J/g} \cdot {}^\circ\text{C})(37 - 3)^\circ\text{C} = 1.4 \times 10^5 \text{ J} = \mathbf{1.4 \times 10^2 \text{ kJ}}$$

(b) We need to calculate both the heat needed to melt the snow and also the heat needed to heat liquid water form 0°C to 37°C (normal body temperature).

The heat needed to melt the snow is:

$$(8.0 \times 10^2 \text{ g}) \times \frac{1 \text{ mol}}{18.02 \text{ g}} \times \frac{6.01 \text{ kJ}}{1 \text{ mol}} = 2.7 \times 10^2 \text{ kJ}$$

The heat needed to raise the temperature of the water from 0°C to 37°C is:

$$q = ms\Delta t = (8.0 \times 10^2 \text{ g})(4.184 \text{ J/g} \cdot {}^\circ\text{C})(37 - 0)^\circ\text{C} = 1.2 \times 10^5 \text{ J} = 1.2 \times 10^2 \text{ kJ}$$

The total heat lost by your body is:

$$(2.7 \times 10^2 \text{ kJ}) + (1.2 \times 10^2 \text{ kJ}) = \mathbf{3.9 \times 10^2 \text{ kJ}}$$

6.110 **(a)** $\Delta H° = \Delta H_f°(F^-) + \Delta H_f°(H_2O) - [\Delta H_f°(HF) + \Delta H_f°(OH^-)]$

$\Delta H° = [(1)(-329.1 \text{ kJ/mol}) + (1)(-285.8 \text{ kJ/mol})] - [(1)(-320.1 \text{ kJ/mol}) + (1)(-229.6 \text{ kJ/mol})$

$\mathbf{\Delta H° = -65.2 \text{ kJ/mol}}$

(b) We can add the equation given in part (a) to that given in part (b) to end up with the equation we are interested in.

$$HF(aq) + \cancel{OH^-(aq)} \longrightarrow F^-(aq) + \cancel{H_2O}(l) \qquad \Delta H° = -65.2 \text{ kJ/mol}$$
$$\cancel{H_2O}(l) \longrightarrow H^+(aq) + \cancel{OH^-(aq)} \qquad \Delta H° = +56.2 \text{ kJ/mol}$$
$$\mathbf{HF(aq) \longrightarrow H^+(aq) + F^-(aq) \qquad \Delta H° = -9.0 \text{ kJ/mol}}$$

6.112 The equation we are interested in is the formation of CO from its elements.

$$C(\text{graphite}) + \tfrac{1}{2}O_2(g) \longrightarrow CO(g) \qquad \Delta H° = ?$$

Try to add the given equations together to end up with the equation above.

$$C(\text{graphite}) + O_2(g) \longrightarrow \cancel{CO_2}(g) \qquad \Delta H° = -393.5 \text{ kJ/mol}$$
$$\cancel{CO_2}(g) \longrightarrow CO(g) + \tfrac{1}{2}\cancel{O_2}(g) \qquad \Delta H° = +283.0 \text{ kJ/mol}$$
$$\mathbf{C(\text{graphite}) + \tfrac{1}{2}O_2(g) \longrightarrow CO(g) \qquad \Delta H° = -110.5 \text{ kJ/mol}}$$

We cannot obtain $\Delta H_f°$ for CO directly, because burning graphite in oxygen will form both CO and CO_2.

6.114 **(a)** mass = 0.0010 kg

Potential energy = mgh

$= (0.0010 \text{ kg})(9.8 \text{ m/s}^2)(51 \text{ m})$

Potential energy = 0.50 J

(b) Kinetic energy $= \frac{1}{2}mu^2 = 0.50 \text{ J}$

$\frac{1}{2}(0.0010 \text{ kg})u^2 = 0.50 \text{ J}$

$u^2 = 1.0 \times 10^3 \text{ m}^2/\text{s}^2$

$\mathbf{u = 32 \text{ m/s}}$

(c) $q = ms\Delta t$

$0.50\cancel{J} = (1.0\cancel{g})(4.184\cancel{J}/\cancel{g}°C)\Delta t$

$\mathbf{\Delta t = 0.12°C}$

6.116 The reaction we are interested in is the formation of ethanol from its elements.

$$2C(graphite) + \frac{1}{2}O_2(g) + 3H_2(g) \longrightarrow C_2H_5OH(l)$$

Along with the reaction for the combustion of ethanol, we can add other reactions together to end up with the above reaction.

Reversing the reaction representing the combustion of ethanol gives:

$$2CO_2(g) + 3H_2O(l) \longrightarrow C_2H_5OH(l) + 3O_2(g) \qquad \Delta H° = +1367.4 \text{ kJ/mol}$$

We need to add equations to add C (graphite) and remove H_2O from the reactants side of the equation. We write:

$$2CO_2(g) + 3H_2O(l) \longrightarrow C_2H_5OH(l) + 3O_2(g) \qquad \Delta H° = +1367.4 \text{ kJ/mol}$$

$$2C(graphite) + 2O_2(g) \longrightarrow 2CO_2(g) \qquad \Delta H° = 2(-393.5 \text{ kJ/mol})$$

$$3H_2(g) + \frac{3}{2}O_2(g) \longrightarrow 3H_2O(l) \qquad \Delta H° = 3(-285.8 \text{ kJ/mol})$$

$$\mathbf{2C(graphite) + \frac{1}{2}O_2(g) + 3H_2(g) \longrightarrow C_2H_5OH(l) \qquad \Delta H_f^\circ = -277.0 \text{ kJ/mol}}$$

6.118 Heat gained by ice = Heat lost by the soft drink

$$m_{ice} \times 334 \text{ J/g} = -m_{sd}s_{sd}\Delta t$$

$$m_{ice} \times 334 \text{ J/g} = -(361 \text{ g})(4.184 \text{ J/g·°C})(0 - 23)\text{°C}$$

$$\mathbf{m_{ice} = 104 \text{ g}}$$

6.120 From Chapter 5, we saw that the kinetic energy (or internal energy) of 1 mole of a gas is $\frac{3}{2}RT$. For 1 mole of an ideal gas, $PV = RT$. We can write:

$$\text{internal energy} = \frac{3}{2}RT = \frac{3}{2}PV$$

$$= \frac{3}{2}(1.2 \times 10^5 \text{ Pa})(5.5 \times 10^3 \text{ m}^3)$$

$$= 9.9 \times 10^8 \text{ Pa·m}^3$$

$$1 \text{ Pa·m}^3 = 1 \frac{\text{N}}{\text{m}^2}\text{m}^3 = 1 \text{ N·m} = 1 \text{ J}$$

Therefore, the internal energy is $\mathbf{9.9 \times 10^8}$ **J**.

The final temperature of the copper metal can be calculated. (10 tons = 9.07×10^6 g)

$$q = m_{Cu}s_{Cu}\Delta t$$

$$9.9 \times 10^8 \text{ J} = (9.07 \times 10^6 \text{ g})(0.385 \text{ J/g°C})(t_f - 21\text{°C})$$

$$(3.49 \times 10^6)t_f = 1.06 \times 10^9$$

$$\mathbf{t_f = 304\text{°C}}$$

6.122 **(a)** $CaC_2(s) + 2H_2O(l) \longrightarrow Ca(OH)_2(s) + C_2H_2(g)$

(b) The reaction for the combustion of acetylene is:

$$2C_2H_2(g) + 5O_2(g) \longrightarrow 4CO_2(g) + 2H_2O(l)$$

We can calculate the enthalpy change for this reaction from standard enthalpy of formation values given in Appendix 3 of the text.

$$\Delta H^\circ_{rxn} = [4\Delta H^\circ_f(CO_2) + 2\Delta H^\circ_f(H_2O)] - [2\Delta H^\circ_f(C_2H_2) + 5\Delta H^\circ_f(O_2)]$$

$$\Delta H^\circ_{rxn} = [(4)(-393.5 \text{ kJ/mol}) + (2)(-285.8 \text{ kJ/mol})] - [(2)(226.6 \text{ kJ/mol}) + (5)(0)]$$

$$\Delta H^\circ_{rxn} = -2599 \text{ kJ/mol}$$

Looking at the balanced equation, this is the amount of heat released when two moles of C_2H_2 are reacted. The problem asks for the amount of heat that can be obtained starting with 74.6 g of CaC_2. From this amount of CaC_2, we can calculate the moles of C_2H_2 produced.

$$74.6 \text{ g } CaC_2 \times \frac{1 \text{ mol } CaC_2}{64.10 \text{ g } CaC_2} \times \frac{1 \text{ mol } C_2H_2}{1 \text{ mol } CaC_2} = 1.16 \text{ mol } C_2H_2$$

Now, we can use the ΔH°_{rxn} calculated above as a conversion factor to determine the amount of heat obtained when 1.16 moles of C_2H_2 are burned.

$$1.16 \text{ mol } C_2H_2 \times \frac{2599 \text{ kJ}}{2 \text{ mol } C_2H_2} = \mathbf{1.51 \times 10^3 \text{ kJ}}$$

6.124 When 1.034 g of naphthalene are burned, 41.56 kJ of heat are evolved. Let's convert this to the amount of heat evolved on a molar basis. The molar mass of naphthalene is 128.2 g/mol.

$$q = \frac{-41.56 \text{ kJ}}{1.034 \text{ g } C_{10}H_8} \times \frac{128.2 \text{ g } C_{10}H_8}{1 \text{ mol } C_{10}H_8} = -5153 \text{ kJ/mol}$$

q has a negative sign because this is an exothermic reaction.

This reaction is run at constant volume ($\Delta V = 0$); therefore, no work will result from the change.

$$w = -P\Delta V = 0$$

From Equation (6.4) of the text, it follows that the change in energy is equal to the heat change.

$$\mathbf{\Delta E} = q + w = q_V = \mathbf{-5153 \text{ kJ/mol}}$$

To calculate ΔH, we rearrange Equation (6.10) of the text.

$$\Delta E = \Delta H - RT\Delta n$$

$$\Delta H = \Delta E + RT\Delta n$$

To calculate ΔH, Δn must be determined, which is the difference in moles of *gas* products and moles of *gas* reactants. Looking at the balanced equation for the combustion of naphthalene:

$C_{10}H_8(s) + 12O_2(g) \rightarrow 10CO_2(g) + 4H_2O(l)$

$$\Delta n = 10 - 12 = -2$$

$$\Delta H = \Delta E + RT\Delta n$$

$$\Delta H = -5153 \text{ kJ/mol} + (8.314 \text{ J/mol}\cdot\text{K})(298 \text{ K})(-2) \times \frac{1 \text{ kJ}}{1000 \text{ J}}$$

$$\mathbf{\Delta H = -5158 \text{ kJ/mol}}$$

Is ΔH equal to q_p in this case?

6.126 We know that $\Delta E = q + w$. $\Delta H = q$, and $w = -P\Delta V = -RT\Delta n$. Using thermodynamic data in Appendix 3 of the text, we can calculate ΔH.

$$2H_2(g) + O_2(g) \rightarrow 2H_2O(l),\ \Delta H = 2(-285.8 \text{ kJ/mol}) = -571.6 \text{ kJ/mol}$$

Next, we calculate w. The change in moles of gas (Δn) equals -3.

$$w = -P\Delta V = -RT\Delta n$$

$$w = -(8.314 \text{ J/mol}\cdot\text{K})(298 \text{ K})(-3) = +7.43 \times 10^3 \text{ J/mol} = 7.43 \text{ kJ/mol}$$

$$\Delta E = q + w$$

$$\mathbf{\Delta E} = -571.6 \text{ kJ/mol} + 7.43 \text{ kJ/mol} = \mathbf{-564.2 \text{ kJ/mol}}$$

Can you explain why ΔE is smaller (in magnitude) than ΔH?

6.128 First, we calculate ΔH for the combustion of 1 mole of glucose using data in Appendix 3 of the text. We can then calculate the heat produced in the calorimeter. Using the heat produced along with ΔH for the combustion of 1 mole of glucose will allow us to calculate the mass of glucose in the sample. Finally, the mass % of glucose in the sample can be calculated.

$$C_6H_{12}O_6(s) + 6O_2(g) \rightarrow 6CO_2(g) + 6H_2O(l)$$

$$\Delta H^\circ_{rxn} = (6)(-393.5 \text{ kJ/mol}) + (6)(-285.8 \text{ kJ/mol}) - (1)(-1274.5 \text{ kJ/mol}) = -2801.3 \text{ kJ/mol}$$

The heat produced in the calorimeter is:

$$(3.134^\circ\text{C})(19.65 \text{ kJ/}^\circ\text{C}) = 61.58 \text{ kJ}$$

Let x equal the mass of glucose in the sample:

$$x \text{ g glucose} \times \frac{1 \text{ mol glucose}}{180.2 \text{ g glucose}} \times \frac{2801.3 \text{ kJ}}{1 \text{ mol glucose}} = 61.58 \text{ kJ}$$

$$x = 3.961 \text{ g}$$

$$\% \text{ glucose} = \frac{3.961 \text{ g}}{4.117 \text{ g}} \times 100\% = \mathbf{96.21\%}$$

6.130 **(a)** From the mass of CO_2 produced, we can calculate the moles of carbon in the compound. From the mass of H_2O produced, we can calculate the moles of hydrogen in the compound.

$$1.419\ \text{g CO}_2 \times \frac{1\ \text{mol CO}_2}{44.01\ \text{g CO}_2} \times \frac{1\ \text{mol C}}{1\ \text{mol CO}_2} = 0.03224\ \text{mol C}$$

$$0.290\ \text{g H}_2\text{O} \times \frac{1\ \text{mol H}_2\text{O}}{18.02\ \text{g H}_2\text{O}} \times \frac{2\ \text{mol H}}{1\ \text{mol H}_2\text{O}} = 0.03219\ \text{mol H}$$

The mole ratio between C and H is 1:1, so the empirical formula is **CH**.

(b) The empirical molar mass of CH is 13.02 g/mol.

$$\frac{\text{molar mass}}{\text{empirical molar mass}} = \frac{76\ \text{g}}{13.02\ \text{g}} = 5.8 \approx 6$$

Therefore, the molecular formula is C_6H_6, and the hydrocarbon is benzene. The combustion reaction is:

$$2C_6H_6(l) + 15O_2(g) \rightarrow 12CO_2(g) + 6H_2O(l)$$

17.55 kJ of heat is released when 0.4196 g of the hydrocarbon undergoes combustion. We can now calculate the enthalpy of combustion ($\Delta H^\circ_{\text{rxn}}$) for the above reaction in units of kJ/mol. Then, from the enthalpy of combustion, we can calculate the enthalpy of formation of C_6H_6.

$$\frac{-17.55\ \text{kJ}}{0.4196\ \text{g C}_6\text{H}_6} \times \frac{78.11\ \text{g C}_6\text{H}_6}{1\ \text{mol C}_6\text{H}_6} \times 2\ \text{mol C}_6\text{H}_6 = -6534\ \text{kJ/mol}$$

$$\Delta H^\circ_{\text{rxn}} = (12)\Delta H^\circ_f(\text{CO}_2) + (6)\Delta H^\circ_f(\text{H}_2\text{O}) - (2)\Delta H^\circ_f(\text{C}_6\text{H}_6)$$

$$-6534\ \text{kJ/mol} = (12)(-393.5\ \text{kJ/mol}) + (6)(-285.8\ \text{kJ/mol}) - (2)\Delta H^\circ_f(\text{C}_6\text{H}_6)$$

$$\Delta H^\circ_f(\text{C}_6\text{H}_6) = \mathbf{49\ kJ/mol}$$

6.132 Begin by using Equation (6.20) of the text, $\Delta H_{\text{soln}} = U + \Delta H_{\text{hydr}}$, where U is the lattice energy.

(1) $Na^+(g) + Cl^-(g) \rightarrow Na^+(aq) + Cl^-(aq)$ $\Delta H_{\text{hydr}} = (4.0 - 788)\ \text{kJ/mol} = -784.0\ \text{kJ/mol}$

(2) $Na^+(g) + I^-(g) \rightarrow Na^+(aq) + I^-(aq)$ $\Delta H_{\text{hydr}} = (-5.1 - 686)\ \text{kJ/mol} = -691.1\ \text{kJ/mol}$

(3) $K^+(g) + Cl^-(g) \rightarrow K^+(aq) + Cl^-(aq)$ $\Delta H_{\text{hydr}} = (17.2 - 699)\ \text{kJ/mol} = -681.8\ \text{kJ/mol}$

Adding together equation (2) and (3) and then subtracting equation (1) gives the equation for the hydration of KI.

(2)	$Na^+(g) + I^-(g) \rightarrow Na^+(aq) + I^-(aq)$	$\Delta H = -691.1\ \text{kJ/mol}$
(3)	$K^+(g) + Cl^-(g) \rightarrow K^+(aq) + Cl^-(aq)$	$\Delta H = -681.8\ \text{kJ/mol}$
(1)	$Na^+(aq) + Cl^-(aq) \rightarrow Na^+(g) + Cl^-(g)$	$\Delta H = +784.0\ \text{kJ/mol}$
	$K^+(g) + I^-(g) \rightarrow K^+(aq) + I^-(aq)$	$\Delta H = -588.9\ \text{kJ/mol}$

We combine this last result with the given value of the lattice energy to arrive at the heat of solution of KI.

$$\Delta H_{\text{soln}} = U + \Delta H_{\text{hydr}} = (632\ \text{kJ/mol} - 588.9\ \text{kJ/mol}) = \mathbf{43\ kJ/mol}$$

CHAPTER 7
QUANTUM THEORY AND THE ELECTRONIC STRUCTURE OF ATOMS

PROBLEM-SOLVING STRATEGIES AND TUTORIAL SOLUTIONS

TYPES OF PROBLEMS

Problem Type 1: Calculating the Frequency and Wavelength of an Electromagnetic Wave.

Problem Type 2: Calculating the Energy of a Photon.

Problem Type 3: Calculating the Energy, Wavelength, or Frequency in the Emission Spectrum of a Hydrogen Atom.

Problem Type 4: The de Broglie Equation: Calculating the Wavelengths of Particles.

Problem Type 5: Quantum Numbers.
(a) Labeling an atomic orbital.
(b) Counting the number of orbitals associated with a principal quantum number.
(c) Assigning quantum numbers to an electron.
(d) Counting the number of electrons in a principal level.

Problem Type 6: Writing Electron Configurations and Orbital Diagrams.

PROBLEM TYPE 1: CALCULATING THE FREQUENCY AND WAVELENGTH OF AN ELECTROMAGNETIC WAVE

All types of *electromagnetic radiation* move through a vacuum at a speed of about 3.00×10^8 m/s, which is called the speed of light (c). Speed is an important property of a wave traveling through space and is equal to the product of the wavelength and the frequency of the wave. For electromagnetic waves

$$c = \lambda\nu \qquad (7.1)$$

Equation (7.1) can be rearranged as necessary to solve for either the wavelength (λ) or the frequency (ν).

EXAMPLE 7.1

A certain AM radio station broadcasts at a frequency of 6.00×10^2 kHz. What is the wavelength of these radio waves in meters?

Strategy: We are given the frequency of an electromagnetic wave and asked to calculate the wavelength. Rearranging Equation (7.1) gives:

$$c = \lambda\nu$$

$$\lambda = \frac{c}{\nu}$$

Solution: Because the speed of light has units of m/s, we must convert the frequency from units of kHz to Hz (s^{-1})

$$(6.00 \times 10^2 \text{ kHz}) \times \frac{1000 \text{ Hz}}{1 \text{ kHz}} = 6.00 \times 10^5 \text{ Hz} = 6.00 \times 10^5 \text{ s}^{-1}$$

Substituting in the frequency and the speed of light constant, the wavelength is:

$$\lambda = \frac{c}{\nu} = \frac{3.00 \times 10^8 \frac{\text{m}}{\text{s}}}{6.00 \times 10^5 \frac{1}{\text{s}}} = \mathbf{5.00 \times 10^2 \text{ m}}$$

Check: Look at Figure 7.4 of the text to confirm that this wavelength corresponds to a radio wave.

EXAMPLE 7.2

What is the frequency of light that has a wavelength of 665 nm?

Strategy: We are given the wavelength of an electromagnetic wave and asked to calculate the frequency. Rearranging Equation (7.1):

$$c = \lambda\nu$$

$$\nu = \frac{c}{\lambda}$$

Solution: Since the speed of light has units of m/s, we must convert the wavelength from units of nm to m.

$$665 \text{ nm} \times \frac{1 \times 10^{-9} \text{ m}}{1 \text{ nm}} = 6.65 \times 10^{-7} \text{ m}$$

Substituting in the wavelength and the speed of light (3.00×10^8 m/s), the frequency is:

$$\nu = \frac{c}{\lambda} = \frac{3.00 \times 10^8 \frac{\text{m}}{\text{s}}}{6.65 \times 10^{-7} \text{ m}} = \mathbf{4.51 \times 10^{14} \text{ s}^{-1}} = \mathbf{4.51 \times 10^{14} \text{ Hz}}$$

PRACTICE EXERCISE

1. Domestic microwave ovens generate microwaves with a frequency of 2.450 GHz. What is the wavelength of this microwave radiation?

Text Problems: 7.8, 7.12, **7.16**

PROBLEM TYPE 2: CALCULATING THE ENERGY OF A PHOTON

Max Planck said that atoms and molecules could emit (or absorb) energy only in discrete quantities. Planck gave the name *quantum* to the smallest quantity of energy that can be emitted (or absorbed) in the form of electromagnetic radiation. The energy E of a single quantum of energy is given by

$$E = h\nu \qquad (7.2, \text{text})$$

where,

h is Planck's constant = 6.63×10^{-34} J·s
ν is the frequency of radiation

EXAMPLE 7.3

The yellow light given off by a sodium vapor lamp has a wavelength of 589 nm. What is the energy of a single photon of this radiation?

Strategy: We are given the wavelength of an electromagnetic wave and asked to calculate its energy. Equation (7.2) of the text relates the energy and frequency of an electromagnetic wave.

$$E = h\nu$$

The relationship between frequency and wavelength is:

$$\nu = \frac{c}{\lambda}$$

Substituting for the frequency gives,

$$E = \frac{hc}{\lambda}$$

Solution: Because the speed of light is in units of m/s, we must convert the wavelength from units of nm to m.

$$589 \text{ nm} \times \frac{1 \times 10^{-9} \text{ m}}{1 \text{ nm}} = 5.89 \times 10^{-7} \text{ m}$$

Substituting in Planck's constant, the speed of light constant, and the wavelength, the energy is:

$$\boldsymbol{E} = \frac{hc}{\lambda} = \frac{(6.63 \times 10^{-34} \text{ J} \cdot \text{s})\left(3.00 \times 10^{8} \frac{\text{m}}{\text{s}}\right)}{5.89 \times 10^{-7} \text{ m}} = \mathbf{3.38 \times 10^{-19}\ J}$$

Check: We expect the energy of a single photon to be a very small energy as calculated above, 3.38×10^{-19} J.

PRACTICE EXERCISE

2. The red line in the spectrum of lithium occurs at 670.8 nm. What is the energy of a photon of this light? What is the energy of 1 mole of these photons?

3. The light-sensitive compound in most photographic films is silver bromide (AgBr). When the film is exposed, assume that the light energy absorbed dissociates the molecule into atoms. (The actual process is more complex.) If the energy of dissociation of AgBr is 1.00×10^{2} kJ/mol, find the wavelength of light that is just able to dissociate AgBr.

Text Problems: 7.16, 7.18, 7.20

PROBLEM TYPE 3: CALCULATING THE ENERGY, WAVELENGTH, OR FREQUENCY IN THE EMISSION SPECTRUM OF A HYDROGEN ATOM

Using arguments based on electrostatic interaction and Newton's laws of motion, Neils Bohr showed that the energies that the electron in the hydrogen atom can possess are given by:

$$E_{\text{n}} = -R_{\text{H}}\left(\frac{1}{n^2}\right) \qquad \text{(7.5, text)}$$

where,

R_{H} is the Rydberg constant = 2.18×10^{-18} J
n is the principal quantum number that has integer values

During the emission process in a hydrogen atom, an electron initially in an excited state characterized by the principal quantum number n_i drops to a lower energy state characterized by the principal quantum number n_f. This lower energy state may be either another excited state or the ground state. The difference between the energies of the initial and final states is:

$$\Delta E = E_f - E_i$$

Substituting Equation (7.5) of the text into the above equation gives:

$$\Delta E = \left(\frac{-R_H}{n_f^2}\right) - \left(\frac{-R_H}{n_i^2}\right)$$

$$\Delta E = R_H\left(\frac{1}{n_i^2} - \frac{1}{n_f^2}\right)$$

Furthermore, since this transition results in the emission of a photon of frequency ν and energy $h\nu$ (See Problem Type 2), we can write:

$$\Delta E = h\nu = R_H\left(\frac{1}{n_i^2} - \frac{1}{n_f^2}\right) \qquad \text{(7.6, text)}$$

EXAMPLE 7.4

What wavelength of radiation will be emitted during an electron transition from the $n = 5$ state to the $n = 1$ state in the hydrogen atom? What region of the electromagnetic spectrum does this wavelength correspond to?

Strategy: We are given the initial and final states in the emission process. We can calculate the energy of the emitted photon using Equation (7.6) of the text. Then, from this energy, we can solve for the wavelength. The value of Rydberg's constant is 2.18×10^{-18} J.

Solution: From Equation (7.6) we write:

$$\Delta E = R_H\left(\frac{1}{n_i^2} - \frac{1}{n_f^2}\right)$$

$$\Delta E = (2.18 \times 10^{-18}\ \text{J})\left(\frac{1}{5^2} - \frac{1}{1^2}\right)$$

$$\Delta E = -2.09 \times 10^{-18}\ \text{J}$$

The negative sign for ΔE indicates that this is energy associated with an emission process. To calculate the wavelength, we will omit the minus sign for ΔE because the wavelength of the photon must be positive. We know that

$$\Delta E = h\nu$$

We also know that $\nu = \frac{c}{\lambda}$. Substituting into the above equation gives:

$$\Delta E = \frac{hc}{\lambda}$$

Solving the equation algebraically for the wavelength, then substituting in the known values gives:

$$\lambda = \frac{hc}{\Delta E} = \frac{(6.63 \times 10^{-34}\ \text{J}\cdot\text{s})\left(3.00 \times 10^{8}\ \frac{\text{m}}{\text{s}}\right)}{2.09 \times 10^{-18}\ \text{J}} = \mathbf{9.52 \times 10^{-8}\ m}$$

To determine the region of the electromagnetic spectrum that this wavelength corresponds to, we should convert the wavelength from units of meters to nanometers, and then compare the value to Figure 7.4 of the text.

$$(9.52 \times 10^{-8}\ \not{m}) \times \frac{1\ \text{nm}}{1 \times 10^{-9}\ \not{m}} = 95.2\ \text{nm}$$

Checking Figure 7.4, we see that the ultraviolet region of the spectrum is centered at a wavelength of 10 nm. Therefore, this emission is in the ultraviolet (UV) region of the electromagnetic spectrum.

PRACTICE EXERCISE

4. A hydrogen emission line in the ultraviolet region of the spectrum at 95.2 nm corresponds to a transition from a higher energy level n_i to the $n = 1$ level. What is the value of n_i for the higher energy level?

Text Problems: 7.30, **7.32**, 7.34

PROBLEM TYPE 4: THE DE BROGLIE EQUATION: CALCULATING THE WAVELENGTHS OF PARTICLES

Albert Einstein showed that light (electromagnetic radiation) can possess particle like properties. De Broglie reasoned that if waves can behave like particles, then particles can exhibit wave properties. De Broglie deduced that the particle and wave properties are related by the expression:

$$\lambda = \frac{h}{mu} \qquad (7.8,\ \text{text})$$

where,

λ is the wavelength associated with the moving particle
m is the mass of the particle
u is the velocity of the particle

Equation (7.8) of the text implies that a particle in motion can be treated as a wave, and a wave can exhibit the properties of a particle.

EXAMPLE 7.5

When an atom of Th-232 undergoes radioactive decay, an alpha particle, which has a mass of 4.0 amu, is ejected from the Th nucleus with a velocity of 1.4×10^7 m/s. What is the de Broglie wavelength of the alpha particle?

Strategy: We are given the mass and the velocity of the alpha particle and asked to calculate the wavelength. We need the de Broglie equation, which is Equation (7.8) of the text. Note that because the units of Planck's constant are J·s, m must be in kg and u must be in m/s (1 J = 1 kg·m^2/s^2).

Solution: Because Planck's constant has units of J·s, and 1 J = 1 kg·m^2/s^2, the mass of the alpha particle must be expressed in kilograms. Since one particle has a mass of 4.0 amu, a mole of alpha particles will have a mass of 4.0 g. A reasonable strategy to complete the conversion is:

g/mol → kg/mol → kg/particle

$$\frac{4.0\ \not{g}}{1\ \not{mol}} \times \frac{1\ \text{kg}}{1000\ \not{g}} \times \frac{1\ \not{mol}}{6.022 \times 10^{23}\ \text{particles}} = 6.6 \times 10^{-27}\ \text{kg/particle}$$

Substitute the known quantities into Equation (7.8) to solve for the wavelength.

$$\lambda = \frac{h}{mu} = \frac{(6.63 \times 10^{-34}\ \not{J}\cdot\not{s}) \times \left(\frac{1\ \not{kg}\cdot m^{\not{2}}/\not{s}^{\not{2}}}{1\ \not{J}}\right)}{(6.6 \times 10^{-27}\ \not{kg})\left(1.4 \times 10^{7}\ \frac{\not{m}}{\not{s}}\right)} = \mathbf{7.2 \times 10^{-15}\ m}$$

The wavelength is smaller than the diameter of the thorium nucleus, which is about 2×10^{-14} m. Is this what you would expect?

PRACTICE EXERCISE

5. The average kinetic energy of a neutron at 25°C is 6.2×10^{-21} J. What is the de Broglie wavelength of an average neutron? The mass of a neutron is 1.008 amu. (**Hint:** kinetic energy = $\frac{1}{2}mu^2$)

Text Problems: 7.40, 7.42

PROBLEM TYPE 5: QUANTUM NUMBERS

See Section 7.6 of your text for a complete discussion. In quantum mechanics, three quantum numbers are required to describe the distribution of electrons in atoms.

(1) **The principal quantum number (*n*).** In a hydrogen atom, the value of *n* determines the energy of an orbital. It can have integral values 1, 2, 3, and so forth.

(2) **The angular momentum quantum number (*l*)** tells us the "shape" of the orbitals. *l* has possible integral values from 0 to $(n - 1)$. The value of *l* is generally designated by the letters *s*, *p*, *d*, ..., as follows:

l	0	1	2	3	4	5
Name of orbital	*s*	*p*	*d*	*f*	*g*	*h*

(3) **The magnetic quantum number (m_l)** describes the orientation of the orbital in space. For a certain value of *l*, there are $(2l + 1)$ integral values of m_l, as follows:

$$-l, (-l + 1), \ldots 0, \ldots (+l - 1), +l$$

The number of m_l values indicates the number of orbitals in a subshell with a particular *l* value.

Finally, there is a fourth quantum number that tells us the spin of the electron. The **electron spin quantum number (m_s)** has values of +1/2 or −1/2, which correspond to the two spinning motions of the electron.

A. Labeling an atomic orbital

Strategy: To "label" an atomic orbital, you need to specify the three quantum numbers (n, l, m_l) that give information about the distribution of electrons in orbitals. Remember, m_s tells us the spin of the electron, which tells us nothing about the orbital.

EXAMPLE 7.6

List the values of *n*, *l*, and m_l for orbitals in the 2*p* subshell.

Solution: The number given in the designation of the subshell is the principal quantum number, so in this case $\boldsymbol{n} = \mathbf{2}$. For *p* orbitals, $\boldsymbol{l} = \mathbf{1}$. m_l can have integer values from $-l$ to $+l$. Therefore, $\boldsymbol{m_l}$ can be **−1**, **0**, and **+1**. (The three values for m_l correspond to the three *p* orbitals.)

PRACTICE EXERCISE

6. List the values of *n*, *l*, and m_l for orbitals in the 4*f* subshell.

Text Problems: 7.56, 7.62

B. Counting the number of orbitals associated with a principal quantum number

Strategy: To work this type of problem, you must take into account the energy level (*n*), the types of orbitals in that energy level (*l*), and the number of orbitals in a subshell with a particular *l* value (m_l).

EXAMPLE 7.7

What is the total number of orbitals associated with the principal quantum number $n = 2$?

Solution: For $n = 2$, there are only two possible values of *l*, 0 and 1. Thus, there is one 2*s* orbital, and there are three 2*p* orbitals. (For $l = 1$, there are three possible m_l values, −1, 0, and +1.)

Therefore, the total number of orbitals in the $n = 2$ energy level is $1 + 3 = \mathbf{4}$.

Tip: The total number of orbitals with a given *n* value is n^2. For Example 7.7, the total number of orbitals in the $n = 2$ level equals $2^2 = \mathbf{4}$.

PRACTICE EXERCISE

7. What is the total number of orbitals associated with the principal quantum number $n = 4$?

Text Problem: 7.62

C. Assigning quantum numbers to an electron

Strategy: In assigning quantum numbers to an electron, you need to specify all four quantum numbers. In most cases, there will be more than one possible set of quantum numbers that can designate an electron.

EXAMPLE 7.8

List the different ways to write the four quantum numbers that designate an electron in a 4*s* orbital.

Solution: To begin with, we know that the principal quantum number *n* is 4, and the angular momentum quantum number *l* is 0 (*s* orbital). For $l = 0$, there is only one possible value for m_l, also 0. Since the electron spin quantum number m_s can be either +1/2 or −1/2, we conclude that there are two possible ways to designate the electron:

(4, 0, 0, +1/2) (4, 0, 0, −1/2)

PRACTICE EXERCISE

8. List the different ways to write the four quantum numbers that designate an electron in a 3*d* orbital.

Text Problems: 7.56, 7.58

D. Counting the number of electrons in a principal level

Strategy: To work this type of problem, you need to know that the number of orbitals with a particular l value is $(2l + 1)$. Also, each orbital can accommodate two electrons.

EXAMPLE 7.9

What is the maximum number of electrons that can be present in the principal level for which $n = 4$?

Solution: When $n = 4$, $l = 0, 1, 2$, and 3. The number of orbitals for each l value is given by

Value of l	Number of orbitals $(2l + 1)$
0	1
1	3
2	5
3	7

The total number of orbitals in the principal level $n = 4$ is sixteen. Since each orbital can accommodate two electrons, the maximum number of electrons that can reside in the orbitals is $2 \times 16 =$ **32**.

> **Tip:** The above result can be generalized by the formula $2n^2$. For Example 7.9, we have $n = 4$, so $2(4)^2 =$ **32**.

PRACTICE EXERCISE

9. What is the maximum number of electrons that can be present in the principal level for which $n = 2$?

Text Problem: 7.64

PROBLEM TYPE 6: WRITING ELECTRON CONFIGURATIONS AND ORBITAL DIAGRAMS

The electron configuration of an atom tells us how the electrons are distributed among the various atomic orbitals. To write electron configurations, you should follow the four rules or guidelines given below.

(1) The electron configurations of all elements except hydrogen and helium are represented by a *noble gas core*, which shows (in brackets) the noble gas element that most nearly precedes the element being considered. The noble gas core is followed by the electron configurations of filled or partially filled subshells in the outermost shells.

(2) *The Aufbau or "building up" principle* states that electrons are added to atomic orbitals starting with the lowest energy orbital and "building up" to higher energy orbitals.

(3) In many-electron atoms, the subshells are filled in the order shown in Figure 7.24 of the text.

(4) Each orbital can hold only *two* electrons.

The electron configuration can be represented in a more detailed manner called an ***orbital diagram*** that shows the spin of the electron. For orbital diagrams, you need to follow two additional rules given below.

(5) The *Pauli exclusion principle* states that no two electrons in an atom can have the same four quantum numbers. This means that electrons occupying the same orbital *cannot* have the same spin.

(6) *Hund's rule* states that the most stable arrangement of electrons in subshells is the one with the greatest number of parallel spins, without violating the Pauli exclusion principle.

EXAMPLE 7.10

Write the ground-state electron configuration and the orbital diagram for selenium.

Strategy: How many electrons are in the Se atom (Z = 34)? We start with $n = 1$ and proceed to fill orbitals in the order shown in Figure 7.23 of the text. Remember that any given orbital can hold at most 2 electrons. However, don't forget about degenerate orbitals. Starting with $n = 2$, there are three p orbitals of equal energy, corresponding to $m_l = -1, 0, 1$. Starting with $n = 3$, there are five d orbitals of equal energy, corresponding to $m_l = -2, -1, 0, 1, 2$. We can place electrons in the orbitals according to the Pauli exclusion principle and Hund's rule. The task is simplified if we use the noble gas core preceding Se for the inner electrons.

Solution: Selenium has 34 electrons. The noble gas core in this case is [Ar]. (Ar is the noble gas in the period preceding germanium.) [Ar] represents $1s^22s^22p^63s^23p^6$. This core accounts for 18 electrons, which leaves 16 electrons to place.

See Figure 7.23 of your text to check the order of filling subshells past the Ar noble gas core. You should find that the order of filling is 4s, 3d, 4p. There are 16 remaining electrons to distribute among these orbitals. The 4s orbital can hold two electrons. Each of the five 3d orbitals can hold two electrons for a total of *10* electrons. This leaves four electrons to place in the 4p orbitals.

The electrons configuration for Se is:

$$\textbf{[Ar]}\ 4s^2 3d^{10} 4p^4$$

To write an *orbital diagram*, we must also specify the spin of the electrons. The 4s and 3d orbitals are filled, so according to the Pauli exclusion principle, the paired electrons in the 4s orbital and in each of the 3d orbitals *must* have opposite spins.

$$[\text{Ar}]\ \underset{4s^2}{\underline{\uparrow\downarrow}} \qquad \underset{3d^{10}}{\underline{\uparrow\downarrow}\ \underline{\uparrow\downarrow}\ \underline{\uparrow\downarrow}\ \underline{\uparrow\downarrow}\ \underline{\uparrow\downarrow}}$$

Now, let's deal with the 4p electrons. Hund's rule states that the most stable arrangement of electrons in subshells is the one with the greatest number of parallel spins. In other words, we want to keep electrons unpaired if possible with parallel spins. Since there are three p orbitals, three of the p electrons can be placed individually in each of the p subshells with parallel spins.

$$\underset{4p^3}{\underline{\uparrow}\ \underline{\uparrow}\ \underline{\uparrow}}$$

Finally, the fourth p electron must be paired up in one of the 4p orbitals. The complete orbital diagram is:

$$[\text{Ar}]\ \underset{4s^2}{\underline{\uparrow\downarrow}} \qquad \underset{3d^{10}}{\underline{\uparrow\downarrow}\ \underline{\uparrow\downarrow}\ \underline{\uparrow\downarrow}\ \underline{\uparrow\downarrow}\ \underline{\uparrow\downarrow}} \qquad \underset{4p^4}{\underline{\uparrow\downarrow}\ \underline{\uparrow}\ \underline{\uparrow}}$$

PRACTICE EXERCISE

10. Write the electron configuration and the orbital diagram for iron (Fe).

Text Problems: 7.76, 7.78, **7.90**, 7.92

ANSWERS TO PRACTICE EXERCISES

1. $\lambda = 0.1224$ m

2. $E = 2.97 \times 10^{-19}$ J
$E = 1.79 \times 10^{5}$ J/mol

3. $\lambda = 1.20 \times 10^{-6}$ m $= 1.20 \times 10^{3}$ nm

4. $n_i = 5$

5. $\lambda = 1.5 \times 10^{-10}$ m

6. $n = 4$, $l = 3$, $m_l = -3, -2, -1, 0, 1, 2, 3$

7. Total number of orbitals $= n^2 = 4^2 = 16$

8.
(3, 2, −2, +1/2) (3, 2, −2, −1/2)
(3, 2, −1, +1/2) (3, 2, −1, −1/2)
(3, 2, 0, +1/2) (3, 2, 0, −1/2)
(3, 2, +1, +1/2) (3, 2, +1, −1/2)
(3, 2, +2, +1/2) (3, 2, +2, −1/2)

9. 8 electrons

10. $[Ar]4s^23d^6$

[Ar] $\underline{\uparrow\downarrow}$ $\underline{\uparrow\downarrow}$ $\underline{\uparrow}$ $\underline{\uparrow}$ $\underline{\uparrow}$ $\underline{\uparrow}$

$4s^2$ $3d^6$

SOLUTIONS TO SELECTED TEXT PROBLEMS

7.8 Calculating the Frequency and Wavelength of an Electromagnetic Wave, Problem Type 1.

(a)
Strategy: We are given the wavelength of an electromagnetic wave and asked to calculate the frequency. Rearranging Equation (7.1) of the text and replacing u with c (the speed of light) gives:

$$\nu = \frac{c}{\lambda}$$

Solution: Because the speed of light is given in meters per second, it is convenient to first convert wavelength to units of meters. Recall that 1 nm = 1×10^{-9} m (see Table 1.3 of the text). We write:

$$456\ \text{nm} \times \frac{1 \times 10^{-9}\ \text{m}}{1\ \text{nm}} = 456 \times 10^{-9}\ \text{m} = 4.56 \times 10^{-7}\ \text{m}$$

Substituting in the wavelength and the speed of light (3.00×10^{8} m/s), the frequency is:

$$\nu = \frac{c}{\lambda} = \frac{3.00 \times 10^{8}\ \frac{\text{m}}{\text{s}}}{4.56 \times 10^{-7}\ \text{m}} = \mathbf{6.58 \times 10^{14}\ s^{-1}} \text{ or } \mathbf{6.58 \times 10^{14}\ Hz}$$

Check: The answer shows that 6.58×10^{14} waves pass a fixed point every second. This very high frequency is in accordance with the very high speed of light.

(b)
Strategy: We are given the frequency of an electromagnetic wave and asked to calculate the wavelength. Rearranging Equation (7.1) of the text and replacing u with c (the speed of light) gives:

$$\lambda = \frac{c}{\nu}$$

Solution: Substituting in the frequency and the speed of light (3.00×10^{8} m/s) into the above equation, the wavelength is:

$$\lambda = \frac{c}{\nu} = \frac{3.00 \times 10^{8}\ \frac{\text{m}}{\text{s}}}{2.45 \times 10^{9}\ \frac{1}{\text{s}}} = 0.122\ \text{m}$$

The problem asks for the wavelength in units of nanometers. Recall that 1 nm = 1×10^{-9} m.

$$\lambda = 0.122\ \text{m} \times \frac{1\ \text{nm}}{1 \times 10^{-9}\ \text{m}} = \mathbf{1.22 \times 10^{8}\ nm}$$

7.10 A radio wave is an electromagnetic wave, which travels at the speed of light. The speed of light is in units of m/s, so let's convert distance from units of miles to meters. (28 million mi = 2.8×10^{7} mi)

$$?\ \text{distance (m)} = (2.8 \times 10^{7}\ \text{mi}) \times \frac{1.61\ \text{km}}{1\ \text{mi}} \times \frac{1000\ \text{m}}{1\ \text{km}} = 4.5 \times 10^{10}\ \text{m}$$

Now, we can use the speed of light as a conversion factor to convert from meters to seconds ($c = 3.00 \times 10^8$ m/s).

$$\textbf{? min} = (4.5 \times 10^{10}\ \text{m}) \times \frac{1\ \text{s}}{3.00 \times 10^8\ \text{m}} = 1.5 \times 10^2\ \text{s} = \textbf{2.5 min}$$

7.12 The wavelength is:

$$\lambda = \frac{1\ \text{m}}{1{,}650{,}763.73\ \text{wavelengths}} = 6.05780211 \times 10^{-7}\ \text{m}$$

$$\nu = \frac{c}{\lambda} = \frac{3.00 \times 10^8\ \text{m/s}}{6.05780211 \times 10^{-7}\ \text{m}} = \mathbf{4.95 \times 10^{14}\ s^{-1}}$$

7.16 Calculating the wavelength of electromagnetic radiation and calculating the energy of a photon, Problem Types 1 and 2.

(a)
Strategy: We are given the frequency of an electromagnetic wave and asked to calculate the wavelength. Rearranging Equation (7.1) of the text and replacing u with c (the speed of light) gives:

$$\lambda = \frac{c}{\nu}$$

Solution: Substituting in the frequency and the speed of light (3.00×10^8 m/s) into the above equation, the wavelength is:

$$\lambda = \frac{3.00 \times 10^8\ \frac{\text{m}}{\text{s}}}{7.5 \times 10^{14}\ \frac{1}{\text{s}}} = 4.0 \times 10^{-7}\ \text{m} = \mathbf{4.0 \times 10^2\ nm}$$

Check: The wavelength of 400 nm calculated is in the blue region of the visible spectrum as expected.

(b)
Strategy: We are given the frequency of an electromagnetic wave and asked to calculate its energy. Equation (7.2) of the text relates the energy and frequency of an electromagnetic wave.

$$E = h\nu$$

Solution: Substituting in the frequency and Planck's constant (6.63×10^{-34} J·s) into the above equation, the energy of a single photon associated with this frequency is:

$$E = h\nu = (6.63 \times 10^{-34}\ \text{J}\cdot\text{s})\left(7.5 \times 10^{14}\ \frac{1}{\text{s}}\right) = \mathbf{5.0 \times 10^{-19}\ J}$$

Check: We expect the energy of a single photon to be a very small energy as calculated above, 5.0×10^{-19} J.

7.18 The energy given in this problem is for *1 mole* of photons. To apply $E = h\nu$, we must divide the energy by Avogadro's number. The energy of one photon is:

$$E = \frac{1.0 \times 10^3\ \text{kJ}}{1\ \text{mol}} \times \frac{1\ \text{mol}}{6.022 \times 10^{23}\ \text{photons}} \times \frac{1000\ \text{J}}{1\ \text{kJ}} = 1.7 \times 10^{-18}\ \text{J/photon}$$

The wavelength of this photon can be found using the relationship, $E = \dfrac{hc}{\lambda}$.

$$\lambda = \frac{hc}{E} = \frac{(6.63 \times 10^{-34}\ \text{J}\cdot\text{s})\left(3.00 \times 10^{8}\ \frac{\text{m}}{\text{s}}\right)}{1.7 \times 10^{-18}\ \text{J}} = 1.2 \times 10^{-7}\ \text{m} \times \frac{1\ \text{nm}}{1 \times 10^{-9}\ \text{m}} = \mathbf{1.2 \times 10^{2}\ nm}$$

The radiation is in the **ultraviolet** region (see Figure 7.4 of the text).

7.20 **(a)** $\lambda = \dfrac{c}{\nu}$

$$\lambda = \frac{3.00 \times 10^{8}\ \frac{\text{m}}{\text{s}}}{8.11 \times 10^{14}\ \frac{1}{\text{s}}} = 3.70 \times 10^{-7}\ \text{m} = \mathbf{3.70 \times 10^{2}\ nm}$$

(b) Checking Figure 7.4 of the text, you should find that the visible region of the spectrum runs from 400 to 700 nm. 370 nm is in the **ultraviolet** region of the spectrum.

(c) $E = h\nu$. Substitute the frequency (ν) into this equation to solve for the energy of one quantum associated with this frequency.

$$E = h\nu = (6.63 \times 10^{-34}\ \text{J}\cdot\text{s})\left(8.11 \times 10^{14}\ \frac{1}{\text{s}}\right) = \mathbf{5.38 \times 10^{-19}\ J}$$

7.26 The emitted light could be analyzed by passing it through a prism.

7.28 Excited atoms of the chemical elements emit the same characteristic frequencies or lines in a terrestrial laboratory, in the sun, or in a star many light-years distant from earth.

7.30 We use more accurate values of h and c for this problem.

$$E = \frac{hc}{\lambda} = \frac{(6.6256 \times 10^{-34}\ \text{J}\cdot\text{s})(2.998 \times 10^{8}\ \text{m/s})}{656.3 \times 10^{-9}\ \text{m}} = \mathbf{3.027 \times 10^{-19}\ J}$$

7.32 Calculating the Energy, Wavelength, or Frequency in the Emission Spectrum of a Hydrogen Atom, Problem Type 3.

Strategy: We are given the initial and final states in the emission process. We can calculate the energy of the emitted photon using Equation (7.6) of the text. Then, from this energy, we can solve for the frequency of the photon, and from the frequency we can solve for the wavelength. The value of Rydberg's constant is 2.18×10^{-18} J.

Solution: From Equation (7.6) we write:

$$\Delta E = R_{\text{H}}\left(\frac{1}{n_{\text{i}}^{2}} - \frac{1}{n_{\text{f}}^{2}}\right)$$

$$\Delta E = (2.18 \times 10^{-18}\ \text{J})\left(\frac{1}{4^{2}} - \frac{1}{2^{2}}\right)$$

$$\Delta E = -4.09 \times 10^{-19}\ \text{J}$$

The negative sign for ΔE indicates that this is energy associated with an emission process. To calculate the frequency, we will omit the minus sign for ΔE because the frequency of the photon must be positive. We know that

$$\Delta E = h\nu$$

Rearranging the equation and substituting in the known values,

$$\nu = \frac{\Delta E}{h} = \frac{(4.09 \times 10^{-19}\ \text{J})}{(6.63 \times 10^{-34}\ \text{J}\cdot\text{s})} = \mathbf{6.17 \times 10^{14}\ s^{-1}} \text{ or } \mathbf{6.17 \times 10^{14}\ Hz}$$

We also know that $\lambda = \frac{c}{\nu}$. Substituting the frequency calculated above into this equation gives:

$$\lambda = \frac{3.00 \times 10^{8}\ \frac{\text{m}}{\text{s}}}{\left(6.17 \times 10^{14}\ \frac{1}{\text{s}}\right)} = 4.86 \times 10^{-7}\ \text{m} = \mathbf{486\ nm}$$

Check: This wavelength is in the visible region of the electromagnetic region (see Figure 7.4 of the text). This is consistent with the fact that because $n_i = 4$ and $n_f = 2$, this transition gives rise to a spectral line in the Balmer series (see Figure 7.6 of the text).

7.34 $$\Delta E = R_H\left(\frac{1}{n_i^2} - \frac{1}{n_f^2}\right)$$

n_f is given in the problem and R_H is a constant, but we need to calculate ΔE. The photon energy is:

$$E = \frac{hc}{\lambda} = \frac{(6.63 \times 10^{-34}\ \text{J}\cdot\text{s})(3.00 \times 10^{8}\ \text{m/s})}{434 \times 10^{-9}\ \text{m}} = 4.58 \times 10^{-19}\ \text{J}$$

Since this is an emission process, the energy change ΔE must be negative, or -4.58×10^{-19} J.

Substitute ΔE into the following equation, and solve for n_i.

$$\Delta E = R_H\left(\frac{1}{n_i^2} - \frac{1}{n_f^2}\right)$$

$$-4.58 \times 10^{-19}\ \text{J} = (2.18 \times 10^{-18}\ \text{J})\left(\frac{1}{n_i^2} - \frac{1}{2^2}\right)$$

$$\frac{1}{n_i^2} = \left(\frac{-4.58 \times 10^{-19}\ \text{J}}{2.18 \times 10^{-18}\ \text{J}}\right) + \frac{1}{2^2} = -0.210 + 0.250 = 0.040$$

$$\boldsymbol{n_i} = \frac{1}{\sqrt{0.040}} = \mathbf{5}$$

7.40 The de Broglie Equation: Calculating the Wavelengths of Particles, Problem Type 4.

Strategy: We are given the mass and the speed of the proton and asked to calculate the wavelength. We need the de Broglie equation, which is Equation (7.8) of the text. Note that because the units of Planck's constant are J·s, m must be in kg and u must be in m/s ($1\text{ J} = 1\text{ kg}\cdot\text{m}^2/\text{s}^2$).

Solution: Using Equation (7.8) we write:

$$\lambda = \frac{h}{mu}$$

$$\lambda = \frac{h}{mu} = \frac{\left(6.63\times10^{-34}\,\frac{\text{kg}\cdot\text{m}^2}{\text{s}^2}\cdot\text{s}\right)}{(1.673\times10^{-27}\text{ kg})(2.90\times10^{8}\text{ m/s})} = 1.37\times10^{-15}\text{ m}$$

The problem asks to express the wavelength in nanometers.

$$\lambda = (1.37\times10^{-15}\text{ m})\times\frac{1\text{ nm}}{1\times10^{-9}\text{ m}} = \mathbf{1.37\times10^{-6}\text{ nm}}$$

7.42 First, we convert mph to m/s.

$$\frac{35\text{ mi}}{1\text{ h}}\times\frac{1.61\text{ km}}{1\text{ mi}}\times\frac{1000\text{ m}}{1\text{ km}}\times\frac{1\text{ h}}{3600\text{ s}} = 16\text{ m/s}$$

$$\lambda = \frac{h}{mu} = \frac{\left(6.63\times10^{-34}\,\frac{\text{kg}\cdot\text{m}^2}{\text{s}^2}\cdot\text{s}\right)}{(2.5\times10^{-3}\text{ kg})(16\text{ m/s})} = 1.7\times10^{-32}\text{ m} = \mathbf{1.7\times10^{-23}\text{ nm}}$$

7.56 Quantum Numbers, Problem Type 5.

Strategy: What are the relationships among n, l, and m_l?

Solution: We are given the principal quantum number, $n = 3$. The possible l values range from 0 to $(n - 1)$. Thus, there are three possible values of l: 0, 1, and 2, corresponding to the s, p, and d orbitals, respectively. The values of m_l can vary from $-l$ to l. The values of m_l for each l value are:

$l = 0$: $m_l = 0$ $l = 1$: $m_l = -1, 0, 1$ $l = 2$: $m_l = -2, -1, 0, 1, 2$

7.58 **(a)** The number given in the designation of the subshell is the principal quantum number, so in this case $\boldsymbol{n = 3}$. For s orbitals, $\boldsymbol{l = 0}$. m_l can have integer values from $-l$ to $+l$, therefore, $\boldsymbol{m_l = 0}$. The electron spin quantum number, $\boldsymbol{m_s}$, can be either **+1/2** or **−1/2**.

Following the same reasoning as part **(a)**

(b) $4p$: $n = 4$; $l = 1$; $m_l = -1, 0, 1$; $m_s = +1/2, -1/2$

(c) $3d$: $n = 3$; $l = 2$; $m_l = -2, -1, 0, 1, 2$; $m_s = +1/2, -1/2$

7.60 The two orbitals are identical in size, shape, and energy. They differ only in their orientation with respect to each other.

Can you assign a specific value of the magnetic quantum number to these orbitals? What are the allowed values of the magnetic quantum number for the $2p$ subshell?

7.62 For $n = 6$, the allowed values of l are 0, 1, 2, 3, 4, and 5 [$l = 0$ to $(n - 1)$, integer values]. These l values correspond to the $6s$, $6p$, $6d$, $6f$, $6g$, and $6h$ subshells. These subshells each have 1, 3, 5, 7, 9, and 11 orbitals, respectively (number of orbitals = $2l + 1$).

7.64

n value	orbital sum	total number of electrons
1	1	2
2	$1 + 3 = 4$	8
3	$1 + 3 + 5 = 9$	18
4	$1 + 3 + 5 + 7 = 16$	32
5	$1 + 3 + 5 + 7 + 9 = 25$	50
6	$1 + 3 + 5 + 7 + 9 + 11 = 36$	72

In each case the total number of orbitals is just the square of the n value (n^2). The total number of electrons is $\mathbf{2n^2}$.

7.66 The electron configurations for the elements are

(a) N: $1s^22s^22p^3$ There are three p-type electrons.

(b) Si: $1s^22s^22p^63s^23p^2$ There are six s-type electrons.

(c) S: $1s^22s^22p^63s^23p^4$ There are no d-type electrons.

7.68 In the many-electron atom, the $3p$ orbital electrons are more effectively shielded by the inner electrons of the atom (that is, the $1s$, $2s$, and $2p$ electrons) than the $3s$ electrons. The $3s$ orbital is said to be more "penetrating" than the $3p$ and $3d$ orbitals. In the hydrogen atom there is only one electron, so the $3s$, $3p$, and $3d$ orbitals have the same energy.

7.70 **(a)** $2s < 2p$ **(b)** $3p < 3d$ **(c)** $3s < 4s$ **(d)** $4d < 5f$

7.76 For aluminum, there are not enough electrons in the $2p$ subshell. (The $2p$ subshell holds six electrons.) The number of electrons (13) is correct. The electron configuration should be $1s^22s^22p^63s^23p^1$. The configuration shown might be an excited state of an aluminum atom.

For boron, there are too many electrons. (Boron only has five electrons.) The electron configuration should be $1s^22s^22p^1$. What would be the electric charge of a boron ion with the electron arrangement given in the problem?

For fluorine, there are also too many electrons. (Fluorine only has nine electrons.) The configuration shown is that of the F^- ion. The correct electron configuration is $1s^22s^22p^5$.

7.78 You should write the electron configurations for each of these elements to answer this question. In some cases, an orbital diagram may be helpful.

B: $[He]2s^22p^1$ (1 unpaired electron)
P: $[Ne]3s^23p^3$ (3 unpaired electrons)
Ne: (0 unpaired electrons, Why?)
Sc: $[Ar]4s^23d^1$ (1 unpaired electron)

Mn: $[Ar]4s^23d^5$ (5 unpaired electrons)
Kr: (0 unpaired electrons)
Cd: $[Kr]5s^24d^{10}$ (0 unpaired electrons)
Pb: $[Xe]6s^24f^{14}5d^{10}6p^2$ (2 unpaired electrons)
Se: $[Ar]4s^23d^{10}4p^4$ (2 unpaired electrons)
Fe: $[Ar]4s^23d^6$ (4 unpaired electrons)
I: $[Kr]5s^24d^{10}5p^5$ (1 unpaired electron)

7.88 The ground state electron configuration of Tc is: $[Kr]5s^24d^5$.

7.90 Writing Electron Configurations, Problem Type 6.

Strategy: How many electrons are in the Ge atom (Z = 32)? We start with $n = 1$ and proceed to fill orbitals in the order shown in Figure 7.23 of the text. Remember that any given orbital can hold at most 2 electrons. However, don't forget about degenerate orbitals. Starting with $n = 2$, there are three p orbitals of equal energy, corresponding to $m_l = -1, 0, 1$. Starting with $n = 3$, there are five d orbitals of equal energy, corresponding to $m_l = -2, -1, 0, 1, 2$. We can place electrons in the orbitals according to the Pauli exclusion principle and Hund's rule. The task is simplified if we use the noble gas core preceding Ge for the inner electrons.

Solution: Germanium has 32 electrons. The noble gas core in this case is [Ar]. (Ar is the noble gas in the period preceding germanium.) [Ar] represents $1s^22s^22p^63s^23p^6$. This core accounts for 18 electrons, which leaves 14 electrons to place.

See Figure 7.23 of your text to check the order of filling subshells past the Ar noble gas core. You should find that the order of filling is 4*s*, 3*d*, 4*p*. There are 14 remaining electrons to distribute among these orbitals. The 4*s* orbital can hold two electrons. Each of the five 3*d* orbitals can hold two electrons for a total of *10* electrons. This leaves two electrons to place in the 4*p* orbitals.

The electrons configuration for Ge is:

$$[Ar]4s^23d^{10}4p^2$$

You should follow the same reasoning for the remaining atoms.

Fe: $[Ar]4s^23d^6$
W: $[Xe]6s^24f^{14}5d^4$
Zn: $[Ar]4s^23d^{10}$
Tl: $[Xe]6s^24f^{14}5d^{10}6p^1$
Ni: $[Ar]4s^23d^8$

7.92

$3s^2$	$3p^3$	$3s^2$	$3p^4$	$3s^2$	$3p^5$
↑↓	↑ ↑ ↑	↑↓	↑↓ ↑ ↑	↑↓	↑↓ ↑↓ ↑

S^+ (5 valence electrons)
3 unpaired electrons

S (6 valence electrons)
2 unpaired electrons

S^- (7 valence electrons)
1 unpaired electron

S^+ has the most unpaired electrons

7.94 Part **(b)** is correct in the view of contemporary quantum theory. Bohr's explanation of emission and absorption line spectra appears to have universal validity. Parts **(a)** and **(c)** are artifacts of Bohr's early planetary model of the hydrogen atom and are *not* considered to be valid today.

7.96 **(a)** With $n = 2$, there are n^2 orbitals $= 2^2 = 4$. $m_s = +1/2$, specifies 1 electron per orbital, for a total of **4 electrons**.

(b) $n = 4$ and $m_l = +1$, specifies one orbital in each subshell with $l = 1, 2,$ or 3 (i.e., a $4p$, $4d$, and $4f$ orbital). Each of the three orbitals holds 2 electrons for a total of **6 electrons**.

(c) If $n = 3$ and $l = 2$, m_l has the values 2, 1, 0, −1, or −2. Each of the five orbitals can hold 2 electrons for a total of **10 electrons** (2 e^- in each of the five $3d$ orbitals).

(d) If $n = 2$ and $l = 0$, then m_l can only be zero. $m_s = -1/2$ specifies 1 electron in this orbital for a total of **1 electron** (one e^- in the $2s$ orbital).

(e) $n = 4$, $l = 3$ and $m_l = -2$, specifies one $4f$ orbital. This orbital can hold **2 electrons**.

7.98 The wave properties of electrons are used in the operation of an electron microscope.

7.100 **(a)** First convert 100 mph to units of m/s.

$$\frac{100\ \text{mi}}{1\ \text{h}} \times \frac{1\ \text{h}}{3600\ \text{s}} \times \frac{1.609\ \text{km}}{1\ \text{mi}} \times \frac{1000\ \text{m}}{1\ \text{km}} = 44.7\ \text{m/s}$$

Using the de Broglie equation:

$$\lambda = \frac{h}{mu} = \frac{\left(6.63 \times 10^{-34}\ \frac{\text{kg}\cdot\text{m}^2}{\text{s}^2}\cdot\text{s}\right)}{(0.141\ \text{kg})(44.7\ \text{m/s})} = 1.05 \times 10^{-34}\ \text{m} = \mathbf{1.05 \times 10^{-25}\ nm}$$

(b) The average mass of a hydrogen atom is:

$$\frac{1.008\ \text{g}}{1\ \text{mol}} \times \frac{1\ \text{mol}}{6.022 \times 10^{23}\ \text{atoms}} = 1.674 \times 10^{-24}\ \text{g/H atom} = 1.674 \times 10^{-27}\ \text{kg}$$

$$\lambda = \frac{h}{mu} = \frac{\left(6.63 \times 10^{-34}\ \frac{\text{kg}\cdot\text{m}^2}{\text{s}^2}\cdot\text{s}\right)}{(1.674 \times 10^{-27}\ \text{kg})(44.7\ \text{m/s})} = 8.86 \times 10^{-9}\ \text{m} = \mathbf{8.86\ nm}$$

7.102 **(a)** First, we can calculate the energy of a single photon with a wavelength of 633 nm.

$$E = \frac{hc}{\lambda} = \frac{(6.63 \times 10^{-34}\ \text{J}\cdot\text{s})(3.00 \times 10^{8}\ \text{m/s})}{633 \times 10^{-9}\ \text{m}} = 3.14 \times 10^{-19}\ \text{J}$$

The number of photons produced in a 0.376 J pulse is:

$$0.376\ \text{J} \times \frac{1\ \text{photon}}{3.14 \times 10^{-19}\ \text{J}} = \mathbf{1.20 \times 10^{18}\ photons}$$

(b) Since a 1 W = 1 J/s, the power delivered per a 1.00×10^{-9} s pulse is:

$$\frac{0.376\ \text{J}}{1.00 \times 10^{-9}\ \text{s}} = 3.76 \times 10^{8}\ \text{J/s} = \mathbf{3.76 \times 10^{8}\ W}$$

Compare this with the power delivered by a 100-W light bulb!

7.104 First, let's find the energy needed to photodissociate one water molecule.

$$\frac{285.8\text{ kJ}}{1\text{ mol}} \times \frac{1\text{ mol}}{6.022 \times 10^{23}\text{ molecules}} = 4.746 \times 10^{-22}\text{ kJ/molecule} = 4.746 \times 10^{-19}\text{ J/molecule}$$

The maximum wavelength of a photon that would provide the above energy is:

$$\lambda = \frac{hc}{E} = \frac{(6.63 \times 10^{-34}\text{ J}\cdot\text{s})(3.00 \times 10^{8}\text{ m/s})}{4.746 \times 10^{-19}\text{ J}} = \mathbf{4.19 \times 10^{-7}\text{ m}} = \mathbf{419\text{ nm}}$$

This wavelength is in the visible region of the electromagnetic spectrum. Since water is continuously being struck by visible radiation *without* decomposition, it seems unlikely that photodissociation of water by this method is feasible.

7.106 Since 1 W = 1 J/s, the energy output of the light bulb in 1 second is 75 J. The actual energy converted to visible light is 15 percent of this value or 11 J.

First, we need to calculate the energy of one 550 nm photon. Then, we can determine how many photons are needed to provide 11 J of energy.

The energy of one 550 nm photon is:

$$E = \frac{hc}{\lambda} = \frac{(6.63 \times 10^{-34}\text{ J}\cdot\text{s})(3.00 \times 10^{8}\text{ m/s})}{550 \times 10^{-9}\text{ m}} = 3.62 \times 10^{-19}\text{ J/photon}$$

The number of photons needed to produce 11 J of energy is:

$$11\text{ J} \times \frac{1\text{ photon}}{3.62 \times 10^{-19}\text{ J}} = \mathbf{3.0 \times 10^{19}\text{ photons}}$$

7.108 The Balmer series corresponds to transitions to the $n = 2$ level.

For He^{+}:

$$\Delta E = R_{He^{+}}\left(\frac{1}{n_i^2} - \frac{1}{n_f^2}\right) \qquad \lambda = \frac{hc}{\Delta E} = \frac{(6.63 \times 10^{-34}\text{ J}\cdot\text{s})(3.00 \times 10^{8}\text{ m/s})}{\Delta E}$$

For the transition, $n = 3 \rightarrow 2$

$$\Delta E = (8.72 \times 10^{-18}\text{ J})\left(\frac{1}{3^2} - \frac{1}{2^2}\right) = -1.21 \times 10^{-18}\text{ J} \qquad \lambda = \frac{1.99 \times 10^{-25}\text{ J}\cdot\text{m}}{1.21 \times 10^{-18}\text{ J}} = 1.64 \times 10^{-7}\text{ m} = \mathbf{164\text{ nm}}$$

For the transition, $n = 4 \rightarrow 2$, $\Delta E = -1.64 \times 10^{-18}$ J $\lambda =$ **121 nm**

For the transition, $n = 5 \rightarrow 2$, $\Delta E = -1.83 \times 10^{-18}$ J $\lambda =$ **109 nm**

For the transition, $n = 6 \rightarrow 2$, $\Delta E = -1.94 \times 10^{-18}$ J $\lambda =$ **103 nm**

For H, the calculations are identical to those above, except the Rydberg constant for H is 2.18×10^{-18} J.

For the transition, $n = 3 \rightarrow 2$, $\Delta E = -3.03 \times 10^{-19}$ J $\lambda =$ **657 nm**

For the transition, $n = 4 \rightarrow 2$, $\Delta E = -4.09 \times 10^{-19}$ J $\lambda =$ **487 nm**

For the transition, $n = 5 \rightarrow 2$, $\Delta E = -4.58 \times 10^{-19}$ J $\quad\quad$ λ = **434 nm**

For the transition, $n = 6 \rightarrow 2$, $\Delta E = -4.84 \times 10^{-19}$ J $\quad\quad$ λ = **411 nm**

All the Balmer transitions for He^+ are in the ultraviolet region; whereas, the transitions for H are all in the visible region. Note the negative sign for energy indicating that a photon has been emitted.

7.110 First, we need to calculate the energy of one 600 nm photon. Then, we can determine how many photons are needed to provide 4.0×10^{-17} J of energy.

The energy of one 600 nm photon is:

$$E = \frac{hc}{\lambda} = \frac{(6.63 \times 10^{-34}\ \text{J}\cdot\text{s})(3.00 \times 10^{8}\ \text{m/s})}{600 \times 10^{-9}\ \text{m}} = 3.32 \times 10^{-19}\ \text{J/photon}$$

The number of photons needed to produce 4.0×10^{-17} J of energy is:

$$(4.0 \times 10^{-17}\ \text{J}) \times \frac{1\ \text{photon}}{3.32 \times 10^{-19}\ \text{J}} = \mathbf{1.2 \times 10^{2}\ photons}$$

7.112 A "blue" photon (shorter wavelength) is higher energy than a "yellow" photon. For the same amount of energy delivered to the metal surface, there must be fewer "blue" photons than "yellow" photons. Thus, the yellow light would eject more electrons since there are more "yellow" photons. Since the "blue" photons are of higher energy, blue light will eject electrons with greater kinetic energy.

7.114 The excited atoms are still neutral, so the total number of electrons is the same as the atomic number of the element.

(a) He (2 electrons), $1s^2$

(b) N (7 electrons), $1s^2 2s^2 2p^3$

(c) Na (11 electrons), $1s^2 2s^2 2p^6 3s^1$

(d) As (33 electrons), $[Ar]4s^2 3d^{10} 4p^3$

(e) Cl (17 electrons), $[Ne]3s^2 3p^5$

7.116 Rutherford and his coworkers might have discovered the wave properties of electrons.

7.118 The wavelength of a He atom can be calculated using the de Broglie equation. First, we need to calculate the root-mean-square speed using Equation (5.16) from the text.

$$u_{\text{rms}} = \sqrt{\frac{3\left(8.314 \frac{\text{J}}{\text{K}\cdot\text{mol}}\right)(273 + 20)\text{K}}{4.003 \times 10^{-3}\ \text{kg/mol}}} = 1.35 \times 10^{3}\ \text{m/s}$$

To calculate the wavelength, we also need the mass of a He atom in kg.

$$\frac{4.003 \times 10^{-3}\ \text{kg He}}{1\ \text{mol He}} \times \frac{1\ \text{mol He}}{6.022 \times 10^{23}\ \text{He atoms}} = 6.647 \times 10^{-27}\ \text{kg/atom}$$

Finally, the wavelength of a He atom is:

$$\lambda = \frac{h}{mu} = \frac{(6.63 \times 10^{-34}\ \text{J}\cdot\text{s})}{(6.647 \times 10^{-27}\ \text{kg})(1.35 \times 10^{3}\ \text{m/s})} = \mathbf{7.39 \times 10^{-11}\ m} = \mathbf{7.39 \times 10^{-2}\ nm}$$

7.120 **(a)** **False.** $n = 2$ is the first excited state.

(b) **False.** In the $n = 4$ state, the electron is (on average) further from the nucleus and hence easier to remove.

(c) **True.**

(d) **False.** The $n = 4$ to $n = 1$ transition is a higher energy transition, which corresponds to a *shorter* wavelength.

(e) **True.**

7.122 We use Heisenberg's uncertainty principle with the equality sign to calculate the minimum uncertainty.

$$\Delta x \Delta p = \frac{h}{4\pi}$$

The momentum (p) is equal to the mass times the velocity.

$$p = mu \qquad \text{or} \qquad \Delta p = m\Delta u$$

We can write:

$$\Delta p = m\Delta u = \frac{h}{4\pi \Delta x}$$

Finally, the uncertainty in the velocity of the oxygen molecule is:

$$\boldsymbol{\Delta u} = \frac{h}{4\pi m \Delta x} = \frac{(6.63 \times 10^{-34}\ \text{J}\cdot\text{s})}{4\pi(5.3 \times 10^{-26}\ \text{kg})(5.0 \times 10^{-5}\ \text{m})} = \mathbf{2.0 \times 10^{-5}\ m/s}$$

7.124 The Pauli exclusion principle states that no two electrons in an atom can have the same four quantum numbers. In other words, only two electrons may exist in the same atomic orbital, and these electrons must have opposite spins. **(a)** and **(f)** violate the Pauli exclusion principle.

Hund's rule states that the most stable arrangement of electrons in subshells is the one with the greatest number of parallel spins. **(b)**, **(d)**, and **(e)** violate Hund's rule.

7.126 As an estimate, we can equate the energy for ionization ($Fe^{13+} \rightarrow Fe^{14+}$) to the average kinetic energy $\left(\frac{3}{2}RT\right)$ of the ions.

$$\frac{3.5 \times 10^4\ \text{kJ}}{1\ \text{mol}} \times \frac{1000\ \text{J}}{1\ \text{kJ}} = 3.5 \times 10^7\ \text{J}$$

$$IE = \frac{3}{2}RT$$

$$3.5 \times 10^7\ \text{J/mol} = \frac{3}{2}(8.314\ \text{J/mol}\cdot\text{K})T$$

$$\boldsymbol{T = 2.8 \times 10^6\ K}$$

The actual temperature can be, and most probably is, higher than this.

7.128 Looking at the de Broglie equation $\lambda = \dfrac{h}{mu}$, the mass of an N_2 molecule (in kg) and the velocity of an N_2 molecule (in m/s) is needed to calculate the de Broglie wavelength of N_2.

First, calculate the root-mean-square velocity of N_2.

$\mathcal{M}(N_2) = 28.02 \text{ g/mol} = 0.02802 \text{ kg/mol}$

$$u_{rms}(N_2) = \sqrt{\frac{(3)\left(8.314 \dfrac{\text{J}}{\text{mol}\cdot\text{K}}\right)(300 \text{ K})}{\left(0.02802 \dfrac{\text{kg}}{\text{mol}}\right)}} = 516.8 \text{ m/s}$$

Second, calculate the mass of one N_2 molecule in kilograms.

$$\frac{28.02 \text{ g } N_2}{1 \text{ mol } N_2} \times \frac{1 \text{ mol } N_2}{6.022 \times 10^{23} \text{ } N_2 \text{ molecules}} \times \frac{1 \text{ kg}}{1000 \text{ g}} = 4.653 \times 10^{-26} \text{ kg/molecule}$$

Now, substitute the mass of an N_2 molecule and the root-mean-square velocity into the de Broglie equation to solve for the de Broglie wavelength of an N_2 molecule.

$$\lambda = \frac{h}{mu} = \frac{(6.63 \times 10^{-34} \text{ J}\cdot\text{s})}{(4.653 \times 10^{-26} \text{ kg})(516.8 \text{ m/s})} = \mathbf{2.76 \times 10^{-11} \text{ m}}$$

7.130 The kinetic energy acquired by the electrons is equal to the voltage times the charge on the electron. After calculating the kinetic energy, we can calculate the velocity of the electrons ($KE = 1/2mu^2$). Finally, we can calculate the wavelength associated with the electrons using the de Broglie equation.

$$KE = (5.00 \times 10^3 \text{ V}) \times \frac{1.602 \times 10^{-19} \text{ J}}{1 \text{ V}} = 8.01 \times 10^{-16} \text{ J}$$

We can now calculate the velocity of the electrons.

$$KE = \frac{1}{2}mu^2$$

$$8.01 \times 10^{-16} \text{ J} = \frac{1}{2}(9.1094 \times 10^{-31} \text{ kg})u^2$$

$$u = 4.19 \times 10^7 \text{ m/s}$$

Finally, we can calculate the wavelength associated with the electrons using the de Broglie equation.

$$\lambda = \frac{h}{mu}$$

$$\lambda = \frac{(6.63 \times 10^{-34} \text{ J}\cdot\text{s})}{(9.1094 \times 10^{-31} \text{ kg})(4.19 \times 10^7 \text{ m/s})} = \mathbf{1.74 \times 10^{-11} \text{ m} = 17.4 \text{ pm}}$$

7.132 The energy given in the problem is the energy of 1 mole of gamma rays. We need to convert this to the energy of one gamma ray, then we can calculate the wavelength and frequency of this gamma ray.

$$\frac{1.29 \times 10^{11}\ \text{J}}{1\ \text{mol}} \times \frac{1\ \text{mol}}{6.022 \times 10^{23}\ \text{gamma rays}} = 2.14 \times 10^{-13}\ \text{J/gamma ray}$$

Now, we can calculate the wavelength and frequency from this energy.

$$E = \frac{hc}{\lambda}$$

$$\lambda = \frac{hc}{E} = \frac{(6.63 \times 10^{-34}\ \text{J}\cdot\text{s})(3.00 \times 10^{8}\ \text{m/s})}{2.14 \times 10^{-13}\ \text{J}} = \mathbf{9.29 \times 10^{-13}\ m} = \mathbf{0.929\ pm}$$

and

$$E = h\nu$$

$$\nu = \frac{E}{h} = \frac{2.14 \times 10^{-13}\ \text{J}}{6.63 \times 10^{-34}\ \text{J}\cdot\text{s}} = \mathbf{3.23 \times 10^{20}\ s^{-1}}$$

7.134 **(a)** Line A corresponds to the longest wavelength or lowest energy transition, which is the 3 → 2 transition. Therefore, line B corresponds to the 4 → 2 transition, and line C corresponds to the 5 → 2 transition.

(b) We can derive an equation for the energy change (ΔE) for an electronic transition.

$$E_f = -R_H Z^2\left(\frac{1}{n_f^2}\right) \quad \text{and} \quad E_i = -R_H Z^2\left(\frac{1}{n_i^2}\right)$$

$$\Delta E = E_f - E_i = -R_H Z^2\left(\frac{1}{n_f^2}\right) - \left(-R_H Z^2\left(\frac{1}{n_i^2}\right)\right)$$

$$\Delta E = R_H Z^2\left(\frac{1}{n_i^2} - \frac{1}{n_f^2}\right)$$

Line C corresponds to the 5 → 2 transition. The energy change associated with this transition can be calculated from the wavelength (27.1 nm).

$$E = \frac{hc}{\lambda} = \frac{(6.63 \times 10^{-34}\ \text{J}\cdot\text{s})(3.00 \times 10^{8}\ \text{m/s})}{(27.1 \times 10^{-9}\ \text{m})} = 7.34 \times 10^{-18}\ \text{J}$$

For the 5 → 2 transition, we now know ΔE, n_i, n_f, and R_H ($R_H = 2.18 \times 10^{-18}$ J). Since this transition corresponds to an emission process, energy is released and ΔE is negative. ($\Delta E = -7.34 \times 10^{-18}$ J). We can now substitute these values into the equation above to solve for Z.

$$\Delta E = R_H Z^2\left(\frac{1}{n_i^2} - \frac{1}{n_f^2}\right)$$

$$-7.34 \times 10^{-18}\ \text{J} = (2.18 \times 10^{-18}\ \text{J})Z^2\left(\frac{1}{5^2} - \frac{1}{2^2}\right)$$

$$-7.34 \times 10^{-18}\ \text{J} = (-4.58 \times 10^{-19})Z^2$$

$$Z^2 = 16.0$$

$$Z = 4$$

Z must be an integer because it represents the atomic number of the parent atom.

Now, knowing the value of Z, we can substitute in n_i and n_f for the $3 \rightarrow 2$ (Line A) and the $4 \rightarrow 2$ (Line B) transitions to solve for ΔE. We can then calculate the wavelength from the energy.

For Line A ($3 \rightarrow 2$)

$$\Delta E = R_H Z^2 \left(\frac{1}{n_i^2} - \frac{1}{n_f^2} \right) = (2.18 \times 10^{-18}\ \text{J})(4)^2 \left(\frac{1}{3^2} - \frac{1}{2^2} \right)$$

$$\Delta E = -4.84 \times 10^{-18}\ \text{J}$$

$$\lambda = \frac{hc}{E} = \frac{(6.63 \times 10^{-34}\ \text{J} \cdot \text{s})(3.00 \times 10^8\ \text{m/s})}{(4.84 \times 10^{-18}\ \text{J})} = \mathbf{4.11 \times 10^{-8}\ m = 41.1\ nm}$$

For Line B ($4 \rightarrow 2$)

$$\Delta E = R_H Z^2 \left(\frac{1}{n_i^2} - \frac{1}{n_f^2} \right) = (2.18 \times 10^{-18}\ \text{J})(4)^2 \left(\frac{1}{4^2} - \frac{1}{2^2} \right)$$

$$\Delta E = -6.54 \times 10^{-18}\ \text{J}$$

$$\lambda = \frac{hc}{E} = \frac{(6.63 \times 10^{-34}\ \text{J} \cdot \text{s})(3.00 \times 10^8\ \text{m/s})}{(6.54 \times 10^{-18}\ \text{J})} = \mathbf{3.04 \times 10^{-8}\ m = 30.4\ nm}$$

(c) The value of the final energy state is $n_f = \infty$. Use the equation derived in part (b) to solve for ΔE.

$$\Delta E = R_H Z^2 \left(\frac{1}{n_i^2} - \frac{1}{n_f^2} \right) = (2.18 \times 10^{-18}\ \text{J})(4)^2 \left(\frac{1}{4^2} - \frac{1}{\infty^2} \right)$$

$$\Delta E = \mathbf{2.18 \times 10^{-18}\ J}$$

(d) As we move to higher energy levels in an atom or ion, the energy levels get closer together. See Figure 7.11 of the text, which represents the energy levels for the hydrogen atom. Transitions from higher energy levels to the $n = 2$ level will be very close in energy and hence will have similar wavelengths. The lines are so close together that they overlap, forming a continuum. The continuum shows that the electron has been removed from the ion, and we no longer have quantized energy levels associated with the electron. In other words, the energy of the electron can now vary continuously.

7.136 To calculate the energy to remove at electron from the $n = 1$ state and the $n = 5$ state in the Li^{2+} ion, we use the equation derived in Problem 7.134 (b).

$$\Delta E = R_H Z^2 \left(\frac{1}{n_i^2} - \frac{1}{n_f^2} \right)$$

For $n_i = 1$, $n_f = \infty$, and $Z = 3$, we have:

$$\Delta E = (2.18 \times 10^{-18}\ \text{J})(3)^2 \left(\frac{1}{1^2} - \frac{1}{\infty^2} \right) = \mathbf{1.96 \times 10^{-17}\ J}$$

For $n_i = 5$, $n_f = \infty$, and Z = 3, we have:

$$\Delta E = (2.18 \times 10^{-18}\ \text{J})(3)^2\left(\frac{1}{5^2} - \frac{1}{\infty^2}\right) = \mathbf{7.85 \times 10^{-19}\ J}$$

To calculate the wavelength of the emitted photon in the electronic transition from $n = 5$ to $n = 1$, we first calculate ΔE and then calculate the wavelength.

$$\Delta E = R_H Z^2\left(\frac{1}{n_i^2} - \frac{1}{n_f^2}\right) = (2.18 \times 10^{-18}\ \text{J})(3)^2\left(\frac{1}{5^2} - \frac{1}{1^2}\right) = -1.88 \times 10^{-17}\ \text{J}$$

We ignore the minus sign for ΔE in calculating λ.

$$\lambda = \frac{hc}{\Delta E} = \frac{(6.63 \times 10^{-34}\ \text{J}\cdot\text{s})(3.00 \times 10^8\ \text{m/s})}{1.88 \times 10^{-17}\ \text{J}}$$

$$\lambda = \mathbf{1.06 \times 10^{-8}\ m = 10.6\ nm}$$

7.138 We calculate W (the energy needed to remove an electron from the metal) at a wavelength of 351 nm. Once W is known, we can then calculate the velocity of an ejected electron using light with a wavelength of 313 nm.

First, we convert wavelength to frequency.

$$\nu = \frac{c}{\lambda} = \frac{3.00 \times 10^8\ \text{m/s}}{351 \times 10^{-9}\ \text{m}} = 8.55 \times 10^{14}\ \text{s}^{-1}$$

$$h\nu = \text{W} + \frac{1}{2}m_e u^2$$

$$(6.63 \times 10^{-34}\ \text{J}\cdot\text{s})(8.55 \times 10^{14}\ \text{s}^{-1}) = \text{W} + \frac{1}{2}(9.1094 \times 10^{-31}\ \text{kg})(0\ \text{m/s})^2$$

$$\text{W} = 5.67 \times 10^{-19}\ \text{J}$$

Next, we convert a wavelength of 313 nm to frequency, and then calculate the velocity of the ejected electron.

$$\nu = \frac{c}{\lambda} = \frac{3.00 \times 10^8\ \text{m/s}}{313 \times 10^{-9}\ \text{m}} = 9.58 \times 10^{14}\ \text{s}^{-1}$$

$$h\nu = \text{W} + \frac{1}{2}m_e u^2$$

$$(6.63 \times 10^{-34}\ \text{J}\cdot\text{s})(9.58 \times 10^{14}\ \text{s}^{-1}) = (5.67 \times 10^{-19}\ \text{J}) + \frac{1}{2}(9.1094 \times 10^{-31}\ \text{kg})u^2$$

$$6.82 \times 10^{-20} = (4.5547 \times 10^{-31})u^2$$

$$\boldsymbol{u} = \mathbf{3.87 \times 10^5\ m/s}$$

7.140 **(a)** We note that the maximum solar radiation centers around 500 nm. Thus, over billions of years, organisms have adjusted their development to capture energy at or near this wavelength. The two most notable cases are photosynthesis and vision.

(b) Astronomers record blackbody radiation curves from stars and compare them with those obtained from objects at different temperatures in the laboratory. Because the shape of the curve and the wavelength corresponding to the maximum depend on the temperature of an object, astronomers can reliably determine the temperature at the surface of a star from the closest matching curve and wavelength.

CHAPTER 8
PERIODIC RELATIONSHIPS AMONG THE ELEMENTS

PROBLEM-SOLVING STRATEGIES AND TUTORIAL SOLUTIONS

TYPES OF PROBLEMS

Problem Type 1: Writing an Electron Configuration and Identifying an Element (see Problem Type 7.6).

Problem Type 2: Electron Configurations of Cations and Anions.

Problem Type 3: Comparing the Sizes of Atoms.

Problem Type 4: Comparing the Sizes of Ions.

Problem Type 5: Comparing the Ionization Energies of Elements.

Problem Type 6: Electron Affinity.

PROBLEM TYPE 1: WRITING AN ELECTRON CONFIGURATION AND IDENTIFYING AN ELEMENT

See Problem Type 7.6 for information on writing electron configurations. When examining the electron configurations of the elements in a particular group, a clear pattern emerges. Elements in the same group have the same electron configuration of their *outer* electrons. The outer electrons of an atom, which are those involved in chemical bonding, are called **valence electrons**. Having the same number of valence electrons accounts for the similarities in chemical behavior among the elements within each of these groups.

In regards to identifying an element, the text considers a number of items. First, according to the type of subshell being filled, the elements can be divided into categories--the representative elements, the noble gases, the transition metals, the lanthanides, and the actinides.

> The **representative elements** are the elements in Groups 1A through 7A, all of which have *incompletely* filled *s* or *p* subshells of the highest principal quantum number.
>
> The **noble gases**, with the exception of helium, all have a *completely* filled *p* subshell.
>
> The **transition metals** are the elements in Groups 1B and 3B through 8B that have *incompletely* filled *d* subshells, or readily produce cations with *incompletely* filled *d* subshells.
>
> The **lanthanides** and **actinides** are sometimes called *f*-block transition elements because they have *incompletely* filled *f* subshells.

Second, you might be able to classify the element as a metal, nonmetal, or metalloid. However, sometimes you need to be careful in making this classification. For example in Group 4A, carbon is a nonmetal, silicon and germanium are metalloids, and tin and lead are metals.

Third, the problem might ask whether the element is paramagnetic or diamagnetic. In Chapter 7, you learned that an element that contains unpaired electrons is *paramagnetic*, and an element in which all electrons are paired is *diamagnetic*.

EXAMPLE 8.1

A neutral atom of a certain element has 19 electrons. Without consulting a periodic table, answer the following questions: (a) What is the electron configuration of the element? (b) How should the element be classified? (c) Are the atoms of this element diamagnetic or paramagnetic?

Strategy: **(a)** We refer to the building-up principle discussed in Section 7.9 of the text. We start writing the electron configuration with principal quantum number $n = 1$ and continue upward in energy until all electrons are accounted for. **(b)** What are the electron configuration characteristics of representative elements, transition elements, and noble gases? **(c)** Examine the pairing scheme of the electrons in the outermost shell. What determines whether an element is diamagnetic or paramagnetic?

Solution:

(a) We know that for $n = 1$, we have a 1*s* orbital (2 electrons). For $n = 2$, we have a 2*s* orbital (2 electrons) and three 2*p* orbitals (6 electrons). For $n = 3$, we have a 3*s* orbital (2 electrons) and three 3*p* orbitals (6 electrons). The number of electrons left to place is 19 – 18 = 1. This electron is placed in the 4*s* orbital. The electron configuration is $1s^22s^22p^63s^23p^64s^1$ or $[Ar]4s^1$.

(b) Because the 4*s* subshell is not completely filled, this is a *representative element.* Without consulting a periodic table, you might know that the alkali metal family has one valence electron in the *s* subshell. You could then further classify this element as an *alkali metal.*

(c) There is *one* unpaired electron in the *s* subshell. Therefore, the atoms of this element are paramagnetic.

Check: For (b), note that a transition metal possesses an incompletely filled *d* subshell, and a noble gas has a completely filled outer-shell. For (c), recall that if the atoms of an element contain an odd number of electrons, the element must be paramagnetic.

PRACTICE EXERCISE

1. A neutral atom of a certain element has 36 electrons. Without consulting a periodic table, answer the following questions: (a) What is the electron configuration of the element? (b) How should the element be classified? (c) Are the atoms of this element diamagnetic or paramagnetic?

Text Problems: 8.20, 8.22, 8.24

PROBLEM TYPE 2: ELECTRON CONFIGURATIONS OF CATIONS AND ANIONS

Writing an electron configuration for a cation or anion requires only a slight extension of the method used for neutral atoms. Let's group the ions in two categories.

(1) Ions Derived from Representative Elements

(a) In the formation of a **cation** from the neutral atom of a representative element, one or more electrons are removed from the highest occupied *n* shell. The number of electrons removed is equal to the charge of the ion.

Examples: Following are the electron configurations of some neutral atoms and their corresponding cations.

Mg: $[Ne]3s^2$	Mg^{2+}: [Ne]
K: $[Ar]4s^1$	K^+: [Ar]

Note: Each ion has a stable noble gas electron configuration.

(b) In the formation of an **anion** from the neutral atom of a representative element, one or more electrons are added to the highest partially filled *n* shell. The number of electrons added is equal to the charge of the ion.

Examples: Following are the electron configurations of some neutral atoms and their corresponding anions.

O: $[He]2s^22p^4$ O^{2-}: [Ne]

Cl: $[Ne]3s^23p^5$ Cl^-: [Ar]

Note: Each ion has a stable noble gas electron configuration.

(2) Cations Derived from Transition Metals

In a neutral transition metal atom, the *ns* orbital is filled before the $(n-1)d$ orbitals. See Figure 7.23 of the text. However, in a transition metal ion, the $(n-1)d$ orbitals are more stable than the *ns* orbital. Hence, when a cation is formed from an atom of a transition metal, electrons are *always* removed first from the *ns* orbital and then from the $(n-1)d$ orbitals if necessary. See Section 8.2 of the text for a more complete discussion.

Examples: Following are the electron configurations of some neutral transition metals and their corresponding cations.

Fe: $[Ar]4s^23d^6$ Fe^{2+}: $[Ar]3d^6$

Fe: $[Ar]4s^23d^6$ Fe^{3+}: $[Ar]3d^5$

Mn: $[Ar]4s^23d^5$ Mn^{7+}: [Ar]

PRACTICE EXERCISE

2. Write electron configurations for the following ions: N^{3-}, Ba^{2+}, Zn^{2+}, and, V^{5+}.

Text Problems: 8.26, **8.28**, 8.30, 8.32

PROBLEM TYPE 3: COMPARING THE SIZES OF ATOMS

The general periodic trends in atomic size are:

(1) Moving from left to right across a row (period) of the periodic table, the atomic radius ***decreases*** due to an increase in effective nuclear charge.

(2) Moving down a column (group) of the periodic table, the atomic radius ***increases*** since the orbital size increases with increasing principal quantum number.

For a more detailed discussion, see Section 8.3 of the text.

EXAMPLE 8.2
Which one of the following has the smallest atomic radius? (a) Li, (b) Na, (c) Be, (d) Mg

Strategy: What are the trends in atomic radii in a periodic group and in a particular period. Which of the above elements are in the same group and which are in the same period?

Solution: Recall that the general periodic trends in atomic size are:

(1) Moving from left to right across a row (period) of the periodic table, the atomic radius ***decreases*** due to an increase in effective nuclear charge.

(2) Moving down a column (group) of the periodic table, the atomic radius ***increases*** since the orbital size increases with increasing principal quantum number.

Atomic radii *increase* going down a group of the periodic table; therefore, Li atoms have a smaller radius than Na atoms, and Be atoms have a smaller radius than Mg atoms. We have narrowed our choices to Li and Be. Atomic radii decrease when moving from left to right across a row of the periodic table. Thus, **Be atoms** are smaller than Li atoms as well as the other choices given.

Check: See Figure 8.5 of the text to confirm that the answer is correct.

PRACTICE EXERCISE

3. Which atom should have the largest atomic radius? (a) Br, (b) Cl, (c) Se, (d) Ge, (e) C.

Text Problems: 8.38, 8.40, 8.42

PROBLEM TYPE 4: COMPARING THE SIZES OF IONS

When a neutral atom is converted to an ion, we expect a change in size. When forming an *anion* from an atom, its size (or radius) *increases*, because the nuclear charge remains the same but the repulsion resulting from the additional electron(s) enlarges the domain of the electron cloud. Conversely, when forming a *cation* from an atom, its size *decreases*, because removing one or more electrons reduces electron-electron repulsion but the nuclear charge remains the same. Thus, the electron cloud shrinks.

The general periodic trends in ionic size are:

(1) Similar to atomic size, ionic radii increase when moving down a column (group) of the periodic table.

(2) The next trend only applies to **isoelectronic ions**. Isoelectronic ions have the same number of electrons and the same electron configuration.

- **(a)** Cations are smaller than anions. For example, Mg^{2+} is smaller than O^{2-}. Both ions have the same number of electrons (10), but Mg (Z = 12) has more protons then O (Z = 8). The larger effective nuclear charge of Mg^{2+} results in a smaller radius.
- **(b)** Considering only *isoelectronic cations*, the radius of a tripositive ion is smaller than the radius of a dipositive ion, which is smaller than the radius of a unipositive ion. For example, Al^{3+} is smaller than Mg^{2+}, which is smaller than Na^+. Each of the cations has 10 electrons, but Al^{3+} has 13 protons, Mg^{2+} has 12 protons, and Na^+ has 11 protons. As the effective nuclear charge increases, the ionic radius decreases.
- **(c)** Considering only *isoelectronic anions*, the radius increases as we go from an ion with a uninegative charge, to one with a dinegative charge, and so on. For example, N^{3-} is larger than O^{2-}, which is larger than F^-. Each of the anions has 10 electrons, but N^{3-} has 7 protons, O^{2-} has 8 protons, and F^- has 9 protons. As the effective nuclear charge increases, ionic radius decreases.

EXAMPLE 8.3

In each of the following pairs, choose the ion with the *larger* ionic radius: (a) K^+ and Na^+; (b) K^+ and Ca^{2+}; (c) K^+ and Cl^-.

Strategy: In comparing ionic radii, it is useful to classify the ions into three categories: (1) isoelectronic ions, (2) ions that carry the same charges and are generated from atoms of the same periodic group, and (3) ions that carry different charges but are generated from the same atom. In case (1), ions carrying a greater negative charge are always larger; in case (2), ions from atoms having a greater atomic number are always larger; in case (3), ions have a smaller positive charge are always larger.

Solution:

(a) K^+ and Na^+ are in the same group of the periodic table (Group 1A, alkali metals). As you proceed down a group of the periodic table, ionic radii increase. This occurs because orbital size increases with increasing principal quantum number. Therefore, $\mathbf{K^+}$ has the larger ionic radius.

(b) K^+ and Ca^{2+} are isoelectronic cations. Both ions have 18 electrons, but Ca^{2+} has one more proton than K^+. The greater effective nuclear charge of Ca^{2+}, pulls the 18 electrons more strongly toward the nucleus. Therefore, $\mathbf{K^+}$ has the larger ionic radius. For isoelectronic cations, the radius of a dipositive ion is smaller than the radius of a unipositive ion.

(c) K^+ and Cl^- are isoelectronic species. Both ions have 18 electrons, but K^+ has the greater effective nuclear charge, with 19 protons compared to 17 protons for Cl^-. Therefore, $\mathbf{Cl^-}$ has the larger ionic radius.

PRACTICE EXERCISE

4. Which of the following has the largest radius: Na^+, Mg^{2+}, Al^{3+}, S^{2-}, or Cl^-?

Text Problems: 8.44, 8.46

PROBLEM TYPE 5: COMPARING THE IONIZATION ENERGIES OF ELEMENTS

Ionization energy is the minimum energy required to remove an electron from a gaseous atom in its ground state. An equation that represents this process is

$$\text{energy} + X(g) \longrightarrow X^+(g) + e^-$$

where,

X represents a gaseous atom of any element

e^- is an electron

The more "tightly" the electron is held in the atom, the more difficult it will be to remove the electron. Hence, the more "tightly" the electron is held, the higher the ionization energy.

The general periodic trends for ionization energy are:

(1) Moving from left to right across a row of the periodic table, the ionization energy ***increases*** due to an increase in effective nuclear charge. As the effective nuclear charge increases, the electrons will be held more "tightly" by the nucleus, making them more difficult to remove.

(2) Moving down a column (group) of the periodic table, the ionization energy ***decreases***. As the principal quantum number n increases, so does the average distance of a valence electron from the nucleus. A greater separation between the electron and the nucleus results in a weaker attraction; hence, a valence electron becomes increasingly easier to remove as we proceed down a group of the periodic table.

As with most periodic trends, there are exceptions. For a more detailed discussion, see Section 8.4 of the text.

EXAMPLE 8.4

Which of the following has the highest ionization energy: K, Br, Cl, or S?

Strategy: The ionization energy increases from left to right across a row, and it decreases moving down a column or family of the periodic table.

Solution: Moving across rows of the periodic table, the ionization energy of Cl is greater than for S, and the ionization energy of Br is greater than for K. We have narrowed our choices to Cl and Br. Moving down a group of the periodic table, the ionization decreases. Therefore, **Cl** has a higher ionization energy than Br as well as the other choices given.

PRACTICE EXERCISE

5. Based on periodic trends, which of the following elements has the greatest ionization energy: Cl, K, S, Se, or Br?

Text Problems: 8.52, 8.54, **8.56**, 8.58

PROBLEM TYPE 6: ELECTRON AFFINITY

Electron affinity is the energy change when an electron is accepted by an atom in the gaseous state. An equation that represents this process is:

$$X(g) + e^- \longrightarrow X^-(g)$$

where,

X represents a gaseous atom of any element

e^- is an electron

A positive electron affinity signifies that energy is liberated when an electron is added to an atom. The more positive the electron affinity, the greater the tendency of the atom to accept an electron. Electron affinity is positive if the reaction is exothermic and negative if the reaction is endothermic.

The general periodic trends for electron affinity are:

(1) The tendency to accept electrons increases (that is, electron affinity values become more positive) as we move from left to right across a period.

(2) The electron affinities of metals are generally less than those of nonmetals.

(3) Electron affinity values vary little within a given group.

EXAMPLE 8.5
Explain why the electron affinities of the halogens are all positive.

The outer-shell electron configuration of the halogens is ns^2np^5. For the process

$$X(g) + e^- \longrightarrow X^-(g)$$

where X denotes a member of the halogen family, the accepted electron would fill the outer shell, giving the halogen ion a stable noble gas electron configuration. This is a favorable process; consequently, halogens have a strong tendency to accept an extra electron (i.e., a highly positive electron affinity).

Text Problems: **8.62**, 8.64

ANSWERS TO PRACTICE EXERCISES

1. **(a)** $1s^22s^22p^63s^23p^64s^23d^{10}4p^6$
   (b) Since the 4*p* subshell is completely filled, this is a *noble gas*. All noble gases are *nonmetals*.
   (c) Since the outer shell is completely filled, there are *no* unpaired electrons. Therefore, the atoms of this element are diamagnetic.

2. N^{3-}: [Ne] Ba^{2+}: [Xe] Zn^{2+}: $[Ar]3d^{10}$ V^{5+}: [Ar]

3. **Ge** has the largest atomic radius of the group.

4. The ions are isoelectronic. $\mathbf{S^{2-}}$ has the largest radius.

5. **Cl**

SOLUTIONS TO SELECTED TEXT PROBLEMS

8.20 Writing an Electron Configuration and Identifying an Element, Problem Type 1.

Strategy: (a) We refer to the building-up principle discussed in Section 7.9 of the text. We start writing the electron configuration with principal quantum number $n = 1$ and continue upward in energy until all electrons are accounted for. **(b)** What are the electron configuration characteristics of representative elements, transition elements, and noble gases? **(c)** Examine the pairing scheme of the electrons in the outermost shell. What determines whether an element is diamagnetic or paramagnetic?

Solution:

(a) We know that for $n = 1$, we have a 1*s* orbital (2 electrons). For $n = 2$, we have a 2*s* orbital (2 electrons) and three 2*p* orbitals (6 electrons). For $n = 3$, we have a 3*s* orbital (2 electrons). The number of electrons left to place is 17 − 12 = 5. These five electrons are placed in the 3*p* orbitals. The electron configuration is $1s^22s^22p^63s^23p^5$ or $[Ne]3s^23p^5$.

(b) Because the 3*p* subshell is not completely filled, this is a *representative element*. Without consulting a periodic table, you might know that the halogen family has seven valence electrons. You could then further classify this element as a *halogen*. In addition, all halogens are *nonmetals*.

(c) If you were to write an orbital diagram for this electron configuration, you would see that there is *one* unpaired electron in the *p* subshell. Remember, the three 3*p* orbitals can hold a total of six electrons. Therefore, the atoms of this element are paramagnetic.

Check: For (b), note that a transition metal possesses an incompletely filled *d* subshell, and a noble gas has a completely filled outer-shell. For (c), recall that if the atoms of an element contain an odd number of electrons, the element must be paramagnetic.

8.22 Elements that have the same number of valence electrons will have similarities in chemical behavior. Looking at the periodic table, elements with the same number of valence electrons are in the same group. Therefore, the pairs that would represent similar chemical properties of their atoms are:

(a) and **(d)** **(b)** and **(e)** **(c)** and **(f)**.

8.24 **(a)** Group 1A **(b)** Group 5A **(c)** Group 8A **(d)** Group 8B

Identify the elements.

8.26 You should realize that the metal ion in question is a transition metal ion because it has five electrons in the 3*d* subshell. Remember that in a transition metal ion, the $(n-1)d$ orbitals are more stable than the *ns* orbital. Hence, when a cation is formed from an atom of a transition metal, electrons are *always* removed first from the *ns* orbital and then from the $(n-1)d$ orbitals if necessary. Since the metal ion has a +3 charge, three electrons have been removed. Since the 4*s* subshell is less stable than the 3*d*, two electrons would have been lost from the 4*s* and one electron from the 3*d*. Therefore, the electron configuration of the neutral atom is $[Ar]4s^23d^6$. This is the electron configuration of iron. Thus, the metal is **iron**.

8.28 Electron Configurations of Cations and Anions, Problem Type 2.

Strategy: In the formation of a **cation** from the neutral atom of a representative element, one or more electrons are *removed* from the highest occupied *n* shell. In the formation of an **anion** from the neutral atom of a representative element, one or more electrons are *added* to the highest partially filled *n* shell. Representative elements typically gain or lose electrons to achieve a stable noble gas electron configuration.

When a cation is formed from an atom of a transition metal, electrons are *always* removed first from the *ns* orbital and then from the $(n-1)d$ orbitals if necessary.

Solution:

(a) [Ne]

(b) same as (a). Do you see why?

(c) [Ar]

(d) Same as (c). Do you see why?

(e) Same as (c)

(f) $[Ar]3d^6$. Why isn't it $[Ar]4s^23d^4$?

(g) $[Ar]3d^9$. Why not $[Ar]4s^23d^7$?

(h) $[Ar]3d^{10}$. Why not $[Ar]4s^23d^8$?

8.30 **(a)** Cr^{3+} **(b)** Sc^{3+} **(c)** Rh^{3+} **(d)** Ir^{3+}

8.32 Isoelectronic means that the species have the same number of electrons and the same electron configuration.

Be^{2+} and He (2 e^-) F^- and N^{3-} (10 e^-) Fe^{2+} and Co^{3+} (24 e^-) S^{2-} and Ar (18 e^-)

8.38 Comparing the Sizes of Atoms, Problem Type 3.

Strategy: What are the trends in atomic radii in a periodic group and in a particular period. Which of the above elements are in the same group and which are in the same period?

Solution: Recall that the general periodic trends in atomic size are:

(1) Moving from left to right across a row (period) of the periodic table, the atomic radius ***decreases*** due to an increase in effective nuclear charge.

(2) Moving down a column (group) of the periodic table, the atomic radius ***increases*** since the orbital size increases with increasing principal quantum number.

The atoms that we are considering are all in the same period of the periodic table. Hence, the atom furthest to the left in the row will have the largest atomic radius, and the atom furthest to the right in the row will have the smallest atomic radius. Arranged in order of decreasing atomic radius, we have:

$$\text{Na} > \text{Mg} > \text{Al} > \text{P} > \text{Cl}$$

Check: See Figure 8.5 of the text to confirm that the above is the correct order of decreasing atomic radius.

8.40 **Fluorine** is the smallest atom in Group 7A. Atomic radius increases moving down a group since the orbital size increases with increasing principal quantum number, *n*.

8.42 The atomic radius is largely determined by how strongly the outer-shell electrons are held by the nucleus. The larger the effective nuclear charge, the more strongly the electrons are held and the smaller the atomic radius. For the second period, the atomic radius of Li is largest because the 2*s* electron is well shielded by the filled 1*s* shell. The effective nuclear charge that the outermost electrons feel increases across the period as a result of incomplete shielding by electrons in the same shell. Consequently, the orbital containing the electrons is compressed and the atomic radius decreases.

8.44 Comparing the Sizes of Ions, Problem Type 4.

Strategy: In comparing ionic radii, it is useful to classify the ions into three categories: (1) isoelectronic ions, (2) ions that carry the same charges and are generated from atoms of the same periodic group, and (3) ions that carry different charges but are generated from the same atom. In case (1), ions carrying a greater negative charge are always larger; in case (2), ions from atoms having a greater atomic number are always larger; in case (3), ions have a smaller positive charge are always larger.

Solution: The ions listed are all isoelectronic. They each have ten electrons. The ion with the fewest protons will have the largest ionic radius, and the ion with the most protons will have the smallest ionic radius. The effective nuclear charge increases with increasing number of protons. The electrons are attracted more strongly by the nucleus, decreasing the ionic radius. N^{3-} has only 7 protons resulting in the smallest attraction exerted by the nucleus on the 10 electrons. N^{3-} is the largest ion of the group. Mg^{2+} has 12 protons resulting in the largest attraction exerted by the nucleus on the 10 electrons. Mg^{2+} is the smallest ion of the group. The order of increasing atomic radius is:

$$\mathbf{Mg^{2+} < Na^{+} < F^{-} < O^{2-} < N^{3-}}$$

8.46 Both selenium and tellurium are Group 6A elements. Since atomic radius increases going down a column in the periodic table, it follows that $\mathbf{Te^{2-}}$ must be larger than Se^{2-}.

8.48 We assume the approximate boiling point of argon is the mean of the boiling points of neon and krypton, based on its position in the periodic table being between Ne and Kr in Group 8A.

$$\textbf{b.p.} = \frac{-245.9°\text{C} + (-152.9°\text{C})}{2} = \mathbf{-199.4°C}$$

The actual boiling point of argon is −185.7°C.

8.52 The general periodic trend for first ionization energy is that it increases across a period (row) of the periodic table and it decreases down a group (column). Of the choices, K will have the smallest ionization energy. Ca, just to the right of K, will have a higher first ionization energy. Moving to the right across the periodic table, the ionization energies will continue to increase as we move to P. Continuing across to Cl and moving up the halogen group, F will have a higher ionization energy than P. Finally, Ne is to the right of F in period two, thus it will have a higher ionization energy. The correct order of increasing first ionization energy is:

$$\mathbf{K < Ca < P < F < Ne}$$

You can check the above answer by looking up the first ionization energies for these elements in Table 8.2 of the text.

8.54 The Group 3A elements (such as Al) all have a single electron in the outermost *p* subshell, which is well shielded from the nuclear charge by the inner electrons and the ns^2 electrons. Therefore, less energy is needed to remove a single *p* electron than to remove a paired *s* electron from the same principal energy level (such as for Mg).

8.56 Comparing the Ionization Energies of Elements, Problem Type 5.

Strategy: Removal of the outermost electron requires less energy if it is shielded by a filled inner shell.

Solution: The lone electron in the 3*s* orbital will be much easier to remove. This lone electron is shielded from the nuclear charge by the filled inner shell. Therefore, the ionization energy of 496 kJ/mol is paired with the electron configuration $1s^22s^22p^63s^1$.

A noble gas electron configuration, such as $1s^22s^22p^6$, is a very stable configuration, making it extremely difficult to remove an electron. The 2*p* electron is not as effectively shielded by electrons in the same energy level. The high ionization energy of 2080 kJ/mol would be associated with the element having this noble gas electron configuration.

Check: Compare this answer to the data in Table 8.2. The electron configuration of $1s^22s^22p^63s^1$ corresponds to a Na atom, and the electron configuration of $1s^22s^22p^6$ corresponds to a Ne atom.

8.58 The atomic number of mercury is 80. We carry an extra significant figure throughout this calculation to avoid rounding errors.

$$\Delta E = (2.18 \times 10^{-18}\text{ J})(80^2)\left(\frac{1}{1^2} - \frac{1}{\infty^2}\right) = 1.395 \times 10^{-14}\text{ J/ion}$$

$$\Delta E = \frac{1.395 \times 10^{-14}\text{ J}}{1\text{ ion}} \times \frac{6.022 \times 10^{23}\text{ ions}}{1\text{ mol}} \times \frac{1\text{ kJ}}{1000\text{ J}} = \mathbf{8.40 \times 10^6\ kJ/mol}$$

8.62 Electron Affinity, Problem Type 6.

Strategy: What are the trends in electron affinity in a periodic group and in a particular period. Which of the above elements are in the same group and which are in the same period?

Solution: One of the general periodic trends for electron affinity is that the tendency to accept electrons increases (that is, electron affinity values become more positive) as we move from left to right across a period. However, this trend does not include the noble gases. We know that noble gases are extremely stable, and they do not want to gain or lose electrons.

Based on the above periodic trend, **Cl** would be expected to have the highest electron affinity. Addition of an electron to Cl forms Cl^-, which has a stable noble gas electron configuration.

8.64 Alkali metals have a valence electron configuration of ns^1 so they can accept another electron in the ns orbital. On the other hand, alkaline earth metals have a valence electron configuration of ns^2. Alkaline earth metals have little tendency to accept another electron, as it would have to go into a higher energy *p* orbital.

8.68 Since ionization energies decrease going down a column in the periodic table, francium should have the lowest first ionization energy of all the alkali metals. As a result, Fr should be the most reactive of all the Group 1A elements toward water and oxygen. The reaction with oxygen would probably be similar to that of K, Rb, or Cs.

What would you expect the formula of the oxide to be? The chloride?

8.70 The Group 1B elements are much less reactive than the Group 1A elements. The 1B elements are more stable because they have much higher ionization energies resulting from incomplete shielding of the nuclear charge by the inner *d* electrons. The ns^1 electron of a Group 1A element is shielded from the nucleus more effectively by the completely filled noble gas core. Consequently, the outer *s* electrons of 1B elements are more strongly attracted by the nucleus.

8.72 **(a)** Lithium oxide is a basic oxide. It reacts with water to form the metal hydroxide:

$$Li_2O(s) + H_2O(l) \longrightarrow 2LiOH(aq)$$

(b) Calcium oxide is a basic oxide. It reacts with water to form the metal hydroxide:

$$CaO(s) + H_2O(l) \longrightarrow Ca(OH)_2(aq)$$

(c) Sulfur trioxide is an acidic oxide. It reacts with water to form sulfuric acid:

$$SO_3(g) + H_2O(l) \longrightarrow H_2SO_4(aq)$$

8.74 As we move down a column, the metallic character of the elements increases. Since magnesium and barium are both Group 2A elements, we expect barium to be more metallic than magnesium and **BaO** to be more basic than MgO.

8.76 **(a)** bromine **(b)** nitrogen **(c)** rubidium **(d)** magnesium

8.78 This is an isoelectronic series with ten electrons in each species. The nuclear charge interacting with these ten electrons ranges from +8 for oxygen to +12 for magnesium. Therefore the +12 charge in Mg^{2+} will draw in the ten electrons more tightly than the +11 charge in Na^+, than the +9 charge in F^-, than the +8 charge in O^{2-}. Recall that the largest species will be the *easiest* to ionize.

(a) increasing ionic radius: $Mg^{2+} < Na^+ < F^- < O^{2-}$

(b) increasing ionization energy: $O^{2-} < F^- < Na^+ < Mg^{2+}$

8.80 According to the *Handbook of Chemistry and Physics* (1966-67 edition), potassium metal has a melting point of 63.6°C, bromine is a reddish brown liquid with a melting point of −7.2°C, and potassium bromide (KBr) is a colorless solid with a melting point of 730°C. **M** is **potassium** (K) and **X** is **bromine** (Br).

8.82 O^+ and N   Ar and S^{2-} Ne and N^{3-} Zn and As^{3+} Cs^+ and Xe

8.84 **(a)** and **(d)**

8.86 Fluorine is a yellow-green gas that attacks glass; chlorine is a pale yellow gas; bromine is a fuming red liquid; and iodine is a dark, metallic-looking solid.

8.88 Fluorine

8.90 H^- and He are isoelectronic species with two electrons. Since H^- has only one proton compared to two protons for He, the nucleus of H^- will attract the two electrons less strongly compared to He. Therefore, $\mathbf{H^-}$ is larger.

8.92

Oxide	Name	Property
Li_2O	lithium oxide	basic
BeO	beryllium oxide	amphoteric
B_2O_3	boron oxide	acidic
CO_2	carbon dioxide	acidic
N_2O_5	dinitrogen pentoxide	acidic

Note that only the highest oxidation states are considered.

8.94 In its chemistry, hydrogen can behave like an alkali metal (H^+) and like a halogen (H^-). H^+ is a single proton.

8.96 Replacing Z in the equation given in Problem 8.57 with (Z − σ) gives:

$$E_n = (2.18 \times 10^{-18}\ \text{J})(Z - \sigma)^2\left(\frac{1}{n^2}\right)$$

For helium, the atomic number (Z) is 2, and in the ground state, its two electrons are in the first energy level, so $n = 1$. Substitute Z, n, and the first ionization energy into the above equation to solve for σ.

$$E_1 = 3.94 \times 10^{-18}\ \text{J} = (2.18 \times 10^{-18}\ \text{J})(2-\sigma)^2\left(\frac{1}{1^2}\right)$$

$$(2-\sigma)^2 = \frac{3.94 \times 10^{-18}\ \text{J}}{2.18 \times 10^{-18}\ \text{J}}$$

$$2-\sigma = \sqrt{1.81}$$

$$\sigma = 2 - 1.35 = \mathbf{0.65}$$

8.98 The percentage of volume occupied by K^+ compared to K is:

$$\frac{\text{volume of K}^+\ \text{ion}}{\text{volume of K atom}} \times 100\% = \frac{\frac{4}{3}\pi(133\ \text{pm})^3}{\frac{4}{3}\pi(216\ \text{pm})^3} \times 100\% = 23.3\%$$

Therefore, there is a decrease in volume of (100 – 23.3)% = **76.7%** when K^+ is formed from K.

8.100 Rearrange the given equation to solve for ionization energy.

$$IE = h\nu - \frac{1}{2}mu^2$$

or,

$$IE = \frac{hc}{\lambda} - KE$$

The kinetic energy of the ejected electron is given in the problem. Substitute h, c, and λ into the above equation to solve for the ionization energy.

$$IE = \frac{(6.63 \times 10^{-34}\ \text{J}\cdot\text{s})(3.00 \times 10^8\ \text{m/s})}{162 \times 10^{-9}\ \text{m}} - (5.34 \times 10^{-19}\ \text{J})$$

$$\boldsymbol{IE = 6.94 \times 10^{-19}\ \text{J}}$$

We might also want to express the ionization energy in kJ/mol.

$$\frac{6.94 \times 10^{-19}\ \text{J}}{1\ \text{photon}} \times \frac{6.022 \times 10^{23}\ \text{photons}}{1\ \text{mol}} \times \frac{1\ \text{kJ}}{1000\ \text{J}} = \mathbf{418\ kJ/mol}$$

To ensure that the ejected electron is the valence electron, UV light of the *longest* wavelength (lowest energy) should be used that can still eject electrons.

8.102 We want to determine the second ionization energy of lithium.

$$Li^+ \longrightarrow Li^{2+} + e^- \qquad I_2 = ?$$

The equation given in Problem 8.57 allows us to determine the third ionization energy for Li. Knowing the total energy needed to remove all three electrons from Li, we can calculate the second ionization energy by difference.

Energy needed to remove three electrons $= I_1 + I_2 + I_3$

First, let's calculate I_3. For Li, Z = 3, and $n = 1$ because the third electron will come from the 1s orbital.

$$I_3 = \Delta E = E_\infty - E_3$$

$$I_3 = -(2.18 \times 10^{-18}\ \text{J})(3)^2\left(\frac{1}{\infty^2}\right) + (2.18 \times 10^{-18}\ \text{J})(3)^2\left(\frac{1}{1^2}\right)$$

$$I_3 = +1.96 \times 10^{-17}\ \text{J}$$

Converting to units of kJ/mol:

$$I_3 = (1.96 \times 10^{-17}\ \text{J}) \times \frac{6.022 \times 10^{23}\ \text{ions}}{1\ \text{mol}} = 1.18 \times 10^7\ \text{J/mol} = 1.18 \times 10^4\ \text{kJ/mol}$$

Energy needed to remove three electrons $= I_1 + I_2 + I_3$

$$1.96 \times 10^4\ \text{kJ/mol} = 520\ \text{kJ/mol} + I_2 + (1.18 \times 10^4\ \text{kJ/mol})$$

$$\mathbf{I_2 = 7.28 \times 10^3\ kJ/mol}$$

8.104 X must belong to Group 4A; it is probably **Sn** or **Pb** because it is not a very reactive metal (it is certainly not reactive like an alkali metal).

Y is a nonmetal since it does *not* conduct electricity. Since it is a light yellow solid, it is probably **phosphorus** (Group 5A).

Z is an **alkali metal** since it reacts with air to form a basic oxide or peroxide.

8.106 $\text{Na} \longrightarrow \text{Na}^+ + e^- \qquad I_1 = 495.9\ \text{kJ/mol}$

This equation is the reverse of the electron affinity for Na^+. Therefore, the electron affinity of Na^+ is **+495.9 kJ/mol**. Note that the electron affinity is positive, indicating that energy is liberated when an electron is added to an atom or ion. You should expect this since we are adding an electron to a positive ion.

8.108 The reaction representing the electron affinity of chlorine is:

$$\text{Cl}(g) + e^- \longrightarrow \text{Cl}^-(g) \qquad \Delta H° = +349\ \text{kJ/mol}$$

It follows that the energy needed for the reverse process is also +349 kJ/mol.

$$\text{Cl}^-(g) + h\nu \longrightarrow \text{Cl}(g) + e^- \qquad \Delta H° = +349\ \text{kJ/mol}$$

The energy above is the energy of one mole of photons. We need to convert to the energy of one photon in order to calculate the wavelength of the photon.

$$\frac{349\ \text{kJ}}{1\ \text{mol photons}} \times \frac{1\ \text{mol photons}}{6.022 \times 10^{23}\ \text{photons}} \times \frac{1000\ \text{J}}{1\ \text{kJ}} = 5.80 \times 10^{-19}\ \text{J/photon}$$

Now, we can calculate the wavelength of a photon with this energy.

$$\lambda = \frac{hc}{E} = \frac{(6.63 \times 10^{-34}\ \text{J}\cdot\text{s})(3.00 \times 10^8\ \text{m/s})}{5.80 \times 10^{-19}\ \text{J}} = 3.43 \times 10^{-7}\ \text{m} = \mathbf{343\ nm}$$

The radiation is in the **ultraviolet** region of the electromagnetic spectrum.

8.110 The equation that we want to calculate the energy change for is:

$$Na(s) \longrightarrow Na^+(g) + e^- \qquad \Delta H° = ?$$

Can we take information given in the problem and other knowledge to end up with the above equation? This is a Hess's law problem (see Chapter 6).

In the problem we are given:	$Na(s) \longrightarrow Na(g)$	$\Delta H° = 108.4$ kJ/mol
We also know the ionization energy of Na (g).	$Na(g) \longrightarrow Na^+(g) + e^-$	$\Delta H° = 495.9$ kJ/mol
Adding the two equations:	$Na(s) \longrightarrow Na^+(g) + e^-$	$\boldsymbol{\Delta H° = 604.3}$ **kJ/mol**

8.112 The electron configuration of titanium is: $[Ar]4s^23d^2$. Titanium has four valence electrons, so the maximum oxidation number it is likely to have in a compound is +4. The compounds followed by the oxidation state of titanium are: K_3TiF_6, +3; $K_2Ti_2O_5$, +4; $TiCl_3$, +3; K_2TiO_4, +6; and K_2TiF_6, +4. $\mathbf{K_2TiO_4}$ is unlikely to exist because of the oxidation state of Ti of +6. Titanium in an oxidation state greater than +4 is unlikely because of the very high ionization energies needed to remove the fifth and sixth electrons.

8.114 The unbalanced ionic equation is: $MnF_6^{2-} + SbF_5 \longrightarrow SbF_6^- + MnF_3 + F_2$

In this redox reaction, Mn^{4+} is reduced to Mn^{3+}, and F^- from both MnF_6^{2-} and SbF_5 is oxidized to F_2.

We can simplify the half-reactions.

$$Mn^{4+} \xrightarrow{\text{reduction}} Mn^{3+}$$

$$F^- \xrightarrow{\text{oxidation}} F_2$$

Balancing the two half-reactions:

$$Mn^{4+} + e^- \longrightarrow Mn^{3+}$$

$$2F^- \longrightarrow F_2 + 2e^-$$

Adding the two half-reactions:

$$2Mn^{4+} + 2F^- \longrightarrow 2Mn^{3+} + F_2$$

We can now reconstruct the complete balanced equation. In the balanced equation, we have 2 moles of Mn ions and 1 mole of F_2 on the products side.

$$\mathbf{2}K_2MnF_6 + SbF_5 \longrightarrow KSbF_6 + \mathbf{2}MnF_3 + \mathbf{1}F_2$$

We can now balance the remainder of the equation by inspection. Notice that there are 4 moles of K^+ on the left, but only 1 mole of K^+ on the right. The balanced equation is:

$$\mathbf{2K_2MnF_6 + 4SbF_5 \longrightarrow 4KSbF_6 + 2MnF_3 + F_2}$$

8.116 To work this problem, assume that the oxidation number of oxygen is −2.

Oxidation number	Chemical formula
+1	N_2O
+2	NO
+3	N_2O_3
+4	NO_2, N_2O_4
+5	N_2O_5

8.118 The larger the effective nuclear charge, the more tightly held are the electrons. Thus, the atomic radius will be small, and the ionization energy will be large. The quantities show an opposite periodic trend.

8.120 We assume that the m.p. and b.p. of bromine will be between those of chlorine and iodine.

Taking the average of the melting points and boiling points:

$$\textbf{m.p.} = \frac{-101.0°C + 113.5°C}{2} = \textbf{6.3°C} \qquad \text{(Handbook: −7.2°C)}$$

$$\textbf{b.p.} = \frac{-34.6°C + 184.4°C}{2} = \textbf{74.9°C} \qquad \text{(Handbook: 58.8 °C)}$$

The estimated values do not agree very closely with the actual values because $Cl_2(g)$, $Br_2(l)$, and $I_2(s)$ are in different physical states. If you were to perform the same calculations for the noble gases, your calculations would be much closer to the actual values.

8.122 The heat generated from the radioactive decay can break bonds; therefore, few radon compounds exist.

8.124 **(a)** It was determined that the periodic table was based on atomic number, not atomic mass.

(b) Argon:

(0.00337 × 35.9675 amu) + (0.00063 × 37.9627 amu) + (0.9960 × 39.9624 amu) = **39.95 amu**

Potassium:

(0.93258 × 38.9637 amu) + (0.000117 × 39.9640 amu) + (0.0673 × 40.9618 amu) = **39.10 amu**

8.126 $Z = 119$

Electron configuration: $[Rn]7s^2 5f^{14} 6d^{10} 7p^6 8s^1$

8.128 There is a large jump from the second to the third ionization energy, indicating a change in the principal quantum number n. In other words, the third electron removed is an inner, noble gas core electron, which is difficult to remove. Therefore, the element is in **Group 2A**.

8.130 **(a)** SiH_4, GeH_4, SnH_4, PbH_4

(b) Metallic character increases going down a family of the periodic table. Therefore, RbH would be more ionic than NaH.

(c) Since Ra is in Group 2A, we would expect the reaction to be the same as other alkaline earth metals with water.

$$Ra(s) + 2H_2O(l) \rightarrow Ra(OH)_2(aq) + H_2(g)$$

(d) Beryllium (diagonal relationship)

8.132 The importance and usefulness of the periodic table lie in the fact that we can use our understanding of the general properties and trends within a group or a period to predict with considerable accuracy the properties of any element, even though the element may be unfamiliar to us. For example, elements in the same group or family have the same valence electron configurations. Due to the same number of valence electrons occupying similar orbitals, elements in the same family have similar chemical properties. In addition, trends in properties such as ionization energy, atomic radius, electron affinity, and metallic character can be predicted based on an element's position in the periodic table. Ionization energy typically increases across a period of the period table and decreases down a group. Atomic radius typically decreases across a period and increases down a group. Electron affinity typically increases across a period and decreases down a group.

Metallic character typically decreases across a period and increases down a group. The periodic table is an extremely useful tool for a scientist. Without having to look in a reference book for a particular element's properties, one can look at its position in the periodic table and make educated predictions as to its many properties such as those mentioned above.

8.134 The first statement that an allotropic form of the element is a colorless crystalline solid, might lead you to think about diamond, a form of carbon. When carbon is reacted with excess oxygen, the colorless gas, carbon dioxide is produced.

$$C(s) + O_2(g) \rightarrow CO_2(g)$$

When $CO_2(g)$ is dissolved in water, carbonic acid is produced.

$$CO_2(g) + H_2O(l) \rightarrow H_2CO_3(aq)$$

Element X is most likely carbon, choice **(c)**.

8.136 The ionization energy of 412 kJ/mol represents the energy difference between the ground state and the dissociation limit, whereas the ionization energy of 126 kJ/mol represents the energy difference between the first excited state and the dissociation limit. Therefore, the energy difference between the ground state and the excited state is:

$$\Delta E = (412 - 126)\text{ kJ/mol} = 286\text{ kJ/mol}$$

The energy of light emitted in a transition from the first excited state to the ground state is therefore 286 kJ/mol. We first convert this energy to units of J/photon, and then we can calculate the wavelength of light emitted in this electronic transition.

$$E = \frac{286 \times 10^3\text{ J}}{1\text{ mol}} \times \frac{1\text{ mol}}{6.022 \times 10^{23}\text{ photons}} = 4.75 \times 10^{-19}\text{ J/photon}$$

$$\lambda = \frac{hc}{E} = \frac{(6.63 \times 10^{-34}\text{ J}\cdot\text{s})(3.00 \times 10^8\text{ m/s})}{4.75 \times 10^{-19}\text{ J}} = \mathbf{4.19 \times 10^{-7}\text{ m} = 419\text{ nm}}$$

8.138 In He, r is greater than that in H. Also, the shielding in He makes Z_{eff} less than 2. Therefore, $I_1(\text{He}) < 2I(\text{H})$. In He^+, there is only one electron so there is no shielding. The greater attraction between the nucleus and the lone electron reduces r to less than the r of hydrogen. Therefore, $I_2(\text{He}) > 2I(\text{H})$.

8.140 We rearrange the equation given in the problem to solve for Z_{eff}.

$$Z_{eff} = n\sqrt{\frac{I_1}{1312\text{ kJ/mol}}}$$

Li: $$Z_{eff} = (2)\sqrt{\frac{520\text{ kJ/mol}}{1312\text{ kJ/mol}}} = 1.26$$

Na: $$Z_{eff} = (3)\sqrt{\frac{495.9\text{ kJ/mol}}{1312\text{ kJ/mol}}} = 1.84$$

K: $$Z_{\text{eff}} = (4)\sqrt{\frac{418.7\text{ kJ/mol}}{1312\text{ kJ/mol}}} = 2.26$$

As we move down a group, Z_{eff} increases. This is what we would expect because shells with larger n values are less effective at shielding the outer electrons from the nuclear charge.

Li: $$\frac{Z_{\text{eff}}}{n} = \frac{1.26}{2} = 0.630$$

Na: $$\frac{Z_{\text{eff}}}{n} = \frac{1.84}{3} = 0.613$$

K: $$\frac{Z_{\text{eff}}}{n} = \frac{2.26}{4} = 0.565$$

The Z_{eff}/n values are fairly constant, meaning that the screening per shell is about the same.

CHAPTER 9
CHEMICAL BONDING I:
BASIC CONCEPTS

PROBLEM-SOLVING STRATEGIES AND TUTORIAL SOLUTIONS

TYPES OF PROBLEMS

Problem Type 1: Classifying Chemical Bonds.

Problem Type 2: Calculating the Lattice Energy of Ionic Compounds.

Problem Type 3: Writing Lewis Structures.

Problem Type 4: Formal Charges.
(a) Assigning formal charges.
(b) Choosing the most plausible Lewis structure based on formal charges.

Problem Type 5: Drawing Resonance Structures.

Problem Type 6: Exceptions to the Octet Rule.
(a) The incomplete octet.
(b) Odd-electron molecules.
(c) The expanded octet.

Problem Type 7: Using Bond enthalpies to Estimate the Enthalpy of a Reaction.

PROBLEM TYPE 1: CLASSIFYING CHEMICAL BONDS

We can classify bonds as three different types: ionic, polar covalent, or covalent.

In a **covalent bond**, the electron pair of the bond is shared *equally* by the two atoms.

In a **polar covalent bond**, the electron pair of the bond is shared *unequally* by the two atoms. The electrons spend more time in the vicinity of one atom than the other.

Covalent bonds (polar or nonpolar) are typically formed between two nonmetals.

In an **ionic bond**, the electron or electrons are nearly completely transferred from one atom to another. Ionic bonds are typically formed between a metal cation and a nonmetal anion or a metal cation and a polyatomic anion.

A property that helps us distinguish a nonpolar covalent bond from a polar covalent bond is **electronegativity**, the ability of an atom to attract toward itself the electrons in a chemical bond. Elements with high electronegativities have a greater tendency to attract electrons than elements with low electronegativities. Linus Pauling devised a method for calculating *relative* electronegativities of most elements. These values are shown in Figure 9.5 of the text.

Only atoms of the same element, which have the same electronegativity, can be joined by a pure *covalent bond.* Atoms of elements with similar electronegativities tend to form *polar covalent bonds* with each other because the shift in electron density is usually small. There is no sharp distinction between a polar bond and an ionic bond, but the following rule is helpful in distinguishing them. An *ionic bond* forms when the electronegativity difference between the two bonding atoms is 2.0 or more. This rule applies to most but not all ionic compounds.

EXAMPLE 9.1

Classify the following bonds as ionic, polar covalent, or covalent: (a) the bond in KCl, (b) the OH bond in H_2O, and (c) the OO bond in oxygen gas (O_2).

Strategy: We can look up electronegativity values in Figure 9.5 of the text. The amount of ionic character is based on the electronegativity difference between the two atoms. The larger the electronegativity difference, the greater the ionic character.

Solution:

(a) In Figure 9.5 of the text, we see that the electronegativity difference between K and Cl is 2.2, above the 2.0 guideline. Therefore, the bond between K and Cl is ionic. Remember that an ionic bond is typically formed between a metal cation and a nonmetal anion or a metal cation and a polyatomic anion.

(b) The electronegativity difference between O and H is 1.4, which is appreciable but not large enough to qualify H_2O as an ionic compound. Therefore, the bond between O and H is polar covalent. Recall that polar covalent compounds are typically formed between two different nonmetals.

(c) The two O atoms are identical in every respect. Therefore, the bond between them is purely covalent.

PRACTICE EXERCISE

1. Classify the following bonds as ionic, polar covalent, or covalent: (a) the NH bond in NH_3, (b) the OO bond in hydrogen peroxide (H_2O_2), and (c) the bond in NaF.

Text Problems: 9.20, **9.36**, 9.38, 9.40

PROBLEM TYPE 2: CALCULATING THE LATTICE ENERGY OF IONIC COMPOUNDS

See Section 9.3 of the text for a complete discussion of lattice energy. Example 9.2 below will illustrate the method used to calculate the lattice energy of an ionic compound.

EXAMPLE 9.2

Calculate the lattice energy of magnesium oxide, MgO, given that the enthalpy of sublimation of Mg is 150 kJ/mol and the electron affinity of O^- is −780 kJ/mol.

Strategy: We want to calculate the enthalpy change corresponding to the following reaction:

$$Mg^{2+}(g) + O^{2-}(g) \longrightarrow MgO(s) \qquad \Delta H° = ?$$

This reaction is the reverse of the reaction for the lattice energy. If we can calculate $\Delta H°$ for the above reaction, we can determine the *lattice energy*. Starting with Mg(s) and $O_2(g)$, we can follow two different pathways to the product, MgO(s). The enthalpy changes for the two pathways will be equal.

Pathway 1: This is the overall reaction, which is the ΔH_f° of MgO(s). You can look up the appropriate value in Appendix 3 of the text.

$$Mg(s) + O_2(g) \longrightarrow MgO(s) \qquad \Delta H° = -601.8 \text{ kJ/mol}$$

Pathway 2: This is the indirect pathway. Using a series of steps, we can form the product MgO(s). The last step in the process will be:

$$Mg^{2+}(g) + O^{2-}(g) \longrightarrow MgO(s)$$

This reaction is the reverse of the reaction for the lattice energy. Knowing the enthalpy changes for the other steps, we can calculate the enthalpy change for the step above.

Solution:

Step 1: Convert solid magnesium to magnesium vapor.

$$Mg(s) \longrightarrow Mg(g) \qquad \Delta H_1^\circ = 150 \text{ kJ/mol}$$

This is the energy needed to sublime Mg(*s*).

Step 2: Dissociate ½ mole of O_2 gas into separate gaseous O atoms.

$$\tfrac{1}{2}O_2(g) \longrightarrow O(g) \qquad \Delta H_2^\circ = \left(\tfrac{1}{2}\right)(498.7 \text{ kJ/mol}) = 249.4 \text{ kJ/mol}$$

498.7 kJ is the amount of energy needed to break a mole of O=O bonds. Here we are breaking the bonds in half a mole of O_2. You can find this bond enthalpy in Table 9.4 of the text.

Step 3: Ionize 1 mole of gaseous Mg atoms.

$$Mg(g) \longrightarrow Mg^{2+}(g) + 2e^- \qquad \Delta H_3^\circ = 738.1 \text{ kJ/mol} + 1450 \text{ kJ/mol} = 2188 \text{ kJ/mol}$$

This process corresponds to the first and second ionization energies of Mg. See Table 8.2 of the text for the ionization energies.

Step 4: Add 2 moles of electrons to 1 mole of gaseous O atoms.

$$O(g) + e^- \longrightarrow O^-(g) \qquad \Delta H_{4a}^\circ = -142 \text{ kJ/mol}$$

$$O^-(g) + e^- \longrightarrow O^{2-}(aq) \qquad \Delta H_{4b}^\circ = 780 \text{ kJ/mol}$$

This process corresponds to the opposite of the electron affinity value of O and the electron affinity value of O^-. The electron affinity of O is given in Table 8.3 of the text.

Step 5: Combine 1 mole of gaseous Mg^{2+} and 1 mole of gaseous O^{2-} to form 1 mole of solid MgO.

$$Mg^{2+}(g) + O^{2-}(g) \longrightarrow MgO(s) \qquad \Delta H_5^\circ = ?$$

This is the enthalpy change that we wish to calculate because it is the opposite of the lattice energy. Changing the sign of the value that we calculate will give us the lattice energy.

Step 6: The enthalpy change for the two pathways is the same. We can write:

$$\Delta H^\circ(\text{pathway 1}) = \Delta H^\circ(\text{pathway 2})$$

$$\Delta H_f^\circ[MgO(s)] = \Delta H_1^\circ + \Delta H_2^\circ + \Delta H_3^\circ + \Delta H_{4a}^\circ + \Delta H_{4b}^\circ + \Delta H_5^\circ$$

Substituting the values from above:

$$-601.8 \text{ kJ/mol} = (150 + 249.4 + 2188 - 142 + 780)\text{ kJ/mol} + \Delta H_5^\circ$$

$$\Delta H_5^\circ = -3827 \text{ kJ/mol}$$

This enthalpy change corresponds to the reaction that is the reverse of the lattice energy. Therefore, the lattice energy will have the opposite sign of ΔH_5°.

$$MgO(s) \longrightarrow Mg^{2+}(g) + O^{2-}(g) \qquad \textbf{lattice energy} = \mathbf{+3827 \text{ kJ/mol}}$$

PRACTICE EXERCISE

2. Given that the enthalpy of sublimation of sodium (Na) is 108 kJ/mol and ΔH_f° [NaF(*s*)] = −570 kJ/mol, calculate the lattice energy of NaF(*s*). See Tables 8.2, 8.3, and 9.4 of the text for other data.

Text Problem: 9.26

PROBLEM TYPE 3: WRITING LEWIS STRUCTURES

The general rules for writing Lewis structures are given below. For more detailed rules, see Section 9.6 of the text.

1. Write the skeletal structure of the compound, using chemical symbols and placing bonded atoms next to one another. In general, the least electronegative atom occupies the central position. Hydrogen and fluorine usually occupy the terminal (end) positions in the Lewis structure.
2. Count the number of valence electrons present. For polyatomic anions, add the number of negative charges to that total. For polyatomic cations, subtract the number of positive charges from the number of valence electrons.
3. Draw a single covalent bond between the central atom and each of the surrounding atoms. Complete the octets of the atoms bonded to the central atom with lone pairs. The total number of electrons to be used is that determined in step 2.
4. Sometimes there will not be enough electrons to satisfy the octet rule of the central atom by placing lone pairs. Try adding double or triple bonds between the surrounding atoms and the central atom, using the lone pairs from the surrounding atoms.

EXAMPLE 9.3

Write the Lewis structure for SO_2.

Strategy: We follow the procedure for drawing Lewis structures outlined in Section 9.6 of the text.

Solution:

Step 1: Sulfur is less electronegative than oxygen, so it occupies the central position. The skeletal structure is:

O S O

Step 2: The outer-shell electron configurations of O and S are $2s^2 2p^4$ and $3s^2 3p^4$, respectively. Thus, there are

$$6 + (2 \times 6) = 18 \text{ valence electrons}$$

Tip: For the representative elements (Group 1A – 7A), the number of valence electrons is equal to the group number. For example, both O and S are in Group 6A; thus, they both have 6 valence electrons.

Step 3: We draw a single covalent bond between S and each O, and then complete the octets for the O atoms. We place the remaining two electrons on S.

:Ö—S̈—Ö:

Step 4: The octet rule is satisfied for the oxygen atoms; however, the S atom does *not* satisfy the octet rule. Let's try making a sulfur-to-oxygen double bond by moving a lone pair from one of the O atoms to form another bond with S.

:Ö—S̈=Ö

Now, the octet rule is also satisfied for the S atom.

Check: As a final check, we verify that there are 18 valence electrons in the Lewis structure of SO_2.

EXAMPLE 9.4

Write the Lewis structure for NO_3^-.

Strategy: We follow the procedure for drawing Lewis structures outlined in Section 9.6 of the text.

Solution:

Step 1: Nitrogen is less electronegative than oxygen, so it occupies the central position. The skeletal structure is:

```
     O

O    N    O
```

Step 2: Nitrate is a polyatomic anion, so we must add the negative charge to the number of valence electrons. The outer-shell electron configurations of O (Group 4A) and N (Group 5A) are $2s^22p^4$ and $2s^22p^3$, respectively. Thus, there are

$$5 + (3 \times 6) + 1 = 24 \text{ valence electrons}$$

Step 3: We draw a single covalent bond between N and each O, and then complete the octets for the O atoms.

```
       ..
      :O:
       |
  ..   |   ..
 :O————N————O:
  ..       ..
```

Step 4: The octet rule is satisfied for the oxygen atoms; however, the N atom does *not* satisfy the octet rule. Let's try making a nitrogen-to-oxygen double bond by moving a lone pair from one of the O atoms to form another bond with N.

```
 [                 ]−
 [      :O:        ]
 [       ||        ]
 [  ..   ||   ..   ]
 [ :O————N————O:   ]
 [  ..        ..   ]
```

Now, the octet rule is also satisfied for the N atom.

Check: As a final check, we verify that there are 24 valence electrons in the Lewis structure of NO_3^-.

Also notice that we draw a bracket with the charge of the polyatomic ion around the Lewis structure. This is to distinguish an ion from a neutral molecule.

PRACTICE EXERCISES

3. Write the Lewis structure for $AsCl_3$.

4. Write the Lewis structure for cyanide ion, CN^-.

Text Problems: 9.44, **9.46**, 9.48

PROBLEM TYPE 4: FORMAL CHARGES

By comparing the number of electrons in an isolated atom (valence electrons) with the number of electrons that are associated with the same atom in a Lewis structure, we can determine the distribution of electrons in the molecule and draw the most plausible Lewis structure. This difference between the valence electrons in an isolated atom and the number of electrons assigned to that atom in a Lewis structure is called the atom's **formal charge**.

To assign the number of electrons on an atom in a Lewis structure, we proceed as:

- All the atom's nonbonding electrons are assigned to the atom.
- Half of the bonding electrons are assigned to the atom. For example, a single bond is a two-electron bond. One electron from the bond would be assigned to the given atom.

When you write formal charges, the following rules are helpful.

- For neutral molecules, the sum of the formal charges must add up to zero.
- For cations, the sum of the formal charges must equal the positive charge.
- For anions, the sum of the formal charges must equal the negative charge.

A. Assigning formal charges

EXAMPLE 9.5
Assign formal charges to the atoms in the following Lewis structures.

(a) :C≡O:

(b) :Ö—S̈=Ö (with lone pairs: O has three pairs, S has one pair, second O has two pairs)

Strategy: Use the approach discussed above to assign formal charges.

Solution:

(a) The formal charge on each atom can be calculated using the following approach:

:C≡O:

	C	O
Valence e^-	4	6
e^- assigned to atom	5	5
Difference (formal charge)	1−	1+

Some chemists do not approve of this structure for CO because it places a positive formal charge on the more electronegative oxygen atom.

(b) The formal charge on each atom can be calculated using the following approach:

:Ö—S̈=Ö

	O	S	O
Valence e^-	6	6	6
e^- assigned to atom	7	5	6
Difference (formal charge)	1−	1+	0

PRACTICE EXERCISE

5. Assign formal charges to the atoms in the following Lewis structures.

(a) $:\ddot{N}=N=\ddot{O}:$

(b) $[:\ddot{\underset{..}{O}}-H]^{-}$

Text Problems: 9.54, 9.56

B. Choosing the most plausible Lewis structure based on formal charges

The following guidelines show how to use formal charges to select a plausible Lewis structure for a given compound.

- For neutral molecules, a Lewis structure in which there are no formal charges is preferable to one in which formal charges are present.
- Lewis structures with large formal charges are less plausible than those with small formal charges.
- When comparing two structures with similar magnitudes of formal charges, the most plausible structure is the one in which negative formal charges are placed on the more electronegative atoms.

EXAMPLE 9.6

Two possible Lewis structures for BF_3 are shown below. Which is the more reasonable structure in terms of formal charges?

(a) :F̈–B–F̈: with :F̈: bonded to B

(b) :F̈–B=F̈ with :F̈: bonded to B

The formal charge on each atom in (a) can be calculated using the following approach:

B atom:	formal charge = 3 valence e^- − 3 e^- assigned to the atom = 0
F atoms:	formal charge = 7 valence e^- − 7 e^- assigned to the atom = 0

The formal charge on each atom in (b) can be calculated using the following approach:

B atom:	formal charge = 3 valence e^- − 4 e^- assigned to the atom = 1−
F atom (double bond):	formal charge = 7 valence e^- − 6 e^- assigned to the atom = 1+
F atoms (single bond):	formal charge = 7 valence e^- − 7 e^- assigned to the atom = 0

The rule used to establish the most plausible structure is: For neutral molecules, a Lewis structure in which there are no formal charges is preferable to one in which formal charges are present. Thus, structure (a) is preferred over structure (b). We could also rule out structure (b) as the most plausible structure because there is a positive formal charge on the very electronegative F atom.

PRACTICE EXERCISE

6. Consider the following Lewis structures for the sulfate ion. Assign formal charges to each atom in the structure, and then determine which structure is more reasonable based on formal charges?

```
      ..              2-          ..              2-
   [ :O: ]                     [ :O: ]
  ..  |  ..                   ..  |  ..
 :O—S—O:                      O=S=O
  ..  |  ..                   ..  |  ..
     :O:                         :O:
      ..                          ..
     (a)                         (b)
```

Text Problem: 9.62

PROBLEM TYPE 5: DRAWING RESONANCE STRUCTURES

The Lewis structure for SO_3 is shown below.

```
     :O:
  ..  ||  ..
 :O—S—O:
  ..      ..
```

We can draw two more equivalent Lewis structures with the double bond between S and a different oxygen atom.

```
      ..               ..
     :O:              :O:
  ..  |  ..        ..  |  ..
  O=S—O:          :O—S=O
  ..      ..       ..      ..
```

Which is the correct structure? Let's consider experimental data. We would expect the S–O bond to be longer than the S=O bond because double bonds are known to be shorter than single bonds. Yet experimental evidence shows that all three sulfur-to-oxygen bonds are equal in length. Therefore, none of the three structures shown accurately represents the molecule. We resolve this conflict by using all *three* Lewis structures to represent SO_3.

```
                         ..                 ..
     :O:                :O:                :O:
  ..  ||  ..         ..  |  ..          ..  |  ..
 :O—S—O:   ⟷        O=S—O:   ⟷        :O—S=O
  ..      ..         ..      ..         ..      ..
```

Each of the three structures is called a **resonance structure**. A resonance structure is one of two or more Lewis structures for a single molecule that cannot be described fully with only one Lewis structure. The symbol ⟷ indicates that the structures shown are resonance structures.

EXAMPLE 9.7

We drew the Lewis structure for nitrate ion, NO_3^-, in Example 9.4. However, experimental evidence shows that all N–O bonds are equivalent. Draw resonance structures to indicate the equivalence of the N–O bonds.

Strategy: We follow the procedure for drawing Lewis structures outlined in Section 9.6 of the text. After we complete the Lewis structure, we draw the resonance structures.

Solution: The Lewis structure drawn in Example 9.4 shown with formal charges is:

```
        :O:
         ||+
 -:Ö—N—Ö:-
```

This structure, while a correct Lewis structure, does not show the equivalence of all three N–O bonds. Three contributing resonance structures can be drawn.

```
   :Ö:-                :O:                :Ö:-
    |+                  ||+                 |+
Ö=N—Ö:-   ⟷   -:Ö—N—Ö:-   ⟷   -:Ö—N=Ö
```

Resonance does not mean that the nitrate ion shifts quickly back and forth from one resonance structure to the other. Keep in mind that *none* of the resonance structures adequately represents the actual molecule, which has its own unique, stable structure. The actual structure is an average or hybrid of the above three structures. Resonance is a human invention, designed to address the limitations in these simple bonding models.

PRACTICE EXERCISE

7. Draw all the resonance structures for N_2O. The skeletal structure is N–N–O.

Text Problems: 9.52, 9.54, 9.56

PROBLEM TYPE 6: EXCEPTIONS TO THE OCTET RULE

A. The incomplete octet

In some compounds the number of electrons surrounding the central atom in a stable molecule is fewer than eight. **Be**, **B**, and **Al** tend to form compounds in which they are surrounded by fewer than eight electrons.

EXAMPLE 9.8

Draw the Lewis structure for GaI_3.

Strategy: We follow the procedure outlined in Section 9.6 of the text for drawing Lewis structures.

Solution:

Step 1: Gallium is less electronegative than iodine, so it occupies the central position. The skeletal structure is:

```
    I
I  Ga  I
```

Step 2: The outer-shell electron configurations of Ga (Group IIIA) and I (Group VIIA) are $4s^2 4p^1$ and $5s^2 5p^5$, respectively. Thus, there are

$$3 + (3 \times 7) = 24 \text{ valence electrons}$$

Step 3: We draw a single covalent bond between Ga and each I, and then complete the octets for the I atoms.

```
       :Ï:
        |
  :Ï—Ga—Ï:
```

The octet rule is satisfied for the iodine atoms; however, the Ga atom does *not* satisfy the octet rule. A resonance structure with a double bond between Ga and I can be drawn that satisfies the octet rule for Ga. However, the properties of GaI_3 are more consistent with a Lewis structure in which there are single bonds between Ga and each I, as shown above.

Based on formal charges, is the structure shown above more plausible than the structure that contains one double bond between Ga and an I atom?

PRACTICE EXERCISE

8. Write the Lewis structure for BCl_3.

Text Problems: 9.62, 9.66

B. Odd-electron molecules

Some molecules contain an **odd** number of electrons. To satisfy the octet rule, we need an even number of electrons. Therefore, the octet rule cannot be satisfied in a molecule that has an odd number of electrons. When drawing a Lewis structure, if an atom has fewer than eight electrons, make sure that it is the least electronegative atom in the compound.

EXAMPLE 9.9

Draw the Lewis structure for NO_2 (all bonds are equivalent).

Strategy: We follow the procedure outlined in Section 9.6 of the text for drawing Lewis structures.

Solution:

Step 1: Nitrogen is less electronegative than oxygen, so it occupies the central position. The skeletal structure is:

O N O

Step 2: The outer-shell electron configurations of N (Group 5A) and O (Group 6A) are $2s^22p^3$ and $2s^22p^4$, respectively. Thus, there are

$$5 + (2 \times 6) = 17 \text{ valence electrons}$$

Step 3: We have an odd number of valence electrons, so the octet rule cannot be satisfied for at least one of the atoms in the molecule. Either nitrogen or oxygen in the structure will have fewer than eight electrons, because a second-row element cannot exceed an octet of electrons. Since N is less electronegative than O, nitrogen should be electron deficient. We draw a single covalent bond between N and each O, and then complete the octets for the O atoms.

:Ö—N—Ö:

Step 4: We have one electron left to place on the molecule. Placing the electron on N only gives five electrons around N. However, if we make a nitrogen-to-oxygen double bond by moving a lone pair from one of the O atoms to form another bond with N, nitrogen will be surrounded by seven electrons.

:Ö—Ṅ=Ö

We could also draw resonance structures in which oxygen is electron deficient with seven electrons. However, the above structure is the most plausible since the least electronegative element is electron deficient.

C. The expanded octet

A number of compounds contain more than eight electrons around an atom. These **expanded octets** only occur for atoms of elements in and beyond the third period of the periodic table. Atoms from the third period on can accommodate more than eight electrons because they also have *d* orbitals that can be used in bonding.

EXAMPLE 9.10

Draw the Lewis structure for ClF_3.

Strategy: We follow the procedure outlined in Section 9.6 of the text for drawing Lewis structures.

Solution:

Step 1: Chlorine is less electronegative than fluorine, so it occupies the central position. The skeletal structure is:

```
     F

  F  Cl  F
```

Step 2: The outer-shell electron configurations of F and Cl are $2s^22p^5$ and $3s^23p^5$, respectively. Thus, there are

$$7 + (3 \times 7) = 28 \text{ valence electrons}$$

Step 3: We draw a single covalent bond between Cl and each F, and then complete the octets for the F atoms.

```
        ..
       :F:
   ..   |   ..
  :F — Cl — F:
   ..       ..
```

Step 4: At this point, we still have two electron pairs (4 e^-) to place. Fluorine cannot exceed an octet of electrons, so the electrons must be placed as lone pairs on chlorine. The correct Lewis structure is:

```
        ..
       :F:
   ..   |   ..
  :F — Cl — F:
   ..  .. ..  ..
```

Chlorine can exceed an octet of electrons. An expanded octet can occur for atoms of elements in and beyond the third period of the periodic table.

Check: As a final check, we verify that there are 28 valence electrons in the Lewis structure of ClF_3.

PRACTICE EXERCISE

9. Write the Lewis structure for PCl_5.

Text Problem: 9.64

PROBLEM TYPE 7: USING BOND ENTHALPIES TO ESTIMATE THE ENTHALPY OF A REACTION

A quantitative measure of the stability of a molecule is its **bond enthalpy**, which is the enthalpy change required to break a particular bond in one mole of gaseous molecules. Table 9.4 of the text lists the average bond enthalpies of a number of bonds found in polyatomic molecules, as well as the bond enthalpies of several diatomic molecules.

In many cases, it is possible to predict the approximate enthalpy of reaction by using the average bond enthalpies. Energy *is always required* to break chemical bonds and chemical bond formation is always accompanied by a *release of energy*. We can estimate the enthalpy of reaction by counting the total number of bonds broken and formed in the reaction and recording all the corresponding energy changes. The enthalpy of reaction in the *gas phase* is given by:

$$\Delta H° = \text{total energy input} - \text{total energy released}$$

$$\Delta H° = \Sigma\text{BE(reactants)} - \Sigma\text{BE(products)}$$

Where,

BE is the average bond enthalpy
Σ represents summation

If the total energy input is greater than the total energy released, $\Delta H°$ *is positive* and the reaction is *endothermic*. On the other hand, if more energy is released than absorbed, $\Delta H°$ *is negative* and the reaction is *exothermic*.

EXAMPLE 9.11

Use average bond enthalpies to estimate $\Delta H°$ for the following reaction:

$$Cl_2(g) + I_2(g) \longrightarrow 2ICl(g)$$

The average I–Cl bond enthalpy is 210 kJ.

Strategy: Keep in mind that bond breaking is an energy absorbing (endothermic) process and bond making is an energy releasing (exothermic) process. Therefore, the overall energy change is the difference between these two opposing processes, as described in Equation (9.3) of the text.

Solution:

$$\Delta H° = \Sigma\text{BE(reactants)} - \Sigma\text{BE(products)}$$

$$\Delta H° = \text{BE(Cl–Cl)} + \text{BE(I–I)} - 2\text{BE(I–Cl)}$$

$$= 242.7 \text{ kJ/mol} + 151.0 \text{ kJ/mol} - 2(210 \text{ kJ/mol})$$

$$\mathbf{\Delta H° = -26 \text{ kJ/mol}}$$

Is the reaction exothermic or endothermic?

PRACTICE EXERCISE

10. Estimate the enthalpy change for the following reaction:

$$N_2(g) + O_2(g) \longrightarrow 2NO(g)$$

Hint: NO has a double bond [BE(N=O) = 630 kJ/mol]

Text Problems: 9.70, 9.72

ANSWERS TO PRACTICE EXERCISES

1. **(a)** polar covalent **(b)** covalent **(c)** ionic

2. lattice energy (NaF) = 919 kJ/mol

3. $AsCl_3$ Lewis structure: :C̤̈l–Äs–C̤̈l: with a third :C̈l: bonded above As (each Cl with three lone pairs; As with one lone pair)

4. $[:C{\equiv}N:]^-$

5. **(a)** Formal charge (left N) = 5 valence e^- – 6 e^- assigned to the atom = 1–

Formal charge (middle N) = 5 valence e^- – 4 e^- assigned to the atom = 1+

Formal charge (O) = 6 valence e^- – 6 e^- assigned to the atom = 0

(b) Formal charge (O) = 6 valence e^- – 7 e^- assigned to the atom = 1–

Formal charge (H) = 1 valence electron – 1 e^- assigned to the atom = 0

6. (b) is the more plausible structure based on formal charges. There is a large positive (+2) formal charge on the S atom in structure (a). The formal charge on the S atom in structure (b) is zero.

7. $\overset{-}{\text{:}\ddot{N}}=\overset{+}{N}=\ddot{O}\text{:} \longleftrightarrow \text{:}N\equiv\overset{+}{N}-\overset{-}{\text{:}\ddot{O}\text{:}} \longleftrightarrow \overset{2-}{\text{:}\ddot{N}}-\overset{+}{N}\equiv\overset{+}{O}\text{:}$

8. $\text{:}\ddot{Cl}-B(-\text{:}\ddot{Cl}\text{:})-\ddot{Cl}\text{:}$ (BCl_3, each Cl with three lone pairs)

9. The lone pairs have been left off the chlorine atoms.

PCl_5: P bonded to five Cl atoms (Cl–P–Cl with Cl above and two Cl below)

10. $\Delta H° = 180$ kJ/mol

SOLUTIONS TO SELECTED TEXT PROBLEMS

9.16 **(a)** RbI, rubidium iodide **(b)** Cs_2SO_4, cesium sulfate

(c) Sr_3N_2, strontium nitride **(d)** Al_2S_3, aluminum sulfide

9.18 The Lewis representations for the reactions are:

(a) $\dot{\text{Sr}}\cdot + \cdot\ddot{\text{Se}}\cdot \longrightarrow Sr^{2+}\ :\ddot{\text{Se}}:^{2-}$

(b) $\dot{\text{Ca}}\cdot + 2\ \text{H}\cdot \longrightarrow Ca^{2+}\ 2\text{H}:^{-}$

(c) $3\text{Li}\cdot + \cdot\dot{\ddot{\text{N}}}\cdot \longrightarrow 3Li^{+}\ :\ddot{\text{N}}:^{3-}$

(d) $2\dot{\text{Al}}\cdot + 3\cdot\ddot{\text{S}}\cdot \longrightarrow 2Al^{3+}\ 3:\ddot{\text{S}}:^{2-}$

9.20 **(a)** Covalent (BF_3, boron trifluoride) **(b)** ionic (KBr, potassium bromide)

9.26

(1) $Ca(s) \rightarrow Ca(g)$	$\Delta H_1^\circ = 121$ kJ/mol
(2) $Cl_2(g) \rightarrow 2Cl(g)$	$\Delta H_2^\circ = 242.8$ kJ/mol
(3) $Ca(g) \rightarrow Ca^+(g) + e^-$	$\Delta H_3^{\circ'} = 589.5$ kJ/mol
$Ca^+(g) \rightarrow Ca^{2+}(g) + e^-$	$\Delta H_3^{\circ''} = 1145$ kJ/mol
(4) $2[Cl(g) + e^- \rightarrow Cl^-(g)]$	$\Delta H_4^\circ = 2(-349\text{ kJ/mol}) = -698$ kJ/mol
(5) $Ca^{2+}(g) + 2Cl^-(g) \rightarrow CaCl_2(s)$	$\Delta H_5^\circ = ?$
$Ca(s) + Cl_2(g) \rightarrow CaCl_2(s)$	$\Delta H_{overall}^\circ = -795$ kJ/mol

Thus we write:

$$\Delta H_{overall}^\circ = \Delta H_1^\circ + \Delta H_2^\circ + \Delta H_3^{\circ'} + \Delta H_3^{\circ''} + \Delta H_4^\circ + \Delta H_5^\circ$$

$$\Delta H_5^\circ = (-795 - 121 - 242.8 - 589.5 - 1145 + 698)\text{kJ/mol} = -2195\text{ kJ/mol}$$

The lattice energy is represented by the reverse of equation (5); therefore, the lattice energy is **+2195 kJ/mol**.

9.36 Classifying Chemical Bonds, Problem Type 1.

Strategy: We can look up electronegativity values in Figure 9.5 of the text. The amount of ionic character is based on the electronegativity difference between the two atoms. The larger the electronegativity difference, the greater the ionic character.

Solution: Let ΔEN = electronegativity difference. The bonds arranged in order of increasing ionic character are:

C–H (ΔEN = 0.4) < Br–H (ΔEN = 0.7) < F–H (ΔEN = 1.9) < Li–Cl (ΔEN = 2.0)

< Na–Cl (ΔEN = 2.1) < K–F (ΔEN = 3.2)

9.38 The order of increasing ionic character is:

Cl–Cl (zero difference in electronegativity) < Br–Cl (difference 0.2) < Si–C (difference 0.7) < Cs–F (difference 3.3).

9.40
(a) The two silicon atoms are the same. The bond is covalent.
(b) The electronegativity difference between Cl and Si is 3.0 – 1.8 = 1.2. The bond is polar covalent.
(c) The electronegativity difference between F and Ca is 4.0 – 1.0 = 3.0. The bond is ionic.
(d) The electronegativity difference between N and H is 3.0 – 2.1 = 0.9. The bond is polar covalent.

9.44

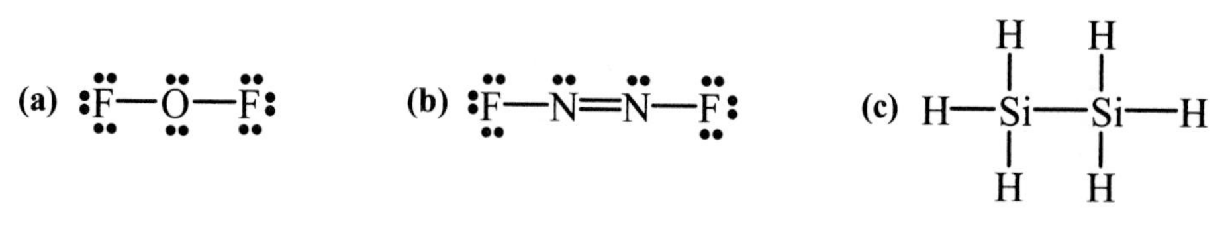

(d) $^{-}$:Ö—H

(e) H—C(H)(:Cl:)—C(=:O:)—Ö:$^{-}$

(f) H—C(H)(H)—N^{+}(H)(H)—H

9.46 Writing Lewis Structures, Problem Type 3.

Strategy: We follow the procedure for drawing Lewis structures outlined in Section 9.6 of the text.

Solution:

(a)
Step 1: It is obvious that the skeletal structure is: O O

Step 2: The outer-shell electron configuration of O is $2s^2 2p^4$. Also, we must add the negative charges to the number of valence electrons, Thus, there are

$$(2 \times 6) + 2 = 14 \text{ valence electrons}$$

Step 3: We draw a single covalent bond between each O, and then attempt to complete the octets for the O atoms.

$^{-}$:Ö—Ö:$^{-}$

Because this structure satisfies the octet rule for both oxygen atoms, step 4 outlined in the text is not required.

Check: As a final check, we verify that there are 14 valence electrons in the Lewis structure of O_2^{-}.

Follow the same procedure as part (a) for parts (b), (c), and (d). The appropriate Lewis structures are:

(b) $^{-}$:C≡C:$^{-}$ (c) :N≡O:$^{+}$ (d) H—N^{+}(H)(H)—H

9.48 **(a)** Neither oxygen atom has a complete octet. The left-most hydrogen atom is forming two bonds (4 e^-). Hydrogen can only be surrounded by at most two electrons.

(b) The correct structure is:

```
    H :O:
    |  ||  ..
H—C—C—O—H
    |      ..
    H
```

Do the two structures have the same number of electrons? Is the octet rule satisfied for all atoms other than hydrogen, which should have a duet of electrons?

9.52 Drawing Resonance Structures, Problem Type 5.

Strategy: We follow the procedure for drawing Lewis structures outlined in Section 9.6 of the text. After we complete the Lewis structure, we draw the resonance structures.

Solution: Following the procedure in Section 9.6 of the text, we come up with the following Lewis structure for ClO_3^-.

```
        :O:
         ||+
 -..          ..-
:O—Cl—O:
  ..    ..    ..
```

We can draw two more equivalent Lewis structures with the double bond between Cl and a different oxygen atom.

The resonance structures with formal charges are as follows:

```
    ..-                       :O:                        ..-
   :O:                         ||+                      :O:
    |+  ..-              -..         ..-          -..    |+
O=Cl—O:      ←——→   :O—Cl—O:    ←——→   :O—Cl=O
..   ..   ..            ..   ..   ..             ..   ..   ..
```

9.54 The structures of the most important resonance forms are:

```
      H                       H
      |   +  -                -|   +
H—C=N=N:   ←——→   H—C—N≡N:
           ..                ..
```

9.56 Three reasonable resonance structures with the formal charges indicated are

```
-..  +    ..         +   ..-       2-..  +   +
 N=N=O  ←→ :N≡N—O:  ←→  :N—N≡O:
 ..       ..               ..             ..
```

9.62 The incomplete octet, Problem Type 6A.

Strategy: We follow the procedure outlined in Section 9.6 of the text for drawing Lewis structures. We assign formal charges as discussed in Section 9.7 of the text.

Solution: Drawing the structure with single bonds between Be and each of the Cl atoms, the octet rule for Be is *not* satisfied. The Lewis structure is:

```
 ..          ..
:Cl—Be—Cl:
 ..          ..
```

An octet of electrons on Be can only be formed by making two double bonds as shown below:

$$\overset{+}{\ddot{\mathrm{Cl}}}{=}\overset{2-}{\mathrm{Be}}{=}\overset{+}{\ddot{\mathrm{Cl}}}$$

This places a high negative formal charge on Be and positive formal charges on the Cl atoms. This structure distributes the formal charges counter to the electronegativities of the elements. It is not a plausible Lewis structure.

9.64 The outer electron configuration of antimony is $5s^2 5p^3$. The Lewis structure is shown below. All five valence electrons are shared in the five covalent bonds. The octet rule is not obeyed. (The electrons on the chlorine atoms have been omitted for clarity.)

Cl
|
Cl—Sb—Cl
Cl / \ Cl

Can Sb have an expanded octet?

9.66 The reaction can be represented as:

$$:\ddot{\mathrm{Cl}}-\underset{\underset{:\underset{\cdot\cdot}{\mathrm{Cl}}:}{|}}{\mathrm{Al}}-\ddot{\mathrm{Cl}}: \quad + \quad :\ddot{\mathrm{Cl}}:^{-} \longrightarrow :\ddot{\mathrm{Cl}}-\overset{\overset{:\ddot{\mathrm{Cl}}:}{|}}{\underset{\underset{:\underset{\cdot\cdot}{\mathrm{Cl}}:}{|}}{\mathrm{Al}^{-}}}-\ddot{\mathrm{Cl}}:$$

The new bond formed is called a **coordinate covalent bond.**

9.70 This problem is similar to Problem Type 7, Using Bond Enthalpies to Estimate the Enthalpy of a Reaction.

Strategy: Keep in mind that bond breaking is an energy absorbing (endothermic) process and bond making is an energy releasing (exothermic) process. Therefore, the overall energy change is the difference between these two opposing processes, as described in Equation (9.3) of the text.

Solution: There are two oxygen-to-oxygen bonds in ozone. We will represent these bonds as O–O. However, these bonds might not be true oxygen-to-oxygen single bonds. Using Equation (9.3) of the text, we write:

$$\Delta H° = \Sigma BE(\text{reactants}) - \Sigma BE(\text{products})$$

$$\Delta H° = BE(\text{O=O}) - 2BE(\text{O–O})$$

In the problem, we are given $\Delta H°$ for the reaction, and we can look up the O=O bond enthalpy in Table 9.4 of the text. Solving for the average bond enthalpy in ozone,

$$-2BE(\text{O–O}) = \Delta H° - BE(\text{O=O})$$

$$BE(\text{O}-\text{O}) = \frac{BE(\text{O=O}) - \Delta H°}{2} = \frac{498.7 \text{ kJ/mol} + 107.2 \text{ kJ/mol}}{2} = \mathbf{303.0 \text{ kJ/mol}}$$

Considering the resonance structures for ozone, is it expected that the O–O bond enthalpy in ozone is between the single O–O bond enthalpy (142 kJ) and the double O=O bond enthalpy (498.7 kJ)?

9.72 **(a)**

Bonds Broken	*Number Broken*	*Bond Enthalpy* (kJ/mol)	*Enthalpy Change* (kJ)
C − H	12	414	4968
C − C	2	347	694
O = O	7	498.7	3491

Bonds Formed	*Number Formed*	*Bond Enthalpy* (kJ/mol)	*Enthalpy Change* (kJ)
C = O	8	799	6392
O − H	12	460	5520

ΔH° = total energy input − total energy released

$$= (4968 + 694 + 3491) - (6392 + 5520) = \mathbf{-2759\ kJ/mol}$$

(b) $\Delta H^\circ = 4\Delta H_f^\circ(CO_2) + 6\Delta H_f^\circ(H_2O) - [2\Delta H_f^\circ(C_2H_6) + 7\Delta H_f^\circ(O_2)]$

$$\Delta H^\circ = (4)(-393.5\text{ kJ/mol}) + (6)(-241.8\text{ kJ/mol}) - [(2)(-84.7\text{ kJ/mol}) + (7)(0)] = \mathbf{-2855\ kJ/mol}$$

The answers for part (a) and (b) are different, because *average* bond enthalpies are used for part (a).

9.74 Typically, ionic compounds are composed of a metal cation and a nonmetal anion. RbCl and KO_2 are ionic compounds.

Typically, covalent compounds are composed of two nonmetals. PF_5, BrF_3, and CI_4 are covalent compounds.

9.76 Recall that you can classify bonds as ionic or covalent based on electronegativity difference.

The melting points (°C) are shown in parentheses following the formulas.

Ionic:	NaF (993)	MgF_2 (1261)	AlF_3 (1291)	
Covalent:	SiF_4 (−90.2)	PF_5 (−83)	SF_6 (−121)	ClF_3 (−83)

Is there any correlation between ionic character and melting point?

9.78 KF is an ionic compound. It is a solid at room temperature made up of K^+ and F^- ions. It has a high melting point, and it is a strong electrolyte. Benzene, C_6H_6, is a covalent compound that exists as discrete molecules. It is a liquid at room temperature. It has a low melting point, is insoluble in water, and is a nonelectrolyte.

9.80 The resonance structures are:

$$^{-}\ddot{\underset{\cdot\cdot}{N}}=\overset{+}{N}=\ddot{\underset{\cdot\cdot}{N}}^{-} \longleftrightarrow :N\equiv\overset{+}{N}-\ddot{\underset{\cdot\cdot}{N}}:^{2-} \longleftrightarrow {}^{2-}:\ddot{\underset{\cdot\cdot}{N}}-\overset{+}{N}\equiv N:$$

Which is the most plausible structure based on a formal charge argument?

9.82 **(a)** An example of an aluminum species that satisfies the octet rule is the anion $AlCl_4^-$. The Lewis dot structure is drawn in Problem 9.66.

(b) An example of an aluminum species containing an expanded octet is anion AlF_6^{3-}. (How many pairs of electrons surround the central atom?)

(c) An aluminum species that has an incomplete octet is the compound $AlCl_3$. The dot structure is given in Problem 9.66.

9.84 CF_2 would be very unstable because carbon does not have an octet. (How many electrons does it have?)

LiO_2 would not be stable because the lattice energy between Li^+ and superoxide O_2^- would be too low to stabilize the solid.

$CsCl_2$ requires a Cs^{2+} cation. The second ionization energy is too large to be compensated by the increase in lattice energy.

PI_5 appears to be a reasonable species (compared to PF_5 in Example 9.10 of the text). However, the iodine atoms are too large to have five of them "fit" around a single P atom.

9.86 **(a)** false **(b)** true **(c)** false **(d)** false

For question (c), what is an example of a second-period species that violates the octet rule?

9.88 The formation of CH_4 from its elements is:

$$C(s) + 2H_2(g) \longrightarrow CH_4(g)$$

The reaction could take place in two steps:

Step 1: $C(s) + 2H_2(g) \longrightarrow C(g) + 4H(g)$ $\quad \Delta H^\circ_{rxn} = (716 + 872.8)\text{kJ/mol} = 1589 \text{ kJ/mol}$

Step 2: $C(g) + 4H(g) \longrightarrow CH_4(g)$ $\quad \Delta H^\circ_{rxn} \approx -4 \times (\text{bond energy of C} - \text{H bond})$

$$= -4 \times 414 \text{ kJ/mol} = -1656 \text{ kJ/mol}$$

Therefore, $\Delta H^\circ_f(CH_4)$ would be approximately the sum of the enthalpy changes for the two steps. See Section 6.6 of the text (Hess's law).

$$\Delta H^\circ_f(CH_4) = \Delta H^\circ_{rxn}(1) + \Delta H^\circ_{rxn}(2)$$

$$\Delta H^\circ_f(CH_4) = (1589 - 1656)\text{kJ/mol} = \mathbf{-67 \text{ kJ/mol}}$$

The actual value of $\Delta H^\circ_f(CH_4) = -74.85$ kJ/mol.

9.90 Only $\mathbf{N_2}$ has a triple bond. Therefore, it has the shortest bond length.

9.92 To be isoelectronic, molecules must have the same number and arrangement of valence electrons. NH_4^+ and CH_4 are isoelectronic (8 valence electrons), as are CO and N_2 (10 valence electrons), as are $B_3N_3H_6$ and C_6H_6 (30 valence electrons). Draw Lewis structures to convince yourself that the electron arrangements are the same in each isoelectronic pair.

9.94 The reaction can be represented as:

$$H-\ddot{N}:^- (-H) + H-\ddot{O}: (-H) \longrightarrow H-\ddot{N}-H (-H) + :\ddot{O}-H^-$$

9.96 The central iodine atom in I_3^- has *ten* electrons surrounding it: two bonding pairs and three lone pairs. The central iodine has an expanded octet. Elements in the second period such as fluorine cannot have an expanded octet as would be required for F_3^-.

9.98 The skeletal structure is:

```
     H
   H C N C O
     H
```

The number of valence electron is: $(1 \times 3) + (2 \times 4) + 5 + 6 = 22$ valence electrons

We can draw two resonance structures for methyl isocyanate.

```
    H                               H
    |   ..       ..                 |   +        ..  -
H — C — N = C = O     <——>     H — C — N ≡ C — O :
    |            ..                 |            ..
    H                               H
```

9.100 **(a)** This is a very good resonance form; there are no formal charges and each atom satisfies the octet rule.

(b) This is a second choice after (a) because of the positive formal charge on the oxygen (high electronegativity).

(c) This is a poor choice for several reasons. The formal charges are placed counter to the electronegativities of C and O, the oxygen atom does not have an octet, and there is no bond between that oxygen and carbon!

(d) This is a mediocre choice because of the large formal charge and lack of an octet on carbon.

9.102 The nonbonding electron pairs around Cl and F are omitted for simplicity.

```
     Cl          Cl          F          F  H
     |           |           |          |  |
 F — C — Cl  F — C — F   H — C — F  F — C — C — F
     |           |           |          |  |
     Cl          Cl          Cl         F  F
```

9.104 **(a)** Using Equation (9.3) of the text,

$$\Delta H = \Sigma BE(\text{reactants}) - \Sigma BE(\text{products})$$

$$\boldsymbol{\Delta H} = [(436.4 + 151.0) - 2(298.3)] = \mathbf{-9.2\ kJ/mol}$$

(b) Using Equation (6.18) of the text,

$$\Delta H^\circ = 2\Delta H_f^\circ[\text{HI}(g)] - \{\Delta H_f^\circ[\text{H}_2(g)] + \Delta H_f^\circ[\text{I}_2(g)]\}$$

$$\boldsymbol{\Delta H^\circ} = (2)(25.9\text{ kJ/mol}) - [(0) + (1)(61.0\text{ kJ/mol})] = \mathbf{-9.2\ kJ/mol}$$

9.106 The Lewis structures are:

(a) $:\overset{-}{C}\equiv\overset{+}{O}:$ **(b)** $:N\equiv\overset{+}{O}:$ **(c)** $:\overset{-}{C}\equiv N:$ **(d)** $:N\equiv N:$

9.108 True. Each noble gas atom already has completely filled *ns* and *np* subshells.

9.110 **(a)** The bond enthalpy of F_2^- is the energy required to break up F_2^- into an F atom and an F^- ion.

$$F_2^-(g) \longrightarrow F(g) + F^-(g)$$

We can arrange the equations given in the problem so that they add up to the above equation. See Section 6.6 of the text (Hess's law).

$$\begin{array}{lll} F_2^-(g) \longrightarrow \cancel{F_2}(g) + \cancel{e^-} & & \Delta H^\circ = 290 \text{ kJ/mol} \\ \cancel{F_2}(g) \longrightarrow \cancel{2}F(g) & & \Delta H^\circ = 156.9 \text{ kJ/mol} \\ \cancel{F}(g) + \cancel{e^-} \longrightarrow F^-(g) & & \Delta H^\circ = -333 \text{ kJ/mol} \\ \hline F_2^-(g) \longrightarrow F(g) + F^-(g) & & \end{array}$$

The bond enthalpy of F_2^- is the sum of the enthalpies of reaction.

$$\boldsymbol{BE(F_2^-)} = [290 + 156.9 + (-333 \text{ kJ})]\text{kJ/mol} = \mathbf{114\ kJ/mol}$$

(b) The bond in F_2^- is weaker (114 kJ/mol) than the bond in F_2 (156.9 kJ/mol), because the extra electron increases repulsion between the F atoms.

9.112 In **(a)** there is a lone pair on the C atom and the negative formal charge is on the less electronegative C atom.

9.114 **(a)** $:\dot{N}=\ddot{O}$ ⟷ $^-:\ddot{N}=\dot{O}^+$

The first structure is the most important. Both N and O have formal charges of zero. In the second structure, the more electronegative oxygen atom has a formal charge of +1. Having a positive formal charge on an highly electronegative atom is not favorable. In addition, both structures leave one atom with an incomplete octet. This cannot be avoided due to the odd number of electrons.

(b) It is not possible to draw a structure with a triple bond between N and O.

$:N\equiv\dot{O}$

Any structure drawn with a triple bond will lead to an expanded octet. Elements in the second row of the period table cannot exceed the octet rule.

9.116 The OCOO structure violates the octet rule (expanded octet). The structure shown below satisfies the octet rule with 22 valence electrons. However, CO_3^{2-} has 24 valence electrons. Adding two more electrons to the structure would cause at least one atom to exceed the octet rule.

$\ddot{O}=C=\ddot{O}-\ddot{O}:$

9.118

Al_2Cl_6 Lewis structure: two Al atoms, each bonded to two terminal :Cl: atoms, bridged by two Cl atoms (Cl → Al arrows).

The arrows indicate coordinate covalent bonds.

9.120 There are four C–H bonds in CH_4, so the average bond enthalpy of a C–H bond is:

$$\frac{1656 \text{ kJ/mol}}{4} = 414 \text{ kJ/mol}$$

The Lewis structure of propane is:

```
     H   H   H
     |   |   |
  H—C—C—C—H
     |   |   |
     H   H   H
```

There are eight C–H bonds and two C–C bonds. We write:

$$8(\text{C–H}) + 2(\text{C–C}) = 4006 \text{ kJ/mol}$$

$$8(414 \text{ kJ/mol}) + 2(\text{C–C}) = 4006 \text{ kJ/mol}$$

$$2(\text{C–C}) = 694 \text{ kJ/mol}$$

So, the average bond enthalpy of a C–C bond is: $\frac{694}{2}$ kJ/mol = **347 kJ/mol**

9.122

(a)

```
  H    H
  |    |
  C == C
  |    |
  H    Cl
```

(b)

```
     H    H    H    H    H    H
     |    |    |    |    |    |
  ——C——C——C——C——C——C——
     |    |    |    |    |    |
     H    Cl   H    Cl   H    Cl
```

(c) In the formation of poly(vinyl chloride) form vinyl chloride, for every C=C double bond broken, 2 C–C single bonds are formed. No other bonds are broken or formed. The energy changes for 1 mole of vinyl chloride reacted are:

total energy input (breaking C=C bonds) = 620 kJ

total energy released (forming C–C bonds) = 2 × 347 kJ = 694 kJ

ΔH° = 620 kJ – 694 kJ = –74 kJ

The negative sign shows that this is an exothermic reaction. To find the total heat released when 1.0×10^3 kg of vinyl chloride react, we proceed as follows:

$$\text{heat released} = (1.0 \times 10^6 \text{ g } C_2H_3Cl) \times \frac{1 \text{ mol } C_2H_3Cl}{62.49 \text{ g } C_2H_3Cl} \times \frac{-74 \text{ kJ}}{1 \text{ mol } C_2H_3Cl} = \mathbf{-1.2 \times 10^6 \text{ kJ}}$$

9.124 $\text{EN(O)} = \frac{1314 + 141}{2} = 727.5$ $\quad\text{EN(F)} = \frac{1680 + 328}{2} = 1004$ $\quad\text{EN(Cl)} = \frac{1251 + 349}{2} = 800$

Using Mulliken's definition, the electronegativity of chlorine is greater than that of oxygen, and fluorine is still the most electronegative element. We can convert to the Pauling scale by dividing each of the above by 230 kJ/mol.

$\text{EN(O)} = \frac{727.5}{230} = \mathbf{3.16}$ $\quad\text{EN(F)} = \frac{1004}{230} = \mathbf{4.37}$ $\quad\text{EN(Cl)} = \frac{800}{230} = \mathbf{3.48}$

These values compare to the Pauling values for oxygen of 3.5, fluorine of 4.0, and chlorine of 3.0.

9.126 (1) You could determine the magnetic properties of the solid. An Mg^+O^- solid would be paramagnetic while $Mg^{2+}O^{2-}$ solid is diamagnetic.

(2) You could determine the lattice energy of the solid. Mg^+O^- would have a lattice energy similar to Na^+Cl^-. This lattice energy is much lower than the lattice energy of $Mg^{2+}O^{2-}$.

9.128 We can arrange the equations for the lattice energy of KCl, ionization energy of K, and electron affinity of Cl, to end up with the desired equation.

$$\cancel{K^+}(g) + \cancel{Cl^-}(g) \rightarrow KCl(s) \qquad \Delta H° = -699 \text{ kJ/mol (equation for lattice energy of KCl, reversed)}$$
$$K(g) \rightarrow \cancel{K^+}(g) + \cancel{e^-} \qquad \Delta H° = 418.7 \text{ kJ/mol (ionization energy of K)}$$
$$Cl(g) + \cancel{e^-} \rightarrow \cancel{Cl^-}(g) \qquad \Delta H° = -349 \text{ kJ/mol (electron affinity of Cl)}$$
$$K(g) + Cl(g) \rightarrow KCl(s) \qquad \mathbf{\Delta H°} = (-699 + 418.7 + -349) \text{ kJ/mol} = \mathbf{-629 \text{ kJ/mol}}$$

9.130 From Table 9.4 of the text, we can find the bond enthalpies of C–N and C=N. The average can be calculated, and then the maximum wavelength associated with this enthalpy can be calculated.

The average bond enthalpy for C–N and C=N is:

$$\frac{(276 + 615)\text{ kJ/mol}}{2} = 446 \text{ kJ/mol}$$

We need to convert this to units of J/bond before the maximum wavelength to break the bond can be calculated. Because there is only 1 CN bond per molecule, there is Avogadro's number of bonds in 1 mole of the amide group.

$$\frac{446 \cancel{\text{kJ}}}{1 \cancel{\text{mol}}} \times \frac{1 \cancel{\text{mol}}}{6.022 \times 10^{23} \text{ bonds}} \times \frac{1000 \text{ J}}{1 \cancel{\text{kJ}}} = 7.41 \times 10^{-19} \text{ J/bond}$$

The maximum wavelength of light needed to break the bond is:

$$\lambda_{max} = \frac{hc}{E} = \frac{(6.63 \times 10^{-34} \cancel{\text{J}}\cdot\cancel{\text{s}})(3.00 \times 10^8 \text{ m/}\cancel{\text{s}})}{7.41 \times 10^{-19} \cancel{\text{J}}} = \mathbf{2.68 \times 10^{-7} \text{ m} = 268 \text{ nm}}$$

9.132 **(a)** We divide the equation given in the problem by 4 to come up with the equation for the decomposition of *1 mole* of nitroglycerin.

$$C_3H_5N_3O_9(l) \rightarrow 3CO_2(g) + \tfrac{5}{2}H_2O(g) + \tfrac{3}{2}N_2(g) + \tfrac{1}{4}O_2(g)$$

We calculate $\Delta H°$ using Equation (6.18) and the enthalpy of formation values from Appendix 3 of the text.

$$\Delta H°_{rxn} = \Sigma n\Delta H°_f(\text{products}) - \Sigma m\Delta H°_f(\text{reactants})$$

$$\mathbf{\Delta H°_{rxn}} = (3)(-395.5 \text{ kJ/mol}) + \left(\tfrac{5}{2}\right)(-241.8 \text{ kJ/mol}) - (1)(-371.1 \text{ kJ/mol}) = \mathbf{-1413.9 \text{ kJ/mol}}$$

Next, we calculate $\Delta H°$ using bond enthalpy values from Table 9.4 of the text.

Bonds Broken	*Number Broken*	*Bond Enthalpy* (kJ/mol)	*Enthalpy Change* (kJ/mol)
C–H	5	414	2070
C–C	2	347	694
C–O	3	351	1053
N–O	6	176	1056
N=O	3	607	1821

Bonds Formed	*Number Formed*	*Bond Enthalpy* (kJ/mol)	*Enthalpy Change* (kJ/mol)
C=O	6	799	4794
O–H	(5/2)(2) = 5	460	2300
N≡N	1.5	941.4	1412.1
O=O	0.25	498.7	124.7

From Equation (9.3) of the text:

$$\Delta H° = \Sigma BE(\text{reactants}) - \Sigma BE(\text{products})$$

$$\boldsymbol{\Delta H°} = (6694 - 8630.8)\text{ kJ/mol} = \mathbf{-1937\ kJ/mol}$$

The $\Delta H°$ values do not agree exactly because average bond enthalpies are used, and nitroglycerin is a liquid (strictly, the bond enthalpy values are for gases).

(b) One mole of nitroglycerin generates, (3 + 2.5 + 1.5 + 0.25) = 7.25 moles of gas. One mole of an ideal gas occupies a volume of 22.41 L at STP.

$$7.25\text{ mol} \times \frac{22.41\text{ L}}{1\text{ mol}} = \mathbf{162\ L}$$

(c) We calculate the pressure exerted by 7.25 moles of gas occupying a volume of 162 L at a temperature of 3000 K.

$$\boldsymbol{P} = \frac{nRT}{V} = \frac{(7.25\text{ mol})(0.0821\text{ L}\cdot\text{atm/mol}\cdot\text{K})(3000\text{ K})}{162\text{ L}} = \mathbf{11.0\ atm}$$

CHAPTER 10
CHEMICAL BONDING II: MOLECULAR GEOMETRY AND HYBRIDIZATION OF ATOMIC ORBITALS

PROBLEM-SOLVING STRATEGIES AND TUTORIAL SOLUTIONS

TYPES OF PROBLEMS

Problem Type 1: Molecular Geometry.
- **(a)** Molecules in which the central atom has *no* lone pairs.
- **(b)** Molecules in which the central atom has one or more lone pairs.
- **(c)** Geometry of molecules with more than one central atom.

Problem Type 2: Predicting Dipole Moments.

Problem Type 3: Hybridization of Atomic Orbitals.
- **(a)** Hybridization of *s* and *p* orbitals.
- **(b)** Hybridization of *s*, *p*, and *d* orbitals.
- **(c)** Hybridization in molecules containing double and triple bonds.

Problem Type 4: Molecular Orbital Diagrams.

PROBLEM TYPE 1: MOLECULAR GEOMETRY

The model that we are going to use to study molecular geometry is called the **valence-shell electron-pair repulsion (VSEPR) model**. It accounts for the geometric arrangement of electron pairs around a central atom in terms of the repulsion between electron pairs. The geometry that a molecule ultimately adopts *minimizes* electron-pair repulsion.

Guidelines for Applying the VSEPR Model

1. Write the Lewis structure of the molecule.
2. Only consider the electron pairs around the *central atom* (the atom that is bonded to more than one other atom). Count the number of electron pairs around the central atom (bonding pairs and lone pairs). For counting purposes, treat double and triple bonds as though they were single bonds. Refer to Table 10.1 of the text to predict the overall arrangement of the electron pairs.
3. Use Tables 10.1 and 10.2 of the text to predict the *geometry* of the molecule.
4. In predicting bond angles, note that a lone pair repels another lone pair or a bonding pair more strongly than a bonding pair repels another bonding pair. There is no easy way to predict bond angles accurately when the central atom possesses one or more lone pairs.

A. Molecules in which the central atom has *no* lone pairs

For simplicity, we will only consider molecules that contain atoms of two elements, A and B, where A is the central atom. These molecules have the general formula AB_x, where x is an integer 2, 3, ... In most cases, x is between 2 and 6.

Table 10.1 of the text shows five possible arrangements of electron pairs around the central atom A. Remember, these arrangements are adopted because electron-pair repulsions are minimized.

Below is a condensed version of Table 10.1 from the text.

Arrangement of electron pairs around a central atom (A) in a molecule, and geometry of molecules if the central atom has *no* lone pairs.

Number of electron pairs around central atom	Arrangement of electron pairs	Molecular geometry
2	Linear, 180°	Linear, AB_2
3	Trigonal planar, 120°	Trigonal planar, AB_3
4	Tetrahedral, 109.5°	Tetrahedral, AB_4
5	Trigonal bipyramidal, 120°, 90°	Trigonal bipyramidal, AB_5
6	Octahedral, 90°	Octahedral, AB_6

Note: Since there are *no* lone pairs around the central atom in all the examples shown above, the molecular geometry is *always* the same as the arrangement of electron pairs around the central atom.

EXAMPLE 10.1

Use the VSEPR model to predict the geometry of the following molecules and ions: (a) $HgCl_2$, (b) $SnCl_4$, (c) NO_3^-, (d) PF_5.

Strategy: The sequence of steps in determining molecular geometry is as follows:

draw Lewis structure ⟶ find arrangement of electrons pairs ⟶ find arrangement of bonding pairs ⟶ determine geometry based on bonding pairs

Solution:

(a)

Step 1: Write the Lewis structure of the molecule (see Chapter 9).

:C̤̈l—Hg—C̤̈l:

Step 2: Count the number of electron pairs around the central atom. There are two electron pairs around Hg.

Step 3: Since there are two electron pairs around Hg, the electron-pair arrangement that minimizes repulsion is **linear**.

In addition, since there are no lone pairs around the central atom, the geometry is also **linear** (AB_2).

(b)

Step 1: Write the Lewis structure of the molecule.

```
        ..
       :Cl:
  ..    |    ..
 :Cl — Sn — Cl:
  ..    |    ..
       :Cl:
        ..
```

Step 2: Count the number of electron pairs around the central atom. There are four electron pairs around Sn.

Step 3: Since there are four electron pairs around Sn, the electron-pair arrangement that minimizes repulsion is **tetrahedral**.

In addition, since there are no lone pairs around the central atom, the geometry is also **tetrahedral** (AB_4).

(c)
Step 1: Write the Lewis structure of the molecule.

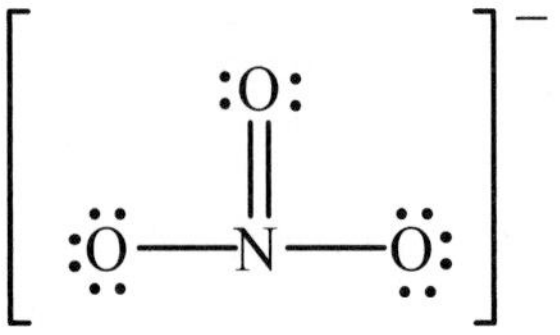

Step 2: Count the number of electron pairs around the central atom. There are three electron pairs around N. Remember, for VSEPR purposes, a double or triple bond counts the same as a single bond.

Step 3: Since there are three electron pairs around N, the electron-pair arrangement that minimizes repulsion is **trigonal planar**.

In addition, since there are no lone pairs around the central atom, the geometry is also **trigonal planar** (AB_3).

(d)
Step 1: Write the Lewis structure of the molecule.

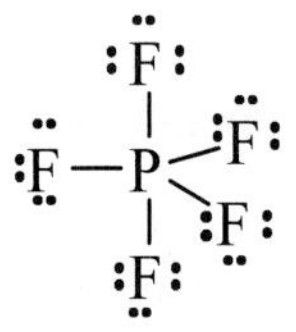

Step 2: Count the number of electron pairs around the central atom. There are five electron pairs around P.

Since there are five electron pairs around P, the electron-pair arrangement that minimizes repulsion is **trigonal bipyramidal**.

In addition, since there are no lone pairs around the central atom, the geometry is also **trigonal bipyramidal** (AB_5).

PRACTICE EXERCISE

1. Use the VSEPR model to predict the geometry of the following molecules: (a) CO_2, (b) CCl_4,(c) SO_3, (d) SF_6.

Text Problems: 10.8, 10.10, 10.12, 10.14

B. Molecules in which the central atom has one or more lone pairs

If lone pairs are present on the central atom, the overall arrangement of electron pairs is ***not*** the same as the geometry of the molecule.

To keep track of the total number of bonding pairs and lone pairs, we will designate molecules with lone pairs as AB_xE_y, where A is the central atom, B is a surrounding atom, and E is a lone pair on the central atom, A. Both x and y are integers: $x = 2, 3, \ldots,$ and $y = 1, 2, \ldots.$

When working with molecules that have lone pairs on the central atom, remember to count all electron pairs on the central atom, both bonding pairs and lone pairs. The total number of electron pairs around the central atom determines the electron arrangement around the central atom. See Table 10.1. However, the molecular geometry will not be the same as this electron arrangement. Use Table 10.2 of the text to determine the molecular geometry, or a better option is to build a model. Gum drops and toothpicks make an effective and inexpensive model kit. And besides, you can eat your model when you are finished. (Not the toothpicks!)

Below is a condensed version of Table 10.2 of the text.

Geometry of simple molecules and ions in which the central atom has one or more lone pairs.

Class of molecule	Total # of electron pairs on central atom	Number of bonding pairs	Arrangement of electron pairs	Number of lone pairs	Molecular geometry
AB_2E	3	2	Trigonal planar	1	Bent
AB_3E	4	3	Tetrahedral	1	Trigonal pyramid
AB_2E_2	4	2	Tetrahedral	2	Bent
AB_4E	5	4	Trigonal bipyramidal	1	Distorted tetrahedron
AB_3E_2	5	3	Trigonal bipyramidal	2	T-shaped
AB_2E_3	5	2	Trigonal bipyramidal	3	Linear
AB_5E	6	5	Octahedral	1	Square pyramidal
AB_4E_2	6	4	Octahedral	2	Square planar

Note: The arrangement of 3 electron-pairs is always trigonal planar; the arrangement of 4 electron-pairs is tetrahedral; the arrangement of 5 electron-pairs is trigonal bipyramidal; the arrangement of 6 electron-pairs is octahedral. However, if there are any lone pairs on the central atom, the molecular geometry will be different from the electron arrangement. **Build models!**

EXAMPLE 10.2

Use the VSEPR model to predict the geometry of the following molecules: (a) O_3, (b) XeF_2, (c) IF_5.

Strategy: The sequence of steps in determining molecular geometry is as follows:

draw Lewis structure $\longrightarrow$ find arrangement of electrons pairs $\longrightarrow$ find arrangement of bonding pairs $\longrightarrow$ determine geometry based on bonding pairs

Solution:

(a)

Step 1: Write the Lewis structure of the molecule (see Chapter 9).

$$\ddot{\underset{\cdot\cdot}{O}}=\ddot{O}-\ddot{\underset{\cdot\cdot}{O}}:$$

Step 2: Count the number of electron pairs around the central atom. There are three electron pairs around the central oxygen. Remember, for VSEPR purposes, a double or triple bond counts the same as a single bond.

Step 3: Since there are three electron pairs around the central O, the electron-pair arrangement that minimizes repulsion is **trigonal planar**.

However, there is one lone pair of electrons around the central atom. Consulting a model or Table 10.2 of the text, you should find that the geometry of the molecule is **bent** (AB_2E).

(b)

Step 1: Write the Lewis structure of the molecule.

```
 ..   .. ..   ..
:F — Xe — F:
 ..   ..      ..
```

Step 2: Count the number of electron pairs around the central atom. There are five electron pairs around xenon.

Step 3: Since there are five electron pairs around xenon, the electron-pair arrangement that minimizes repulsion is **trigonal bipyramidal**.

However, there are three lone pairs of electrons around the central atom. Consulting a model or Table 10.2 of the text, you should find that the geometry of the molecule is **linear** (AB_2E_3).

(c)

Step 1: Write the Lewis structure of the molecule.

```
         ..
        :F:
  ..     |  ..
 :F — :I < :F:
  ..     | :F:
        :F:  ..
         ..
```

Step 2: Count the number of electron pairs around the central atom. There are six electron pairs around iodine.

Step 3: Since there are six electron pairs around iodine, the electron-pair arrangement that minimizes repulsion is **octahedral**.

However, there is one lone pair of electrons around the central atom. Consulting a model or Table 10.2 of the text, you should find that the geometry of the molecule is **square pyramidal** (AB_5E).

PRACTICE EXERCISE

2. Use the VSEPR model to predict the geometry of the following molecules: (a) SO_2, (b) XeF_4, (c) SF_4.

Text Problems: 10.10, 10.12, 10.14

C. Geometry of molecules with more than one central atom

So far, we have discussed the geometry of molecules having only one central atom. (The term "central atom" means an atom that is not a terminal atom in a polyatomic molecule.) Many molecules will have more than one "central atom". In these cases, you have to apply the VSEPR method presented in parts A and B above to each of the central atoms.

EXAMPLE 10.3

Use the VSEPR model to predict the geometry of C_2H_6.

Step 1: Write the Lewis structure of the molecule. Hydrogens must be in terminal positions. The two carbons must be the "central atoms".

```
    H   H
    |   |
H — C — C — H
    |   |
    H   H
```

Step 2: Count the number of electron pairs around the "central atoms". There are four electron pairs around each carbon.

Step 3: Since there are four electron pairs around each carbon, the electron-pair arrangement around each carbon that minimizes repulsion is **tetrahedral**.

In addition, since there are no lone pairs around the central atoms, their geometries are also **tetrahedral**.

PRACTICE EXERCISE

3. Use the VSEPR model to predict the geometry of C_2H_2.

PROBLEM TYPE 2: PREDICTING DIPOLE MOMENTS

To determine if a molecule has a dipole moment (a measure of electrical charge separation in a molecule) you must consider two factors.

1. Are the *bonds* in the molecule polar?

A bond will be polar if there is a difference in electronegativity between the two atoms comprising the bond (see Section 9.5 of the text). For example, in an O–H bond, there is a shift in electron density from H to O because O is more electronegative than H. The shift in electron density is symbolized by placing a crossed arrow ($\leftarrow\!\!\!+$) above the bond to indicate the direction of the shift in electron density. This is called a *bond* moment. For example:

$$\overset{\longleftarrow\!\!+}{\text{O} - \text{H}}$$

The consequent charge separation can be represented as:

$$\overset{\delta^-}{\text{O}} - \overset{\delta^+}{\text{H}}$$

where δ (delta) represents a partial charge.

2. Is the *molecule* polar?

Diatomic molecules containing atoms of the same element do not have dipole moments and so are **nonpolar molecules**. There is no difference in electronegativity since the two elements are the same.

However, most molecules will have polar bonds (bond moments). If a molecule has polar bonds, does this mean that it is a polar molecule? Not necessarily. The *bond moment* is a vector quantity, which means that it has both a magnitude and direction. The measured *dipole moment* of the molecule is equal to the vector sum of the bond moments. For example, in CO_2, the two bond moments are equal in magnitude and opposite in direction. The sum or resultant dipole moment will be *zero*. Hence, CO_2 is a **nonpolar molecule**.

$$\overset{\longleftarrow\!\!+}{\ddot{\underset{\cdot\cdot}{\text{O}}}}\!=\!\text{C}\!=\!\overset{+\!\!\longrightarrow}{\ddot{\underset{\cdot\cdot}{\text{O}}}}$$

In summary, to determine if a molecule is **polar** (i.e., does the molecule have a dipole moment), you must sum the individual bond moments to determine if there is a resultant dipole moment.

EXAMPLE 10.4

Predict whether each of the following molecules has a dipole moment: (a) CCl_4 and (b) $CHCl_3$.

Strategy: Keep in mind that the dipole moment of a molecule depends on both the difference in electronegativities of the elements present and its geometry. A molecule can have polar bonds (if the bonded atoms have different electronegativities), but it may not possess a dipole moment if it has a highly symmetrical geometry.

Solution:

(a) Write the Lewis structure for the molecule. The lone pairs on Cl have been omitted.

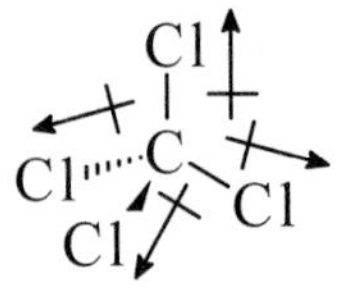

Shown on the Lewis structure are the bond moments. Chlorine is more electronegative than C, so the arrows indicate the shift in electron density toward Cl. However, these polar bonds are arranged in a symmetric tetrahedral fashion about the central C atom. The sum or resultant dipole moment is *zero*. CCl_4 is a **nonpolar molecule**.

(b) Write the Lewis structure for the molecule. The lone pairs on Cl have been omitted.

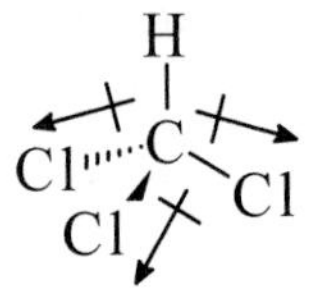

Shown on the Lewis structure are the bond moments. Chlorine is more electronegative than C, so the arrows indicate the shift in electron density toward Cl. The electronegativity difference between C and H is very small, so a C–H bond is essentially nonpolar. In this case, the three bond moments partially reinforce each other. Thus, $CHCl_3$ has a dipole moment and is therefore a **polar molecule**.

PRACTICE EXERCISE

4. Predict whether each of the following molecules has a dipole moment: (a) CO, (b) BCl_3, and (c) XeF_4.

Text Problems: 10.20, 10.22, **10.24**

PROBLEM TYPE 3: HYBRIDIZATION OF ATOMIC ORBITALS

The **VSEPR** model is very powerful considering its simplicity; however, it does not explain chemical bond formation in any detail. In the 1930s, **valence bond (VB) theory** was introduced to account for chemical bond formation. **VB** theory describes covalent bonding as the overlapping of atomic orbitals. This means that the orbitals share a common region in space.

VB theory uses a concept called **hybridization** to explain covalent bonding. Hybridization is the mixing of atomic orbitals in an atom (usually a central atom) to generate a set of new atomic orbitals, called *hybrid orbitals*. Hybrid orbitals are atomic orbitals obtained when two or more nonequivalent orbitals of the same atom combine. The hybrid orbitals are used to form covalent bonds.

A. Hybridization of *s* and *p* orbitals

1. *sp* hybrid orbitals

Let's consider the central atom Be in the BeH_2 molecule. Be has a ground state electron configuration of $1s^22s^2$. Only valence electrons are involved in bonding, so an orbital diagram of the valence electrons is:

$$\underset{2s}{\uparrow\downarrow} \quad \underset{2p}{\_\ \_\ \_}$$

With this ground-state electron configuration, Be cannot form bonds with H because its valence electrons are paired in a 2*s* orbital.

To explain the bonding, first an electron is promoted from the 2*s* orbital to a 2*p* orbital.

$$\underset{2s}{\uparrow} \quad \underset{2p}{\uparrow\ \_\ \_}$$

Now, we have two different orbitals that can bond to the two hydrogens, which would result in two nonequivalent Be–H bonds. However, experimental evidence shows that there are two equivalent Be–H bonds.

This is where hybridization comes in. By mixing the 2*s* orbital with one of the 2*p* orbitals, we can generate two equivalent *sp* hybrid orbitals.

$$\underset{sp \text{ hybrid orbitals}}{\uparrow\ \uparrow} \quad \underset{\text{empty } p \text{ orbitals}}{\_\ \_}$$

Figure 10.10 of the text shows the shape and orientation of the *sp* hybrid orbitals. These two hybrid orbitals lie along the same line so that the angle between them is 180° (linear arrangement). Each of the BeH bonds is formed by the overlap of a Be *sp* hybrid and a H 1*s* orbital. The resulting BeH_2 molecule has a linear geometry.

2. sp^2 hybrid orbitals

Following the same type of argument for *sp* hybrids, sp^2 hybrid orbitals are formed by mixing an *s* orbital with two *p* orbitals. Three equivalent sp^2 hybrid orbitals are formed.

$$\underset{s}{\uparrow} \quad \underset{p}{\uparrow\ \uparrow\ \_}$$

Mixing an *s* with two *p* orbitals gives:

$$\underset{sp^2 \text{ hybrid orbitals}}{\uparrow\ \uparrow\ \uparrow} \quad \underset{\text{empty } p \text{ orbital}}{\_}$$

Figure 10.12 of the text shows the shape and orientation of the sp^2 hybrid orbitals. These three hybrid orbitals lie in a plane with an angle between any two hybrids of 120°. The three sp^2 hybrid orbitals are arranged in a trigonal planar fashion.

3. sp^3 hybrid orbitals

Again, following the same argument as above, sp^3 hybrid orbitals are formed by mixing an *s* orbital with three *p* orbitals. Four equivalent sp^3 hybrid orbitals are formed.

↑ (s) ↑ ↑ ↑ (p)

Mixing an *s* with three *p* orbitals gives:

↑ ↑ ↑ ↑
sp^3 hybrid orbitals

Figure 10.7 of the text shows the shape and orientation of the sp^3 hybrid orbitals. These four equivalent hybrid orbitals are directed toward the four corners of a regular tetrahedron with 109.5° bond angles.

Summarizing,

Type of hybrid	No. of equivalent hybrid orbitals	Arrangement of hybrid orbitals	No. of empty *p* orbitals
sp	2	linear, 180°	2
sp^2	3	trigonal planar, 120°	1
sp^3	4	tetrahedral, 109.5°	0

EXAMPLE 10.5

Determine the hybridization of the central (underlined) atom in each of the following molecules: (a) $\underline{C}Cl_4$ and (b) $\underline{B}Cl_3$.

Strategy: The steps for determining the hybridization of the central atom in a molecule are:

draw Lewis Structure of the molecule ⟶ use VSEPR to determine the electron pair arrangement surrounding the central atom (Table 10.1 of the text) ⟶ use Table 10.4 of the text to determine the hybridization state of the central atom

Solution:

(a) Write the Lewis structure of the molecule. The lone pairs on the chlorines have been omitted.

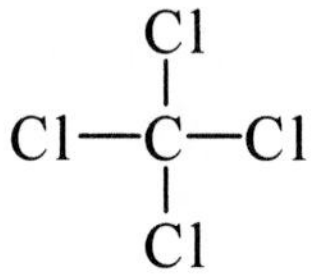

```
     Cl
     |
Cl—C—Cl
     |
     Cl
```

Count the number of electron pairs around the central atom. Since there are four electron pairs around C, the electron arrangement that minimizes electron-pair repulsion is **tetrahedral.**

We conclude that C is $\boldsymbol{sp^3}$ **hybridized** because it has the electron arrangement of four sp^3 hybrid orbitals.

(b) Write the Lewis structure of the molecule. The lone pairs on the chlorines have been omitted.

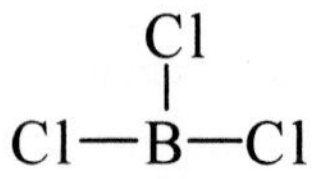

```
     Cl
     |
Cl—B—Cl
```

Count the number of electron pairs around the central atom. Since there are three electron pairs around B, the electron arrangement that minimizes electron-pair repulsion is **trigonal planar.**

We conclude that B is $\boldsymbol{sp^2}$ **hybridized** because it has the electron arrangement of three sp^2 hybrid orbitals.

PRACTICE EXERCISE

5. Determine the hybridization of the central (underlined) atom in each of the following molecules or ions: (a) $\underline{C}O_2$ and (b) $\underline{C}O_3^{2-}$.

Text Problems: 10.34, 10.36, 10.38, **10.40**

B. Hybridization of *s*, *p*, and *d* orbitals

1. sp^3d hybrid orbitals

We use the same approach that we used for hybridizing *s* and *p* orbitals, but now we are also mixing in a *d* orbital. The sp^3d hybrid orbitals are formed by mixing an *s* orbital, three *p* orbitals, and a *d* orbital. Five equivalent sp^3d hybrid orbitals are formed.

↑ (*s*) ↑ ↑ ↑ (*p*) ↑ __ __ __ __ (*d*)

Mixing an *s* orbital, three *p* orbitals, and a *d* orbital gives:

↑ ↑ ↑ ↑ ↑ (sp^3d hybrid orbitals) __ __ __ __ (empty *d* orbitals)

Table 10.4 of the text shows the shape and orientation of the sp^3d hybrid orbitals. These five equivalent hybrid orbitals are directed toward the five corners of a trigonal bipyramid with bond angles of 120° and 90°.

2. sp^3d^2 hybrid orbitals

We use the same approach that we used above, but now we are mixing in two *d* orbitals. The sp^3d^2 hybrid orbitals are formed by mixing an *s* orbital, three *p* orbitals, and two *d* orbitals. Six equivalent sp^3d^2 hybrid orbitals are formed.

↑ (*s*) ↑ ↑ ↑ (*p*) ↑ ↑ __ __ __ (*d*)

Mixing an *s* orbital, three *p* orbitals, and two *d* orbitals gives:

↑ ↑ ↑ ↑ ↑ ↑ (sp^3d^2 hybrid orbitals) __ __ __ (empty *d* orbitals)

Table 10.4 of the text shows the shape and orientation of the sp^3d^2 hybrid orbitals. These six equivalent hybrid orbitals are directed toward the six corners of an octahedron with bond angles of 90°.

Summarizing,

Type of hybrid	No. of equivalent hybrid orbitals	Arrangement of hybrid orbitals
sp^3d	5	trigonal bipyramid, 120°, 90°
sp^3d^2	6	octahedral, 90°

EXAMPLE 10.6

Describe the hybridization of xenon in xenon tetrafluoride (XeF_4).

Strategy: The steps for determining the hybridization of the central atom in a molecule are:

draw Lewis Structure of the molecule → use VSEPR to determine the electron pair arrangement surrounding the central atom (Table 10.1 of the text) → use Table 10.4 of the text to determine the hybridization state of the central atom

Solution: Write the Lewis structure for XeF_4. The lone pairs of electrons on F have been omitted.

```
      F
      |•
  F —•Xe— F
      |
      F
```

Count the number of electron pairs around the central atom. Since there are six electron pairs around Xe, the electron arrangement that minimizes electron-pair repulsions is **octahedral**.

We conclude that Xe is sp^3d^2 **hybridized** because it has the electron arrangement of six sp^3d^2 hybrid orbitals.

> **Tip:** It is important to use the electron arrangement to determine the hybridization of the central atom and not the geometry. The two lone pairs on Xe are occupying two of the hybrid orbitals, so the lone pairs must be included to determine the correct hybridization.

PRACTICE EXERCISE

6. Describe the hybridization of sulfur in sulfur tetrafluoride, SF_4.

Text Problem: 10.42

C. Hybridization in molecules containing double or triple bonds

We can determine the hybridization of molecules containing double or triple bonds in the same manner as molecules with single bonds. Furthermore, we would like to determine which orbitals overlap to form the double or triple bond.

In a double or triple bond, there are two types of covalent bonds formed. One involves end-to-end overlap of orbitals in which the electron density is concentrated between the nuclei of the bonding atoms. A bond of this type is called a **sigma bond (σ bond)**. The second type involves side-to-side overlap of orbitals in which electron density is concentrated above and below the plane of the nuclei of the bonding atoms. This type of bond is called a **pi bond (π bond)**. For the molecules we will be considering, a π bond is formed from the side-to-side overlap of two *p* orbitals.

EXAMPLE 10.7

Describe the bonding in carbon dioxide, CO_2.

Strategy: The steps for determining the hybridization of the central atom in a molecule are:

draw Lewis Structure of the molecule → use VSEPR to determine the electron pair arrangement surrounding the central atom (Table 10.1 of the text) → use Table 10.4 of the text to determine the hybridization state of the central atom

Solution: Write the Lewis structure for CO_2.

```
 ••     ••
 O==C==O
 ••     ••
```

Count the number of electron pairs around the central atom. Since there are two electron pairs around C, the electron arrangement that minimizes electron-pair repulsions is **linear**.

We conclude that C is ***sp* hybridized** because it has the electron arrangement of two *sp* hybrid orbitals.

Next, count the number of electron pairs around each oxygen atom. Since there are three electron pairs around each O, the electron arrangement that minimizes repulsions is **trigonal planar**.

We conclude that each O is **sp^2 hybridized** because it has the electron arrangement of three sp^2 hybrid orbitals.

Describing the bonding in CO_2, each of the *sp* orbitals of the C atom forms a sigma bond with an sp^2 hybrid on each of the O atoms. Carbon has two "pure" *p* orbitals that did not mix with the *s* orbital. Each of these *p* orbitals of the C atom forms a pi bond by overlapping in a side-to-side fashion with a *p* orbital on each of the oxygen atoms. Each double bond is composed of one σ bond and one π bond. Finally, the two lone pairs on each O atom are placed in its two remaining sp^2 orbitals.

> **Tip:** A double bond is typically composed of one σ bond and one π bond. A triple bond is composed of one σ bond and two π bonds.

PRACTICE EXERCISE

7. Describe the bonding in a nitrogen molecule, N_2.

Text Problems: 10.38, **10.40**

PROBLEM TYPE 4: MOLECULAR ORBITAL DIAGRAMS

When working molecular orbital problems, realize that this is another bonding model that is different from the other models we have encountered. So far, we have looked at the Lewis model and valence bond theory. We will focus on molecular orbital diagrams for homonuclear diatomic molecules of second-period elements. Examples include N_2, O_2, and F_2.

A. Writing electron configurations

For homonuclear diatomic molecules of second-period elements, the types of molecular orbitals used in bonding will be similar. The order of filling molecular orbitals for Li_2, B_2, C_2, and N_2 is:

$\sigma^{\star}_{2p_x}$ ____

$\pi^{\star}_{2p_y}$, $\pi^{\star}_{2p_z}$ ____ ____

σ_{2p_x} ____

π_{2p_y}, π_{2p_z} ____ ____

$\sigma^{\star}_{2s}$ ____

σ_{2s} ____

$\sigma^{\star}_{1s}$ ____

σ_{1s} ____

The order of filling molecular orbitals for O_2 and F_2 is:

$\sigma^{\star}_{2p_x}$	___	
$\pi^{\star}_{2p_y}$, $\pi^{\star}_{2p_z}$	___	___
π_{2p_y}, π_{2p_z}	___	___
σ_{2p_x}	___	
$\sigma^{\star}_{2s}$	___	
σ_{2s}	___	
$\sigma^{\star}_{1s}$	___	
σ_{1s}	___	

Note that for O_2 and F_2, the σ_{2p_x} is lower in energy than the π_{2p_y}, π_{2p_z} molecular orbitals.

EXAMPLE 10.8

Write the electron configuration for the ion, O_2^-.

Strategy: Count the number of electrons in the ion.

Each oxygen has 8 electrons, plus we need to add one electron for the negative charge. The total number of electrons in the ion is 17.

Place the electrons in molecular orbitals following this convention.

1. Build up from the lowest energy molecular orbital to higher energy orbitals.
2. Each molecular orbital can hold a maximum of two electrons.
3. Follow Hund's rule: the most stable arrangement of electrons in molecular orbitals with equal energy is the one with the greatest number of parallel spins.
4. When electrons are paired in a molecular orbital, they must have opposite spin.

Solution: Placing 17 electrons following the above rules gives:

$\sigma^{\star}_{2p_x}$	___	
$\pi^{\star}_{2p_y}$, $\pi^{\star}_{2p_z}$	↑↓	↑
π_{2p_y}, π_{2p_z}	↑↓	↑↓
σ_{2p_x}	↑↓	
$\sigma^{\star}_{2s}$	↑↓	
σ_{2s}	↑↓	
$\sigma^{\star}_{1s}$	↑↓	
σ_{1s}	↑↓	

The electron configuration is: $(\sigma_{1s})^2(\sigma^{\star}_{1s})^2(\sigma_{2s})^2(\sigma^{\star}_{2s})^2(\sigma_{2p_x})^2(\pi_{2p_y})^2(\pi_{2p_z})^2(\pi^{\star}_{2p_y})^2(\pi^{\star}_{2p_z})^1$

PRACTICE EXERCISE

8. Write the electron configuration for C_2^+.

Text Problems: 10.52, 10.54, 10.56, 10.58, 10.60

B. Calculating bond order

Placing electrons in a "bonding" molecular orbital yields a stable covalent bond, whereas placing electrons in an "antibonding" molecular orbital results in an unstable bond. We can evaluate the stability of molecules or ions by calculating their **bond order**. The bond order indicates the strength of the bond; the greater the bond order, the stronger the bond. We can calculate bond order as follows:

$$\text{bond order} = \frac{1}{2}\begin{pmatrix}\text{number of electrons} & - & \text{number of electrons} \\ \text{in bonding MOs} & & \text{in antibonding MOs}\end{pmatrix}$$

EXAMPLE 10.9

Determine the bond order of O_2^-.

Referring to Example 10.8, the bond order is:

$$\textbf{bond order} = \frac{1}{2}(10 - 7) = \textbf{1.5}$$

The bond order of O_2 is 2. This indicates that O_2 is more stable than O_2^-. Is this what you would expect?

PRACTICE EXERCISE

9. Determine the bond order of C_2^+.

Text Problems: 10.52, 10.54, 10.56, 10.58

C. Determining the magnetic character of a molecule or ion

In Section 7.8 of the text, the terms "paramagnetic" and "diamagnetic" were discussed. A *paramagnetic* substance contains unpaired electrons and is *attracted* by an external magnetic field. Any substance with an *odd* number of electrons must be paramagnetic, because we need an even number of electrons for complete pairing. In a *diamagnetic* substance, all the electron spins are paired. Diamagnetic substances are slightly *repelled* by an external magnetic field.

Substances containing an even number of electrons may be either diamagnetic or paramagnetic. A molecular orbital diagram can be helpful in determining the magnetic character of a molecule or ion with an even number of electrons. O_2 is an example of a molecule with an even number of electrons (16) that is paramagnetic (see Table 10.5 of the text).

EXAMPLE 10.10

Determine the magnetic character of O_2^-.

Since O_2^- has an odd number of electrons (17), it is paramagnetic. We could also write the molecular orbital diagram for O_2^- to determine its magnetic character. Looking at the MO diagram for O_2^- in Example 10.8, we see that there is a single unpaired electron in the $\pi^{\star}_{2p_z}$ molecular orbital. Hence, O_2^- is paramagnetic.

PRACTICE EXERCISE

10. Determine the magnetic character of C_2^+.

Text Problems: 10.56, 10.58

ANSWERS TO PRACTICE EXERCISES

1. **(a)** linear **(b)** tetrahedral **(c)** trigonal planar **(d)** octahedral

2. **(a)** bent **(b)** square planar **(c)** distorted tetrahedron (seesaw)

3. The electron arrangement around each C that minimizes electron-pair repulsion is linear. Since there are no lone pairs around each C, the geometry is also **linear**.

4. **(a)** Yes, the molecule is polar.
   (b) No, the molecule is nonpolar.
   (c) No, the molecule is nonpolar.

5. **(a)** *sp* **(b)** sp^2

6. When you draw the Lewis structure for SF_4, you will find five electron pairs around the central atom, S (four bonding pairs and one lone pair). The electron arrangement that minimizes electron-pair repulsion is trigonal bipyramidal. You should conclude that S is dsp^3 hybridized because it has the electron arrangement of five dsp^3 hybrid orbitals.

7. :N≡N:

   The structure of N_2 is **linear**. We conclude that each N is ***sp* hybridized** because it has the electron arrangement of two *sp* hybrid orbitals. An *sp* orbital of one N atom overlaps with an *sp* orbital on the other N to form a sigma bond. The other *sp* orbital of each N contains the lone pair of electrons. Each nitrogen atom has two "pure" *p* orbitals that did not mix with the *s* orbital. The two *p* orbitals on one N can form two pi bonds by overlapping side-to-side with the two *p* orbitals on the other N atom. The triple bond is composed of one σ bond and two π bonds.

8. $(\sigma_{1s})^2(\sigma_{1s}^\star)^2(\sigma_{2s})^2(\sigma_{2s}^\star)^2(\pi_{2p_y})^2(\pi_{2p_z})^1$

9. bond order = 1.5

10. C_2^+ is paramagnetic.

SOLUTIONS TO SELECTED TEXT PROBLEMS

10.8 Molecular Geometry, Problem Type 1.

Strategy: The sequence of steps in determining molecular geometry is as follows:

draw Lewis structure ⟶ find arrangement of electrons pairs ⟶ find arrangement of bonding pairs ⟶ determine geometry based on bonding pairs

Solution:

	Lewis structure	Electron pairs on central atom	Electron arrangement	Lone pairs	Geometry
(a)	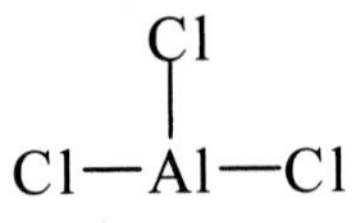 Cl—Al(—Cl)—Cl	3	trigonal planar	0	trigonal planar, AB_3
(b)	Cl—Zn—Cl	2	linear	0	linear, AB_2
(c)	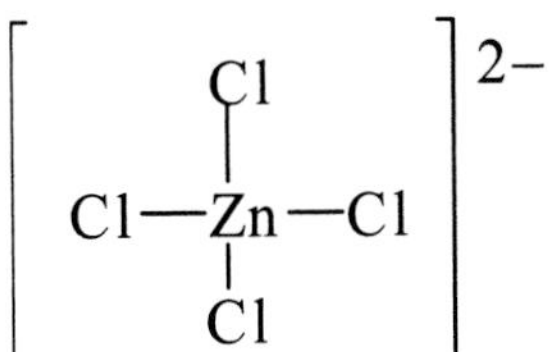 $[ZnCl_4]^{2-}$: Cl—Zn(—Cl)(—Cl)—Cl	4	tetrahedral	0	tetrahedral, AB_4

10.10 We use the following sequence of steps to determine the geometry of the molecules.

draw Lewis structure ⟶ find arrangement of electrons pairs ⟶ find arrangement of bonding pairs ⟶ determine geometry based on bonding pairs

(a) Looking at the Lewis structure we find 4 pairs of electrons around the central atom. The electron pair arrangement is tetrahedral. Since there are no lone pairs on the central atom, the geometry is also **tetrahedral**.

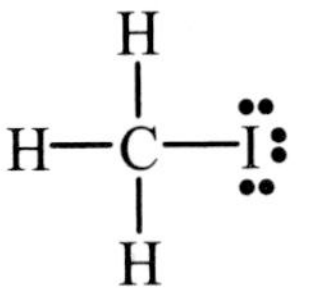

H
|
H—C—Ï:
|
H

(b) Looking at the Lewis structure we find 5 pairs of electrons around the central atom. The electron pair arrangement is trigonal bipyramidal. There are two lone pairs on the central atom, which occupy positions in the trigonal plane. The geometry is **t-shaped**.

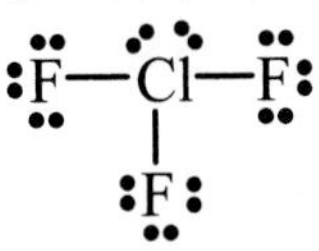

:F̤̈—C̈l—F̤̈:
|
:F̤:

(c) Looking at the Lewis structure we find 4 pairs of electrons around the central atom. The electron pair arrangement is tetrahedral. There are two lone pairs on the central atom. The geometry is **bent**.

H—S—H

(d) Looking at the Lewis structure, there are 3 VSEPR pairs of electrons around the central atom. Recall that a double bond counts as one VSEPR pair. The electron pair arrangement is trigonal planar. Since there are no lone pairs on the central atom, the geometry is also **trigonal planar**.

:O:
||
:O—S—O:

(e) Looking at the Lewis structure, there are 4 pairs of electrons around the central atom. The electron pair arrangement is tetrahedral. Since there are no lone pairs on the central atom, the geometry is also **tetrahedral**.

$\left[\begin{array}{c} :O: \\ | \\ :O—S—O: \\ | \\ :O: \end{array}\right]^{2-}$

10.12

(a)	AB_4	tetrahedral	**(f)**	AB_4	tetrahedral
(b)	AB_2E_2	bent	**(g)**	AB_5	trigonal bipyramidal
(c)	AB_3	trigonal planar	**(h)**	AB_3E	trigonal pyramidal
(d)	AB_2E_3	linear	**(i)**	AB_4	tetrahedral
(e)	AB_4E_2	square planar			

10.14 Only molecules with four bonds to the central atom and no lone pairs are tetrahedral (AB_4).

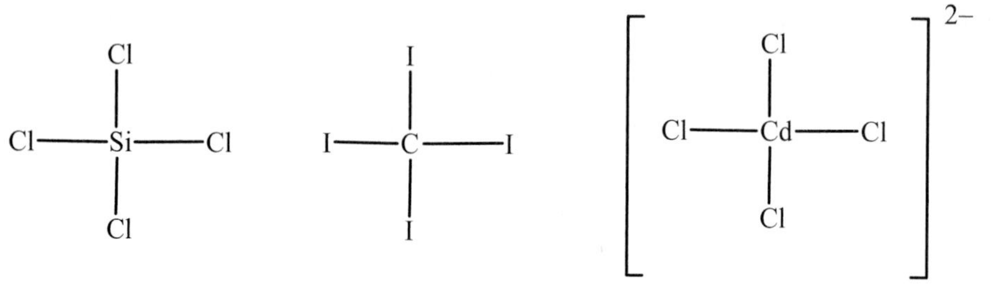

What are the Lewis structures and shapes for XeF_4 and SeF_4?

10.20 The electronegativity of the halogens decreases from F to I. Thus, the polarity of the H–X bond (where X denotes a halogen atom) also decreases from HF to HI. This difference in electronegativity accounts for the decrease in dipole moment.

10.22 Draw the Lewis structures. Both molecules are linear (AB_2). In CS_2, the two C–S bond moments are equal in magnitude and opposite in direction. The sum or resultant dipole moment will be *zero*. Hence, CS_2 is a nonpolar molecule. Even though OCS is linear, the C–O and C–S bond moments are not exactly equal, and there will be a small net dipole moment. Hence, OCS has a **larger** dipole moment than CS_2 (zero).

10.24 Predicting Dipole Moments, Problem Type 2.

Strategy: Keep in mind that the dipole moment of a molecule depends on both the difference in electronegativities of the elements present and its geometry. A molecule can have polar bonds (if the bonded atoms have different electronegativities), but it may not possess a dipole moment if it has a highly symmetrical geometry.

Solution: Each vertex of the hexagonal structure of benzene represents the location of a C atom. Around the ring, there is no difference in electronegativity between C atoms, so the only bonds we need to consider are the polar C–Cl bonds.

The molecules shown in **(b)** and **(d)** are nonpolar. Due to the high symmetry of the molecules and the equal magnitude of the bond moments, the bond moments in each molecule cancel one another. The resultant dipole moment will be *zero*. For the molecules shown in **(a)** and **(c),** the bond moments do not cancel and there will be net dipole moments. The dipole moment of the molecule in **(a)** is larger than that in **(c)**, because in **(a)** all the bond moments point in the same relative direction, reinforcing each other (see Lewis structure below). Therefore, the order of increasing dipole moments is:

$$\textbf{(b) = (d) < (c) < (a).}$$

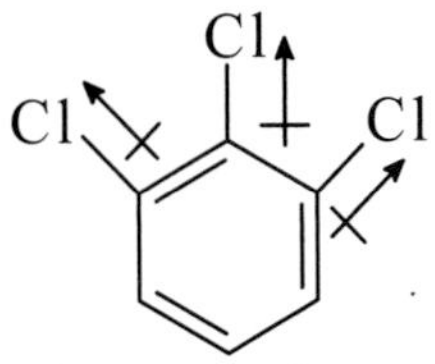

Cl Cl Cl

(a)

10.34 Hybridization of Atomic Orbitals, Problem Type 3.

Strategy: The steps for determining the hybridization of the central atom in a molecule are:

draw Lewis Structure of the molecule → use VSEPR to determine the electron pair arrangement surrounding the central atom (Table 10.1 of the text) → use Table 10.4 of the text to determine the hybridization state of the central atom

Solution:

(a) Write the Lewis structure of the molecule.

$$\begin{array}{c} \text{H} \\ | \\ \text{H}-\text{Si}-\text{H} \\ | \\ \text{H} \end{array}$$

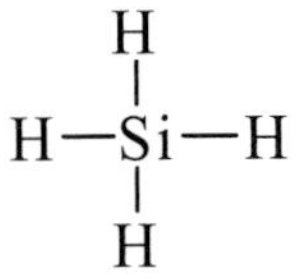

Count the number of electron pairs around the central atom. Since there are four electron pairs around Si, the electron arrangement that minimizes electron-pair repulsion is **tetrahedral**.

We conclude that Si is ***sp***^{3} **hybridized** because it has the electron arrangement of four *sp*^{3} hybrid orbitals.

(b) Write the Lewis structure of the molecule.

```
    H    H
    |    |
H—Si—Si—H
    |    |
    H    H
```

Count the number of electron pairs around the "central atoms". Since there are four electron pairs around each Si, the electron arrangement that minimizes electron-pair repulsion for each Si is **tetrahedral**.

We conclude that each Si is ***sp*^{3} hybridized** because it has the electron arrangement of four *sp*^{3} hybrid orbitals.

10.36 Draw the Lewis structures. Before the reaction, boron is *sp*^{2} hybridized (trigonal planar electron arrangement) in BF_3 and nitrogen is *sp*^{3} hybridized (tetrahedral electron arrangement) in NH_3. After the reaction, boron and nitrogen are both *sp*^{3} hybridized (tetrahedral electron arrangement).

10.38 **(a)** Each carbon has four bond pairs and no lone pairs and therefore has a tetrahedral electron pair arrangement. This implies *sp*^{3} hybrid orbitals.

```
    H    H
    |    |
H—C——C—H
    |    |
    H    H
```

(b) The left-most carbon is tetrahedral and therefore has *sp*^{3} hybrid orbitals. The two carbon atoms connected by the double bond are trigonal planar with *sp*^{2} hybrid orbitals.

```
    H    H    H
    |    |    |
H—C——C═C—H
    |
    H
```

(c) Carbons 1 and 4 have *sp*^{3} hybrid orbitals. Carbons 2 and 3 have *sp* hybrid orbitals.

```
    H                H
    |                |
H—C1——C2≡C3——C4——OH
    |                |
    H                H
```

(d) The left-most carbon is tetrahedral (*sp*^{3} hybrid orbitals). The carbon connected to oxygen is trigonal planar (why?) and has *sp*^{2} hybrid orbitals.

```
    H    H
    |    |
H—C——C═O
    |
    H
```

(e) The left-most carbon is tetrahedral (sp^3 hybrid orbitals). The other carbon is trigonal planar with sp^2 hybridized orbitals.

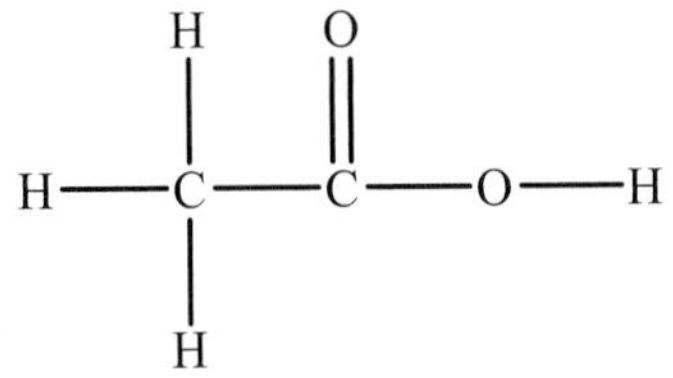

```
      H    O
      |    ||
H —— C —— C —— O —— H
      |
      H
```

10.40 Hybridization of Atomic Orbitals, Problem Type 3.

Strategy: The steps for determining the hybridization of the central atom in a molecule are:

draw Lewis Structure of the molecule → use VSEPR to determine the electron pair arrangement surrounding the central atom (Table 10.1 of the text) → use Table 10.4 of the text to determine the hybridization state of the central atom

Solution:

Write the Lewis structure of the molecule. Several resonance forms with formal charges are shown.

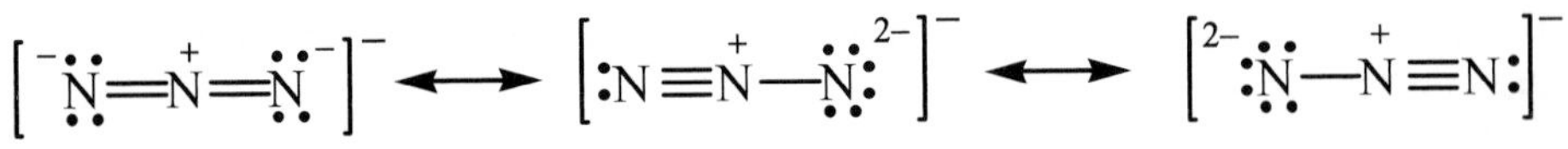

$$[\,^{-}\ddot{N}{=}\overset{+}{N}{=}\ddot{N}^{-}\,]^{-} \longleftrightarrow [\,:N{\equiv}\overset{+}{N}{-}\ddot{N}:^{2-}\,]^{-} \longleftrightarrow [\,^{2-}:\ddot{N}{-}\overset{+}{N}{\equiv}N:\,]^{-}$$

Count the number of electron pairs around the central atom. Since there are two electron pairs around N, the electron arrangement that minimizes electron-pair repulsion is **linear** (AB_2). Remember, for VSEPR purposes a multiple bond counts the same as a single bond.

We conclude that N is ***sp* hybridized** because it has the electron arrangement of two *sp* hybrid orbitals.

10.42 Hybridization of Atomic Orbitals, Problem Type 3.

Strategy: The steps for determining the hybridization of the central atom in a molecule are:

draw Lewis Structure of the molecule → use VSEPR to determine the electron pair arrangement surrounding the central atom (Table 10.1 of the text) → use Table 10.4 of the text to determine the hybridization state of the central atom

Solution:

Write the Lewis structure of the molecule.

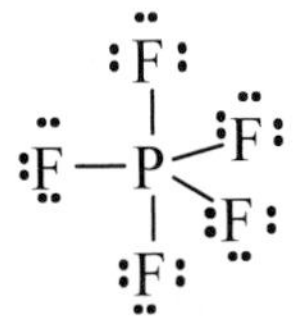

PF_5 (Lewis structure: P bonded to five F atoms, each F with three lone pairs)

Count the number of electron pairs around the central atom. Since there are five electron pairs around P, the electron arrangement that minimizes electron-pair repulsion is **trigonal bipyramidal** (AB_5).

We conclude that P is **sp^3d hybridized** because it has the electron arrangement of five sp^3d hybrid orbitals.

10.44 A single bond is usually a sigma bond, a double bond is usually a sigma bond and a pi bond, and a triple bond is always a sigma bond and two pi bonds. Therefore, there are **nine pi bonds** and **nine sigma bonds** in the molecule.

10.50 In order for the two hydrogen atoms to combine to form a H_2 molecule, the electrons must have opposite spins. Furthermore, the combined energy of the two atoms must not be too great. Otherwise, the H_2 molecule will possess too much energy and will break apart into two hydrogen atoms.

10.52 The electron configurations are listed. Refer to Table 10.5 of the text for the molecular orbital diagram.

Li_2: $(\sigma_{1s})^2(\sigma_{1s}^{\star})^2(\sigma_{2s})^2$ bond order $= 1$

Li_2^+: $(\sigma_{1s})^2(\sigma_{1s}^{\star})^2(\sigma_{2s})^1$ bond order $= \frac{1}{2}$

Li_2^-: $(\sigma_{1s})^2(\sigma_{1s}^{\star})^2(\sigma_{2s})^2(\sigma_{2s}^{\star})^1$ bond order $= \frac{1}{2}$

Order of increasing stability: $\mathbf{Li_2^-} = \mathbf{Li_2^+} < \mathbf{Li_2}$

In reality, Li_2^+ is more stable than Li_2^- because there is less electrostatic repulsion in Li_2^+.

10.54 See Table 10.5 of the text. Removing an electron from B_2 (bond order = 1) gives B_2^+, which has a bond order of (1/2). Therefore, $\mathbf{B_2^+}$ has a weaker and longer bond than B_2.

10.56 In both the Lewis structure and the molecular orbital energy level diagram (Table 10.5 of the text), the oxygen molecule has a double bond (bond order = 2). The principal difference is that the molecular orbital treatment predicts that the molecule will have two unpaired electrons (paramagnetic). Experimentally this is found to be true.

10.58 We refer to Table 10.5 of the text.

O_2 has a bond order of 2 and is paramagnetic (two unpaired electrons).

O_2^+ has a bond order of 2.5 and is paramagnetic (one unpaired electron).

O_2^- has a bond order of 1.5 and is paramagnetic (one unpaired electron).

O_2^{2-} has a bond order of 1 and is diamagnetic.

Based on molecular orbital theory, the stability of these molecules increases as follows:

$$\mathbf{O_2^{2-} < O_2^- < O_2 < O_2^+}$$

10.60 As discussed in the text (see Table 10.5), the single bond in B_2 is a pi bond (the electrons are in a pi *bonding* molecular orbital) and the double bond in C_2 is made up of two pi bonds (the electrons are in the pi *bonding* molecular orbitals).

10.64 The symbol on the left shows the pi bond delocalized over the entire molecule. The symbol on the right shows only one of the two resonance structures of benzene; it is an incomplete representation.

10.66 **(a)** Two Lewis resonance forms are shown below. Formal charges different than zero are indicated.

$$\ddot{O}=\overset{\overset{:\ddot{F}:}{|}}{\underset{+}{N}}-\ddot{O}:^{-} \longleftrightarrow {}^{-}:\ddot{O}-\overset{\overset{:\ddot{F}:}{|}}{\underset{+}{N}}=\ddot{O}$$

(b) There are no lone pairs on the nitrogen atom; it should have a trigonal planar electron pair arrangement and therefore use sp^2 hybrid orbitals.

(c) The bonding consists of sigma bonds joining the nitrogen atom to the fluorine and oxygen atoms. In addition there is a pi molecular orbital delocalized over the N and O atoms. Is nitryl fluoride isoelectronic with the carbonate ion?

10.68 The Lewis structures of ozone are:

$$\ddot{O}=\ddot{O}-\ddot{O}: \longleftrightarrow :\ddot{O}-\ddot{O}=\ddot{O}$$

The central oxygen atom is sp^2 hybridized (AB_2E). The unhybridized $2p_z$ orbital on the central oxygen overlaps with the $2p_z$ orbitals on the two end atoms.

10.70 Molecular Geometry, Problem Type 1.

Strategy: The sequence of steps in determining molecular geometry is as follows:

draw Lewis structure ⟶ find arrangement of electrons pairs ⟶ find arrangement of bonding pairs ⟶ determine geometry based on bonding pairs

Solution:

Write the Lewis structure of the molecule.

$$:\ddot{Br}-Hg-\ddot{Br}:$$

Count the number of electron pairs around the central atom. There are two electron pairs around Hg.

Since there are two electron pairs around Hg, the electron-pair arrangement that minimizes electron-pair repulsion is **linear**.

In addition, since there are no lone pairs around the central atom, the geometry is also **linear** (AB_2).

You could establish the geometry of $HgBr_2$ by measuring its dipole moment. If mercury(II) bromide were bent, it would have a measurable dipole moment. Experimentally, it has no dipole moment and therefore must be linear.

10.72 According to valence bond theory, a pi bond is formed through the side-to-side overlap of a pair of p orbitals. As atomic size increases, the distance between atoms is too large for p orbitals to overlap effectively in a side-to-side fashion. If two orbitals overlap poorly, that is, they share very little space in common, then the resulting bond will be very weak. This situation applies in the case of pi bonds between silicon atoms as well as between any other elements not found in the second period. It is usually far more energetically favorable for silicon, or any other heavy element, to form two single (sigma) bonds to two other atoms than to form a double bond (sigma + pi) to only one other atom.

10.74 The Lewis structures and VSEPR geometries of these species are shown below. The three nonbonding pairs of electrons on each fluorine atom have been omitted for simplicity.

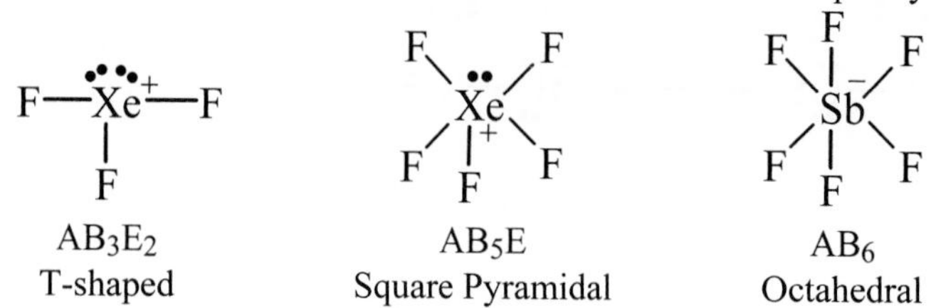

AB_3E_2 T-shaped | AB_5E Square Pyramidal | AB_6 Octahedral

10.76 To predict the bond angles for the molecules, you would have to draw the Lewis structure and determine the geometry using the VSEPR model. From the geometry, you can predict the bond angles.

(a) $BeCl_2$: AB_2 type, 180° (linear).

(b) BCl_3: AB_3 type, 120° (trigonal planar).

(c) CCl_4: AB_4 type, 109.5° (tetrahedral).

(d) CH_3Cl: AB_4 type, 109.5° (tetrahedral with a possible slight distortion resulting from the different sizes of the chlorine and hydrogen atoms).

(e) Hg_2Cl_2: Each mercury atom is of the AB_2 type. The entire molecule is linear, 180° bond angles.

(f) $SnCl_2$: AB_2E type, roughly 120° (bent).

(g) H_2O_2: The atom arrangement is HOOH. Each oxygen atom is of the AB_2E_2 type and the H–O–O angles will be roughly 109.5°.

(h) SnH_4: AB_4 type, 109.5° (tetrahedral).

10.78 Since arsenic and phosphorus are both in the same group of the periodic table, this problem is exactly like Problem 10.42. AsF_5 is an AB_5 type molecule, so the geometry is trigonal bipyramidal. We conclude that As is ***sp^3d* hybridized** because it has the electron arrangement of five sp^3d hybrid orbitals.

10.80 Only ICl_2^- and $CdBr_2$ will be linear. The rest are bent.

10.82 **(a)** Hybridization of Atomic Orbitals, Problem Type 3.

Strategy: The steps for determining the hybridization of the central atom in a molecule are:

draw Lewis Structure of the molecule → use VSEPR to determine the electron pair arrangement surrounding the central atom (Table 10.1 of the text) → use Table 10.4 of the text to determine the hybridization state of the central atom

Solution:

The geometry around each nitrogen is identical. To complete an octet of electrons around N, you must add a lone pair of electrons. Count the number of electron pairs around N. There are three electron pairs around each N.

Since there are three electron pairs around N, the electron-pair arrangement that minimizes electron-pair repulsion is **trigonal planar**.

We conclude that each N is sp^2 **hybridized** because it has the electron arrangement of three sp^2 hybrid orbitals.

(b) Predicting Dipole Moments, Problem Type 2.

Strategy: Keep in mind that the dipole moment of a molecule depends on both the difference in electronegativities of the elements present and its geometry. A molecule can have polar bonds (if the bonded atoms have different electronegativities), but it may not possess a dipole moment if it has a highly symmetrical geometry.

Solution: An N–F bond is polar because F is more electronegative than N. The structure on the right has a dipole moment because the two N–F bond moments do not cancel each other out and so the molecule has a net dipole moment. On the other hand, the two N–F bond moments in the left-hand structure cancel. The sum or resultant dipole moment will be *zero*.

10.84 In 1,2-dichloroethane, the two C atoms are joined by a sigma bond. Rotation about a sigma bond does not destroy the bond, and the bond is therefore free (or relatively free) to rotate. Thus, all angles are permitted and the molecule is nonpolar because the C–Cl bond moments cancel each other because of the averaging effect brought about by rotation. In *cis*-dichloroethylene the two C–Cl bonds are locked in position. The π bond between the C atoms prevents rotation (in order to rotate, the π bond must be broken, using an energy source such as light or heat). Therefore, there is no rotation about the C=C in *cis*-dichloroethylene, and the molecule is polar.

10.86 O_3, CO, CO_2, NO_2, N_2O, CH_4, and $CFCl_3$ are greenhouse gases.

10.88 The Lewis structure is:

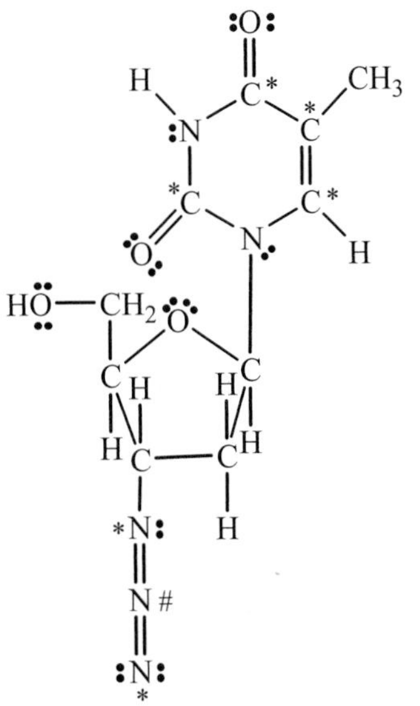

```
              :O:
               ||
      H       C*       CH3
        \   /    \ *  /
         :N        C
          |        ||
         *C        C*
        //  \    /    \
       :O:    N:        H
HO—CH2  :O:    |
   |   /       |
   C  H      H C
   |\|        |/|
   H C———————C H
     |        |
   *N:        H
     ||
     N #
     ||
    :N:
     *
```

The carbon atoms and nitrogen atoms marked with an asterisk (C∗ and N∗) are sp^2 hybridized; unmarked carbon atoms and nitrogen atoms are sp^3 hybridized; and the nitrogen atom marked with (#) is *sp* hybridized.

10.90 C has no *d* orbitals but Si does (3*d*). Thus, H_2O molecules can add to Si in hydrolysis (valence-shell expansion).

10.92 The carbons are in sp^2 hybridization states. The nitrogens are in the sp^3 hybridization state, except for the ring nitrogen double-bonded to a carbon that is sp^2 hybridized. The oxygen atom is sp^2 hybridized.

10.94 **(a)** Use a conventional oven. A microwave oven would not cook the meat from the outside toward the center (it penetrates).

(b) Polar molecules absorb microwaves and would interfere with the operation of radar.

(c) Too much water vapor (polar molecules) absorbed the microwaves, interfering with the operation of radar.

10.96 The smaller size of F compared to Cl results in a shorter F–F bond than a Cl–Cl bond. The closer proximity of the lone pairs of electrons on the F atoms results in greater electron-electron repulsions that weaken the bond.

10.98 $1\text{ D} = 3.336 \times 10^{-30}\text{ C·m}$

electronic charge (e) $= 1.6022 \times 10^{-19}\text{ C}$

$$\frac{\mu}{ed} \times 100\% = \frac{1.92\ \text{D} \times \dfrac{3.336 \times 10^{-30}\ \text{C·m}}{1\ \text{D}}}{(1.6022 \times 10^{-19}\ \text{C}) \times (91.7 \times 10^{-12}\ \text{m})} \times 100\% = \mathbf{43.6\%}\ \text{ionic character}$$

10.100 The second and third vibrational motions are responsible for CO_2 to behave as a greenhouse gas. CO_2 is a nonpolar molecule. The second and third vibrational motions, create a changing dipole moment. The first vibration, a symmetric stretch, does *not* create a dipole moment. Since CO, NO_2, and N_2O are all polar molecules, they will also act as greenhouse gases.

10.102 **(a)** A σ bond is formed by orbitals overlapping end-to-end. Rotation will not break this end-to-end overlap. A π bond is formed by the sideways overlapping of orbitals. The two 90° rotations (180° total) will break and then reform the pi bond, thereby converting *cis*-dichloroethylene to *trans*-dichloroethylene.

(b) The pi bond is weaker because of the lesser extent of sideways orbital overlap, compared to the end-to-end overlap in a sigma bond.

(c) The bond enthalpy is given in the unit, kJ/mol. To find the longest wavelength of light needed to bring about the conversion from *cis* to *trans*, we need the energy to break a pi bond in a single molecule. We convert from kJ/mol to J/molecule.

$$\frac{270\ \text{kJ}}{1\ \text{mol}} \times \frac{1\ \text{mol}}{6.022 \times 10^{23}\ \text{molecules}} = 4.48 \times 10^{-22}\ \text{kJ/molecule} = 4.48 \times 10^{-19}\ \text{J/molecule}$$

Now that we have the energy needed to cause the conversion from *cis* to *trans* in one molecule, we can calculate the wavelength from this energy.

$$E = \frac{hc}{\lambda}$$

$$\lambda = \frac{hc}{E} = \frac{(6.63 \times 10^{-34}\ \text{J·s})(3.00 \times 10^{8}\ \text{m/s})}{4.48 \times 10^{-19}\ \text{J}}$$

$$\lambda = \mathbf{4.44 \times 10^{-7}\ m = 444\ nm}$$

10.104 In each case, we examine the molecular orbital that is occupied by the valence electrons of the molecule to see if it is a bonding or antibonding molecular orbital. If the electron is in a bonding molecular orbital, it is more stable than an electron in an atomic orbital (1*s* or 2*p* atomic orbital) and thus will have a higher ionization energy compared to the lone atom. On the other hand, if the electron is in an antibonding molecular orbital, it is less stable than an electron in an atomic orbital (1*s* or 2*p* atomic orbital) and thus will have a lower ionization energy compared to the lone atom. Refer to Table 10.5 of the text.

(a) H_2 **(b)** N_2 **(c)** O **(d)** F

10.106 **(a)** Looking at the electronic configuration for N_2 shown in Table 10.5 of the text, we write the electronic configuration for P_2.

$$[Ne_2](\sigma_{3s})^2(\sigma_{3s}^\star)^2(\pi_{3p_y})^2(\pi_{3p_z})^2(\sigma_{3p_x})^2$$

(b) Past the Ne_2 core configuration, there are 8 bonding electrons and 2 antibonding electrons. The bond order is:

$$\textbf{bond order} = \tfrac{1}{2}(8-2) = \mathbf{3}$$

(c) All the electrons in the electronic configuration are paired. P_2 is **diamagnetic**.

10.108 The Lewis structure shows 4 pairs of electrons on the two oxygen atoms. From Table 10.5 of the text, we see that these 8 valence electrons are placed in the σ_{2p_x}, π_{2p_y}, π_{2p_z}, $\pi_{2p_y}^\star$, and $\pi_{2p_z}^\star$ orbitals. For all the electrons to be paired, energy is needed to flip the spin in one of the antibonding molecular orbitals ($\pi_{2p_y}^\star$ or $\pi_{2p_z}^\star$). According to Hund's rule, this arrangement is less stable than the ground-state configuration shown in Table 10.5, and hence the Lewis structure shown actually corresponds to an excited state of the oxygen molecule.

10.110 **(a)** Although the O atoms are sp^3 hybridized, they are locked in a planar structure by the benzene rings. The molecule is symmetrical and therefore does not possess a dipole moment.

(b) 20 σ bonds and 6 π bonds.

10.112 **(a)** $:\overset{-}{C}\equiv\overset{+}{O}:$

This is the only reasonable Lewis structure for CO. The electronegativity difference suggests that electron density should concentrate on the O atom, but assigning formal charges places a negative charge on the C atom. Therefore, we expect CO to have a small dipole moment.

(b) The Lewis structure shows a triple bond. The molecular orbital description gives a bond order of 3, just like N_2 (see Table 10.5 of the text).

(c) Normally, we would expect the more electronegative atom to bond with the metal ion. In this case, however, the small dipole moment suggests that the C atom may form a stronger bond with Fe^{2+} than O. More elaborate analysis of the orbitals involved shows that this is indeed the case.

CHAPTER 11
INTERMOLECULAR FORCES AND LIQUIDS AND SOLIDS

PROBLEM-SOLVING STRATEGIES AND TUTORIAL SOLUTIONS

TYPES OF PROBLEMS

Problem Type 1: Identifying Intermolecular Forces.

Problem Type 2: Identifying Hydrogen Bonds.

Problem Type 3: Counting the Number of Atoms in a Unit Cell.

Problem Type 4: Calculating Density from Crystal Structure and Atomic Radius.

PROBLEM TYPE 1: IDENTIFYING INTERMOLECULAR FORCES

Intermolecular forces are attractive forces between molecules. Intermolecular forces account for the existence of the condensed states of matter--liquids and solids. To understand the properties of condensed matter, you must understand the different types of intermolecular forces.

(1) **Dipole-dipole forces:** These are attractive forces that act between polar molecules, that is, between molecules that possess dipole moments (see Section 11.2 of the text). The larger the dipole moments, the greater the attractive force.

(2) **Ion-dipole forces:** These are attractive forces that occur between an ion (either a cation or an anion) and a polar molecule (see Figure 11.2 of the text). The strength of this interaction depends on the charge and size of the ion and on the magnitude of the dipole moment and the size of the molecule.

(3) **Dispersion forces:** These are attractive forces that arise as a result of temporary dipoles induced in the atoms or molecules. This is the only type of intermolecular force that exists between nonpolar atoms or molecules. The likelihood of a dipole moment being induced depends on the polarizability of the atom or molecule. *Polarizability* is the ease with which the electron distribution in the atom or molecule can be distorted. Dispersion forces usually increase with molar mass.

For a complete discussion of intermolecular forces, see Section 11.2 of the text.

EXAMPLE 11.1
Indicate all the different types of intermolecular forces that exist in each of the following substances: (a) $CCl_4(l)$ and (b) $HBr(l)$.

Strategy: Classify the species into three categories: ionic, polar (possessing a dipole moment), and nonpolar. Keep in mind that dispersion forces exist between *all* species.

Solution:

(a) CCl_4 is nonpolar. The only type of intermolecular forces present in nonpolar molecules is *dispersion forces*.

(b) HBr is a polar molecule. The types of intermolecular forces present are *dipole-dipole* and *dispersion forces*. There is no hydrogen bonding in HBr. The Br atom is too large and is not electronegative enough.

PRACTICE EXERCISE

1. Which of the following substances should have the strongest intermolecular attractive forces: N_2, Ar, F_2, or Cl_2?

2. The dipole moment (μ) in HCl is 1.03 D, and in HCN it is 2.99 D. Which substance should have the higher boiling point?

Text Problems: 11.8, 11.10, 11.16, **11.18**

PROBLEM TYPE 2: IDENTIFYING HYDROGEN BONDS

The boiling points of NH_3, H_2O, and HF are much higher than expected if the boiling point is solely based on molar mass (see Figure 11.6 of the text). The high boiling points in these compounds are due to extensive *hydrogen bonding*. The hydrogen bond is a special type of dipole-dipole interaction between the hydrogen atom in a polar bond, such as N–H, O–H, or F–H, and an electronegative O, N, or F atom. This interaction is written

$$\text{A–H}\cdots\text{B} \qquad \text{or} \qquad \text{A–H}\cdots\text{A}$$

where A and B represent O, N, or F. A–H is one molecule or part of a molecule and B is a part of another molecule. The dotted line represents the hydrogen bond.

The average energy of a hydrogen bond is quite large for a dipole-dipole interaction (up to 40 kJ/mol). Thus, hydrogen bonds are a powerful force in determining the structures and properties of many compounds.

EXAMPLE 11.2

Predict whether hydrogen bonding intermolecular attractions are present in the following substances: (a) $CH_3OH(l)$ and (b) $CH_3CH_2OCH_2CH_3(l)$.

Strategy: If a molecule contains an N–H, O–H, or F–H bond it can form intermolecular hydrogen bonds. A hydrogen bond is a particularly strong dipole-dipole intermolecular attraction.

Solution:

(a) CH_3OH (methanol) is polar and has a hydrogen atom bound to an oxygen atom. Hydrogen bonding intermolecular forces are present between CH_3OH molecules. Dipole-dipole and dispersion forces are also present.

(b) $CH_3CH_2OCH_2CH_3$ (diethyl ether) is polar and does contain both oxygen atoms and hydrogen atoms. However, all the hydrogen atoms are bonded to carbon, not oxygen. There is no hydrogen bonding in diethyl ether, because carbon is not electronegative enough. The intermolecular attractive forces present are dipole-dipole and dispersion forces.

PRACTICE EXERCISE

3. Which member of each pair has the stronger intermolecular forces of attraction:
(a) H_2O or H_2S, (b) HCl or HF, and (c) NH_3 or PH_3?

Text Problems: 11.12, **11.14, 11.18**, 11.20

PROBLEM TYPE 3: COUNTING THE NUMBER OF ATOMS IN A UNIT CELL

To solve this problem, you must realize that because every unit cell in a crystalline solid is adjacent to other unit cells, most of a cell's atoms are shared by neighboring cells. In cubic cells, each corner atom is shared by eight unit cells [see Figure 11.19(a) of the text]. A face-centered atom is shared by two unit cells [See Figure 11.19(c) of the text].

A center atom in a unit cell is solely contained by that unit cell and is not shared. An edge atom is shared by four unit cells. Table 11.1 summarizes this information.

TABLE 11.1

Type of atom	Amount of atom contained in unit cell
Corner	1/8
Edge	1/4
Face-centered	1/2
Center	1

EXAMPLE 11.3

If atoms of a solid occupy a face-centered cubic lattice, how many atoms are there per unit cell?

Strategy: Recall that a corner atom is shared with 8 unit cells and therefore only 1/8 of corner atom is within a given unit cell. Also recall that a face atom is shared with 2 unit cells and therefore 1/2 of a face atom is within a given unit cell. See Figure 11.19 of the text.

Solution: In a face-centered cubic unit cell, there are atoms at each of the eight corners, and there is one atom in each of the six faces.

(8 corner atoms)(1/8 atom per corner) + (6 face-centered atoms)(1/2 atom per face) = **4 atoms/unit cell**

PRACTICE EXERCISE

4. Atoms of polonium (Po) occupy a simple cubic lattice. How many Po atoms are there per unit cell?

Text Problems: 11.38, **11.44**

PROBLEM TYPE 4: CALCULATING DENSITY FROM CRYSTAL STRUCTURE AND ATOMIC RADIUS

Step 1: To solve this type of problem, you must know how many atoms are contained in the different types of cubic unit cells. Table 11.2 summarizes the number of atoms per unit cell.

TABLE 11.2

Type of cubic unit cell	Number of atoms/unit cell
Simple cubic	1
Body-centered cubic	2
Face-centered cubic	4

Step 2: You can look up the relationship between edge length (a) and atomic radius (r). See Figure 11.22 of the text. Table 11.3 summarizes these relationships.

TABLE 11.3

Type of cubic unit cell	Relationship between a and r
Simple cubic	$a = 2r$
Body-centered cubic	$a = \frac{4}{\sqrt{3}}r$
Face-centered cubic	$a = \sqrt{8}\,r$

Step 3: Density = mass/volume. The volume of a cube is equal to the edge length cubed.

$$V = a^3$$

You will know how many atoms are in a unit cell (*Step 1* above). To calculate the mass, you need to convert from atoms/unit cell to grams/unit cell. A reasonable strategy would be

$$\frac{\text{atoms}}{\text{unit cell}} \rightarrow \frac{\text{mol}}{\text{unit cell}} \rightarrow \frac{\text{grams}}{\text{unit cell}}$$

Step 4: Substitute the mass and volume into the density equation to solve for the density.

EXAMPLE 11.4

Nickel crystallizes in a face-centered cubic lattice with an edge length of 352 pm. Calculate the density of nickel.

Strategy: To calculate the density, we need to find the mass and the volume of the unit cell. The volume of the unit cell can be calculated from the edge length. Because nickel crystallizes in a face-centered cubic lattice, there are four Ni atoms per unit cell. We can convert from 4 atoms/unit cell to grams/unit cell.

Solution: Density = mass/volume. The volume of a cube is equal to the edge length cubed.

$$V = a^3$$

$$V = (325\ \text{pm})^3 = 3.43 \times 10^7\ \text{pm}^3$$

We convert pm^3 to cm^3 because the density of a solid is typically expressed in units of g/cm^3.

$$(3.43 \times 10^7\ \text{pm}^3) \times \left(\frac{1 \times 10^{-12}\ \text{m}}{1\ \text{pm}}\right)^3 \times \left(\frac{1\ \text{cm}}{1 \times 10^{-2}\ \text{m}}\right)^3 = 3.43 \times 10^{-23}\ \text{cm}^3$$

There are four atoms/unit cell. To calculate the mass, you need to convert from atoms/unit cell to grams/unit cell. A reasonable strategy would be:

atoms/unit cell → mol/unit cell → grams/unit cell

$$\frac{4\ \text{Ni atoms}}{\text{unit cell}} \times \frac{1\ \text{mol Ni}}{6.022 \times 10^{23}\ \text{Ni atoms}} \times \frac{58.71\ \text{g Ni}}{1\ \text{mol Ni}} = \frac{3.900 \times 10^{-22}\ \text{g Ni}}{1\ \text{unit cell}}$$

Substitute the mass and volume into the density equation to solve for the density.

$$d = \frac{m}{V} = \frac{3.900 \times 10^{-22}\ \text{g Ni}}{3.43 \times 10^{-23}\ \text{cm}^3} = \mathbf{11.4\ g/cm^3}$$

PRACTICE EXERCISE

5. Potassium crystallizes in a body-centered cubic lattice and has a density of $0.856\ \text{g/cm}^3$ at 25°C.
   (a) How many atoms are there per unit cell?
   (b) What is the length of an edge of the unit cell?

Text Problems: 11.40, 11.42

ANSWERS TO PRACTICE EXERCISES

1. Cl_2 **2.** HCN **3.** **(a)** H_2O **(b)** HF **(c)** NH_3

4. 1 atom/unit cell **5.** **(a)** 2 atoms/unit cell **(b)** $a = 5.34 \times 10^{-8}\ \text{cm} = 534\ \text{pm}$

SOLUTIONS TO SELECTED TEXT PROBLEMS

11.8 Identifying Intermolecular Forces, Problem Type 1.

Strategy: Classify the species into three categories: ionic, polar (possessing a dipole moment), and nonpolar. Keep in mind that dispersion forces exist between *all* species.

Solution: The three molecules are essentially nonpolar. There is little difference in electronegativity between carbon and hydrogen. Thus, the only type of intermolecular attraction in these molecules is dispersion forces. Other factors being equal, the molecule with the greater number of electrons will exert greater intermolecular attractions. By looking at the molecular formulas you can predict that the order of increasing boiling points will be $CH_4 < C_3H_8 < C_4H_{10}$.

On a very cold day, propane and butane would be liquids (boiling points −44.5°C and −0.5°C, respectively); only **methane** would still be a gas (boiling point −161.6°C).

11.10 **(a)** Benzene (C_6H_6) molecules are nonpolar. Only dispersion forces will be present.

(b) Chloroform (CH_3Cl) molecules are polar (why?). Dispersion and dipole-dipole forces will be present.

(c) Phosphorus trifluoride (PF_3) molecules are polar. Dispersion and dipole-dipole forces will be present.

(d) Sodium chloride (NaCl) is an ionic compound. Ion-ion (and dispersion) forces will be present.

(e) Carbon disulfide (CS_2) molecules are nonpolar. Only dispersion forces will be present.

11.12 In this problem you must identify the species capable of hydrogen bonding among themselves, not with water. In order for a molecule to be capable of hydrogen bonding with another molecule like itself, it must have at least one hydrogen atom bonded to N, O, or F. Of the choices, only **(e)** CH_3COOH (acetic acid) shows this structural feature. The others cannot form hydrogen bonds among themselves.

11.14 Identifying Hydrogen Bonds, Problem Type 2.

Strategy: The molecule with the stronger intermolecular forces will have the higher boiling point. If a molecule contains an N–H, O–H, or F–H bond it can form intermolecular hydrogen bonds. A hydrogen bond is a particularly strong dipole-dipole intermolecular attraction.

Solution: 1-butanol has the higher boiling point because the molecules can form hydrogen bonds with each other (It contains an O–H bond). Diethyl ether molecules do contain both oxygen atoms and hydrogen atoms. However, all the hydrogen atoms are bonded to carbon, not oxygen. There is no hydrogen bonding in diethyl ether, because carbon is not electronegative enough.

11.16 **(a)** Xe: it has more electrons and therefore stronger dispersion forces.

(b) CS_2: it has more electrons (both molecules nonpolar) and therefore stronger dispersion forces.

(c) Cl_2: it has more electrons (both molecules nonpolar) and therefore stronger dispersion forces.

(d) LiF: it is an ionic compound, and the ion-ion attractions are much stronger than the dispersion forces between F_2 molecules.

(e) NH_3: it can form hydrogen bonds and PH_3 cannot.

11.18 Identifying Intermolecular Forces and Hydrogen Bonding, Problem Types 1 and 2.

Strategy: Classify the species into three categories: ionic, polar (possessing a dipole moment), and nonpolar. Also look for molecules that contain an N–H, O–H, or F–H bond, which are capable of forming intermolecular hydrogen bonds. Keep in mind that dispersion forces exist between *all* species.

Solution:

(a) Water has O–H bonds. Therefore, water molecules can form hydrogen bonds. The attractive forces that must be overcome are hydrogen bonding and dispersion forces.

(b) Bromine (Br_2) molecules are nonpolar. Only dispersion forces must be overcome.

(c) Iodine (I_2) molecules are nonpolar. Only dispersion forces must be overcome.

(d) In this case, the F–F bond must be broken. This is an *intra*molecular force between two F atoms, not an *inter*molecular force between F_2 molecules. The attractive forces of the covalent bond must be overcome.

11.20 The lower melting compound (shown below) can form hydrogen bonds only with itself (*intra*molecular hydrogen bonds), as shown in the figure. Such bonds do not contribute to *inter*molecular attraction and do not help raise the melting point of the compound. The other compound can form *inter*molecular hydrogen bonds; therefore, it will take a higher temperature to provide molecules of the liquid with enough kinetic energy to overcome these attractive forces to escape into the gas phase.

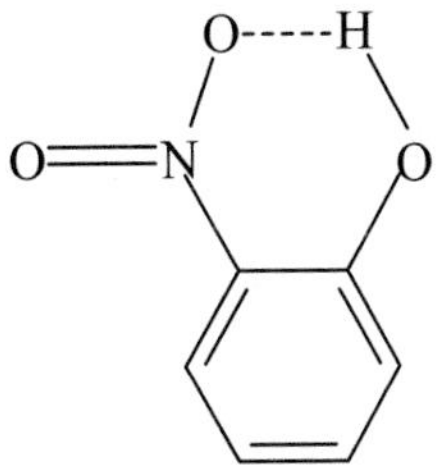

11.32 Ethylene glycol has two –OH groups, allowing it to exert strong intermolecular forces through hydrogen bonding. Its viscosity should fall between ethanol (1 OH group) and glycerol (3 OH groups).

11.38 A corner sphere is shared equally among eight unit cells, so only one-eighth of each corner sphere "belongs" to any one unit cell. A face-centered sphere is divided equally between the two unit cells sharing the face. A body-centered sphere belongs entirely to its own unit cell.

In a *simple cubic cell* there are eight corner spheres. One-eighth of each belongs to the individual cell giving a total of **one** whole sphere per cell. In a *body-centered cubic cell*, there are eight corner spheres and one body-center sphere giving a total of **two** spheres per unit cell (one from the corners and one from the body-center). In a *face-center* sphere, there are eight corner spheres and six face-centered spheres (six faces). The total number of spheres would be **four**: one from the corners and three from the faces.

11.40 Similar to Problem Type 4, Calculating the Density from Crystal Structure and Atomic Radius.

Strategy: The problem gives a generous hint. First, we need to calculate the volume (in cm^3) occupied by 1 mole of Ba atoms. Next, we calculate the volume that a Ba atom occupies. Once we have these two pieces of information, we can multiply them together to end up with the number of Ba atoms per mole of Ba.

$$\frac{\text{number of Ba atoms}}{\cancel{\text{cm}^3}} \times \frac{\cancel{\text{cm}^3}}{1 \text{ mol Ba}} = \frac{\text{number of Ba atoms}}{1 \text{ mol Ba}}$$

Solution: The volume that contains one mole of barium atoms can be calculated from the density using the following strategy:

$$\frac{\text{volume}}{\text{mass of Ba}} \rightarrow \frac{\text{volume}}{\text{mol Ba}}$$

$$\frac{1\text{ cm}^3}{3.50\text{ g Ba}} \times \frac{137.3\text{ g Ba}}{1\text{ mol Ba}} = \frac{39.23\text{ cm}^3}{1\text{ mol Ba}}$$

We carry an extra significant figure in this calculation to limit rounding errors. Next, the volume that contains two barium atoms is the volume of the body-centered cubic unit cell. Some of this volume is empty space because packing is only 68.0 percent efficient. But, this will not affect our calculation.

$$V = a^3$$

Let's also convert to cm^3.

$$V = (502\text{ pm})^3 \times \left(\frac{1 \times 10^{-12}\text{ m}}{1\text{ pm}}\right)^3 \times \left(\frac{1\text{ cm}}{0.01\text{ m}}\right)^3 = \frac{1.265 \times 10^{-22}\text{ cm}^3}{2\text{ Ba atoms}}$$

We can now calculate the number of barium atoms in one mole using the strategy presented above.

$$\frac{\text{number of Ba atoms}}{\text{cm}^3} \times \frac{\text{cm}^3}{1\text{ mol Ba}} = \frac{\text{number of Ba atoms}}{1\text{ mol Ba}}$$

$$\frac{2\text{ Ba atoms}}{1.265 \times 10^{-22}\text{ cm}^3} \times \frac{39.23\text{ cm}^3}{1\text{ mol Ba}} = \mathbf{6.20 \times 10^{23}\text{ atoms/mol}}$$

This is close to Avogadro's number, 6.022×10^{23} particles/mol.

11.42 The mass of the unit cell is the mass in grams of two europium atoms.

$$m = \frac{2\text{ Eu atoms}}{1\text{ unit cell}} \times \frac{1\text{ mol Eu}}{6.022 \times 10^{23}\text{ Eu atoms}} \times \frac{152.0\text{ g Eu}}{1\text{ mol Eu}} = 5.048 \times 10^{-22}\text{ g Eu/unit cell}$$

$$V = \frac{5.048 \times 10^{-22}\text{ g}}{1\text{ unit cell}} \times \frac{1\text{ cm}^3}{5.26\text{ g}} = 9.60 \times 10^{-23}\text{ cm}^3\text{/unit cell}$$

The edge length (a) is:

$$\boldsymbol{a} = V^{1/3} = (9.60 \times 10^{-23}\text{ cm}^3)^{1/3} = 4.58 \times 10^{-8}\text{ cm} = \mathbf{458\text{ pm}}$$

11.44 Similar to Problem Type 4, Counting the Number of Atoms in a Unit Cell.

Strategy: Recall that a corner atom is shared with 8 unit cells and therefore only 1/8 of corner atom is within a given unit cell. Also recall that a face atom is shared with 2 unit cells and therefore 1/2 of a face atom is within a given unit cell. See Figure 11.19 of the text.

Solution: In a face-centered cubic unit cell, there are atoms at each of the eight corners, and there is one atom in each of the six faces. Only one-half of each face-centered atom and one-eighth of each corner atom belongs to the unit cell.

X atoms/unit cell = (8 corner atoms)(1/8 atom per corner) = 1 X atom/unit cell

Y atoms/unit cell = (6 face-centered atoms)(1/2 atom per face) = 3 Y atoms/unit cell

The unit cell is the smallest repeating unit in the crystal; therefore, the empirical formula is $\mathbf{XY_3}$.

11.48 Rearranging the Bragg equation, we have:

$$\lambda = \frac{2d\sin\theta}{n} = \frac{2(282\text{ pm})(\sin 23.0^\circ)}{1} = 220\text{ pm} = \mathbf{0.220\text{ nm}}$$

11.52 See Table 11.4 of the text. The properties listed are those of a **molecular solid**.

11.54 In a molecular crystal the lattice points are occupied by molecules. Of the solids listed, the ones that are composed of molecules are Se_8, HBr, CO_2, P_4O_6, and SiH_4. In covalent crystals, atoms are held together in an extensive three-dimensional network entirely by covalent bonds. Of the solids listed, the ones that are composed of atoms held together by covalent bonds are Si and C.

11.56 In diamond, each carbon atom is covalently bonded to four other carbon atoms. Because these bonds are strong and uniform, diamond is a very hard substance. In graphite, the carbon atoms in each layer are linked by strong bonds, but the layers are bound by weak dispersion forces. As a result, graphite may be cleaved easily between layers and is not hard.

In graphite, all atoms are sp^2 hybridized; each atom is covalently bonded to three other atoms. The remaining unhybridized $2p$ orbital is used in pi bonding forming a delocalized molecular orbital. The electrons are free to move around in this extensively delocalized molecular orbital making graphite a good conductor of electricity in directions along the planes of carbon atoms.

11.78 *Step 1:* Warming ice to the melting point.

$$q_1 = ms\Delta t = (866\text{ g H}_2\text{O})(2.03\text{ J/g}^\circ\text{C})[0-(-10)]^\circ\text{C} = 17.6\text{ kJ}$$

Step 2: Converting ice at the melting point to liquid water at 0°C. (See Table 11.8 of the text for the heat of fusion of water.)

$$q_2 = 866\text{ g H}_2\text{O} \times \frac{1\text{ mol}}{18.02\text{ g H}_2\text{O}} \times \frac{6.01\text{ kJ}}{1\text{ mol}} = 289\text{ kJ}$$

Step 3: Heating water from 0°C to 100°C.

$$q_3 = ms\Delta t = (866\text{ g H}_2\text{O})(4.184\text{ J/g}^\circ\text{C})[(100-0)^\circ\text{C}] = 362\text{ kJ}$$

Step 4: Converting water at 100°C to steam at 100°C. (See Table 11.6 of the text for the heat of vaporization of water.)

$$q_4 = 866\text{ g H}_2\text{O} \times \frac{1\text{ mol}}{18.02\text{ g H}_2\text{O}} \times \frac{40.79\text{ kJ}}{1\text{ mol}} = 1.96\times10^3\text{ kJ}$$

Step 5: Heating steam from 100°C to 126°C.

$$q_5 = ms\Delta t = (866\text{ g H}_2\text{O})(1.99\text{ J/g}^\circ\text{C})[(126-100)^\circ\text{C}] = 44.8\text{ kJ}$$

$$\mathbf{q_{total}} = q_1 + q_2 + q_3 + q_4 + q_5 = \mathbf{2.67\times10^3\text{ kJ}}$$

How would you set up and work this problem if you were computing the heat lost in cooling steam from 126°C to ice at −10°C?

11.80 $\Delta H_{vap} = \Delta H_{sub} - \Delta H_{fus} = 62.30\text{ kJ/mol} - 15.27\text{ kJ/mol} = \mathbf{47.03\text{ kJ/mol}}$

11.82 Two phase changes occur in this process. First, the liquid is turned to solid (freezing), then the solid ice is turned to gas (sublimation).

11.84 When steam condenses to liquid water at 100°C, it releases a large amount of heat equal to the enthalpy of vaporization. Thus steam at 100°C exposes one to more heat than an equal amount of water at 100°C.

11.86 We can use a modified form of the Clausius-Clapeyron equation to solve this problem. See Equation (11.5) in the text.

$P_1 = 40.1\text{ mmHg}$ $\quad$ $\mathbf{P_2 = ?}$

$T_1 = 7.6°C = 280.6\text{ K}$ $\quad$ $T_2 = 60.6°C = 333.6\text{ K}$

$$\ln\frac{P_1}{P_2} = \frac{\Delta H_{vap}}{R}\left(\frac{1}{T_2} - \frac{1}{T_1}\right)$$

$$\ln\frac{40.1}{P_2} = \frac{31000\text{ J/mol}}{8.314\text{ J/K}\cdot\text{mol}}\left(\frac{1}{333.6\text{ K}} - \frac{1}{280.6\text{ K}}\right)$$

$$\ln\frac{40.1}{P_2} = -2.11$$

Taking the antilog of both sides, we have:

$$\frac{40.1}{P_2} = 0.121$$

$$\mathbf{P_2 = 331\text{ mmHg}}$$

11.88 Using Equation (11.5) of the text:

$$\ln\frac{P_1}{P_2} = \frac{\Delta H_{vap}}{R}\left(\frac{1}{T_2} - \frac{1}{T_1}\right)$$

$$\ln\left(\frac{1}{2}\right) = \left(\frac{\Delta H_{vap}}{8.314\text{ J/K}\cdot\text{mol}}\right)\left(\frac{1}{368\text{ K}} - \frac{1}{358\text{ K}}\right) = \Delta H_{vap}\left(\frac{-7.59\times10^{-5}}{8.314\text{ J/mol}}\right)$$

$$\Delta H_{vap} = 7.59\times10^{4}\text{ J/mol} = \mathbf{75.9\text{ kJ/mol}}$$

11.92 Initially, the ice melts because of the increase in pressure. As the wire sinks into the ice, the water above the wire refreezes. Eventually the wire actually moves completely through the ice block without cutting it in half.

11.94 Region labels: The region containing point A is the solid region. The region containing point B is the liquid region. The region containing point C is the gas region.

(a) Raising the temperature at constant pressure beginning at A implies starting with solid ice and warming until melting occurs. If the warming continued, the liquid water would eventually boil and change to steam. Further warming would increase the temperature of the steam.

(b) At point C water is in the gas phase. Cooling without changing the pressure would eventually result in the formation of solid ice. Liquid water would never form.

(c) At B the water is in the liquid phase. Lowering the pressure without changing the temperature would eventually result in boiling and conversion to water in the gas phase.

11.96 **(a)** A low surface tension means the attraction between molecules making up the surface is weak. Water has a high surface tension; water bugs could not "walk" on the surface of a liquid with a low surface tension.

(b) A low critical temperature means a gas is very difficult to liquefy by cooling. This is the result of weak intermolecular attractions. Helium has the lowest known critical temperature (5.3 K).

(c) A low boiling point means weak intermolecular attractions. It takes little energy to separate the particles. All ionic compounds have extremely high boiling points.

(d) A low vapor pressure means it is difficult to remove molecules from the liquid phase because of high intermolecular attractions. Substances with low vapor pressures have high boiling points (why?).

Thus, only choice **(d)** indicates strong intermolecular forces in a liquid. The other choices indicate weak intermolecular forces in a liquid.

11.98 The properties of hardness, high melting point, poor conductivity, and so on, could place boron in either the ionic or covalent categories. However, boron atoms will not alternately form positive and negative ions to achieve an ionic crystal. The structure is **covalent** because the units are single boron atoms.

11.100 **CCl_4**. Generally, the larger the number of electrons and the more diffuse the electron cloud in an atom or a molecule, the greater its polarizability. Recall that polarizability is the ease with which the electron distribution in an atom or molecule can be distorted.

11.102 The vapor pressure of mercury (as well as all other substances) is 760 mmHg at its normal boiling point.

11.104 It has reached the critical point; the point of critical temperature (T_c) and critical pressure (P_c).

11.106 Crystalline SiO_2. Its regular structure results in a more efficient packing.

11.108 **(a)** **False**. Permanent dipoles are usually much stronger than temporary dipoles.

(b) **False**. The hydrogen atom must be bonded to N, O, or F.

(c) **True**.

(d) **False**. The magnitude of the attraction depends on both the ion charge and the polarizability of the neutral atom or molecule.

11.110 Sublimation temperature is –78°C or 195 K at a pressure of 1 atm.

$$\ln\frac{P_1}{P_2} = \frac{\Delta H_{\text{sub}}}{R}\left(\frac{1}{T_2} - \frac{1}{T_1}\right)$$

$$\ln\frac{1}{P_2} = \frac{25.9 \times 10^3\ \text{J/mol}}{8.314\ \text{J/mol}\cdot\text{K}}\left(\frac{1}{150\ \text{K}} - \frac{1}{195\ \text{K}}\right)$$

$$\ln\frac{1}{P_2} = 4.79$$

Taking the antiln of both sides gives:

$$P_2 = \mathbf{8.3 \times 10^{-3}\ atm}$$

11.112 **(a)** K_2S: Ionic forces are much stronger than the dipole-dipole forces in $(CH_3)_3N$.

(b) Br_2: Both molecules are nonpolar; but Br_2 has more electrons. (The boiling point of Br_2 is 50°C and that of C_4H_{10} is −0.5°C.)

11.114 CH_4 is a tetrahedral, nonpolar molecule that can only exert weak dispersion type attractive forces. SO_2 is bent (why?) and possesses a dipole moment, which gives rise to stronger dipole-dipole attractions. Sulfur dioxide will have a larger value of "*a*" in the van der Waals equation (*a* is a measure of the strength of the interparticle attraction) and will behave less like an ideal gas than methane.

11.116 The standard enthalpy change for the formation of gaseous iodine from solid iodine is simply the difference between the standard enthalpies of formation of the products and the reactants in the equation:

$$I_2(s) \rightarrow I_2(g)$$

$$\Delta H_{vap} = \Delta H_f^\circ[I_2(g)] - \Delta H_f^\circ[I_2(s)] = 62.4\text{ kJ/mol} - 0\text{ kJ/mol} = \mathbf{62.4\ kJ/mol}$$

11.118 Smaller ions have more concentrated charges (charge densities) and are more effective in ion-dipole interaction. The greater the ion-dipole interaction, the larger is the heat of hydration.

11.120 **(a)** For the process: $Br_2(l) \rightarrow Br_2(g)$

$$\boldsymbol{\Delta H°} = \Delta H_f^\circ[Br_2(g)] - \Delta H_f^\circ[Br_2(l)] = (1)(30.7\text{ kJ/mol}) - 0 = \mathbf{30.7\ kJ/mol}$$

(b) For the process: $Br_2(g) \rightarrow 2Br(g)$

$\Delta H° = $ **192.5 kJ/mol** (from Table 9.4 of the text)

As expected, the bond enthalpy represented in part (b) is much greater than the energy of vaporization represented in part (a). It requires more energy to break the bond than to vaporize the molecule.

11.122 **(a)** Decreases **(b)** No change **(c)** No change

11.124 $CaCO_3(s) \rightarrow CaO(s) + CO_2(g)$

Three phases (two solid and one gas). $CaCO_3$ and CaO constitute two separate solid phases because they are separated by well-defined boundaries.

11.126 SiO_2 has an extensive three-dimensional structure. CO_2 exists as discrete molecules. It will take much more energy to break the strong network covalent bonds of SiO_2; therefore, SiO_2 has a much higher boiling point than CO_2.

11.128 The moles of water vapor can be calculated using the ideal gas equation.

$$n = \frac{PV}{RT} = \frac{\left(187.5\ \text{mmHg} \times \frac{1\ \text{atm}}{760\ \text{mmHg}}\right)(5.00\ \text{L})}{\left(0.0821\frac{\text{L}\cdot\text{atm}}{\text{mol}\cdot\text{K}}\right)(338\ \text{K})} = 0.0445\ \text{mol}$$

$$\text{mass of water vapor} = 0.0445\ \text{mol} \times 18.02\ \text{g/mol} = 0.802\ \text{g}$$

Now, we can calculate the percentage of the 1.20 g sample of water that is vapor.

$$\textbf{\% of H}_2\textbf{O vaporized} = \frac{0.802\ \text{g}}{1.20\ \text{g}} \times 100\% = \mathbf{66.8\%}$$

11.130 The packing efficiency is: $\frac{\text{volume of atoms in unit cell}}{\text{volume of unit cell}} \times 100\%$

An atom is assumed to be spherical, so the volume of an atom is $(4/3)\pi r^3$. The volume of a cubic unit cell is a^3 (a is the length of the cube edge). The packing efficiencies are calculated below:

(a) Simple cubic cell: cell edge (a) = $2r$

$$\text{Packing efficiency} = \frac{\left(\frac{4\pi r^3}{3}\right) \times 100\%}{(2r)^3} = \frac{4\pi r^3 \times 100\%}{24r^3} = \frac{\pi}{6} \times 100\% = \mathbf{52.4\%}$$

(b) Body-centered cubic cell: cell edge $= \frac{4r}{\sqrt{3}}$

$$\text{Packing efficiency} = \frac{2 \times \left(\frac{4\pi r^3}{3}\right) \times 100\%}{\left(\frac{4r}{\sqrt{3}}\right)^3} = \frac{2 \times \left(\frac{4\pi r^3}{3}\right) \times 100\%}{\left(\frac{64r^3}{3\sqrt{3}}\right)} = \frac{2\pi\sqrt{3}}{16} \times 100\% = \mathbf{68.0\%}$$

Remember, there are two atoms per body-centered cubic unit cell.

(c) Face-centered cubic cell: cell edge $= \sqrt{8}r$

$$\text{Packing efficiency} = \frac{4 \times \left(\frac{4\pi r^3}{3}\right) \times 100\%}{\left(\sqrt{8}r\right)^3} = \frac{\left(\frac{16\pi r^3}{3}\right) \times 100\%}{8r^3\sqrt{8}} = \frac{2\pi}{3\sqrt{8}} \times 100\% = \mathbf{74.0\%}$$

Remember, there are four atoms per face-centered cubic unit cell.

11.132 For a face-centered cubic unit cell, the length of an edge (a) is given by:

$$a = \sqrt{8}r$$

$$a = \sqrt{8}\,(191\,\text{pm}) = 5.40 \times 10^2\ \text{pm}$$

The volume of a cube equals the edge length cubed (a^3).

$$V = a^3 = (5.40 \times 10^2 \text{ pm})^3 \times \left(\frac{1 \times 10^{-12} \text{ m}}{1 \text{ pm}}\right)^3 \times \left(\frac{1 \text{ cm}}{1 \times 10^{-2} \text{ m}}\right)^3 = 1.57 \times 10^{-22} \text{ cm}^3$$

Now that we have the volume of the unit cell, we need to calculate the mass of the unit cell in order to calculate the density of Ar. The number of atoms in one face centered cubic unit cell is four.

$$m = \frac{4 \text{ atoms}}{1 \text{ unit cell}} \times \frac{1 \text{ mol}}{6.022 \times 10^{23} \text{ atoms}} \times \frac{39.95 \text{ g}}{1 \text{ mol}} = \frac{2.65 \times 10^{-22} \text{ g}}{1 \text{ unit cell}}$$

$$d = \frac{m}{V} = \frac{2.65 \times 10^{-22} \text{ g}}{1.57 \times 10^{-22} \text{ cm}^3} = \mathbf{1.69 \text{ g/cm}^3}$$

11.134 **(a)** Two triple points: Diamond/graphite/liquid and graphite/liquid/vapor.

(b) Diamond.

(c) Apply high pressure at high temperature.

11.136 The cane is made of many molecules held together by intermolecular forces. The forces are strong and the molecules are packed tightly. Thus, when the handle is raised, all the molecules are raised because they are held together.

11.138 When the tungsten filament inside the bulb is heated to a high temperature (about 3000°C), the tungsten sublimes (solid → gas phase transition) and then it condenses on the inside walls of the bulb. The inert, pressurized Ar gas retards sublimation and oxidation of the tungsten filament.

11.140 The fuel source for the Bunsen burner is most likely methane gas. When methane burns in air, carbon dioxide and water are produced.

$$CH_4(g) + 2O_2(g) \rightarrow CO_2(g) + 2H_2O(g)$$

The water vapor produced during the combustion condenses to liquid water when it comes in contact with the outside of the cold beaker.

11.142 First, we need to calculate the volume (in cm^3) occupied by 1 mole of Fe atoms. Next, we calculate the volume that a Fe atom occupies. Once we have these two pieces of information, we can multiply them together to end up with the number of Fe atoms per mole of Fe.

$$\frac{\text{number of Fe atoms}}{\text{cm}^3} \times \frac{\text{cm}^3}{1 \text{ mol Fe}} = \frac{\text{number of Fe atoms}}{1 \text{ mol Fe}}$$

The volume that contains one mole of iron atoms can be calculated from the density using the following strategy:

$$\frac{\text{volume}}{\text{mass of Fe}} \rightarrow \frac{\text{volume}}{\text{mol Fe}}$$

$$\frac{1 \text{ cm}^3}{7.874 \text{ g Fe}} \times \frac{55.85 \text{ g Fe}}{1 \text{ mol Fe}} = \frac{7.093 \text{ cm}^3}{1 \text{ mol Fe}}$$

Next, the volume that contains two iron atoms is the volume of the body-centered cubic unit cell. Some of this volume is empty space because packing is only 68.0 percent efficient. But, this will not affect our calculation.

$$V = a^3$$

Let's also convert to cm^3.

$$V = (286.7\ \text{pm})^3 \times \left(\frac{1 \times 10^{-12}\ \text{m}}{1\ \text{pm}}\right)^3 \times \left(\frac{1\ \text{cm}}{0.01\ \text{m}}\right)^3 = \frac{2.357 \times 10^{-23}\ \text{cm}^3}{2\ \text{Fe atoms}}$$

We can now calculate the number of iron atoms in one mole using the strategy presented above.

$$\frac{\text{number of Fe atoms}}{\text{cm}^3} \times \frac{\text{cm}^3}{1\ \text{mol Fe}} = \frac{\text{number of Fe atoms}}{1\ \text{mol Fe}}$$

$$\frac{2\ \text{Fe atoms}}{2.357 \times 10^{-23}\ \text{cm}^3} \times \frac{7.093\ \text{cm}^3}{1\ \text{mol Ba}} = \mathbf{6.019 \times 10^{23}\ Fe\ atoms/mol}$$

The small difference between the above number and 6.022×10^{23} is the result of rounding off and using rounded values for density and other constants.

11.144 Figure 11.29 of the text shows that all alkali metals have a body-centered cubic structure. Figure 11.22 of the text gives the equation for the radius of a body-centered cubic unit cell.

$$r = \frac{\sqrt{3}a}{4}, \text{ where } a \text{ is the edge length.}$$

Because $V = a^3$, if we can determine the volume of the unit cell (V), then we can calculate a and r.

Using the ideal gas equation, we can determine the moles of metal in the sample.

$$n = \frac{PV}{RT} = \frac{\left(19.2\ \text{mmHg} \times \frac{1\ \text{atm}}{760\ \text{mmHg}}\right)(0.843\ \text{L})}{\left(0.0821 \frac{\text{L} \cdot \text{atm}}{\text{mol} \cdot \text{K}}\right)(1235\ \text{K})} = 2.10 \times 10^{-4}\ \text{mol}$$

Next, we calculate the volume of the cube, and then convert to the volume of one unit cell.

$$\text{Vol. of cube} = (0.171\ \text{cm})^3 = 5.00 \times 10^{-3}\ \text{cm}^3$$

This is the volume of 2.10×10^{-4} mole. We convert from volume/mole to volume/unit cell.

$$\text{Vol. of unit cell} = \frac{5.00 \times 10^{-3}\ \text{cm}^3}{2.10 \times 10^{-4}\ \text{mol}} \times \frac{1\ \text{mol}}{6.022 \times 10^{23}\ \text{atoms}} \times \frac{2\ \text{atoms}}{1\ \text{unit cell}} = 7.91 \times 10^{-23}\ \text{cm}^3/\text{unit cell}$$

Recall that there are 2 atoms in a body-centered cubic unit cell.

Next, we can calculate the edge length (a) from the volume of the unit cell.

$$a = \sqrt[3]{V} = \sqrt[3]{7.91 \times 10^{-23}\ \text{cm}^3} = 4.29 \times 10^{-8}\ \text{cm}$$

Finally, we can calculate the radius of the alkali metal.

$$r = \frac{\sqrt{3}a}{4} = \frac{\sqrt{3}(4.29 \times 10^{-8}\text{ cm})}{4} = 1.86 \times 10^{-8}\text{ cm} = \mathbf{186\ pm}$$

Checking Figure 8.5 of the text, we conclude that the metal is **sodium, Na**.

To calculate the density of the metal, we need the mass and volume of the unit cell. The volume of the unit cell has been calculated (7.91×10^{-23} cm^3/unit cell). The mass of the unit cell is

$$2\text{ Na atoms} \times \frac{22.99\text{ amu}}{1\text{ Na atom}} \times \frac{1\text{ g}}{6.022 \times 10^{23}\text{ amu}} = 7.635 \times 10^{-23}\text{ g}$$

$$d = \frac{m}{V} = \frac{7.635 \times 10^{-23}\text{ g}}{7.91 \times 10^{-23}\text{ cm}^3} = \mathbf{0.965\ g/cm^3}$$

The density value also matches that of sodium.

11.146 The original diagram shows that as heat is supplied to the water, its temperature rises. At the boiling point (represented by the horizontal line), water is converted to steam. Beyond this point the temperature of the steam rises above 100°C.

Choice (a) is eliminated because it shows no change from the original diagram even though the mass of water is doubled.

Choice (b) is eliminated because the rate of heating is greater than that for the original system. Also, it shows water boiling at a higher temperature, which is not possible.

Choice (c) is eliminated because it shows that water now boils at a temperature below 100°C, which is not possible.

Choice **(d)** therefore represents what actually happens. The heat supplied is enough to bring the water to its boiling point, but not raise the temperature of the steam.

CHAPTER 12
PHYSICAL PROPERTIES OF SOLUTIONS

PROBLEM-SOLVING STRATEGIES AND TUTORIAL SOLUTIONS

TYPES OF PROBLEMS

Problem Type 1: Predicting Solubility Based on Intermolecular Forces.

Problem Type 2: Types of Concentration Units.
- **(a)** Percent by mass.
- **(b)** Molarity.
- **(c)** Molality.

Problem Type 3: Converting between Concentration Units.
- **(a)** Converting molality to molarity.
- **(b)** Converting molarity to molality.
- **(c)** Converting percent by mass to molality.

Problem Type 4: Effect of Pressure on Solubility: Henry's Law.

Problem Type 5: Colligative Properties of Nonelectrolytes.
- **(a)** Vapor-pressure lowering, Raoult's law.
- **(b)** Boiling-point elevation.
- **(c)** Freezing-point depression.
- **(d)** Osmotic pressure.

Problem Type 6: Determining Molar Mass Using Colligative Properties.
- **(a)** Calculating molar mass from freezing-point depression.
- **(b)** Calculating molar mass from osmotic pressure.

Problem Type 7: Colligative Properties of Electrolytes.

PROBLEM TYPE 1: PREDICTING SOLUBILITY BASED ON INTERMOLECULAR FORCES

Two substances with intermolecular forces of similar type and magnitude are likely to be soluble in each other. The saying "*like dissolves like*" will help you predict the solubility of a substance in a solvent.

EXAMPLE 12.1

Which of the following would be a better solvent for molecular $I_2(s)$: CCl_4 or H_2O?

Strategy: In predicting solubility, remember the saying: Like dissolves like. A nonpolar solute will dissolve in a nonpolar solvent; ionic compounds will generally dissolve in polar solvents due to favorable ion-dipole interactions; solutes that can form hydrogen bonds with a solvent will have high solubility in the solvent.

Solution: I_2 is a nonpolar molecule (see Chapter 10). Using the like-dissolves-like rule, I_2 will be more soluble in the nonpolar solvent, CCl_4, than in the polar solvent, H_2O.

PRACTICE EXERCISE

1. In which solvent will NaBr be more soluble, benzene (C_6H_6) or water?

Text Problems: 12.10, 12.12

PROBLEM TYPE 2: TYPES OF CONCENTRATION UNITS

We will focus on three of the most common units of concentration: percent by mass, molarity, and molality.

A. Percent by mass

The percent by mass is defined as:

$$\textbf{percent by mass of solute} = \frac{\text{mass of solute}}{\text{mass of solute + mass of solvent}} \times 100\% \qquad (12.1, \text{text})$$

$$= \frac{\text{mass of solute}}{\text{mass of soln}} \times 100\%$$

The percent by mass has no units because it is a ratio of two similar quantities.

EXAMPLE 12.2

The dehydrated form of Epsom salt is magnesium sulfate. What is the percent $MgSO_4$ by mass in a solution made from 16.0 g $MgSO_4$ and 100 mL of H_2O at 25°C? The density of water at 25°C is 0.997 g/mL.

Strategy: We are given the mass of the solute, and the volume and density of the solvent. The mass of solvent can be calculated from its density and volume, and then we can use Equation (12.1) of the text to solve for the mass percent of $MgSO_4$ in the solution.

Solution:

Calculate the mass of 100 mL of water using the density of water as a conversion factor.

$$? \text{ g of water} = 100 \text{ mL } H_2O \times \frac{0.997 \text{ g } H_2O}{1 \text{ mL } H_2O} = 99.7 \text{ g } H_2O$$

Substitute the mass of solute and the mass of solvent into Equation (12.1) to calculate the percent by mass of $MgSO_4$.

$$\textbf{percent by mass MgSO}_4 = \frac{\text{mass of solute}}{\text{mass of solute + mass of solvent}} \times 100\%$$

$$= \frac{16.0 \text{ g}}{16.0 \text{ g} + 99.7 \text{ g}} \times 100\% = \mathbf{13.8\%}$$

PRACTICE EXERCISE

2. An aqueous solution contains 167 g $CuSO_4$ in 820 mL of solution. The density of the solution is 1.195 g/mL. Calculate the percent $CuSO_4$ by mass in the solution.

Text Problem: 12.16

B. Molarity (*M*)

We have already defined molarity in Chapter 4. Molarity is defined as

$$\text{molarity} = \frac{\text{moles of solute}}{\text{liters of soln}}$$

Molarity has units of mol/L.

EXAMPLE 12.3

What is the molarity of the $MgSO_4$ solution made in Example 12.2? Assume that the density remains unchanged upon addition of $MgSO_4$ to the water.

Strategy: To calculate molarity, we need moles of solute and volume of solution. Moles of solute can be calculated from the mass of solute. To find the volume of the solution, we add together the mass of the solute and solvent and then use the density to convert to volume.

Solution: Calculate the number of moles $MgSO_4$ in 16.0 g. Use the molar mass of $MgSO_4$ as a conversion factor.

$$? \text{ mol } MgSO_4 = 16.0 \text{ g } MgSO_4 \times \frac{1 \text{ mol } MgSO_4}{120.38 \text{ g } MgSO_4} = 0.133 \text{ mol } MgSO_4$$

Calculate the volume of the solution. The total mass of the solution is:

$$16.0 \text{ g } MgSO_4 + \left(100 \text{ mL } H_2O \times \frac{0.997 \text{ g } H_2O}{1 \text{ mL } H_2O}\right) = 115.7 \text{ g}$$

Assuming that the density of the solution is the same as that of water, we can calculate the volume of the solution as follows:

$$\text{Volume of solution} = 115.7 \text{ g} \times \frac{1 \text{ mL}}{0.997 \text{ g}} = 116.0 \text{ mL soln}$$

We calculate the molarity of the solution by dividing the moles of solute by volume of solution (in L).

$$\textbf{molarity} = \frac{\text{moles of solute}}{\text{liters of soln}} = \frac{0.133 \text{ mol}}{0.116 \text{ L}} = \mathbf{1.15}\ \boldsymbol{M}$$

PRACTICE EXERCISE

3. An aqueous solution contains 167 g $CuSO_4$ in 820 mL of solution. The density of the solution is 1.195 g/mL. Calculate the molarity of the solution.

C. Molality (*m*)

Molality is defined as the number of moles of solute per mass of solvent (in kg).

$$\text{molality} = \frac{\text{moles of solute}}{\text{mass of solvent (kg)}} \qquad (12.2, \text{text})$$

EXAMPLE 12.4

What is the molality of the $MgSO_4$ solution made in Example 12.2?

Strategy: To calculate molality, we need moles of solute and mass of solvent (water) in kg. From Example 12.3, we know the moles of solute (0.133 mole $MgSO_4$), and from Example 12.2, we know the mass of water (99.7 g). We will repeat these calculations below, and then solve for molality.

Solution: Calculate the number of moles $MgSO_4$ in 16.0 g. Use the molar mass of $MgSO_4$ as a conversion factor.

$$? \text{ mol } MgSO_4 = 16.0 \text{ g } MgSO_4 \times \frac{1 \text{ mol } MgSO_4}{120.38 \text{ g } MgSO_4} = 0.133 \text{ mol } MgSO_4$$

Calculate the mass (in kg) of 100 mL of H_2O. Use the density of water as a conversion factor.

$$? \text{ mass of } H_2O = 100 \text{ mL } H_2O \times \frac{0.997 \text{ g } H_2O}{1 \text{ mL } H_2O} \times \frac{1 \text{ kg}}{1000 \text{ g}} = 0.0997 \text{ kg } H_2O$$

Calculate the molality by substituting the mol of solute and the mass of solvent (in kg) into Equation (12.2) of the text.

$$\textbf{molality} = \frac{\text{moles of solute}}{\text{mass of solvent (kg)}} = \frac{0.133 \text{ mol}}{0.0997 \text{ kg}} = \mathbf{1.33}\ \boldsymbol{m}$$

PRACTICE EXERCISE

4. An aqueous solution contains 167 g $CuSO_4$ in 820 mL of solution. The density of the solution is 1.195 g/mL. Calculate the molality of the solution.

Text Problems: 12.18, 12.20

PROBLEM TYPE 3: CONVERTING BETWEEN CONCENTRATION UNITS

There are advantages and disadvantages to each type of concentration unit. An advantage of molality and percent by mass is that the concentration is temperature independent. On the other hand, molarity changes with temperature, because solution volume typically increases with increasing temperature. However, the advantage of molarity is that it is generally easier to measure the volume of solution than to weigh the solvent or solution.

A. Converting molality to molarity

EXAMPLE 12.5

Calculate the molarity of a 2.44 *m* NaCl solution given that its density is 1.089 g/mL.

Strategy: To calculate molarity, we need moles of solute and volume of solution. The moles of solute can be obtained directly from the molality. 2.44 *m* means 2.44 moles of solute per 1 kg (1000 g) of solvent. Next, we need to determine the mass of the solution, and then we can use the density to convert to volume of solution. To calculate the mass of the solution, you must calculate the mass of the solute and then add that to the mass of water (1 kg = 1000 g).

Solution: Use the molar mass of the solute as a conversion factor to convert from moles of solute to mass of solute.

$$? \text{ mass of solute} = 2.44 \text{ mol NaCl} \times \frac{58.44 \text{ g NaCl}}{1 \text{ mol NaCl}} = 143 \text{ g NaCl}$$

$$? \text{ mass of solution} = \text{mass of solute} + \text{mass of solvent}$$

$$= 143 \text{ g} + 1000 \text{ g} = 1143 \text{ g}$$

From the mass of the solution, we can calculate the volume of solution using the solution density as a conversion factor.

$$? \text{ volume of solution} = 1143 \text{ g} \times \frac{1 \text{ mL}}{1.089 \text{ g}} = 1.050 \times 10^3 \text{ mL} = 1.050 \text{ L}$$

Dividing moles of solute (2.44 moles) by liters of solution gives the molarity of the solution.

$$\textbf{molarity} = \frac{\text{mol of solute}}{\text{L of soln}} = \frac{2.44 \text{ mol}}{1.050 \text{ L}} = \mathbf{2.32}\ \boldsymbol{M}$$

PRACTICE EXERCISE

5. Concentrated hydrochloric acid is 15.7 *m*. Calculate the molarity of concentrated HCl given that its density is 1.18 g/mL.

Text Problem: 12.22

B. Converting molarity to molality

EXAMPLE 12.6
Calculate the molality of a 2.55 *M* NaCl solution given that its density is 1.089 g/mL.

Strategy: To calculate molality, we need moles of solute and mass of solvent (in kg). The moles of solute can be obtained directly from the molarity. 2.55 *M* means 2.55 moles of solute per 1 L of solution. Next, we can determine the mass of the solution from the density and volume of solution. Subtracting off the mass of the solute from the mass of the solution will give the mass of solvent.

Solution: From the volume of solution, we can calculate the mass of the solution using the solution density as a conversion factor. Remember that 2.55 *M* means 2.55 moles of solute per 1 L (1000 mL) of solution.

$$?\text{ mass of solution} = 1000\text{ mL soln} \times \frac{1.089\text{ g}}{1\text{ mL soln}} = 1089\text{ g}$$

To calculate the mass of the *solvent* (water), you need to subtract the mass due to the *solute* from the mass of solution calculated. You can calculate the mass of the solute from the moles of solute using molar mass as a conversion factor.

$$?\text{ mass of solute} = 2.55\text{ mol NaCl} \times \frac{58.44\text{ g NaCl}}{1\text{ mol NaCl}} = 149\text{ g NaCl}$$

$$?\text{ mass of solvent} = \text{mass of soln} - \text{mass of solute}$$

$$= 1089\text{ g} - 149\text{ g} = 9.40 \times 10^2\text{ g} = 0.940\text{ kg solvent}$$

The moles of solute are given in the molarity (2.55 mol). Divide moles of solute by mass of solvent in kg to calculate the molality of the solution.

$$\textbf{molality} = \frac{\text{mol of solute}}{\text{kg of solvent}} = \frac{2.55\text{ mol}}{0.940\text{ kg}} = \mathbf{2.71}\ \boldsymbol{m}$$

PRACTICE EXERCISE

6. Concentrated sulfuric acid, H_2SO_4 is 18.0 *M*. Calculate the molality of concentrated H_2SO_4 given that its density is 1.83 g/mL.

C. Converting percent by mass to molality

EXAMPLE 12.7
Concentrated hydrochloric acid is 36.5 percent HCl by mass. Its density is 1.18 g/mL. Calculate the molality of concentrated HCl.

Strategy: In solving this type of problem, it is convenient to assume that we start with 100.0 grams of the solution. If the mass of hydrochloric acid is 36.5% of 100.0 g, or 36.5 g, the percent by mass of water must be 100.0% – 36.5% = 63.5%. The mass of water in 100.0 g of solution would be 63.5 g. From the definition of molality, we need to find moles of solute (hydrochloric acid) and kilograms of solvent (water).

Solution: Since the definition of molality is

$$\text{molality} = \frac{\text{moles of solute}}{\text{mass of solvent (kg)}}$$

we first convert 36.5 g HCl to moles of HCl using its molar mass, then we convert 63.5 g of H_2O to units of kilograms.

$$36.5 \cancel{g}\ \text{HCl} \times \frac{1 \text{ mol HCl}}{36.46 \cancel{g}\ \text{HCl}} = 1.00 \text{ mol HCl}$$

$$63.5 \cancel{g}\ H_2O \times \frac{1 \text{ kg}}{1000 \cancel{g}} = 0.0635 \text{ kg } H_2O$$

Lastly, we divide moles of solute by mass of solvent in kg to calculate the molality of the solution.

$$\textbf{molality} = \frac{\text{mol of solute}}{\text{kg of solvent}} = \frac{1.00 \text{ mol}}{0.0635 \text{ kg}} = \mathbf{15.7}\ \boldsymbol{m}$$

PRACTICE EXERCISE

7. What is the molality of a 3.0 percent hydrogen peroxide (H_2O_2) aqueous solution? The density of the solution is 1.0 g/mL.

Text Problem: 12.22

PROBLEM TYPE 4: EFFECT OF PRESSURE ON SOLUBILITY: HENRY'S LAW

Solubility is defined as the maximum amount of a solute that will dissolve in a given quantity of solvent at a specific temperature. For all practical purposes, external pressure has no influence on the solubilities of liquids and solids, but it does greatly affect the solubility of gases. There is a quantitative relationship between gas solubility and pressure called **Henry's law**, which states that the solubility of a gas in a liquid is proportional to the pressure of the gas over the solution:

$$c = kP$$

where,

c is the molar concentration (mol/L) of the dissolved gas.
P is the pressure, in atmospheres, of the gas over the solution.
k is a constant for a given gas that depends only on the temperature. k has units of mol/L·atm.

As the pressure of the gas over the solution increases, more gas molecules strike the surface of the liquid increasing the number of gas molecules that dissolve in the liquid. For further discussion, see Section 12.5 of the text.

EXAMPLE 12.8

What is the concentration of O_2 at 25°C in water that is saturated with *air* at an atmospheric pressure of 645 mmHg? The Henry's law constant (k) for oxygen is 3.5×10^{-4} mol/L·atm. Assume that the mole fraction of oxygen in air is 0.209.

Strategy: We want to calculate the molar concentration (c) of O_2 in water. We can use Henry's law, $c_{O_2} = kP_{O_2}$. k is given in the problem. To calculate the molar concentration, we must first calculate the partial pressure of oxygen in air. The partial pressure of O_2 can be found using Dalton's law of partial pressures (see Chapter 5).

Solution:

$$P_{O_2} = X_{O_2}P_T = (0.209)(645 \text{ mmHg}) \times \frac{1 \text{ atm}}{760 \text{ mmHg}} = 0.177 \text{ atm}$$

Substitute the partial pressure of O_2 and k into the Henry's law expression to solve for the molar concentration of O_2 in water.

$$c_{O_2} = kP_{O_2} = (3.5 \times 10^{-4} \text{ mol/L} \cdot \text{atm})(0.177 \text{ atm}) = \mathbf{6.2 \times 10^{-5} \text{ mol/L}}$$

PRACTICE EXERCISE

8. The Henry's law constant for CO is 9.73×10^{-4} mol/L·atm at 25°C. What is the concentration of dissolved CO in water if the partial pressure of CO in the air is 0.015 mmHg?

Text Problems: 12.36, **12.38**

PROBLEM TYPE 5: COLLIGATIVE PROPERTIES OF NONELECTROLYTES

Several important properties of solutions depend on the number of solute particles in solution and not on the nature of the solute particles. These properties are called **colligative properties**.

A. Vapor pressure lowering, Raoult's law

If a solute is nonvolatile, the vapor pressure of its solution is always less than that of the pure solvent. Raoult's law quantifies this relationship by stating that the partial pressure of a solvent over a solution, P_1, is given by the vapor pressure of the pure solvent, P_1°, times the mole fraction of the solvent in the solution, X_1:

$$P_1 = X_1P_1^\circ \qquad (12.4, \text{text})$$

If the solution contains only one solute, $X_1 = 1 - X_2$, where X_2 is the mole fraction of the solute. Substituting for X_1 in Equation (12.4) of the text gives:

$$P_1 = (1 - X_2)P_1^\circ$$

$$P_1 = P_1^\circ - X_2P_1^\circ$$

$$P_1^\circ - P_1 = \Delta P = X_2P_1^\circ \qquad (12.5, \text{text})$$

Equation (12.5) of the text shows that a decrease in vapor pressure, ΔP, is directly proportional to the concentration of the solute in solution, X_2.

EXAMPLE 12.9

Calculate the vapor pressure of an aqueous solution at 30°C made from 1.00×10^2 g of sucrose ($C_{12}H_{22}O_{11}$) and 1.00×10^2 g of water. The vapor pressure of pure water at 30°C is 31.8 mmHg.

Strategy: Equation (12.4) of the text gives a relationship between the vapor pressure of the solvent over a solution and the mole fraction of the solvent. If we can calculate the mole fraction of the solvent, we can calculate the vapor pressure of the solution.

Solution:

$$\text{mol water} = (1.00 \times 10^2 \text{ g water}) \times \frac{1 \text{ mol water}}{18.02 \text{ g water}} = 5.55 \text{ mol}$$

$$\text{mol sucrose} = (1.00 \times 10^2 \cancel{\text{g}}\ \text{sucrose}) \times \frac{1 \text{ mol sucrose}}{342.3 \cancel{\text{g}}\ \text{sucrose}} = 0.292 \text{ mol}$$

$$X_{\text{water}} = \frac{\text{mol}_{\text{water}}}{\text{mol}_{\text{water}} + \text{mol}_{\text{sucrose}}} = \frac{5.55 \text{ mol}}{5.55 \text{ mol} + 0.292 \text{ mol}} = 0.950$$

Substitute X_{water} and the vapor pressure of pure water into Equation (12.4) to solve for the vapor pressure of the solution.

$$P_{\text{soln}} = X_{\text{water}} P^\circ_{\text{water}} = (0.950)(31.8 \text{ mmHg}) = \mathbf{30.2\ mmHg}$$

Tip: You could also have solved this problem using Equation (12.5). You could calculate the change in vapor pressure (ΔP) from the mole fraction of solute. Then, you could calculate the vapor pressure of the solution by subtracting ΔP from the vapor pressure of the pure solvent (water).

PRACTICE EXERCISE

9. At 25°C, the vapor pressure of pure water is 23.76 mmHg and that of an aqueous sucrose ($C_{12}H_{22}O_{11}$) is 23.28 mmHg. Calculate the molality of the solution.

Text Problems: 12.52, 12.54, 12.56

B. Boiling-point elevation

The **boiling point** of a solution is the temperature at which its vapor pressure equals the external atmospheric pressure (see Section 11.8 of the text). We just saw in Part (a) that a nonvolatile solute always decreases the vapor pressure of the solution relative to the pure solvent. Consequently, the boiling point of the solution is *higher* than the pure solvent, because more energy in the form of heat must be added to raise the vapor pressure of the solution to the external atmospheric pressure. The change in boiling point is proportional to the concentration of solute.

$$\Delta T_b = K_b m \qquad (12.6, \text{text})$$

where,

$\Delta T_b = T_b - T_b^\circ$ (where T_b is the boiling point of the solution and T_b° is the boiling point of the pure solvent)

m is the molal concentration of the solute

K_b is the molal boiling-point elevation constant of the solvent with units of °C/m

Do you know why molality is used for the concentration instead of molarity? Because we are dealing with a system that is not kept at constant temperature, we cannot express the concentration in molarity because molarity changes with temperature.

EXAMPLE 12.10

What is the boiling point of an "antifreeze/coolant" solution made from a 50-50 mixture (by volume) of ethylene glycol, $C_2H_6O_2$ and water? Assume the density of water is 1.00 g/mL and the density of ethylene glycol is 1.11 g/mL.

Strategy: Using Equation (12.6) of the text, we can calculate the change in boiling point, ΔT_b, by first calculating the molality of the solution and then multiplying by K_b for water (see Table 12.2 of the text).

Solution: For simplicity, assume that you have 100.0 mL of solution. Since the mixture is 50-50 by volume, there are 50.0 mL of ethylene glycol and 50.0 mL of water. To calculate the molality of the solution, you need the moles of solute (ethylene glycol) and the mass of solvent (water) in kg.

$$\text{mol of ethylene glycol} = 50.0 \text{ mL } C_2H_6O_2 \times \frac{1.11 \text{ g } C_2H_6O_2}{1 \text{ mL } C_2H_6O_2} \times \frac{1 \text{ mol } C_2H_6O_2}{62.07 \text{ g } C_2H_6O_2} = 0.894 \text{ mol}$$

$$\text{mass of water} = 50.0 \text{ mL } H_2O \times \frac{1.00 \text{ g } H_2O}{1 \text{ mL } H_2O} \times \frac{1 \text{ kg}}{1000 \text{ g}} = 0.0500 \text{ kg}$$

$$\text{molality} = \frac{\text{mol solute}}{\text{kg solvent}} = \frac{0.894 \text{ mol}}{0.0500 \text{ kg}} = 17.9\ m$$

Substitute the molality of the solution and K_b into Equation (12.6) to solve for the change in boiling point. Then, add the change in boiling point to the normal boiling point of water (100.0°C) to calculate the boiling point of the solution.

$$\Delta T_b = K_b m = (0.52°\text{C}/m)(17.9\ m) = 9.3°\text{C}$$

$$\textbf{b.p. of soln} = 100.0°\text{C} + 9.3°\text{C} = \mathbf{109.3°C}$$

PRACTICE EXERCISE

10. What is the boiling point of an aqueous solution of a nonvolatile solute that freezes at −3.0°C?

C. Freezing-point depression

The **freezing point** of a liquid (or the melting point of a solid) is the temperature at which the solid and liquid phases coexist in equilibrium. It might be easier to understand freezing-point depression by looking at the opposite of freezing, melting. To melt a solid, the intermolecular forces holding the solid molecules together must be overcome. Adding another solid substance to a pure solid disrupts the intermolecular forces of the formerly pure solid. Hence, it is easier to overcome the intermolecular forces and the mixture melts at a lower temperature than the pure solid. The melting point (or the freezing point) is depressed. The depression of freezing point can be represented by the following equation.

$$\Delta T_f = K_f m \qquad (12.7, \text{text})$$

where,

$\Delta T_f = T_f^\circ - T_f$ (where T_f is the freezing point of the solution and T_f° is the freezing point of the pure solvent)

m is the molal concentration of the solute

K_f is the molal freezing-point depression constant with units of °C/m

EXAMPLE 12.11

How many grams of isopropyl alcohol, C_3H_7OH, should be added to 1.0 L of water to give a solution that will not freeze above −16°C? Assume the density of water is 1.00 g/mL.

Strategy: We want to lower the freezing point of 1.0 L of water by 16°C. Using Equation (12.7), we can calculate the molality of the solution needed. Then, from the molality, we can calculate the grams of isopropyl alcohol needed.

Solution: Rearrange Equation (12.7) of the text to solve for the molality. You can look up K_f in Table 12.2 of the text.

$$m = \frac{\Delta T_f}{K_f} = \frac{16°\text{C}}{1.86°\text{C}/m} = 8.6\ m$$

8.6 m means that the solution contains 8.6 mol of solute per 1 kg of solvent. Assume that the density of water is 1.0 g/mL; thus, 1 kg (1000 g) of water has a volume of 1 L. Convert 8.6 mol of isopropyl alcohol to grams of isopropyl alcohol using the molar mass as a conversion factor.

$$\textbf{? g isopropyl alcohol} = 8.6 \text{ mol } C_3H_7OH \times \frac{60.09 \text{ g } C_3H_7OH}{1 \text{ mol } C_3H_7OH} = \mathbf{5.2 \times 10^2\ g}$$

PRACTICE EXERCISE

11. Benzene melts at 5.50°C. When 2.11 g of naphthalene, $C_{10}H_8$, is added to 100 g of benzene, the solution freezes at 4.65°C. Calculate the freezing-point depression constant (K_f) for benzene.

Text Problems: 12.58, 12.62

D. Osmotic pressure

Osmosis is the net movement of solvent molecules through a semipermeable membrane from a pure solvent or from a dilute solution to a more concentrated solution. The **osmotic pressure (π)** of a solution is the pressure required to stop osmosis. The osmotic pressure of a solution is given by:

$$\pi = MRT \qquad (12.8, \text{text})$$

where,

M is the molarity of the solution
R is the gas constant (0.0821 L·atm/K·mol)
T is the absolute temperature in K

EXAMPLE 12.12

The average osmotic pressure of seawater is about 30.0 atm at 25°C. Calculate the molar concentration of an aqueous solution of urea (NH_2CONH_2) that is isotonic with seawater.

Strategy: A solution of urea that is isotonic with seawater must have the same osmotic pressure, 30.0 atm. Solve Equation (12.8) of the text algebraically for the molar concentration. Then, substitute π, R, and T (in K) into the equation to solve for the molar concentration of the urea solution.

Solution:

$$\textbf{molarity} = \frac{\pi}{RT} = \frac{30.0\ \cancel{\text{atm}}}{298\ \cancel{\text{K}}} \times \frac{\text{mol}\cdot\cancel{\text{K}}}{0.0821\ \text{L}\cdot\cancel{\text{atm}}} = \mathbf{1.23}\ \boldsymbol{M}$$

PRACTICE EXERCISE

12. The walls of red blood cells are semipermeable membranes, and the solution of NaCl within those walls exerts an osmotic pressure of 7.82 atm at 37°C. What concentration of NaCl must a *surrounding* solution have so that this pressure is balanced and cell rupture (hemolysis) is prevented?

PROBLEM TYPE 6: DETERMINING MOLAR MASS USING COLLIGATIVE PROPERTIES

Any of the four colligative properties discussed in Problem Type 5 can be used to calculate the molar mass of the solute. However, in practice, only freezing-point depression and osmotic pressure are used because they show the most pronounced changes.

A. Calculating molar mass from freezing-point depression

You can solve this type of problem by first calculating the molality of the solute using Equation (12.7) of the text and then calculating the molar mass from the molality.

EXAMPLE 12.13

Benzene has a normal freezing point of 5.51°C. The addition of 1.25 g of an unknown compound to 85.0 g of benzene produces a solution with a freezing point of 4.52°C. What is the molar mass of the unknown compound?

Strategy: We are asked to calculate the molar mass of the unknown. Grams of the compound are given in the problem, so we need to solve for moles of compound.

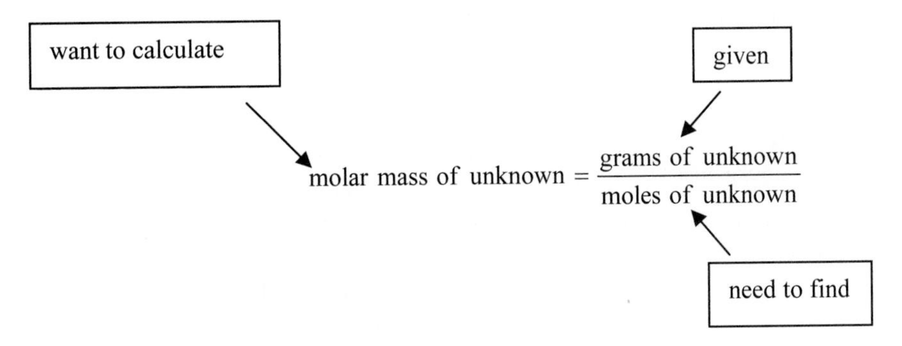

From the freezing point depression, we can calculate the molality of the solution. Then, from the molality, we can determine the number of moles in 1.25 g of the unknown compound.

Solution: Solve Equation (12.7) algebraically for molality (m), then substitute ΔT_f and K_f into the equation to calculate the molality.

$$\Delta T_f = 5.51°C - 4.52°C = 0.99°C$$

$$m = \frac{\Delta T_f}{K_f} = \frac{0.99°\cancel{C}}{5.12°\cancel{C}/m} = 0.19\ m$$

Multiplying the molality by the mass of solvent (in kg) gives moles of unknown solute. Then, dividing the mass of solute (in g) by the moles of solute, gives the molar mass of the unknown solute.

$$?\ \text{mol of unknown solute} = \frac{0.19\ \text{mol solute}}{1\ \cancel{\text{kg}}\ \text{benzene}} \times 0.085\ \cancel{\text{kg}}\ \text{benzene} = 0.016\ \text{mol solute}$$

$$\textbf{molar mass of unknown} = \frac{1.25\ \text{g}}{0.016\ \text{mol}} = \mathbf{78\ g/mol}$$

PRACTICE EXERCISE

13. When 48 g of glucose (a nonelectrolyte) is dissolved in 500 g of H_2O, the solution has a freezing point of −0.94°C. What is the molar mass of glucose?

14. In the course of research, a chemist isolates a new compound. An elemental analysis shows the following: C, 50.7 percent; H, 4.25 percent; O, 45.1 percent. If 5.01 g of the compound is dissolved in 100 g of water, a solution with a freezing point of −0.65°C is produced. What is the molecular formula of the compound?

Text Problems: 12.60, 12.64

B. Calculating the molar mass from osmotic pressure

You can solve this type of problem by first calculating the molarity of the solution using Equation (12.8) of the text and then calculating the molar mass from the molarity.

EXAMPLE 12.14

30.0 g of sucrose is dissolved in water making 1.00 × 10^2 mL of solution. The solution has an osmotic pressure of 20.8 atm at 16.0°C. What is the molar mass of sucrose?

Strategy: We are asked to calculate the molar mass of the sucrose. Grams of the sucrose are given in the problem, so we need to solve for moles of sucrose.

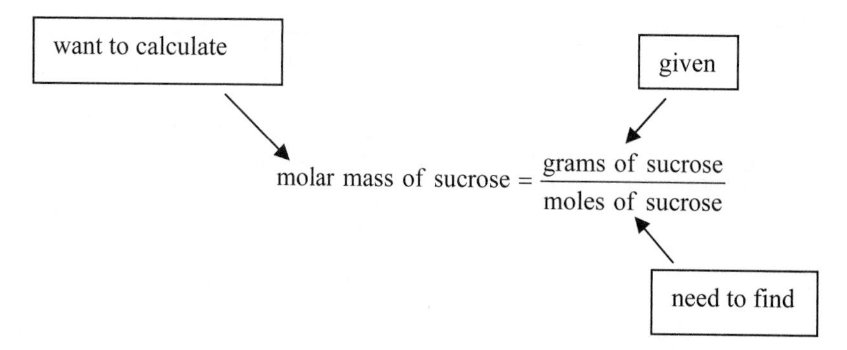

From the osmotic pressure of the solution, we can calculate the molarity of the solution. Then, from the molarity, we can determine the number of moles in 30.0 g of sucrose. What units should we use for π and temperature?

Solution: First, we calculate the molarity using Equation (12.8) of the text.

$$\pi = MRT$$

$$\text{molarity} = \frac{\pi}{RT} = \frac{20.8\ \text{atm}}{289\ \text{K}} \times \frac{\text{mol} \cdot \text{K}}{0.0821\ \text{L} \cdot \text{atm}} = 0.877\ M$$

Multiplying the molarity by the volume of solution (in L) gives moles of solute (sucrose).

$$?\ \text{mol of sucrose} = (0.877\ \text{mol/L})(0.100\ \text{L}) = 0.0877\ \text{mol sucrose}$$

Lastly, dividing the mass of sucrose (in g) by the moles of sucrose, gives the molar mass of sucrose.

$$\textbf{molar mass of sucrose} = \frac{30.0\ \text{g sucrose}}{0.0877\ \text{mol sucrose}} = \mathbf{342\ g/mol}$$

PRACTICE EXERCISE

15. Peruvian Indians use a dart poison from root extracts called curare. It is a nonelectrolyte. The osmotic pressure at 20.0°C of an aqueous solution containing 0.200 g of curare in 1.00×10^2 mL of solution is 56.2 mmHg. Calculate the molar mass of curare.

Text Problems: 12.66, 12.68

PROBLEM TYPE 7: COLLIGATIVE PROPERTIES OF ELECTROLYTES

Electrolytes dissociate into ions in solution, so this requires us to take a slightly different approach than that used for the colligative properties of nonelectrolytes. Remember, it is the number of solute particles that determines the colligative properties of a solution. To account for the dissociation of an electrolyte into ions, the equations for colligative properties must be modified as follows:

$$\Delta T_b = iK_b m \qquad (12.10,\ \text{text})$$

$$\Delta T_f = iK_f m \qquad (12.11,\ \text{text})$$

$$\pi = iMRT \qquad (12.12,\ \text{text})$$

where,

i is the van't Hoff factor which is defined as:

$$i = \frac{\text{actual number of particles in soln after dissociation}}{\text{number of formula units initially dissolved in soln}}$$

Thus, i should be 1 for all nonelectrolytes. For strong electrolytes such as KCl and $BaSO_4$, i should be 2, and for strong electrolytes such as $CaCl_2$ and Na_2CO_3, i should be 3. In reality, the colligative properties of *electrolyte* solutions are usually smaller than anticipated. At higher concentrations, electrostatic forces come into play, drawing cations and anions together. The ion pairs that are formed decrease the number of solute particles in solution. Table 12.3 of the text lists the van't Hoff factor (i) for various solutes.

EXAMPLE 12.15

Calculate the value of *i* for an electrolyte that should dissociate into 2 ions, if a 1.0 *m* aqueous solution of the electrolyte freezes at –3.28°C. Why is the value of *i* less than 2?

Strategy: Solve Equation (12.11) of the text algebraically for i, then substitute the values of ΔT_f, K_f, and m into the equation to solve for i.

Solution:

$\Delta T_f = 3.28°C$

$$i = \frac{\Delta T_f}{K_f m} = \frac{3.28\cancel{°C}}{(1.86\cancel{°C}/\cancel{m})(1.0\cancel{m})} = \mathbf{1.76}$$

The value if i is less than 2 because ion pairing reduces the number of particles in solution. Remember, colligative properties depend only on the number of solute particles in solution and *not* on the nature of the solute particles.

PRACTICE EXERCISE

16. Arrange the following aqueous solutions in order of increasing boiling points: 0.100 m ethanol, 0.050 m $Ca(NO_3)_2$, 0.100 m NaBr, 0.050 m HCl.

Text Problems: 12.74, 12.76, 12.78, **12.80**, 12.82

ANSWERS TO PRACTICE EXERCISES

1. Remember that "like dissolves like". Therefore, the ionic solid NaBr will be more soluble in the polar solvent, $\mathbf{H_2O}$.

2. 17.0 percent $CuSO_4$

3. 1.28 M $CuSO_4$

4. 1.29 m $CuSO_4$

5. 11.8 M HCl

6. $m = 3 \times 10^2\ m$

7. 0.91 m H_2O_2

8. $1.9 \times 10^{-8}\ M$ CO

9. $m = 1.1\ m$

10. b.p. of soln = 100.84°C

11. K_f (benzene) = 5.15°C/m

12. 0.15 M NaCl

13. molar mass = 1.9×10^2 g/mol

14. $C_6H_6O_4$

15. molar mass of curare = 6.50×10^2 g/mol

16. ethanol ≈ HCl < $Ca(NO_3)_2$ < NaBr

SOLUTIONS TO SELECTED TEXT PROBLEMS

12.10 Predicting Solubility Based on Intermolecular Forces, Problem Type 1.

Strategy: In predicting solubility, remember the saying: Like dissolves like. A nonpolar solute will dissolve in a nonpolar solvent; ionic compounds will generally dissolve in polar solvents due to favorable ion-dipole interactions; solutes that can form hydrogen bonds with a solvent will have high solubility in the solvent.

Solution: Strong hydrogen bonding (dipole-dipole attraction) is the principal intermolecular attraction in liquid ethanol, but in liquid cyclohexane the intermolecular forces are dispersion forces because cyclohexane is nonpolar. Cyclohexane cannot form hydrogen bonds with ethanol, and therefore cannot attract ethanol molecules strongly enough to form a solution.

12.12 The longer the C–C chain, the more the molecule "looks like" a hydrocarbon and the less important the –OH group becomes. Hence, as the C–C chain length increases, the molecule becomes less polar. Since "like dissolves like", as the molecules become more nonpolar, the solubility in polar water decreases. The –OH group of the alcohols can form strong hydrogen bonds with water molecules, but this property decreases as the chain length increases.

12.16 Percent by Mass Calculation, Problem Type 2.

Strategy: We are given the percent by mass of the solute and the mass of the solute. We can use Equation (12.1) of the text to solve for the mass of the solvent (water).

Solution:

(a) The percent by mass is defined as

$$\text{percent by mass of solute} = \frac{\text{mass of solute}}{\text{mass of solute} + \text{mass of solvent}} \times 100\%$$

Substituting in the percent by mass of solute and the mass of solute, we can solve for the mass of solvent (water).

$$16.2\% = \frac{5.00\text{ g urea}}{5.00\text{ g urea} + \text{mass of water}} \times 100\%$$

$$(0.162)(\text{mass of water}) = 5.00\text{ g} - (0.162)(5.00\text{g})$$

$$\textbf{mass of water} = \textbf{25.9 g}$$

(b) Similar to part (a),

$$1.5\% = \frac{26.2\text{ g MgCl}_2}{26.2\text{ g MgCl}_2 + \text{mass of water}} \times 100\%$$

$$\textbf{mass of water} = \mathbf{1.72 \times 10^3\text{ g}}$$

12.18
$$\text{molality} = \frac{\text{moles of solute}}{\text{mass of solvent (kg)}}$$

(a)
$$\text{mass of 1 L soln} = 1000\ \cancel{\text{mL}} \times \frac{1.08\text{ g}}{1\ \cancel{\text{mL}}} = 1080\text{ g}$$

$$\text{mass of water} = 1080 \text{ g} - \left(2.50 \text{ mol NaCl} \times \frac{58.44 \text{ g NaCl}}{1 \text{ mol NaCl}}\right) = 934 \text{ g} = 0.934 \text{ kg}$$

$$m = \frac{2.50 \text{ mol NaCl}}{0.934 \text{ kg } H_2O} = \mathbf{2.68}\ \boldsymbol{m}$$

(b) 100 g of the solution contains 48.2 g KBr and 51.8 g H_2O.

$$\text{mol of KBr} = 48.2 \text{ g KBr} \times \frac{1 \text{ mol KBr}}{119.0 \text{ g KBr}} = 0.405 \text{ mol KBr}$$

$$\text{mass of } H_2O \text{ (in kg)} = 51.8 \text{ g } H_2O \times \frac{1 \text{ kg}}{1000 \text{ g}} = 0.0518 \text{ kg } H_2O$$

$$m = \frac{0.405 \text{ mol KBr}}{0.0518 \text{ kg } H_2O} = \mathbf{7.82}\ \boldsymbol{m}$$

12.20 Let's assume that we have 1.0 L of a 0.010 *M* solution.

Assuming a solution density of 1.0 g/mL, the mass of 1.0 L (1000 mL) of the solution is 1000 g or 1.0×10^3 g.

The mass of 0.010 mole of urea is:

$$0.010 \text{ mol urea} \times \frac{60.06 \text{ g urea}}{1 \text{ mol urea}} = 0.60 \text{ g urea}$$

The mass of the solvent is:

$$\text{(solution mass)} - \text{(solute mass)} = (1.0 \times 10^3 \text{ g}) - (0.60 \text{ g}) = 1.0 \times 10^3 \text{ g} = 1.0 \text{ kg}$$

$$m = \frac{\text{moles solute}}{\text{mass solvent}} = \frac{0.010 \text{ mol}}{1.0 \text{ kg}} = \mathbf{0.010}\ \boldsymbol{m}$$

12.22 Converting between Concentration Units, Problem Type 3.

(a) Converting mass percent to molality.

Strategy: In solving this type of problem, it is convenient to assume that we start with 100.0 grams of the solution. If the mass of sulfuric acid is 98.0% of 100.0 g, or 98.0 g, the percent by mass of water must be 100.0% – 98.0% = 2.0%. The mass of water in 100.0 g of solution would be 2.0 g. From the definition of molality, we need to find moles of solute (sulfuric acid) and kilograms of solvent (water).

Solution: Since the definition of molality is

$$\text{molality} = \frac{\text{moles of solute}}{\text{mass of solvent (kg)}}$$

we first convert 98.0 g H_2SO_4 to moles of H_2SO_4 using its molar mass, then we convert 2.0 g of H_2O to units of kilograms.

$$98.0 \text{ g } H_2SO_4 \times \frac{1 \text{ mol } H_2SO_4}{98.09 \text{ g } H_2SO_4} = 0.999 \text{ mol } H_2SO_4$$

$$2.0\ \text{g}\ H_2O \times \frac{1\ \text{kg}}{1000\ \text{g}} = 2.0 \times 10^{-3}\ \text{kg}\ H_2O$$

Lastly, we divide moles of solute by mass of solvent in kg to calculate the molality of the solution.

$$m = \frac{\text{mol of solute}}{\text{kg of solvent}} = \frac{0.999\ \text{mol}}{2.0 \times 10^{-3}\ \text{kg}} = \mathbf{5.0 \times 10^2\ m}$$

(b) Converting molality to molarity.

Strategy: From part (a), we know the moles of solute (0.999 mole H_2SO_4) and the mass of the solution (100.0 g). To solve for molarity, we need the volume of the solution, which we can calculate from its mass and density.

Solution: First, we use the solution density as a conversion factor to convert to volume of solution.

$$?\ \text{volume of solution} = 100.0\ \text{g} \times \frac{1\ \text{mL}}{1.83\ \text{g}} = 54.6\ \text{mL} = 0.0546\ \text{L}$$

Since we already know moles of solute from part (a), 0.999 mole H_2SO_4, we divide moles of solute by liters of solution to calculate the molarity of the solution.

$$M = \frac{\text{mol of solute}}{\text{L of soln}} = \frac{0.999\ \text{mol}}{0.0546\ \text{L}} = \mathbf{18.3\ \mathit{M}}$$

12.24 Assume 100.0 g of solution.

(a) The mass of ethanol in the solution is $0.100 \times 100.0\ \text{g} = 10.0\ \text{g}$. The mass of the water is $100.0\ \text{g} - 10.0\ \text{g} = 90.0\ \text{g} = 0.0900\ \text{kg}$. The amount of ethanol in moles is:

$$10.0\ \text{g ethanol} \times \frac{1\ \text{mol}}{46.07\ \text{g}} = 0.217\ \text{mol ethanol}$$

$$m = \frac{\text{mol solute}}{\text{kg solvent}} = \frac{0.217\ \text{mol}}{0.0900\ \text{kg}} = \mathbf{2.41\ \mathit{m}}$$

(b) The volume of the solution is:

$$100.0\ \text{g} \times \frac{1\ \text{mL}}{0.984\ \text{g}} = 102\ \text{mL} = 0.102\ \text{L}$$

The amount of ethanol in moles is 0.217 mole [part (a)].

$$M = \frac{\text{mol solute}}{\text{liters of soln}} = \frac{0.217\ \text{mol}}{0.102\ \text{L}} = \mathbf{2.13\ \mathit{M}}$$

(c) **Solution volume** $= 0.125\ \text{mol} \times \frac{1\ \text{L}}{2.13\ \text{mol}} = \mathbf{0.0587\ L = 58.7\ mL}$

12.28 At 75°C, 155 g of KNO_3 dissolves in 100 g of water to form 255 g of solution. When cooled to 25°C, only 38.0 g of KNO_3 remain dissolved. This means that (155 – 38.0) g = 117 g of KNO_3 will crystallize. The amount of KNO_3 formed when 100 g of saturated solution at 75°C is cooled to 25°C can be found by a simple unit conversion.

$$100\ \text{g saturated soln} \times \frac{117\ \text{g KNO}_3\ \text{crystallized}}{255\ \text{g saturated soln}} = \mathbf{45.9\ g\ KNO_3}$$

12.36 According to Henry's law, the solubility of a gas in a liquid increases as the pressure increases ($c = kP$). The soft drink tastes flat at the bottom of the mine because the carbon dioxide pressure is greater and the dissolved gas is not released from the solution. As the miner goes up in the elevator, the atmospheric carbon dioxide pressure decreases and dissolved gas is released from his stomach.

12.38 Effect of Pressure on Solubility, Problem Type 4.

Strategy: The given solubility allows us to calculate Henry's law constant (k), which can then be used to determine the concentration of N_2 at 4.0 atm. We can then compare the solubilities of N_2 in blood under normal pressure (0.80 atm) and under a greater pressure that a deep-sea diver might experience (4.0 atm) to determine the moles of N_2 released when the diver returns to the surface. From the moles of N_2 released, we can calculate the volume of N_2 released.

Solution: First, calculate the Henry's law constant, k, using the concentration of N_2 in blood at 0.80 atm.

$$k = \frac{c}{P}$$

$$k = \frac{5.6 \times 10^{-4}\ \text{mol/L}}{0.80\ \text{atm}} = 7.0 \times 10^{-4}\ \text{mol/L} \cdot \text{atm}$$

Next, we can calculate the concentration of N_2 in blood at 4.0 atm using k calculated above.

$$c = kP$$

$$c = (7.0 \times 10^{-4}\ \text{mol/L} \cdot \text{atm})(4.0\ \text{atm}) = 2.8 \times 10^{-3}\ \text{mol/L}$$

From each of the concentrations of N_2 in blood, we can calculate the number of moles of N_2 dissolved by multiplying by the total blood volume of 5.0 L. Then, we can calculate the number of moles of N_2 released when the diver returns to the surface.

The number of moles of N_2 in 5.0 L of blood at 0.80 atm is:

$$(5.6 \times 10^{-4}\ \text{mol/L})(5.0\ \text{L}) = 2.8 \times 10^{-3}\ \text{mol}$$

The number of moles of N_2 in 5.0 L of blood at 4.0 atm is:

$$(2.8 \times 10^{-3}\ \text{mol/L})(5.0\ \text{L}) = 1.4 \times 10^{-2}\ \text{mol}$$

The amount of N_2 released in moles when the diver returns to the surface is:

$$(1.4 \times 10^{-2}\ \text{mol}) - (2.8 \times 10^{-3}\ \text{mol}) = 1.1 \times 10^{-2}\ \text{mol}$$

Finally, we can now calculate the volume of N_2 released using the ideal gas equation. The total pressure pushing on the N_2 that is released is atmospheric pressure (1 atm).

The volume of N_2 released is:

$$V_{N_2} = \frac{nRT}{P}$$

$$V_{N_2} = \frac{(1.1 \times 10^{-2} \text{ mol})(273 + 37)\text{K}}{(1.0 \text{ atm})} \times \frac{0.0821 \text{ L}\cdot\text{atm}}{\text{mol}\cdot\text{K}} = \mathbf{0.28\ L}$$

12.52 Vapor-pressure lowering, Raoult's law, Problem Type 5A.

Strategy: From the vapor pressure of water at 20°C and the change in vapor pressure for the solution (2.0 mmHg), we can solve for the mole fraction of sucrose using Equation (12.5) of the text. From the mole fraction of sucrose, we can solve for moles of sucrose. Lastly, we convert form moles to grams of sucrose.

Solution: Using Equation (12.5) of the text, we can calculate the mole fraction of sucrose that causes a 2.0 mmHg drop in vapor pressure.

$$\Delta P = X_2 P_1^\circ$$

$$\Delta P = X_{\text{sucrose}} P_{\text{water}}^\circ$$

$$X_{\text{sucrose}} = \frac{\Delta P}{P_{\text{water}}^\circ} = \frac{2.0 \text{ mmHg}}{17.5 \text{ mmHg}} = 0.11$$

From the definition of mole fraction, we can calculate moles of sucrose.

$$X_{\text{sucrose}} = \frac{n_{\text{sucrose}}}{n_{\text{water}} + n_{\text{sucrose}}}$$

$$\text{moles of water} = 552 \text{ g} \times \frac{1 \text{ mol}}{18.02 \text{ g}} = 30.6 \text{ mol } H_2O$$

$$X_{\text{sucrose}} = 0.11 = \frac{n_{\text{sucrose}}}{30.6 + n_{\text{sucrose}}}$$

$$n_{\text{sucrose}} = 3.8 \text{ mol sucrose}$$

Using the molar mass of sucrose as a conversion factor, we can calculate the mass of sucrose.

$$\textbf{mass of sucrose} = 3.8 \text{ mol sucrose} \times \frac{342.3 \text{ g sucrose}}{1 \text{ mol sucrose}} = \mathbf{1.3 \times 10^3\ g\ sucrose}$$

12.54 For any solution the sum of the mole fractions of the components is always 1.00, so the mole fraction of 1−propanol is 0.700. The partial pressures are:

$$P_{\text{ethanol}} = X_{\text{ethanol}} \times P_{\text{ethanol}}^\circ = (0.300)(100 \text{ mmHg}) = \mathbf{30.0\ mmHg}$$

$$P_{\text{1-propanol}} = X_{\text{1-propanol}} \times P_{\text{1-propanol}}^\circ = (0.700)(37.6 \text{ mmHg}) = \mathbf{26.3\ mmHg}$$

Is the vapor phase richer in one of the components than the solution? Which component? Should this always be true for ideal solutions?

12.56 This problem is very similar to Problem 12.52.

$$\Delta P = X_{\text{urea}} P^{\circ}_{\text{water}}$$

$$2.50 \text{ mmHg} = X_{\text{urea}}(31.8 \text{ mmHg})$$

$$X_{\text{urea}} = 0.0786$$

The number of moles of water is:

$$n_{\text{water}} = 450 \text{ g H}_2\text{O} \times \frac{1 \text{ mol H}_2\text{O}}{18.02 \text{ g H}_2\text{O}} = 25.0 \text{ mol H}_2\text{O}$$

$$X_{\text{urea}} = \frac{n_{\text{urea}}}{n_{\text{water}} + n_{\text{urea}}}$$

$$0.0786 = \frac{n_{\text{urea}}}{25.0 + n_{\text{urea}}}$$

$$n_{\text{urea}} = 2.13 \text{ mol}$$

$$\textbf{mass of urea} = 2.13 \text{ mol urea} \times \frac{60.06 \text{ g urea}}{1 \text{ mol urea}} = \textbf{128 g of urea}$$

12.58

$$m = \frac{\Delta T_{\text{f}}}{K_{\text{f}}} = \frac{1.1^{\circ}\text{C}}{1.86^{\circ}\text{C}/m} = \mathbf{0.59}\ \boldsymbol{m}$$

12.60 This is a combination of Problem Type 4B, Chapter 3, and Problem Type 6A, Chapter 12.

METHOD 1:

Strategy: First, we can determine the empirical formula from mass percent data. Then, we can determine the molar mass from the freezing-point depression. Finally, from the empirical formula and the molar mass, we can find the molecular formula.

Solution: If we assume that we have 100 g of the compound, then each percentage can be converted directly to grams. In this sample, there will be 40.0 g of C, 6.7 g of H, and 53.3 g of O. Because the subscripts in the formula represent a mole ratio, we need to convert the grams of each element to moles. The conversion factor needed is the molar mass of each element. Let n represent the number of moles of each element so that

$$n_{\text{C}} = 40.0 \text{ g C} \times \frac{1 \text{ mol C}}{12.01 \text{ g C}} = 3.33 \text{ mol C}$$

$$n_{\text{H}} = 6.7 \text{ g H} \times \frac{1 \text{ mol H}}{1.008 \text{ g H}} = 6.6 \text{ mol H}$$

$$n_{\text{O}} = 53.3 \text{ g O} \times \frac{1 \text{ mol O}}{16.00 \text{ g O}} = 3.33 \text{ mol O}$$

Thus, we arrive at the formula $C_{3.33}H_{6.6}O_{3.3}$, which gives the identity and the ratios of atoms present. However, chemical formulas are written with whole numbers. Try to convert to whole numbers by dividing all the subscripts by the smallest subscript.

$$\text{C}: \frac{3.33}{3.33} = 1.00 \qquad \text{H}: \frac{6.6}{3.33} = 2.0 \qquad \text{O}: \frac{3.33}{3.33} = 1.00$$

This gives us the empirical, CH_2O.

Now, we can use the freezing point data to determine the molar mass. First, calculate the molality of the solution.

$$m = \frac{\Delta T_f}{K_f} = \frac{1.56\text{°C}}{8.00\text{°C}/m} = 0.195\ m$$

Multiplying the molality by the mass of solvent (in kg) gives moles of unknown solute. Then, dividing the mass of solute (in g) by the moles of solute, gives the molar mass of the unknown solute.

$$?\text{ mol of unknown solute} = \frac{0.195\text{ mol solute}}{1\text{ kg diphenyl}} \times 0.0278\text{ kg diphenyl}$$

$$= 0.00542\text{ mol solute}$$

$$\textbf{molar mass of unknown} = \frac{0.650\text{ g}}{0.00542\text{ mol}} = \mathbf{1.20 \times 10^2\ g/mol}$$

Finally, we compare the empirical molar mass to the molar mass above.

$$\text{empirical molar mass} = 12.01\text{ g} + 2(1.008\text{ g}) + 16.00\text{ g} = 30.03\text{ g/mol}$$

The number of (CH_2O) units present in the molecular formula is:

$$\frac{\text{molar mass}}{\text{empirical molar mass}} = \frac{1.20 \times 10^2\text{ g}}{30.03\text{ g}} = 4.00$$

Thus, there are four CH_2O units in each molecule of the compound, so the molecular formula is $(CH_2O)_4$, or $\mathbf{C_4H_8O_4}$.

METHOD 2:

Strategy: As in Method 1, we determine the molar mass of the unknown from the freezing point data. Once the molar mass is known, we can multiply the mass % of each element (converted to a decimal) by the molar mass to convert to grams of each element. From the grams of each element, the moles of each element can be determined and hence the mole ratio in which the elements combine.

Solution: We use the freezing point data to determine the molar mass. First, calculate the molality of the solution.

$$m = \frac{\Delta T_f}{K_f} = \frac{1.56\text{°C}}{8.00\text{°C}/m} = 0.195\ m$$

Multiplying the molality by the mass of solvent (in kg) gives moles of unknown solute. Then, dividing the mass of solute (in g) by the moles of solute, gives the molar mass of the unknown solute.

$$?\text{ mol of unknown solute} = \frac{0.195\text{ mol solute}}{1\text{ kg diphenyl}} \times 0.0278\text{ kg diphenyl}$$

$$= 0.00542\text{ mol solute}$$

$$\textbf{molar mass of unknown} = \frac{0.650\text{ g}}{0.00542\text{ mol}} = \mathbf{1.20 \times 10^2\ g/mol}$$

Next, we multiply the mass % (converted to a decimal) of each element by the molar mass to convert to grams of each element. Then, we use the molar mass to convert to moles of each element.

$$n_C = (0.400) \times (1.20 \times 10^2 \text{ g}) \times \frac{1 \text{ mol C}}{12.01 \text{ g C}} = 4.00 \text{ mol C}$$

$$n_H = (0.067) \times (1.20 \times 10^2 \text{ g}) \times \frac{1 \text{ mol H}}{1.008 \text{ g H}} = 7.98 \text{ mol H}$$

$$n_O = (0.533) \times (1.20 \times 10^2 \text{ g}) \times \frac{1 \text{ mol O}}{16.00 \text{ g O}} = 4.00 \text{ mol O}$$

Since we used the molar mass to calculate the moles of each element present in the compound, this method directly gives the molecular formula. The formula is $\mathbf{C_4H_8O_4}$.

12.62 We first find the number of moles of gas using the ideal gas equation.

$$n = \frac{PV}{RT} = \frac{\left(748 \text{ mmHg} \times \dfrac{1 \text{ atm}}{760 \text{ mmHg}}\right)(4.00 \text{ L})}{(27 + 273) \text{ K}} \times \frac{\text{mol} \cdot \text{K}}{0.0821 \text{ L} \cdot \text{atm}} = 0.160 \text{ mol}$$

$$\text{molality} = \frac{0.160 \text{ mol}}{0.0580 \text{ kg benzene}} = 2.76\ m$$

$$\Delta T_f = K_f m = (5.12°\text{C}/m)(2.76\ m) = 14.1°\text{C}$$

freezing point = 5.5°C − 14.1°C = **−8.6°C**

12.64 First, from the freezing point depression we can calculate the molality of the solution. See Table 12.2 of the text for the normal freezing point and K_f value for benzene.

$$\Delta T_f = (5.5 - 4.3)°\text{C} = 1.2°\text{C}$$

$$m = \frac{\Delta T_f}{K_f} = \frac{1.2°\text{C}}{5.12°\text{C}/m} = 0.23\ m$$

Multiplying the molality by the mass of solvent (in kg) gives moles of unknown solute. Then, dividing the mass of solute (in g) by the moles of solute, gives the molar mass of the unknown solute.

$$\text{? mol of unknown solute} = \frac{0.23 \text{ mol solute}}{1 \text{ kg benzene}} \times 0.0250 \text{ kg benzene}$$

$$= 0.0058 \text{ mol solute}$$

$$\textbf{molar mass of unknown} = \frac{2.50 \text{ g}}{0.0058 \text{ mol}} = \mathbf{4.3 \times 10^2 \text{ g/mol}}$$

The empirical molar mass of C_6H_5P is 108.1 g/mol. Therefore, the molecular formula is $(C_6H_5P)_4$ or $\mathbf{C_{24}H_{20}P_4}$.

12.66 Calculating molar mass from osmotic pressure, Problem Type 6B.

Strategy: We are asked to calculate the molar mass of the polymer. Grams of the polymer are given in the problem, so we need to solve for moles of polymer.

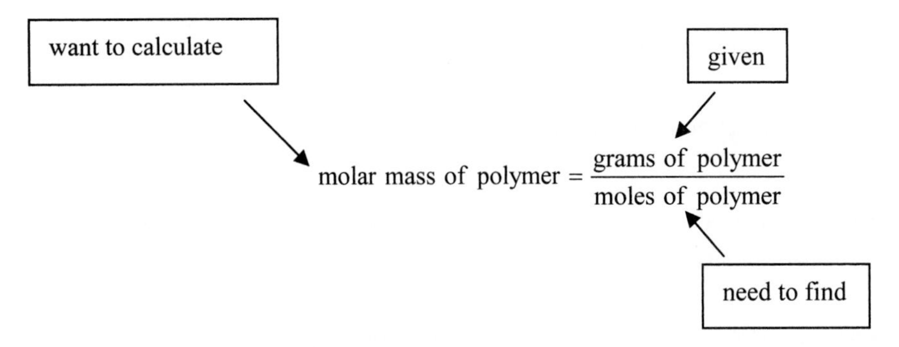

From the osmotic pressure of the solution, we can calculate the molarity of the solution. Then, from the molarity, we can determine the number of moles in 0.8330 g of the polymer. What units should we use for π and temperature?

Solution: First, we calculate the molarity using Equation (12.8) of the text.

$$\pi = MRT$$

$$M = \frac{\pi}{RT} = \frac{\left(5.20\ \text{mmHg} \times \dfrac{1\ \text{atm}}{760\ \text{mmHg}}\right)}{298\ \text{K}} \times \frac{\text{mol}\cdot\text{K}}{0.0821\ \text{L}\cdot\text{atm}} = 2.80 \times 10^{-4}\ M$$

Multiplying the molarity by the volume of solution (in L) gives moles of solute (polymer).

$$?\ \text{mol of polymer} = (2.80 \times 10^{-4}\ \text{mol/L})(0.170\ \text{L}) = 4.76 \times 10^{-5}\ \text{mol polymer}$$

Lastly, dividing the mass of polymer (in g) by the moles of polymer, gives the molar mass of the polymer.

$$\textbf{molar mass of polymer} = \frac{0.8330\ \text{g polymer}}{4.76 \times 10^{-5}\ \text{mol polymer}} = \mathbf{1.75 \times 10^{4}\ g/mol}$$

12.68 We use the osmotic pressure data to determine the molarity.

$$M = \frac{\pi}{RT} = \frac{4.61\ \text{atm}}{(20 + 273)\ \text{K}} \times \frac{\text{mol}\cdot\text{K}}{0.0821\ \text{L}\cdot\text{atm}} = 0.192\ \text{mol/L}$$

Next we use the density and the solution mass to find the volume of the solution.

$$\text{mass of soln} = 6.85\ \text{g} + 100.0\ \text{g} = 106.9\ \text{g soln}$$

$$\text{volume of soln} = 106.9\ \text{g soln} \times \frac{1\ \text{mL}}{1.024\ \text{g}} = 104.4\ \text{mL} = 0.1044\ \text{L}$$

Multiplying the molarity by the volume (in L) gives moles of solute (carbohydrate).

$$\text{mol of solute} = M \times \text{L} = (0.192\ \text{mol/L})(0.1044\ \text{L}) = 0.0200\ \text{mol solute}$$

Finally, dividing mass of carbohydrate by moles of carbohydrate gives the molar mass of the carbohydrate.

$$\textbf{molar mass} = \frac{6.85\text{ g carbohydrate}}{0.0200\text{ mol carbohydrate}} = \textbf{343 g/mol}$$

12.74 Boiling point, vapor pressure, and osmotic pressure all depend on particle concentration. Therefore, these solutions also have the same boiling point, osmotic pressure, and vapor pressure.

12.76 The freezing point will be depressed most by the solution that contains the most solute particles. You should try to classify each solute as a strong electrolyte, a weak electrolyte, or a nonelectrolyte. All three solutions have the same concentration, so comparing the solutions is straightforward. HCl is a strong electrolyte, so under ideal conditions it will completely dissociate into two particles per molecule. The concentration of particles will be 1.00 *m*. Acetic acid is a weak electrolyte, so it will only dissociate to a small extent. The concentration of particles will be greater than 0.50 *m*, but less than 1.00 *m*. Glucose is a nonelectrolyte, so glucose molecules remain as glucose molecules in solution. The concentration of particles will be 0.50 *m*. For these solutions, the order in which the freezing points become *lower* is:

$$0.50\ m\text{ glucose} > 0.50\ m\text{ acetic acid} > 0.50\ m\text{ HCl}$$

In other words, the HCl solution will have the lowest freezing point (greatest freezing point depression).

12.78 Using Equation (12.5) of the text, we can find the mole fraction of the NaCl. We use subscript 1 for H_2O and subscript 2 for NaCl.

$$\Delta P = X_2 P_1^\circ$$

$$X_2 = \frac{\Delta P}{P_1^\circ}$$

$$X_2 = \frac{23.76\text{ mmHg} - 22.98\text{ mmHg}}{23.76\text{ mmHg}} = 0.03283$$

Let's assume that we have 1000 g (1 kg) of water as the solvent, because the definition of molality is moles of solute per kg of solvent. We can find the number of moles of particles dissolved in the water using the definition of mole fraction.

$$X_2 = \frac{n_2}{n_1 + n_2}$$

$$n_1 = 1000\text{ g }H_2O \times \frac{1\text{ mol }H_2O}{18.02\text{ g }H_2O} = 55.49\text{ mol }H_2O$$

$$\frac{n_2}{55.49 + n_2} = 0.03283$$

$$n_2 = 1.884\text{ mol}$$

Since NaCl dissociates to form two particles (ions), the number of moles of NaCl is half of the above result.

$$\text{Moles NaCl} = 1.884\text{ mol particles} \times \frac{1\text{ mol NaCl}}{2\text{ mol particles}} = 0.9420\text{ mol}$$

The molality of the solution is:

$$\frac{0.9420 \text{ mol}}{1.000 \text{ kg}} = \mathbf{0.9420}\ \boldsymbol{m}$$

12.80 Colligative Properties of Electrolytes, Problem Type 7.

Strategy: We want to calculate the osmotic pressure of a NaCl solution. Since NaCl is a strong electrolyte, i in the van't Hoff equation is 2.

$$\pi = iMRT$$

Since, R is a constant and T is given, we need to first solve for the molarity of the solution in order to calculate the osmotic pressure (π). If we assume a given volume of solution, we can then use the density of the solution to determine the mass of the solution. The solution is 0.86% by mass NaCl, so we can find grams of NaCl in the solution.

Solution: To calculate molarity, let's assume that we have 1.000 L of solution (1.000×10^3 mL). We can use the solution density as a conversion factor to calculate the mass of 1.000×10^3 mL of solution.

$$(1.000 \times 10^3 \text{ mL soln}) \times \frac{1.005 \text{ g soln}}{1 \text{ mL soln}} = 1005 \text{ g of soln}$$

Since the solution is 0.86% by mass NaCl, the mass of NaCl in the solution is:

$$1005 \text{ g} \times \frac{0.86\%}{100\%} = 8.6 \text{ g NaCl}$$

The molarity of the solution is:

$$\frac{8.6 \text{ g NaCl}}{1.000 \text{ L}} \times \frac{1 \text{ mol NaCl}}{58.44 \text{ g NaCl}} = 0.15\ M$$

Since NaCl is a strong electrolyte, we assume that the van't Hoff factor is 2. Substituting i, M, R, and T into the equation for osmotic pressure gives:

$$\pi = iMRT = (2)\left(\frac{0.15 \text{ mol}}{\text{L}}\right)\left(\frac{0.0821 \text{ L}\cdot\text{atm}}{\text{mol}\cdot\text{K}}\right)(310 \text{ K}) = \mathbf{7.6\ atm}$$

12.82 From Table 12.3 of the text, $i = 1.3$

$$\pi = iMRT$$

$$\pi = (1.3)\left(\frac{0.0500 \text{ mol}}{\text{L}}\right)\left(\frac{0.0821 \text{ L}\cdot\text{atm}}{\text{mol}\cdot\text{K}}\right)(298 \text{ K})$$

$$\pi = \mathbf{1.6\ atm}$$

12.86 At constant temperature, the osmotic pressure of a solution is proportional to the molarity. When equal volumes of the two solutions are mixed, the molarity will just be the mean of the molarities of the two solutions (assuming additive volumes). Since the osmotic pressure is proportional to the molarity, the osmotic pressure of the solution will be the mean of the osmotic pressure of the two solutions.

$$\pi = \frac{2.4 \text{ atm} + 4.6 \text{ atm}}{2} = \mathbf{3.5\ atm}$$

12.88 **(a)** We use Equation (12.4) of the text to calculate the vapor pressure of each component.

$$P_1 = X_1 P_1^\circ$$

First, you must calculate the mole fraction of each component.

$$X_A = \frac{n_A}{n_A + n_B} = \frac{1.00\text{ mol}}{1.00\text{ mol} + 1.00\text{ mol}} = 0.500$$

Similarly,

$$X_B = 0.500$$

Substitute the mole fraction calculated above and the vapor pressure of the pure solvent into Equation (12.4) to calculate the vapor pressure of each component of the solution.

$$P_A = X_A P_A^\circ = (0.500)(76\text{ mmHg}) = 38\text{ mmHg}$$

$$P_B = X_B P_B^\circ = (0.500)(132\text{ mmHg}) = 66\text{ mmHg}$$

The total vapor pressure is the sum of the vapor pressures of the two components.

$$\boldsymbol{P_{\textbf{Total}}} = P_A + P_B = 38\text{ mmHg} + 66\text{ mmHg} = \mathbf{104\ mmHg}$$

(b) This problem is solved similarly to part (a).

$$X_A = \frac{n_A}{n_A + n_B} = \frac{2.00\text{ mol}}{2.00\text{ mol} + 5.00\text{ mol}} = 0.286$$

Similarly,

$$X_B = 0.714$$

$$P_A = X_A P_A^\circ = (0.286)(76\text{ mmHg}) = 22\text{ mmHg}$$

$$P_B = X_B P_B^\circ = (0.714)(132\text{ mmHg}) = 94\text{ mmHg}$$

$$\boldsymbol{P_{\textbf{Total}}} = P_A + P_B = 22\text{ mmHg} + 94\text{ mmHg} = \mathbf{116\ mmHg}$$

12.90 From the osmotic pressure, you can calculate the molarity of the solution.

$$M = \frac{\pi}{RT} = \frac{\left(30.3\text{ mmHg} \times \dfrac{1\text{ atm}}{760\text{ mmHg}}\right)}{308\text{ K}} \times \frac{\text{mol}\cdot\text{K}}{0.0821\text{ L}\cdot\text{atm}} = 1.58 \times 10^{-3}\text{ mol/L}$$

Multiplying molarity by the volume of solution in liters gives the moles of solute.

$$(1.58 \times 10^{-3}\text{ mol solute/L soln}) \times (0.262\text{ L soln}) = 4.14 \times 10^{-4}\text{ mol solute}$$

Divide the grams of solute by the moles of solute to calculate the molar mass.

$$\textbf{molar mass of solute} = \frac{1.22\text{ g}}{4.14 \times 10^{-4}\text{ mol}} = \mathbf{2.95 \times 10^3\ g/mol}$$

12.92 Solve Equation (12.7) of the text algebraically for molality (m), then substitute ΔT_f and K_f into the equation to calculate the molality. You can find the normal freezing point for benzene and K_f for benzene in Table 12.2 of the text.

$$\Delta T_f = 5.5°C - 3.9°C = 1.6°C$$

$$m = \frac{\Delta T_f}{K_f} = \frac{1.6°\cancel{C}}{5.12°\cancel{C}/m} = 0.31\ m$$

Multiplying the molality by the mass of solvent (in kg) gives moles of unknown solute. Then, dividing the mass of solute (in g) by the moles of solute, gives the molar mass of the unknown solute.

$$?\text{ mol of unknown solute} = \frac{0.31\text{ mol solute}}{1\ \cancel{\text{kg}}\text{ benzene}} \times (8.0 \times 10^{-3}\ \cancel{\text{kg}}\text{ benzene})$$

$$= 2.5 \times 10^{-3}\text{ mol solute}$$

$$\text{molar mass of unknown} = \frac{0.50\text{ g}}{2.5 \times 10^{-3}\text{ mol}} = 2.0 \times 10^{2}\text{ g/mol}$$

The molar mass of cocaine $C_{17}H_{21}NO_4$ = 303 g/mol, so the compound is not cocaine. We assume in our analysis that the compound is a pure, monomeric, nonelectrolyte.

12.94 The molality of the solution assuming $AlCl_3$ to be a nonelectrolyte is:

$$\text{mol AlCl}_3 = 1.00\ \cancel{\text{g}}\text{ AlCl}_3 \times \frac{1\text{ mol AlCl}_3}{133.3\ \cancel{\text{g}}\text{ AlCl}_3} = 0.00750\text{ mol AlCl}_3$$

$$m = \frac{0.00750\text{ mol}}{0.0500\text{ kg}} = 0.150\ m$$

The molality calculated with Equation (12.7) of the text is:

$$m = \frac{\Delta T_f}{K_f} = \frac{1.11°\cancel{C}}{1.86°\cancel{C}/m} = 0.597\ m$$

The ratio $\frac{0.597\ m}{0.150\ m}$ is 4. Thus each $AlCl_3$ dissociates as follows:

$$\mathbf{AlCl_3(\mathit{s}) \rightarrow Al^{3+}(\mathit{aq}) + 3Cl^{-}(\mathit{aq})}$$

12.96 First, we tabulate the concentration of all of the ions. Notice that the chloride concentration comes from more than one source.

$MgCl_2$:	If $[MgCl_2] = 0.054\ M$,	$[Mg^{2+}] = 0.054\ M$	$[Cl^-] = 2 \times 0.054\ M$
Na_2SO_4:	if $[Na_2SO_4] = 0.051\ M$,	$[Na^+] = 2 \times 0.051\ M$	$[SO_4^{2-}] = 0.051\ M$
$CaCl_2$:	if $[CaCl_2] = 0.010\ M$,	$[Ca^{2+}] = 0.010\ M$	$[Cl^-] = 2 \times 0.010\ M$
$NaHCO_3$:	if $[NaHCO_3] = 0.0020\ M$	$[Na^+] = 0.0020\ M$	$[HCO_3^-] = 0.0020\ M$
KCl:	if $[KCl] = 0.0090\ M$	$[K^+] = 0.0090\ M$	$[Cl^-] = 0.0090\ M$

The subtotal of chloride ion concentration is:

$$[Cl^-] = (2 \times 0.0540) + (2 \times 0.010) + (0.0090) = 0.137\ M$$

Since the required $[Cl^-]$ is 2.60 M, the difference (2.6 – 0.137 = 2.46 M) must come from NaCl.

The subtotal of sodium ion concentration is:

$$[Na^+] = (2 \times 0.051) + (0.0020) = 0.104\ M$$

Since the required $[Na^+]$ is 2.56 M, the difference (2.56 – 0.104 = 2.46 M) must come from NaCl.

Now, calculating the mass of the compounds required:

NaCl: $$2.46\ \text{mol} \times \frac{58.44\ \text{g NaCl}}{1\ \text{mol NaCl}} = \mathbf{143.8\ g}$$

$MgCl_2$: $$0.054\ \text{mol} \times \frac{95.21\ \text{g } MgCl_2}{1\ \text{mol } MgCl_2} = \mathbf{5.14\ g}$$

Na_2SO_4: $$0.051\ \text{mol} \times \frac{142.1\ \text{g } Na_2SO_4}{1\ \text{mol } Na_2SO_4} = \mathbf{7.25\ g}$$

$CaCl_2$: $$0.010\ \text{mol} \times \frac{111.0\ \text{g } CaCl_2}{1\ \text{mol } CaCl_2} = \mathbf{1.11\ g}$$

KCl: $$0.0090\ \text{mol} \times \frac{74.55\ \text{g KCl}}{1\ \text{mol KCl}} = \mathbf{0.67\ g}$$

$NaHCO_3$: $$0.0020\ \text{mol} \times \frac{84.01\ \text{g } NaHCO_3}{1\ \text{mol } NaHCO_3} = \mathbf{0.17\ g}$$

12.98 Solution A: Let molar mass be $\mathcal{M}$.

$$\Delta P = X_A P_A^\circ$$

$$(760 - 754.5) = X_A(760)$$

$$X_A = 7.237 \times 10^{-3}$$

$$n = \frac{\text{mass}}{\text{molar mass}}$$

$$X_A = \frac{n_A}{n_A + n_{water}} = \frac{5.00/\mathcal{M}}{5.00/\mathcal{M} + 100/18.02} = 7.237 \times 10^{-3}$$

$$\mathcal{M} = \mathbf{124\ g/mol}$$

Solution B: Let molar mass be $\mathcal{M}$

$$\Delta P = X_B P_B^\circ$$

$$X_B = 7.237 \times 10^{-3}$$

$$n = \frac{\text{mass}}{\text{molar mass}}$$

$$X_B = \frac{n_B}{n_B + n_{\text{benzene}}} = \frac{2.31/\mathcal{M}}{2.31/\mathcal{M} + 100/78.11} = 7.237 \times 10^{-3}$$

$$\mathcal{M} = \mathbf{248\ g/mol}$$

The molar mass in benzene is about twice that in water. This suggests some sort of dimerization is occurring in a nonpolar solvent such as benzene.

12.100 As the chain becomes longer, the alcohols become more like hydrocarbons (nonpolar) in their properties. The alcohol with five carbons (*n*-pentanol) would be the best solvent for iodine (a) and *n*-pentane (c) (why?). Methanol (CH_3OH) is the most water like and is the best solvent for an ionic solid like KBr.

12.102 $I_2 - H_2O$: Dipole - induced dipole.

$I_3^- - H_2O$: Ion - dipole. Stronger interaction causes more I_2 to be converted to I_3^-.

12.104 **(a)** If the membrane is permeable to all the ions and to the water, the result will be the same as just removing the membrane. You will have two solutions of equal NaCl concentration.

(b) This part is tricky. The movement of one ion but not the other would result in one side of the apparatus acquiring a positive electric charge and the other side becoming equally negative. This has never been known to happen, so we must conclude that migrating ions always drag other ions of the opposite charge with them. In this hypothetical situation only water would move through the membrane from the dilute to the more concentrated side.

(c) This is the classic osmosis situation. Water would move through the membrane from the dilute to the concentrated side.

12.106 First, we calculate the number of moles of HCl in 100 g of solution.

$$n_{\text{HCl}} = 100\ \text{g soln} \times \frac{37.7\ \text{g HCl}}{100\ \text{g soln}} \times \frac{1\ \text{mol HCl}}{36.46\ \text{g HCl}} = 1.03\ \text{mol HCl}$$

Next, we calculate the volume of 100 g of solution.

$$V = 100\ \text{g} \times \frac{1\ \text{mL}}{1.19\ \text{g}} \times \frac{1\ \text{L}}{1000\ \text{mL}} = 0.0840\ \text{L}$$

Finally, the molarity of the solution is:

$$\frac{1.03\ \text{mol}}{0.0840\ \text{L}} = \mathbf{12.3}\ \boldsymbol{M}$$

12.108 Let the mass of NaCl be x g. Then, the mass of sucrose is $(10.2 - x)$g.

We know that the equation representing the osmotic pressure is:

$$\pi = MRT$$

π, R, and T are given. Using this equation and the definition of molarity, we can calculate the percentage of NaCl in the mixture.

$$\text{molarity} = \frac{\text{mol solute}}{\text{L soln}}$$

Remember that NaCl dissociates into two ions in solution; therefore, we multiply the moles of NaCl by two.

$$\text{mol solute} = 2\left(x\text{ g NaCl} \times \frac{1\text{ mol NaCl}}{58.44\text{ g NaCl}}\right) + \left((10.2 - x)\text{g sucrose} \times \frac{1\text{ mol sucrose}}{342.3\text{ g sucrose}}\right)$$

$$\text{mol solute} = 0.03422x + 0.02980 - 0.002921x$$

$$\text{mol solute} = 0.03130x + 0.02980$$

$$\text{Molarity of solution} = \frac{\text{mol solute}}{\text{L soln}} = \frac{(0.03130x + 0.02980)\text{ mol}}{0.250\text{ L}}$$

Substitute molarity into the equation for osmotic pressure to solve for x.

$$\pi = MRT$$

$$7.32\text{ atm} = \left(\frac{(0.03130x + 0.02980)\text{ mol}}{0.250\text{ L}}\right)\left(0.0821\frac{\text{L}\cdot\text{atm}}{\text{mol}\cdot\text{K}}\right)(296\text{ K})$$

$$0.0753 = 0.03130x + 0.02980$$

$$x = 1.45\text{ g} = \text{mass of NaCl}$$

$$\textbf{Mass \% NaCl} = \frac{1.45\text{ g}}{10.2\text{ g}} \times 100\% = \textbf{14.2\%}$$

12.110 **(a)** Solubility decreases with increasing lattice energy.

(b) Ionic compounds are more soluble in a polar solvent.

(c) Solubility increases with enthalpy of hydration of the cation and anion.

12.112

$$\text{molality} = \frac{98.0\text{ g } H_2SO_4 \times \dfrac{1\text{ mol } H_2SO_4}{98.09\text{ g } H_2SO_4}}{2.0\text{ g } H_2O \times \dfrac{1\text{ kg } H_2O}{1000\text{ g } H_2O}} = \mathbf{5.0 \times 10^2}\ \boldsymbol{m}$$

We can calculate the density of sulfuric acid from the molarity.

$$\text{molarity} = 18\ M = \frac{18\text{ mol } H_2SO_4}{1\text{ L soln}}$$

The 18 mol of H_2SO_4 has a mass of:

$$18\text{ mol } H_2SO_4 \times \frac{98.0\text{ g } H_2SO_4}{1\text{ mol } H_2SO_4} = 1.8 \times 10^3\text{ g } H_2SO_4$$

$$1\text{ L} = 1000\text{ mL}$$

$$\textbf{density} = \frac{\text{mass } H_2SO_4}{\text{volume}} = \frac{1.8 \times 10^3\text{ g}}{1000\text{ mL}} = \textbf{1.80 g/mL}$$

12.114 $P_A = X_A P_A^\circ$

$P_{ethanol} = (0.62)(108\ \text{mmHg}) = 67.0\ \text{mmHg}$

$P_{1\text{-propanol}} = (0.38)(40.0\ \text{mmHg}) = 15.2\ \text{mmHg}$

In the vapor phase:

$$X_{ethanol} = \frac{67.0}{67.0 + 15.2} = \mathbf{0.815}$$

12.116 NH_3 can form hydrogen bonds with water; NCl_3 cannot. (Like dissolves like.)

12.118 We can calculate the molality of the solution from the freezing point depression.

$$\Delta T_f = K_f m$$

$$0.203 = 1.86\ m$$

$$m = \frac{0.203}{1.86} = 0.109\ m$$

The molality of the original solution was 0.106 *m*. Some of the solution has ionized to H^+ and CH_3COO^-.

	CH_3COOH $\rightleftharpoons$	CH_3COO^- +	H^+
Initial	0.106 *m*	0	0
Change	$-x$	$+x$	$+x$
Equil.	$0.106\ m - x$	x	x

At equilibrium, the total concentration of species in solution is 0.109 *m*.

$$(0.106 - x) + 2x = 0.109\ m$$

$$x = 0.003\ m$$

The percentage of acid that has undergone ionization is:

$$\frac{0.003\ m}{0.106\ m} \times 100\% = \mathbf{3\%}$$

12.120 First, we can calculate the molality of the solution from the freezing point depression.

$$\Delta T_f = (5.12)m$$

$$(5.5 - 3.5) = (5.12)m$$

$$m = 0.39$$

Next, from the definition of molality, we can calculate the moles of solute.

$$m = \frac{\text{mol solute}}{\text{kg solvent}}$$

$$0.39\ m = \frac{\text{mol solute}}{80 \times 10^{-3}\ \text{kg benzene}}$$

$$\text{mol solute} = 0.031\ \text{mol}$$

The molar mass ($\mathcal{M}$) of the solute is:

$$\frac{3.8\text{ g}}{0.031\text{ mol}} = \mathbf{1.2 \times 10^2\ g/mol}$$

The molar mass of CH_3COOH is 60.05 g/mol. Since the molar mass of the solute calculated from the freezing point depression is twice this value, the structure of the solute most likely is a dimer that is held together by hydrogen bonds.

$$H_3C-C(=O\cdots H-O)(-O-H\cdots O=)C-CH_3$$

A dimer

12.122 **(a)** $\Delta T_f = K_f m$

$2 = (1.86)(m)$

$molality = \mathbf{1.1\ m}$

This concentration is too high and is *not* a reasonable physiological concentration.

(b) Although the protein is present in low concentrations, it can prevent the formation of ice crystals.

12.124 As the water freezes, dissolved minerals in the water precipitate from solution. The minerals refract light and create an opaque appearance.

12.126 To solve for the molality of the solution, we need the moles of solute (urea) and the kilograms of solvent (water). If we assume that we have 1 mole of water, we know the mass of water. Using the change in vapor pressure, we can solve for the mole fraction of urea and then the moles of urea. Using Equation (12.5) of the text, we solve for the mole fraction of urea.

$\Delta P = 23.76\text{ mmHg} - 22.98\text{ mmHg} = 0.78\text{ mmHg}$

$$\Delta P = X_2 P_1^\circ = X_{\text{urea}} P_{\text{water}}^\circ$$

$$X_{\text{urea}} = \frac{\Delta P}{P_{\text{water}}^\circ} = \frac{0.78\text{ mmHg}}{23.76\text{ mmHg}} = 0.033$$

Assuming that we have 1 mole of water, we can now solve for moles of urea.

$$X_{\text{urea}} = \frac{\text{mol urea}}{\text{mol urea} + \text{mol water}}$$

$$0.033 = \frac{n_{\text{urea}}}{n_{\text{urea}} + 1}$$

$$0.033 n_{\text{urea}} + 0.033 = n_{\text{urea}}$$

$$0.033 = 0.967 n_{\text{urea}}$$

$$n_{\text{urea}} = 0.034\text{ mol}$$

1 mole of water has a mass of 18.02 g or 0.01802 kg. We now know the moles of solute (urea) and the kilograms of solvent (water), so we can solve for the molality of the solution.

$$\boldsymbol{m} = \frac{\text{mol solute}}{\text{kg solvent}} = \frac{0.034\text{ mol}}{0.01802\text{ kg}} = \mathbf{1.9\ m}$$

12.128 **(a)** The solution is prepared by mixing equal masses of A and B. Let's assume that we have 100 grams of each component. We can convert to moles of each substance and then solve for the mole fraction of each component.

Since the molar mass of A is 100 g/mol, we have 1.00 mole of A. The moles of B are:

$$100\ \text{g B} \times \frac{1\ \text{mol B}}{110\ \text{g B}} = 0.909\ \text{mol B}$$

The mole fraction of A is:

$$\boldsymbol{X}_\mathbf{A} = \frac{n_\text{A}}{n_\text{A} + n_\text{B}} = \frac{1}{1 + 0.909} = \mathbf{0.524}$$

Since this is a two component solution, the mole fraction of B is: $\boldsymbol{X}_\mathbf{B} = 1 - 0.524 = \mathbf{0.476}$

(b) We can use Equation (12.4) of the text and the mole fractions calculated in part (a) to calculate the partial pressures of A and B over the solution.

$$\boldsymbol{P}_\mathbf{A} = X_\text{A}P_\text{A}^\circ = (0.524)(95\ \text{mmHg}) = \mathbf{50\ mmHg}$$

$$\boldsymbol{P}_\mathbf{B} = X_\text{B}P_\text{B}^\circ = (0.476)(42\ \text{mmHg}) = \mathbf{20\ mmHg}$$

(c) Recall that pressure of a gas is directly proportional to moles of gas ($P \propto n$). The ratio of the partial pressures calculated in part (b) is 50 : 20, and therefore the ratio of moles will also be 50 : 20. Let's assume that we have 50 moles of A and 20 moles of B. We can solve for the mole fraction of each component and then solve for the vapor pressures using Equation (12.4) of the text.

The mole fraction of A is:

$$\boldsymbol{X}_\mathbf{A} = \frac{n_\text{A}}{n_\text{A} + n_\text{B}} = \frac{50}{50 + 20} = \mathbf{0.71}$$

Since this is a two component solution, the mole fraction of B is: $\boldsymbol{X}_\mathbf{B} = 1 - 0.71 = \mathbf{0.29}$

The vapor pressures of each component above the solution are:

$$\boldsymbol{P}_\mathbf{A} = X_\text{A}P_\text{A}^\circ = (0.71)(95\ \text{mmHg}) = \mathbf{67\ mmHg}$$

$$\boldsymbol{P}_\mathbf{B} = X_\text{B}P_\text{B}^\circ = (0.29)(42\ \text{mmHg}) = \mathbf{12\ mmHg}$$

12.130 To calculate the mole fraction of urea in the solutions, we need the moles of urea and the moles of water. The number of moles of urea in each beaker is:

$$\text{moles urea (1)} = \frac{0.10\ \text{mol}}{1\ \text{L}} \times 0.050\ \text{L} = 0.0050\ \text{mol}$$

$$\text{moles urea (2)} = \frac{0.20\ \text{mol}}{1\ \text{L}} \times 0.050\ \text{L} = 0.010\ \text{mol}$$

The number of moles of water in each beaker initially is:

$$\text{moles water} = 50\ \text{mL} \times \frac{1\ \text{g}}{1\ \text{mL}} \times \frac{1\ \text{mol}}{18.02\ \text{g}} = 2.8\ \text{mol}$$

The mole fraction of urea in each beaker initially is:

$$X_1 = \frac{0.0050 \text{ mol}}{0.0050 \text{ mol} + 2.8 \text{ mol}} = 1.8 \times 10^{-3}$$

$$X_2 = \frac{0.010 \text{ mol}}{0.010 \text{ mol} + 2.8 \text{ mol}} = 3.6 \times 10^{-3}$$

Equilibrium is attained by the transfer of water (via water vapor) from the less concentrated solution to the more concentrated one until the mole fractions of urea are equal. At this point, the mole fractions of water in each beaker are also equal, and Raoult's law implies that the vapor pressures of the water over each beaker are the same. Thus, there is no more net transfer of solvent between beakers. Let y be the number of moles of water transferred to reach equilibrium.

$$X_1 \text{ (equil.)} = X_2 \text{ (equil.)}$$

$$\frac{0.0050 \text{ mol}}{0.0050 \text{ mol} + 2.8 \text{ mol} - y} = \frac{0.010 \text{ mol}}{0.010 \text{ mol} + 2.8 \text{ mol} + y}$$

$$0.014 + 0.0050y = 0.028 - 0.010y$$

$$y = 0.93$$

The mole fraction of urea at equilibrium is:

$$\frac{0.010 \text{ mol}}{0.010 \text{ mol} + 2.8 \text{ mol} + 0.93 \text{ mol}} = \mathbf{2.7 \times 10^{-3}}$$

This solution to the problem assumes that the volume of water left in the bell jar as vapor is negligible compared to the volumes of the solutions. It is interesting to note that at equilibrium, 16.8 mL of water has been transferred from one beaker to the other.

12.132 Starting with $n = kP$ and substituting into the ideal gas equation ($PV = nRT$), we find:

$$PV = (kP)RT$$

$$V = kRT$$

This equation shows that the volume of a gas that dissolves in a given amount of solvent is dependent on the *temperature*, not the pressure of the gas.

12.134 To calculate the freezing point of the solution, we need the solution molality and the freezing-point depression constant for water (see Table 12.2 of the text). We can first calculate the molarity of the solution using Equation (12.8) of the text: $\pi = MRT$. The solution molality can then be determined from the molarity.

$$M = \frac{\pi}{RT} = \frac{10.50 \text{ atm}}{(0.0821 \text{ L}\cdot\text{atm/mol}\cdot\text{K})(298 \text{ K})} = 0.429\ M$$

Let's assume that we have 1 L (1000 mL) of solution. The mass of 1000 mL of solution is:

$$\frac{1.16 \text{ g}}{1 \text{ mL}} \times 1000 \text{ mL} = 1160 \text{ g soln}$$

The mass of the solvent (H_2O) is:

$$\text{mass } H_2O = \text{mass soln} - \text{mass solute}$$

$$\text{mass } H_2O = 1160 \text{ g} - \left(0.429 \text{ mol glucose} \times \frac{180.2 \text{ g glucose}}{1 \text{ mol glucose}}\right) = 1083 \text{ g} = 1.083 \text{ kg}$$

The molality of the solution is:

$$\text{molality} = \frac{\text{mol solute}}{\text{kg solvent}} = \frac{0.429 \text{ mol}}{1.083 \text{ kg}} = 0.396\ m$$

The freezing point depression is:

$$\Delta T_f = K_f m = (1.86°\text{C}/m)(0.396\ m) = 0.737°\text{C}$$

The solution will freeze at 0°C – 0.737°C = **–0.737°C**

CHAPTER 13
CHEMICAL KINETICS

PROBLEM-SOLVING STRATEGIES AND TUTORIAL SOLUTIONS

TYPES OF PROBLEMS

Problem Type 1: Writing Rate Expressions.

Problem Type 2: Determining the Rate Law of a Reaction.

Problem Type 3: First-Order Reactions.
- **(a)** Analyzing a first-order reaction.
- **(b)** Determining the half-life of a first-order reaction.
- **(c)** Analyzing first-order kinetics.

Problem Type 4: Second-Order Reactions.
- **(a)** Analyzing a second-order reaction.
- **(b)** Determining the half-life of a second-order reaction.
- **(c)** Analyzing second-order kinetics.

Problem Type 5: The Arrhenius Equation.
- **(a)** Applying the Arrhenius equation.
- **(b)** Applying a modified form of the Arrhenius equation that relates the rate constants at two different temperatures.

Problem Type 6: Studying Reaction Mechanisms.

PROBLEM TYPE 1: WRITING RATE EXPRESSIONS

As a chemical reaction proceeds, the concentrations of reactants and products change with time. As the reaction

$$A + B \longrightarrow C$$

progresses, the concentration of C increases. The rate can be expressed as the change in concentration of C during the time interval Δt.

$$\text{rate} = \frac{\Delta[C]}{\Delta t}$$

For a specific reaction, we need to take into account the stoichiometry of the balanced equation. For example, let's express the rate of the following reaction in terms of the concentrations of the individual reactants and products.

$$2NO(g) + O_2(g) \longrightarrow 2NO_2(g)$$

Notice from the balanced equation that the concentration of NO will decrease twice as fast as that of O_2. We can write the rate as

$$\text{rate} = -\frac{1}{2}\frac{\Delta[NO]}{\Delta t} \quad \text{or} \quad \text{rate} = -\frac{\Delta[O_2]}{\Delta t} \quad \text{or} \quad \text{rate} = \frac{1}{2}\frac{\Delta[NO_2]}{\Delta t}$$

Division of each concentration by the coefficient from the balanced equation makes all of the above rates equal. Notice also that a negative sign is inserted before terms involving reactants. The Δ[NO] is a negative quantity because the concentration of NO *decreases* with time. Therefore, multiplying Δ[NO] by a negative sign makes the rate of reaction a positive quantity.

For a general reaction:

$$aA + bB \longrightarrow cC + dD$$

the rate of reaction can be expressed in terms of any reactant or product.

$$\text{rate} = -\frac{1}{a}\frac{\Delta[A]}{\Delta t} = -\frac{1}{b}\frac{\Delta[B]}{\Delta t} = \frac{1}{c}\frac{\Delta[C]}{\Delta t} = \frac{1}{d}\frac{\Delta[D]}{\Delta t}$$

EXAMPLE 13.1

Write expressions for the rate of the following reaction in terms of each of the reactants and products.

$$2N_2O_5(g) \longrightarrow 4NO_2(g) + O_2(g)$$

Strategy: The rate is defined as the change in concentration of a reactant or product with time. Each "change in concentration" term is divided by the corresponding stoichiometric coefficient. Terms involving reactants are preceded by a minus sign.

Solution:

$$\textbf{Rate} = -\frac{1}{2}\frac{\Delta[N_2O_5]}{\Delta t} = \frac{1}{4}\frac{\Delta[NO_2]}{\Delta t} = \frac{\Delta[O_2]}{\Delta t}$$

PRACTICE EXERCISE

1. Oxygen gas can be formed by the decomposition of nitrogen monoxide (nitric oxide).

$$2NO(g) \longrightarrow O_2(g) + N_2(g)$$

If the rate of formation of O_2 is 0.054 *M*/s, what is the rate of change of the NO concentration?

Text Problems: 13.6, **13.8**

PROBLEM TYPE 2: DETERMINING THE RATE LAW OF A REACTION

The **rate law** is an expression relating the rate of a reaction to the rate constant and the concentrations of reactants. For a general reaction of the type

$$aA + bB \longrightarrow cC + dD$$

the rate law takes the form

$$\text{rate} = k[A]^x[B]^y$$

The term k is the **rate constant**, a constant of proportionality between the reaction rate and the concentrations of the reactants. The sum of the powers to which all reactant concentrations appearing in the rate law are raised is called the overall **reaction order**. In the rate law expression shown above, the overall reaction order is given by $x + y$.

k, x, and y must be determined experimentally. One method to determine x and y is to keep the concentrations of all but one reactant fixed. Then, the rate of reaction is measured as a function of the concentration of the one reactant whose concentration is varied. Any variation in rate is due solely to the variation in this reactant's concentration.

The example below will show you how to determine the rate law and how to calculate the rate constant.

EXAMPLE 13.2
The following rate data were collected for the reaction

$$\mathbf{2NO + 2H_2 \longrightarrow N_2 + 2H_2O}$$

(a) Determine the rate law.
(b) Calculate the rate constant.

Experiment	[NO] (M)	$[H_2]$ (M)	$\Delta[N_2]/\Delta t$ (M/h)
1	0.60	0.15	0.076
2	0.60	0.30	0.15
3	0.60	0.60	0.30
4	1.20	0.60	1.21

Strategy: We are given a set of concentrations and rate data and asked to determine the rate law and the rate constant. We assume that the rate law takes the form

$$\text{rate} = k[\text{NO}]^x[\text{H}_2]^y$$

How do we use the data to determine x and y? Once the orders of the reactants are known, we can calculate k for any set of rate and concentrations.

Solution:

(a) Experiments 1 and 2 show that when we double the concentration of H_2 at constant concentration of NO, the rate doubles. Taking the ratio of the rates from these two experiments:

$$\frac{\text{rate}_2}{\text{rate}_1} = \frac{0.15\ M/\text{h}}{0.076\ M/\text{h}} \approx 2 = \frac{k(0.60)^x(0.30)^y}{k(0.60)^x(0.15)^y}$$

Therefore,

$$\frac{(0.30)^y}{(0.15)^y} = 2^y = 2$$

or, $y = 1$. That is, the reaction is first-order in H_2. Experiments 3 and 4 indicate that doubling [NO] at constant $[H_2]$ quadruples the rate. Here we write the ratio as:

$$\frac{\text{rate}_4}{\text{rate}_3} = \frac{1.21\ M/\text{h}}{0.30\ M/\text{h}} \approx 4 = \frac{k(1.20)^x(0.60)^y}{k(0.60)^x(0.60)^y}$$

Therefore,

$$\frac{(1.20)^x}{(0.60)^x} = 2^x = 4$$

or, $x = 2$. That is, the reaction is second-order in NO. Hence, the rate law is given by:

$$\mathbf{rate} = \boldsymbol{k}\mathbf{[NO]^2[H_2]}$$

(b) The rate constant k can be calculated using the data from any one of the experiments. Rearranging the rate law and using the first set of data, we find:

$$k = \frac{\text{rate}}{[\text{NO}]^2[\text{H}_2]}$$

$$\boldsymbol{k} = \frac{0.076\ \cancel{M}/\text{h}}{(0.60\ M)^2(0.15\ \cancel{M})} = \mathbf{1.4}\ \boldsymbol{M}^{\mathbf{-2}}\mathbf{h^{-1}}$$

PRACTICE EXERCISE

2. The following experimental data were obtained for the reaction

$$2A + B \longrightarrow \text{products}$$

What is the rate law for this reaction?

Experiment	$[A]_0$ (M)	$[B]_0$ (M)	Rate (M/s)
1	0.40	0.20	5.6×10^{-3}
2	0.80	0.20	5.5×10^{-3}
3	0.40	0.40	22.3×10^{-3}

Text Problems: 13.16, **13.18**, 13.20, **13.52 a,b**

PROBLEM TYPE 3: FIRST-ORDER REACTIONS

One of the most widely encountered kinetic forms is the first-order rate equation. Consider the reaction,

$$A \longrightarrow \text{products}$$

For a first-order reaction, the exponent of [A] in the rate law is 1. We can write:

$$\text{rate} = k[A]$$

We also know that

$$\text{rate} = \frac{-\Delta[A]}{\Delta t}$$

Combining the two equations gives:

$$k[A] = \frac{-\Delta[A]}{\Delta t}$$

Using calculus, we can integrate both sides of the above equation to give:

$$\ln\frac{[A]_t}{[A]_0} = -kt \qquad (13.3, \text{text})$$

where ln is the natural logarithm, and $[A]_0$ and $[A]_t$ are the concentrations of A at times $t = 0$ and $t = t$, respectively.

A. Analyzing a first-order reaction

EXAMPLE 13.3

Methyl isocyanide undergoes a first-order isomerization to from methyl cyanide.

$$CH_3NC(g) \longrightarrow CH_3CN(g)$$

The reaction was studied at 199°C. The initial concentration of CH_3NC was 0.0258 mol/L and after 11.4 min, analysis showed the concentration of the product to be 1.30×10^{-3} mol/L.

(a) What is the first-order rate constant?

(b) How long will it take for 90.0 percent of the CH_3NC to react?

(a)
Strategy: The relationship between the concentration of a reactant at different times in a first-order reaction is given by Equations (13.3) and (13.4) of the text. From the initial concentration and the concentration at time = 11.4 min, the first-order rate constant can be determined.

Solution: We can calculate $[CH_3NC]_t$ by realizing that the amount of product formed equals the amount of reactant lost due to the 1:1 mole ratio between reactant and product. Thus,

$$[CH_3NC]_t = [CH_3NC]_0 - (1.30 \times 10^{-3}\ M)$$
$$= 0.0258\ M - (1.30 \times 10^{-3}\ M) = 0.0245\ M$$

Using Equation (13.3) of the text, we plug in the concentrations and time to solve for k.

$$\ln\frac{[CH_3NC]_t}{[CH_3NC]_0} = -kt$$

$$\ln\frac{0.0245\ M}{0.0258\ M} = -k(11.4\ \text{min})$$

$$\boldsymbol{k} = -\frac{\ln(0.9496)}{11.4\ \text{min}} = \mathbf{4.54 \times 10^{-3}\ min^{-1}}$$

(b)
Strategy: The relationship between the concentration of a reactant at different times in a first-order reaction is given by Equations (13.3) and (13.4) of the text. We are asked to determine the time required for 90% of CH_3NC to react. If we initially have 100% of the compound and 90% has reacted, then what is left must be (100% – 90%), or 10%. Thus, the ratio of the percentages will be equal to the ratio of the actual concentrations; that is, $[A]_t/[A]_0$ = 10%/100%, or 0.10/1.00.

Solution: The time required for 90% of CH_3NC to react can be found using Equation (13.3) of the text.

$$\ln\frac{[CH_3NC]_t}{[CH_3NC]_0} = -kt$$

$$\ln\frac{(0.10)}{(1.00)} = -(4.54 \times 10^{-3}\ \text{min}^{-1})t$$

$$\boldsymbol{t} = -\frac{\ln(0.10)}{4.54 \times 10^{-3}\ \text{min}^{-1}} = \mathbf{507\ min}$$

PRACTICE EXERCISE

3. The hydrolysis of sucrose ($C_{12}H_{22}O_{11}$) yields the simple sugars glucose ($C_6H_{12}O_6$) and fructose ($C_6H_{12}O_6$), which happen to be isomers.

$$C_{12}H_{22}O_{11} + H_2O \longrightarrow C_6H_{12}O_6 + C_6H_{12}O_6$$

The reaction is first-order in glucose concentration.

$$\text{rate} = k[C_{12}H_{22}O_{11}]$$

At 27°C, the rate constant is $2.1 \times 10^{-6}\ s^{-1}$. Starting with a sucrose solution with a concentration of 0.10 M at 27°C, what would the concentration of sucrose be 24 hours later? (The solution is maintained at 27°C.)

Text Problem: 13.28

B. Determining the half-life of a first-order reaction

The **half-life** of a reaction, $t_{\frac{1}{2}}$, is the time required for the concentration of a reactant to decrease to half of its initial concentration. From Equation (13.3) of the text, we can write

$$\ln\frac{[A]_0}{[A]_t} = kt$$

From the definition of half-life, when $t = t_{\frac{1}{2}}$

$$[A]_t = \frac{[A]_0}{2}$$

Substituting gives:

$$\ln\frac{[A]_0}{\frac{[A]_0}{2}} = kt_{\frac{1}{2}}$$

$$\ln 2 = kt_{\frac{1}{2}}$$

$$t_{\frac{1}{2}} = \frac{\ln 2}{k} = \frac{0.693}{k} \qquad \text{(13.6, text)}$$

Equation (13.6) of the text shows that the half-life of a first-order reaction is *independent* of the initial concentration of the reactant. Thus, it would take the same time for the concentration of the reactant to decrease from 1.0 M to 0.50 M as it would to decrease from 0.10 M to 0.050 M.

The half-life can also be used to determine the rate constant of a first-order reaction.

EXAMPLE 13.4

Methyl isocyanide undergoes a first-order isomerization to form methyl cyanide.

$$CH_3NC(g) \longrightarrow CH_3CN(g)$$

The reaction was studied at 199°C. The initial concentration of CH_3NC was 0.0258 mol/L and after 11.4 min, analysis showed the concentration of the product to be 1.30×10^{-3} mol/L. Using the rate constant calculated in Example 13.3, calculate the half-life of methyl isocyanide.

Strategy: To calculate the half-life of a first-order reaction, we use Equation (13.6) of the text. The rate constant for this reaction was determined in Example 13.3 ($k = 4.54 \times 10^{-3}$ min^{-1}).

Solution: For a first-order reaction, we only need the rate constant to calculate the half-life. From Equation (13.6),

$$t_{\frac{1}{2}} = \frac{0.693}{k} = \frac{0.693}{4.54 \times 10^{-3}\ \text{min}^{-1}} = \mathbf{153\ min}$$

PRACTICE EXERCISE

4. The hydrolysis of sucrose ($C_{12}H_{22}O_{11}$) yields the simple sugars glucose ($C_6H_{12}O_6$) and fructose ($C_6H_{12}O_6$), which happen to be isomers.

$$C_{12}H_{22}O_{11} + H_2O \longrightarrow C_6H_{12}O_6 + C_6H_{12}O_6$$

The reaction is first-order in sucrose concentration.

$$\text{rate} = k[C_{12}H_{22}O_{11}]$$

At 27°C, the rate constant is 2.1×10^{-6} s^{-1}. What is the half-life of sucrose at 27°C?

Text Problem: 13.28

C. Analyzing first-order kinetics

From Equation (13.3),

$$\ln\frac{[A]_t}{[A]_0} = -kt \qquad (13.3, text)$$

$$\ln[A]_t - \ln[A]_0 = -kt$$

or

$$\ln[A]_t = -kt + \ln[A]_0 \qquad (13.4, text)$$

Equation (13.4) has the form of a linear equation.

$$y = mx + b$$

where,

m is the slope of the line
b is the y-intercept

A plot of ln[A] versus t (y vs. x) gives a straight line with a slope of $-k$ (m). Thus, we can calculate the rate constant k from the slope of the plot.

EXAMPLE 13.5
At 500 K, butadiene gas converts to cyclobutene gas:

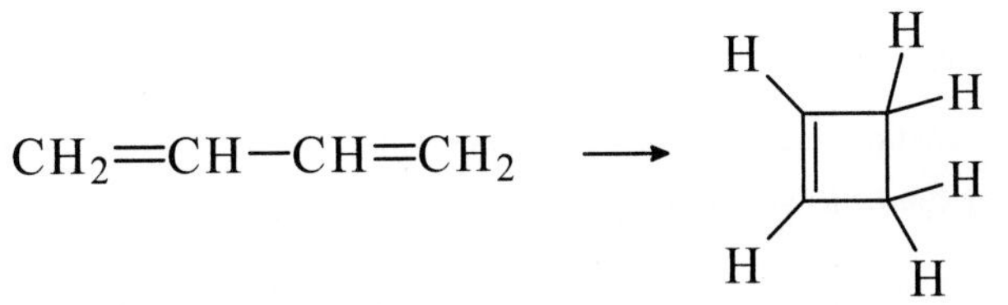

$$CH_2{=}CH{-}CH{=}CH_2 \longrightarrow C_4H_6 \text{ (cyclobutene)}$$

Given the following data, is the reaction first-order in butadiene concentration?

Time from start (s)	Concentration of butadiene (M)
195	0.0162
604	0.0147
1246	0.0129
2180	0.0110
4140	0.0084
8135	0.0057

Strategy: If the reaction is first-order in butadiene, then a plot of ln[butadiene] versus t will be a straight line.

Solution:

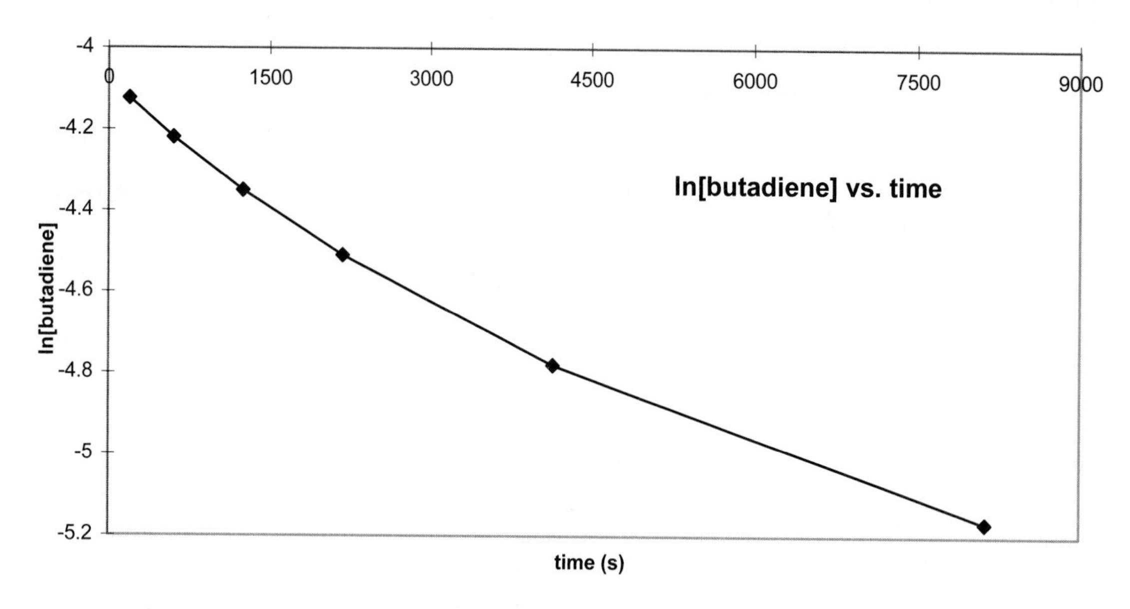

As you can see, a plot of ln[butadiene] versus t does *not* give a straight line. Hence, the reaction is *not* first-order in butadiene.

PRACTICE EXERCISE

5. In a certain experiment, the rate of hydrogen peroxide decomposition,

$$2H_2O_2 \rightarrow 2H_2O + O_2$$

is followed by titration against a potassium permanganate solution. At regular intervals, an equal volume of H_2O_2 is withdrawn to give the following data:

Time (min)	0	10.0	20.0	30.0
Volume of $KMnO_4$ used (mL)	22.8	13.8	8.25	5.00

Confirm that the reaction is first-order in hydrogen peroxide and calculate the rate constant.

Hint: The concentration of hydrogen peroxide is directly proportional to the volume (in mL) of $KMnO_4$ used in each titration.

Text Problem: 13.22

PROBLEM TYPE 4: SECOND-ORDER REACTIONS

Second-order reactions are also encountered quite often. Consider the reaction,

$$A \longrightarrow \text{products}$$

For a second-order reaction, the exponent of [A] in the rate law is 2. We can write:

$$\text{rate} = k[A]^2$$

We also know that

$$\text{rate} = \frac{-\Delta[A]}{\Delta t}$$

Combining the two equations gives:

$$k[\mathrm{A}]^2 = \frac{-\Delta[\mathrm{A}]}{\Delta t}$$

Using calculus, we can integrate both sides of the above equation to give:

$$\frac{1}{[\mathrm{A}]_t} = kt + \frac{1}{[\mathrm{A}]_0} \quad \text{(13.7, text)}$$

where $[\mathrm{A}]_0$ and $[\mathrm{A}]_t$ are the concentrations of A at times $t = 0$ and $t = t$, respectively.

A. Analyzing a second-order reaction

EXAMPLE 13.6
At 500 K, butadiene gas converts to cyclobutene gas:

$$CH_2{=}CH{-}CH{=}CH_2 \longrightarrow C_4H_6 \text{ (cyclobutene)}$$

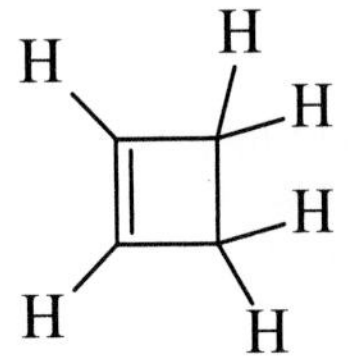

At 500 K, the rate constant for the reaction is 0.0143/*M*·s. If the initial concentration of butadiene is 0.272 *M*, calculate the concentration of butadiene after 30.0 min.

Strategy: The relationship between the concentration of a reactant at different times in a second-order reaction is given by Equation (13.7) of the text. From the initial concentration and the rate constant, the concentration at $t = 30$ min can be determined.

Solution: Using Equation (13.7) of the text, we plug in the initial concentration, k, and t, to solve for the concentration at $t = 30$ min.

$$\frac{1}{[\text{butadiene}]_t} = kt + \frac{1}{[\text{butadiene}]_0}$$

$$\frac{1}{[\text{butadiene}]_t} = (0.0143\ /M\cdot\text{s})(1.80\times10^3\ \text{s}) + \frac{1}{0.272\ M}$$

$$\frac{1}{[\text{butadiene}]_t} = 29.4\ M^{-1}$$

$$\textbf{[butadiene]} = \mathbf{0.0340}\ \boldsymbol{M}$$

PRACTICE EXERCISE

6. For the reaction shown in Example 13.6 above, how long will it take for the concentration of butadiene to decrease from its initial concentration of 0.272 *M* to 0.100 *M*?

Text Problem: 13.30

B. Determining the half-life of a second-order reaction

The **half-life** of a reaction, $t_{\frac{1}{2}}$, is the time required for the concentration of a reactant to decrease to half of its initial concentration. Starting with Equation (13.7) of the text,

$$\frac{1}{[A]_t} = kt + \frac{1}{[A]_0} \qquad (13.7, \text{text})$$

and the definition of half-life, $t = t_{\frac{1}{2}}$, we write:

$$[A]_t = \frac{[A]_0}{2}$$

Substituting into Equation (13.7) gives:

$$\frac{1}{\frac{[A]_0}{2}} = kt_{\frac{1}{2}} + \frac{1}{[A]_0}$$

$$\frac{2}{[A]_0} - \frac{1}{[A]_0} = kt_{\frac{1}{2}}$$

$$t_{\frac{1}{2}} = \frac{1}{k[A]_0} \qquad (13.8, \text{text})$$

Equation (13.8) of the text shows that the half-life of a second-order reaction is *dependent* on the initial concentration of the reactant, unlike the half-life of a first-order reaction.

EXAMPLE 13.7
The following reaction follows second-order kinetics:

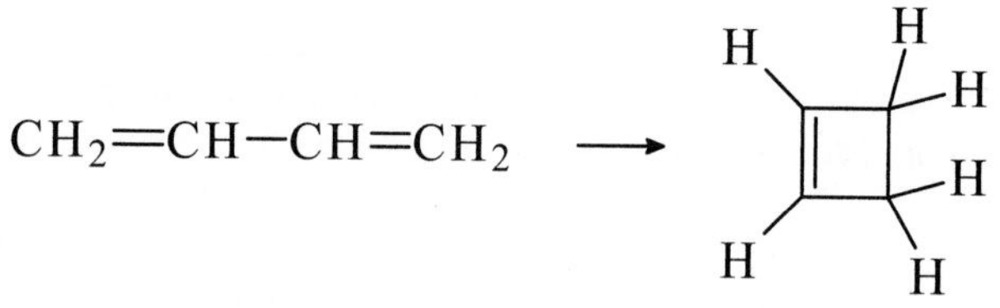

$$CH_2{=}CH{-}CH{=}CH_2 \longrightarrow C_4H_6 \text{ (cyclobutene)}$$

butadiene(g) cyclobutene(g)

At 500 K, the rate constant is 0.0143/*M*·s. The initial concentration of butadiene is 0.272 *M*. What is the half-life for this reaction?

Strategy: To calculate the half-life of a second-order reaction, we use Equation (13.8) of the text. The initial concentration and the rate constant are needed to solve for the half-life.

Solution: For a second-order reaction, we need the rate constant and the initial concentration to calculate the half-life. From Equation (13.8),

$$t_{\frac{1}{2}} = \frac{1}{k[A]_0} = \frac{1}{(0.0143\ /M\cdot s)(0.272\ M)} = \mathbf{257\ s}$$

PRACTICE EXERCISE

7. For the reaction shown in Example 13.7 above, the half-life of the reaction is determined to be 66.6 s at 500 K. What is the initial concentration of butadiene? (k = 0.0143 /m·s at 500 K)

C. Analyzing second-order kinetics

Equation (13.7) of the text has the form of a linear equation.

$$\frac{1}{[A]_t} = kt + \frac{1}{[A]_0} \qquad (13.7, \text{text})$$

$$y = mx + b$$

where,

m is the slope of the line
b is the y-intercept

A plot of $\frac{1}{[A]}$ versus t (y vs. x) gives a straight line with a slope of k (m). Thus, we can calculate the rate constant k from the slope of the plot.

EXAMPLE 13.8
At 500 K, butadiene gas converts to cyclobutene gas:

$$CH_2{=}CH{-}CH{=}CH_2 \longrightarrow \text{cyclobutene}$$

H H H H H H

Given the following data, is the reaction second-order in butadiene concentration?

Time from start (s)	Concentration of butadiene (*M*)
195	0.0162
604	0.0147
1246	0.0129
2180	0.0110
4140	0.0084
8135	0.0057

Strategy: If the reaction is second-order in butadiene, then a plot of $\frac{1}{[\text{butadiene}]}$ versus t will be a straight line.

Solution:

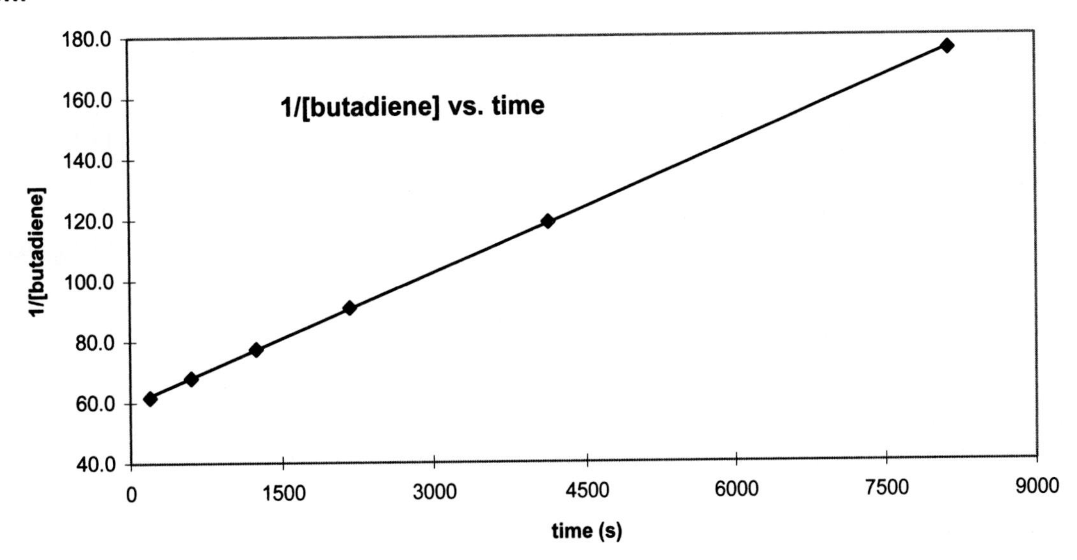

As you can see, a plot of $\frac{1}{[\text{butadiene}]}$ versus t *does* give a straight line. Hence, the reaction is second-order in butadiene.

PRACTICE EXERCISE

8. If a plot of $\frac{1}{[A]}$ versus time produces a straight line, which of the following must be *true*?

a. The reaction is first-order in A.
b. The reaction is second-order in A.
c. The reaction is first-order in two reactants.
d. The rate of the reaction does not depend on the concentration of A.
e. None of the above

Text Problem: 13.22

PROBLEM TYPE 5: THE ARRHENIUS EQUATION

The dependence of the rate constant of a reaction on temperature can be expressed by the following equation, called the **Arrhenius equation**.

$$k = Ae^{-E_a/RT}$$

where, E_a is the activation energy of the reaction (in kJ/mol)
R is the gas constant (8.314 J/mol·K)
T is the absolute temperature (in K)
e is the base of the natural logarithm scale (see Appendix 4)

The quantity A represents the collision frequency and is called the *frequency factor*. It can be treated as a constant for a given reacting system over a fairly wide temperature range. The Arrhenius equation shows that the rate constant is directly proportional to A. Therefore, as the number of collisions increase, the rate increases.

The minus sign associated with the exponent E_a/RT indicates that the rate constant decreases with increasing activation energy and increases with increasing temperature.

The Arrhenius equation can be expressed in a more useful form by taking the natural logarithm of both sides.

$$\ln k = \ln\left[Ae^{-E_a/RT}\right]$$

$$\ln k = \ln A - \frac{E_a}{RT} \qquad (13.12, \text{text})$$

or

$$\ln k = \left(-\frac{E_a}{R}\right)\left(\frac{1}{T}\right) + \ln A \qquad (13.13, \text{text})$$

Equation (13.13) of the text has the form of the linear equation

$$y = mx + b$$

where,

m is the slope of the line
b is the y-intercept

A plot of $\ln [k]$ versus $\frac{1}{T}$ (y vs. x) gives a straight line with a slope of $\frac{-E_a}{R}$ (m). Thus, we can calculate the activation energy (E_a) from the slope of the plot.

A. Applying the Arrhenius Equation

EXAMPLE 13.9
Variation of the rate constant with temperature for the reaction

$$NO + O_3 \rightarrow NO_2 + O_2$$

is given in the following table. Determine graphically the activation energy for the reaction.

Temperature (K)	k ($M^{-1}s^{-1}$)
283	9.30×10^6
293	1.08×10^7
303	1.25×10^7
313	1.43×10^7
323	1.62×10^7

Strategy: A plot of ln [k] versus $\frac{1}{T}$ should give a straight line with a slope of $\frac{-E_a}{R}$. Thus, we can calculate the activation energy (E_a) from the slope of the plot.

Solution: From the given data we obtain:

$1/T$ (K^{-1})	ln k
3.53×10^{-3}	16.05
3.41×10^{-3}	16.20
3.30×10^{-3}	16.34
3.19×10^{-3}	16.48
3.10×10^{-3}	16.60

These data, when plotted, yield the graph shown below.

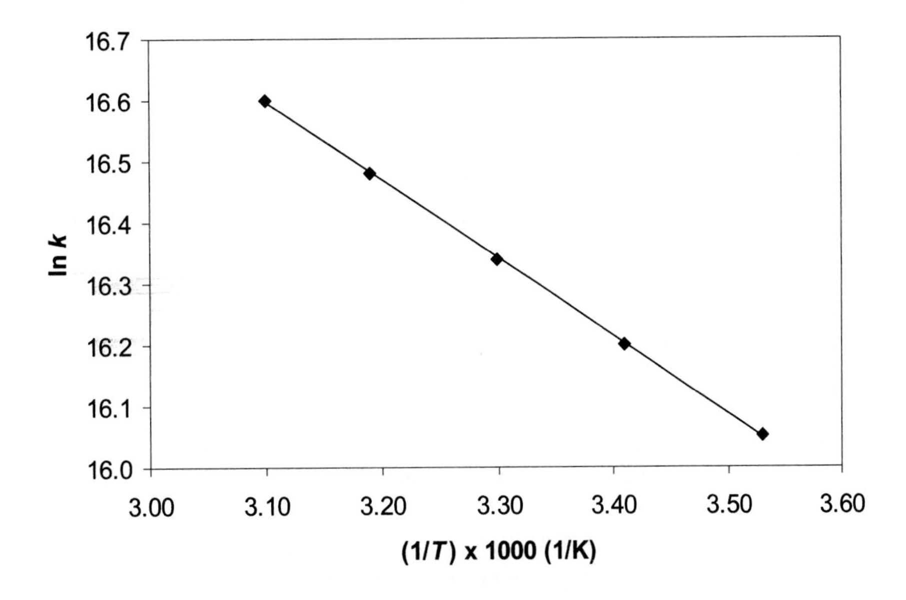

Calculating the slope from the first and last points gives:

$$\text{slope} = \frac{16.05 - 16.60}{(3.53 \times 10^{-3}) - (3.10 \times 10^{-3})} = -1.28 \times 10^3 \text{ K}$$

The slope is equal to $-E_a/R$ or

$$E_a = -\text{slope} \times R$$

$$\boldsymbol{E_a} = -(-1.28 \times 10^3 \cancel{\text{K}})(8.314 \text{ J/mol}\cdot\cancel{\text{K}}) = \mathbf{1.06 \times 10^4 \text{ J/mol}} = \mathbf{10.6 \text{ kJ/mol}}$$

B. Applying a modified form of the Arrhenius equation that relates the rate constants at two different temperatures

Starting with Equation (13.12) of the text, we can write,

$$\ln k_1 = \ln A - \frac{E_a}{RT_1}$$

and

$$\ln k_2 = \ln A - \frac{E_a}{RT_2}$$

Subtracting $\ln k_2$ from $\ln k_1$ gives:

$$\ln k_1 - \ln k_2 = -\frac{E_a}{RT_1} + \frac{E_a}{RT_2}$$

$$\ln \frac{k_1}{k_2} = \frac{E_a}{R}\left(\frac{1}{T_2} - \frac{1}{T_1}\right)$$

or

$$\ln \frac{k_1}{k_2} = \frac{E_a}{R}\left(\frac{T_1 - T_2}{T_1 T_2}\right) \quad \text{(13.14, text)}$$

Using Equation (13.14) of the text, we can calculate the activation energy if we know the rate constants at two temperatures or find the rate constant at another temperature if the activation energy is known.

EXAMPLE 13.10
For the reaction

$$\mathbf{NO + O_3 \longrightarrow NO_2 + O_2}$$

the following rate constants have been obtained. Calculate the activation energy for this reaction.

Temperature (°C)	k ($M^{-1}s^{-1}$)
10.0	9.3×10^6
30.0	1.25×10^7

Strategy: A modified form of the Arrhenius equation relates two rate constants at two different temperatures [see Equation (13.14) of the text]. The activation energy can be calculated using this equation.

Solution: The data are: $T_1 = 10°C = 283$ K, $k_1 = 9.3 \times 10^6\ M^{-1}s^{-1}$, $T_2 = 30°C = 303$ K, $k_2 = 1.25 \times 10^7\ M^{-1}s^{-1}$. Recall that $R = 8.314$ J/mol·K. Substituting into Equation (13.14) of the text,

$$\ln\frac{k_1}{k_2} = \frac{E_a}{R}\left(\frac{T_1 - T_2}{T_1T_2}\right) \qquad (13.14,\ \text{text})$$

$$\ln\left(\frac{9.3 \times 10^6\ M^{-1}\text{s}^{-1}}{1.25 \times 10^7\ M^{-1}\text{s}^{-1}}\right) = \frac{E_a}{8.314\ \text{J/mol}\cdot\text{K}}\left(\frac{283\ \text{K} - 303\ \text{K}}{(283\ \text{K})(303\ \text{K})}\right)$$

$$\ln(0.744)(8.314\ \text{J/mol}\cdot\text{K}) = E_a\left(\frac{-20\ \cancel{\text{K}}}{8.57 \times 10^4\ \text{K}^{\cancel{2}}}\right)$$

$$(-0.296)\left(8.314\frac{\text{J}}{\text{mol}\cdot\cancel{\text{K}}}\right) = E_a\left(-2.33 \times 10^{-4}\frac{1}{\cancel{\text{K}}}\right)$$

$$\mathbf{E_a = 1.06 \times 10^4\ J/mol = 10.6\ kJ/mol}$$

Compare the answer in this example with that obtained in Example 13.9.

PRACTICE EXERCISE

9. At 300 K, the rate constant is $1.5 \times 10^{-5}\ M^{-1}\text{s}^{-1}$ for the following reaction:

$$2\text{NOCl} \longrightarrow 2\text{NO} + \text{Cl}_2$$

The activation energy is 90.2 kJ/mol. Calculate the value of the rate constant at 310 K.

Text Problems: 13.38, 13.40, 13.42

PROBLEM TYPE 6: STUDYING REACTION MECHANISMS

For a complete discussion of reaction mechanisms, see Section 13.5 of the text. Experimental studies of reaction mechanisms begin with the collection of data (rate measurements). Next the data are analyzed to determine the rate constant and the order of reaction, so that the rate law can be written. Finally, a plausible mechanism is suggested for the reaction in terms of elementary steps. This sequence of steps is summarized in Figure 13.20 of the text.

The elementary steps of the proposed mechanism must satisfy two requirements:

- The sum of the elementary steps must give the overall balanced equation for the reaction.
- The **rate-determining step**, which is the slowest step in the sequence of steps leading to product formation, should predict the same rate law as is determined experimentally.

EXAMPLE 13.11

The rate law for the substitution of NH_3 for H_2O in the following reaction is first order in $Ni(H_2O)_6^{2+}$ and zero order in NH_3.

$$\mathbf{Ni(H_2O)_6^{2+}(\mathit{aq}) + NH_3(\mathit{aq}) \longrightarrow Ni(H_2O)_5(NH_3)^{2+}(\mathit{aq}) + H_2O(\mathit{l})}$$

Show that the following mechanism is consistent with the experimental rate law.

$$\text{Ni(H}_2\text{O)}_6^{2+}(aq) \longrightarrow \text{Ni(H}_2\text{O)}_5^{2+}(aq) + \text{H}_2\text{O}(l) \qquad \text{(slow)}$$

$$\text{Ni(H}_2\text{O)}_5^{2+}(aq) + \text{NH}_3(aq) \longrightarrow \text{Ni(H}_2\text{O)}_5(\text{NH}_3)^{2+}(aq) \qquad \text{(fast)}$$

Strategy: Do the elementary steps add to give the overall balanced equation? Does the rate law written from the rate-determining step match the experimentally determined rate law?

Solution: First, does the sum of the elementary steps give the overall balanced equation. Yes, the sum of the steps gives:

$$Ni(H_2O)_6^{2+}(aq) + NH_3(aq) \longrightarrow Ni(H_2O)_5(NH_3)^{2+}(aq) + H_2O(l)$$

Next, the reaction was experimentally determined to be first order in $Ni(H_2O)_6^{2+}$ and zero order in NH_3. We can write the rate law for the reaction.

$$\text{rate} = k[Ni(H_2O)_6^{2+}]$$

Does the rate law match the rate law of the proposed mechanism? We can write the rate law from the rate determining step of the proposed mechanism. Step 1 is the slow step, so we can write:

$$\text{rate} = k[Ni(H_2O)_6^{2+}]$$

which matches the experimentally determined rate law.

The elementary steps of the proposed mechanism satisfy the two requirements outlined above; therefore, it is a valid mechanism.

PRACTICE EXERCISE

10. The reaction of nitric oxide and chlorine,

$$2NO(g) + Cl_2(g) \rightarrow 2NOCl(g)$$

has been proposed to proceed by the following two-step mechanism:

$$NO(g) + Cl_2(g) \rightarrow NOCl_2(g)$$

$$NO(g) + NOCl_2(g) \rightarrow 2NOCl(g)$$

What rate law is predicted if the first step of the proposed mechanism is the rate-determining step?

Text Problems: 13.52c, 13.54, 13.62

ANSWERS TO PRACTICE EXERCISES

1. Rate of change of NO = −0.11 M/s

2. Rate = $k[B]^2$

3. [sucrose] = 0.083 M after 24 h

4. $t_{\frac{1}{2}} = 3.3 \times 10^5$ s

5. The slope of the straight line plot equals $-k$. You should find $\boldsymbol{k} = \mathbf{0.0504\ min^{-1}}$.

6. 442 s

7. 1.05 M

8. **(b)**

9. $k = 4.8 \times 10^{-5}\ M^{-1}s^{-1}$

10. rate = $k[NO][Cl_2]$

SOLUTIONS TO SELECTED TEXT PROBLEMS

13.6 **(a)** $$\text{rate} = -\frac{1}{2}\frac{\Delta[H_2]}{\Delta t} = -\frac{\Delta[O_2]}{\Delta t} = \frac{1}{2}\frac{\Delta[H_2O]}{\Delta t}$$

(b) $$\text{rate} = -\frac{1}{4}\frac{\Delta[NH_3]}{\Delta t} = -\frac{1}{5}\frac{\Delta[O_2]}{\Delta t} = \frac{1}{4}\frac{\Delta[NO]}{\Delta t} = \frac{1}{6}\frac{\Delta[H_2O]}{\Delta t}$$

13.8 Writing Rate Expressions, Problem Type 1.

Strategy: The rate is defined as the change in concentration of a reactant or product with time. Each "change in concentration" term is divided by the corresponding stoichiometric coefficient. Terms involving reactants are preceded by a minus sign.

$$\text{rate} = -\frac{\Delta[N_2]}{\Delta t} = -\frac{1}{3}\frac{\Delta[H_2]}{\Delta t} = \frac{1}{2}\frac{\Delta[NH_3]}{\Delta t}$$

Solution:

(a) If hydrogen is reacting at the rate of -0.074 M/s, the rate at which ammonia is being formed is

$$\frac{1}{2}\frac{\Delta[NH_3]}{\Delta t} = -\frac{1}{3}\frac{\Delta[H_2]}{\Delta t}$$

or

$$\frac{\Delta[NH_3]}{\Delta t} = -\frac{2}{3}\frac{\Delta[H_2]}{\Delta t}$$

$$\frac{\Delta[NH_3]}{\Delta t} = -\frac{2}{3}(-0.074\ M/\text{s}) = \mathbf{0.049\ \textit{M}/s}$$

(b) The rate at which nitrogen is reacting must be:

$$\frac{\Delta[N_2]}{\Delta t} = \frac{1}{3}\frac{\Delta[H_2]}{\Delta t} = \frac{1}{3}(-0.074\ M/\text{s}) = \mathbf{-0.025\ \textit{M}/s}$$

Will the rate at which ammonia forms always be twice the rate of reaction of nitrogen, or is this true only at the instant described in this problem?

13.16 Assume the rate law has the form:

$$\text{rate} = k[F_2]^x[ClO_2]^y$$

To determine the order of the reaction with respect to F_2, find two experiments in which the $[ClO_2]$ is held constant. Compare the data from experiments 1 and 3. When the concentration of F_2 is doubled, the reaction rate doubles. Thus, the reaction is *first-order* in F_2.

To determine the order with respect to ClO_2, compare experiments 1 and 2. When the ClO_2 concentration is quadrupled, the reaction rate quadruples. Thus, the reaction is *first-order* in ClO_2.

The rate law is:

$$\text{rate} = k[F_2][ClO_2]$$

The value of k can be found using the data from any of the experiments. If we take the numbers from the second experiment we have:

$$k = \frac{\text{rate}}{[F_2][ClO_2]} = \frac{4.8 \times 10^{-3}\ \cancel{M}/s}{(0.10\ \cancel{M})(0.040\ M)} = 1.2\ M^{-1}s^{-1}$$

Verify that the same value of k can be obtained from the other sets of data.

Since we now know the rate law and the value of the rate constant, we can calculate the rate at any concentration of reactants.

$$\textbf{rate} = k[F_2][ClO_2] = (1.2\ \cancel{M}^{-1}s^{-1})(0.010\ \cancel{M})(0.020\ M) = \mathbf{2.4 \times 10^{-4}\ M/s}$$

13.18 Determining the Rate Law of a Reaction, Problem Type 2.

Strategy: We are given a set of concentrations and rate data and asked to determine the order of the reaction and the initial rate for specific concentrations of X and Y. To determine the order of the reaction, we need to find the rate law for the reaction. We assume that the rate law takes the form

$$\text{rate} = k[X]^x[Y]^y$$

How do we use the data to determine x and y? Once the orders of the reactants are known, we can calculate k for any set of rate and concentrations. Finally, the rate law enables us to calculate the rate at any concentrations of X and Y.

Solution:

(a) Experiments 2 and 5 show that when we double the concentration of X at constant concentration of Y, the rate quadruples. Taking the ratio of the rates from these two experiments

$$\frac{\text{rate}_5}{\text{rate}_2} = \frac{0.509\ M/s}{0.127\ M/s} \approx 4 = \frac{k(0.40)^x(0.30)^y}{k(0.20)^x(0.30)^y}$$

Therefore,

$$\frac{(0.40)^x}{(0.20)^x} = 2^x = 4$$

or, $x = 2$. That is, the reaction is second order in X. Experiments 2 and 4 indicate that doubling [Y] at constant [X] doubles the rate. Here we write the ratio as

$$\frac{\text{rate}_4}{\text{rate}_2} = \frac{0.254\ M/s}{0.127\ M/s} = 2 = \frac{k(0.20)^x(0.60)^y}{k(0.20)^x(0.30)^y}$$

Therefore,

$$\frac{(0.60)^y}{(0.30)^y} = 2^y = 2$$

or, $y = 1$. That is, the reaction is first order in Y. Hence, the rate law is given by:

$$\text{rate} = k[X]^2[Y]$$

The order of the reaction is $(2 + 1) = \mathbf{3}$. The reaction is *3rd-order*.

(b) The rate constant k can be calculated using the data from any one of the experiments. Rearranging the rate law and using the first set of data, we find:

$$k = \frac{\text{rate}}{[X]^2[Y]} = \frac{0.053\ \cancel{M}/s}{(0.10\ M)^2(0.50\ \cancel{M})} = 10.6\ M^{-2}s^{-1}$$

Next, using the known rate constant and substituting the concentrations of X and Y into the rate law, we can calculate the initial rate of disappearance of X.

$$\textbf{rate} = (10.6\ M^{-2}s^{-1})(0.30\ M)^2(0.40\ M) = \mathbf{0.38\ M/s}$$

13.20 **(a)** For a reaction first-order in A,

$$\text{Rate} = k[A]$$

$$1.6 \times 10^{-2}\ \cancel{M}/s = k(0.35\ \cancel{M})$$

$$\mathbf{k = 0.046\ s^{-1}}$$

(b) For a reaction second-order in A,

$$\text{Rate} = k[A]^2$$

$$1.6 \times 10^{-2}\ M/s = k(0.35\ M)^2$$

$$\mathbf{k = 0.13\ /M{\cdot}s}$$

13.22 Let P_0 be the pressure of $ClCO_2CCl_3$ at $t = 0$, and let x be the decrease in pressure after time t. Note that from the coefficients in the balanced equation that the loss of 1 atmosphere of $ClCO_2CCl_3$ results in the formation of two atmospheres of $COCl_2$. We write:

$$ClCO_2CCl_3 \rightarrow 2COCl_2$$

Time	$[ClCO_2CCl_3]$	$[COCl_2]$
$t = 0$	P_0	0
$t = t$	$P_0 - x$	$2x$

Thus the change (increase) in pressure (ΔP) is $2x - x = x$. We have:

t(s)	P (mmHg)	$\Delta P = x$	$P_{ClCO_2CCl_3}$	$\ln P_{ClCO_2CCl_3}$	$\frac{1}{P_{ClCO_2CCl_3}}$
0	15.76	0.00	15.76	2.757	0.0635
181	18.88	3.12	12.64	2.537	0.0791
513	22.79	7.03	8.73	2.167	0.115
1164	27.08	11.32	4.44	1.491	0.225

If the reaction is first order, then a plot of $\ln P_{ClCO_2CCl_3}$ vs. t would be linear. If the reaction is second order, a plot of $1/P_{ClCO_2CCl_3}$ vs. t would be linear. The two plots are shown below.

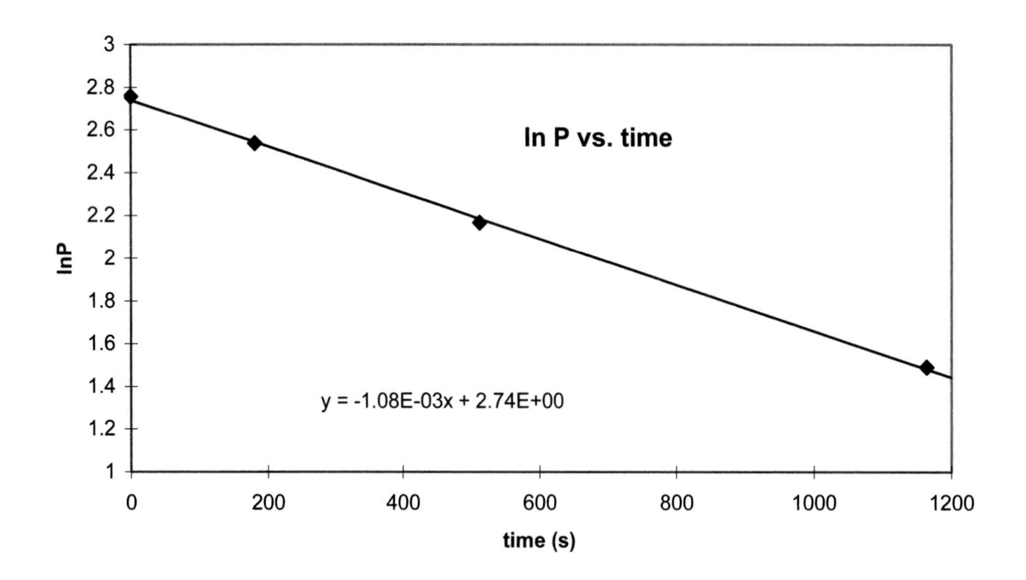

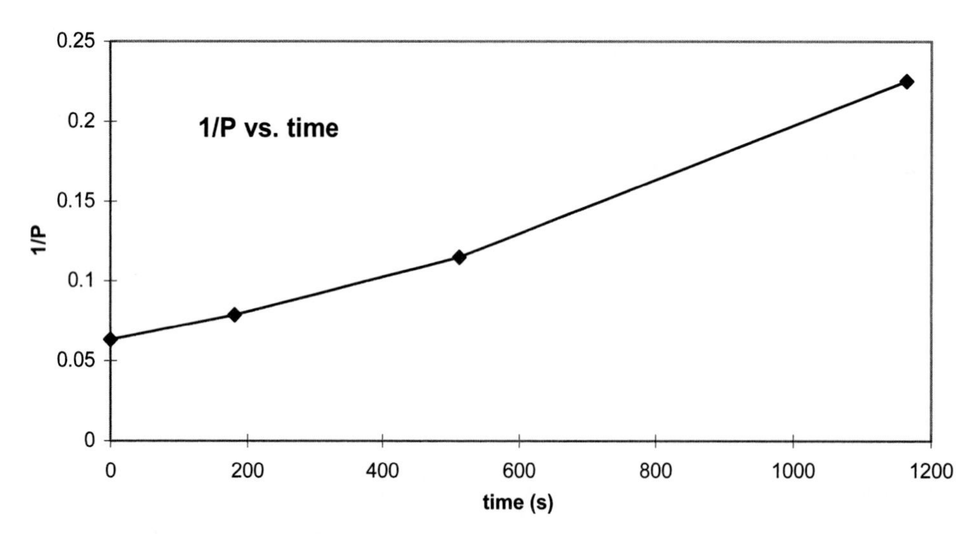

From the graphs we see that the reaction must be **first-order**. For a first-order reaction, the slope is equal to $-k$. The equation of the line is given on the graph. The rate constant is: $\mathbf{k = 1.08 \times 10^{-3}\ s^{-1}}$.

13.28 Analyzing first-order kinetics, Problem Type 3A, and determining the half-life of a first-order reaction, Problem Type 3B.

(a)
Strategy: To calculate the rate constant, k, from the half-life of a first-order reaction, we use Equation (13.6) of the text.

Solution: For a first-order reaction, we only need the half-life to calculate the rate constant. From Equation (13.6)

$$k = \frac{0.693}{t_{\frac{1}{2}}}$$

$$k = \frac{0.693}{35.0 \text{ s}} = \mathbf{0.0198\ s^{-1}}$$

(b)

Strategy: The relationship between the concentration of a reactant at different times in a first-order reaction is given by Equations (13.3) and (13.4) of the text. We are asked to determine the time required for 95% of the phosphine to decompose. If we initially have 100% of the compound and 95% has reacted, then what is left must be (100% – 95%), or 5%. Thus, the ratio of the percentages will be equal to the ratio of the actual concentrations; that is, $[A]_t/[A]_0 = 5\%/100\%$, or 0.05/1.00.

Solution: The time required for 95% of the phosphine to decompose can be found using Equation (13.3) of the text.

$$\ln\frac{[A]_t}{[A]_0} = -kt$$

$$\ln\frac{(0.05)}{(1.00)} = -(0.0198 \text{ s}^{-1})t$$

$$t = -\frac{\ln(0.0500)}{0.0198 \text{ s}^{-1}} = \mathbf{151\ s}$$

13.30

$$\frac{1}{[A]} = \frac{1}{[A]_0} + kt$$

$$\frac{1}{0.28} = \frac{1}{0.62} + 0.54t$$

$$t = \mathbf{3.6\ s}$$

13.38 Applying a modified form of the Arrhenius equation, Problem Type 5B.

Strategy: A modified form of the Arrhenius equation relates two rate constants at two different temperatures [see Equation (13.14) of the text]. Make sure the units of R and E_a are consistent. Since the rate of the reaction at 250°C is 1.50×10^3 times faster than the rate at 150°C, the ratio of the rate constants, k, is also 1.50×10^3 : 1, because rate and rate constant are directly proportional.

Solution: The data are: T_1 = 250°C = 523 K, T_2 = 150°C = 423 K, and $k_1/k_2 = 1.50 \times 10^3$. Substituting into Equation (13.14) of the text,

$$\ln\frac{k_1}{k_2} = \frac{E_a}{R}\left(\frac{T_1 - T_2}{T_1 T_2}\right)$$

$$\ln(1.50 \times 10^3) = \frac{E_a}{8.314 \text{ J/mol}\cdot\text{K}}\left(\frac{523 \text{ K} - 423 \text{ K}}{(523 \text{ K})(423 \text{ K})}\right)$$

$$7.31 = \frac{E_a}{8.314 \dfrac{\text{J}}{\text{mol}\cdot\text{K}}}\left(4.52 \times 10^{-4} \frac{1}{\text{K}}\right)$$

$$E_a = \mathbf{1.35 \times 10^5\ J/mol = 135\ kJ/mol}$$

13.40 Use a modified form of the Arrhenius equation to calculate the temperature at which the rate constant is $8.80 \times 10^{-4}\ s^{-1}$. We carry an extra significant figure throughout this calculation to minimize rounding errors.

$$\ln\frac{k_1}{k_2} = \frac{E_a}{R}\left(\frac{1}{T_2} - \frac{1}{T_1}\right)$$

$$\ln\left(\frac{4.60 \times 10^{-4}\ \text{s}^{-1}}{8.80 \times 10^{-4}\ \text{s}^{-1}}\right) = \frac{1.04 \times 10^{5}\ \text{J/mol}}{8.314\ \text{J/mol}\cdot\text{K}}\left(\frac{1}{T_2} - \frac{1}{623\ \text{K}}\right)$$

$$\ln(0.5227) = (1.251 \times 10^{4}\ \text{K})\left(\frac{1}{T_2} - \frac{1}{623\ \text{K}}\right)$$

$$-0.6487 + 20.08 = \frac{1.251 \times 10^{4}\ \text{K}}{T_2}$$

$$19.43T_2 = 1.251 \times 10^{4}\ \text{K}$$

$$\mathbf{T_2 = 644\ K = 371°C}$$

13.42 Since the ratio of rates is equal to the ratio of rate constants, we can write:

$$\ln\frac{\text{rate}_1}{\text{rate}_2} = \ln\frac{k_1}{k_2}$$

$$\ln\frac{k_1}{k_2} = \ln\left(\frac{2.0 \times 10^{2}}{39.6}\right) = \frac{E_a}{8.314\ \text{J/K}\cdot\text{mol}}\left(\frac{(300\ \text{K} - 278\ \text{K})}{(300\ \text{K})(278\ \text{K})}\right)$$

$$\mathbf{E_a = 5.10 \times 10^{4}\ J/mol = 51.0\ kJ/mol}$$

13.52 **(a)** Determining the Rate Law of a Reaction, Problem Type 2.

Strategy: We are given information as to how the concentrations of X_2, Y, and Z affect the rate of the reaction and are asked to determine the rate law. We assume that the rate law takes the form

$$\text{rate} = k[X_2]^x[Y]^y[Z]^z$$

How do we use the information to determine x, y, and z?

Solution: Since the reaction rate doubles when the X_2 concentration is doubled, the reaction is first-order in X. The reaction rate triples when the concentration of Y is tripled, so the reaction is also first-order in Y. The concentration of Z has no effect on the rate, so the reaction is zero-order in Z.

The rate law is:

$$\mathbf{rate = k[X_2][Y]}$$

(b) If a change in the concentration of Z has no effect on the rate, the concentration of Z is not a term in the rate law. This implies that Z does not participate in the rate-determining step of the reaction mechanism.

(c) Studying Reaction Mechanisms, Problem Type 6.

Strategy: The rate law, determined in part (a), shows that the slow step involves reaction of a molecule of X_2 with a molecule of Y. Since Z is not present in the rate law, it does not take part in the slow step and must appear in a fast step at a later time. (If the fast step involving Z happened before the rate-determining step, the rate law would involve Z in a more complex way.)

Solution: A mechanism that is consistent with the rate law could be:

$$X_2 + Y \longrightarrow XY + X \qquad \text{(slow)}$$

$$X + Z \longrightarrow XZ \qquad \text{(fast)}$$

The rate law only tells us about the slow step. Other mechanisms with different subsequent fast steps are possible. Try to invent one.

Check: The rate law written from the rate-determining step in the proposed mechanism matches the rate law determined in part (a). Also, the two elementary steps add to the overall balanced equation given in the problem.

13.54 The experimentally determined rate law is first order in H_2 and second order in NO. In Mechanism I the slow step is bimolecular and the rate law would be:

$$\text{rate} = k[H_2][NO]$$

Mechanism I can be discarded.

The rate-determining step in Mechanism II involves the simultaneous collision of two NO molecules with one H_2 molecule. The rate law would be:

$$\text{rate} = k[H_2][NO]^2$$

Mechanism II is a possibility.

In Mechanism III we assume the forward and reverse reactions in the first fast step are in dynamic equilibrium, so their rates are equal:

$$k_f[NO]^2 = k_r[N_2O_2]$$

The slow step is bimolecular and involves collision of a hydrogen molecule with a molecule of N_2O_2. The rate would be:

$$\text{rate} = k_2[H_2][N_2O_2]$$

If we solve the dynamic equilibrium equation of the first step for $[N_2O_2]$ and substitute into the above equation, we have the rate law:

$$\text{rate} = \frac{k_2 k_f}{k_r}[H_2][NO]^2 = k[H_2][NO]^2$$

Mechanism III is also a possibility. Can you suggest an experiment that might help to decide between the two mechanisms?

13.62 The rate-determining step involves the breakdown of ES to E and P. The rate law for this step is:

$$\text{rate} = k_2[ES]$$

In the first elementary step, the intermediate ES is in equilibrium with E and S. The equilibrium relationship is:

$$\frac{[ES]}{[E][S]} = \frac{k_1}{k_{-1}}$$

or

$$[\text{ES}] = \frac{k_1}{k_{-1}}[\text{E}][\text{S}]$$

Substitute [ES] into the rate law expression.

$$\textbf{rate} = k_2[\text{ES}] = \frac{k_1 k_2}{k_{-1}}[\mathbf{E}][\mathbf{S}]$$

13.64 Temperature, energy of activation, concentration of reactants, and a catalyst.

13.66 First, calculate the radius of the 10.0 cm^3 sphere.

$$V = \frac{4}{3}\pi r^3$$

$$10.0\ \text{cm}^3 = \frac{4}{3}\pi r^3$$

$$r = 1.34\ \text{cm}$$

The surface area of the sphere is:

$$\textbf{area} = 4\pi r^2 = 4\pi(1.34\ \text{cm})^2 = \mathbf{22.6\ cm^2}$$

Next, calculate the radius of the 1.25 cm^3 sphere.

$$V = \frac{4}{3}\pi r^3$$

$$1.25\ \text{cm}^3 = \frac{4}{3}\pi r^3$$

$$r = 0.668\ \text{cm}$$

The surface area of one sphere is:

$$\text{area} = 4\pi r^2 = 4\pi(0.668\ \text{cm})^2 = 5.61\ \text{cm}^2$$

$$\textbf{The total area of 8 spheres} = 5.61\ \text{cm}^2 \times 8 = \mathbf{44.9\ cm^2}$$

Obviously, the surface area of the eight spheres (44.9 cm^2) is greater than that of one larger sphere (22.6 cm^2). A greater surface area promotes the catalyzed reaction more effectively.

It can be dangerous to work in grain elevators, because the large surface area of the grain dust can result in a violent explosion.

13.68 The overall rate law is of the general form: $\text{rate} = k[H_2]^x[\text{NO}]^y$

(a) Comparing Experiment #1 and Experiment #2, we see that the concentration of NO is constant and the concentration of H_2 has decreased by one-half. The initial rate has also decreased by one-half. Therefore, the initial rate is directly proportional to the concentration of H_2; $x = 1$.

Comparing Experiment #1 and Experiment #3, we see that the concentration of H_2 is constant and the concentration of NO has decreased by one-half. The initial rate has decreased by one-fourth. Therefore, the initial rate is proportional to the squared concentration of NO; $y = 2$.

The overall rate law is: rate $= k[H_2][NO]^2$, and the order of the reaction is $1 + 2 =$ **3.**

(b) Using Experiment #1 to calculate the rate constant,

$$\text{rate} = k[H_2][NO]^2$$

$$k = \frac{\text{rate}}{[H_2][NO]^2}$$

$$\boldsymbol{k} = \frac{2.4 \times 10^{-6}\ M/\text{s}}{(0.010\ M)(0.025\ M)^2} = \mathbf{0.38\ /M^2 \cdot s}$$

(c) Consulting the rate law, we assume that the slow step in the reaction mechanism will probably involve one H_2 molecule and two NO molecules. Additionally the hint tells us that O atoms are an intermediate.

$H_2 + 2NO \rightarrow N_2 + H_2O + O$	slow step
$O + H_2 \rightarrow H_2O$	fast step
$2H_2 + 2NO \rightarrow N_2 + 2H_2O$	

13.70 If water is also the solvent in this reaction, it is present in vast excess over the other reactants and products. Throughout the course of the reaction, the concentration of the water will not change by a measurable amount. As a result, the reaction rate will not appear to depend on the concentration of water.

13.72 Since the reaction is first order in both A and B, then we can write the rate law expression:

$$\text{rate} = k[A][B]$$

Substituting in the values for the rate, [A], and [B]:

$$4.1 \times 10^{-4}\ M/\text{s} = k(1.6 \times 10^{-2})(2.4 \times 10^{-3})$$

$$\boldsymbol{k} = \mathbf{10.7\ /M \cdot s}$$

Knowing that the overall reaction was second order, could you have predicted the units for k?

13.74 Recall that the pressure of a gas is directly proportional to the number of moles of gas. This comes from the ideal gas equation.

$$P = \frac{nRT}{V}$$

The balanced equation is:

$$2N_2O(g) \longrightarrow 2N_2(g) + O_2(g)$$

From the stoichiometry of the balanced equation, for every one mole of N_2O that decomposes, one mole of N_2 and 0.5 moles of O_2 will be formed. Let's assume that we had 2 moles of N_2O at $t = 0$. After one half-life there will be one mole of N_2O remaining and one mole of N_2 and 0.5 moles of O_2 will be formed. The total number of moles of gas after one half-life will be:

$$n_T = n_{N_2O} + n_{N_2} + n_{O_2} = 1 \text{ mol} + 1 \text{ mol} + 0.5 \text{ mol} = 2.5 \text{ mol}$$

At $t = 0$, there were 2 mol of gas. Now, at $t_{\frac{1}{2}}$, there are 2.5 mol of gas. Since the pressure of a gas is directly proportional to the number of moles of gas, we can write:

$$\frac{2.10 \text{ atm}}{2 \cancel{\text{mol}} \text{ gas } (t = 0)} \times 2.5 \cancel{\text{mol}} \text{ gas}\left(\text{at } t_{\frac{1}{2}}\right) = \textbf{2.63 atm after one half-life}$$

13.76 The rate expression for a third order reaction is:

$$\text{rate} = -\frac{\Delta[\text{A}]}{\Delta t} = k[\text{A}]^3$$

The units for the rate law are:

$$\frac{M}{\text{s}} = kM^3$$

$$\boldsymbol{k = M^{-2}\text{s}^{-1}}$$

13.78 Both compounds, A and B, decompose by first-order kinetics. Therefore, we can write a first-order rate equation for A and also one for B.

$$\ln\frac{[\text{A}]_t}{[\text{A}]_0} = -k_\text{A}t \qquad \ln\frac{[\text{B}]_t}{[\text{B}]_0} = -k_\text{B}t$$

$$\frac{[\text{A}]_t}{[\text{A}]_0} = e^{-k_\text{A}t} \qquad \frac{[\text{B}]_t}{[\text{B}]_0} = e^{-k_\text{B}t}$$

$$[\text{A}]_t = [\text{A}]_0 e^{-k_\text{A}t} \qquad [\text{B}]_t = [\text{B}]_0 e^{-k_\text{B}t}$$

We can calculate each of the rate constants, k_A and k_B, from their respective half-lives.

$$k_\text{A} = \frac{0.693}{50.0 \text{ min}} = 0.0139 \text{ min}^{-1} \qquad k_\text{B} = \frac{0.693}{18.0 \text{ min}} = 0.0385 \text{ min}^{-1}$$

The initial concentration of A and B are equal. $[\text{A}]_0 = [\text{B}]_0$. Therefore, from the first-order rate equations, we can write:

$$\frac{[\text{A}]_t}{[\text{B}]_t} = 4 = \frac{[\text{A}]_0 e^{-k_\text{A}t}}{[\text{B}]_0 e^{-k_\text{B}t}} = \frac{e^{-k_\text{A}t}}{e^{-k_\text{B}t}} = e^{(k_\text{B} - k_\text{A})t} = e^{(0.0385 - 0.0139)t}$$

$$4 = e^{0.0246t}$$

$$\ln 4 = 0.0246t$$

$$\boldsymbol{t = 56.4 \textbf{ min}}$$

13.80 **(a)** Changing the concentration of a reactant has no effect on k.

(b) If a reaction is run in a solvent other than in the gas phase, then the reaction mechanism will probably change and will thus change k.

(c) Doubling the pressure simply changes the concentration. No effect on k, as in (a).

(d) The rate constant k changes with temperature.

(e) A catalyst changes the reaction mechanism and therefore changes k.

13.82 Mathematically, the amount left after ten half–lives is:

$$\left(\frac{1}{2}\right)^{10} = \mathbf{9.8 \times 10^{-4}}$$

13.84 The net ionic equation is:

$$Zn(s) + 2H^+(aq) \longrightarrow Zn^{2+}(aq) + H_2(g)$$

(a) Changing from the same mass of granulated zinc to powdered zinc **increases** the rate because the surface area of the zinc (and thus its concentration) has increased.

(b) Decreasing the mass of zinc (in the same granulated form) will **decrease** the rate because the total surface area of zinc has decreased.

(c) The concentration of protons has decreased in changing from the strong acid (hydrochloric) to the weak acid (acetic); the rate will **decrease.**

(d) An increase in temperature will **increase** the rate constant k; therefore, the rate of reaction increases.

13.86 If the reaction is 35.5% complete, the amount of A remaining is 64.5%. The ratio of $[A]_t/[A]_0$ is 64.5%/100% or 0.645/1.00. Using the first-order integrated rate law, Equation (13.3) of the text, we have

$$\ln\frac{[A]_t}{[A]_0} = -kt$$

$$\ln\frac{0.645}{1.00} = -k(4.90\text{ min})$$

$$-0.439 = -k(4.90\text{ min})$$

$$\boldsymbol{k} = \mathbf{0.0896\ min^{-1}}$$

13.88 The first-order rate equation can be arranged to take the form of a straight line.

$$\ln[A] = -kt + \ln[A]_0$$

If a reaction obeys first-order kinetics, a plot of ln[A] vs. t will be a straight line with a slope of $-k$.

The slope of a plot of $\ln[N_2O_5]$ vs. t is $-6.18 \times 10^{-4}\text{ min}^{-1}$. Thus,

$$k = 6.18 \times 10^{-4}\text{ min}^{-1}$$

The equation for the half-life of a first-order reaction is:

$$t_{\frac{1}{2}} = \frac{0.693}{k}$$

$$\boldsymbol{t_{\frac{1}{2}}} = \frac{0.693}{6.18 \times 10^{-4}\text{ min}^{-1}} = \mathbf{1.12 \times 10^3\ min}$$

13.90 **(a)** In the two-step mechanism the rate-determining step is the collision of a hydrogen molecule with two iodine atoms. If visible light increases the concentration of iodine atoms, then the rate must increase. If the true rate-determining step were the collision of a hydrogen molecule with an iodine molecule (the one-step mechanism), then the visible light would have no effect (it might even slow the reaction by depleting the number of available iodine molecules).

(b) To split hydrogen molecules into atoms, one needs ultraviolet light of much higher energy.

13.92 **(a)** We can write the rate law for an elementary step directly from the stoichiometry of the balanced reaction. In this rate-determining elementary step three molecules must collide simultaneously (one X and two Y's). This makes the reaction termolecular, and consequently the rate law must be third order: first order in X and second order in Y.

The rate law is:

$$\textbf{rate} = \boldsymbol{k[\mathrm{X}][\mathrm{Y}]^2}$$

(b) The value of the rate constant can be found by solving algebraically for k.

$$k = \frac{\text{rate}}{[\mathrm{X}][\mathrm{Y}]^2} = \frac{3.8 \times 10^{-3}\ \cancel{M}/\mathrm{s}}{(0.26\ \cancel{M})(0.88\ M)^2} = \mathbf{1.9 \times 10^{-2}}\ \boldsymbol{M^{-2}\mathrm{s}^{-1}}$$

Could you write the rate law if the reaction shown were the overall balanced equation and not an elementary step?

13.94

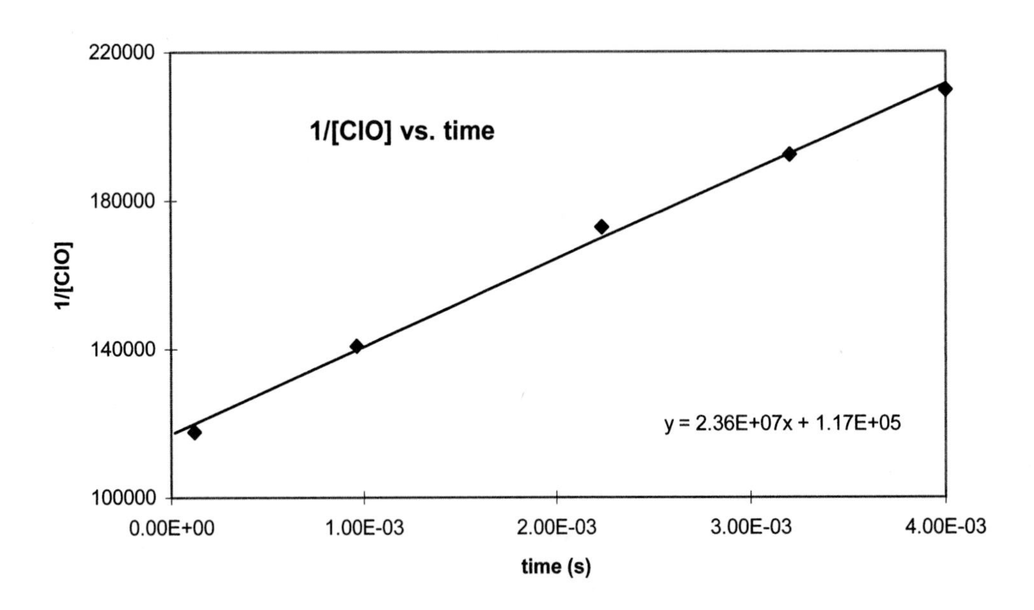

Reaction is **second-order** because a plot of 1/[ClO] vs. time is a straight line. The slope of the line equals the rate constant, k.

k = Slope = $\mathbf{2.4 \times 10^7}\ \boldsymbol{/M\cdot s}$

13.96 During the first five minutes or so the engine is relatively cold, so the exhaust gases will not fully react with the components of the catalytic converter. Remember, for almost all reactions, the rate of reaction increases with temperature.

13.98 **(a)** E_a has a large value.

(b) $E_a \approx 0$. Orientation factor is not important.

13.100 First, solve for the rate constant, k, from the half-life of the decay.

$$t_{\frac{1}{2}} = 2.44 \times 10^5 \text{ yr} = \frac{0.693}{k}$$

$$k = \frac{0.693}{2.44 \times 10^5 \text{ yr}} = 2.84 \times 10^{-6} \text{ yr}^{-1}$$

Now, we can calculate the time for the plutonium to decay from 5.0×10^2 g to 1.0×10^2 g using the equation for a first-order reaction relating concentration and time.

$$\ln\frac{[\text{A}]_t}{[\text{A}]_0} = -kt$$

$$\ln\frac{1.0 \times 10^2}{5.0 \times 10^2} = -(2.84 \times 10^{-6} \text{ yr}^{-1})t$$

$$-1.61 = -(2.84 \times 10^{-6} \text{ yr}^{-1})t$$

$$\boldsymbol{t = 5.7 \times 10^5 \text{ yr}}$$

13.102 **(a)** Catalyst: Mn^{2+}; intermediate: Mn^{3+}

First step is rate-determining.

(b) Without the catalyst, the reaction would be a termolecular one involving 3 cations! (Tl^+ and two Ce^{4+}). The reaction would be slow.

(c) The catalyst is a homogeneous catalyst because it has the same phase (aqueous) as the reactants.

13.104 Initially, the number of moles of gas in terms of the volume is:

$$n = \frac{PV}{RT} = \frac{(0.350 \cancel{\text{atm}})V}{\left(0.0821\frac{\text{L}\cdot\cancel{\text{atm}}}{\text{mol}\cdot\cancel{\text{K}}}\right)(450 + 273)\cancel{\text{K}}} = 5.90 \times 10^{-3}\, V$$

We can calculate the concentration of dimethyl ether from the following equation.

$$\ln\frac{[(CH_3)_2O]_t}{[(CH_3)_2O]_0} = -kt$$

$$\frac{[(CH_3)_2O]_t}{[(CH_3)_2O]_0} = e^{-kt}$$

Since, the volume is held constant, it will cancel out of the equation. The concentration of dimethyl ether after 8.0 minutes (480 s) is:

$$[(CH_3)_2O]_t = \left(\frac{5.90 \times 10^{-3}\, V}{V}\right) e^{-\left(3.2 \times 10^{-4}\frac{1}{\cancel{s}}\right)(480\cancel{s})}$$

$$[(CH_3)_2O]_t = 5.06 \times 10^{-3}\ M$$

After 8.0 min, the concentration of $(CH_3)_2O$ has decreased by $(5.90 \times 10^{-3} - 5.06 \times 10^{-3})M$ or $8.4 \times 10^{-4}\ M$. Since three moles of product form for each mole of dimethyl ether that reacts, the concentrations of the products are $(3)(8.4 \times 10^{-4}\ M) = 2.5 \times 10^{-3}\ M$.

The pressure of the system after 8.0 minutes is:

$$P = \frac{nRT}{V} = \left(\frac{n}{V}\right)RT = MRT$$

$$P = [(5.06 \times 10^{-3}) + (2.5 \times 10^{-3})]\cancel{M} \times (0.0821\ \cancel{L}\cdot\text{atm}/\cancel{\text{mol}}\cdot\cancel{K})(723\ \cancel{K})$$

$$\mathbf{P = 0.45\ atm}$$

13.106 **(a)** $\frac{\Delta[B]}{\Delta t} = k_1[A] - k_2[B]$

(b) If, $\frac{\Delta[B]}{\Delta t} = 0$

Then, from part (a) of this problem:

$$k_1[A] = k_2[B]$$

$$\mathbf{[B] = \frac{k_1}{k_2}[A]}$$

13.108 **(a)** The first-order rate constant can be determined from the half-life.

$$t_{\frac{1}{2}} = \frac{0.693}{k}$$

$$k = \frac{0.693}{t_{\frac{1}{2}}} = \frac{0.693}{28.1\ \text{yr}} = \mathbf{0.0247\ yr^{-1}}$$

(b) See Problem 13.82. Mathematically, the amount left after ten half–lives is:

$$\left(\frac{1}{2}\right)^{10} = \mathbf{9.8 \times 10^{-4}}$$

(c) If 99.0% has disappeared, then 1.0% remains. The ratio of $[A]_t/[A]_0$ is 1.0%/100% or 0.010/1.00. Substitute into the first-order integrated rate law, Equation (13.3) of the text, to determine the time.

$$\ln\frac{[A]_t}{[A]_0} = -kt$$

$$\ln\frac{0.010}{1.0} = -(0.0247\ \text{yr}^{-1})t$$

$$-4.6 = -(0.0247\ \text{yr}^{-1})t$$

$$\mathbf{t = 186\ yr}$$

13.110 **(a)** There are three elementary steps: A → B, B → C, and C → D.

(b) There are two intermediates: B and C.

(c) The third step, C → D, is rate determining because it has the largest activation energy.

(d) The overall reaction is exothermic.

13.112 Let $k_{cat} = k_{uncat}$

Then,

$$Ae^{\frac{-E_a(\text{cat})}{RT_1}} = Ae^{\frac{-E_a(\text{uncat})}{RT_2}}$$

Since the frequency factor is the same, we can write:

$$e^{\frac{-E_a(\text{cat})}{RT_1}} = e^{\frac{-E_a(\text{uncat})}{RT_2}}$$

Taking the natural log (*ln*) of both sides of the equation gives:

$$\frac{-E_a(\text{cat})}{RT_1} = \frac{-E_a(\text{uncat})}{RT_2}$$

or,

$$\frac{E_a(\text{cat})}{T_1} = \frac{E_a(\text{uncat})}{T_2}$$

Substituting in the given values:

$$\frac{7.0\ \text{kJ/mol}}{293\ \text{K}} = \frac{42\ \text{kJ/mol}}{T_2}$$

$$\mathbf{T_2 = 1.8 \times 10^3\ K}$$

This temperature is much too high to be practical.

13.114 **(a)** The rate law for the reaction is:

$$\text{rate} = k[\text{Hb}][\text{O}_2]$$

We are given the rate constant and the concentration of Hb and O_2, so we can substitute in these quantities to solve for rate.

$$\text{rate} = (2.1 \times 10^6\ /M\cdot\text{s})(8.0 \times 10^{-6}\ M)(1.5 \times 10^{-6}\ M)$$

$$\mathbf{rate = 2.5 \times 10^{-5}\ \mathit{M}/s}$$

(b) If HbO_2 is being formed at the rate of 2.5×10^{-5} *M*/s, then O_2 is being consumed at the same rate, **2.5×10^{-5} *M*/s**. Note the 1:1 mole ratio between O_2 and HbO_2.

(c) The rate of formation of HbO_2 increases, but the concentration of Hb remains the same. Assuming that temperature is constant, we can use the same rate constant as in part (a). We substitute rate, [Hb], and the rate constant into the rate law to solve for O_2 concentration.

$$\text{rate} = k[\text{Hb}][\text{O}_2]$$

$$1.4 \times 10^{-4}\ M/\text{s} = (2.1 \times 10^{6}\ /M\cdot\text{s})(8.0 \times 10^{-6}\ M)[\text{O}_2]$$

$$\mathbf{[O_2] = 8.3 \times 10^{-6}\ \mathit{M}}$$

13.116 $t_{\frac{1}{2}} \propto \dfrac{1}{[\text{A}]_0^{n-1}}$

$$t_{\frac{1}{2}} = C\frac{1}{[\text{A}]_0^{n-1}}$$, where C is a proportionality constant.

Substituting in for zero, first, and second-order reactions gives:

$n = 0$ $\quad t_{\frac{1}{2}} = C\dfrac{1}{[\text{A}]_0^{-1}} = C[\text{A}]_0$

$n = 1$ $\quad t_{\frac{1}{2}} = C\dfrac{1}{[\text{A}]_0^{0}} = C$

$n = 2$ $\quad t_{\frac{1}{2}} = C\dfrac{1}{[\text{A}]_0}$

Compare these results with those in Table 13.3 of the text. What is C in each case?

13.118 (a) The units of the rate constant show the reaction to be second-order, meaning the rate law is most likely:

$$\text{Rate} = k[\text{H}_2][\text{I}_2]$$

We can use the ideal gas equation to solve for the concentrations of H_2 and I_2. We can then solve for the initial rate in terms of H_2 and I_2 and then convert to the initial rate of formation of HI. We carry an extra significant figure throughout this calculation to minimize rounding errors.

$$n = \frac{PV}{RT}$$

$$\frac{n}{V} = M = \frac{P}{RT}$$

Since the total pressure is 1658 mmHg and there are equimolar amounts of H_2 and I_2 in the vessel, the partial pressure of each gas is 829 mmHg.

$$[\text{H}_2] = [\text{I}_2] = \frac{\left(829\ \text{mmHg} \times \dfrac{1\ \text{atm}}{760\ \text{mmHg}}\right)}{\left(0.0821\dfrac{\text{L}\cdot\text{atm}}{\text{mol}\cdot\text{K}}\right)(400 + 273)\text{K}} = 0.01974\ M$$

Let's convert the units of the rate constant to $/M{\cdot}$min, and then we can substitute into the rate law to solve for rate.

$$k = 2.42 \times 10^{-2}\frac{1}{M \cdot \text{s}} \times \frac{60\ \text{s}}{1\ \text{min}} = 1.452\frac{1}{M \cdot \text{min}}$$

$$\text{Rate} = k[\text{H}_2][\text{I}_2]$$

$$\text{Rate} = \left(1.452\frac{1}{M \cdot \text{min}}\right)(0.01974\ M)(0.01974\ M) = 5.658 \times 10^{-4}\ M/\text{min}$$

We know that,

$$\text{Rate} = \frac{1}{2}\frac{\Delta[\text{HI}]}{\Delta t}$$

or

$$\frac{\Delta[\text{HI}]}{\Delta t} = 2 \times \text{Rate} = (2)(5.658 \times 10^{-4}\ M/\text{min}) = \mathbf{1.13 \times 10^{-3}\ M/min}$$

(b) We can use the second-order integrated rate law to calculate the concentration of H_2 after 10.0 minutes. We can then substitute this concentration back into the rate law to solve for rate.

$$\frac{1}{[\text{H}_2]_t} = kt + \frac{1}{[\text{H}_2]_0}$$

$$\frac{1}{[\text{H}_2]_t} = \left(1.452\frac{1}{M \cdot \text{min}}\right)(10.0\ \text{min}) + \frac{1}{0.01974\ M}$$

$$[\text{H}_2]_t = 0.01534\ M$$

We can now substitute this concentration back into the rate law to solve for rate. The concentration of I_2 after 10.0 minutes will also equal 0.01534 *M*.

$$\text{Rate} = k[\text{H}_2][\text{I}_2]$$

$$\text{Rate} = \left(1.452\frac{1}{M \cdot \text{min}}\right)(0.01534\ M)(0.01534\ M) = 3.417 \times 10^{-4}\ M/\text{min}$$

We know that,

$$\text{Rate} = \frac{1}{2}\frac{\Delta[\text{HI}]}{\Delta t}$$

or

$$\frac{\Delta[\text{HI}]}{\Delta t} = 2 \times \text{Rate} = (2)(3.417 \times 10^{-4}\ M/\text{min}) = \mathbf{6.83 \times 10^{-4}\ M/min}$$

The concentration of HI after 10.0 minutes is:

$$[\text{HI}]_t = ([\text{H}_2]_0 - [\text{H}_2]_t) \times 2$$

$$\mathbf{[HI]_t} = (0.01974\ M - 0.01534\ M) \times 2 = \mathbf{8.8 \times 10^{-3}\ M}$$

13.120 The half-life is related to the initial concentration of A by

$$t_{\frac{1}{2}} \propto \frac{1}{[\mathrm{A}]_0^{n-1}}$$

According to the data given, the half-life doubled when $[\mathrm{A}]_0$ was halved. This is only possible if the half-life is inversely proportional to $[\mathrm{A}]_0$. Substituting $n = 2$ into the above equation gives:

$$t_{\frac{1}{2}} \propto \frac{1}{[\mathrm{A}]_0}$$

Looking at this equation, it is clear that if $[\mathrm{A}]_0$ is halved, the half-life would double. The reaction is **second-order**.

We use Equation (13.8) of the text to calculate the rate constant.

$$t_{\frac{1}{2}} = \frac{1}{k[\mathrm{A}]_0}$$

$$\boldsymbol{k} = \frac{1}{[\mathrm{A}]_0 t_{\frac{1}{2}}} = \frac{1}{(1.20\ M)(2.0\ \text{min})} = \mathbf{0.42\ /\boldsymbol{M} \cdot min}$$

13.122 From Equation (13.14) of the text,

$$\ln\frac{k_1}{k_2} = \frac{E_a}{R}\left(\frac{1}{T_2} - \frac{1}{T_1}\right)$$

$$\ln\left(\frac{k_1}{k_2}\right) = \frac{2.4 \times 10^5\ \text{J/mol}}{8.314\ \text{J/mol}\cdot\text{K}}\left(\frac{1}{606\ \text{K}} - \frac{1}{600\ \text{K}}\right)$$

$$\ln\left(\frac{k_1}{k_2}\right) = -0.48$$

$$\frac{k_2}{k_1} = e^{0.48} = 1.6$$

The rate constant at 606 K is 1.6 times greater than that at 600 K. This is a **60%** increase in the rate constant for a 1% increase in temperature! The result shows the profound effect of an exponential dependence. In general, the larger the E_a, the greater the influence of T on k.

CHAPTER 14
CHEMICAL EQUILIBRIUM

PROBLEM-SOLVING STRATEGIES AND TUTORIAL SOLUTIONS

TYPES OF PROBLEMS

Problem Type 1: Homogeneous Equilibria.
- **(a)** Writing expressions for K_c and K_P.
- **(b)** Calculating equilibrium partial pressures.
- **(c)** Converting between K_c and K_P.

Problem Type 2: Heterogeneous Equilibria, Calculating K_P and K_c.

Problem Type 3: The Form of K and the Equilibrium Equation.

Problem Type 4: Using the Reaction Quotient (Q_c) to Predict the Direction of a Reaction.

Problem Type 5: Calculating Equilibrium Concentrations.

Problem Type 6: Factors that Affect Chemical Equilibrium.
- **(a)** Changes in concentration.
- **(b)** Changes in pressure and volume.
- **(c)** Changes in temperature.
- **(d)** The effect of a catalyst.

PROBLEM TYPE 1: HOMOGENEOUS EQUILIBRIA

Let's start by considering the following reversible reaction.

$$a\text{A} + b\text{B} \rightleftharpoons c\text{C} + d\text{D}$$

where *a, b, c,* and *d* are the stoichiometric coefficients for the reacting species A, B, C, and D. The equilibrium constant for the reaction at a particular temperature is:

$$K = \frac{[\text{C}]^c[\text{D}]^d}{[\text{A}]^a[\text{B}]^b}$$

This equation is the mathematical form of the **law of mass action**. It relates the concentrations of reactants and products at equilibrium in terms of a quantity called the **equilibrium constant (*K*)**.

The magnitude of the equilibrium constant, K, is important. If the equilibrium constant is much greater than one ($K >> 1$), we say that the equilibrium lies to the right and favors the products. Conversely, if $K << 1$, the equilibrium lies to the left and favors the reactants.

A. Writing expressions for K_c and K_P

The term **homogeneous equilibrium** applies to reactions in which all reacting species are in the same phase. An example of a homogeneous gas-phase equilibrium is the reaction between sulfur dioxide and oxygen to form sulfur trioxide.

$$2SO_2(g) + O_2(g) \rightleftharpoons 2SO_3(g)$$

The equilibrium constant as given in Equation (14.2) of the text is:

$$K_c = \frac{[SO_3]^2}{[SO_2]^2[O_2]}$$

The subscript c of K_c denotes that the concentrations of all species are expressed in mol/L.

The concentrations of reactants and products in gas-phase reactions can also be expressed in terms of their partial pressures. Starting with the ideal gas equation, we can write:

$$P = \left(\frac{n}{V}\right)RT$$

At constant temperature, the pressure (P) of a gas is directly related to the concentration of the gas in mol/L, $\left(\frac{n}{V}\right)$.

Thus, for the equilibrium shown above, we can write:

$$K_P = \frac{P_{SO_3}^2}{P_{SO_2}^2 P_{O_2}}$$

where P_{SO_3}, P_{SO_2}, and P_{O_2} are the equilibrium partial pressures (in atmospheres) of SO_3, SO_2, and O_2, respectively. The subscript p of K_P indicates the equilibrium concentrations are expressed in terms of pressure.

EXAMPLE 14.1
Write the equilibrium constant expression for the following reversible reaction:

$$4NH_3(g) + 5O_2(g) \rightleftharpoons 4NO(g) + 6H_2O(g)$$

Strategy: Remember that the concentration of each component in the equilibrium constant expression is raised to a power equal to its coefficient in the balanced equation.

Solution:

$$K_c = \frac{[NO]^4[H_2O]^6}{[NH_3]^4[O_2]^5}$$

You could also write the equilibrium constant expression in terms of the partial pressures of the gaseous components.

$$K_P = \frac{P_{NO}^4 P_{H_2O}^6}{P_{NH_3}^4 P_{O_2}^5}$$

PRACTICE EXERCISE

1. Write the equilibrium constant expression for the following reaction.

$$2N_2O(g) + 3O_2(g) \rightleftharpoons 2N_2O_4(g)$$

Text Problem: See Review Question 14.8

B. Calculating equilibrium partial pressures

EXAMPLE 14.2

At 400°C, $K_P = 64$ for the following reaction:

$$H_2(g) + I_2(g) \rightleftharpoons 2HI(g)$$

At equilibrium, the partial pressures of H_2 and I_2 in a closed container are 0.20 atm and 0.50 atm respectively. What is the partial pressure of HI in the mixture?

Strategy: Write K_P in terms of the partial pressures of the reacting species, and then substitute the known values into the equation to solve for the unknown value.

Solution:

$$K_P = \frac{P_{HI}^2}{P_{H_2}P_{I_2}}$$

Solve the above equation algebraically for the unknown partial pressure. Then, substitute the known values into the equation to solve for the unknown.

$$P_{HI}^2 = K_P P_{H_2}P_{I_2}$$

$$P_{HI} = \sqrt{K_P P_{H_2}P_{I_2}}$$

$$\mathbf{P_{HI}} = \sqrt{(64)(0.20)(0.50)} = \mathbf{2.5\ atm}$$

PRACTICE EXERCISE

2. Consider the following reaction at a temperature of 250°C.

$$PCl_5(g) \rightleftharpoons PCl_3(g) + Cl_2(g)$$

At equilibrium, the partial pressures of PCl_5, PCl_3, and Cl_2 are 0.0704 atm, 0.340 atm, and 0.218 atm, respectively. What is the value of K_P?

C. Converting K_c to K_P

In general, K_c is not equal to K_P, since the partial pressures of reactants and products are not equal to their concentrations expressed in mol/L. A simple relationship between K_P and K_c can be derived. See Section 14.2 of the text for the derivation. The relationship is:

$$K_P = K_c(RT)^{\Delta n}$$

where,

Δn = (mol of gaseous product in balanced equation) – (mol of gaseous reactants in balanced eq.)

If we express pressure in atmospheres, the gas constant R is given by 0.0821 L·atm/mol·K. We can write the relationship between K_P and K_c as

$$K_P = K_c(0.0821\,T)^{\Delta n} \qquad (14.5, \text{text})$$

In general, $K_P \neq K_c$ except in the special case when $\Delta n = 0$. In this case, Equation (14.5) of the text can be written as:

$$K_P = K_c(0.0821\,T)^0$$

and

$$K_P = K_c$$

EXAMPLE 14.3

In the decomposition of carbon dioxide at 2000°C,

$$2CO_2(g) \rightleftharpoons 2CO(g) + O_2(g)$$

$K_P = 1.2 \times 10^{-4}$. Calculate K_c for this reaction.

Strategy: The relationship between K_c and K_P is given by Equation (14.5) of the text. What is the change in the number of moles of gases from reactant to product? Recall that

$$\Delta n = \text{moles of gaseous products} - \text{moles of gaseous reactants}$$

What unit of temperature should we use?

Solution: The relationship between K_c and K_P is given by Equation (14.5) of the text.

$$K_P = K_c(0.0821\,T)^{\Delta n}$$

Rearrange the equation relating K_P and K_c, solving for K_c.

$$K_c = \frac{K_P}{(0.0821\,T)^{\Delta n}}$$

Substitute the given values into the above equation to solve for K_c. Temperature must be in units of Kelvin. **Note:** In the balanced equation, there are a total of three moles of gaseous products and two moles of gaseous reactant, so $\Delta n = 3 - 2$.

$$\mathbf{K_c} = \frac{K_P}{(0.0821\,T)^{\Delta n}} = \frac{1.2 \times 10^{-4}}{[(0.0821)(2273\text{K})]^{(3-2)}} = \mathbf{6.4 \times 10^{-7}}$$

PRACTICE EXERCISE

3. At 400°C, $K_P = 64$ for the following reaction:

$$H_2(g) + I_2(g) \rightleftharpoons 2HI(g)$$

What is the value of K_c for this reaction?

Text Problems: 14.18, 14.22

PROBLEM TYPE 2: HETEROGENEOUS EQUILIBRIA, CALCULATING K_P AND K_c

Whenever a reaction involves reactants and products that exist in different phases, it is called a *heterogeneous reaction*. If the reaction is carried out in a closed container, a **heterogeneous equilibrium** will result. For example, when steam is brought into contact with charcoal in a closed container, the following equilibrium is established:

$$C(s) + H_2O(g) \rightleftharpoons H_2(g) + CO(g)$$

At equilibrium, we might write the equilibrium constant as:

$$K_c' = \frac{[H_2][CO]}{[C][H_2O]}$$

However, the "concentration" of a solid, like its density, is an intensive property and thus does not depend on the amount of substance present. For this reason, the term [C] is a constant and can be combined with the equilibrium constant. We can simplify the equation by writing:

$$[C]K_c' = K_c = \frac{[H_2][CO]}{[H_2O]}$$

Keep in mind that the value of K_c does not depend on how much carbon is present, as long as some amount is present at equilibrium.

The argument presented above for solids also applies to pure liquids. Thus, if a liquid is a reactant or product, we can treat its concentration as a constant and omit it from the equilibrium constant expression.

EXAMPLE 14.4

What are the values of K_P and K_c at 1000°C for the reaction

$$CaCO_3(s) \rightleftharpoons CaO(s) + CO_2(g)$$

if the pressure of CO_2 in equilibrium with $CaCO_3$ and CaO is 3.87 atm?

Strategy: Because they are constant quantities, the concentrations of solids and liquids do not appear in the equilibrium constant expressions for heterogeneous systems. The only species that enters into the equilibrium constant expression is CO_2. After solving for K_P, we can calculate K_c from K_P.

Solution: As stated above, if a solid or a liquid is a reactant or product, we can treat its concentration as a constant and omit it from the equilibrium constant expression. Thus, enough information is given to calculate K_P for this heterogeneous equilibrium.

$$K_P = P_{CO_2} = 3.87$$

To calculate K_c, rearrange the equation relating K_P and K_c, solving for K_c. Then, substitute the given values into the equation to obtain the answer.

$$K_P = K_c(0.0821\,T)^{\Delta n}$$

$$K_c = \frac{K_P}{(0.0821\,T)^{\Delta n}}$$

$$\mathbf{K_c} = \frac{3.87}{[(0.0821)(1273\text{K})]^{(1-0)}} = \mathbf{0.0370}$$

Text Problems: 14.22, **14.24**

PROBLEM TYPE 3: THE FORM OF *K* AND THE EQUILIBRIUM EQUATION

The equilibrium constant expression and its value depend on how an equation is balanced. Often an equation can be balanced with more than one set of coefficients. For example,

$$2SO_2(g) + O_2(g) \rightleftharpoons 2SO_3(g) \qquad K_c = \frac{[SO_3]^2}{[SO_2]^2[O_2]}$$

and

$$SO_2(g) + \frac{1}{2}O_2(g) \rightleftharpoons SO_3(g) \qquad K_c' = \frac{[SO_3]}{[SO_2][O_2]^{\frac{1}{2}}}$$

Is there a relationship between the equilibrium constants for the two reactions? Note that K_c' is the square root of the equilibrium constant for the first reaction, K_c.

$$K_c' = \frac{[SO_3]}{[SO_2][O_2]^{\frac{1}{2}}} = \sqrt{\frac{[SO_3]^2}{[SO_2]^2[O_2]}} = \sqrt{K_c}$$

The general relationship is that you raise K_c to the power by which the equation was multiplied. To come up with the second balanced equation above, we had to multiply the first equation by 1/2. Thus,

$$K_c' = (K_c)^{\frac{1}{2}} = \sqrt{K_c}$$

What if the reaction is written in the reverse direction?

$$2SO_3(g) \rightleftharpoons 2SO_2(g) + O_2(g) \qquad K_c'' = \frac{[SO_2]^2[O_2]}{[SO_3]^2}$$

Is there a relationship between K_c'' and K_c? By inspection, you should find that K_c'' is the reciprocal of K_c for the forward reaction.

$$K_c'' = \frac{[SO_2]^2[O_2]}{[SO_3]^2} = \frac{1}{K_c}$$

Remember, always use the K_c expression and value that are consistent with the way in which the balanced equation is written.

EXAMPLE 14.5

For the reaction, $2HBr(g) \rightleftharpoons H_2(g) + Br_2(g)$, $K_P = 1.4 \times 10^{-5}$ at 700 K. What are the values of K_P for the following reactions at the same temperature?

(a) $4HBr(g) \rightleftharpoons 2H_2(g) + 2Br_2(g)$

(b) $H_2(g) + Br_2(g) \rightleftharpoons 2HBr(g)$

(a) In Equation (a), the original equation has been multiplied by *two*. The general relationship is that you raise K to the power by which the equation was multiplied by. Thus,

$$K_P' = (K_P)^2 = (1.4 \times 10^{-5})^2 = \mathbf{2.0 \times 10^{-10}}$$

(b) Equation (b) is the reverse of the original equation. By inspection, you should find that K_P'' is the reciprocal of K_P.

$$K_P'' = \frac{1}{K_P} = \frac{1}{1.4 \times 10^{-5}} = \mathbf{7.1 \times 10^4}$$

PRACTICE EXERCISE

4. For the reaction, $2HBr(g) \rightleftharpoons H_2(g) + Br_2(g)$, $K_P = 1.4 \times 10^{-5}$ at 700 K. What is the value of K_P for the following reaction at the same temperature?

$$HBr(g) \rightleftharpoons \frac{1}{2}H_2(g) + \frac{1}{2}Br_2(g)$$

Text Problems: 14.30, 14.32

PROBLEM TYPE 4: USING THE REACTION QUOTIENT (*Q*) TO PREDICT THE DIRECTION OF A REACTION

Equilibrium constants provide useful information about chemical reaction systems. For instance, equilibrium constants can be used to predict the direction in which a reaction will proceed to establish equilibrium.

The reaction quotient, Q_c, is a useful aid in predicting whether or not a reaction system is at equilibrium. Again, consider the reaction

$$2SO_2(g) + O_2(g) \rightleftharpoons 2SO_3(g)$$

The reaction quotient is:

$$Q_c = \frac{[SO_3]_0^2}{[SO_2]_0^2[O_2]_0}$$

You should notice that Q_c has the same algebraic form as K_c. However, the concentrations are not necessarily equilibrium concentrations. We will call them initial concentrations, represented by a subscript 0 after the square brackets, $[\ ]_0$. Substituting initial concentrations into the reaction quotient, gives a value for Q_c. In order to predict whether the system is at equilibrium, the magnitude of Q_c must be compared with that of K_c.

If $Q_c = K_c$, The initial concentrations are equilibrium concentrations. The system is at equilibrium.

If $Q_c > K_c$, The ratio of initial concentrations of products to reactants is too large. To reach equilibrium, products must be converted to reactants. The system proceeds from right to left (consuming products, forming reactants) to reach equilibrium.

If $Q_c < K_c$, The ratio of initial concentrations of products to reactants is too small. To reach equilibrium, reactants must be converted to products. The system proceeds from left to right (consuming reactants, forming products) to reach equilibrium.

EXAMPLE 14.6

At a certain temperature, the reaction: $CO(g) + Cl_2(g) \rightleftharpoons COCl_2(g)$, has an equilibrium constant, $K_c = 13.8$. Is the following mixture an equilibrium mixture? If not, in which direction (right or left) will the reaction proceed to reach equilibrium?

$$[CO]_0 = 2.5\ M,\ [Cl_2]_0 = 1.2\ M,\ \text{and}\ [COCl_2]_0 = 5.0\ M$$

Strategy: We are given the initial concentrations of the gases, so we can calculate the reaction quotient (Q_c). How does a comparison of Q_c with K_c enable us to determine if the system is at equilibrium or, if not, in which direction the net reaction will proceed to reach equilibrium?

Solution: Recall that for a system to be at equilibrium, $Q_c = K_c$. Substitute the given concentrations into the equation for the reaction quotient to calculate Q_c.

$$Q_c = \frac{[COCl_2]_0}{[CO]_0[Cl_2]_0} = \frac{5.0}{(2.5)(1.2)} = 1.7$$

Compare Q_c to K_c. Since $Q_c < K_c$, the ratio of initial concentrations of products to reactants is too small. To reach equilibrium, reactants must be converted to products. The system proceeds from **left to right** (consuming reactants, forming products) to reach equilibrium.

PRACTICE EXERCISE

5. Given the reaction,

$$N_2(g) + O_2(g) \rightleftharpoons 2NO(g) \qquad K_c = 2.5 \times 10^{-3} \text{ at } 2130°C$$

decide whether the following mixture is at equilibrium or if a net forward or reverse reaction will occur. $[NO] = 0.0050\ M$, $[O_2] = 0.25\ M$, and $[N_2] = 0.020\ M$

Text Problem: 14.40

PROBLEM TYPE 5: CALCULATING EQUILIBRIUM CONCENTRATIONS

The expected concentrations at equilibrium can be calculated from a knowledge of the initial concentrations and the equilibrium constant. In these types of problems, it will be very helpful to recall that

equilibrium concentration = initial concentration ± the change due to reaction.

The next *two* examples illustrate this important type of calculation.

EXAMPLE 14.7

A 0.25 mole sample of N_2O_4 dissociates and comes to equilibrium in a 1.5 L flask at 100°C. The reaction is

$$N_2O_4(g) \rightleftharpoons 2NO_2(g) \qquad K_c = 0.36 \text{ at } 100°C$$

What are the equilibrium concentrations of NO_2 and N_2O_4?

Strategy: We are given the initial amount of N_2O_4 (in moles) in a vessel of known volume (in liters), so we can calculate its molar concentration. Because initially no NO_2 molecules are present, the system could not be at equilibrium. Therefore, some N_2O_4 will dissociate to form NO_2 molecules until equilibrium is established.

Solution: We follow the procedure outlined in Section 14.4 of the text to calculate the equilibrium concentrations.

Step 1: The initial concentration of N_2O_4 is 0.25 mol/1.5 L = 0.17 *M*. The stoichiometry of the problem shows 1 mole of N_2O_4 dissociating to 2 moles of NO_2 molecules. Let x be the amount (in mol/L) of N_2O_4 dissociated. It follows that the equilibrium concentration of NO_2 molecules must be $2x$. We summarize the changes in concentrations as follows:

	$N_2O_4(g)$	$\rightleftharpoons$	$2NO_2(g)$
Initial (*M*):	0.17		0
Change (*M*):	$-x$		$+2x$
Equilibrium (*M*):	$0.17 - x$		$2x$

We call this type of table an **ICE** table for **I**nitial, **C**hange, and **E**quilibrium.

Step 2: Write the equilibrium constant expression in terms of the equilibrium concentrations. Knowing the value of the equilibrium constant, solve for x.

$$K_c = \frac{[NO_2]^2}{[N_2O_4]}$$

$$0.36 = \frac{(2x)^2}{0.17 - x}$$

$$0.061 - 0.36x = 4x^2$$

$$4x^2 + 0.36x - 0.061 = 0$$

The above equation is a quadratic equation of the form $ax^2 + bx + c = 0$. The solution for a quadratic equation is:

$$x = \frac{-b \pm \sqrt{b^2 - 4ac}}{2a}$$

Here, we have a = 4, b = 0.36, and c = −0.061. Substituting into the above equation,

$$x = \frac{-0.36 \pm \sqrt{(0.36)^2 - 4(4)(-0.061)}}{2(4)}$$

$$x = \frac{-0.36 \pm 1.05}{8}$$

$$x = 0.086\ M \quad \text{or} \quad x = -0.18\ M$$

The second solution is physically impossible because you cannot have a negative concentration. The first solution is the correct answer.

> **Tip:** In solving a quadratic equation of this type, one answer is always physically impossible, so the choice of which value to use for x is easy to make.

Step 3: Having solved for x, calculate the equilibrium concentrations of all species.

$$[N_2O_4] = (0.17 - 0.086)M = \mathbf{0.08\ M}$$

$$[NO_2] = 2(0.086\ M) = \mathbf{0.17\ M}$$

EXAMPLE 14.8

A 1.00 L vessel initially contains 0.776 mol of SO_3 (*g*) at 1100 K. What is the value of K_c for the following reaction if 0.520 mol of SO_3 remain at equilibrium?

$$2SO_3(g) \rightleftharpoons 2SO_2(g) + O_2(g)$$

Strategy: If we can calculate the equilibrium concentrations for all species, we can calculate the equilibrium constant K_c. Because the initial amount of SO_3 and the amount of SO_3 at equilibrium are given, the equilibrium concentration of SO_3 can be calculated. The equilibrium concentrations of other species can be determined from the amount of SO_3 reacted.

Solution:

Step 1: In this problem, we are given both the initial and equilibrium concentrations of SO_3. Recalling that

equilibrium concentration = initial concentration ± the change due to reaction

we can calculate the change in concentration of SO_3 due to reaction. Let's call this change, $2x$, because of the coefficient of 2 for SO_3 in the balanced equation.

change in concentration of SO_3 = $2x$ = $0.776\ M - 0.520\ M = 0.256\ M$

$2x = 0.256,\ x = 0.128\ M$

Complete a table that lists the initial concentrations, the change in concentrations, and the equilibrium concentrations.

	$2SO_3(g)$	$\rightleftharpoons$	$2SO_2(g)$	+	$O_2(g)$
Initial (M):	0.776		0		0
Change (M):	$-2x = -0.256$		$+2x$		$+x$
Equilibrium (M):	0.520		$2x$		x

So, at equilibrium,

$$[SO_2] = 2x = 0.256\ M$$
$$[O_2] = x = 0.128\ M$$

Tip: You probably could have come up with the equilibrium concentrations of SO_2 and O_2 without the use of a table. However, a table is a simple way to keep all the data organized.

Step 2: Substitute the equilibrium concentrations into the equilibrium constant expression to solve for K_c.

$$K_c = \frac{[SO_2]^2[O_2]}{[SO_3]^2}$$

$$\mathbf{K_c} = \frac{(0.256)^2(0.128)}{(0.520)^2} = \mathbf{0.0310}$$

PRACTICE EXERCISE

6. For the reaction,

$$N_2(g) + O_2(g) \rightleftharpoons 2NO(g) \qquad K_P = 3.80 \times 10^{-4} \text{ at } 2000°C$$

what equilibrium pressures of N_2, O_2, and NO will result when a 10.0 L reactor vessel is filled with 2.00 atm of N_2 and 0.400 atm of O_2 and the reaction is allowed to come to equilibrium?

7. Initially a 1.0 L vessel contains 10.0 mol of NO and 6.0 mol of O_2 at a certain temperature.

$$2NO(g) + O_2(g) \rightleftharpoons 2NO_2(g)$$

At equilibrium, the vessel contains 8.8 mol of NO_2. Determine the value of K_c at this temperature.

Text Problems: 14.42, 14.44, 14.46, 14.48

PROBLEM TYPE 6: FACTORS THAT AFFECT CHEMICAL EQUILIBRIUM

What effect does a change in concentration, pressure, volume, or temperature have on a system at equilibrium? This question can be answered qualitatively by using **Le Châtelier's principle**. It states that when an external stress is applied to a system at equilibrium, the system adjusts in such a way that the stress is partially offset. The word "stress" here means a change in concentration, pressure, volume, or temperature that removes a system from the equilibrium state.

A. Changes in concentration

EXAMPLE 14.9
For the following reaction at equilibrium in a closed container,

$$2NaHCO_3(s) \rightleftharpoons Na_2CO_3(s) + H_2O(g) + CO_2(g)$$

state the effects (increase, decrease, or no change) of the following stresses on the number of moles of sodium carbonate, Na_2CO_3. Note that Na_2CO_3 is a solid (this is a heterogeneous equilibrium); its concentration will remain constant, but its amount can change.

(a) Removing $CO_2(g)$.
(b) Adding $H_2O(g)$.
(c) Adding $NaHCO_3(s)$.

Strategy: (a) The stress is the removal of CO_2 gas. How will the system adjust to partially offset the stress? (b) The stress is the addition of H_2O gas. How will the system adjust to partially offset the stress? (c) The stress is the addition of $NaHCO_3(s)$. Do pure solids or pure liquids enter into the equilibrium constant expression?

Solution: Applying Le Châtelier's principle,

(a) If CO_2 concentration is lowered, the system will react to offset the change. That is, a shift to the right will replace some of the removed CO_2. Moles of Na_2CO_3 will **increase**.

(b) The system will respond to the stress of added H_2O by shifting to the left to remove some of the water. Moles of Na_2CO_3 will **decrease**.

(c) The position of a heterogeneous equilibrium does not depend on the amounts of pure solids or liquids present. Remember that the concentrations of pure solids and liquids are constant and thus do not enter into the equilibrium constant expression. Hence, there is no shift in the equilibrium, and the amount of Na_2CO_3 is **unchanged**.

B. Changes in pressure and volume

The pressure of a system of gases in chemical equilibrium can be increased by decreasing the available volume. The ideal gas equation shows this inverse relationship between pressure and volume.

$$PV = nRT$$

A decrease in volume causes the concentrations of all *gas-phase* components to increase. Remember that pressure is directly proportional to the concentration of a gas.

$$P = \left(\frac{n}{V}\right)RT$$

The stress caused by the increased pressure will be partially offset by a net reaction that will lower the total pressure. In other words, the system will adjust by lowering the total concentration of gas molecules to reestablish equilibrium. Again, consider the following equilibrium:

$$2SO_2(g) + O_2(g) \rightleftharpoons 2SO_3(g)$$

When the molecules of the above gases are compressed into a smaller volume, the total pressure increases and hence the total concentration increases (this is a stress). A net forward reaction (shift to the right) will bring the system to a new state of equilibrium, in which the *total concentration* of all molecules will be lowered somewhat. This partially offsets the initial stress on the system. The total concentration is lowered somewhat because when 2 moles of SO_2 react with 1 mole of O_2 (a total of 3 moles), only 2 moles of SO_3 are produced.

In general, an increase in pressure (decrease in volume) favors the net reaction that decreases the total number of moles of gases. On the other hand, a decrease in pressure (increase in volume) favors the net reaction that increases the total number of moles of gases.

EXAMPLE 14.10
For the following reaction at equilibrium in a closed container,

$$2NaHCO_3(s) \rightleftharpoons Na_2CO_3(s) + H_2O(g) + CO_2(g)$$

state the effect (increase, decrease, or no change) of increasing the volume of the container on the number of moles of sodium carbonate, Na_2CO_3. Note that Na_2CO_3 is a solid (this is a heterogeneous equilibrium); its concentration will remain constant, but its amount can change.

Strategy: A change in pressure can affect only the volume of a gas, but not that of a solid or liquid because solids and liquids are much less compressible. The stress applied is an decrease in pressure (increase in volume). According to Le Châtelier's principle, the system will adjust to partially offset this stress. In other words, the system will adjust to increase the pressure. This can be achieved by shifting to the side of the equation that has more moles of gas. Recall that pressure is directly proportional to moles of gas: $PV = nRT$ so $P \propto n$.

Solution: Looking at the above reaction, there are zero moles of gas on the reactants' side, and two moles of gas on the products' side. The pressure will be increased by shifting to the right (more moles of gas).

Because the reaction shifts to the right to establish equilibrium, the amount of Na_2CO_3 will **increase**.

C. Changes in temperature

A change in concentration, pressure, or volume may alter the equilibrium position, but it does not change the value of the equilibrium constant. However, a change in temperature can alter the equilibrium constant.

To decide how a temperature stress will affect a system at equilibrium, you must look at whether the reaction is endothermic or exothermic. The following reaction is endothermic (positive ΔH). Endothermic reactions absorb heat from the surroundings; therefore, we can think of heat as a reactant.

$$\text{heat} + PCl_5(g) \rightleftharpoons PCl_3(g) + Cl_2(g) \qquad \Delta H° = 92.9 \text{ kJ/mol}$$

What happens if we heat the above system at equilibrium? We have added heat (a stress). The system will shift in the direction that will partially offset this stress by removing some heat. Shifting to the right will remove heat. PCl_5 dissociates into PCl_3 and Cl_2 molecules. Consequently, the equilibrium constant for the above reaction, given by

$$K_c = \frac{[PCl_3][Cl_2]}{[PCl_5]}$$

increases with temperature.

In general, a temperature *increase* favors an *endothermic* reaction, and a temperature *decrease* favors an *exothermic* reaction.

EXAMPLE 14.11
For the following reaction at equilibrium in a closed container,

$$2NaHCO_3(s) \rightleftharpoons Na_2CO_3(s) + H_2O(g) + CO_2(g) \qquad \Delta H = 128 \text{ kJ/mol}$$

state the effect (increase, decrease, or no change) of decreasing the temperature on the number of moles of sodium carbonate, Na_2CO_3. Note that Na_2CO_3 is a solid (this is a heterogeneous equilibrium); its concentration will remain constant, but its amount can change.

Strategy: What does the sign of $\Delta H°$ indicate about the heat change (endothermic or exothermic) for the forward reaction?

Solution: The stress applied is heat removed from the system. Note that the reaction is endothermic ($\Delta H° > 0$). Endothermic reactions absorb heat from the surroundings; therefore, we can think of heat as a reactant.

$$\text{heat} + 2NaHCO_3(s) \rightleftharpoons Na_2CO_3(s) + H_2O(g) + CO_2(g)$$

As heat is removed, the system will shift to replace some of the heat by shifting to the left. Thus, the reverse reaction is favored, and the amount of Na_2CO_3 will **decrease**.

PRACTICE EXERCISE

8. For the decomposition of calcium carbonate

$$CaCO_3(s) \rightleftharpoons CaO(s) + CO_2(g) \qquad \Delta H° = 175 \text{ kJ/mol}$$

how will the amount (not concentration) of $CaCO_3(s)$ change with the following stresses to the system at equilibrium?

(a) $CO_2(g)$ is removed.
(b) $CaO(s)$ is added.
(c) The temperature is raised.
(d) The volume of the container is decreased.

Text Problems: 14.54, **14.56, 14.58**, 14.60, 14.62

D. The effect of a catalyst

A catalyst increases the rate at which a reaction occurs. For a reversible reaction, a catalyst affects the rate in the forward and reverse directions to the same extent. Therefore, the presence of a catalyst does not alter the equilibrium constant, nor does it shift the position of an equilibrium system.

ANSWERS TO PRACTICE EXERCISES

1. If concentration is expressed in mol/L, $K_c = \dfrac{[N_2O_4]^2}{[N_2O]^2[O_2]^3}$

or, in terms of partial pressures, $K_P = \dfrac{P^2_{N_2O_4}}{P^2_{N_2O}P^3_{O_2}}$

2. $K_P = 1.05$ **3.** $K_P = K_c = 64$ **4.** $K'_P = 3.7 \times 10^{-3}$

5. Since $Q_c > K_c$, the ratio of initial concentrations of products to reactants is too large. To reach equilibrium, products must be converted to reactants. The system proceeds from **right to left** (consuming products, forming reactants) to reach equilibrium.

6. $P_{NO_2} = 1.74 \times 10^{-2}$ atm

$P_{N_2} = 1.99$ atm

$P_{O_2} = 0.391$ atm

7. $K_c = 34$

8. **(a)** The amount of $CaCO_3$ will **decrease**.

(b) The amount of $CaCO_3$ will **not change**.

(c) The amount of $CaCO_3$ will **decrease**.

(d) The amount of $CaCO_3$ will **increase**.

SOLUTIONS TO SELECTED TEXT PROBLEMS

14.14 Note that we are comparing similar reactions at equilibrium – two reactants producing one product, all with coefficients of one in the balanced equation.

(a) The reaction, $A + C \rightleftharpoons AC$ has the largest equilibrium constant. Of the three diagrams, there is the most product present at equilibrium.

(b) The reaction, $A + D \rightleftharpoons AD$ has the smallest equilibrium constant. Of the three diagrams, there is the least amount of product present at equilibrium.

14.16 The problem states that the system is at equilibrium, so we simply substitute the equilibrium concentrations into the equilibrium constant expression to calculate K_c.

Step 1: Calculate the concentrations of the components in units of mol/L. The molarities can be calculated by simply dividing the number of moles by the volume of the flask.

$$[H_2] = \frac{2.50 \text{ mol}}{12.0 \text{ L}} = 0.208\ M$$

$$[S_2] = \frac{1.35 \times 10^{-5} \text{ mol}}{12.0 \text{ L}} = 1.13 \times 10^{-6}\ M$$

$$[H_2S] = \frac{8.70 \text{ mol}}{12.0 \text{ L}} = 0.725\ M$$

Step 2: Once the molarities are known, K_c can be found by substituting the molarities into the equilibrium constant expression.

$$\boldsymbol{K_c} = \frac{[H_2S]^2}{[H_2]^2[S_2]} = \frac{(0.725)^2}{(0.208)^2(1.13 \times 10^{-6})} = \mathbf{1.08 \times 10^7}$$

If you forget to convert moles to moles/liter, will you get a different answer? Under what circumstances will the two answers be the same?

14.18 Converting between K_c and K_P, Problem Type 1C.

Strategy: The relationship between K_c and K_P is given by Equation (14.5) of the text. What is the change in the number of moles of gases from reactant to product? Recall that

$$\Delta n = \text{moles of gaseous products} - \text{moles of gaseous reactants}$$

What unit of temperature should we use?

Solution: The relationship between K_c and K_P is given by Equation (14.5) of the text.

$$K_P = K_c(0.0821\ T)^{\Delta n}$$

Rearrange the equation relating K_P and K_c, solving for K_c.

$$K_c = \frac{K_P}{(0.0821T)^{\Delta n}}$$

Because $T = 623$ K and $\Delta n = 3 - 2 = 1$, we have:

$$\boldsymbol{K_c} = \frac{K_P}{(0.0821T)^{\Delta n}} = \frac{1.8 \times 10^{-5}}{(0.0821)(623\text{ K})} = \mathbf{3.5 \times 10^{-7}}$$

14.20 The equilibrium constant expressions are:

(a) $$K_c = \frac{[NH_3]^2}{[N_2][H_2]^3}$$

(b) $$K_c = \frac{[NH_3]}{[N_2]^{\frac{1}{2}}[H_2]^{\frac{3}{2}}}$$

Substituting the given equilibrium concentration gives:

(a) $$\boldsymbol{K_c} = \frac{(0.25)^2}{(0.11)(1.91)^3} = \mathbf{0.082}$$

(b) $$\boldsymbol{K_c} = \frac{(0.25)}{(0.11)^{\frac{1}{2}}(1.91)^{\frac{3}{2}}} = \mathbf{0.29}$$

Is there a relationship between the K_c values from parts (a) and (b)?

14.22 Because pure solids do not enter into an equilibrium constant expression, we can calculate K_P directly from the pressure that is due solely to $CO_2(g)$.

$$\boldsymbol{K_P} = P_{CO_2} = \mathbf{0.105}$$

Now, we can convert K_P to K_c using the following equation.

$$K_P = K_c(0.0821\,T)^{\Delta n}$$

$$K_c = \frac{K_P}{(0.0821T)^{\Delta n}}$$

$$\boldsymbol{K_c} = \frac{0.105}{(0.0821 \times 623)^{(1-0)}} = \mathbf{2.05 \times 10^{-3}}$$

14.24 Heterogeneous Equilibria, Calculating K_P. Problem Type 2.

Strategy: Because they are constant quantities, the concentrations of solids and liquids do not appear in the equilibrium constant expressions for heterogeneous systems. The total pressure at equilibrium that is given is due to both NH_3 and CO_2. Note that for every 1 atm of CO_2 produced, 2 atm of NH_3 will be produced due to the stoichiometry of the balanced equation. Using this ratio, we can calculate the partial pressures of NH_3 and CO_2 at equilibrium.

Solution: The equilibrium constant expression for the reaction is

$$K_P = P_{NH_3}^2 P_{CO_2}$$

The total pressure in the flask (0.363 atm) is a sum of the partial pressures of NH_3 and CO_2.

$$P_T = P_{NH_3} + P_{CO_2} = 0.363\text{ atm}$$

Let the partial pressure of $CO_2 = x$. From the stoichiometry of the balanced equation, you should find that $P_{NH_3} = 2P_{CO_2}$. Therefore, the partial pressure of $NH_3 = 2x$. Substituting into the equation for total pressure gives:

$$P_T = P_{NH_3} + P_{CO_2} = 2x + x = 3x$$

$$3x = 0.363\text{ atm}$$

$$x = P_{CO_2} = 0.121\text{ atm}$$

$$P_{NH_3} = 2x = 0.242\text{ atm}$$

Substitute the equilibrium pressures into the equilibrium constant expression to solve for K_P.

$$\mathbf{K_P} = P_{NH_3}^2 P_{CO_2} = (0.242)^2(0.121) = \mathbf{7.09 \times 10^{-3}}$$

14.26 If the CO pressure at equilibrium is 0.497 atm, the balanced equation requires the chlorine pressure to have the same value. The initial pressure of phosgene gas can be found from the ideal gas equation.

$$P = \frac{nRT}{V} = \frac{(3.00 \times 10^{-2}\text{ mol})(0.0821\text{ L}\cdot\text{atm/mol}\cdot\text{K})(800\text{ K})}{(1.50\text{ L})} = 1.31\text{ atm}$$

Since there is a 1:1 mole ratio between phosgene and CO, the partial pressure of CO formed (0.497 atm) equals the partial pressure of phosgene reacted. The phosgene pressure at equilibrium is:

	$CO(g)$ +	$Cl_2(g)$ $\rightleftharpoons$	$COCl_2(g)$
Initial (atm):	0	0	1.31
Change (atm):	+0.497	+0.497	−0.497
Equilibrium (atm):	0.497	0.497	0.81

The value of K_P is then found by substitution.

$$\mathbf{K_P} = \frac{P_{COCl_2}}{P_{CO}P_{Cl_2}} = \frac{0.81}{(0.497)^2} = \mathbf{3.3}$$

14.28 In this problem, you are asked to calculate K_c.

Step 1: Calculate the initial concentration of NOCl. We carry an extra significant figure throughout this calculation to minimize rounding errors.

$$[NOCl]_0 = \frac{2.50\text{ mol}}{1.50\text{ L}} = 1.667\ M$$

Step 2: Let's represent the change in concentration of NOCl as $-2x$. Setting up a table:

	$2NOCl(g)$ $\rightleftharpoons$	$2NO(g)$ +	$Cl_2(g)$
Initial (M):	1.667	0	0
Change (M):	$-2x$	$+2x$	$+x$
Equilibrium (M):	$1.667 - 2x$	$2x$	x

If 28.0 percent of the NOCl has dissociated at equilibrium, the amount reacted is:

$$(0.280)(1.667\ M) = 0.4668\ M$$

In the table above, we have represented the amount of NOCl that reacts as $2x$. Therefore,

$$2x = 0.4668\ M$$

$$x = 0.2334\ M$$

The equilibrium concentrations of NOCl, NO, and Cl_2 are:

$$[NOCl] = (1.67 - 2x)M = (1.667 - 0.4668)M = 1.200\ M$$
$$[NO] = 2x = 0.4668\ M$$
$$[Cl_2] = x = 0.2334\ M$$

Step 3: The equilibrium constant K_c can be calculated by substituting the above concentrations into the equilibrium constant expression.

$$\boldsymbol{K_c} = \frac{[NO]^2[Cl_2]}{[NOCl]^2} = \frac{(0.4668)^2(0.2334)}{(1.200)^2} = \mathbf{0.0353}$$

14.30 $K = K'K''$

$$K = (6.5 \times 10^{-2})(6.1 \times 10^{-5})$$

$$\boldsymbol{K} = \mathbf{4.0 \times 10^{-6}}$$

14.32 To obtain $2SO_2$ as a reactant in the final equation, we must reverse the first equation and multiply by two.

For the equilibrium, $2SO_2(g) \rightleftharpoons 2S(s) + 2O_2(g)$

$$K_c''' = \left(\frac{1}{K_c'}\right)^2 = \left(\frac{1}{4.2 \times 10^{52}}\right)^2 = 5.7 \times 10^{-106}$$

Now we can add the above equation to the second equation to obtain the final equation. Since we add the two equations, the equilibrium constant is the product of the equilibrium constants for the two reactions.

$$2SO_2(g) \rightleftharpoons 2S(s) + 2O_2(g) \qquad K_c''' = 5.7 \times 10^{-106}$$

$$2S(s) + 3O_2(g) \rightleftharpoons 2SO_3(g) \qquad K_c'' = 9.8 \times 10^{128}$$

$$2SO_2(g) + O_2(g) \rightleftharpoons 2SO_3(g) \qquad \boldsymbol{K_c} = K_c''' \times K_c'' = \mathbf{5.6 \times 10^{23}}$$

14.36 At equilibrium, the value of K_c is equal to the ratio of the forward rate constant to the rate constant for the reverse reaction.

$$K_c = \frac{k_f}{k_r} = \frac{k_f}{5.1 \times 10^{-2}} = 12.6$$

$$k_f = (12.6)(5.1 \times 10^{-2}) = 0.64$$

The forward reaction is third order, so the units of k_f must be:

$$\text{rate} = k_f[A]^2[B]$$

$$k_f = \frac{\text{rate}}{(\text{concentration})^3} = \frac{M/\text{s}}{M^3} = 1/M^2\cdot\text{s}$$

$$\mathbf{k_f = 0.64\ /M^2\cdot s}$$

14.40 Using the Reaction Quotient (Q_c) to Predict the Direction of a Reaction, Problem Type 4.

Strategy: We are given the initial concentrations of the gases, so we can calculate the reaction quotient (Q_c). How does a comparison of Q_c with K_c enable us to determine if the system is at equilibrium or, if not, in which direction the net reaction will proceed to reach equilibrium?

Solution: Recall that for a system to be at equilibrium, $Q_c = K_c$. Substitute the given concentrations into the equation for the reaction quotient to calculate Q_c.

$$Q_c = \frac{[NH_3]_0^2}{[N_2]_0[H_2]_0^3} = \frac{[0.48]^2}{[0.60][0.76]^3} = 0.87$$

Comparing Q_c to K_c, we find that $Q_c < K_c$ $(0.87 < 1.2)$. The ratio of initial concentrations of products to reactants is too small. To reach equilibrium, reactants must be converted to products. The system proceeds from left to right (consuming reactants, forming products) to reach equilibrium.

Therefore, **$[NH_3]$ will increase** and **$[N_2]$ and $[H_2]$ will decrease** at equilibrium.

14.42 Calculating Equilibrium Concentrations, Problem Type 5.

Strategy: The equilibrium constant K_P is given, and we start with pure NO_2. The partial pressure of O_2 at equilibrium is 0.25 atm. From the stoichiometry of the reaction, we can determine the partial pressure of NO at equilibrium. Knowing K_P and the partial pressures of both O_2 and NO, we can solve for the partial pressure of NO_2.

Solution: Since the reaction started with only pure NO_2, the equilibrium concentration of NO must be twice the equilibrium concentration of O_2, due to the 2:1 mole ratio of the balanced equation. Therefore, the equilibrium partial pressure of **NO** is $(2 \times 0.25\text{ atm}) = \mathbf{0.50\ atm}$.

We can find the equilibrium NO_2 pressure by rearranging the equilibrium constant expression, then substituting in the known values.

$$K_P = \frac{P_{NO}^2 P_{O_2}}{P_{NO_2}^2}$$

$$\mathbf{P_{NO_2}} = \sqrt{\frac{P_{NO}^2 P_{O_2}}{K_P}} = \sqrt{\frac{(0.50)^2(0.25)}{158}} = \mathbf{0.020\ atm}$$

14.44 Calculating Equilibrium Concentrations, Problem Type 5.

Strategy: We are given the initial amount of I_2 (in moles) in a vessel of known volume (in liters), so we can calculate its molar concentration. Because initially no I atoms are present, the system could not be at equilibrium. Therefore, some I_2 will dissociate to form I atoms until equilibrium is established.

Solution: We follow the procedure outlined in Section 14.4 of the text to calculate the equilibrium concentrations.

Step 1: The initial concentration of I_2 is 0.0456 mol/2.30 L = 0.0198 *M*. The stoichiometry of the problem shows 1 mole of I_2 dissociating to 2 moles of I atoms. Let x be the amount (in mol/L) of I_2 dissociated. It follows that the equilibrium concentration of I atoms must be $2x$. We summarize the changes in concentrations as follows:

	$I_2(g)$	$\rightleftharpoons$	$2I(g)$
Initial (M):	0.0198		0.000
Change (M):	$-x$		$+2x$
Equilibrium (M):	$(0.0198 - x)$		$2x$

Step 2: Write the equilibrium constant expression in terms of the equilibrium concentrations. Knowing the value of the equilibrium constant, solve for x.

$$K_c = \frac{[I]^2}{[I_2]} = \frac{(2x)^2}{(0.0198 - x)} = 3.80 \times 10^{-5}$$

$$4x^2 + (3.80 \times 10^{-5})x - (7.52 \times 10^{-7}) = 0$$

The above equation is a quadratic equation of the form $ax^2 + bx + c = 0$. The solution for a quadratic equation is

$$x = \frac{-b \pm \sqrt{b^2 - 4ac}}{2a}$$

Here, we have a = 4, b = 3.80×10^{-5}, and c = -7.52×10^{-7}. Substituting into the above equation,

$$x = \frac{(-3.80 \times 10^{-5}) \pm \sqrt{(3.80 \times 10^{-5})^2 - 4(4)(-7.52 \times 10^{-7})}}{2(4)}$$

$$x = \frac{(-3.80 \times 10^{-5}) \pm (3.47 \times 10^{-3})}{8}$$

$$x = 4.29 \times 10^{-4}\,M \quad \text{or} \quad x = -4.39 \times 10^{-4}\,M$$

The second solution is physically impossible because you cannot have a negative concentration. The first solution is the correct answer.

Step 3: Having solved for x, calculate the equilibrium concentrations of all species.

$$[\mathbf{I}] = 2x = (2)(4.29 \times 10^{-4}\,M) = \mathbf{8.58 \times 10^{-4}\,M}$$

$$[\mathbf{I_2}] = (0.0198 - x) = [0.0198 - (4.29 \times 10^{-4})]\,M = \mathbf{0.0194\,M}$$

Tip: We could have simplified this problem by assuming that x was small compared to 0.0198. We could then assume that $0.0198 - x \approx 0.0198$. By making this assumption, we could have avoided solving a quadratic equation.

14.46 **(a)** The equilibrium constant, K_c, can be found by simple substitution.

$$\boldsymbol{K_c} = \frac{[H_2O][CO]}{[CO_2][H_2]} = \frac{(0.040)(0.050)}{(0.086)(0.045)} = \mathbf{0.52}$$

(b) The magnitude of the reaction quotient Q_c for the system after the concentration of CO_2 becomes 0.50 mol/L, but before equilibrium is reestablished, is:

$$Q_c = \frac{(0.040)(0.050)}{(0.50)(0.045)} = 0.089$$

The value of Q_c is smaller than K_c; therefore, the system will shift to the right, increasing the concentrations of CO and H_2O and decreasing the concentrations of CO_2 and H_2. Let x be the depletion in the concentration of CO_2 at equilibrium. The stoichiometry of the balanced equation then requires that the decrease in the concentration of H_2 must also be x, and that the concentration increases of CO and H_2O be equal to x as well. The changes in the original concentrations are shown in the table.

	CO_2	+	H_2	$\rightleftharpoons$	CO	+	H_2O
Initial (M):	0.50		0.045		0.050		0.040
Change (M):	$-x$		$-x$		$+x$		$+x$
Equilibrium (M):	$(0.50 - x)$		$(0.045 - x)$		$(0.050 + x)$		$(0.040 + x)$

The equilibrium constant expression is:

$$K_c = \frac{[H_2O][CO]}{[CO_2][H_2]} = \frac{(0.040 + x)(0.050 + x)}{(0.50 - x)(0.045 - x)} = 0.52$$

$$0.52(x^2 - 0.545x + 0.0225) = x^2 + 0.090x + 0.0020$$

$$0.48x^2 + 0.373x - (9.7 \times 10^{-3}) = 0$$

The positive root of the equation is $x = 0.025$.

The equilibrium concentrations are:

$$\mathbf{[CO_2]} = (0.50 - 0.025)\ M = \mathbf{0.48\ \mathit{M}}$$
$$\mathbf{[H_2]} = (0.045 - 0.025)\ M = \mathbf{0.020\ \mathit{M}}$$
$$\mathbf{[CO]} = (0.050 + 0.025)\ M = \mathbf{0.075\ \mathit{M}}$$
$$\mathbf{[H_2O]} = (0.040 + 0.025)\ M = \mathbf{0.065\ \mathit{M}}$$

14.48 The initial concentrations are $[H_2] = 0.80\text{ mol}/5.0\text{ L} = 0.16\ M$ and $[CO_2] = 0.80\text{ mol}/5.0\text{ L} = 0.16\ M$.

	$H_2(g)$	+	$CO_2(g)$	$\rightleftharpoons$	$H_2O(g)$	+	$CO(g)$
Initial (M):	0.16		0.16		0.00		0.00
Change (M):	$-x$		$-x$		$+x$		$+x$
Equilibrium (M):	$0.16 - x$		$0.16 - x$		x		x

$$K_c = \frac{[H_2O][CO]}{[H_2][CO_2]} = 4.2 = \frac{x^2}{(0.16 - x)^2}$$

Taking the square root of both sides, we obtain:

$$\frac{x}{0.16 - x} = 2.0$$

$$x = 0.11\ M$$

The equilibrium concentrations are:

$$[\mathbf{H_2}] = [\mathbf{CO_2}] = (0.16 - 0.11)\ M = \mathbf{0.05\ M}$$

$$[\mathbf{H_2O}] = [\mathbf{CO}] = \mathbf{0.11\ M}$$

14.54 **(a)** Removal of $CO_2(g)$ from the system would shift the position of equilibrium to the **right**.

(b) Addition of more solid Na_2CO_3 would have **no effect**. $[Na_2CO_3]$ does not appear in the equilibrium constant expression.

(c) Removal of some of the solid $NaHCO_3$ would have **no effect**. Same reason as (b).

14.56 Factors that Affect Chemical Equilibrium, Problem Type 6.

Strategy: A change in pressure can affect only the volume of a gas, but not that of a solid or liquid because solids and liquids are much less compressible. The stress applied is an increase in pressure. According to Le Châtelier's principle, the system will adjust to partially offset this stress. In other words, the system will adjust to decrease the pressure. This can be achieved by shifting to the side of the equation that has fewer moles of gas. Recall that pressure is directly proportional to moles of gas: $PV = nRT$ so $P \propto n$.

Solution:

(a) Changes in pressure ordinarily do not affect the concentrations of reacting species in condensed phases because liquids and solids are virtually incompressible. Pressure change should have **no effect** on this system.

(b) Same situation as (a).

(c) Only the product is in the gas phase. An increase in pressure should favor the reaction that decreases the total number of moles of gas. The equilibrium should shift to the **left**, that is, the amount of B should decrease and that of A should increase.

(d) In this equation there are equal moles of gaseous reactants and products. A shift in either direction will have no effect on the total number of moles of gas present. There will be **no change** when the pressure is increased.

(e) A shift in the direction of the reverse reaction (**left**) will have the result of decreasing the total number of moles of gas present.

14.58 Factors that Affect Chemical Equilibrium, Problem Type 6.

Strategy: (a) What does the sign of $\Delta H°$ indicate about the heat change (endothermic or exothermic) for the forward reaction? (b) The stress is the addition of Cl_2 gas. How will the system adjust to partially offset the stress? (c) The stress is the removal of PCl_3 gas. How will the system adjust to partially offset the stress? (d) The stress is an increase in pressure. The system will adjust to decrease the pressure. Remember, pressure is directly proportional to moles of gas. (e) What is the function of a catalyst? How does it affect a reacting system not at equilibrium? at equilibrium?

Solution:

(a) The stress applied is the heat added to the system. Note that the reaction is endothermic ($\Delta H^\circ > 0$). Endothermic reactions absorb heat from the surroundings; therefore, we can think of heat as a reactant.

$$\text{heat} + PCl_5(g) \rightleftharpoons PCl_3(g) + Cl_2(g)$$

The system will adjust to remove some of the added heat by undergoing a decomposition reaction (from **left to right**)

(b) The stress is the addition of Cl_2 gas. The system will shift in the direction to remove some of the added Cl_2. The system shifts from **right to left** until equilibrium is reestablished.

(c) The stress is the removal of PCl_3 gas. The system will shift to replace some of the PCl_3 that was removed. The system shifts from **left to right** until equilibrium is reestablished.

(d) The stress applied is an increase in pressure. The system will adjust to remove the stress by decreasing the pressure. Recall that pressure is directly proportional to the number of moles of gas. In the balanced equation we see 1 mole of gas on the reactants side and 2 moles of gas on the products side. The pressure can be decreased by shifting to the side with the fewer moles of gas. The system will shift from **right to left** to reestablish equilibrium.

(e) The function of a catalyst is to increase the rate of a reaction. If a catalyst is added to the reacting system not at equilibrium, the system will reach equilibrium faster than if left undisturbed. If a system is already at equilibrium, as in this case, the addition of a catalyst will not affect either the concentrations of reactant and product, or the equilibrium constant.

14.60 There will be no change in the pressures. A catalyst has no effect on the position of the equilibrium.

14.62 For this system, $K_P = [CO_2]$.

This means that to remain at equilibrium, the pressure of carbon dioxide must stay at a fixed value as long as the temperature remains the same.

(a) If the volume is increased, the pressure of CO_2 will drop (Boyle's law, pressure and volume are inversely proportional). Some $CaCO_3$ will break down to form more CO_2 and CaO. (**Shift right**)

(b) Assuming that the amount of added solid CaO is not so large that the volume of the system is altered significantly, there should be **no change** at all. If a huge amount of CaO were added, this would have the effect of reducing the volume of the container. What would happen then?

(c) Assuming that the amount of $CaCO_3$ removed doesn't alter the container volume significantly, there should be **no change**. Removing a huge amount of $CaCO_3$ will have the effect of increasing the container volume. The result in that case will be the same as in part (a).

(d) The pressure of CO_2 will be greater and will exceed the value of K_P. Some CO_2 will combine with CaO to form more $CaCO_3$. (**Shift left**)

(e) Carbon dioxide combines with aqueous NaOH according to the equation

$$CO_2(g) + NaOH(aq) \rightarrow NaHCO_3(aq)$$

This will have the effect of reducing the CO_2 pressure and causing more $CaCO_3$ to break down to CO_2 and CaO. (**Shift right**)

(f) Carbon dioxide does not react with hydrochloric acid, but $CaCO_3$ does.

$$CaCO_3(s) + 2HCl(aq) \rightarrow CaCl_2(aq) + CO_2(g) + H_2O(l)$$

The CO_2 produced by the action of the acid will combine with CaO as discussed in (d) above. **(Shift left)**

(g) This is a decomposition reaction. Decomposition reactions are endothermic. Increasing the temperature will favor this reaction and produce more CO_2 and CaO. **(Shift right)**

14.64 **(a)** Since the total pressure is 1.00 atm, the sum of the partial pressures of NO and Cl_2 is

$$1.00 \text{ atm} - \text{partial pressure of NOCl} = 1.00 \text{ atm} - 0.64 \text{ atm} = 0.36 \text{ atm}$$

The stoichiometry of the reaction requires that the partial pressure of NO be twice that of Cl_2. Hence, the partial pressure of NO is **0.24 atm** and the partial pressure of Cl_2 is **0.12 atm**.

(b) The equilibrium constant K_P is found by substituting the partial pressures calculated in part (a) into the equilibrium constant expression.

$$\boldsymbol{K_P} = \frac{P_{NO}^2 P_{Cl_2}}{P_{NOCl}^2} = \frac{(0.24)^2(0.12)}{(0.64)^2} = \mathbf{0.017}$$

14.66 The equilibrium expression for this system is given by:

$$K_P = P_{CO_2}P_{H_2O}$$

(a) In a closed vessel the decomposition will stop when the product of the partial pressures of CO_2 and H_2O equals K_P. Adding more sodium bicarbonate will have **no effect**.

(b) In an open vessel, $CO_2(g)$ and $H_2O(g)$ will escape from the vessel, and the partial pressures of CO_2 and H_2O will never become large enough for their product to equal K_P. Therefore, equilibrium will never be established. Adding more sodium bicarbonate will result in the production of **more CO_2 and H_2O.**

14.68 **(a)** The equation that relates K_P and K_c is:

$$K_P = K_c(0.0821\,T)^{\Delta n}$$

For this reaction, $\Delta n = 3 - 2 = 1$

$$\boldsymbol{K_c} = \frac{K_P}{(0.0821T)} = \frac{2 \times 10^{-42}}{(0.0821 \times 298)} = \mathbf{8 \times 10^{-44}}$$

(b) Because of a very large activation energy, the reaction of hydrogen with oxygen is infinitely slow without a catalyst or an initiator. The action of a single spark on a mixture of these gases results in the explosive formation of water.

14.70 **(a)** Calculate the value of K_P by substituting the equilibrium partial pressures into the equilibrium constant expression.

$$\boldsymbol{K_P} = \frac{P_B}{P_A^2} = \frac{(0.60)}{(0.60)^2} = \mathbf{1.7}$$

(b) The total pressure is the sum of the partial pressures for the two gaseous components, A and B. We can write:

$$P_A + P_B = 1.5 \text{ atm}$$

and

$$P_B = 1.5 - P_A$$

Substituting into the expression for K_P gives:

$$K_P = \frac{(1.5 - P_A)}{P_A^2} = 1.7$$

$$1.7P_A^2 + P_A - 1.5 = 0$$

Solving the quadratic equation, we obtain:

$$\mathbf{P_A = 0.69 \text{ atm}}$$

and by difference,

$$\mathbf{P_B = 0.81 \text{ atm}}$$

Check that substituting these equilibrium concentrations into the equilibrium constant expression gives the equilibrium constant calculated in part (a).

$$K_P = \frac{P_B}{P_A^2} = \frac{0.81}{(0.69)^2} = 1.7$$

14.72 Total number of moles of gas is:

$$0.020 + 0.040 + 0.96 = 1.02 \text{ mol of gas}$$

You can calculate the partial pressure of each gaseous component from the mole fraction and the total pressure.

$$P_{NO} = X_{NO}P_T = \frac{0.040}{1.02} \times 0.20 = 0.0078 \text{ atm}$$

$$P_{O_2} = X_{O_2}P_T = \frac{0.020}{1.02} \times 0.20 = 0.0039 \text{ atm}$$

$$P_{NO_2} = X_{NO_2}P_T = \frac{0.96}{1.02} \times 0.20 = 0.19 \text{ atm}$$

Calculate K_P by substituting the partial pressures into the equilibrium constant expression.

$$\mathbf{K_P} = \frac{P_{NO_2}^2}{P_{NO}^2 P_{O_2}} = \frac{(0.19)^2}{(0.0078)^2(0.0039)} = \mathbf{1.5 \times 10^5}$$

14.74 Set up a table that contains the initial concentrations, the change in concentrations, and the equilibrium concentration. Assume that the vessel has a volume of 1 L.

	H_2	+	Cl_2	$\rightleftharpoons$	2HCl
Initial (*M*):	0.47		0		3.59
Change (*M*):	$+x$		$+x$		$-2x$
Equilibrium (*M*):	$(0.47 + x)$		x		$(3.59 - 2x)$

Substitute the equilibrium concentrations into the equilibrium constant expression, then solve for x. Since $\Delta n = 0$, $K_c = K_P$.

$$K_c = \frac{[HCl]^2}{[H_2][Cl_2]} = \frac{(3.59 - 2x)^2}{(0.47 + x)x} = 193$$

Solving the quadratic equation,

$$x = 0.10$$

Having solved for x, calculate the equilibrium concentrations of all species.

$$[H_2] = 0.57\ M \qquad [Cl_2] = 0.10\ M \qquad [HCl] = 3.39\ M$$

Since we assumed that the vessel had a volume of 1 L, the above molarities also correspond to the number of moles of each component.

From the mole fraction of each component and the total pressure, we can calculate the partial pressure of each component.

$$\text{Total number of moles} = 0.57 + 0.10 + 3.39 = 4.06\ \text{mol}$$

$$P_{H_2} = \frac{0.57}{4.06} \times 2.00 = \mathbf{0.28\ atm}$$

$$P_{Cl_2} = \frac{0.10}{4.06} \times 2.00 = \mathbf{0.049\ atm}$$

$$P_{HCl} = \frac{3.39}{4.06} \times 2.00 = \mathbf{1.67\ atm}$$

14.76 This is a difficult problem. Express the equilibrium number of moles in terms of the initial moles and the change in number of moles (x). Next, calculate the mole fraction of each component. Using the mole fraction, you should come up with a relationship between partial pressure and total pressure for each component. Substitute the partial pressures into the equilibrium constant expression to solve for the total pressure, P_T.

The reaction is:

	N_2	+	$3\ H_2$	$\rightleftharpoons$	$2\ NH_3$
Initial (mol):	1		3		0
Change (mol):	$-x$		$-3x$		$2x$
Equilibrium (mol):	$(1 - x)$		$(3 - 3x)$		$2x$

$$\text{Mole fraction of } NH_3 = \frac{\text{mol of } NH_3}{\text{total number of moles}}$$

$$X_{NH_3} = \frac{2x}{(1 - x) + (3 - 3x) + 2x} = \frac{2x}{4 - 2x}$$

$$0.21 = \frac{2x}{4 - 2x}$$

$$x = 0.35\ \text{mol}$$

Substituting x into the following mole fraction equations, the mole fractions of N_2 and H_2 can be calculated.

$$X_{N_2} = \frac{1-x}{4-2x} = \frac{1-0.35}{4-2(0.35)} = 0.20$$

$$X_{H_2} = \frac{3-3x}{4-2x} = \frac{3-3(0.35)}{4-2(0.35)} = 0.59$$

The partial pressures of each component are equal to the mole fraction multiplied by the total pressure.

$$P_{NH_3} = 0.21P_T \qquad P_{N_2} = 0.20P_T \qquad P_{H_2} = 0.59P_T$$

Substitute the partial pressures above (in terms of P_T) into the equilibrium constant expression, and solve for P_T.

$$K_P = \frac{P_{NH_3}^2}{P_{H_2}^3 P_{N_2}}$$

$$4.31 \times 10^{-4} = \frac{(0.21)^2 P_T^2}{(0.59P_T)^3(0.20P_T)}$$

$$4.31 \times 10^{-4} = \frac{1.07}{P_T^{\ 2}}$$

$$\mathbf{P_T = 5.0 \times 10^1\ atm}$$

14.78 We carry an additional significant figure throughout this calculation to minimize rounding errors. The initial molarity of SO_2Cl_2 is:

$$[SO_2Cl_2] = \frac{6.75\ \text{g}\ SO_2Cl_2 \times \dfrac{1\ \text{mol}\ SO_2Cl_2}{135.0\ \text{g}\ SO_2Cl_2}}{2.00\ \text{L}} = 0.02500\ M$$

The concentration of SO_2 at equilibrium is:

$$[SO_2] = \frac{0.0345\ \text{mol}}{2.00\ \text{L}} = 0.01725\ M$$

Since there is a 1:1 mole ratio between SO_2 and SO_2Cl_2, the concentration of SO_2 at equilibrium (0.01725 M) equals the concentration of SO_2Cl_2 reacted. The concentrations of SO_2Cl_2 and Cl_2 at equilibrium are:

	$SO_2Cl_2(g)$ $\rightleftharpoons$	$SO_2(g)$ +	$Cl_2(g)$
Initial (M):	0.02500	0	0
Change (M):	−0.01725	+0.01725	+0.01725
Equilibrium (M):	0.00775	0.01725	0.01725

Substitute the equilibrium concentrations into the equilibrium constant expression to calculate K_c.

$$\mathbf{K_c} = \frac{[SO_2][Cl_2]}{[SO_2Cl_2]} = \frac{(0.01725)(0.01725)}{(0.00775)} = \mathbf{3.84 \times 10^{-2}}$$

14.80 $I_2(g) \rightleftharpoons 2I(g)$

Assuming 1 mole of I_2 is present originally and α moles reacts, at equilibrium: $[I_2] = 1 - \alpha$, $[I] = 2\alpha$. The total number of moles present in the system $= (1 - \alpha) + 2\alpha = 1 + \alpha$. From Problem 14.109(a) in the text, we know that K_P is equal to:

$$K_P = \frac{4\alpha^2}{1-\alpha^2}P \qquad (1)$$

If there were no dissociation, then the pressure would be:

$$P = \frac{nRT}{V} = \frac{\left(1.00\ \text{g} \times \frac{1\ \text{mol}}{253.8\ \text{g}}\right)\left(0.0821\frac{\text{L}\cdot\text{atm}}{\text{mol}\cdot\text{K}}\right)(1473\ \text{K})}{0.500\ \text{L}} = 0.953\ \text{atm}$$

$$\frac{\text{observed pressure}}{\text{calculated pressure}} = \frac{1.51\ \text{atm}}{0.953\ \text{atm}} = \frac{1+\alpha}{1}$$

$$\alpha = 0.584$$

Substituting in equation (1) above:

$$\boldsymbol{K_P} = \frac{4\alpha^2}{1-\alpha^2}P = \frac{(4)(0.584)^2}{1-(0.584)^2} \times 1.51 = \mathbf{3.13}$$

14.82 According to the ideal gas law, pressure is directly proportional to the concentration of a gas in mol/L if the reaction is at constant volume and temperature. Therefore, pressure may be used as a concentration unit. The reaction is:

	N_2	+	$3H_2$	$\rightleftharpoons$	$2NH_3$
Initial (atm):	0.862		0.373		0
Change (atm):	$-x$		$-3x$		$+2x$
Equilibrium (atm):	$(0.862 - x)$		$(0.373 - 3x)$		$2x$

$$K_P = \frac{P_{NH_3}^2}{P_{H_2}^3 P_{N_2}}$$

$$4.31 \times 10^{-4} = \frac{(2x)^2}{(0.373-3x)^3(0.862-x)}$$

At this point, we need to make two assumptions that $3x$ is very small compared to 0.373 and that x is very small compared to 0.862. Hence,

$$0.373 - 3x \approx 0.373$$

and

$$0.862 - x \approx 0.862$$

$$4.31 \times 10^{-4} \approx \frac{(2x)^2}{(0.373)^3(0.862)}$$

Solving for x.

$$x = 2.20 \times 10^{-3}\ \text{atm}$$

The equilibrium pressures are:

$$P_{N_2} = [0.862 - (2.20 \times 10^{-3})]\text{atm} = \mathbf{0.860\ atm}$$

$$P_{H_2} = [0.373 - (3)(2.20 \times 10^{-3})]\text{atm} = \mathbf{0.366\ atm}$$

$$P_{NH_3} = (2)(2.20 \times 10^{-3}\ \text{atm}) = \mathbf{4.40 \times 10^{-3}\ atm}$$

Was the assumption valid that we made above? Typically, the assumption is considered valid if x is less than 5 percent of the number that we said it was very small compared to. Is this the case?

14.84 **(a)** The equation is:

	fructose	⇌	glucose
Initial (M):	0.244		0
Change (M):	−0.131		+0.131
Equilibrium (M):	0.113		0.131

Calculating the equilibrium constant,

$$\boldsymbol{K_c} = \frac{[\text{glucose}]}{[\text{fructose}]} = \frac{0.131}{0.113} = \mathbf{1.16}$$

(b) $$\textbf{Percent converted} = \frac{\text{amount of fructose converted}}{\text{original amount of fructose}} \times 100\%$$

$$= \frac{0.131}{0.244} \times 100\% = \mathbf{53.7\%}$$

14.86 **(a)** There is only one gas phase component, O_2. The equilibrium constant is simply

$$\boldsymbol{K_P} = P_{O_2} = \mathbf{0.49\ atm}$$

(b) From the ideal gas equation, we can calculate the moles of O_2 produced by the decomposition of CuO.

$$n_{O_2} = \frac{PV}{RT} = \frac{(0.49\ \text{atm})(2.0\ \text{L})}{(0.0821\ \text{L}\cdot\text{atm/K}\cdot\text{mol})(1297\ \text{K})} = 9.2 \times 10^{-3}\ \text{mol}\ O_2$$

From the balanced equation,

$$(9.2 \times 10^{-3}\ \text{mol}\ O_2) \times \frac{4\ \text{mol CuO}}{1\ \text{mol}\ O_2} = 3.7 \times 10^{-2}\ \text{mol CuO decomposed}$$

$$\textbf{Fraction of CuO decomposed} = \frac{\text{amount of CuO lost}}{\text{original amount of CuO}}$$

$$= \frac{3.7 \times 10^{-2}\ \text{mol}}{0.16\ \text{mol}} = \mathbf{0.23}$$

(c) If a 1.0 mol sample were used, the pressure of oxygen would still be the same (0.49 atm) and it would be due to the same quantity of O_2. Remember, a pure solid does not affect the equilibrium position. The moles of CuO lost would still be 3.7×10^{-2} mol. Thus the fraction decomposed would be:

$$\frac{0.037}{1.0} = \mathbf{0.037}$$

(d) If the number of moles of CuO were less than 3.7×10^{-2} mol, the equilibrium could not be established because the pressure of O_2 would be less than 0.49 atm. Therefore, the smallest number of moles of CuO needed to establish equilibrium must be slightly greater than 3.7×10^{-2} mol.

14.88 We first must find the initial concentrations of all the species in the system.

$$[H_2]_0 = \frac{0.714 \text{ mol}}{2.40 \text{ L}} = 0.298\ M$$

$$[I_2]_0 = \frac{0.984 \text{ mol}}{2.40 \text{ L}} = 0.410\ M$$

$$[HI]_0 = \frac{0.886 \text{ mol}}{2.40 \text{ L}} = 0.369\ M$$

Calculate the reaction quotient by substituting the initial concentrations into the appropriate equation.

$$Q_c = \frac{[HI]_0^2}{[H_2]_0[I_2]_0} = \frac{(0.369)^2}{(0.298)(0.410)} = 1.11$$

We find that Q_c is less than K_c. The equilibrium will shift to the right, decreasing the concentrations of H_2 and I_2 and increasing the concentration of HI.

We set up the usual table. Let x be the decrease in concentration of H_2 and I_2.

	H_2	+	I_2	$\rightleftharpoons$	2 HI
Initial (M):	0.298		0.410		0.369
Change (M):	$-x$		$-x$		$+2x$
Equilibrium (M):	$(0.298 - x)$		$(0.410 - x)$		$(0.369 + 2x)$

The equilibrium constant expression is:

$$K_c = \frac{[HI]^2}{[H_2][I_2]} = \frac{(0.369 + 2x)^2}{(0.298 - x)(0.410 - x)} = 54.3$$

This becomes the quadratic equation

$$50.3x^2 - 39.9x + 6.48 = 0$$

The smaller root is $x = 0.228\ M$. (The larger root is physically impossible.)

Having solved for x, calculate the equilibrium concentrations.

$$\mathbf{[H_2]} = (0.298 - 0.228)\ M = \mathbf{0.070\ M}$$
$$\mathbf{[I_2]} = (0.410 - 0.228)\ M = \mathbf{0.182\ M}$$
$$\mathbf{[HI]} = [0.369 + 2(0.228)]\ M = \mathbf{0.825\ M}$$

14.90 The gas cannot be (a) because the color became lighter with heating. Heating (a) to 150°C would produce some HBr, which is colorless and would lighten rather than darken the gas.

The gas cannot be (b) because Br_2 doesn't dissociate into Br atoms at 150°C, so the color shouldn't change.

The gas must be **(c)**. From 25°C to 150°C, heating causes N_2O_4 to dissociate into NO_2, thus darkening the color (NO_2 is a brown gas).

$$N_2O_4(g) \rightarrow 2NO_2(g)$$

Above 150°C, the NO_2 breaks up into colorless NO and O_2.

$$2NO_2(g) \rightarrow 2NO(g) + O_2(g)$$

An increase in pressure shifts the equilibrium back to the left, forming NO_2, thus darkening the gas again.

$$2NO(g) + O_2(g) \rightarrow 2NO_2(g)$$

14.92 Given the following: $K_c = \dfrac{[NH_3]^2}{[N_2][H_2]^3} = 1.2$

(a) Temperature must have units of Kelvin.

$$K_P = K_c(0.0821\,T)^{\Delta n}$$

$$\boldsymbol{K_P} = (1.2)(0.0821 \times 648)^{(2-4)} = \mathbf{4.2 \times 10^{-4}}$$

(b) Recalling that,

$$K_{\text{forward}} = \frac{1}{K_{\text{reverse}}}$$

Therefore,

$$\boldsymbol{K_c'} = \frac{1}{1.2} = \mathbf{0.83}$$

(c) Since the equation

$$\tfrac{1}{2}N_2(g) + \tfrac{3}{2}H_2(g) \rightleftharpoons NH_3(g)$$

is equivalent to

$$\tfrac{1}{2}[N_2(g) + 3H_2(g) \rightleftharpoons 2NH_3(g)]$$

then, K_c' for the reaction:

$$\tfrac{1}{2}N_2(g) + \tfrac{3}{2}H_2(g) \rightleftharpoons NH_3(g)$$

equals $(K_c)^{\frac{1}{2}}$ for the reaction:

$$N_2(g) + 3H_2(g) \rightleftharpoons 2NH_3(g)$$

Thus,

$$\boldsymbol{K_c'} = (K_c)^{\frac{1}{2}} = \sqrt{1.2} = \mathbf{1.1}$$

(d) For K_P in part (b):

$$\boldsymbol{K_P} = (0.83)(0.0821 \times 648)^{+2} = \mathbf{2.3 \times 10^3}$$

and for K_P in part (c):

$$\boldsymbol{K_P} = (1.1)(0.0821 \times 648)^{-1} = \mathbf{0.021}$$

14.94 The vapor pressure of water is equivalent to saying the partial pressure of $H_2O(g)$.

$$\boldsymbol{K_P} = P_{H_2O} = \mathbf{0.0231}$$

$$\boldsymbol{K_c} = \frac{K_p}{(0.0821T)^{\Delta n}} = \frac{0.0231}{(0.0821 \times 293)^1} = \mathbf{9.60 \times 10^{-4}}$$

14.96 We can calculate the average molar mass of the gaseous mixture from the density.

$$\mathcal{M} = \frac{dRT}{P}$$

Let $\overline{\mathcal{M}}$ be the average molar mass of NO_2 and N_2O_4. The above equation becomes:

$$\overline{\mathcal{M}} = \frac{dRT}{P} = \frac{(2.3\text{ g/L})(0.0821\text{ L}\cdot\text{atm/K}\cdot\text{mol})(347\text{ K})}{1.3\text{ atm}}$$

$$\overline{\mathcal{M}} = 50.4\text{ g/mol}$$

The average molar mass is equal to the sum of the molar masses of each component times the respective mole fractions. Setting this up, we can calculate the mole fraction of each component.

$$\overline{\mathcal{M}} = X_{NO_2}\mathcal{M}_{NO_2} + X_{N_2O_4}\mathcal{M}_{N_2O_4} = 50.4\text{ g/mol}$$

$$X_{NO_2}(46.01\text{ g/mol}) + (1 - X_{NO_2})(92.01\text{ g/mol}) = 50.4\text{ g/mol}$$

$$X_{NO_2} = 0.905$$

We can now calculate the partial pressure of NO_2 from the mole fraction and the total pressure.

$$P_{NO_2} = X_{NO_2}P_T$$

$$\boldsymbol{P_{NO_2}} = (0.905)(1.3\text{ atm}) = 1.18\text{ atm} = \mathbf{1.2\text{ atm}}$$

We can calculate the partial pressure of N_2O_4 by difference.

$$P_{N_2O_4} = P_T - P_{NO_2}$$

$$\boldsymbol{P_{N_2O_4}} = (1.3 - 1.18)\text{atm} = \mathbf{0.12\text{ atm}}$$

Finally, we can calculate K_P for the dissociation of N_2O_4.

$$\boldsymbol{K_P} = \frac{P_{NO_2}^2}{P_{N_2O_4}} = \frac{(1.2)^2}{0.12} = \mathbf{12}$$

14.98 **(a)** shifts to right **(b)** shifts to right **(c)** no change **(d)** no change **(e)** no change **(f)** shifts to left

14.100 The equilibrium is: $N_2O_4(g) \rightleftharpoons 2NO_2(g)$

$$K_P = \frac{(P_{NO_2})^2}{P_{N_2O_4}} = \frac{0.15^2}{0.20} = 0.113$$

Volume is doubled so pressure is halved. Let's calculate Q_P and compare it to K_P.

$$Q_P = \frac{\left(\frac{0.15}{2}\right)^2}{\left(\frac{0.20}{2}\right)} = 0.0563 < K_P$$

Equilibrium will shift to the right. Some N_2O_4 will react, and some NO_2 will be formed. Let x = amount of N_2O_4 reacted.

	$N_2O_4(g)$ $\rightleftharpoons$	$2NO_2(g)$
Initial (atm):	0.10	0.075
Change (atm):	$-x$	$+2x$
Equilibrium (atm):	$0.10 - x$	$0.075 + 2x$

Substitute into the K_P expression to solve for x.

$$K_P = 0.113 = \frac{(0.075 + 2x)^2}{0.10 - x}$$

$$4x^2 + 0.413x - 5.68 \times 10^{-3} = 0$$

$$x = 0.0123$$

At equilibrium:

$$P_{NO_2} = 0.075 + 2(0.0123) = 0.0996 \approx \mathbf{0.100\ atm}$$

$$P_{N_2O_4} = 0.10 - 0.0123 = \mathbf{0.09\ atm}$$

Check:

$$K_P = \frac{(0.100)^2}{0.09} = 0.111 \qquad \text{close enough to } 0.113$$

14.102 **(a)** Molar mass of PCl_5 = 208.2 g/mol

$$\boldsymbol{P} = \frac{nRT}{V} = \frac{\left(2.50\ \text{g} \times \frac{1\ \text{mol}}{208.2\ \text{g}}\right)\left(0.0821\frac{\text{L}\cdot\text{atm}}{\text{mol}\cdot\text{K}}\right)(523\ \text{K})}{0.500\ \text{L}} = \mathbf{1.03\ atm}$$

(b)

	PCl_5 $\rightleftharpoons$	PCl_3 +	Cl_2
Initial (atm)	1.03	0	0
Change (atm)	$-x$	$+x$	$+x$
Equilibrium (atm)	$1.03 - x$	x	x

$$K_P = 1.05 = \frac{x^2}{1.03 - x}$$

$$x^2 + 1.05x - 1.08 = 0$$

$$x = 0.639$$

At equilibrium:

$$P_{PCl_5} = 1.03 - 0.639 = \mathbf{0.39\ atm}$$

(c) $P_T = (1.03 - x) + x + x = 1.03 + 0.639 = \mathbf{1.67\ atm}$

(d) $\dfrac{0.639\text{ atm}}{1.03\text{ atm}} = \mathbf{0.620}$

14.104 **(a)** $K_P = P_{Hg} = 0.0020\text{ mmHg} = 2.6 \times 10^{-6}\text{ atm} = \mathbf{2.6 \times 10^{-6}}$ (equil. constants are expressed without units)

$$K_c = \frac{K_P}{(0.0821T)^{\Delta n}} = \frac{2.6 \times 10^{-6}}{(0.0821 \times 299)^1} = \mathbf{1.1 \times 10^{-7}}$$

(b) Volume of lab $= (6.1\text{ m})(5.3\text{ m})(3.1\text{ m}) = 100\text{ m}^3$

$[Hg] = K_c$

$$\textbf{Total mass of Hg vapor} = \frac{1.1 \times 10^{-7}\text{ mol}}{1\text{ L}} \times \frac{200.6\text{ g}}{1\text{ mol}} \times \frac{1\text{ L}}{1000\text{ cm}^3} \times \left(\frac{1\text{ cm}}{0.01\text{ m}}\right)^3 \times 100\text{ m}^3 = \mathbf{2.2\ g}$$

The concentration of mercury vapor in the room is:

$$\frac{2.2\text{ g}}{100\text{ m}^3} = 0.022\text{ g/m}^3 = \mathbf{22\ mg/m^3}$$

Yes! This concentration exceeds the safety limit of 0.05 mg/m^3. Better clean up the spill!

14.106 There is a temporary dynamic equilibrium between the melting ice cubes and the freezing of water between the ice cubes.

14.108 First, let's calculate the initial concentration of ammonia.

$$[NH_3] = \frac{14.6\text{ g} \times \dfrac{1\text{ mol }NH_3}{17.03\text{ g }NH_3}}{4.00\text{ L}} = 0.214\ M$$

Let's set up a table to represent the equilibrium concentrations. We represent the amount of NH_3 that reacts as $2x$.

	$2NH_3(g)$ $\rightleftharpoons$	$N_2(g)$ +	$3H_2(g)$
Initial (M):	0.214	0	0
Change (M):	$-2x$	$+x$	$+3x$
Equilibrium (M):	$0.214 - 2x$	x	$3x$

Substitute into the equilibrium constant expression to solve for x.

$$K_c = \frac{[N_2][H_2]^3}{[NH_3]^2}$$

$$0.83 = \frac{(x)(3x)^3}{(0.214 - 2x)^2} = \frac{27x^4}{(0.214 - 2x)^2}$$

Taking the square root of both sides of the equation gives:

$$0.91 = \frac{5.20x^2}{0.214 - 2x}$$

Rearranging,

$$5.20x^2 + 1.82x - 0.195 = 0$$

Solving the quadratic equation gives the solutions:

$$x = 0.086\ M \text{ and } x = -0.44\ M$$

The positive root is the correct answer. The equilibrium concentrations are:

$$\mathbf{[NH_3]} = 0.214 - 2(0.086) = \mathbf{0.042}\ \boldsymbol{M}$$
$$\mathbf{[N_2]} = \mathbf{0.086}\ \boldsymbol{M}$$
$$\mathbf{[H_2]} = 3(0.086) = \mathbf{0.26}\ \boldsymbol{M}$$

14.110 To determine $\Delta H°$, we need to plot $\ln K_P$ versus $1/T$ (y vs. x).

$\ln K_P$	$1/T$
4.93	0.00167
1.63	0.00143
−0.83	0.00125
−2.77	0.00111
−4.34	0.00100

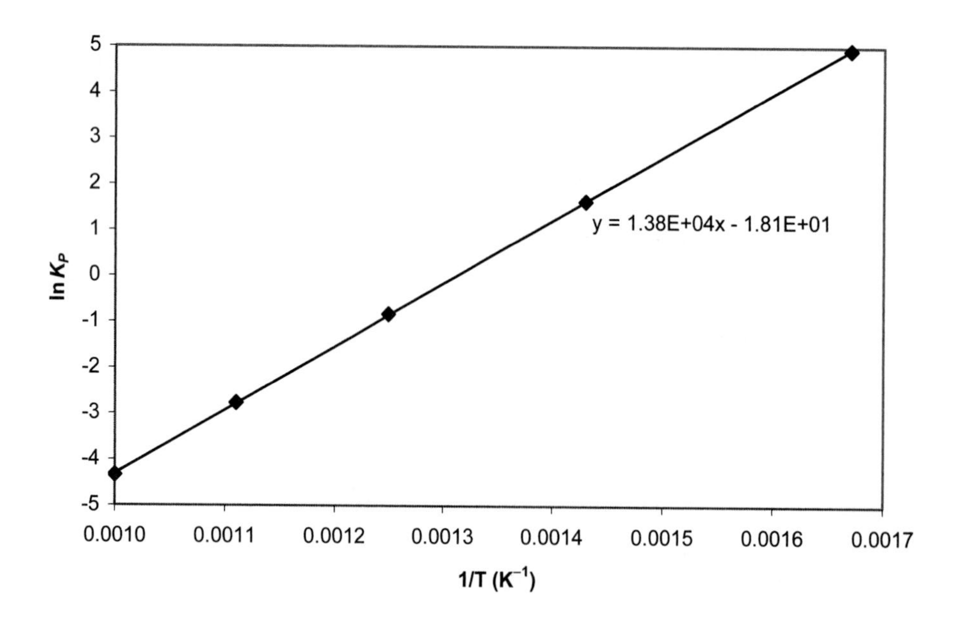

The slope of the plot equals $-\Delta H°/R$.

$$1.38 \times 10^4\ \text{K} = -\frac{\Delta H°}{8.314\ \text{J/mol}\cdot\text{K}}$$

$$\Delta H° = \mathbf{-1.15 \times 10^5\ J/mol} = \mathbf{-115\ kJ/mol}$$

14.112 Initially, the pressure of SO_2Cl_2 is 9.00 atm. The pressure is held constant, so after the reaction reaches equilibrium, $P_{SO_2Cl_2} + P_{SO_2} + P_{Cl_2} = 9.00\ \text{atm}$. The amount (pressure) of SO_2Cl_2 reacted must equal the pressure of SO_2 and Cl_2 produced for the pressure to remain constant. If we let $P_{SO_2} + P_{Cl_2} = x$, then the pressure of SO_2Cl_2 reacted must be $2x$. We set up a table showing the initial pressures, the change in pressures, and the equilibrium pressures.

	$SO_2Cl_2(g)$	$\rightleftharpoons$	$SO_2(g)$	+	$Cl_2(g)$
Initial (atm):	9.00		0		0
Change (atm):	$-2x$		$+x$		$+x$
Equilibrium (atm):	$9.00 - 2x$		x		x

Again, note that the change in pressure for SO_2Cl_2 ($-2x$) does not match the stoichiometry of the reaction, because we are expressing changes in pressure. The total pressure is kept at 9.00 atm throughout.

$$K_P = \frac{P_{SO_2}P_{Cl_2}}{P_{SO_2Cl_2}}$$

$$2.05 = \frac{(x)(x)}{9.00 - 2x}$$

$$x^2 + 4.10x - 18.45 = 0$$

Solving the quadratic equation, $x = 2.71$ atm. At equilibrium,

$$P_{SO_2} = P_{Cl_2} = x = \mathbf{2.71\ atm}$$

$$P_{SO_2Cl_2} = 9.00 - 2(2.71) = \mathbf{3.58\ atm}$$

14.114 We start with a table.

	A_2	+	B_2	$\rightleftharpoons$	2AB
Initial (mol):	1		3		0
Change (mol):	$-\frac{x}{2}$		$-\frac{x}{2}$		$+x$
Equilibrium (mol):	$1 - \frac{x}{2}$		$3 - \frac{x}{2}$		x

After the addition of 2 moles of A,

	A_2	+	B_2	$\rightleftharpoons$	2AB
Initial (mol):	$3 - \frac{x}{2}$		$3 - \frac{x}{2}$		x
Change (mol):	$-\frac{x}{2}$		$-\frac{x}{2}$		$+x$
Equilibrium (mol):	$3 - x$		$3 - x$		$2x$

We write two different equilibrium constants expressions for the two tables.

$$K = \frac{[AB]^2}{[A_2][B_2]}$$

$$K = \frac{x^2}{\left(1 - \frac{x}{2}\right)\left(3 - \frac{x}{2}\right)} \quad \text{and} \quad K = \frac{(2x)^2}{(3 - x)(3 - x)}$$

We equate the equilibrium constant expressions and solve for x.

$$\frac{x^2}{\left(1 - \frac{x}{2}\right)\left(3 - \frac{x}{2}\right)} = \frac{(2x)^2}{(3 - x)(3 - x)}$$

$$\frac{1}{\frac{1}{4}(x^2 - 8x + 12)} = \frac{4}{x^2 - 6x + 9}$$

$$-6x + 9 = -8x + 12$$

$$x = 1.5$$

We substitute x back into one of the equilibrium constant expressions to solve for K.

$$\boldsymbol{K} = \frac{(2x)^2}{(3 - x)(3 - x)} = \frac{(3)^2}{(1.5)(1.5)} = \mathbf{4.0}$$

Substitute x into the other equilibrium constant expression to see if you obtain the same value for K. Note that we used moles rather than molarity for the concentrations, because the volume, V, cancels in the equilibrium constant expressions.

CHAPTER 15
ACIDS AND BASES

PROBLEM-SOLVING STRATEGIES AND TUTORIAL SOLUTIONS

TYPES OF PROBLEMS

Problem Type 1: Identifying Conjugate Acid-Base Pairs.

Problem Type 2: The Ion-Product Constant (K_w), Calculating $[H^+]$ from $[OH^-]$.

Problem Type 3: pH Calculations.
- **(a)** Calculating pH from $[H^+]$.
- **(b)** Calculating $[H^+]$ from pH.

Problem Type 4: Calculating the pH of a Strong Acid and a Strong Base.

Problem Type 5: Weak Acids.
- **(a)** Ionization of a weak monoprotic acid.
- **(b)** Determining K_a from a pH measurement.

Problem Type 6: Calculating the pH of a Diprotic Acid.

Problem Type 7: Calculating the pH of a Weak Base.

Problem Type 8: Predicting the Acid-Base Properties of Salt Solutions.

PROBLEM TYPE 1: IDENTIFYING CONJUGATE ACID-BASE PAIRS

A **conjugate acid-base pair** can be defined as *an acid and its conjugate base or a base and its conjugate acid.* The conjugate base of a Brønsted acid is the species that remains when *one* proton has been removed from the acid. Conversely, a conjugate acid results from the addition of *one* proton to a Brønsted base.

EXAMPLE 15.1
Consider the reaction

$$HSO_4^-(aq) + HCO_3^-(aq) \rightleftharpoons SO_4^{2-}(aq) + H_2CO_3(aq)$$

(a) Identify the acids and bases for the forward and reverse reactions.
(b) Identify the conjugate acid-base pairs.

Strategy: (a) An acid is a proton donor and a base is a proton acceptor. (b) Remember that a conjugate base always has one fewer H atom and one more negative charge (or one fewer positive charge) than the formula of the corresponding acid.

Solution:

(a) In the forward reaction, HSO_4^- is the proton donor, which makes it an acid. The proton acceptor, HCO_3^-, is a base. In the reverse reaction, the proton donor (acid) is H_2CO_3, and SO_4^{2-} is the proton acceptor (base).

(b) $HSO_4^-(aq) + HCO_3^-(aq) \rightleftharpoons SO_4^{2-}(aq) + H_2CO_3(aq)$

acid$_1$ base$_2$ base$_1$ acid$_2$

The subscripts 1 and 2 designate the two conjugate acid-base pairs.

PRACTICE EXERCISE

1. Identify the conjugate bases of the following acids:

(a) CH_3COOH (b) H_2S (c) HSO_3^- (d) $HClO$

Text Problems: 15.4, 15.6, 15.8

PROBLEM TYPE 2: THE ION-PRODUCT CONSTANT (K_w), CALCULATING [H$^+$] FROM [OH$^-$]

Water is a very weak electrolyte and therefore a poor conductor of electricity, but it does ionize to a small extent:

$$H_2O(l) \rightleftharpoons H^+(aq) + OH^-(aq)$$

We can write the equilibrium constant for the autoionization of water as

$$K_c = \frac{[H^+][OH^-]}{[H_2O]}$$

Since a very small fraction of water molecules are ionized, the concentration of water, $[H_2O]$, remains virtually unchanged. Therefore, we assume that the concentration of water is constant, and we write:

$$K_c[H_2O] = K_w = [H^+][OH^-] \qquad (15.3,\ \text{text})$$

In pure water at 25°C, the concentrations of H^+ and OH^- ions are equal and found to be $[H^+] = [OH^-] = 1.0 \times 10^{-7}\ M$. Substituting these concentrations into Equation (15.3) of the text,

$$K_w = [H^+][OH^-] = (1 \times 10^{-7})(1 \times 10^{-7}) = 1 \times 10^{-14}$$

Thus, knowing K_w and one of the concentrations, either $[H^+]$ or $[OH^-]$, we can easily calculate the other concentration.

EXAMPLE 15.2

The OH^- ion concentration in a certain solution at 25°C is 5.0×10^{-5} *M*. What is the H^+ concentration?

Strategy: We are given the concentration of OH^- ions and asked to calculate $[H^+]$. The relationship between $[H^+]$ and $[OH^-]$ in water or an aqueous solution is given by the ion-product of water, K_w [Equation (15.3) of the text].

Solution: The ion product of water is applicable to all aqueous solutions. At 25°C,

$$K_w = 1.0 \times 10^{-14} = [H^+][OH^-]$$

Rearrange the equation to solve for $[H^+]$.

$$\mathbf{[H^+]} = \frac{1.0 \times 10^{-14}}{[OH^-]} = \frac{1.0 \times 10^{-14}}{5.0 \times 10^{-5}} = \mathbf{2.0 \times 10^{-10}}\ \boldsymbol{M}$$

PRACTICE EXERCISE

2. The H^+ concentration in a certain solution is 6.6×10^{-4} *M*. What is the OH^- concentration?

Text Problems: 15.16, 15.18, **15.20c**

PROBLEM TYPE 3: pH CALCULATIONS

Because the concentrations of H^+ and OH^- ions in aqueous solutions are frequently very small numbers making them inconvenient to work with, the Danish biochemist Soren Sorensen in 1909 proposed a more practical measure called pH. The **pH** of a solution is defined as *the negative log of the hydrogen ion concentration* (in moles per liter).

$$\text{pH} = -\log[\text{H}^+]$$

The pOH of a solution is defined in a similar manner.

$$\text{pOH} = -\log[\text{OH}^-]$$

A useful relationship between pH and pOH is:

$$\text{pH} + \text{pOH} = 14.$$

See Section 15.3 of the text for a complete discussion.

A. Calculating pH from $[H^+]$

EXAMPLE 15.3

The $[H^+]$ of a solution is 0.015 *M*. What is the pH of the solution?

Strategy: The relationship between $[H^+]$ and pH is: $\text{pH} = -\log[\text{H}^+]$. The hydrogen ion concentration is given in the problem.

Solution: Substitute the H^+ concentration into the above equation to calculate the pH of the solution.

$$\textbf{pH} = -\log[\text{H}^+] = -\log[0.015] = \mathbf{1.82}$$

PRACTICE EXERCISE

3. The OH^- concentration of a certain ammonia solution is 7.2×10^{-4} *M*. What is the pOH and pH?

Text Problem: 15.18

B. Calculating $[H^+]$ from pH or pOH

EXAMPLE 15.4

What is the H^+ concentration in a solution with a pOH of 3.9?

Strategy: Here we are given the pOH of a solution and asked to calculate $[H^+]$. First, the pH can be calculated from the pOH. Because pH is defined as $\text{pH} = -\log[\text{H}^+]$, we can solve for $[H^+]$ by taking the antilog of the pH; that is, $[\text{H}^+] = 10^{-\text{pH}}$.

Solution: Recall that pH + pOH = 14.00. Since pOH is given in the problem, we can calculate the pH.

$$\text{pH} = 14.00 - \text{pOH}$$

$$\text{pH} = 14.00 - 3.9 = 10.1$$

The H^+ concentration can now be determined from the pH.

$$\text{pH} = -\log[H^+]$$

or,

$$-\text{pH} = \log[H^+]$$

Taking the antilog of both sides of the equation,

$$10^{-\text{pH}} = [H^+]$$

$$\mathbf{[H^+]} = 10^{-10.1} = \mathbf{8 \times 10^{-11}}\ \boldsymbol{M}$$

PRACTICE EXERCISE

4. The pH of a solution is 4.45. What is the H^+ concentration?

Text Problem: 15.20 a,b

PROBLEM TYPE 4: CALCULATING THE pH OF A STRONG ACID AND A STRONG BASE

Strong acids and strong bases are strong electrolytes that are assumed to ionize completely in water. For example, consider the strong acid, hydrochloric acid [HCl(*aq*)]. We assume that it ionizes completely in water.

$$HCl(aq) \rightarrow H^+(aq) + Cl^-(aq)$$

Since strong acids ionize completely in water, we can easily calculate the $[H^+]$ in solution. For example, consider a 0.10 *M* HCl solution. Let's set up a table to determine the $[H^+]$ of the solution.

	$HCl(aq)$	$\rightarrow$	$H^+(aq)$	+	$Cl^-(aq)$
initial conc.	0.10 *M*		0		0
conc. after ionization	0		0.10 *M*		0.10 *M*

Since there is a one:one mole ratio between HCl and H^+, the H^+ concentration after ionization equals the initial concentration of HCl.

Note that the above reaction can be written more accurately as:

$$HCl(aq) + H_2O(l) \rightarrow H_3O^+(aq) + Cl^-(aq)$$

HCl is an acid and donates a proton (H^+) to the weak base, H_2O. H^+ in aqueous solution is really shorthand notation for H_3O^+.

You need to know the six strong acids and the six strong bases. Half the battle in pH calculations is recognizing the type of species in solution (i.e., strong acid, strong base, weak acid, weak base, or salt).

The six strong acids

$HClO_4$	perchloric acid
HI	hydroiodic acid
HBr	hydrobromic acid
HCl	hydrochloric acid
H_2SO_4	sulfuric acid
HNO_3	nitric acid

The six **strong bases** include the five alkali metal hydroxides (LiOH, NaOH, KOH, RbOH, and CsOH), plus barium hydroxide, $Ba(OH)_2$.

EXAMPLE 15.5

Calculate the pH of a 0.10 *M* $Ba(OH)_2$ solution.

Strategy: You must recognize that barium hydroxide is one of the six strong bases. Strong bases ionize completely in solution.

Solution: $Ba(OH)_2$ is a strong base; each $Ba(OH)_2$ unit produces two OH^- ions. The changes in concentrations of all species can be represented as follows:

	$Ba(OH)_2(aq)$ →	$Ba^{2+}(aq)$ +	$2OH^-(aq)$
Initial (*M*):	0.10	0	0
Change (*M*):	−0.10	+0.10	+2(0.10)
Final (*M*):	0	0.10	0.20

Because there is a 2:1 mole ratio between OH^- and $Ba(OH)_2$, the OH^- concentration is double the $Ba(OH)_2$ concentration.

Calculate the pOH from the OH^- concentration.

$$\text{pOH} = -\log[\text{OH}^-]$$

$$\text{pOH} = -\log[0.20] = 0.70$$

Use the relationship, pH + pOH = 14.0 to calculate the pH of the $Ba(OH)_2$ solution.

$$\textbf{pH} = 14.00 - 0.70 = \mathbf{13.30}$$

PRACTICE EXERCISE

5. What is the pH of a 0.025 *M* HCl solution?

Text Problems: 15.24, 15.26

PROBLEM TYPE 5: WEAK ACIDS

Most acids ionize only to a limited extent in water. Such acids are classified as **weak acids**. For example, HNO_2 (nitrous acid), is a weak acid. Its ionization in water is represented by

$$HNO_2(aq) + H_2O(l) \rightleftharpoons H_3O^+(aq) + NO_2^-(aq)$$

or simply

$$HNO_2(aq) \rightleftharpoons H^+(aq) + NO_2^-(aq)$$

The equilibrium constant for this acid ionization is called the **acid ionization constant, K_a**. It is given by:

$$K_a = \frac{[H^+][NO_2^-]}{[HNO_2]}$$

In general,

$$K_a = \frac{[H^+][A^-]}{[HA]}$$

where,

A^- is the conjugate base of the weak acid, HA.

A. Ionization of a weak monoprotic acid

This problem is very similar to equilibrium calculations in Chapter 14. Try to think of this problem as just another equilibrium calculation. The only difference is that in Chapter 14, we dealt mostly with gas phase reactions. Now, we are dealing with weak acids in aqueous solution. Some problems will ask you to calculate the equilibrium concentrations of all species, others will ask for the pH of the solution.

EXAMPLE 15.6

Calculate the pH of a 0.20 *M* acetic acid, CH_3COOH, solution.

Strategy: First, recognize that acetic acid is a weak acid. It is not one of the six strong acids, so it must be a weak acid. Recall that a weak acid only partially ionizes in water. We are given the initial quantity of a weak acid (CH_3COOH) and asked to calculate the pH. In order to calculate pH, we need to determine the H^+ concentration. In determining the H^+ concentration, we ignore the ionization of H_2O as a source of H^+, so the major source of H^+ ions is the acid. We follow the procedure outlined in Section 15.5 of the text.

Solution:

Step 1: We ignore water's contribution to $[H^+]$. We consider CH_3COOH as the only source of H^+ ions.

Step 2: Letting x be the equilibrium concentration of H^+ and CH_3COO^- ions in mol/L, we summarize:

	$CH_3COOH(aq)$ $\rightleftharpoons$	$H^+(aq)$ +	$CH_3COO^-(aq)$
Initial (*M*):	0.20	0	0
Change (*M*):	$-x$	$+x$	$+x$
Equilibrium (*M*):	$0.20 - x$	x	x

Step 3: Write the ionization constant expression in terms of the equilibrium concentrations. Knowing the value of the equilibrium constant (K_a), solve for x.

$$K_a = \frac{[H^+][CH_3COO^-]}{[CH_3COOH]}$$

You can look up the K_a value for acetic acid in Table 15.3 of the text.

$$1.8 \times 10^{-5} = \frac{(x)(x)}{(0.20 - x)}$$

At this point, we can make an assumption that x is very small compared to 0.20. Hence,

$$0.20 - x \approx 0.20$$

Oftentimes, assumptions such as these are valid if K is very small. A very small value of K means that a very small amount of reactants go to products. Hence, x is small. If we did not make this assumption, we would have to solve a quadratic equation.

$$1.8 \times 10^{-5} \approx \frac{(x)(x)}{0.20}$$

Solving for x.

$$x = 1.9 \times 10^{-3}\ M = [H^+]$$

Step 4: Having solved for the $[H^+]$, calculate the pH of the solution.

$$\mathbf{pH} = -\log[H^+] = -\log(1.9 \times 10^{-3}) = \mathbf{2.72}$$

Check: Checking the validity of the assumption,

$$\frac{1.9 \times 10^{-3}}{0.20} \times 100\% = 0.95\% < 5\%$$

The assumption was valid.

PRACTICE EXERCISE

6. Calculate the pH of a 0.50 M nitrous acid (HNO_2) solution. The K_a value for HNO_2 is 4.5×10^{-4}.

Text Problems: 15.44, 15.46, 15.48, 15.50

B. Determining K_a from a pH measurement

This problem is the reverse of Example 15.6 above. From the pH of the solution, you can calculate the $[H^+]$. Then from the $[H^+]$, you can calculate the equilibrium concentrations of the other species in solution. Substitute these equilibrium concentrations into the ionization constant expression to solve for K_a.

EXAMPLE 15.7

The pH of a 0.10 M solution of a weak monoprotic acid is 5.15. What is the K_a of the acid?

Strategy: Weak acids only partially ionize in water.

$$HA(aq) \rightleftharpoons H^+(aq) + A^-(aq)$$

Note that the concentration of the weak acid given refers to the initial concentration before ionization has started. The pH of the solution, on the other hand, refers to the situation at equilibrium. To calculate K_a, we need to know the concentrations of all three species, [HA], $[H^+]$, and $[A^-]$ at equilibrium. We ignore the ionization of water as a source of H^+ ions.

Solution: We proceed as follows.

Step 1: The major species in solution are HA, H^+, and the conjugate acid A^-.

Step 2: First, we need to calculate the hydrogen ion concentration from the pH value.

$$pH = -\log[H^+]$$

or,

$$-pH = \log[H^+]$$

Taking the antilog of both sides of the equation,

$$10^{-pH} = [H^+]$$

$$[H^+] = 10^{-5.15} = 7.1 \times 10^{-6}\ M$$

Step 3: If the concentration of H^+ is 7.1×10^{-6} M at equilibrium, that must mean that 7.1×10^{-6} M of the acid ionized. We summarize the changes.

	$HA(aq)$	$\rightleftharpoons$	$H^+(aq)$	+	$A^-(aq)$
Initial (M):	0.10		0		0
Change (M):	-7.1×10^{-6}		$+7.1 \times 10^{-6}$		$+7.1 \times 10^{-6}$
Equilibrium (M):	$0.10 - (7.1 \times 10^{-6})$		7.1×10^{-6}		7.1×10^{-6}

Step 4: Substitute the equilibrium concentrations into the ionization constant expression to solve for K_a.

$$K_a = \frac{[H^+][A^-]}{[HA]}$$

$$\mathbf{K_a} = \frac{(7.1 \times 10^{-6})^2}{(0.10)} = \mathbf{5.0 \times 10^{-10}}$$

PRACTICE EXERCISE

7. The pH of a 0.50 M monoprotic weak acid is 2.24. What is the K_a of the acid?

Text Problem: 15.54

PROBLEM TYPE 6: CALCULATING THE pH OF A DIPROTIC ACID

Diprotic acids may yield more than one hydrogen ion per molecule. These acids ionize in a stepwise manner, that is, they lose one proton at a time. An ionization constant expression can be written for each ionization stage. Consequently, two equilibrium constant expressions must be used to calculate the concentrations of species in the acid solution. For the generic diprotic acid, H_2A, we can write:

$$H_2A(aq) \rightleftharpoons H^+(aq) + HA^-(aq) \qquad K_{a_1} = \frac{[H^+][HA^-]}{[H_2A]}$$

$$HA^-(aq) \rightleftharpoons H^+(aq) + A^{2-}(aq) \qquad K_{a_2} = \frac{[H^+][A^{2-}]}{[HA^-]}$$

Note that the conjugate base in the first ionization becomes the acid in the second ionization.

EXAMPLE 15.8

Calculate the concentrations of H_2A, HA^-, H^+, and A^{2-} in a 1.0 M H_2A solution. The first and second ionization constants for H_2A are 1.3×10^{-2} and 6.3×10^{-8}, respectively.

Strategy: Determining the pH of a diprotic acid in aqueous solution is more involved than for a monoprotic acid. H_2A is a weak acid, and the conjugate base produced in the first ionization (HA^-) is also a weak acid. We follow the procedure for determining the pH of a weak acid for both stages.

Solution: We proceed according to the following steps.

Step 1: Complete a table showing the concentrations for the first ionization stage.

	$H_2A(aq)$	$\rightleftharpoons$	$H^+(aq)$	+	$HA^-(aq)$
Initial (M):	1.0		0		0
Change (M):	$-x$		$+x$		$+x$
Equilibrium (M):	$1.0 - x$		x		x

Step 2: Write the ionization constant expression for K_{a_1}. Then, solve for x.

$$K_{a_1} = \frac{[H^+][HA^-]}{[H_2A]}$$

$$1.3 \times 10^{-2} = \frac{(x)(x)}{(1.0 - x)}$$

Since K_{a_1} is quite large, we cannot make the assumption

$$1.0 - x \approx 1.0$$

Therefore, we must solve a quadratic equation.

$$x^2 + 0.013x - 0.013 = 0$$

The above equation is a quadratic equation of the form $ax^2 + bx + c = 0$. The solution for a quadratic equation is:

$$x = \frac{-b \pm \sqrt{b^2 - 4ac}}{2a}$$

Here, we have a = 1, b = 0.013, and c = −0.013. Substituting into the above equation,

$$x = \frac{-0.013 \pm \sqrt{(0.013)^2 - 4(1)(-0.013)}}{2(1)}$$

$$x = \frac{-0.013 \pm 0.23}{2}$$

$$x = 0.11\ M \quad \text{or} \quad x = -0.12\ M$$

The second solution is physically impossible because you cannot have a negative concentration. The first solution is the correct answer.

Step 3: Having solved for x, calculate the concentrations when the equilibrium for the first stage of ionization is reached.

Because $K_{a_1} >> K_{a_2}$, we assume that essentially all the H^+ comes from the first ionization stage. Hence,

$$\mathbf{[H^+] = [HA^-] = x = 0.11\ M}$$

$$\mathbf{[H_2A]} = 1.0 - x = 1.00 - 0.11 = \mathbf{0.89\ M}$$

Step 4: Now, consider the second stage of ionization. Set up a table showing the concentrations for the second ionization stage. Let y be the change in concentration.

	$HA^-(aq)$	$\rightleftharpoons$	$H^+(aq)$	+	$A^{2-}(aq)$
Initial (M):	0.11		0.11		0
Change (M):	$-y$		$+y$		$+y$
Equilibrium (M):	$0.11 - y$		$0.11 + y$		y

Step 5: Write the ionization constant expression for K_{a_2}. Then, solve for y.

$$K_{a_2} = \frac{[H^+][A^{2-}]}{[HA^-]}$$

$$6.3 \times 10^{-8} = \frac{(0.11 + y)(y)}{(0.11 - y)}$$

Since K_{a_2} is very small, we can make an assumption that y is very small compared to 0.11.

Hence,

$$0.11 \pm y \approx 0.11$$

$$6.3 \times 10^{-8} = \frac{(0.11)(y)}{0.11}$$

Solving for y,

$$\mathbf{y = 6.3 \times 10^{-8}\ M = [A^{2-}]}$$

Check: Checking the validity of the assumption,

$$\frac{6.3 \times 10^{-8}}{0.11} \times 100\% = 5.7 \times 10^{-5}\% < 5\%$$

The assumption was valid.

PRACTICE EXERCISE

8. The first and second ionization constants of H_2CO_3 are 4.2×10^{-7} and 4.8×10^{-11}, respectively. Calculate the concentrations of H^+, HCO_3^-, CO_3^{2-}, and unionized H_2CO_3 in a 0.080 M H_2CO_3 solution.

Text Problems: 15.62, 15.64

PROBLEM TYPE 7: CALCULATING THE pH OF A WEAK BASE

The procedure used to calculate the pH of a weak base solution is essentially the same as the one used for weak acids.

EXAMPLE 15.9

What is the pH of a 0.10 M C_5H_5N (pyridine) solution?

Strategy: First, recognize that pyridine is a weak base. It is an amine. Recall that a weak base only partially ionizes in water. We are given the initial quantity of a weak base (C_5H_5N) and asked to calculate the pH. To calculate pH, we can first determine the pOH from the OH^- concentration and then calculate pH from pOH. In determining the OH^- concentration, we ignore the ionization of H_2O as a source of OH^-, so the major source of OH^- ions is the base. We follow the procedure outlined in Section 15.6 of the text.

Solution:

Step 1: We ignore water's contribution to $[OH^-]$. We consider C_5H_5N as the only source of OH^- ions.

Step 2: Letting x be the equilibrium concentration of OH^- and $C_5H_5NH^+$ ions in mol/L, we summarize:

	$C_5H_5N(aq)$ + $H_2O(l)$ ⇌	$C_5H_5NH^+(aq)$ +	$OH^-(aq)$
Initial (M):	0.10	0	0
Change (M):	$-x$	$+x$	$+x$
Equilibrium (M):	$0.10 - x$	x	x

Step 3: Write the ionization constant expression in terms of the equilibrium concentrations. Knowing the value of the equilibrium constant (K_b), solve for x.

$$K_b = \frac{[C_5H_5NH^+][OH^-]}{[C_5H_5N]}$$

You can look up the K_b value for pyridine in Table 15.4 of the text.

$$1.7 \times 10^{-9} = \frac{(x)(x)}{(0.10 - x)}$$

At this point, we can make an assumption that x is very small compared to 0.10. Hence,

$$0.10 - x \approx 0.10$$

Oftentimes, assumptions such as these are valid if K is very small. A very small value of K means that a very small amount of reactants go to products. Hence, x is small. If we did not make this assumption, we would have to solve a quadratic equation.

$$1.7 \times 10^{-9} \approx \frac{(x)(x)}{0.10}$$

Solving for x,

$$x = 1.3 \times 10^{-5}\ M = [OH^-]$$

Step 4: Having solved for the $[OH^-]$, calculate the pOH of the solution. Then use the relationship, pH + pOH = 14, to solve for the pH of the solution.

$$\text{pOH} = -\log[OH^-] = -\log(1.3 \times 10^{-5}) = 4.89$$

$$\mathbf{pH} = 14.00 - \text{pOH} = 14.00 - 4.89 = \mathbf{9.11}$$

PRACTICE EXERCISE

9. What is the pH of a 2.0 M NH_3 solution?

Text Problem: 15.56

PROBLEM TYPE 8: PREDICTING THE ACID-BASE PROPERTIES OF SALT SOLUTIONS

Salts, when dissolved in water, can produce neutral, acidic, or basic solutions. We need to consider four possibilities.

1. **Salts that produce neutral solutions**. It is generally true that salts containing an alkali metal ion or an alkaline earth metal ion (except Be^{2+}) and the conjugate base of a strong acid (for example, Cl^-, NO_3^-, and ClO_4^-) do not undergo hydrolysis, and thus their solutions are neutral.

2. **Salts that produce basic solutions**. Salts that contain the conjugate base of a weak acid and an alkali metal or alkaline earth metal ion, produce basic solutions. As an example, consider potassium fluoride. The dissociation of KF in water is given by:

$$KF(s) \xrightarrow{H_2O} K^+(aq) + F^-(aq)$$

A hydrated alkali or alkaline earth metal ion has no acidic or basic properties. However, the conjugate base of a weak acid is a weak base (it has an affinity for H^+ ions). F^- is the conjugate base of the weak acid HF. The hydrolysis reaction is given by:

$$F^-(aq) + H_2O(l) \rightleftharpoons HF(aq) + OH^-(aq)$$

Because this reaction produces OH^- ions, the potassium fluoride solution will be basic.

3. **Salts that produce acidic solutions**. Salts that contain the conjugate acid of a weak base and the conjugate base of a strong acid, produce acidic solutions. As an example, consider ammonium iodide. The dissociation of NH_4I in water is given by

$$NH_4I(s) \xrightarrow{H_2O} NH_4^+(aq) + I^-(aq)$$

The conjugate base of a strong acid, such as I^-, does not undergo hydrolysis. However, the conjugate acid of a weak base is a weak acid and does undergo hydrolysis. The reaction is

$$NH_4^+(aq) + H_2O(l) \rightleftharpoons NH_3(aq) + H_3O^+(aq)$$

Since this reaction produces H_3O^+ ions, the ammonium iodide solution will be acidic.

4. **Salts in which both the cation and anion hydrolyze.** In these cases, you must compare the base strength of the anion to the acid strength of the cation. We consider three situations.

- $K_b > K_a$. If K_b for the anion is greater than K_a for the cation, the solution is basic because the anion will hydrolyze to a greater extent than the cation. At equilibrium, there will be more OH^- ions than H^+ ions.

- $K_b < K_a$. Conversely, if K_b for the anion is smaller than K_a for the cation, the solution will be acidic because cation hydrolysis will be more extensive than anion hydrolysis. At equilibrium, there will be more H^+ ions than OH^- ions.

- $K_b \approx K_a$. If K_b is approximately equal to K_a, the solution will be close to neutral.

EXAMPLE 15.10

Is a 0.10 *M* solution of Na_2CO_3 acidic, basic, or neutral?

Strategy: In deciding whether a salt will undergo hydrolysis, ask yourself the following questions: Is the cation a highly charged metal ion or the conjugate acid of a weak base? Is the anion the conjugate base of a weak acid? If yes to either question, then hydrolysis will occur. In cases where both the cation and the anion react with water, the pH of the solution will depend on the relative magnitudes of K_a for the cation and K_b for the anion (see Table 15.7 of the text).

Solution: We first break up the salt into its cation and anion components and then examine the possible reaction of each ion with water. The dissociation of Na_2CO_3 in water is given by

$$Na_2CO_3(s) \xrightarrow{H_2O} 2Na^+(aq) + CO_3^{2-}(aq)$$

A hydrated alkali or alkaline earth metal ion has no acidic or basic properties. However, the conjugate base of a weak acid is a weak base (it has an affinity for H^+ ions). CO_3^{2-} is the conjugate base of the weak acid HCO_3^-.

The hydrolysis reaction is given by

$$CO_3^{2-}(aq) + H_2O(l) \rightleftharpoons HCO_3^-(aq) + OH^-(aq)$$

Because this reaction produces OH^- ions, the sodium carbonate solution will be **basic**.

PRACTICE EXERCISE

10. Predict whether the following aqueous solutions will be acidic, basic, or neutral.

(a) KI
(b) NH_4Cl
(c) CH_3COOK

Text Problems: 15.76, 15.78, 15.80, 15.82

ANSWERS TO PRACTICE EXERCISES

1. **(a)** CH_3COO^- **(b)** HS^- **(c)** SO_3^{2-} **(d)** ClO^-

2. $[OH^-] = 1.5 \times 10^{-11}$ *M*

3. pOH = 3.14
pH = 10.86

4. $[H^+] = 3.5 \times 10^{-5}$ *M*

5. pH = 1.60

6. pH = 1.82

7. $K_a = 6.69 \times 10^{-5}$

8. $[H^+] = [HCO_3^-] = 1.8 \times 10^{-4}$ *M* $[H_2CO_3] = 0.080$ *M* $[CO_3^{2-}] = 4.8 \times 10^{-11}$ *M*

9. pH = 11.8

10. **(a)** neutral **(b)** acidic **(c)** basic

SOLUTIONS TO SELECTED TEXT PROBLEMS

15.4 Recall that the conjugate base of a Brønsted acid is the species that remains when *one* proton has been removed from the acid.

(a) nitrite ion: NO_2^-
(b) hydrogen sulfate ion (also called bisulfate ion): HSO_4^-
(c) hydrogen sulfide ion (also called bisulfide ion): HS^-
(d) cyanide ion: CN^-
(e) formate ion: $HCOO^-$

15.6 The conjugate acid of any base is just the base with a proton added.

(a) H_2S **(b)** H_2CO_3 **(c)** HCO_3^- **(d)** H_3PO_4 **(e)** $H_2PO_4^-$
(f) HPO_4^{2-} **(g)** H_2SO_4 **(h)** HSO_4^- **(i)** HSO_3^-

15.8 The conjugate base of any acid is simply the acid minus one proton.

(a) CH_2ClCOO^- **(b)** IO_4^- **(c)** $H_2PO_4^-$ **(d)** HPO_4^{2-} **(e)** PO_4^{3-}
(f) HSO_4^- **(g)** SO_4^{2-} **(h)** IO_3^- **(i)** SO_3^{2-} **(j)** NH_3
(k) HS^- **(l)** S^{2-} **(m)** OCl^-

15.16 $[OH^-] = 0.62\ M$

$$[H^+] = \frac{K_w}{[OH^-]} = \frac{1.0 \times 10^{-14}}{0.62} = \mathbf{1.6 \times 10^{-14}}\ M$$

15.18 **(a)** $Ba(OH)_2$ is ionic and fully ionized in water. The concentration of the hydroxide ion is $5.6 \times 10^{-4}\ M$ (Why? What is the concentration of Ba^{2+}?) We find the hydrogen ion concentration.

$$[H^+] = \frac{K_w}{[OH^-]} = \frac{1.0 \times 10^{-14}}{5.6 \times 10^{-4}} = 1.8 \times 10^{-11}\ M$$

The pH is then: $\mathbf{pH} = -\log[H^+] = -\log(1.8 \times 10^{-11}) = \mathbf{10.74}$

(b) Nitric acid is a strong acid, so the concentration of hydrogen ion is also $5.2 \times 10^{-4}\ M$. The pH is:

$$\mathbf{pH} = -\log[H^+] = -\log(5.2 \times 10^{-4}) = \mathbf{3.28}$$

15.20 For (a) and (b) we can calculate the H^+ concentration using the equation representing the definition of pH. Problem Type 3B.

Strategy: Here we are given the pH of a solution and asked to calculate $[H^+]$. Because pH is defined as $pH = -\log[H^+]$, we can solve for $[H^+]$ by taking the antilog of the pH; that is, $[H^+] = 10^{-pH}$.

Solution: From Equation (15.4) of the text:

(a) $pH = -\log [H^+] = 5.20$

$\log[H^+] = -5.20$

To calculate $[H^+]$, we need to take the antilog of −5.20.

$$[\mathbf{H^+}] = 10^{-5.20} = \mathbf{6.3 \times 10^{-6}}\ \boldsymbol{M}$$

Check: Because the pH is between 5 and 6, we can expect $[H^+]$ to be between 1×10^{-5} *M* and 1×10^{-6} *M*. Therefore, the answer is reasonable.

(b) $\text{pH} = -\log[H^+] = 16.00$

$$\log[H^+] = -16.00$$

$$[\mathbf{H^+}] = 10^{-16.00} = \mathbf{1.0 \times 10^{-16}}\ \boldsymbol{M}$$

(c) For part (c), it is probably easiest to calculate the $[H^+]$ from the ion product of water, Problem Type 2.

Strategy: We are given the concentration of OH^- ions and asked to calculate $[H^+]$. The relationship between $[H^+]$ and $[OH^-]$ in water or an aqueous solution is given by the ion-product of water, K_w [Equation (15.3) of the text].

Solution: The ion product of water is applicable to all aqueous solutions. At 25°C,

$$K_w = 1.0 \times 10^{-14} = [H^+][OH^-]$$

Rearranging the equation to solve for $[H^+]$, we write

$$[\mathbf{H^+}] = \frac{1.0 \times 10^{-14}}{[OH^-]} = \frac{1.0 \times 10^{-14}}{3.7 \times 10^{-9}} = \mathbf{2.7 \times 10^{-6}}\ \boldsymbol{M}$$

Check: Since the $[OH^-] < 1 \times 10^{-7}$ *M* we expect the $[H^+]$ to be greater than 1×10^{-7} *M*.

15.22 **(a)** acidic **(b)** neutral **(c)** basic

15.24 $$5.50\ \text{mL} \times \frac{1\ \text{L}}{1000\ \text{mL}} \times \frac{0.360\ \text{mol}}{1\ \text{L}} = \mathbf{1.98 \times 10^{-3}\ mol\ KOH}$$

KOH is a strong base and therefore ionizes completely. The OH^- concentration equals the KOH concentration, because there is a 1:1 mole ratio between KOH and OH^-.

$$[OH^-] = 0.360\ M$$

$$\mathbf{pOH} = -\log[OH^-] = \mathbf{0.444}$$

15.26 Molarity of the HCl solution is: $$\frac{18.4\ \text{g HCl} \times \dfrac{1\ \text{mol HCl}}{36.46\ \text{g HCl}}}{662 \times 10^{-3}\ \text{L}} = 0.762\ M$$

$$\mathbf{pH} = -\log(0.762) = \mathbf{0.118}$$

15.32 **(1)** The two steps in the ionization of a weak diprotic acid are:

$$H_2A(aq) + H_2O(l) \rightleftharpoons H_3O^+(aq) + HA^-(aq)$$

$$HA^-(aq) + H_2O(l) \rightleftharpoons H_3O^+(aq) + A^{2-}(aq)$$

The diagram that represents a weak diprotic acid is **(c)**. In this diagram, we only see the first step of the ionization, because HA^- is a much weaker acid than H_2A.

(2) Both **(b)** and **(d)** are chemically implausible situations. Because HA^- is a much weaker acid than H_2A, you would not see a higher concentration of A^{2-} compared to HA^-.

15.34 **(a)** strong base **(b)** weak base **(c)** weak base **(d)** weak base **(e)** strong base

15.36 **(a)** false, they are equal **(b)** true, find the value of log(1.00) on your calculator
(c) true **(d)** false, if the acid is strong, [HA] = 0.00 *M*

15.38 Cl^- is the conjugate base of the strong acid, HCl. It is a negligibly weak base and has no affinity for protons. Therefore, the reaction will *not* proceed from left to right to any measurable extent.

Another way to think about this problem is to consider the possible products of the reaction.

$$CH_3COOH(aq) + Cl^-(aq) \rightarrow HCl(aq) + CH_3COO^-(aq)$$

The favored reaction is the one that proceeds from right to left. HCl is a strong acid and will ionize completely, donating all its protons to the base, CH_3COO^-.

15.44 Ionization of a weak monoprotic acid, Problem Type 5A.

Strategy: Recall that a weak acid only partially ionizes in water. We are given the initial quantity of a weak acid (CH_3COOH) and asked to calculate the concentrations of H^+, CH_3COO^-, and CH_3COOH at equilibrium. First, we need to calculate the initial concentration of CH_3COOH. In determining the H^+ concentration, we ignore the ionization of H_2O as a source of H^+, so the major source of H^+ ions is the acid. We follow the procedure outlined in Section 15.5 of the text.

Solution:
Step 1: Calculate the concentration of acetic acid before ionization.

$$0.0560 \text{ g acetic acid} \times \frac{1 \text{ mol acetic acid}}{60.05 \text{ g acetic acid}} = 9.33 \times 10^{-4} \text{ mol acetic acid}$$

$$\frac{9.33 \times 10^{-4} \text{ mol}}{0.0500 \text{ L soln}} = 0.0187\ M \text{ acetic acid}$$

Step 2: We ignore water's contribution to $[H^+]$. We consider CH_3COOH as the only source of H^+ ions.

Step 3: Letting x be the equilibrium concentration of H^+ and CH_3COO^- ions in mol/L, we summarize:

	$CH_3COOH(aq)$	$\rightleftharpoons$ $H^+(aq)$	+ $CH_3COO^-(aq)$
Initial (*M*):	0.0187	0	0
Change (*M*):	$-x$	$+x$	$+x$
Equilibrium (*M*):	$0.0187 - x$	x	x

Step 4: Write the ionization constant expression in terms of the equilibrium concentrations. Knowing the value of the equilibrium constant (K_a), solve for x. You can look up the K_a value in Table 15.3 of the text.

$$K_a = \frac{[H^+][CH_3COO^-]}{[CH_3COOH]}$$

$$1.8 \times 10^{-5} = \frac{(x)(x)}{(0.0187 - x)}$$

At this point, we can make an assumption that x is very small compared to 0.0187. Hence,

$$0.0187 - x \approx 0.0187$$

$$1.8 \times 10^{-5} = \frac{(x)(x)}{0.0187}$$

$$\mathbf{x = 5.8 \times 10^{-4}\ M = [H^+] = [CH_3COO^-]}$$

$$\mathbf{[CH_3COOH]} = (0.0187 - 5.8 \times 10^{-4})M = \mathbf{0.0181\ M}$$

Check: Testing the validity of the assumption,

$$\frac{5.8 \times 10^{-4}}{0.0187} \times 100\% = 3.1\% < 5\%$$

The assumption is valid.

15.46 A pH of 3.26 corresponds to a $[H^+]$ of 5.5×10^{-4} M. Let the original concentration of formic acid be I. If the concentration of $[H^+]$ is 5.5×10^{-4} M, that means that 5.5×10^{-4} M of HCOOH ionized because of the 1:1 mole ratio between HCOOH and H^+.

	$HCOOH(aq)$	$\rightleftharpoons$	$H^+(aq)$	+	$HCOO^-(aq)$
Initial (M):	I		0		0
Change (M):	-5.5×10^{-4}		$+5.5 \times 10^{-4}$		$+5.5 \times 10^{-4}$
Equilibrium (M):	$I - (5.5 \times 10^{-4})$		5.5×10^{-4}		5.5×10^{-4}

Substitute K_a and the equilibrium concentrations into the ionization constant expression to solve for I.

$$K_a = \frac{[H^+][HCOO^-]}{[HCOOH]}$$

$$1.7 \times 10^{-4} = \frac{(5.5 \times 10^{-4})^2}{x - (5.5 \times 10^{-4})}$$

$$\mathbf{I = [HCOOH] = 2.3 \times 10^{-3}\ M}$$

15.48 Percent ionization is defined as:

$$\text{percent ionization} = \frac{\text{ionized acid concentration at equilibrium}}{\text{initial concentration of acid}} \times 100\%$$

For a monoprotic acid, HA, the concentration of acid that undergoes ionization is equal to the concentration of H^+ ions or the concentration of A^- ions at equilibrium. Thus, we can write:

$$\text{percent ionization} = \frac{[H^+]}{[HA]_0} \times 100\%$$

(a) First, recognize that hydrofluoric acid is a weak acid. It is not one of the six strong acids, so it must be a weak acid.

Step 1: Express the equilibrium concentrations of all species in terms of initial concentrations and a single unknown x, that represents the change in concentration. Let $(-x)$ be the depletion in concentration (mol/L) of HF. From the stoichiometry of the reaction, it follows that the increase in concentration for both H^+ and F^- must be x. Complete a table that lists the initial concentrations, the change in concentrations, and the equilibrium concentrations.

	$HF(aq)$	$\rightleftharpoons$	$H^+(aq)$	+	$F^-(aq)$
Initial (M):	0.60		0		0
Change (M):	$-x$		$+x$		$+x$
Equilibrium (M):	$0.60 - x$		x		x

Step 2: Write the ionization constant expression in terms of the equilibrium concentrations. Knowing the value of the equilibrium constant (K_a), solve for x.

$$K_a = \frac{[H^+][F^-]}{[HF]}$$

You can look up the K_a value for hydrofluoric acid in Table 15.3 of your text.

$$7.1 \times 10^{-4} = \frac{(x)(x)}{(0.60 - x)}$$

At this point, we can make an assumption that x is very small compared to 0.60. Hence,

$$0.60 - x \approx 0.60$$

Oftentimes, assumptions such as these are valid if K is very small. A very small value of K means that a very small amount of reactants go to products. Hence, x is small. If we did not make this assumption, we would have to solve a quadratic equation.

$$7.1 \times 10^{-4} = \frac{(x)(x)}{0.60}$$

Solving for x.

$$x = 0.021\ M = [H^+]$$

Step 3: Having solved for the $[H^+]$, calculate the percent ionization.

$$\textbf{percent ionization} = \frac{[H^+]}{[HF]_0} \times 100\%$$

$$= \frac{0.021\ M}{0.60\ M} \times 100\% = \mathbf{3.5\%}$$

(b) – (c) are worked in a similar manner to part (a). However, as the initial concentration of HF becomes smaller, the assumption that x is very small compared to this concentration will no longer be valid. You must solve a quadratic equation.

(b) $$K_a = \frac{[H^+][F^-]}{[HF]} = \frac{x^2}{(0.0046 - x)} = 7.1 \times 10^{-4}$$

$$x^2 + (7.1 \times 10^{-4})x - (3.3 \times 10^{-6}) = 0$$

$$x = 1.5 \times 10^{-3}\ M$$

$$\textbf{Percent ionization} = \frac{1.5 \times 10^{-3}\ M}{0.0046\ M} \times 100\% = \textbf{33\%}$$

(c) $$K_a = \frac{[H^+][F^-]}{[HF]} = \frac{x^2}{(0.00028 - x)} = 7.1 \times 10^{-4}$$

$$x^2 + (7.1 \times 10^{-4})x - (2.0 \times 10^{-7}) = 0$$

$$x = 2.2 \times 10^{-4}\ M$$

$$\textbf{Percent ionization} = \frac{2.2 \times 10^{-4}\ M}{0.00028\ M} \times 100\% = \textbf{79\%}$$

As the solution becomes more dilute, the percent ionization increases.

15.50 The equilibrium is:

	$C_9H_8O_4(aq)$	$\rightleftharpoons$ $H^+(aq)$	+ $C_9H_7O_4^-(aq)$
Initial (M):	0.20	0	0
Change (M):	$-x$	$+x$	$+x$
Equilibrium (M):	$0.20 - x$	x	x

(a) $$K_a = \frac{[H^+][C_9H_7O_4^-]}{[C_9H_8O_4]}$$

$$3.0 \times 10^{-4} = \frac{x^2}{(0.20 - x)}$$

Assuming $(0.20 - x) \approx 0.20$

$$x = [H^+] = 7.7 \times 10^{-3}\ M$$

$$\textbf{Percent ionization} = \frac{x}{0.20} \times 100\% = \frac{7.7 \times 10^{-3}\ M}{0.20\ M} \times 100\% = \textbf{3.9\%}$$

(b) At pH 1.00 the concentration of hydrogen ion is 0.10 M ($[H^+] = 10^{-pH}$). The extra hydrogen ions will tend to suppress the ionization of the weak acid (LeChâtelier's principle, Section 14.5 of the text). The position of equilibrium is shifted in the direction of the un-ionized acid. Let's set up a table of concentrations with the initial concentration of H^+ equal to 0.10 M.

	$C_9H_8O_4(aq)$	$\rightleftharpoons$ $H^+(aq)$ +	$C_9H_7O_4^-(aq)$
Initial (M):	0.20	0.10	0
Change (M):	$-x$	$+x$	$+x$
Equilibrium (M):	$0.20 - x$	$0.10 + x$	x

$$K_a = \frac{[H^+][C_9H_7O_4^-]}{[C_9H_8O_4]}$$

$$3.0 \times 10^{-4} = \frac{x(0.10 + x)}{(0.20 - x)}$$

Assuming $(0.20 - x) \approx 0.20$ and $(0.10 + x) \approx 0.10$

$$x = 6.0 \times 10^{-4}\ M$$

$$\textbf{Percent ionization} = \frac{x}{0.20} \times 100\% = \frac{6.0 \times 10^{-4}\ M}{0.20\ M} \times 100\% = \mathbf{0.30\%}$$

The high acidity of the gastric juices appears to enhance the rate of absorption of unionized aspirin molecules through the stomach lining. In some cases this can irritate these tissues and cause bleeding.

15.54 Similar to Problem Type 5B, Determining K_a from a pH measurement.

Strategy: Weak bases only partially ionize in water.

$$B(aq) + H_2O(l) \rightleftharpoons BH^+(aq) + OH^-(aq)$$

Note that the concentration of the weak base given refers to the initial concentration before ionization has started. The pH of the solution, on the other hand, refers to the situation at equilibrium. To calculate K_b, we need to know the concentrations of all three species, [B], $[BH^+]$, and $[OH^-]$ at equilibrium. We ignore the ionization of water as a source of OH^- ions.

Solution: We proceed as follows.

Step 1: The major species in solution are B, OH^-, and the conjugate acid BH^+.

Step 2: First, we need to calculate the hydroxide ion concentration from the pH value. Calculate the pOH from the pH. Then, calculate the OH^- concentration from the pOH.

$$\text{pOH} = 14.00 - \text{pH} = 14.00 - 10.66 = 3.34$$

$$\text{pOH} = -\log[OH^-]$$

$$-\text{pOH} = \log[OH^-]$$

Taking the antilog of both sides of the equation,

$$10^{-\text{pOH}} = [OH^-]$$

$$[OH^-] = 10^{-3.34} = 4.6 \times 10^{-4}\ M$$

Step 3: If the concentration of OH^- is 4.6×10^{-4} *M* at equilibrium, that must mean that 4.6×10^{-4} *M* of the base ionized. We summarize the changes.

	$B(aq)$ + $H_2O(l)$ $\rightleftharpoons$	$BH^+(aq)$ +	$OH^-(aq)$
Initial (*M*):	0.30	0	0
Change (*M*):	-4.6×10^{-4}	$+4.6 \times 10^{-4}$	$+4.6 \times 10^{-4}$
Equilibrium (*M*):	$0.30 - (4.6 \times 10^{-4})$	4.6×10^{-4}	4.6×10^{-4}

Step 4: Substitute the equilibrium concentrations into the ionization constant expression to solve for K_b.

$$K_b = \frac{[BH^+][OH^-]}{[B]}$$

$$\mathbf{K_b} = \frac{(4.6 \times 10^{-4})^2}{(0.30)} = \mathbf{7.1 \times 10^{-7}}$$

15.56 The reaction is:

	$NH_3(aq) + H_2O(l) \rightleftharpoons$	$NH_4^+(aq)$ +	$OH^-(aq)$
Initial (*M*):	0.080	0	0
Change (*M*):	$-x$	$+x$	$+x$
Equilibrium (*M*):	$0.080 - x$	x	x

At equilibrium we have:

$$K_a = \frac{[NH_4^+][OH^-]}{[NH_3]}$$

$$1.8 \times 10^{-5} = \frac{x^2}{(0.080 - x)} \approx \frac{x^2}{0.080}$$

$$x = 1.2 \times 10^{-3}\ M$$

$$\textbf{Percent } \mathbf{NH_3} \textbf{ present as } \mathbf{NH_4^+} = \frac{1.2 \times 10^{-3}}{0.080} \times 100\% = \mathbf{1.5\%}$$

15.62 The pH of a 0.040 *M* HCl solution (strong acid) is: pH = −log(0.040) = **1.40**. Follow the procedure for calculating the pH of a diprotic acid, Problem Type 6, to calculate the pH of the sulfuric acid solution.

Strategy: Determining the pH of a diprotic acid in aqueous solution is more involved than for a monoprotic acid. The first stage of ionization for H_2SO_4 goes to completion. We follow the procedure for determining the pH of a strong acid for this stage. The conjugate base produced in the first ionization (HSO_4^-) is a weak acid. We follow the procedure for determining the pH of a weak acid for this stage.

Solution: We proceed according to the following steps.

Step 1: H_2SO_4 is a strong acid. The first ionization stage goes to completion. The ionization of H_2SO_4 is

$$H_2SO_4(aq) \rightarrow H^+(aq) + HSO_4^-(aq)$$

The concentrations of all the species (H_2SO_4, H^+, and HSO_4^-) before and after ionization can be represented as follows.

	$H_2SO_4(aq)$ →	$H^+(aq)$ +	$HSO_4^-(aq)$
Initial (*M*):	0.040	0	0
Change (*M*):	−0.040	+0.040	+0.040
Final (*M*):	0	0.040	0.040

Step 2: Now, consider the second stage of ionization. HSO_4^- is a weak acid. Set up a table showing the concentrations for the second ionization stage. Let x be the change in concentration. Note that the initial concentration of H^+ is 0.040 *M* from the first ionization.

	$HSO_4^-(aq)$ ⇌	$H^+(aq)$ +	$SO_4^{2-}(aq)$
Initial (*M*):	0.040	0.040	0
Change (*M*):	$-x$	$+x$	$+x$
Equilibrium (*M*):	$0.040 - x$	$0.040 + x$	x

Write the ionization constant expression for K_a. Then, solve for x. You can find the K_a value in Table 15.5 of the text.

$$K_a = \frac{[H^+][SO_4^{2-}]}{[HSO_4^-]}$$

$$1.3 \times 10^{-2} = \frac{(0.040 + x)(x)}{(0.040 - x)}$$

Since K_a is quite large, we cannot make the assumptions that

$$0.040 - x \approx 0.040 \quad \text{and} \quad 0.040 + x \approx 0.040$$

Therefore, we must solve a quadratic equation.

$$x^2 + 0.053x - (5.2 \times 10^{-4}) = 0$$

$$x = \frac{-0.053 \pm \sqrt{(0.053)^2 - 4(1)(-5.2 \times 10^{-4})}}{2(1)}$$

$$x = \frac{-0.053 \pm 0.070}{2}$$

$$x = 8.5 \times 10^{-3}\ M \quad \text{or} \quad x = -0.062\ M$$

The second solution is physically impossible because you cannot have a negative concentration. The first solution is the correct answer.

Step 3: Having solved for x, we can calculate the H^+ concentration at equilibrium. We can then calculate the pH from the H^+ concentration.

$$[H^+] = 0.040\ M + x = [0.040 + (8.5 \times 10^{-3})]M = 0.049\ M$$

$$\mathbf{pH} = -\log(0.049) = \mathbf{1.31}$$

Without doing any calculations, could you have known that the pH of the sulfuric acid would be lower (more acidic) than that of the hydrochloric acid?

15.64 For the first stage of ionization:

	$H_2CO_3(aq)$	$\rightleftharpoons$	$H^+(aq)$	+	$HCO_3^-(aq)$
Initial (M):	0.025		0.00		0.00
Change (M):	$-x$		$+x$		$+x$
Equilibrium (M):	$(0.025 - x)$		x		x

$$K_{a_1} = \frac{[H^+][HCO_3^-]}{[H_2CO_3]}$$

$$4.2 \times 10^{-7} = \frac{x^2}{(0.025 - x)} \approx \frac{x^2}{0.025}$$

$$x = 1.0 \times 10^{-4}\ M$$

For the second ionization,

	$HCO_3^-(aq)$	$\rightleftharpoons$	$H^+(aq)$	+	$CO_3^{2-}(aq)$
Initial (M):	1.0×10^{-4}		1.0×10^{-4}		0.00
Change (M):	$-x$		$+x$		$+x$
Equilibrium (M):	$(1.0 \times 10^{-4}) - x$		$(1.0 \times 10^{-4}) + x$		x

$$K_{a_2} = \frac{[H^+][CO_3^{2-}]}{[HCO_3^-]}$$

$$4.8 \times 10^{-11} = \frac{[(1.0 \times 10^{-4}) + x](x)}{(1.0 \times 10^{-4}) - x} \approx \frac{(1.0 \times 10^{-4})(x)}{(1.0 \times 10^{-4})}$$

$$x = 4.8 \times 10^{-11}\ M$$

Since HCO_3^- is a very weak acid, there is little ionization at this stage. Therefore we have:

$$\mathbf{[H^+] = [HCO_3^-] = 1.0 \times 10^{-4}\ M} \text{ and } \mathbf{[CO_3^{2-}]} = x = \mathbf{4.8 \times 10^{-11}\ M}$$

15.68 All the listed pairs are oxoacids that contain different central atoms whose elements are in the same group of the periodic table and have the same oxidation number. In this situation the acid with the most electronegative central atom will be the strongest.

(a) $H_2SO_4 > H_2SeO_4$.

(b) $H_3PO_4 > H_3AsO_4$

15.70 The conjugate bases are $C_6H_5O^-$ from phenol and CH_3O^- from methanol. The $C_6H_5O^-$ is stabilized by resonance:

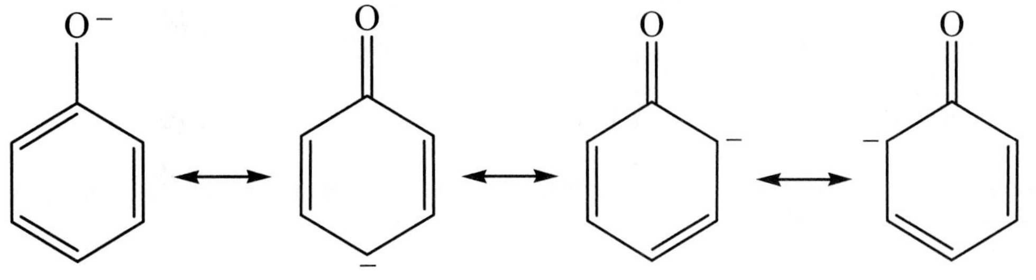

The CH_3O^- ion has no such resonance stabilization. A more stable conjugate base means an increase in the strength of the acid.

15.76 Predicting the Acid-Base Properties of Salt Solutions, Problem Type 8.

Strategy: In deciding whether a salt will undergo hydrolysis, ask yourself the following questions: Is the cation a highly charged metal ion or an ammonium ion? Is the anion the conjugate base of a weak acid? If yes to either question, then hydrolysis will occur. In cases where both the cation and the anion react with water, the pH of the solution will depend on the relative magnitudes of K_a for the cation and K_b for the anion (see Table 15.7 of the text).

Solution: We first break up the salt into its cation and anion components and then examine the possible reaction of each ion with water.

(a) The Na^+ cation does not hydrolyze. The Br^- anion is the conjugate base of the strong acid HBr. Therefore, Br^- will not hydrolyze either, and the solution is **neutral**.

(b) The K^+ cation does not hydrolyze. The SO_3^{2-} anion is the conjugate base of the weak acid HSO_3^- and will hydrolyze to give HSO_3^- and OH^-. The solution will be **basic**.

(c) Both the NH_4^+ and NO_2^- ions will hydrolyze. NH_4^+ is the conjugate acid of the weak base NH_3, and NO_2^- is the conjugate base of the weak acid HNO_2. From Tables 15.3 and 15.4 of the text, we see that the K_a of NH_4^+ (5.6×10^{-10}) is greater than the K_b of NO_2^- (2.2×10^{-11}). Therefore, the solution will be **acidic**.

(d) Cr^{3+} is a small metal cation with a high charge, which hydrolyzes to produce H^+ ions. The NO_3^- anion does not hydrolyze. It is the conjugate base of the strong acid, HNO_3. The solution will be **acidic**.

15.78 There is an inverse relationship between acid strength and conjugate base strength. As acid strength decreases, the proton accepting power of the conjugate base increases. In general the weaker the acid, the stronger the conjugate base. All three of the potassium salts ionize completely to form the conjugate base of the respective acid. The greater the pH, the stronger the conjugate base, and therefore, the weaker the acid.

The order of increasing acid strength is **HZ < HY < HX**.

15.80 The salt ammonium chloride completely ionizes upon dissolution, producing 0.42 M [NH_4^+] and 0.42 M [Cl^-] ions. NH_4^+ will undergo hydrolysis because it is a weak acid (NH_4^+ is the conjugate acid of the weak base, NH_3).

Step 1: Express the equilibrium concentrations of all species in terms of initial concentrations and a single unknown x, that represents the change in concentration. Let $(-x)$ be the depletion in concentration (mol/L) of NH_4^+. From the stoichiometry of the reaction, it follows that the increase in concentration for both H_3O^+ and NH_3 must be x. Complete a table that lists the initial concentrations, the change in concentrations, and the equilibrium concentrations.

	$NH_4^+(aq) + H_2O(l) \rightleftharpoons$	$NH_3(aq)$ +	$H_3O^+(aq)$
Initial (M):	0.42	0.00	0.00
Change (M):	$-x$	$+x$	$+x$
Equilibrium (M):	$(0.42 - x)$	x	x

Step 2: You can calculate the K_a value for NH_4^+ from the K_b value of NH_3. The relationship is

$$K_a \times K_b = K_w$$

or

$$K_a = \frac{K_w}{K_b} = \frac{1.0 \times 10^{-14}}{1.8 \times 10^{-5}} = 5.6 \times 10^{-10}$$

Step 3: Write the ionization constant expression in terms of the equilibrium concentrations. Knowing the value of the equilibrium constant (K_a), solve for x.

$$K_a = \frac{[NH_3][H_3O^+]}{[NH_4^+]}$$

$$5.6 \times 10^{-10} = \frac{x^2}{0.42 - x} \approx \frac{x^2}{0.42}$$

$$x = [H^+] = 1.5 \times 10^{-5}\ M$$

$$\mathbf{pH} = -\log(1.5 \times 10^{-5}) = \mathbf{4.82}$$

Since NH_4Cl is the salt of a weak base (aqueous ammonia) and a strong acid (HCl), we expect the solution to be slightly acidic, which is confirmed by the calculation.

15.82 The acid and base reactions are:

acid: $HPO_4^{2-}(aq) \rightleftharpoons H^+(aq) + PO_4^{3-}(aq)$

base: $HPO_4^{2-}(aq) + H_2O(l) \rightleftharpoons H_2PO_4^-(aq) + OH^-(aq)$

K_a for HPO_4^{2-} is 4.8×10^{-13}. Note that HPO_4^{2-} is the conjugate base of $H_2PO_4^-$, so K_b is 1.6×10^{-7}. Comparing the two K's, we conclude that the monohydrogen phosphate ion is a much stronger proton acceptor (base) than a proton donor (acid). The solution will be **basic**.

15.86 The most basic oxides occur with metal ions having the lowest positive charges (or lowest oxidation numbers).

(a) $Al_2O_3 < BaO < K_2O$ **(b)** $CrO_3 < Cr_2O_3 < CrO$

15.88 $Al(OH)_3$ is an amphoteric hydroxide. The reaction is:

$$Al(OH)_3(s) + OH^-(aq) \rightarrow Al(OH)_4^-(aq)$$

This is a Lewis acid-base reaction. Can you identify the acid and base?

15.92 $AlCl_3$ is a Lewis acid with an incomplete octet of electrons and Cl^- is the Lewis base donating a pair of electrons.

$$AlCl_3 + :\ddot{Cl}:^- \longrightarrow [AlCl_4]^-$$

15.94 By definition Brønsted acids are proton donors, therefore such compounds must contain at least one hydrogen atom. In Problem 15.91, Lewis acids that do not contain hydrogen, and therefore are not Brønsted acids, are CO_2, SO_2, and BCl_3. Can you name others?

15.96 We first find the number of moles of CO_2 produced in the reaction:

$$0.350 \text{ g NaHCO}_3 \times \frac{1 \text{ mol NaHCO}_3}{84.01 \text{ g NaHCO}_3} \times \frac{1 \text{ mol CO}_2}{1 \text{ mol NaHCO}_3} = 4.17 \times 10^{-3} \text{ mol CO}_2$$

$$V_{CO_2} = \frac{n_{CO_2}RT}{P} = \frac{(4.17 \times 10^{-3} \text{ mol})(0.0821 \text{ L}\cdot\text{atm/K}\cdot\text{mol})(37.0 + 273)\text{K}}{(1.00 \text{ atm})} = \mathbf{0.106 \text{ L}}$$

15.98 If we assume that the unknown monoprotic acid is a strong acid that is 100% ionized, then the $[H^+]$ concentration will be 0.0642 *M*.

$$\text{pH} = -\log(0.0642) = 1.19$$

Since the actual pH of the solution is higher, the acid must be a weak acid.

15.100 The reaction of a weak acid with a strong base is driven to completion by the formation of water. Irrespective of whether the strong base is reacting with a strong monoprotic acid or a weak monoprotic acid, the same number of moles of acid is required to react with a constant number of moles of base. Therefore the volume of base required to react with the same concentration of acid solutions (either both weak, both strong, or one strong and one weak) will be the same.

15.102 High oxidation state leads to covalent compounds and low oxidation state leads to ionic compounds. Therefore, CrO is ionic and basic and CrO_3 is covalent and acidic.

15.104 We can write two equilibria that add up to the equilibrium in the problem.

$$CH_3COOH(aq) \rightleftharpoons H^+(aq) + CH_3COO^-(aq) \qquad K_a = \frac{[H^+][CH_3COO^-]}{[CH_3COOH]} = 1.8 \times 10^{-5}$$

$$H^+(aq) + NO_2^-(aq) \rightleftharpoons HNO_2(aq) \qquad K_a' = \frac{1}{K_a(HNO_2)} = \frac{1}{4.5 \times 10^{-4}} = 2.2 \times 10^3$$

$$K_a' = \frac{[HNO_2]}{[H^+][NO_2^-]}$$

$$CH_3COOH(aq) + NO_2^-(aq) \rightleftharpoons CH_3COO^-(aq) + HNO_2(aq) \qquad K = \frac{[CH_3COO^-][HNO_2]}{[CH_3COOH][NO_2^-]} = K_a \times K_a'$$

The equilibrium constant for this sum is the product of the equilibrium constants of the component reactions.

$$\boldsymbol{K} = K_a \times K_a' = (1.8 \times 10^{-5})(2.2 \times 10^3) = \mathbf{4.0 \times 10^{-2}}$$

15.106 In this specific case the K_a of ammonium ion is the same as the K_b of acetate ion [$K_a(NH_4^+) = 5.6 \times 10^{-10}$, $K_b(CH_3COO^-) = 5.6 \times 10^{-10}$]. The two are of exactly (to two significant figures) equal strength. The solution will have **pH 7.00**.

What would the pH be if the concentration were 0.1 *M* in ammonium acetate? 0.4 *M*?

15.108 The fact that fluorine attracts electrons in a molecule more strongly than hydrogen should cause NF_3 to be a poor electron pair donor and a poor base. **NH_3** is the stronger base.

15.110 The autoionization for deuterium-substituted water is: $D_2O \rightleftharpoons D^+ + OD^-$

$$[D^+][OD^-] = 1.35 \times 10^{-15} \qquad (1)$$

(a) The definition of pD is: $\mathbf{pD} = -\log[D^+] = -\log\sqrt{1.35 \times 10^{-15}} = \mathbf{7.43}$

(b) To be acidic, the **pD** must be < **7.43**.

(c) Taking –log of both sides of equation (1) above:

$$-\log[D^+] + -\log[OD^-] = -\log(1.35 \times 10^{-15})$$

$$\mathbf{pD + pOD = 14.87}$$

15.112 First we must calculate the molarity of the trifluoromethane sulfonic acid. (Molar mass = 150.1 g/mol)

$$\text{Molarity} = \frac{0.616\ \cancel{g} \times \frac{1\ \text{mol}}{150.1\ \cancel{g}}}{0.250\ \text{L}} = 0.0164\ M$$

Since trifluoromethane sulfonic acid is a strong acid and is 100% ionized, the $[H^+]$ is 0.0165 M.

$$\mathbf{pH} = -\log(0.0164) = \mathbf{1.79}$$

15.114 The reactions are

$$HF \rightleftharpoons H^+ + F^- \qquad (1)$$

$$F^- + HF \rightleftharpoons HF_2^- \qquad (2)$$

Note that for equation (2), the equilibrium constant is relatively large with a value of 5.2. This means that the equilibrium lies to the right. Applying Le Châtelier's principle, as HF ionizes in the first step, the F^- that is produced is partially removed in the second step. More HF must ionize to compensate for the removal of the F^-, at the same time producing more H^+.

15.116 (a) We must consider both the complete ionization of the strong acid, and the partial ionization of water.

$$HA \longrightarrow H^+ + A^-$$

$$H_2O \rightleftharpoons H^+ + OH^-$$

From the above two equations, the $[H^+]$ in solution is:

$$[H^+] = [A^-] + [OH^-] \qquad (1)$$

We can also write:

$$[H^+][OH^-] = K_w$$

$$[OH^-] = \frac{K_w}{[H^+]}$$

Substituting into Equation (1):

$$[H^+] = [A^-] + \frac{K_w}{[H^+]}$$

$$[H^+]^2 = [A^-][H^+] + K_w$$

$$[H^+]^2 - [A^-][H^+] - K_w = 0$$

Solving a quadratic equation:

$$\mathbf{[H^+]} = \frac{\mathbf{[A^-] \pm \sqrt{[A^-]^2 + 4K_w}}}{\mathbf{2}}$$

(b) For the strong acid, HCl, with a concentration of 1.0×10^{-7} *M*, the $[Cl^-]$ will also be 1.0×10^{-7} *M*.

$$[H^+] = \frac{[Cl^-] \pm \sqrt{[Cl^-]^2 + 4K_w}}{2} = \frac{1 \times 10^{-7} \pm \sqrt{(1 \times 10^{-7})^2 + 4(1 \times 10^{-14})}}{2}$$

$[H^+] = 1.6 \times 10^{-7}\ M$ (or -6.0×10^{-8} *M*, which is impossible)

$\mathbf{pH} = -\log[1.6 \times 10^{-7}] = \mathbf{6.80}$

15.118 The solution for the first step is standard:

	$H_3PO_4(aq)$	$\rightleftharpoons$ $H^+(aq)$	+ $H_2PO_4^-(aq)$
Initial (*M*):	0.100	0.000	0.000
Change (*M*):	$-x$	$+x$	$+x$
Equil. (*M*):	$(0.100 - x)$	x	x

$$K_{a_1} = \frac{[H^+][H_2PO_4^-]}{[H_3PO_4]}$$

$$7.5 \times 10^{-3} = \frac{x^2}{(0.100 - x)}$$

In this case we probably cannot say that $(0.100 - x) \approx 0.100$ due to the magnitude of K_a. We obtain the quadratic equation:

$$x^2 + (7.5 \times 10^{-3})x - (7.5 \times 10^{-4}) = 0$$

The positive root is $x = 0.0239\ M$. We have:

$$[H^+] = [H_2PO_4^-] = 0.0239\ M$$

$$[H_3PO_4] = (0.100 - 0.0239)\ M = 0.076\ M$$

For the second ionization:

	$H_2PO_4^-(aq)$	$\rightleftharpoons$	$H^+(aq)$	+	$HPO_4^{2-}(aq)$
Initial (M):	0.0239		0.0239		0.000
Change (M):	$-y$		$+y$		$+y$
Equil (M):	$(0.0239 - y)$		$(0.0239 + y)$		y

$$K_{a_2} = \frac{[H^+][HPO_4^{2-}]}{[H_2PO_4^-]}$$

$$6.2 \times 10^{-8} = \frac{(0.0239 + y)(y)}{(0.0239 - y)} \approx \frac{(0.0239)(y)}{(0.0239)}$$

$$y = 6.2 \times 10^{-8}\ M.$$

Thus,

$$[H^+] = [H_2PO_4^-] = 0.0239\ M$$

$$[HPO_4^{2-}] = y = 6.2 \times 10^{-8}\ M$$

We set up the problem for the third ionization in the same manner.

	$HPO_4^{2-}(aq)$	$\rightleftharpoons$	$H^+(aq)$	+	$PO_4^{3-}(aq)$
Initial (M):	6.2×10^{-8}		0.0239		0
Change (M):	$-z$		$+z$		$+z$
Equil. (M):	$(6.2 \times 10^{-8}) - z$		$0.0239 + z$		z

$$K_{a_3} = \frac{[H^+][PO_4^{3-}]}{[HPO_4^{2-}]}$$

$$4.8 \times 10^{-13} = \frac{(0.0239 + z)(z)}{(6.2 \times 10^{-8}) - z} \approx \frac{(0.239)(z)}{(6.2 \times 10^{-8})}$$

$$z = 1.2 \times 10^{-18}\ M$$

The equilibrium concentrations are:

$$\mathbf{[H^+] = [H_2PO_4^-] = 0.0239\ M}$$

$$\mathbf{[H_3PO_4] = 0.076\ M}$$

$$\mathbf{[HPO_4^{2-}] = 6.2 \times 10^{-8}\ M}$$

$$\mathbf{[PO_4^{3-}] = 1.2 \times 10^{-18}\ M}$$

15.120 $0.100\ M\ Na_2CO_3 \rightarrow 0.200\ M\ Na^+ + 0.100\ M\ CO_3^{2-}$

First stage:

	$CO_3^{2-}(aq) + H_2O(l)$	$\rightleftharpoons$	$HCO_3^-(aq)$	+	$OH^-(aq)$
Initial (M):	0.100		0		0
Change (M):	$-x$		$+x$		$+x$
Equilibrium (M):	$0.100 - x$		x		x

$$K_1 = \frac{K_w}{K_2} = \frac{1.0 \times 10^{-14}}{4.8 \times 10^{-11}} = 2.1 \times 10^{-4}$$

$$K_1 = \frac{[HCO_3^-][OH^-]}{[CO_3^{2-}]}$$

$$2.1 \times 10^{-4} = \frac{x^2}{0.100 - x} \approx \frac{x^2}{0.100}$$

$$x = 4.6 \times 10^{-3}\ M = [HCO_3^-] = [OH^-]$$

<u>Second stage:</u>

	$HCO_3^-(aq) + H_2O(l) \rightleftharpoons$	$H_2CO_3(aq)$	$+\ OH^-(aq)$
Initial (M):	4.6×10^{-3}	0	4.6×10^{-3}
Change (M):	$-y$	$+y$	$+y$
Equilibrium (M):	$(4.6 \times 10^{-3}) - y$	y	$(4.6 \times 10^{-3}) + y$

$$K_2 = \frac{[H_2CO_3][OH^-]}{[HCO_3^-]}$$

$$2.4 \times 10^{-8} = \frac{y[(4.6 \times 10^{-3}) + y]}{(4.6 \times 10^{-3}) - y} \approx \frac{(y)(4.6 \times 10^{-3})}{(4.6 \times 10^{-3})}$$

$$y = 2.4 \times 10^{-8}\ M$$

At equilibrium:

$$\mathbf{[Na^+] = 0.200\ M}$$

$$\mathbf{[HCO_3^-]} = (4.6 \times 10^{-3})\ M - (2.4 \times 10^{-8})\ M \approx \mathbf{4.6 \times 10^{-3}\ M}$$

$$\mathbf{[H_2CO_3] = 2.4 \times 10^{-8}\ M}$$

$$\mathbf{[OH^-]} = (4.6 \times 10^{-3})\ M + (2.4 \times 10^{-8}]\ M \approx \mathbf{4.6 \times 10^{-3}\ M}$$

$$\mathbf{[H^+]} = \frac{1.0 \times 10^{-14}}{4.6 \times 10^{-3}} = \mathbf{2.2 \times 10^{-12}\ M}$$

15.122 When NaCN is treated with HCl, the following reaction occurs.

$$NaCN + HCl \rightarrow NaCl + HCN$$

HCN is a very weak acid, and only partially ionizes in solution.

$$HCN(aq) \rightleftharpoons H^+(aq) + CN^-(aq)$$

The main species in solution is HCN which has a tendency to escape into the gas phase.

$$HCN(aq) \rightleftharpoons HCN(g)$$

Since the HCN(g) that is produced is a highly poisonous compound, it would be dangerous to treat NaCN with acids without proper ventilation.

15.124 pH = 2.53 = $-\log[H^+]$

$[H^+] = 2.95 \times 10^{-3}\ M$

Since the concentration of H^+ at equilibrium is $2.95 \times 10^{-3}\ M$, that means that $2.95 \times 10^{-3}\ M$ HCOOH ionized. Let' represent the initial concentration of HCOOH as *I*. The equation representing the ionization of formic acid is:

	$HCOOH(aq)$	$\rightleftharpoons$	$H^+(aq)$	+	$HCOO^-(aq)$
Initial (*M*):	I		0		0
Change (*M*):	-2.95×10^{-3}		$+2.95 \times 10^{-3}$		$+2.95 \times 10^{-3}$
Equilibrium (*M*):	$I - (2.95 \times 10^{-3})$		2.95×10^{-3}		2.95×10^{-3}

$$K_a = \frac{[H^+][HCOO^-]}{[HCOOH]}$$

$$1.7 \times 10^{-4} = \frac{(2.95 \times 10^{-3})^2}{I - (2.95 \times 10^{-3})}$$

$$I = 0.054\ M$$

There are 0.054 moles of formic acid in 1000 mL of solution. The mass of formic acid in 100 mL is:

$$100\ \text{mL} \times \frac{0.054\ \text{mol formic acid}}{1000\ \text{mL soln}} \times \frac{46.03\ \text{g formic acid}}{1\ \text{mol formic acid}} = \mathbf{0.25\ g\ formic\ acid}$$

15.126 The balanced equation is: $Mg + 2HCl \rightarrow MgCl_2 + H_2$

$$\text{mol of Mg} = 1.87\ \text{g Mg} \times \frac{1\ \text{mol Mg}}{24.31\ \text{g Mg}} = 0.0769\ \text{mol}$$

From the balanced equation:

mol of HCl required for reaction = 2 × mol Mg = (2)(0.0769 mol) = 0.154 mol HCl

The concentration of HCl:

pH = −0.544, thus $[H^+] = 3.50\ M$

initial mol HCl = $M \times$ Vol (L) = (3.50 *M*)(0.0800 L) = 0.280 mol HCl

Moles of HCl left after reaction:

initial mol HCl − mol HCl reacted = 0.280 mol − 0.154 mol = 0.126 mol HCl

Molarity of HCl left after reaction:

M = mol/L = 0.126 mol/0.080 L = 1.58 *M*

pH = −log(1.58) = **−0.20**

15.128 The important equation is the hydrolysis of NO_2^-: $NO_2^- + H_2O \rightleftharpoons HNO_2 + OH^-$

(a) Addition of HCl will result in the reaction of the H^+ from the HCl with the OH^- that was present in the solution. The OH^- will effectively be removed and the equilibrium will **shift to the right** to compensate (more hydrolysis).

(b) Addition of NaOH is effectively addition of more OH^- which places stress on the right hand side of the equilibrium. The equilibrium will **shift to the left** (less hydrolysis) to compensate for the addition of OH^-.

(c) Addition of NaCl will have **no effect**.

(d) Recall that the percent ionization of a weak acid increases with dilution (see Figure 15.4 of the text). The same is true for weak bases. Thus dilution will cause more hydrolysis, shifting the equilibrium to the **right**.

15.130 In Chapter 11, we found that salts with their formal electrostatic intermolecular attractions had low vapor pressures and thus high boiling points. Ammonia and its derivatives (amines) are molecules with dipole-dipole attractions; as long as the nitrogen has one direct N–H bond, the molecule will have hydrogen bonding. Even so, these molecules will have much higher vapor pressures than ionic species. Thus, if we could convert the neutral ammonia-type molecules into salts, their vapor pressures, and thus associated odors, would decrease. Lemon juice contains acids which can react with neutral ammonia-type (amine) molecules to form ammonium salts.

$$NH_3 + H^+ \rightarrow NH_4^+$$

$$RNH_2 + H^+ \rightarrow RNH_3^+$$

15.132

	$HCOOH$	$\rightleftharpoons$ H^+	+ $HCOO^-$
Initial (M):	0.400	0	0
Change (M):	$-x$	$+x$	$+x$
Equilibrium (M):	$0.400 - x$	x	x

Total concentration of particles in solution: $(0.400 - x) + x + x = 0.400 + x$

Assuming the molarity of the solution is equal to the molality, we can write:

$$\Delta T_f = K_f m$$

$$0.758 = (1.86)(0.400 + x)$$

$$x = 0.00753 = [H^+] = [HCOO^-]$$

$$\boldsymbol{K_a} = \frac{[H^+][HCOO^-]}{[HCOOH]} = \frac{(0.00753)(0.00753)}{0.400 - 0.00753} = \mathbf{1.4 \times 10^{-4}}$$

15.134 $SO_2(g) + H_2O(l) \rightleftharpoons H^+(aq) + HSO_3^-(aq)$

Recall that 0.12 ppm SO_2 would mean 0.12 parts SO_2 per 1 million (10^6) parts of air by volume. The number of particles of SO_2 per volume will be directly related to the pressure.

$$P_{SO_2} = \frac{0.12 \text{ parts } SO_2}{10^6 \text{ parts air}} \text{ atm} = 1.2 \times 10^{-7} \text{ atm}$$

We can now calculate the $[H^+]$ from the equilibrium constant expression.

$$K = \frac{[H^+][HSO_3^-]}{P_{SO_2}}$$

$$1.3 \times 10^{-2} = \frac{x^2}{1.2 \times 10^{-7}}$$

$$x^2 = (1.3 \times 10^{-2})(1.2 \times 10^{-7})$$

$$x = 3.9 \times 10^{-5}\ M = [H^+]$$

$$\mathbf{pH} = -\log(3.9 \times 10^{-5}) = \mathbf{4.40}$$

15.136 In inhaling the smelling salt, some of the powder dissolves in the basic solution. The ammonium ions react with the base as follows:

$$NH_4^+(aq) + OH^-(aq) \rightarrow NH_3(aq) + H_2O$$

It is the pungent odor of ammonia that prevents a person from fainting.

15.138 **(c)** does not represent a Lewis acid-base reaction. In this reaction, the F–F single bond is broken and single bonds are formed between P and each F atom. For a Lewis acid-base reaction, the Lewis acid is an electron-pair acceptor and the Lewis base is an electron-pair donor.

15.140 From the given pH's, we can calculate the $[H^+]$ in each solution.

Solution (1): $[H^+] = 10^{-pH} = 10^{-4.12} = 7.6 \times 10^{-5}\ M$
Solution (2): $[H^+] = 10^{-5.76} = 1.7 \times 10^{-6}\ M$
Solution (3): $[H^+] = 10^{-5.34} = 4.6 \times 10^{-6}\ M$

We are adding solutions (1) and (2) to make solution (3). The volume of solution (2) is 0.528 L. We are going to add a given volume of solution (1) to solution (2). Let's call this volume x. The moles of H^+ in solutions (1) and (2) will equal the moles of H^+ in solution (3).

$$\text{mol } H^+ \text{ soln (1)} + \text{mol } H^+ \text{ soln (2)} = \text{mol } H^+ \text{ soln (3)}$$

Recall that mol = $M \times$ L. We have:

$$(7.6 \times 10^{-5}\text{ mol/L})(x\text{ L}) + (1.7 \times 10^{-6}\text{ mol/L})(0.528\text{ L}) = (4.6 \times 10^{-6}\text{ mol/L})(0.528 + x)\text{L}$$

$$(7.6 \times 10^{-5})x + (9.0 \times 10^{-7}) = (2.4 \times 10^{-6}) + (4.6 \times 10^{-6})x$$

$$(7.1 \times 10^{-5})x = 1.5 \times 10^{-6}$$

$$x = 0.021\text{ L} = \mathbf{21\ mL}$$

15.142 The balanced equations for the two reactions are:

$$MCO_3(s) + 2HCl(aq) \longrightarrow MCl_2(aq) + CO_2(g) + H_2O(l)$$

$$HCl(aq) + NaOH(aq) \longrightarrow NaCl(aq) + H_2O(l)$$

First, let's find the number of moles of excess acid from the reaction with NaOH.

$$0.03280\ \cancel{L} \times \frac{0.588\ \cancel{mol}\ NaOH}{1\ \cancel{L}\ soln} \times \frac{1\ mol\ HCl}{1\ \cancel{mol}\ NaOH} = 0.0193\ mol\ HCl$$

The original number of moles of acid was:

$$0.500\ \cancel{L} \times \frac{0.100\ mol\ HCl}{1\ \cancel{L}\ soln} = 0.0500\ mol\ HCl$$

The amount of hydrochloric acid that reacted with the metal carbonate is:

$$(0.0500\ mol\ HCl) - (0.0193\ mol\ HCl) = 0.0307\ mol\ HCl$$

The mole ratio from the balanced equation is 1 mole MCO_3 : 2 mole HCl. The moles of MCO_3 that reacted are:

$$0.0307\ \cancel{mol}\ HCl \times \frac{1\ mol\ MCO_3}{2\ \cancel{mol}\ HCl} = 0.01535\ mol\ MCO_3$$

We can now determine the molar mass of MCO_3, which will allow us to identify the metal.

$$\text{molar mass } MCO_3 = \frac{1.294\ g\ MCO_3}{0.01535\ mol\ MCO_3} = 84.3\ g/mol$$

We subtract off the mass of CO_3^{2-} to identify the metal.

$$\text{molar mass M} = 84.3\ g/mol - 60.01\ g/mol = 24.3\ g/mol$$

The metal is **magnesium**.

15.144 Because HF is a much stronger acid than HCN, we can assume that the pH is largely determined by the ionization of HF.

	$HF(aq) + H_2O(l) \rightleftharpoons$	$H_3O^+(aq)$	$+ F^-(aq)$
Initial (M):	1.00	0	0
Change (M):	$-x$	$+x$	$+x$
Equilibrium (M):	$1.00 - x$	x	x

$$K_a = \frac{[H_3O^+][F^-]}{[HF]}$$

$$7.1 \times 10^{-4} = \frac{x^2}{1.00 - x} \approx \frac{x^2}{1.00}$$

$$x = 0.027\ M = [H_3O^+]$$

$$\mathbf{pH = 1.57}$$

HCN is a very weak acid, so at equilibrium, $[HCN] \approx 1.00\ M$.

$$K_a = \frac{[H_3O^+][CN^-]}{[HCN]}$$

$$4.9 \times 10^{-10} = \frac{(0.027)[\text{CN}^-]}{1.00}$$

$$\mathbf{[CN^-] = 1.8 \times 10^{-8}\ \mathit{M}}$$

In a 1.00 *M* HCN solution, the concentration of $[\text{CN}^-]$ would be:

	$\text{HCN}(aq) + \text{H}_2\text{O}(l) \rightleftharpoons$	$\text{H}_3\text{O}^+(aq)$ +	$\text{CN}^-(aq)$
Initial (*M*):	1.00	0	0
Change (*M*):	$-x$	$+x$	$+x$
Equilibrium (*M*):	$1.00 - x$	x	x

$$K_a = \frac{[\text{H}_3\text{O}^+][\text{CN}^-]}{[\text{HCN}]}$$

$$4.9 \times 10^{-10} = \frac{x^2}{1.00 - x} \approx \frac{x^2}{1.00}$$

$$x = \mathbf{2.2 \times 10^{-5}\ \mathit{M} = [CN^-]}$$

$[\text{CN}^-]$ is greater in the 1.00 *M* HCN solution compared to the 1.00 *M* HCN/1.00 *M* HF solution. According to LeChâtelier's principle, the high $[\text{H}_3\text{O}^+]$ (from HF) shifts the HCN equilibrium from right to left decreasing the ionization of HCN. The result is a smaller $[\text{CN}^-]$ in the presence of HF.

15.146 The van't Hoff equation allows the calculation of an equilibrium constant at a different temperature if the value of the equilibrium constant at another temperature and $\Delta H°$ for the reaction are known.

$$\ln\frac{K_1}{K_2} = \frac{\Delta H°}{R}\left(\frac{1}{T_2} - \frac{1}{T_1}\right)$$

First, we calculate $\Delta H°$ for the ionization of water using data in Appendix 3 of the text.

$$\text{H}_2\text{O}(l) \rightleftharpoons \text{H}^+(aq) + \text{OH}^-(aq)$$

$$\Delta H° = [\Delta H_f°(\text{H}^+) + \Delta H_f°(\text{OH}^-)] - \Delta H_f°(\text{H}_2\text{O})$$

$$\Delta H° = (0 - 229.94\ \text{kJ/mol}) - (-285.8\ \text{kJ/mol})$$

$$\Delta H° = 55.9\ \text{kJ/mol}$$

We substitute $\Delta H°$ and the equilibrium constant at 25°C (298 K) into the van't Hoff equation to solve for the equilibrium constant at 100°C (373 K).

$$\ln\frac{1.0 \times 10^{-14}}{K_2} = \frac{55.9 \times 10^3\ \text{J/mol}}{8.314\ \text{J/mol}\cdot\text{K}}\left(\frac{1}{373\ \text{K}} - \frac{1}{298\ \text{K}}\right)$$

$$\frac{1.0 \times 10^{-14}}{K_2} = e^{-4.537}$$

$$K_2 = 9.3 \times 10^{-13}$$

We substitute into the equilibrium constant expression for the ionization of water to solve for $[H^+]$ and then pH.

$$K_2 = [H^+][OH^-]$$

$$9.3 \times 10^{-13} = x^2$$

$$x = [H^+] = 9.6 \times 10^{-7}\ M$$

$$\mathbf{pH} = -\log(9.6 \times 10^{-7}) = \mathbf{6.02}$$

Note that the water is **not** acidic at 100°C because $[H^+] = [OH^-]$.

CHAPTER 16
ACID-BASE EQUILIBRIA AND SOLUBILITY EQUILIBRIA

PROBLEM-SOLVING STRATEGIES AND TUTORIAL SOLUTIONS

TYPES OF PROBLEMS

Problem Type 1: Buffers.
- **(a)** Identifying buffer systems.
- **(b)** Calculating the pH of a buffer system.
- **(c)** Preparing a buffer solution with a specific pH.

Problem Type 2: Titrations.
- **(a)** Strong acid–strong base titrations.
- **(b)** Weak acid–strong base titrations.
- **(c)** Strong acid–weak base titrations.

Problem Type 3: Choosing Suitable Acid-Base Indicators.

Problem Type 4: Solubility Equilibria.
- **(a)** Calculating K_{sp} from molar solubility.
- **(b)** Calculating solubility from K_{sp}.
- **(c)** Predicting a precipitation reaction.
- **(d)** The effect of a common ion on solubility.

Problem Type 5: Complex Ion Equilibria and Solubility.

PROBLEM TYPE 1: BUFFERS

A **buffer solution** is a solution of (1) a weak acid or a weak base and (2) its salt; both components must be present. The solution has the ability to resist change in pH upon the addition of small amounts of either acid or base. A buffer resists change in pH, because the weak acid component reacts with small amounts of added base. The weak base component of the buffer reacts with small amounts of added acid.

$$\underset{\text{weak acid of buffer}}{HA(aq)} + \underset{\text{added base}}{OH^-(aq)} \longrightarrow A^-(aq) + H_2O(l)$$

$$\underset{\text{weak base of buffer}}{A^-(aq)} + \underset{\text{added acid}}{H_3O^+(aq)} \longrightarrow HA(aq) + H_2O(l)$$

A. Identifying buffer systems

To identify a buffer system, look for a weak acid and its conjugate base (usually the anion in a soluble salt) or a weak base and its conjugate acid (usually the cation in a soluble salt).

EXAMPLE 16.1

Which of the following solutions are buffer systems: (a) CH_3COONa/CH_3COOH, (b) KNO_3/HNO_3, and (c) NH_3/NH_4Cl?

Strategy: What constitutes a buffer system? Which of the preceding solutions contains a weak acid and its salt (containing the weak conjugate base)? Which of the preceding solutions contains a weak base and its salt (containing the weak conjugate acid)? Why is the conjugate base of a strong acid not able to neutralize an added acid?

Solution: The criteria for a buffer system are that we must have a weak acid and its salt (containing the weak conjugate base) or a weak base and its salt (containing the weak conjugate acid).

(a) CH_3COOH (acetic acid) is a weak acid, and its conjugate base, CH_3COO^- (acetate ion, the anion of the salt CH_3COONa), is a weak base. Therefore, this is a buffer system.

(b) Because HNO_3 is a strong acid, its conjugate base, NO_3^-, is an extremely weak base. This means that NO_3^- will not combine with H^+ in solution to form HNO_3. Thus, the system cannot act as a buffer system.

(c) NH_3 (ammonia) is a weak base and its conjugate acid, NH_4^+ (ammonium ion, the cation of the salt NH_4Cl), is a weak acid. Therefore, this is a buffer system.

PRACTICE EXERCISE

1. Which of the following solutions are buffer systems: (a) KCN/HCN, (b) NaCl/HCl, and (c) KNO_2/HNO_2?

Text Problem: 16.10

B. Calculating the pH of a buffer system

Calculating the pH of a buffer system is similar to calculating the pH of a weak acid solution or the pH of a weak base solution (see Chapter 15). The difference is that the initial concentration of the conjugate of the weak acid or weak base is *not zero*. The initial concentration of the conjugate comes from the salt component of the buffer.

For example, consider a solution that is 1.0 *M* in CH_3COOH (acetic acid) and 1.0 *M* in CH_3COONa (sodium acetate). If this were only a 1.0 *M* acetic acid solution, the initial concentration of the conjugate base (CH_3COO^-) would be *zero*. However, CH_3COONa is a soluble salt that ionizes completely.

	$CH_3COONa(aq)$	$\longrightarrow$	$Na^+(aq)$	+	$CH_3COO^-(aq)$
Initial	1.0 *M*		0		0
After dissociation	0		1.0 *M*		1.0 *M*

The CH_3COO^- ion is present initially and enters into the weak acid equilibrium of acetic acid. Thus, if we look at the weak acid equilibrium for acetic acid, we have:

	$CH_3COOH(aq)$	$\rightleftharpoons$	$H^+(aq)$	+	$CH_3COO^-(aq)$
Initial (*M*):	1.0		0		1.0
Change (*M*):	$-x$		$+x$		$+x$
Equilibrium (*M*):	$1.0 - x$		x		$1.0 + x$

You should notice that the only difference between this equilibrium, and a weak acid equilibrium (see Chapter 15), is that the initial concentration of the conjugate base, CH_3COO^-, is *not zero*, as it would be for a weak acid equilibrium. Remember, a buffer contains both a weak acid and its conjugate base or a weak base and its conjugate acid. The above system is a buffer because it initially contains both the weak acid (CH_3COOH) and its conjugate base (CH_3COO^-). Example 16.2 illustrates how to calculate the pH of a buffer system.

EXAMPLE 16.2

(a) Calculate the pH of a buffer system containing 0.25 *M* HF and 0.50 *M* NaF.

(b) What is the pH of the buffer system after the addition of 0.060 mol of gaseous HCl to 1.00 L of the solution? Assume that the volume of the solution does not change upon addition of the HCl.

Strategy: (a) The pH of a buffer system can be calculated in a similar manner to a weak acid equilibrium problem. The difference is that a common-ion is present in solution. The K_a of HF is 7.1×10^{-4} (see Table 15.3 of the text).
(b) Added acid will react with the base component of the buffer. The concentrations of both buffer components will change. The weak base component will decrease in concentration, and the weak acid component will increase in concentration. Set up a new equilibrium calculation with these concentrations.

Solution:

(a)

Step 1: Recognize that NaF is a soluble salt that will completely dissociate into ions. From the dissociation of NaF, we can calculate the initial concentration of the weak base, F^-.

	$NaF(aq) \longrightarrow$	$Na^+(aq)$ +	$F^-(aq)$
Initial	0.50 *M*	0	0
After dissociation	0	0.50 *M*	0.50 *M*

Step 2: The initial concentration of F^- is 0.50 *M*. We summarize the concentrations of the species at equilibrium as follows:

	$HF(aq) \rightleftharpoons$	$H^+(aq)$ +	$F^-(aq)$
Initial (*M*):	0.25	0	0.50
Change (*M*):	$-x$	$+x$	$+x$
Equilibrium (*M*):	$0.25 - x$	x	$0.50 + x$

Step 3: Write the ionization constant expression in terms of the equilibrium concentrations. Knowing the value of the equilibrium constant (K_a), solve for x.

$$K_a = \frac{[H^+][F^-]}{[HF]}$$

You can look up the K_a value for hydrofluoric acid in Table 15.3 of the text.

$$7.1 \times 10^{-4} = \frac{(x)(0.50 + x)}{(0.25 - x)}$$

At this point, we will make two assumptions.

$$0.25 - x \approx 0.25$$

and

$$0.50 + x \approx 0.50$$

Oftentimes, assumptions such as these are valid if K is very small. A very small value of K means that a very small amount of reactants go to products. Hence, x is small. If we did not make this assumption, we would have to solve a quadratic equation.

$$7.1 \times 10^{-4} \approx \frac{(x)(0.50)}{0.25}$$

Solving for x,

$$x = 3.6 \times 10^{-4}\ M = [H^+]$$

Step 4: Having solved for the $[H^+]$, calculate the pH of the solution.

$$\mathbf{pH} = -\log[H^+] = -\log(3.6 \times 10^{-4}) = \mathbf{3.44}$$

Check: Checking the validity of the assumption,

$$\frac{3.6 \times 10^{-4}}{0.25} \times 100\% = 0.14\% < 5\%$$

The assumption is valid.

(b) We have added 0.060 mol of HCl (strong acid), which will react completely with the weak base component of the buffer. This reaction will change the equilibrium concentrations of both the acid and base components of the equilibrium. Therefore, after the reaction between HCl and the weak base of the buffer, the equilibrium must be reestablished.

Step 1: Write the reaction that occurs between the strong acid, HCl, and the weak base component of the buffer, F^-. After addition of HCl, complete ionization of HCl occurs.

	$HCl(aq)$	$\longrightarrow$	$H^+(aq)$	+	$Cl^-(aq)$
Initial	0.060 mol		0		0
After ionization	0		0.060 mol		0.060 mol

Next, H^+ will react with the weak base component of the buffer, F^-. Since H^+ is the strongest acid that can exist in water, this reaction will be driven to completion.

$$H^+(aq) + F^-(aq) \rightarrow HF(aq)$$

Since we have 1.00 L of buffer solution, the number of moles of F^- in solution is:

$$\text{mol } F^- = \frac{0.50 \text{ mol}}{1 \text{ L}} \times 1.00 \text{ L} = 0.50 \text{ mol}$$

Similarly, the number of moles of HF is 0.25 mol.

We can now calculate the moles of F^- and HF that remain after F^- reacts with the strong acid HCl.

	$H^+(aq)$	+	$F^-(aq)$	$\longrightarrow$	$HF(aq)$
Initial (mol):	0.060		0.50		0.25
Change (mol):	−0.060		−0.060		+0.060
Final (mol):	0		0.44		0.31

Step 2: After the reaction above, HF and F^- are no longer in equilibrium. Equilibrium must be reestablished. The concentrations of HF and F^- are:

$$[HF] = \frac{0.31 \text{ mol}}{1.00 \text{ L}} = 0.31\ M$$

$$[F^-] = \frac{0.44 \text{ mol}}{1.00 \text{ L}} = 0.44\ M$$

Reestablishing equilibrium between HF and F^-, we have:

	$HF(aq)$	$\rightleftharpoons$	$H^+(aq)$	+	$F^-(aq)$
Initial (M):	0.31		0		0.44
Change (M):	$-x$		$+x$		$+x$
Equilibrium (M):	$0.31 - x$		x		$0.44 + x$

Step 3: Write the ionization constant expression in terms of the equilibrium concentrations. Knowing the value of the equilibrium constant (K_a), solve for x.

$$K_a = \frac{[H^+][F^-]}{[HF]}$$

$$7.1 \times 10^{-4} = \frac{(x)(0.44 + x)}{(0.31 - x)}$$

$$7.1 \times 10^{-4} \approx \frac{(x)(0.44)}{0.31}$$

$$x = [H^+] = 5.0 \times 10^{-4}\ M$$

Step 4: Having solved for the $[H^+]$, calculate the pH of the solution.

$$\mathbf{pH} = -\log[H^+] = -\log(5.0 \times 10^{-4}) = \mathbf{3.30}$$

The pH of the solution dropped only 0.15 pH units upon addition of 0.060 mol of the strong acid, HCl. Thus, this buffer solution resisted change in pH.

Check: Does it make sense that the pH of the solution decreased upon the addition of a strong acid to the buffer?

PRACTICE EXERCISE

2. A buffer solution is prepared by mixing 500 mL of 0.600 M CH_3COOH with 500 mL of a 1.00 M CH_3COONa solution. What is the pH of the solution?

3. (a) Calculate the pH of a buffer system that is 0.0600 M HNO_2 and 0.160 M $NaNO_2$.
(b) What is the pH after 2.00 mL of 2.00 M NaOH are added to 1.00 L of this buffer?

Text Problems: 16.12, 16.14, 16.18

C. Preparing a buffer solution with a specific pH

To prepare a buffer solution with a specific pH, we need to consider the Henderson-Hasselbalch equation. For a derivation of this equation, see Section 16.2 of the text.

$$pH = pK_a + \log\frac{[\text{conjugate base}]}{[\text{acid}]}$$

where,

$$pK_a = -\log K_a$$

If the molar concentrations of the acid and its conjugate base are approximately equal, that is,

$$[\text{acid}] \approx [\text{conjugate base}]$$

then,

$$\log\frac{[\text{conjugate base}]}{[\text{acid}]} \approx 0$$

Substituting into the Henderson-Hasselbalch equation gives:

$$\text{pH} \approx \text{p}K_a$$

Thus, to prepare a buffer solution, we should choose a weak acid with a pK_a value close to the desired pH. This choice not only gives the desired pH value for the buffer system, but also ensures that we have comparable amounts of the weak acid and its conjugate base present. Having pH $\approx$ pK_a is a prerequisite for the buffer system to function effectively.

EXAMPLE 16.3

Which of the following mixtures is suitable for making a buffer solution with an optimum pH of about 9.2?

(a) CH_3COONa/CH_3COOH (b) NH_3/NH_4Cl (c) NaF/HF
(d) $NaNO_2/HNO_2$ (e) NaCl/HCl

Strategy: For a buffer to function effectively, the concentration of the acid component must be roughly equal to the conjugate base component. According to Equation (16.4) of the text, when the desired pH is close to the pK_a of the acid, that is, when pH $\approx$ pK_a,

$$\log\frac{[\text{conjugate base}]}{[\text{acid}]} \approx 0$$

or

$$\frac{[\text{conjugate base}]}{[\text{acid}]} \approx 1$$

Solution: To prepare a solution of a desired pH, we should choose a weak acid with a pK_a value close to the desired pH. We can rule out choice (e) immediately, because it contains a strong acid and its conjugate base. This solution is not a buffer. Calculating the pK_a for each acid:

(a) pK_a = 4.74
(b) pK_a = 9.26
(c) pK_a = 3.15
(d) pK_a = 3.35

Thus, the only solution that would make an effective buffer at pH = 9.2 is choice (b), NH_4Cl/NH_3.

Tip: We could have saved some time by recognizing that if the K_a value is less than 1×10^{-7}, the pK_a value will be greater than 7. The only choice that had a K_a value less than 1×10^{-7} was choice (b). Thus, it was the only choice that has a pK_a value greater than 7.

PRACTICE EXERCISE

4. How would you prepare a liter of an HF/F^- buffer at a pH of 2.85?

Text Problem: 16.20

PROBLEM TYPE 2: TITRATIONS

A. Strong acid–strong base titrations

The reaction between a strong acid (HNO_3) and a strong base (KOH) can be written as:

$$HNO_3(aq) + KOH(aq) \longrightarrow KNO_3(aq) + H_2O(l)$$

or in terms of the net ionic equation

$$H^+(aq) + OH^-(aq) \longrightarrow H_2O(l)$$

At the equivalence point of a titration, the $[H^+] = [OH^-]$. Thus, the only species present in solution other than water at the equivalence point is $KNO_3(aq)$. Potassium nitrate is a salt that does not undergo hydrolysis (see Chapter 15); therefore, the pH at the equivalence point of a strong acid–strong base titration is 7.

A typical strong acid–strong base titration problem involves calculating the pH at various points in the titration. Example 16.4 below illustrates this type of problem.

EXAMPLE 16.4

A 20.0 mL sample of 0.0200 *M* HNO_3 is titrated with 0.0100 *M* KOH.

(a) What is the pH at the equivalence point?

(b) How many mL of KOH are required to reach the equivalence point?

(c) What is the pH before any KOH is added?

(d) What will be the pH after 10.0 mL of KOH are added?

(e) What will be the pH after 45.0 mL of KOH are added?

(a) Since this is a strong acid–strong base titration, the pH at the equivalence point is 7.

(b) We worked this type of problem in Chapter 4.

Step 1: In order to have the correct mole ratio to solve the problem, you must start with a balanced chemical equation.

$$HNO_3(aq) + KOH(aq) \longrightarrow KNO_3(aq) + H_2O(l)$$

Step 2: From the molarity and volume of the HNO_3 solution, you can calculate moles of HNO_3. Then, using the mole ratio from the balanced equation above, you can calculate moles of KOH.

$$\text{mol KOH} = 20.0\ \text{mL soln} \times \frac{0.0200\ \text{mol HNO}_3}{1000\ \text{mL soln}} \times \frac{1\ \text{mol KOH}}{1\ \text{mol HNO}_3} = 4.00 \times 10^{-4}\ \text{mol KOH}$$

Step 3: Solve the molarity equation algebraically for liters of solution. Then, substitute in the moles of KOH and molarity of KOH to solve for volume of KOH.

$$M = \frac{\text{moles of solute}}{\text{liters of solution}}$$

$$\text{liters of solution} = \frac{\text{moles of solute}}{M}$$

$$\textbf{volume of KOH} = \frac{4.00 \times 10^{-4}\ \text{mol KOH}}{0.0100\ \text{mol/L}} = \textbf{0.0400 L} = \textbf{40.0 mL}$$

(c) We only have the strong acid, HNO_3, in solution before any KOH is added. Thus, we need to calculate the pH of a strong acid (see Chapter 15).

Step 1: Strong acids ionize completely in solution. Let's write the reaction and set up a table to calculate the H^+ concentration in solution.

	$HNO_3(aq)$	$\longrightarrow$	$H^+(aq)$	+	$NO_3^-(aq)$
initial conc.	0.0200 M		0		0
conc. after ionization	0		0.0200 M		0.0200 M

Since there is a 1:1 mole ratio between H^+ and HNO_3, the H^+ concentration equals the HNO_3 concentration.

Step 2: Calculate the pH from the H^+ concentration.

$$\textbf{pH} = -\log[H^+] = -\log[0.0200] = \textbf{1.70}$$

(d) Any strong base added will react completely with the strong acid in solution. The reaction is:

$$HNO_3(aq) + KOH(aq) \longrightarrow KNO_3(aq) + H_2O(l)$$

Step 1: Calculate the number of moles of HNO_3 in 20.0 mL, and calculate the number of moles of KOH in 10.0 mL.

$$\text{mol } HNO_3 = 20.0 \text{ mL soln} \times \frac{1.00 \text{ L}}{1000 \text{ mL}} \times \frac{0.0200 \text{ mol } HNO_3}{1.00 \text{ L soln}} = 4.00 \times 10^{-4} \text{ mol}$$

$$\text{mol KOH} = 10.0 \text{ mL soln} \times \frac{1.00 \text{ L}}{1000 \text{ mL}} \times \frac{0.0100 \text{ mol KOH}}{1.00 \text{ L soln}} = 1.00 \times 10^{-4} \text{ mol}$$

Step 2: Set up a table showing the number of moles of HNO_3 and KOH before and after the reaction.

	$HNO_3(aq)$	+	$KOH(aq)$	$\longrightarrow$	$KNO_3(aq)$	+	$H_2O(l)$
Initial (mol):	4.00×10^{-4}		1.00×10^{-4}		0		
Change (mol):	-1.00×10^{-4}		-1.00×10^{-4}		$+1.00 \times 10^{-4}$		
Final (mol):	3.00×10^{-4}		0		1.00×10^{-4}		

Thus, 3.00×10^{-4} mol of the strong acid HNO_3 remain after the addition of 10.0 mL of KOH.

Step 3: The total volume of solution is now the sum of the volume of the acid solution and the volume of the base solution.

$$V_{soln} = V_{acid} + V_{base} = 20.0 \text{ mL} + 10.0 \text{ mL} = 30.0 \text{ mL}$$

We can now calculate the concentration of H^+ in solution. Since HNO_3 ionizes completely, the number of moles of H^+ in solution is 3.00×10^{-4} mol. Calculate the molarity by dividing the moles by liters of solution.

30.0 mL = 0.0300 L

$$[H^+] = \frac{3.00 \times 10^{-4} \text{ mol}}{0.0300 \text{ L}} = 0.0100\ M$$

Step 4: Calculate the pH from the H^+ concentration.

$$\textbf{pH} = -\log[H^+] = -\log[0.0100] = \textbf{2.00}$$

(e) Start this part by following the same procedure as in part (d).

Step 1: Calculate the number of moles of KOH in 45.0 mL.

$$\text{mol KOH} = 45.0\ \text{mL soln} \times \frac{1.00\ \text{L}}{1000\ \text{mL}} \times \frac{0.0100\ \text{mol KOH}}{1.00\ \text{L soln}} = 4.50 \times 10^{-4}\ \text{mol}$$

Step 2: Set up a table showing the number of moles of HNO_3 and KOH before and after the reaction.

	$HNO_3(aq)$ +	$KOH(aq)$ ⟶	$KNO_3(aq) + H_2O(l)$
Initial (mol):	4.00×10^{-4}	4.50×10^{-4}	0
Change (mol):	-4.00×10^{-4}	-4.00×10^{-4}	$+4.00 \times 10^{-4}$
Final (mol):	0	5.00×10^{-5}	4.00×10^{-4}

We have passed the equivalence point of the titration. The only species of significance that remains in solution is the strong base, KOH. 5.00×10^{-5} mol of the strong base KOH remain after the addition of 45.0 mL of KOH.

Step 3: The total volume of solution is now the sum of the volume of the acid solution and the volume of the base solution.

$$V_{\text{soln}} = V_{\text{acid}} + V_{\text{base}} = 20.0\ \text{mL} + 45.0\ \text{mL} = 65.0\ \text{mL}$$

We can now calculate the concentration of OH^- in solution. Since KOH ionizes completely, the number of moles of OH^- in solution is 5.00×10^{-5} mol. Calculate the molarity by dividing the moles by liters of solution.

65.0 mL = 0.0650 L

$$[OH^-] = \frac{5.00 \times 10^{-5}\ \text{mol}}{0.0650\ \text{L}} = 7.69 \times 10^{-4}\ M$$

Step 4: Calculate the pOH from the OH^- concentration.

$$\text{pOH} = -\log[OH^-] = -\log[7.69 \times 10^{-4}] = 3.11$$

Step 5: Use the relationship, pH + pOH = 14.00, to calculate the pH of the solution.

$$\textbf{pH} = 14.00 - 3.11 = \mathbf{10.89}$$

PRACTICE EXERCISE

5. Consider the titration of 25.0 mL of 0.250 M KOH with 0.100 M HCl.

(a) What is the pH at the equivalence point?
(b) How many mL of HCl are required to reach the equivalence point?
(c) What is the pH before any HCl is added?
(d) What will be the pH after 15.0 mL of HCl are added?
(e) What will be the pH after 75.0 mL of HCl are added?

Text Problem: 16.24

B. Weak acid–strong base titrations

Consider the neutralization between nitrous acid (a weak acid) and sodium hydroxide (a strong base):

$$HNO_2(aq) + NaOH(aq) \longrightarrow NaNO_2(aq) + H_2O(l)$$

The net ionic equation is:

$$HNO_2(aq) + OH^-(aq) \longrightarrow NO_2^-(aq) + H_2O(l)$$

At the equivalence point of a titration, the $[HNO_2] = [OH^-]$. Thus, the major species present in solution other than water at the equivalence point are $NO_2^-(aq)$ and $Na^+(aq)$. NO_2^- is a weak base; it is the conjugate base of the weak acid, HNO_2 (see Chapter 15). The hydrolysis reaction is given by:

$$NO_2^-(aq) + H_2O(l) \rightleftharpoons HNO_2(aq) + OH^-(aq)$$

Therefore, at the equivalence point, the pH will be *greater than 7* as a result of the excess OH^- formed.

A typical weak acid–strong base titration problem involves calculating the pH at various points in the titration. Example 16.5 below illustrates this type of problem.

EXAMPLE 16.5

Consider the titration of 50.0 mL of 0.100 *M* CH_3COOH with 0.100 *M* NaOH.

(a) How many mL of NaOH are required to reach the equivalence point?
(b) What is the pH before any NaOH is added?
(c) What will be the pH after 25.0 mL of NaOH are added?
(d) What is the pH at the equivalence point?
(e) What will be the pH after 60.0 mL of KOH are added?

(a) We worked this type of problem in Chapter 4 and in Example 16.4 above.

Step 1: In order to have the correct mole ratio to solve the problem, you must start with a balanced chemical equation.

$$CH_3COOH(aq) + NaOH(aq) \longrightarrow CH_3COONa(aq) + H_2O(l)$$

Step 2: We can take a shortcut on this problem by recognizing that both CH_3COOH and NaOH have the same concentration, 0.100 *M*. Since the mole ratio between CH_3COOH and NaOH is 1:1, it will require the same volume of each component to neutralize the other.

Thus, the volume of NaOH required to reach the equivalence point is **50.0 mL**.

(b) We only have the weak acid, CH_3COOH, in solution before any NaOH is added. Thus, we need to calculate the pH of a weak acid (see Chapter 15).

Step 1: Letting x be the equilibrium concentration of H^+ and CH_3COO^- ions in mol/L, we summarize:

	$CH_3COOH(aq)$	$\rightleftharpoons$ $H^+(aq)$	+ $CH_3COO^-(aq)$
Initial (*M*):	0.100	0	0
Change (*M*):	$-x$	$+x$	$+x$
Equilibrium (*M*):	$0.100 - x$	x	x

Step 2: Write the ionization constant expression in terms of the equilibrium concentrations. Knowing the value of the equilibrium constant (K_a), solve for x.

$$K_a = \frac{[H^+][CH_3COO^-]}{[CH_3COOH]}$$

You can look up the K_a value for acetic acid in Table 15.3 of the text.

$$1.8 \times 10^{-5} = \frac{(x)(x)}{(0.100 - x)}$$

At this point, we can make an assumption that x is very small compared to 0.100. Hence,

$$0.100 - x \approx 0.100$$

and,

$$1.8 \times 10^{-5} = \frac{(x)(x)}{0.100}$$

Solving for x.

$$x = 1.34 \times 10^{-3}\ M = [H^+]$$

Step 3: Having solved for the $[H^+]$, calculate the pH of the solution.

$$\mathbf{pH} = -\log[H^+] = -\log(1.34 \times 10^{-3}) = \mathbf{2.87}$$

Check: Checking the validity of the assumption,

$$\frac{1.34 \times 10^{-3}}{0.100} \times 100\% = 1.34\% < 5\%$$

The assumption was valid.

(c) You should recognize that when 25.0 mL of NaOH are added, we are half-way to the equivalence point. For a weak acid-strong base titration or a weak base-strong acid titration, the pH at the half-way point equals pK_a. This relation can be derived from the Henderson-Hasselbalch equation. At the half-way point, the concentration of the acid equals the concentration of its conjugate base, because half of the acid (CH_3COOH) has been neutralized, forming an equal amount of its conjugate base (CH_3COO^-).

$$pH = pK_a + \log\frac{[\text{conjugate base}]}{[\text{acid}]}$$

At the halfway point,

$$[\text{acid}] = [\text{conjugate base}]$$

and,

$$\log\frac{[\text{conjugate base}]}{[\text{acid}]} = 0$$

Substituting into the Henderson-Hasselbalch equation gives:

$$pH = pK_a$$

$$\mathbf{pH} = pK_a = -\log(1.8 \times 10^{-5}) = \mathbf{4.74}$$

(d) At the equivalence point the strong base has completely neutralized the weak acid in solution. The reaction is:

$$CH_3COOH(aq) + NaOH(aq) \longrightarrow CH_3COONa(aq) + H_2O(l)$$

Step 1: The number of moles of acetic acid equals the number of moles of NaOH at the equivalence point. Calculate the number of moles of acetic acid and sodium hydroxide.

$$\text{mol } CH_3COOH = 50.0 \text{ mL soln} \times \frac{1.00 \text{ L}}{1000 \text{ mL}} \times \frac{0.100 \text{ mol } CH_3COOH}{1.00 \text{ L soln}} = 5.00 \times 10^{-3} \text{ mol}$$

$$\text{mol NaOH} = \text{mol } CH_3COOH$$

Step 2: Set up a table showing the number of moles of CH_3COOH and NaOH before and after the reaction.

	$CH_3COOH(aq)$	+ $NaOH(aq)$	$\longrightarrow$ $CH_3COONa(aq)$	+ $H_2O(l)$
Initial (mol):	5.00×10^{-3}	5.00×10^{-3}	0	
Change (mol):	-5.00×10^{-3}	-5.00×10^{-3}	$+5.00 \times 10^{-3}$	
Final (mol):	0	0	5.00×10^{-3}	

The only species in solution of significance at the equivalence point is the salt CH_3COONa. This salt contains the conjugate base (CH_3COO^-) of the weak acid, CH_3COOH. The conjugate base is a weak base. The concentration of CH_3COO^- is

$$[CH_3COO^-] = \frac{5.00 \times 10^{-3} \text{ mol } CH_3COO^-}{0.100 \text{ L soln}} = 0.0500\ M$$

Remember to calculate the total volume of solution by adding the volume of the acid to the volume of the base.

$$50.0 \text{ mL acid} + 50.0 \text{ mL base} = 100.0 \text{ mL soln} = 0.100 \text{ L soln}$$

At this point, this problem just becomes a weak base calculation (see Chapter 15).

Step 3: Letting x be the equilibrium concentration of OH^- and CH_3COOH ions in mol/L, we summarize:

	$CH_3COO^-(aq)$ + $H_2O(l)$	$\rightleftharpoons$ $CH_3COOH(aq)$	+ $OH^-(aq)$
Initial (M):	0.0500	0	0
Change (M):	$-x$	$+x$	$+x$
Equilibrium (M):	$0.0500 - x$	x	x

Step 4: Write the ionization constant expression in terms of the equilibrium concentrations. Knowing the value of the equilibrium constant (K_b), solve for x.

We can calculate K_b from the K_a value of acetic acid.

$$K_b = \frac{K_w}{K_a} = \frac{1.0 \times 10^{-14}}{1.8 \times 10^{-5}} = 5.6 \times 10^{-10}$$

$$K_b = \frac{[CH_3COOH][OH^-]}{[CH_3COO^-]}$$

$$5.6 \times 10^{-10} = \frac{(x)(x)}{(0.0500 - x)}$$

$$5.6 \times 10^{-10} \approx \frac{(x)(x)}{0.0500}$$

Solving for x.

$$x = 5.3 \times 10^{-6}\ M = [OH^-]$$

Step 5: Having solved for the $[OH^-]$, calculate the pOH of the solution. Then use the relationship, pH + pOH = 14, to solve for the pH of the solution.

$$pOH = -\log[OH^-] = -\log(5.3 \times 10^{-6}) = 5.28$$

$$\mathbf{pH} = 14.00 - pOH = 14.00 - 5.28 = \mathbf{8.72}$$

Does it make sense that the pH at the equivalence point is *greater than 7*?

(e) After 60.0 mL of NaOH have been added, we have passed the equivalence point. The reaction is:

$$CH_3COOH(aq) + NaOH(aq) \rightarrow CH_3COONa(aq) + H_2O(l)$$

Step 1: Calculate the number of moles of sodium hydroxide in 60.0 mL of solution.

$$\text{mol NaOH} = 60.0 \text{ mL soln} \times \frac{1.00 \text{ L}}{1000 \text{ mL}} \times \frac{0.100 \text{ mol NaOH}}{1.00 \text{ L soln}} = 6.00 \times 10^{-3} \text{ mol}$$

Step 2: Set up a table showing the number of moles of CH_3COOH and NaOH before and after the reaction.

	$CH_3COOH(aq)$ +	$NaOH(aq)$ →	$CH_3COONa(aq)$ + $H_2O(l)$
Initial (mol):	5.00×10^{-3}	6.00×10^{-3}	0
Change (mol):	-5.00×10^{-3}	-5.00×10^{-3}	$+5.00 \times 10^{-3}$
Final (mol):	0	1.00×10^{-3}	5.00×10^{-3}

The only species in solution of significance past the equivalence point is the strong base NaOH. The salt does contain a weak base, but because NaOH is a strong base, we can assume that all the OH^- comes from NaOH. The concentration of OH^- is:

$$[OH^-] = \frac{1.00 \times 10^{-3} \text{ mol } OH^-}{0.110 \text{ L soln}} = 9.09 \times 10^{-3}\ M$$

Step 3: Having solved for the $[OH^-]$, calculate the pOH of the solution. Then use the relationship, pH + pOH = 14, to solve for the pH of the solution.

$$pOH = -\log[OH^-] = -\log(9.09 \times 10^{-3}) = 2.04$$

$$\mathbf{pH} = 14.00 - pOH = 14.00 - 2.04 = \mathbf{11.96}$$

PRACTICE EXERCISE

6. Consider the titration of 20.0 mL of 0.200 M HNO_2 with 0.100 M NaOH.

(a) How many mL of NaOH are required to reach the equivalence point?
(b) What is the pH before any NaOH is added?
(c) What will be the pH after 20.0 mL of NaOH are added?
(d) What is the pH at the equivalence point?
(e) What will be the pH after 50.0 mL of KOH are added?

Text Problems: 16.26, 16.28, 16.30

C. Strong acid–weak base titrations

Consider the neutralization between ammonia (a weak base) and nitric acid (a strong acid):

$$NH_3(aq) + HNO_3(aq) \rightarrow NH_4NO_3(aq)$$

The net ionic equation is:

$$NH_3(aq) + H^+(aq) \rightarrow NH_4^+(aq)$$

At the equivalence point of a titration, the $[NH_3] = [H^+]$. Thus, the major species present in solution other than water at the equivalence point are $NH_4^+(aq)$ and $NO_3^-(aq)$. NH_4^+ is a weak acid; it is the conjugate acid of the weak base, NH_3 (see Chapter 15). The hydrolysis reaction is given by:

$$NH_4^+(aq) + H_2O(l) \rightleftharpoons NH_3(aq) + H_3O^+(aq)$$

Therefore, at the equivalence point, the pH will be *less than 7* as a result of the excess H_3O^+ ions formed.

A weak base–strong acid titration problem is worked similarly to a weak acid–strong base titration problem (see Example 16.5).

PROBLEM TYPE 3: CHOOSING SUITABLE ACID-BASE INDICATORS

The criterion for choosing an appropriate indicator for a given titration is whether the pH range over which the indicator changes color corresponds with the steep portion of the titration curve. If the indicator does not change color during this portion of the curve, it will not accurately identify the equivalence point. Table 16.1 of the text lists some common acid-base indicators and the pH ranges in which they change color.

Typically, the pH range for an indicator is equal to the pK_a of the indicator, plus or minus one pH unit.

$$\text{pH range} = pK_a \pm 1$$

EXAMPLE 16.6

Will bromophenol blue be a good choice as an indicator for a weak acid–strong base titration?

Strategy: The choice of an indicator for a particular titration is based on the fact that its pH range for color change must overlap the steep portion of the titration curve. Otherwise we cannot use the color change to locate the equivalence point.

Solution: A typical titration curve for a weak acid–strong base titration is shown in Figure 16.5 of the text. The pH at the equivalence point is greater than 7. Checking Table 16.1 of the text, bromophenol blue changes color over the pH range, 3.0–4.6. This indicator would change color before the steep portion of the titration curve for a typical weak acid–strong base titration. Thus, bromophenol blue would *not* be a good choice as an indicator for a weak acid–strong base titration.

PRACTICE EXERCISE

7. Would methyl red be a suitable indicator for a titration that has a pH of 5.0 at the equivalence point?

Text Problems: 16.36, 16.38

PROBLEM TYPE 4: SOLUBILITY EQUILIBRIA

Solubility equilibria typically involve salts of low solubility. For example, calcium phosphate, $[Ca_3(PO_4)_2]$, is practically insoluble in water. However, a very small amount of calcium phosphate will dissolve and dissociate completely into Ca^{2+} and PO_4^{3-} ions. An equilibrium will then be established between undissolved calcium phosphate and the ions in solution. Consider a saturated solution of calcium phosphate that is in contact with solid calcium phosphate. The solubility equilibrium can be represented as

$$Ca_3(PO_4)_2(s) \rightleftharpoons 3Ca^{2+}(aq) + 2PO_4^{3-}(aq)$$

We know from Chapter 14 that for heterogeneous equilibria, the concentration of a solid is a constant. Thus, we can write the equilibrium constant for the dissolution of $Ca_3(PO_4)_2$ as

$$K_{sp} = [Ca^{2+}]^3[PO_4^{3-}]^2$$

where K_{sp} is called the solubility product constant or simply the solubility product. In general, the **solubility product** of a compound is the product of the molar concentrations of the constituent ions, each raised to the power of its stoichiometric coefficient in the balanced equilibrium equation.

A. Calculating K_{sp} from molar solubility

Molar solubility is the number of moles of solute in 1 L of a saturated solution (moles per liter). The concentrations of the ions in the solubility product expression are also molar concentrations. From the molar solubility, we can calculate the molar concentrations of the ions in solution from the stoichiometry of the balanced equilibrium equation.

EXAMPLE 16.7

If the solubility of $Fe(OH)_2$ in water is 7.7×10^{-6} mol/L at a certain temperature, what is its K_{sp} value at that temperature?

Strategy: From the molar solubility, the concentrations of Fe^{2+} and OH^- can be calculated. Then, from these concentrations, we can determine K_{sp}.

Solution: Consider the dissociation of $Fe(OH)_2$ in water. Let s be the molar solubility of $Fe(OH)_2$.

	$Fe(OH)_2(s)$	$\rightleftharpoons$ $Fe^{2+}(aq)$	$+ \; 2OH^-(aq)$
Initial (M):		0	0
Change (M):	$-s$	$+s$	$+2s$
Equilibrium (M):		s	$2s$

$$K_{sp} = [Fe^{2+}][OH^-]^2 = (s)(2s)^2 = 4s^3$$

The molar solubility (s) is given in the problem. Substitute into the equilibrium constant expression to solve for K_{sp}.

$$\mathbf{K_{sp}} = [Fe^{2+}][OH^-]^2 = 4s^3 = 4(7.7 \times 10^{-6})^3 = \mathbf{1.8 \times 10^{-15}}$$

PRACTICE EXERCISE

8. At a certain temperature, the solubility of barium chromate ($BaCrO_4$) is 1.8×10^{-5} mol/L. What is the K_{sp} value at this temperature?

Text Problems: 16.46, 16.48, 16.52

B. Calculating solubility from K_{sp}

As in other equilibrium calculations, the expected concentrations at equilibrium can be calculated from a knowledge of the initial concentrations and the equilibrium constant (see Chapter 14). In these types of problems, it will be very helpful to recall that

$$\text{equilibrium concentration} = \text{initial concentration} \pm \text{the change due to reaction.}$$

Example 16.8 illustrates this important type of calculation.

EXAMPLE 16.8

What is the molar solubility of silver phosphate (Ag_3PO_4) in water? $K_{sp} = 1.8 \times 10^{-18}$.

Strategy: The K_{sp} value for Ag_3PO_4 is given in the problem. Setting up the dissociation equilibrium of Ag_3PO_4 in water, we can solve for the molar solubility, *s*.

Solution: Consider the dissociation of Ag_3PO_4 in water.

	$Ag_3PO_4(s)$	$\rightleftharpoons$	$3Ag^+(aq)$	+	$PO_4^{3-}(aq)$
Initial (M):			0		0
Change (M):	$-s$		$+3s$		$+s$
Equilibrium (M):			$3s$		s

Recall, that the concentration of a pure solid does not enter into an equilibrium constant expression. Therefore, the concentration of Ag_3PO_4 is not important.

Substitute the value of K_{sp} and the concentrations of Ag^+ and PO_4^{3-} in terms of *s* into the solubility product expression to solve for *s*, the molar solubility.

$$K_{sp} = [Ag^+]^3[PO_4^{3-}]$$

$$1.8 \times 10^{-18} = (3s)^3(s)$$

$$1.8 \times 10^{-18} = 27s^4$$

$$s = \text{molar solubility} = \mathbf{1.6 \times 10^{-5}\ mol/L}$$

The molar solubility indicates that 1.6×10^{-5} mole of Ag_3PO_4 will dissolve in 1 L of an aqueous solution.

PRACTICE EXERCISE

9. The K_{sp} value of silver carbonate (Ag_2CO_3) is 8.1×10^{-12} at 25°C. What is the solubility in g/L of silver carbonate in water at 25°C?

Text Problems: 16.50, 16.54

C. Predicting a precipitation reaction

To predict whether a precipitate will form, we must calculate the **ion product, *Q***. We will follow the same procedure outlined in Section 14.4 of the text. The ion product represents the molar concentrations of the ions raised to the power of their stoichiometric coefficients. Thus, for an aqueous solution containing Pb^{2+} and S^{2-} ions at 25°C,

$$Q = [Pb^{2+}]_0[S^{2-}]_0$$

The subscript 0 indicates that these are initial concentrations and do not necessarily correspond to equilibrium concentrations.

There are three possible relationships between Q and K_{sp}.

- $Q = K_{sp}$ $[Pb^{2+}]_0[S^{2-}]_0 = 3.4 \times 10^{-28}$ Since $Q = K_{sp}$, the reaction is already at equilibrium. The solution is saturated.

- $Q < K_{sp}$ $[Pb^{2+}]_0[S^{2-}]_0 < 3.4 \times 10^{-28}$ With $Q < K_{sp}$, the solution is not saturated. More solid could dissolve to produce more ions in solution. No precipitate will form.

- $Q > K_{sp}$ $[Pb^{2+}]_0[S^{2-}]_0 > 3.4 \times 10^{-28}$ With $Q > K_{sp}$, the solution is supersaturated. There are too many ions in solution. Some ions will combine to form PbS, which will precipitate out until the product of the ion concentrations is equal to 3.4×10^{-28}.

EXAMPLE 16.9

Predict whether a precipitate of PbI_2 will form when 200 mL of 0.015 *M* $Pb(NO_3)_2$ and 300 mL of 0.050 *M* NaI are mixed together. K_{sp} (PbI_2) = 1.4×10^{-8}.

Strategy: Under what condition will an ionic compound precipitate from solution? The ions in solution are Pb^{2+}, NO_3^-, Na^+, and I^-. According to the solubility rules listed in Table 4.2 of the text, the only precipitate that can form is PbI_2. From the information given, we can calculate $[Pb^{2+}]_0$ and $[I^-]_0$ because we know the number of moles of ions in the original solutions and the volume of the combined solution. Next, we calculate the reaction quotient Q ($Q = [Pb^{2+}]_0[I^-]_0^2$) and compare the value of Q with K_{sp} of PbI_2 to see if a precipitate will form, that is, if the solution is supersaturated.

Solution: The number of moles of Pb^{2+} present in the original 200 mL of solution is:

$$200 \text{ mL} \times \frac{0.015 \text{ mol Pb}^{2+}}{1000 \text{ mL soln}} = 3.00 \times 10^{-3} \text{ mol Pb}^{2+}$$

The total volume after combining the solutions is 500 mL (0.500 L). The concentration of Pb^{2+} in the 0.500 L volume is:

$$[\text{Pb}^{2+}]_0 = \frac{3.00 \times 10^{-3} \text{ mol}}{0.500 \text{ L}} = 6.00 \times 10^{-3}\ M$$

The number of moles of I^- present in the original 300 mL of solution is:

$$300 \text{ mL} \times \frac{0.050 \text{ mol I}^-}{1000 \text{ mL soln}} = 1.50 \times 10^{-2} \text{ mol I}^-$$

The concentration of I^- in the 0.500 L volume is:

$$[\text{I}^-]_0 = \frac{1.50 \times 10^{-2} \text{ mol}}{0.500 \text{ L}} = 3.00 \times 10^{-2}\ M$$

Now, we must compare Q and K_{sp} from Table 16.2 of the text.

$$\text{PbI}_2(s) \rightleftharpoons \text{Pb}^{2+}(aq) + 2\text{I}^-(aq) \qquad K_{sp} = 1.4 \times 10^{-8}$$

$$Q = [\text{Pb}^{2+}]_0[\text{I}^-]_0^2$$

$$Q = (6.0 \times 10^{-3})(3.00 \times 10^{-2})^2 = 5.4 \times 10^{-6}$$

Comparing Q to K_{sp}, we find the $Q > K_{sp}$. This means that the solution is supersaturated. There are too many ions in solution. Some ions will combine to form PbI_2, which will precipitate out until the product of the ion concentrations is equal to K_{sp}, 1.4×10^{-8}.

PRACTICE EXERCISE

10. Will a precipitate of MgF_2 form when 6.00×10^2 mL of a solution that is 2.0×10^{-4} M in $MgCl_2$ is added to 3.00×10^2 mL of a 1.1×10^{-2} M NaF solution? K_{sp} $(MgF_2) = 6.6 \times 10^{-9}$.

Text Problem: 16.68

D. The effect of a common ion on solubility

Consider the slightly soluble salt copper (I) iodide. We can write its equilibrium reaction in water as follows:

$$\text{CuI}(s) \rightleftharpoons \text{Cu}^+(aq) + \text{I}^-(aq)$$

Now, suppose that instead of dissolving CuI in water, we attempted to dissolve a certain quantity of CuI in a potassium iodide (KI) solution. Would the presence of KI affect the solubility of the copper(I) iodide?

Remember, that KI is a soluble salt, so it will dissociate completely into K^+ and I^- ions.

$$\text{KI}(aq) \rightarrow \text{K}^+(aq) + \text{I}^-(aq)$$

The I^- ions from KI *will affect* the copper(I) iodide equilibrium. See Chapter 14 for a discussion of Le Châtelier's principle. Le Châtelier's principle states that when an external stress is applied to a system at equilibrium, the system adjusts in such a way that the stress is partially offset. In this system, the stress is additional I^- ions in solution (from KI). The CuI equilibrium will shift to the left to remove some of the additional I^- ions, decreasing the solubility of CuI.

In summary, the effect of adding a **common ion**, in this case I^-, is to **decrease** the solubility of the salt (CuI) in solution.

Calculating the solubility of a salt when a common ion is present is very similar to calculating the solubility from K_{sp} discussed earlier in this chapter. The only difference is that the initial concentration of the common ion is *not* zero. The initial concentration of the common ion comes from the soluble salt that is present in solution. Example 16.10 below illustrates this type of problem.

EXAMPLE 16.10

What is the molar solubility of $PbCl_2$ in a 0.50 M NaCl solution? K_{sp} $(PbCl_2) = 2.4 \times 10^{-4}$.

This problem is worked in a similar manner to Example 16.8.

Strategy: This is a common-ion problem. The common ion is Cl^-, which is supplied by both $PbCl_2$ and NaCl. Remember that the presence of a common ion will affect only the solubility of $PbCl_2$, but not the K_{sp} value because it is an equilibrium constant.

Solution: Set up a table to find the equilibrium concentrations in 0.50 M NaCl. NaCl is a soluble salt that ionizes completely giving an initial concentration of $Cl^- = 0.50$ M.

	$PbCl_2(s)$	$\rightleftharpoons$	$Pb^{2+}(aq)$	+	$2Cl^-(aq)$
Initial (M):			0		0.50
Change (M):	$-s$		$+s$		$+2s$
Equilibrium (M):			s		$0.50 + 2s$

Recall that the concentration of a pure solid does not enter into an equilibrium constant expression. Therefore, the concentration of $PbCl_2$ is not important.

$$K_{sp} = [Pb^{2+}][Cl^-]^2$$

$$2.4 \times 10^{-4} = (s)(0.50 + 2s)^2$$

Let's assume that $0.50 >> 2s$. This is a valid assumption because K_{sp} is small.

$$2.4 \times 10^{-4} \approx (s)(0.50)^2$$

$$s = \text{molar solubility} = \mathbf{9.6 \times 10^{-4}\ mol/L}$$

The molar solubility of lead(II) chloride in pure water is 3.9×10^{-2} M. Does it make sense that the molar solubility decreased in 0.50 M NaCl?

Check: Checking the validity of the assumption,

$$\frac{2(9.6 \times 10^{-4})}{0.50} \times 100\% = 0.38\% < 5\%$$

PRACTICE EXERCISE

11. What is the molar solubility of Ag_3PO_4 in 0.20 M $AgNO_3$? K_{sp} $(Ag_3PO_4) = 1.8 \times 10^{-18}$.

Text Problems: 16.60, 16.62

PROBLEM TYPE 5: COMPLEX ION EQUILIBRIA AND SOLUBILITY

We can define a **complex ion** as an ion containing a central metal cation bonded to one or more molecules or ions. Transition metals have a particular tendency to form complex ions. For example, silver chloride (AgCl) is insoluble in water. However, when aqueous ammonia is added to AgCl in water, the silver chloride dissolves to form $Ag(NH_3)_2^+(aq)$ and $Cl^-(aq)$. The equilibrium equation is:

$$Ag^+(aq) + 2NH_3(aq) \rightleftharpoons Ag(NH_3)_2^+(aq)$$

A measure of the tendency of a metal ion to form a particular complex ion is given by the **formation constant K_f** (also called the stability constant). The formation constant is simply the equilibrium constant for complex ion formation. The larger the K_f value, the more stable the complex ion. Table 16.4 of the text lists the formation constants of a number of complex ions.

The formation constant expression for the formation of $Ag(NH_3)_2^+$ can be written as:

$$K_f = \frac{[Ag(NH_3)_2^+]}{[Ag^+][NH_3]^2} = 1.5 \times 10^7$$

The very large value of K_f in this case indicates the great stability of the complex ion $[Ag(NH_3)_2^+]$ in solution. Thus, the insoluble silver chloride dissolves to form the stable complex ion, $Ag(NH_3)_2^+$.

A typical complex ion equilibrium problem involves using the formation constant to calculate the equilibrium concentrations of species in solution. See Example 16.11 below.

EXAMPLE 16.11

Calculate the concentration of free Ag^+ ions in a solution formed by adding 0.20 mol of $AgNO_3$ to 1.0 L of 1.0 *M* NaCN. Assume no volume change upon addition of $AgNO_3$.

Strategy: The addition of $AgNO_3$ to the NaCN solution results in complex ion formation. In solution, Ag^+ ions will complex with CN^- ions. The concentration of Ag^+ will be determined by the following equilibrium

$$Ag^+(aq) + 2CN^-(aq) \rightleftharpoons Ag(CN)_2^-$$

From Table 16.4 of the text, we see that the formation constant (K_f) for this reaction is very large ($K_f = 1.0 \times 10^{21}$). Because K_f is so large, the reaction lies mostly to the right. At equilibrium, the concentration of Ag^+ will be very small. As a good approximation, we can assume that essentially all the dissolved Ag^+ ions end up as $Ag(CN)_2^-$ ions. What is the initial concentration of Ag^+ ions? A very small amount of Ag^+ will be present at equilibrium. Set up the K_f expression for the above equilibrium to solve for $[Ag^+]$.

Solution: The initial concentration of Ag^+ is 0.20 *M*. If we assume that the above equilibrium goes to completion, we can write:

	$Ag^+(aq)$	+ $2CN^-(aq)$	$\rightleftharpoons Ag(CN)_2^-(aq)$
Initial (*M*):	0.20	1.0	0
Change (*M*):	−0.20	−2(0.20)	+0.20
Final (*M*):	0	0.60	0.20

To find the concentration of free Ag^+ at equilibrium, use the formation constant expression.

$$K_f = \frac{[Ag(CN)_2^-]}{[Ag^+][CN^-]^2}$$

Rearranging,

$$[Ag^+] = \frac{[Ag(CN)_2^-]}{K_f[CN^-]^2}$$

Substitute the equilibrium concentrations calculated above into the formation constant expression to calculate the equilibrium concentration of Ag^+.

$$\mathbf{[Ag^+]} = \frac{[Ag(CN)_2^-]}{K_f[CN^-]^2} = \frac{0.20}{(1.0 \times 10^{21})(0.60)^2} = \mathbf{5.6 \times 10^{-22}}\ \boldsymbol{M}$$

This concentration corresponds to only *three* Ag^+ ions per 10 mL of solution!

PRACTICE EXERCISE

12. If 0.0100 mol of $Cu(NO_3)_2$ are dissolved in 1.00 L of 1.00 *M* NH_3, what are the concentrations of Cu^{2+}, $Cu(NH_3)_4^{2+}$, and NH_3 at equilibrium? Assume no volume change upon addition of the $Cu(NO_3)_2$ to the ammonia solution.

Text Problems: 16.72, 16.74, 16.76

ANSWERS TO PRACTICE EXERCISES

1. Both (a) and (c). **2.** pH = 4.96 **3.** **(a)** pH = 3.77 **(b)** pH = 3.81

4. To prepare a pH = 2.85 buffer you would need to use a ratio of $[F^-]$ to [HF] of 0.50, or 1 to 2.

One way to prepare this buffer would be to add 0.50 mol of NaF (21 g) to 1.0 L of 1.0 *M* HF (assuming no change in volume).

5. **(a)** pH = 7 **(b)** 62.5 mL of HCl **(c)** pH = 13.40
(d) pH = 13.08 **(e)** pH = 1.90

6. **(a)** 40.0 mL of NaOH **(b)** pH = 2.02 **(c)** Halfway point, pH = pK_a = 3.35
(d) pH = 8.09 **(e)** pH = 12.16

7. Yes, methyl red would be a suitable indicator.

8. $K_{sp} = 3.2 \times 10^{-10}$ **9.** solubility = 0.035 g/L **10.** No, a precipitate will *not* form.

11. s = molar solubility = 2.3×10^{-16} *M* **12.** $[NH_3]$ = 0.96 *M*
$[Cu(NH_3)_4^{2+}]$ = 0.0100 *M*
$[Cu^{2+}] = 2.4 \times 10^{-16}$ *M*

SOLUTIONS TO SELECTED TEXT PROBLEMS

16.6 **(a)** This is a weak base calculation.

	$NH_3(aq) + H_2O(l) \rightleftharpoons$	$NH_4^+(aq)$	$+ OH^-(aq)$
Initial (*M*):	0.20	0	0
Change (*M*):	$-x$	$+x$	$+x$
Equilibrium (*M*):	$0.20 - x$	x	x

$$K_b = \frac{[NH_4^+][OH^-]}{[NH_3]}$$

$$1.8 \times 10^{-5} = \frac{(x)(x)}{0.20 - x} \approx \frac{x^2}{0.20}$$

$$x = 1.9 \times 10^{-3}\ M = [OH^-]$$

$$\text{pOH} = 2.72$$

$$\mathbf{pH = 11.28}$$

(b) The initial concentration of NH_4^+ is 0.30 *M* from the salt NH_4Cl. We set up a table as in part (a).

	$NH_3(aq) + H_2O(l) \rightleftharpoons$	$NH_4^+(aq)$	$+ OH^-(aq)$
Initial (*M*):	0.20	0.30	0
Change (*M*):	$-x$	$+x$	$+x$
Equilibrium (*M*):	$0.20 - x$	$0.30 + x$	x

$$K_b = \frac{[NH_4^+][OH^-]}{[NH_3]}$$

$$1.8 \times 10^{-5} = \frac{(x)(0.30 + x)}{0.20 - x} \approx \frac{x(0.30)}{0.20}$$

$$x = 1.2 \times 10^{-5}\ M = [OH^-]$$

$$\text{pOH} = 4.92$$

$$\mathbf{pH = 9.08}$$

Alternatively, we could use the Henderson-Hasselbalch equation to solve this problem. Table 15.4 gives the value of K_a for the ammonium ion. Substituting into the Henderson-Hasselbalch equation gives:

$$\text{pH} = \text{p}K_a + \log\frac{[\text{conjugate base}]}{\text{acid}} = -\log(5.6 \times 10^{-10}) + \log\frac{(0.20)}{(0.30)}$$

$$\mathbf{pH} = 9.25 - 0.18 = \mathbf{9.07}$$

Is there any difference in the Henderson-Hasselbalch equation in the cases of a weak acid and its conjugate base and a weak base and its conjugate acid?

16.10 Identifying buffer systems, Problem Type 1A.

Strategy: What constitutes a buffer system? Which of the preceding solutions contains a weak acid and its salt (containing the weak conjugate base)? Which of the preceding solutions contains a weak base and its salt (containing the weak conjugate acid)? Why is the conjugate base of a strong acid not able to neutralize an added acid?

Solution: The criteria for a buffer system are that we must have a weak acid and its salt (containing the weak conjugate base) or a weak base and its salt (containing the weak conjugate acid).

(a) HCN is a weak acid, and its conjugate base, CN^-, is a weak base. Therefore, this is a buffer system.

(b) HSO_4^- is a weak acid, and its conjugate base, SO_4^{2-} is a weak base (see Table 15.5 of the text). Therefore, this is a buffer system.

(c) NH_3 (ammonia) is a weak base, and its conjugate acid, NH_4^+ is a weak acid. Therefore, this is a buffer system.

(d) Because HI is a strong acid, its conjugate base, I^-, is an extremely weak base. This means that the I^- ion will not combine with a H^+ ion in solution to form HI. Thus, this system cannot act as a buffer system.

16.12 Calculating the pH of a buffer system, Problem Type 1B.

Strategy: The pH of a buffer system can be calculated in a similar manner to a weak acid equilibrium problem. The difference is that a common-ion is present in solution. The K_a of CH_3COOH is 1.8×10^{-5} (see Table 15.3 of the text).

Solution:

(a) We summarize the concentrations of the species at equilibrium as follows:

	$CH_3COOH(aq)$	$\rightleftharpoons$	$H^+(aq)$	+	$CH_3COO^-(aq)$
Initial (M):	2.0		0		2.0
Change (M):	$-x$		$+x$		$+x$
Equilibrium (M):	$2.0 - x$		x		$2.0 + x$

$$K_a = \frac{[H^+][CH_3COO^-]}{[CH_3COOH]}$$

$$K_a = \frac{[H^+](2.0 + x)}{(2.0 - x)} \approx \frac{[H^+](2.0)}{2.0}$$

$$K_a = [H^+]$$

Taking the –log of both sides,

$$pK_a = pH$$

Thus, for a buffer system in which the [weak acid] = [weak base],

$$pH = pK_a$$

$$\mathbf{pH} = -\log(1.8 \times 10^{-5}) = \mathbf{4.74}$$

(b) Similar to part (a),

$$\mathbf{pH} = pK_a = \mathbf{4.74}$$

Buffer (a) will be a more effective buffer because the concentrations of acid and base components are ten times higher than those in (b). Thus, buffer (a) can neutralize 10 times more added acid or base compared to buffer (b).

16.14 *Step 1:* Write the equilibrium that occurs between $H_2PO_4^-$ and HPO_4^{2-}. Set up a table relating the initial concentrations, the change in concentration to reach equilibrium, and the equilibrium concentrations.

	$H_2PO_4^-(aq)$	$\rightleftharpoons$ $H^+(aq)$	+ $HPO_4^{2-}(aq)$
Initial (M):	0.15	0	0.10
Change (M):	$-x$	$+x$	$+x$
Equilibrium (M):	$0.15 - x$	x	$0.10 + x$

Step 2: Write the ionization constant expression in terms of the equilibrium concentrations. Knowing the value of the equilibrium constant (K_a), solve for x.

$$K_a = \frac{[H^+][HPO_4^{2-}]}{[H_2PO_4^-]}$$

You can look up the K_a value for dihydrogen phosphate in Table 15.5 of your text.

$$6.2 \times 10^{-8} = \frac{(x)(0.10 + x)}{(0.15 - x)}$$

$$6.2 \times 10^{-8} \approx \frac{(x)(0.10)}{(0.15)}$$

$$x = [H^+] = 9.3 \times 10^{-8}\ M$$

Step 3: Having solved for the $[H^+]$, calculate the pH of the solution.

$$\mathbf{pH} = -\log[H^+] = -\log(9.3 \times 10^{-8}) = \mathbf{7.03}$$

16.16 We can use the Henderson-Hasselbalch equation to calculate the ratio $[HCO_3^-]/[H_2CO_3]$. The Henderson-Hasselbalch equation is:

$$\text{pH} = pK_a + \log\frac{[\text{conjugate base}]}{[\text{acid}]}$$

For the buffer system of interest, HCO_3^- is the conjugate base of the acid, H_2CO_3. We can write:

$$\text{pH} = 7.40 = -\log(4.2 \times 10^{-7}) + \log\frac{[HCO_3^-]}{[H_2CO_3]}$$

$$7.40 = 6.38 + \log\frac{[HCO_3^-]}{[H_2CO_3]}$$

The [conjugate base]/[acid] ratio is:

$$\log\frac{[HCO_3^-]}{[H_2CO_3]} = 7.40 - 6.38 = 1.02$$

$$\frac{\mathbf{[HCO_3^-]}}{\mathbf{[H_2CO_3]}} = 10^{1.02} = \mathbf{1.0 \times 10^1}$$

The buffer should be more effective against an added acid because ten times more base is present compared to acid. Note that a pH of 7.40 is only a two significant figure number (Why?); the final result should only have two significant figures.

16.18 As calculated in Problem 16.12, the pH of this buffer system is equal to pK_a.

$$pH = pK_a = -\log(1.8 \times 10^{-5}) = 4.74$$

(a) The added NaOH will react completely with the acid component of the buffer, CH_3COOH. NaOH ionizes completely; therefore, 0.080 mol of OH^- are added to the buffer.

Step 1: The neutralization reaction is:

	$CH_3COOH(aq)$ +	$OH^-(aq)$ ⟶	$CH_3COO^-(aq)$ + $H_2O(l)$
Initial (mol):	1.00	0.080	1.00
Change (mol):	−0.080	−0.080	+0.080
Final (mol):	0.92	0	1.08

Step 2: Now, the acetic acid equilibrium is reestablished. Since the volume of the solution is 1.00 L, we can convert directly from moles to molar concentration.

	$CH_3COOH(aq)$ ⇌	$H^+(aq)$ +	$CH_3COO^-(aq)$
Initial (*M*):	0.92	0	1.08
Change (*M*):	$-x$	$+x$	$+x$
Equilibrium (*M*):	$0.92 - x$	x	$1.08 + x$

Write the K_a expression, then solve for x.

$$K_a = \frac{[H^+][CH_3COO^-]}{[CH_3COOH]}$$

$$1.8 \times 10^{-5} = \frac{(x)(1.08 + x)}{(0.92 - x)} \approx \frac{x(1.08)}{0.92}$$

$$x = [H^+] = 1.5 \times 10^{-5}\ M$$

Step 3: Having solved for the $[H^+]$, calculate the pH of the solution.

$$\mathbf{pH} = -\log[H^+] = -\log(1.5 \times 10^{-5}) = \mathbf{4.82}$$

The pH of the buffer increased from 4.74 to 4.82 upon addition of 0.080 mol of strong base.

(b) The added acid will react completely with the base component of the buffer, CH_3COO^-. HCl ionizes completely; therefore, 0.12 mol of H^+ ion are added to the buffer

Step 1: The neutralization reaction is:

	$CH_3COO^-(aq)$ +	$H^+(aq)$ $\longrightarrow$	$CH_3COOH(aq)$
Initial (mol):	1.00	0.12	1.00
Change (mol):	−0.12	−0.12	+0.12
Final (mol):	0.88	0	1.12

Step 2: Now, the acetic acid equilibrium is reestablished. Since the volume of the solution is 1.00 L, we can convert directly from moles to molar concentration.

	$CH_3COOH(aq)$ $\rightleftharpoons$	$H^+(aq)$ +	$CH_3COO^-(aq)$
Initial (M):	1.12	0	0.88
Change (M):	$-x$	$+x$	$+x$
Equilibrium (M):	$1.12 - x$	x	$0.88 + x$

Write the K_a expression, then solve for x.

$$K_a = \frac{[H^+][CH_3COO^-]}{[CH_3COOH]}$$

$$1.8 \times 10^{-5} = \frac{(x)(0.88 + x)}{(1.12 - x)} \approx \frac{x(0.88)}{1.12}$$

$$x = [H^+] = 2.3 \times 10^{-5}\ M$$

Step 3: Having solved for the $[H^+]$, calculate the pH of the solution.

$$\textbf{pH} = -\log[H^+] = -\log(2.3 \times 10^{-5}) = \mathbf{4.64}$$

The pH of the buffer decreased from 4.74 to 4.64 upon addition of 0.12 mol of strong acid.

16.20 Preparing a buffer solution with a specific pH, Problem Type 1C.

Strategy: For a buffer to function effectively, the concentration of the acid component must be roughly equal to the conjugate base component. According to Equation (16.4) of the text, when the desired pH is close to the pK_a of the acid, that is, when pH ≈ pK_a,

$$\log\frac{[\text{conjugate base}]}{[\text{acid}]} \approx 0$$

or

$$\frac{[\text{conjugate base}]}{[\text{acid}]} \approx 1$$

Solution: To prepare a solution of a desired pH, we should choose a weak acid with a pK_a value close to the desired pH. Calculating the pK_a for each acid:

For HA, $pK_a = -\log(2.7 \times 10^{-3}) = 2.57$

For HB, $pK_a = -\log(4.4 \times 10^{-6}) = 5.36$

For HC, $\quad pK_a = -\log(2.6 \times 10^{-9}) = 8.59$

The buffer solution with a pK_a closest to the desired pH is HC. Thus, **HC** is the best choice to prepare a buffer solution with pH = 8.60.

16.24 We want to calculate the molar mass of the diprotic acid. The mass of the acid is given in the problem, so we need to find moles of acid in order to calculate its molar mass.

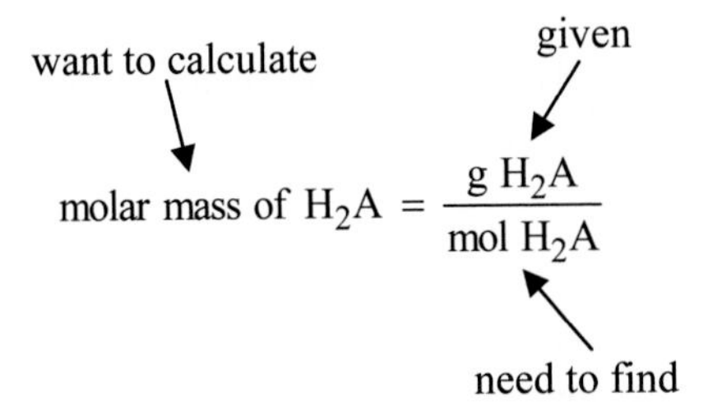

The neutralization reaction is:

$$2KOH(aq) + H_2A(aq) \longrightarrow K_2A(aq) + 2H_2O(l)$$

From the volume and molarity of the base needed to neutralize the acid, we can calculate the number of moles of H_2A reacted.

$$11.1 \text{ mL KOH} \times \frac{1.00 \text{ mol KOH}}{1000 \text{ mL}} \times \frac{1 \text{ mol } H_2A}{2 \text{ mol KOH}} = 5.55 \times 10^{-3} \text{ mol } H_2A$$

We know that 0.500 g of the diprotic acid were reacted (1/10 of the 250 mL was tested). Divide the number of grams by the number of moles to calculate the molar mass.

$$\mathcal{M}\,(\mathbf{H_2A}) = \frac{0.500 \text{ g } H_2A}{5.55 \times 10^{-3} \text{ mol } H_2A} = \mathbf{90.1\ g/mol}$$

16.26 We want to calculate the molarity of the $Ba(OH)_2$ solution. The volume of the solution is given (19.3 mL), so we need to find the moles of $Ba(OH)_2$ to calculate the molarity.

want to calculate → M of $Ba(OH)_2$; need to find → mol $Ba(OH)_2$; given → L of $Ba(OH)_2$ soln

$$M \text{ of } Ba(OH)_2 = \frac{\text{mol } Ba(OH)_2}{\text{L of } Ba(OH)_2 \text{ soln}}$$

The neutralization reaction is:

$$2HCOOH + Ba(OH)_2 \rightarrow (HCOO)_2Ba + 2H_2O$$

From the volume and molarity of HCOOH needed to neutralize $Ba(OH)_2$, we can determine the moles of $Ba(OH)_2$ reacted.

$$20.4 \text{ mL HCOOH} \times \frac{0.883 \text{ mol HCOOH}}{1000 \text{ mL}} \times \frac{1 \text{ mol Ba(OH)}_2}{2 \text{ mol HCOOH}} = 9.01 \times 10^{-3} \text{ mol Ba(OH)}_2$$

The molarity of the $Ba(OH)_2$ solution is:

$$\frac{9.01 \times 10^{-3} \text{ mol Ba(OH)}_2}{19.3 \times 10^{-3} \text{ L}} = \mathbf{0.467\ \mathit{M}}$$

16.28 The resulting solution is not a buffer system. There is excess NaOH and the neutralization is well past the equivalence point.

$$\text{Moles NaOH} = 0.500 \text{ L} \times \frac{0.167 \text{ mol}}{1 \text{ L}} = 0.0835 \text{ mol}$$

$$\text{Moles CH}_3\text{COOH} = 0.500 \text{ L} \times \frac{0.100 \text{ mol}}{1 \text{ L}} = 0.0500 \text{ mol}$$

	$CH_3COOH(aq)$ +	$NaOH(aq)$ →	$CH_3COONa(aq)$ + $H_2O(l)$
Initial (mol):	0.0500	0.0835	0
Change (mol):	−0.0500	−0.0500	+0.0500
Final (mol):	0	0.0335	0.0500

The volume of the resulting solution is 1.00 L (500 mL + 500 mL = 1000 mL).

$$[\mathbf{OH^-}] = \frac{0.0335 \text{ mol}}{1.00 \text{ L}} = \mathbf{0.0335\ \mathit{M}}$$

$$[\mathbf{Na^+}] = \frac{(0.0335 + 0.0500) \text{ mol}}{1.00 \text{ L}} = \mathbf{0.0835\ \mathit{M}}$$

$$[\mathbf{H^+}] = \frac{K_w}{[OH^-]} = \frac{1.0 \times 10^{-14}}{0.0335} = \mathbf{3.0 \times 10^{-13}\ \mathit{M}}$$

$$[\mathbf{CH_3COO^-}] = \frac{0.0500 \text{ mol}}{1.00 \text{ L}} = \mathbf{0.0500\ \mathit{M}}$$

	$CH_3COO^-(aq) + H_2O(l)$ ⇌	$CH_3COOH(aq)$ +	$OH^-(aq)$
Initial (M):	0.0500	0	0.0335
Change (M):	$-x$	$+x$	$+x$
Equilibrium (M):	$0.0500 - x$	x	$0.0335 + x$

$$K_b = \frac{[CH_3COOH][OH^-]}{[CH_3COO^-]}$$

$$5.6 \times 10^{-10} = \frac{(x)(0.0335 + x)}{(0.0500 - x)} \approx \frac{(x)(0.0335)}{(0.0500)}$$

$$x = [\mathbf{CH_3COOH}] = \mathbf{8.4 \times 10^{-10}\ \mathit{M}}$$

16.30 Let's assume we react 1 L of HCOOH with 1 L of NaOH.

	$HCOOH(aq)$ +	$NaOH(aq)$ →	$HCOONa(aq) + H_2O(l)$
Initial (mol):	0.10	0.10	0
Change (mol):	−0.10	−0.10	+0.10
Final (mol):	0	0	0.10

The solution volume has doubled (1 L + 1 L = 2 L). The concentration of HCOONa is:

$$M\ (\text{HCOONa}) = \frac{0.10\ \text{mol}}{2\ \text{L}} = 0.050\ M$$

$HCOO^-(aq)$ is a weak base. The hydrolysis is:

	$HCOO^-(aq) + H_2O(l)$ ⇌	$HCOOH(aq)$ +	$OH^-(aq)$
Initial (M):	0.050	0	0
Change (M):	$-x$	$+x$	$+x$
Equilibrium (M):	$0.050 - x$	x	x

$$K_b = \frac{[\text{HCOOH}][\text{OH}^-]}{[\text{HCOO}^-]}$$

$$5.9 \times 10^{-11} = \frac{x^2}{0.050 - x} \approx \frac{x^2}{0.050}$$

$$x = 1.7 \times 10^{-6}\ M = [\text{OH}^-]$$

$$\text{pOH} = 5.77$$

$$\mathbf{pH = 8.23}$$

16.32 The reaction between NH_3 and HCl is:

$$NH_3(aq) + HCl(aq) \rightarrow NH_4Cl(aq)$$

We see that 1 mole $NH_3 \simeq$ 1 mol HCl. Therefore, at every stage of titration, we can calculate the number of moles of base reacting with acid, and the pH of the solution is determined by the excess base or acid left over. At the equivalence point, however, the neutralization is complete, and the pH of the solution will depend on the extent of the hydrolysis of the salt formed, which is NH_4Cl.

(a) No HCl has been added. This is a weak base calculation.

	$NH_3(aq) + H_2O(l)$ ⇌	$NH_4^+(aq)$ +	$OH^-(aq)$
Initial (M):	0.300	0	0
Change (M):	$-x$	$+x$	$+x$
Equilibrium (M):	$0.300 - x$	x	x

$$K_b = \frac{[\text{NH}_4^+][\text{OH}^-]}{[\text{NH}_3]}$$

$$1.8 \times 10^{-5} = \frac{(x)(x)}{0.300 - x} \approx \frac{x^2}{0.300}$$

$$x = 2.3 \times 10^{-3}\ M = [OH^-]$$

$$pOH = 2.64$$

$$\mathbf{pH = 11.36}$$

(b) The number of moles of NH_3 originally present in 10.0 mL of solution is:

$$10.0\ \text{mL} \times \frac{0.300\ \text{mol } NH_3}{1000\ \text{mL } NH_3\ \text{soln}} = 3.00 \times 10^{-3}\ \text{mol}$$

The number of moles of HCl in 10.0 mL is:

$$10.0\ \text{mL} \times \frac{0.100\ \text{mol HCl}}{1000\ \text{mL HCl soln}} = 1.00 \times 10^{-3}\ \text{mol}$$

We work with moles at this point because when two solutions are mixed, the solution volume increases. As the solution volume increases, molarity will change, but the number of moles will remain the same. The changes in number of moles are summarized.

	$NH_3(aq)$	+	$HCl(aq)$	→	$NH_4Cl(aq)$
Initial (mol):	3.00×10^{-3}		1.00×10^{-3}		0
Change (mol):	-1.00×10^{-3}		-1.00×10^{-3}		$+1.00 \times 10^{-3}$
Final (mol):	2.00×10^{-3}		0		1.00×10^{-3}

At this stage, we have a buffer system made up of NH_3 and NH_4^+ (from the salt, NH_4Cl). We use the Henderson-Hasselbalch equation to calculate the pH.

$$pH = pK_a + \log\frac{[\text{conjugate base}]}{[\text{acid}]}$$

$$pH = -\log(5.6 \times 10^{-10}) + \log\left(\frac{2.00 \times 10^{-3}}{1.00 \times 10^{-3}}\right)$$

$$\mathbf{pH = 9.55}$$

(c) This part is solved similarly to part (b).

The number of moles of HCl in 20.0 mL is:

$$20.0\ \text{mL} \times \frac{0.100\ \text{mol HCl}}{1000\ \text{mL HCl soln}} = 2.00 \times 10^{-3}\ \text{mol}$$

The changes in number of moles are summarized.

	$NH_3(aq)$	+	$HCl(aq)$	→	$NH_4Cl(aq)$
Initial (mol):	3.00×10^{-3}		2.00×10^{-3}		0
Change (mol):	-2.00×10^{-3}		-2.00×10^{-3}		$+2.00 \times 10^{-3}$
Final (mol):	1.00×10^{-3}		0		2.00×10^{-3}

At this stage, we have a buffer system made up of NH_3 and NH_4^+ (from the salt, NH_4Cl). We use the Henderson-Hasselbalch equation to calculate the pH.

$$pH = pK_a + \log\frac{[\text{conjugate base}]}{[\text{acid}]}$$

$$pH = -\log(5.6 \times 10^{-10}) + \log\left(\frac{1.00 \times 10^{-3}}{2.00 \times 10^{-3}}\right)$$

$$\mathbf{pH = 8.95}$$

(d) We have reached the equivalence point of the titration. 3.00×10^{-3} mole of NH_3 reacts with 3.00×10^{-3} mole HCl to produce 3.00×10^{-3} mole of NH_4Cl. The only major species present in solution at the equivalence point is the salt, NH_4Cl, which contains the conjugate acid, NH_4^+. Let's calculate the molarity of NH_4^+. The volume of the solution is: (10.0 mL + 30.0 mL = 40.0 mL = 0.0400 L).

$$M\ (NH_4^+) = \frac{3.00 \times 10^{-3}\ \text{mol}}{0.0400\ \text{L}} = 0.0750\ M$$

We set up the hydrolysis of NH_4^+, which is a weak acid.

	$NH_4^+(aq) + H_2O(l) \rightleftharpoons$	$H_3O^+(aq)$	$+ NH_3(aq)$
Initial (M):	0.0750	0	0
Change (M):	$-x$	$+x$	$+x$
Equilibrium (M):	$0.0750 - x$	x	x

$$K_a = \frac{[H_3O^+][NH_3]}{[NH_4^+]}$$

$$5.6 \times 10^{-10} = \frac{(x)(x)}{0.0750 - x} \approx \frac{x^2}{0.0750}$$

$$x = 6.5 \times 10^{-6}\ M = [H_3O^+]$$

$$\mathbf{pH = 5.19}$$

(e) We have passed the equivalence point of the titration. The excess strong acid, HCl, will determine the pH at this point. The moles of HCl in 40.0 mL are:

$$40.0\ \text{mL} \times \frac{0.100\ \text{mol HCl}}{1000\ \text{mL HCl soln}} = 4.00 \times 10^{-3}\ \text{mol}$$

The changes in number of moles are summarized.

	$NH_3(aq)$ +	$HCl(aq)$ →	$NH_4Cl(aq)$
Initial (mol):	3.00×10^{-3}	4.00×10^{-3}	0
Change (mol):	-3.00×10^{-3}	-3.00×10^{-3}	$+3.00 \times 10^{-3}$
Final (mol):	0	1.00×10^{-3}	3.00×10^{-3}

Let's calculate the molarity of the HCl in solution. The volume of the solution is now 50.0 mL = 0.0500 L.

$$M\ (HCl) = \frac{1.00 \times 10^{-3}\ \text{mol}}{0.0500\ \text{L}} = 0.0200\ M$$

HCl is a strong acid. The pH is:

$$\mathbf{pH} = -\log(0.0200) = \mathbf{1.70}$$

16.36 CO_2 in the air dissolves in the solution:

$$CO_2 + H_2O \rightleftharpoons H_2CO_3$$

The carbonic acid neutralizes the NaOH.

16.38 According to Section 16.5 of the text, when $[HIn] \approx [In^-]$ the indicator color is a mixture of the colors of HIn and In^-. In other words, the indicator color changes at this point. When $[HIn] \approx [In^-]$ we can write:

$$\frac{[In^-]}{[HIn]} = \frac{K_a}{[H^+]} = 1$$

$$[H^+] = K_a = 2.0 \times 10^{-6}$$

$$\mathbf{pH = 5.70}$$

16.46 Calculating K_{sp} from molar solubility, Problem Type 4A.

Strategy: In each part, we can calculate the number of moles of compound dissolved in one liter of solution (the molar solubility). Then, from the molar solubility, s, we can determine K_{sp}.

Solution:

(a) $$\frac{7.3 \times 10^{-2}\ \text{g SrF}_2}{1\ \text{L soln}} \times \frac{1\ \text{mol SrF}_2}{125.6\ \text{g SrF}_2} = 5.8 \times 10^{-4}\ \text{mol/L} = s$$

Consider the dissociation of SrF_2 in water. Let s be the molar solubility of SrF_2.

	$SrF_2(s)$ $\rightleftharpoons$	$Sr^{2+}(aq)$ +	$2F^-(aq)$
Initial (M):		0	0
Change (M):	$-s$	$+s$	$+2s$
Equilibrium (M):		s	$2s$

$$K_{sp} = [Sr^{2+}][F^-]^2 = (s)(2s)^2 = 4s^3$$

The molar solubility (s) was calculated above. Substitute into the equilibrium constant expression to solve for K_{sp}.

$$\boldsymbol{K_{sp}} = [Sr^{2+}][F^-]^2 = 4s^3 = 4(5.8 \times 10^{-4})^3 = \mathbf{7.8 \times 10^{-10}}$$

(b) $$\frac{6.7 \times 10^{-3}\ \text{g Ag}_3\text{PO}_4}{1\ \text{L soln}} \times \frac{1\ \text{mol Ag}_3\text{PO}_4}{418.7\ \text{g Ag}_3\text{PO}_4} = 1.6 \times 10^{-5}\ \text{mol/L} = s$$

(b) is solved in a similar manner to (a)

The equilibrium equation is:

	$Ag_3PO_4(s)$ ⇌	$3Ag^+(aq)$ +	$PO_4^{3-}(aq)$
Initial (M):		0	0
Change (M):	$-s$	$+3s$	$+s$
Equilibrium (M):		$3s$	s

$$K_{sp} = [Ag^+]^3[PO_4^{3-}] = (3s)^3(s) = 27s^4 = 27(1.6 \times 10^{-5})^4 = \mathbf{1.8 \times 10^{-18}}$$

16.48 First, we can convert the solubility of MX in g/L to mol/L.

$$\frac{4.63 \times 10^{-3} \text{ g MX}}{1 \text{ L soln}} \times \frac{1 \text{ mol MX}}{346 \text{ g MX}} = 1.34 \times 10^{-5} \text{ mol/L} = s \text{ (molar solubility)}$$

The equilibrium reaction is:

	$MX(s)$ ⇌	$M^{n+}(aq)$ +	$X^{n-}(aq)$
Initial (M):		0	0
Change (M):	$-s$	$+s$	$+s$
Equilibrium (M):		s	s

$$K_{sp} = [M^{n+}][X^{n-}] = s^2 = (1.34 \times 10^{-5})^2 = \mathbf{1.80 \times 10^{-10}}$$

16.50 Calculating solubility from K_{sp}, Problem Type 4B.

Strategy: We can look up the K_{sp} value of CaF_2 in Table 16.2 of the text. Then, setting up the dissociation equilibrium of CaF_2 in water, we can solve for the molar solubility, s.

Solution: Consider the dissociation of CaF_2 in water.

	$CaF_2(s)$ ⇌	$Ca^{2+}(aq)$ +	$2F^-(aq)$
Initial (M):		0	0
Change (M):	$-s$	$+s$	$+2s$
Equilibrium (M):		s	$2s$

Recall, that the concentration of a pure solid does not enter into an equilibrium constant expression. Therefore, the concentration of CaF_2 is not important.

Substitute the value of K_{sp} and the concentrations of Ca^{2+} and F^- in terms of s into the solubility product expression to solve for s, the molar solubility.

$$K_{sp} = [Ca^{2+}][F^-]^2$$

$$4.0 \times 10^{-11} = (s)(2s)^2$$

$$4.0 \times 10^{-11} = 4s^3$$

$$s = \text{molar solubility} = \mathbf{2.2 \times 10^{-4} \text{ mol/L}}$$

The molar solubility indicates that 2.2×10^{-4} mol of CaF_2 will dissolve in 1 L of an aqueous solution.

16.52 First we can calculate the OH^- concentration from the pH.

$$pOH = 14.00 - pH$$

$$pOH = 14.00 - 9.68 = 4.32$$

$$[OH^-] = 10^{-pOH} = 10^{-4.32} = 4.8 \times 10^{-5}\ M$$

The equilibrium equation is:

$$MOH(s) \rightleftharpoons M^+(aq) + OH^-(aq)$$

From the balanced equation we know that $[M^+] = [OH^-]$

$$\boldsymbol{K_{sp}} = [M^+][OH^-] = (4.8 \times 10^{-5})^2 = \mathbf{2.3 \times 10^{-9}}$$

16.54 The net ionic equation is:

$$Sr^{2+}(aq) + 2F^-(aq) \longrightarrow SrF_2(s)$$

Let's find the limiting reagent in the precipitation reaction.

$$\text{Moles } F^- = 75\ \text{mL} \times \frac{0.060\ \text{mol}}{1000\ \text{mL soln}} = 0.0045\ \text{mol}$$

$$\text{Moles } Sr^{2+} = 25\ \text{mL} \times \frac{0.15\ \text{mol}}{1000\ \text{mL soln}} = 0.0038\ \text{mol}$$

From the stoichiometry of the balanced equation, twice as many moles of F^- are required to react with Sr^{2+}. This would require 0.0076 mol of F^-, but we only have 0.0045 mol. Thus, F^- is the limiting reagent.

Let's assume that the above reaction goes to completion. Then, we will consider the equilibrium that is established when SrF_2 partially dissociates into ions.

	$Sr^{2+}(aq)$ +	$2F^-(aq)$ ⟶	$SrF_2(s)$
Initial (mol):	0.0038	0.0045	0
Change (mol):	−0.00225	−0.0045	+0.00225
Final (mol):	0.00155	0	0.00225

Now, let's establish the equilibrium reaction. The total volume of the solution is 100 mL = 0.100 L. Divide the above moles by 0.100 L to convert to molar concentration.

	$SrF_2(s)$ ⇌	$Sr^{2+}(aq)$ +	$2F^-(aq)$
Initial (M):	0.0225	0.0155	0
Change (M):	$-s$	$+s$	$+2s$
Equilibrium (M):	$0.0225 - s$	$0.0155 + s$	$2s$

Write the solubility product expression, then solve for s.

$$K_{sp} = [Sr^{2+}][F^-]^2$$

$$2.0 \times 10^{-10} = (0.0155 + s)(2s)^2 \approx (0.0155)(2s)^2$$

$$s = 5.7 \times 10^{-5}\ M$$

$$[\mathbf{F^-}] = 2s = \mathbf{1.1 \times 10^{-4}}\ \boldsymbol{M}$$

$$[\mathbf{Sr^{2+}}] = 0.0155 + s = \mathbf{0.016}\ \boldsymbol{M}$$

Both sodium ions and nitrate ions are spectator ions and therefore do not enter into the precipitation reaction.

$$[\mathbf{NO_3^-}] = \frac{2(0.0038)\,\text{mol}}{0.10\ \text{L}} = \mathbf{0.076}\ \boldsymbol{M}$$

$$[\mathbf{Na^+}] = \frac{0.0045\ \text{mol}}{0.10\ \text{L}} = \mathbf{0.045}\ \boldsymbol{M}$$

16.56 For $Fe(OH)_3$, $K_{sp} = 1.1 \times 10^{-36}$. When $[Fe^{3+}] = 0.010\ M$, the $[OH^-]$ value is:

$$K_{sp} = [Fe^{3+}][OH^-]^3$$

or

$$[OH^-] = \left(\frac{K_{sp}}{[Fe^{3+}]}\right)^{\frac{1}{3}}$$

$$[OH^-] = \left(\frac{1.1 \times 10^{-36}}{0.010}\right)^{\frac{1}{3}} = 4.8 \times 10^{-12}\ M$$

This $[OH^-]$ corresponds to a pH of 2.68. In other words, $Fe(OH)_3$ will begin to precipitate from this solution at pH of 2.68.

For $Zn(OH)_2$, $K_{sp} = 1.8 \times 10^{-14}$. When $[Zn^{2+}] = 0.010\ M$, the $[OH^-]$ value is:

$$[OH^-] = \left(\frac{K_{sp}}{[Zn^{2+}]}\right)^{\frac{1}{2}}$$

$$[OH^-] = \left(\frac{1.8 \times 10^{-14}}{0.010}\right)^{\frac{1}{2}} = 1.3 \times 10^{-6}\ M$$

This corresponds to a pH of 8.11. In other words $Zn(OH)_2$ will begin to precipitate from the solution at pH = 8.11. These results show that $Fe(OH)_3$ will precipitate when the pH just exceeds 2.68 and that $Zn(OH)_2$ will precipitate when the pH just exceeds 8.11. Therefore, to selectively remove iron as $Fe(OH)_3$, the pH must be *greater than* **2.68** but *less than* **8.11**.

16.60 The effect of a common ion on solubility, Problem Type 4D.

Strategy: In parts (b) and (c), this is a common-ion problem. In part (b), the common ion is Br^-, which is supplied by both $PbBr_2$ and KBr. Remember that the presence of a common ion will affect only the solubility of $PbBr_2$, but not the K_{sp} value because it is an equilibrium constant. In part (c), the common ion is Pb^{2+}, which is supplied by both $PbBr_2$ and $Pb(NO_3)_2$.

Solution:

(a) Set up a table to find the equilibrium concentrations in pure water.

	$PbBr_2(s)$	$\rightleftharpoons$	$Pb^{2+}(aq)$	+	$2Br^-(aq)$
Initial (M)			0		0
Change (M)	$-s$		$+s$		$+2s$
Equilibrium (M)			s		$2s$

$$K_{sp} = [Pb^{2+}][Br^-]^2$$

$$8.9 \times 10^{-6} = (s)(2s)^2$$

$$s = \text{molar solubility} = \mathbf{0.013}\ \boldsymbol{M}$$

(b) Set up a table to find the equilibrium concentrations in 0.20 M KBr. KBr is a soluble salt that ionizes completely giving an initial concentration of $Br^- = 0.20\ M$.

	$PbBr_2(s)$	$\rightleftharpoons$	$Pb^{2+}(aq)$	+	$2Br^-(aq)$
Initial (M)			0		0.20
Change (M)	$-s$		$+s$		$+2s$
Equilibrium (M)			s		$0.20 + 2s$

$$K_{sp} = [Pb^{2+}][Br^-]^2$$

$$8.9 \times 10^{-6} = (s)(0.20 + 2s)^2$$

$$8.9 \times 10^{-6} \approx (s)(0.20)^2$$

$$s = \text{molar solubility} = \mathbf{2.2 \times 10^{-4}}\ \boldsymbol{M}$$

Thus, the molar solubility of $PbBr_2$ is reduced from 0.013 M to 2.2×10^{-4} M as a result of the common ion (Br^-) effect.

(c) Set up a table to find the equilibrium concentrations in 0.20 M $Pb(NO_3)_2$. $Pb(NO_3)_2$ is a soluble salt that dissociates completely giving an initial concentration of $[Pb^{2+}] = 0.20\ M$.

	$PbBr_2(s)$	$\rightleftharpoons$	$Pb^{2+}(aq)$	+	$2Br^-(aq)$
Initial (M):	0.20		0		
Change (M):	$-s$		$+s$		$+2s$
Equilibrium (M):			$0.20 + s$		$2s$

$$K_{sp} = [Pb^{2+}][Br^-]^2$$

$$8.9 \times 10^{-6} = (0.20 + s)(2s)^2$$

$$8.9 \times 10^{-6} \approx (0.20)(2s)^2$$

$$s = \text{molar solubility} = \mathbf{3.3 \times 10^{-3}}\ \boldsymbol{M}$$

Thus, the molar solubility of $PbBr_2$ is reduced from 0.013 M to 3.3×10^{-3} M as a result of the common ion (Pb^{2+}) effect.

Check: You should also be able to predict the decrease in solubility due to a common-ion using Le Châtelier's principle. Adding Br^- or Pb^{2+} ions shifts the system to the left, thus decreasing the solubility of $PbBr_2$.

16.62 **(a)** The equilibrium reaction is:

	$BaSO_4(s) \rightleftharpoons$	$Ba^{2+}(aq)$ +	$SO_4^{2-}(aq)$
Initial (M):		0	0
Change (M):	$-s$	$+s$	$+s$
Equilibrium (M):		s	s

$$K_{sp} = [Ba^{2+}][SO_4^{2-}]$$

$$1.1 \times 10^{-10} = s^2$$

$$\mathbf{s = 1.0 \times 10^{-5}\ M}$$

The molar solubility of $BaSO_4$ in pure water is 1.0×10^{-5} mol/L.

(b) The initial concentration of SO_4^{2-} is 1.0 M.

	$BaSO_4(s) \rightleftharpoons$	$Ba^{2+}(aq)$ +	$SO_4^{2-}(aq)$
Initial (M):		0	1.0
Change (M):	$-s$	$+s$	$+s$
Equilibrium (M):		s	$1.0 + s$

$$K_{sp} = [Ba^{2+}][SO_4^{2-}]$$

$$1.1 \times 10^{-10} = (s)(1.0 + s) \approx (s)(1.0)$$

$$\mathbf{s = 1.1 \times 10^{-10}\ M}$$

Due to the common ion effect, the molar solubility of $BaSO_4$ decreases to 1.1×10^{-10} mol/L in 1.0 M $SO_4^{2-}(aq)$ compared to 1.0×10^{-5} mol/L in pure water.

16.64 **(b)** $SO_4^{2-}(aq)$ is a weak base
(c) $OH^-(aq)$ is a strong base
(d) $C_2O_4^{2-}(aq)$ is a weak base
(e) $PO_4^{3-}(aq)$ is a weak base.

The solubilities of the above will increase in acidic solution. Only (a), which contains an extremely weak base (I^- is the conjugate base of the strong acid HI) is unaffected by the acid solution.

16.66 From Table 16.2, the value of K_{sp} for iron(II) is 1.6×10^{-14}.

(a) At pH = 8.00, pOH = 14.00 − 8.00 = 6.00, and $[OH^-] = 1.0 \times 10^{-6}\ M$

$$[Fe^{2+}] = \frac{K_{sp}}{[OH^-]^2} = \frac{1.6 \times 10^{-14}}{(1.0 \times 10^{-6})^2} = 0.016\ M$$

The *molar solubility* of iron(II) hydroxide at pH = 8.00 is **0.016 *M***

(b) At pH = 10.00, pOH = 14.00 − 10.00 = 4.00, and $[OH^-] = 1.0 \times 10^{-4}\ M$

$$[Fe^{2+}] = \frac{K_{sp}}{[OH^-]^2} = \frac{1.6 \times 10^{-14}}{(1.0 \times 10^{-4})^2} = 1.6 \times 10^{-6}\ M$$

The *molar solubility* of iron(II) hydroxide at pH = 10.00 is **1.6×10^{-6} *M*.**

16.68 We first determine the effect of the added ammonia. Let's calculate the concentration of NH_3. This is a dilution problem.

$$M_iV_i = M_fV_f$$
$$(0.60\ M)(2.00\ \text{mL}) = M_f(1002\ \text{mL})$$
$$M_f = 0.0012\ M\ NH_3$$

Ammonia is a weak base ($K_b = 1.8 \times 10^{-5}$).

	NH_3 + H_2O	$\rightleftharpoons$	NH_4^+	+ OH^-
Initial (*M*):	0.0012		0	0
Change (*M*):	$-x$		$+x$	$+x$
Equil. (*M*):	$0.0012 - x$		x	x

$$K_b = \frac{[NH_4^+][OH^-]}{[NH_3]}$$

$$1.8 \times 10^{-5} = \frac{x^2}{(0.0012 - x)}$$

Solving the resulting quadratic equation gives $x = 0.00014$, or $[OH^-] = 0.00014\ M$

This is a solution of iron(II) sulfate, which contains Fe^{2+} ions. These Fe^{2+} ions could combine with OH^- to precipitate $Fe(OH)_2$. Therefore, we must use K_{sp} for iron(II) hydroxide. We compute the value of Q_c for this solution.

$$Fe(OH)_2(s) \rightleftharpoons Fe^{2+}(aq) + 2OH^-(aq)$$

$$Q = [Fe^{2+}]_0[OH^-]_0^2 = (1.0 \times 10^{-3})(0.00014)^2 = 2.0 \times 10^{-11}$$

Note that when adding 2.00 mL of NH_3 to 1.0 L of $FeSO_4$, the concentration of $FeSO_4$ will decrease slightly. However, rounding off to 2 significant figures, the concentration of 1.0×10^{-3} *M* does not change. Q is larger than K_{sp} $[Fe(OH)_2] = 1.6 \times 10^{-14}$. The concentrations of the ions in solution are greater than the equilibrium concentrations; the solution is saturated. The system will shift left to reestablish equilibrium; therefore, **a precipitate of $Fe(OH)_2$ will form**.

16.72 Complex Ion Equilibria and Solubility, Problem Type 5.

Strategy: The addition of $Cd(NO_3)_2$ to the NaCN solution results in complex ion formation. In solution, Cd^{2+} ions will complex with CN^- ions. The concentration of Cd^{2+} will be determined by the following equilibrium

$$Cd^{2+}(aq) + 4CN^-(aq) \rightleftharpoons Cd(CN)_4^{2-}$$

From Table 16.4 of the text, we see that the formation constant (K_f) for this reaction is very large ($K_f = 7.1 \times 10^{16}$). Because K_f is so large, the reaction lies mostly to the right. At equilibrium, the concentration of Cd^{2+} will be very small. As a good approximation, we can assume that essentially all the dissolved Cd^{2+} ions end up as $Cd(CN)_4^{2-}$ ions. What is the initial concentration of Cd^{2+} ions? A very small amount of Cd^{2+} will be present at equilibrium. Set up the K_f expression for the above equilibrium to solve for $[Cd^{2+}]$.

Solution: Calculate the initial concentration of Cd^{2+} ions.

$$[Cd^{2+}]_0 = \frac{0.50\ \text{g} \times \frac{1\ \text{mol}\ Cd(NO_3)_2}{236.42\ \text{g}\ Cd(NO_3)_2} \times \frac{1\ \text{mol}\ Cd^{2+}}{1\ \text{mol}\ Cd(NO_3)_2}}{0.50\ \text{L}} = 4.2 \times 10^{-3}\ M$$

If we assume that the above equilibrium goes to completion, we can write

	$Cd^{2+}(aq)$	+	$4CN^-(aq)$	$\longrightarrow$	$Cd(CN)_4^{2-}(aq)$
Initial (M):	4.2×10^{-3}		0.50		0
Change (M):	-4.2×10^{-3}		$-4(4.2 \times 10^{-3})$		$+4.2 \times 10^{-3}$
Final (M):	0		0.48		4.2×10^{-3}

To find the concentration of free Cd^{2+} at equilibrium, use the formation constant expression.

$$K_f = \frac{[Cd(CN)_4^{2-}]}{[Cd^{2+}][CN^-]^4}$$

Rearranging,

$$[Cd^{2+}] = \frac{[Cd(CN)_4^{2-}]}{K_f[CN^-]^4}$$

Substitute the equilibrium concentrations calculated above into the formation constant expression to calculate the equilibrium concentration of Cd^{2+}.

$$\mathbf{[Cd^{2+}]} = \frac{[Cd(CN)_4^{2-}]}{K_f[CN^-]^4} = \frac{4.2 \times 10^{-3}}{(7.1 \times 10^{16})(0.48)^4} = \mathbf{1.1 \times 10^{-18}\ \mathit{M}}$$

$$\mathbf{[CN^-]} = 0.48\ M + 4(1.1 \times 10^{-18}\ M) = \mathbf{0.48\ \mathit{M}}$$

$$\mathbf{[Cd(CN)_4^{2-}]} = (4.2 \times 10^{-3}\ M) - (1.1 \times 10^{-18}) = \mathbf{4.2 \times 10^{-3}\ \mathit{M}}$$

Check: Substitute the equilibrium concentrations calculated into the formation constant expression to calculate K_f. Also, the small value of $[Cd^{2+}]$ at equilibrium, compared to its initial concentration of 4.2×10^{-3} M, certainly justifies our approximation that almost all the Cd^{2+} ions react.

16.74 Silver iodide is only slightly soluble. It dissociates to form a small amount of Ag^+ and I^- ions. The Ag^+ ions then complex with NH_3 in solution to form the complex ion $Ag(NH_3)_2^+$. The balanced equations are:

$$AgI(s) \rightleftharpoons Ag^+(aq) + I^-(aq) \qquad K_{sp} = [Ag^+][I^-] = 8.3 \times 10^{-17}$$

$$Ag^+(aq) + 2NH_3(aq) \rightleftharpoons Ag(NH_3)_2^+(aq) \qquad K_f = \frac{[Ag(NH_3)_2^+]}{[Ag^+][NH_3]^2} = 1.5 \times 10^7$$

Overall: $$AgI(s) + 2NH_3(aq) \rightleftharpoons Ag(NH_3)_2^+(aq) + I^-(aq) \qquad K = K_{sp} \times K_f = 1.2 \times 10^{-9}$$

If s is the molar solubility of AgI then,

	$AgI(s)$ +	$2NH_3(aq)$	$\rightleftharpoons$	$Ag(NH_3)_2^+(aq)$ +	$I^-(aq)$
Initial (M):		1.0		0.0	0.0
Change (M):	$-s$	$-2s$		$+s$	$+s$
Equilibrium (M):		$(1.0 - 2s)$		s	s

Because K_f is large, we can assume all of the silver ions exist as $Ag(NH_3)_2^+$. Thus,

$$[Ag(NH_3)_2^+] = [I^-] = s$$

We can write the equilibrium constant expression for the above reaction, then solve for s.

$$K = 1.2 \times 10^{-9} = \frac{(s)(s)}{(1.0 - 2s)^2} \approx \frac{(s)(s)}{(1.0)^2}$$

$$\mathbf{s = 3.5 \times 10^{-5}\ M}$$

At equilibrium, 3.5×10^{-5} moles of AgI dissolves in 1 L of 1.0 M NH_3 solution.

16.76 **(a)** The equations are as follows:

$$CuI_2(s) \rightleftharpoons Cu^{2+}(aq) + 2I^-(aq)$$

$$\mathbf{Cu^{2+}(aq) + 4NH_3(aq) \rightleftharpoons [Cu(NH_3)_4]^{2+}(aq)}$$

The ammonia combines with the Cu^{2+} ions formed in the first step to form the complex ion $[Cu(NH_3)_4]^{2+}$, effectively removing the Cu^{2+} ions, causing the first equilibrium to shift to the right (resulting in more CuI_2 dissolving).

(b) Similar to part (a):

$$AgBr(s) \rightleftharpoons Ag^+(aq) + Br^-(aq)$$

$$\mathbf{Ag^+(aq) + 2CN^-(aq) \rightleftharpoons [Ag(CN)_2]^-(aq)}$$

(c) Similar to parts (a) and (b).

$$HgCl_2(s) \rightleftharpoons Hg^{2+}(aq) + 2Cl^-(aq)$$

$$\mathbf{Hg^{2+}(aq) + 4Cl^-(aq) \rightleftharpoons [HgCl_4]^{2-}(aq)}$$

16.80 Since some $PbCl_2$ precipitates, the solution is saturated. From Table 16.2, the value of K_{sp} for lead(II) chloride is 2.4×10^{-4}. The equilibrium is:

$$PbCl_2(aq) \rightleftharpoons Pb^{2+}(aq) + 2Cl^-(aq)$$

We can write the solubility product expression for the equilibrium.

$$K_{sp} = [Pb^{2+}][Cl^-]^2$$

K_{sp} and $[Cl^-]$ are known. Solving for the Pb^{2+} concentration,

$$\mathbf{[Pb^{2+}]} = \frac{K_{sp}}{[Cl^-]^2} = \frac{2.4 \times 10^{-4}}{(0.15)^2} = \mathbf{0.011\ M}$$

16.82 Chloride ion will precipitate Ag^+ but not Cu^{2+}. So, dissolve some solid in H_2O and add HCl. If a precipitate forms, the salt was $AgNO_3$. A flame test will also work. Cu^{2+} gives a green flame test.

16.84 We can use the Henderson-Hasselbalch equation to solve for the pH when the indicator is 90% acid / 10% conjugate base and when the indicator is 10% acid / 90% conjugate base.

$$\text{pH} = \text{p}K_a + \log\frac{[\text{conjugate base}]}{[\text{acid}]}$$

Solving for the pH with 90% of the indicator in the HIn form:

$$\text{pH} = 3.46 + \log\frac{[10]}{[90]} = 3.46 - 0.95 = 2.51$$

Next, solving for the pH with 90% of the indicator in the In^- form:

$$\text{pH} = 3.46 + \log\frac{[90]}{[10]} = 3.46 + 0.95 = 4.41$$

Thus the pH range varies from **2.51 to 4.41** as the [HIn] varies from 90% to 10%.

16.86 First, calculate the pH of the 2.00 *M* weak acid (HNO_2) solution before any NaOH is added.

	$HNO_2(aq)$	$\rightleftharpoons$	$H^+(aq)$	+	$NO_2^-(aq)$
Initial (*M*):	2.00		0		0
Change (*M*):	$-x$		$+x$		$+x$
Equilibrium (*M*):	$2.00 - x$		x		x

$$K_a = \frac{[H^+][NO_2^-]}{[HNO_2]}$$

$$4.5 \times 10^{-4} = \frac{x^2}{2.00 - x} \approx \frac{x^2}{2.00}$$

$$x = [H^+] = 0.030\ M$$

$$\text{pH} = -\log(0.030) = 1.52$$

Since the pH after the addition is 1.5 pH units greater, the new pH = 1.52 + 1.50 = 3.02.

From this new pH, we can calculate the $[H^+]$ in solution.

$$[H^+] = 10^{-\text{pH}} = 10^{-3.02} = 9.55 \times 10^{-4}\ M$$

When the NaOH is added, we dilute our original 2.00 M HNO_2 solution to:

$$M_iV_i = M_fV_f$$
$$(2.00\ M)(400\ \text{mL}) = M_f(600\ \text{mL})$$
$$M_f = 1.33\ M$$

Since we have not reached the equivalence point, we have a buffer solution. The reaction between HNO_2 and NaOH is:

$$HNO_2(aq) + NaOH(aq) \longrightarrow NaNO_2(aq) + H_2O(l)$$

Since the mole ratio between HNO_2 and NaOH is 1:1, the decrease in $[HNO_2]$ is the same as the decrease in [NaOH].

We can calculate the decrease in $[HNO_2]$ by setting up the weak acid equilibrium. From the pH of the solution, we know that the $[H^+]$ at equilibrium is 9.55×10^{-4} M.

	$HNO_2(aq)$	$\rightleftharpoons$	$H^+(aq)$	+	$NO_2^-(aq)$
Initial (M):	1.33		0		0
Change (M):	$-x$				$+x$
Equilibrium (M):	$1.33 - x$		9.55×10^{-4}		x

We can calculate x from the equilibrium constant expression.

$$K_a = \frac{[H^+][NO_2^-]}{[HNO_2]}$$

$$4.5 \times 10^{-4} = \frac{(9.55 \times 10^{-4})(x)}{1.33 - x}$$

$$x = 0.426\ M$$

Thus, x is the decrease in $[HNO_2]$ which equals the concentration of added OH^-. However, this is the concentration of NaOH after it has been diluted to 600 mL. We need to correct for the dilution from 200 mL to 600 mL to calculate the concentration of the original NaOH solution.

$$M_iV_i = M_fV_f$$
$$M_i(200\ \text{mL}) = (0.426\ M)(600\ \text{mL})$$
$$\mathbf{[NaOH]} = M_i = \mathbf{1.28\ M}$$

16.88 The resulting solution is not a buffer system. There is excess NaOH and the neutralization is well past the equivalence point.

$$\text{Moles NaOH} = 0.500\ \cancel{L} \times \frac{0.167\ \text{mol}}{1\ \cancel{L}} = 0.0835\ \text{mol}$$

$$\text{Moles HCOOH} = 0.500\ \cancel{L} \times \frac{0.100\ \text{mol}}{1\ \cancel{L}} = 0.0500\ \text{mol}$$

	$HCOOH(aq)$ +	$NaOH(aq)$	$\rightarrow$ $HCOONa(aq)$ + $H_2O(l)$
Initial (mol):	0.0500	0.0835	0
Change (mol):	−0.0500	−0.0500	+0.0500
Final (mol):	0	0.0335	0.0500

The volume of the resulting solution is 1.00 L (500 mL + 500 mL = 1000 mL).

$$[\mathbf{OH^-}] = \frac{0.0335\text{ mol}}{1.00\text{ L}} = \mathbf{0.0335}\ \boldsymbol{M}$$

$$[\mathbf{Na^+}] = \frac{(0.0335 + 0.0500)\text{ mol}}{1.00\text{ L}} = \mathbf{0.0835}\ \boldsymbol{M}$$

$$[\mathbf{H^+}] = \frac{K_w}{[OH^-]} = \frac{1.0\times10^{-14}}{0.0335} = \mathbf{3.0\times10^{-13}}\ \boldsymbol{M}$$

$$[\mathbf{HCOO^-}] = \frac{0.0500\text{ mol}}{1.00\text{ L}} = \mathbf{0.0500}\ \boldsymbol{M}$$

	$HCOO^-(aq) + H_2O(l)$	$\rightleftharpoons$ $HCOOH(aq)$	$+\ OH^-(aq)$
Initial (M):	0.0500	0	0.0335
Change (M):	$-x$	$+x$	$+x$
Equilibrium (M):	$0.0500 - x$	x	$0.0335 + x$

$$K_b = \frac{[HCOOH][OH^-]}{[HCOO^-]}$$

$$5.9\times10^{-11} = \frac{(x)(0.0335 + x)}{(0.0500 - x)} \approx \frac{(x)(0.0335)}{(0.0500)}$$

$$x = [\mathbf{HCOOH}] = \mathbf{8.8\times10^{-11}}\ \boldsymbol{M}$$

16.90 The number of moles of $Ba(OH)_2$ present in the original 50.0 mL of solution is:

$$50.0\ \text{mL} \times \frac{1.00\text{ mol }Ba(OH)_2}{1000\text{ mL soln}} = 0.0500\text{ mol }Ba(OH)_2$$

The number of moles of H_2SO_4 present in the original 86.4 mL of solution, assuming complete dissociation, is:

$$86.4\ \text{mL} \times \frac{0.494\text{ mol }H_2SO_4}{1000\text{ mL soln}} = 0.0427\text{ mol }H_2SO_4$$

The reaction is:

	$Ba(OH)_2(aq)$	$+\ H_2SO_4(aq)$	$\rightarrow\ BaSO_4(s)$	$+\ 2H_2O(l)$
Initial (mol):	0.0500	0.0427	0	
Change (mol):	−0.0427	−0.0427	+0.0427	
Final (mol):	0.0073	0	0.0427	

Thus the mass of $BaSO_4$ formed is:

$$0.0427\text{ mol }BaSO_4 \times \frac{233.4\text{ g }BaSO_4}{1\text{ mol }BaSO_4} = \mathbf{9.97\ g\ BaSO_4}$$

The pH can be calculated from the excess OH^- in solution. First, calculate the molar concentration of OH^-. The total volume of solution is 136.4 mL = 0.1364 L.

$$[OH^-] = \frac{0.0073 \text{ mol Ba(OH)}_2 \times \frac{2 \text{ mol OH}^-}{1 \text{ mol Ba(OH)}_2}}{0.1364 \text{ L}} = 0.11\ M$$

$$pOH = -\log(0.11) = 0.96$$

$$\mathbf{pH} = 14.00 - pOH = 14.00 - 0.96 = \mathbf{13.04}$$

16.92 First, we calculate the molar solubility of $CaCO_3$.

	$CaCO_3(s)$	$\rightleftharpoons$	$Ca^{2+}(aq)$	+	$CO_3^{2-}(aq)$
Initial (M):			0		0
Change (M):	$-s$		$+s$		$+s$
Equil. (M):			s		s

$$K_{sp} = [Ca^{2+}][CO_3^{2-}] = s^2 = 8.7 \times 10^{-9}$$

$$s = 9.3 \times 10^{-5}\ M = 9.3 \times 10^{-5} \text{ mol/L}$$

The moles of $CaCO_3$ in the kettle are:

$$116 \text{ g} \times \frac{1 \text{ mol CaCO}_3}{100.1 \text{ g CaCO}_3} = 1.16 \text{ mol CaCO}_3$$

The volume of distilled water needed to dissolve 1.16 moles of $CaCO_3$ is:

$$1.16 \text{ mol CaCO}_3 \times \frac{1 \text{ L}}{9.3 \times 10^{-5} \text{ mol CaCO}_3} = 1.2 \times 10^4 \text{ L}$$

The number of times the kettle would have to be filled is:

$$(1.2 \times 10^4 \text{ L}) \times \frac{1 \text{ filling}}{2.0 \text{ L}} = \mathbf{6.0 \times 10^3 \text{ fillings}}$$

Note that the very important assumption is made that each time the kettle is filled, the calcium carbonate is allowed to reach equilibrium before the kettle is emptied.

16.94 First we find the molar solubility and then convert moles to grams. The solubility equilibrium for silver carbonate is:

	$Ag_2CO_3(s)$	$\rightleftharpoons$	$2Ag^+(aq)$	+	$CO_3^{2-}(aq)$
Initial (M):			0		0
Change (M):	$-s$		$+2s$		$+s$
Equilibrium (M):			$2s$		s

$$K_{sp} = [Ag^+]^2[CO_3^{2-}] = (2s)^2(s) = 4s^3 = 8.1 \times 10^{-12}$$

$$s = \left(\frac{8.1 \times 10^{-12}}{4}\right)^{\frac{1}{3}} = 1.3 \times 10^{-4}\ M$$

Converting from mol/L to g/L:

$$\frac{1.3 \times 10^{-4} \text{ mol}}{1 \text{ L soln}} \times \frac{275.8 \text{ g}}{1 \text{ mol}} = \mathbf{0.036 \text{ g/L}}$$

16.96 **(a)** To 2.50×10^{-3} mole HCl (that is, 0.0250 L of 0.100 *M* solution) is added 1.00×10^{-3} mole CH_3NH_2 (that is, 0.0100 L of 0.100 *M* solution).

	$HCl(aq)$ +	$CH_3NH_2(aq)$ →	$CH_3NH_3Cl(aq)$
Initial (mol):	2.50×10^{-3}	1.00×10^{-3}	0
Change (mol):	-1.00×10^{-3}	-1.00×10^{-3}	$+1.00 \times 10^{-3}$
Equilibrium (mol):	1.50×10^{-3}	0	1.00×10^{-3}

After the acid-base reaction, we have 1.50×10^{-3} mol of HCl remaining. Since HCl is a strong acid, the $[H^+]$ will come from the HCl. The total solution volume is 35.0 mL = 0.0350 L.

$$[H^+] = \frac{1.50 \times 10^{-3} \text{ mol}}{0.0350 \text{ L}} = 0.0429 \; M$$

pH = 1.37

(b) When a total of 25.0 mL of CH_3NH_2 is added, we reach the equivalence point. That is, 2.50×10^{-3} mol HCl reacts with 2.50×10^{-3} mol CH_3NH_2 to form 2.50×10^{-3} mol CH_3NH_3Cl. Since there is a total of 50.0 mL of solution, the concentration of $CH_3NH_3^+$ is:

$$[CH_3NH_3^+] = \frac{2.50 \times 10^{-3} \text{ mol}}{0.0500 \text{ L}} = 5.00 \times 10^{-2} \; M$$

This is a problem involving the hydrolysis of the weak acid $CH_3NH_3^+$.

	$CH_3NH_3^+(aq)$ ⇌	$H^+(aq)$ +	$CH_3NH_2(aq)$
Initial (*M*):	5.00×10^{-2}	0	0
Change (*M*):	$-x$	$+x$	$+x$
Equilibrium (*M*):	$(5.00 \times 10^{-2}) - x$	x	x

$$K_a = \frac{[CH_3NH_2][H^+]}{[CH_3NH_3^+]}$$

$$2.3 \times 10^{-11} = \frac{x^2}{(5.00 \times 10^{-2}) - x} \approx \frac{x^2}{5.00 \times 10^{-2}}$$

$$1.15 \times 10^{-12} = x^2$$

$$x = 1.07 \times 10^{-6} \; M = [H^+]$$

pH = 5.97

(c) 35.0 mL of 0.100 M CH_3NH_2 (3.50×10^{-3} mol) is added to the 25 mL of 0.100 M HCl (2.50×10^{-3} mol).

	$HCl(aq)$ +	$CH_3NH_2(aq)$ →	$CH_3NH_3Cl(aq)$
Initial (mol):	2.50×10^{-3}	3.50×10^{-3}	0
Change (mol):	-2.50×10^{-3}	-2.50×10^{-3}	$+2.50 \times 10^{-3}$
Equilibrium (mol):	0	1.00×10^{-3}	2.50×10^{-3}

This is a buffer solution. Using the Henderson-Hasselbalch equation:

$$\text{pH} = \text{p}K_a + \log\frac{[\text{conjugate base}]}{[\text{acid}]}$$

$$\textbf{pH} = -\log(2.3 \times 10^{-11}) + \log\frac{(1.00 \times 10^{-3})}{(2.50 \times 10^{-3})} = \textbf{10.24}$$

16.98 The precipitate is HgI_2.

$$Hg^{2+}(aq) + 2I^-(aq) \longrightarrow HgI_2(s)$$

With further addition of I^-, a soluble complex ion is formed and the precipitate redissolves.

$$HgI_2(s) + 2I^-(aq) \longrightarrow HgI_4^{2-}(aq)$$

16.100 We can use the Henderson-Hasselbalch equation to solve for the pH when the indicator is 95% acid / 5% conjugate base and when the indicator is 5% acid / 95% conjugate base.

$$\text{pH} = \text{p}K_a + \log\frac{[\text{conjugate base}]}{[\text{acid}]}$$

Solving for the pH with 95% of the indicator in the HIn form:

$$\text{pH} = 9.10 + \log\frac{[5]}{[95]} = 9.10 - 1.28 = 7.82$$

Next, solving for the pH with 95% of the indicator in the In^- form:

$$\text{pH} = 9.10 + \log\frac{[95]}{[5]} = 9.10 + 1.28 = 10.38$$

Thus the pH range varies from **7.82 to 10.38** as the [HIn] varies from 95% to 5%.

16.102 **(a)** We abbreviate the name of cacodylic acid to CacH. We set up the usual table.

	$CacH(aq)$ ⇌	$Cac^-(aq)$ +	$H^+(aq)$
Initial (M):	0.10	0	0
Change (M):	$-x$	$+x$	$+x$
Equilibrium (M):	$0.10 - x$	x	x

$$K_a = \frac{[H^+][Cac^-]}{[CacH]}$$

$$6.4 \times 10^{-7} = \frac{x^2}{0.10 - x} \approx \frac{x^2}{0.10}$$

$$x = 2.5 \times 10^{-4}\ M = [H^+]$$

$$\textbf{pH} = -\log(2.5 \times 10^{-4}) = \textbf{3.60}$$

(b) We set up a table for the hydrolysis of the anion:

	$Cac^-(aq)$ + $H_2O(l)$ $\rightleftharpoons$	$CacH(aq)$ +	$OH^-(aq)$
Initial (M):	0.15	0	0
Change (M):	$-x$	$+x$	$+x$
Equilibrium (M):	$0.15 - x$	x	x

The ionization constant, K_b, for Cac^- is:

$$K_b = \frac{K_w}{K_a} = \frac{1.0 \times 10^{-14}}{6.4 \times 10^{-7}} = 1.6 \times 10^{-8}$$

$$K_b = \frac{[CacH][OH^-]}{[Cac^-]}$$

$$1.6 \times 10^{-8} = \frac{x^2}{0.15 - x} \approx \frac{x^2}{0.15}$$

$$x = 4.9 \times 10^{-5}\ M$$

$$pOH = -\log(4.9 \times 10^{-5}) = 4.31$$

$$\textbf{pH} = 14.00 - 4.31 = \textbf{9.69}$$

(c) Number of moles of CacH from (a) is:

$$50.0\ \text{mL CacH} \times \frac{0.10\ \text{mol CacH}}{1000\ \text{mL}} = 5.0 \times 10^{-3}\ \text{mol CacH}$$

Number of moles of Cac^- from (b) is:

$$25.0\ \text{mL CacNa} \times \frac{0.15\ \text{mol CacNa}}{1000\ \text{mL}} = 3.8 \times 10^{-3}\ \text{mol CacNa}$$

At this point we have a buffer solution.

$$\textbf{pH} = pK_a + \log\frac{[Cac^-]}{[CacH]} = -\log(6.4 \times 10^{-7}) + \log\frac{3.8 \times 10^{-3}}{5.0 \times 10^{-3}} = \textbf{6.07}$$

16.104 (a) $MCO_3 + 2HCl \rightarrow MCl_2 + H_2O + CO_2$

$HCl + NaOH \rightarrow NaCl + H_2O$

(b) We are given the mass of the metal carbonate, so we need to find moles of the metal carbonate to calculate its molar mass. We can find moles of MCO_3 from the moles of HCl reacted.

$$\text{Moles of HCl reacted with MCO}_3 = \text{Total moles of HCl} - \text{Moles of excess HCl}$$

$$\text{Total moles of HCl} = 20.00 \text{ mL} \times \frac{0.0800 \text{ mol}}{1000 \text{ mL soln}} = 1.60 \times 10^{-3} \text{ mol HCl}$$

$$\text{Moles of excess HCl} = 5.64 \text{ mL} \times \frac{0.1000 \text{ mol}}{1000 \text{ mL soln}} = 5.64 \times 10^{-4} \text{ mol HCl}$$

$$\text{Moles of HCl reacted with MCO}_3 = (1.60 \times 10^{-3} \text{ mol}) - (5.64 \times 10^{-4} \text{ mol}) = 1.04 \times 10^{-3} \text{ mol HCl}$$

$$\text{Moles of MCO}_3 \text{ reacted} = (1.04 \times 10^{-3} \text{ mol HCl}) \times \frac{1 \text{ mol MCO}_3}{2 \text{ mol HCl}} = 5.20 \times 10^{-4} \text{ mol MCO}_3$$

$$\textbf{Molar mass of MCO}_3 = \frac{0.1022 \text{ g}}{5.20 \times 10^{-4} \text{ mol}} = \textbf{197 g/mol}$$

$$\text{Molar mass of CO}_3 = 60.01 \text{ g}$$

$$\text{Molar mass of M} = 197 \text{ g/mol} - 60.01 \text{ g/mol} = 137 \text{ g/mol}$$

The metal, M, is **Ba**!

16.106 The number of moles of NaOH reacted is:

$$15.9 \text{ mL NaOH} \times \frac{0.500 \text{ mol NaOH}}{1000 \text{ mL soln}} = 7.95 \times 10^{-3} \text{ mol NaOH}$$

Since two moles of NaOH combine with one mole of oxalic acid, the number of moles of oxalic acid reacted is 3.98×10^{-3} mol. This is the number of moles of oxalic acid hydrate in 25.0 mL of solution. In 250 mL, the number of moles present is 3.98×10^{-2} mol. Thus the molar mass is:

$$\frac{5.00 \text{ g}}{3.98 \times 10^{-2} \text{ mol}} = 126 \text{ g/mol}$$

From the molecular formula we can write:

$$2(1.008)\text{g} + 2(12.01)\text{g} + 4(16.00)\text{g} + x(18.02)\text{g} = 126 \text{ g}$$

Solving for x:

$$\boldsymbol{x = 2}$$

16.108 **(a)** $\text{pH} = \text{p}K_\text{a} + \log\dfrac{[\text{conjugate base}]}{[\text{acid}]}$

$$8.00 = 9.10 + \log\frac{[\text{ionized}]}{[\text{un-ionized}]}$$

$$\frac{[\text{un-ionized}]}{[\text{ionized}]} = \textbf{12.6} \qquad (1)$$

(b) First, let's calculate the total concentration of the indicator. 2 drops of the indicator are added and each drop is 0.050 mL.

$$2\ \text{drops} \times \frac{0.050\ \text{mL phenolphthalein}}{1\ \text{drop}} = 0.10\ \text{mL phenolphthalein}$$

This 0.10 mL of phenolphthalein of concentration 0.060 M is diluted to 50.0 mL.

$$M_iV_i = M_fV_f$$
$$(0.060\ M)(0.10\ \text{mL}) = M_f(50.0\ \text{mL})$$
$$M_f = 1.2 \times 10^{-4}\ M$$

Using equation (1) above and letting y = [ionized], then [un-ionized] = $(1.2 \times 10^{-4}) - y$.

$$\frac{(1.2 \times 10^{-4}) - y}{y} = 12.6$$

$$\mathbf{y = 8.8 \times 10^{-6}\ M}$$

16.110 **(a)** Add sulfate. Na_2SO_4 is soluble, $BaSO_4$ is not.
(b) Add sulfide. K_2S is soluble, PbS is not
(c) Add iodide. ZnI_2 is soluble, HgI_2 is not.

16.112 The amphoteric oxides cannot be used to prepare buffer solutions because they are insoluble in water.

16.114 The ionized polyphenols have a dark color. In the presence of citric acid from lemon juice, the anions are converted to the lighter-colored acids.

16.116 Assuming the density of water to be 1.00 g/mL, 0.05 g Pb^{2+} per 10^6 g water is equivalent to 5×10^{-5} g Pb^{2+}/L

$$\frac{0.05\ \text{g Pb}^{2+}}{1 \times 10^6\ \text{g H}_2\text{O}} \times \frac{1\ \text{g H}_2\text{O}}{1\ \text{mL H}_2\text{O}} \times \frac{1000\ \text{mL H}_2\text{O}}{1\ \text{L H}_2\text{O}} = 5 \times 10^{-5}\ \text{g Pb}^{2+}/\text{L}$$

	$PbSO_4$	$\rightleftharpoons$	Pb^{2+}	+	SO_4^{2-}
Initial (M):			0		0
Change (M):	$-s$		$+s$		$+s$
Equilibrium (M):			s		s

$$K_{sp} = [Pb^{2+}][SO_4^{2-}]$$
$$1.6 \times 10^{-8} = s^2$$
$$s = 1.3 \times 10^{-4}\ M$$

The solubility of $PbSO_4$ in g/L is:

$$\frac{1.3 \times 10^{-4}\ \text{mol}}{1\ \text{L}} \times \frac{303.3\ \text{g}}{1\ \text{mol}} = 4.0 \times 10^{-2}\ \text{g/L}$$

Yes. The $[Pb^{2+}]$ exceeds the safety limit of 5×10^{-5} g Pb^{2+}/L.

16.118 **(c)** has the highest $[H^+]$

$F^- + SbF_5 \rightarrow SbF_6^-$

Removal of F^- promotes further ionization of HF.

6.120 **(a)** This is a common ion (CO_3^{2-}) problem.

The dissociation of Na_2CO_3 is:

$$Na_2CO_3(s) \xrightarrow{H_2O} 2Na^+(aq) + CO_3^{2-}(aq)$$
$$\qquad\qquad\qquad 2(0.050\ M) \qquad 0.050\ M$$

Let s be the molar solubility of $CaCO_3$ in Na_2CO_3 solution. We summarize the changes as:

$$CaCO_3(s) \rightleftharpoons Ca^{2+}(aq) + CO_3^{2-}(aq)$$

	$Ca^{2+}(aq)$	$CO_3^{2-}(aq)$
Initial (M):	0.00	0.050
Change (M):	$+s$	$+s$
Equil. (M):	$+s$	$0.050 + s$

$$K_{sp} = [Ca^{2+}][CO_3^{2-}]$$

$$8.7 \times 10^{-9} = s(0.050 + s)$$

Since s is small, we can assume that $0.050 + s \approx 0.050$

$$8.7 \times 10^{-9} = 0.050s$$

$$\mathbf{s = 1.7 \times 10^{-7}\ M}$$

Thus, the addition of washing soda to permanent hard water removes most of the Ca^{2+} ions as a result of the common ion effect.

(b) Mg^{2+} is not removed by this procedure, because $MgCO_3$ is fairly soluble ($K_{sp} = 4.0 \times 10^{-5}$).

(c) The K_{sp} for $Ca(OH)_2$ is 8.0×10^{-6}.

$$Ca(OH)_2 \rightleftharpoons Ca^{2+} + 2OH^-$$

At equil.: $\quad s \quad 2s$

$$K_{sp} = 8.0 \times 10^{-6} = [Ca^{2+}][OH^-]^2$$

$$4s^3 = 8.0 \times 10^{-6}$$

$$s = 0.0126\ M$$

$$[OH^-] = 2s = 0.0252\ M$$

$$pOH = -\log(0.0252) = 1.60$$

$$\mathbf{pH = 12.40}$$

(d) The $[OH^-]$ calculated above is 0.0252 *M*. At this rather high concentration of OH^-, most of the Mg^{2+} will be removed as $Mg(OH)_2$. The small amount of Mg^{2+} remaining in solution is due to the following equilibrium:

$$Mg(OH)_2(s) \rightleftharpoons Mg^{2+}(aq) + 2OH^-(aq)$$

$$K_{sp} = [Mg^{2+}][OH^-]^2$$

$$1.2 \times 10^{-11} = [Mg^{2+}](0.0252)^2$$

$$\mathbf{[Mg^{2+}] = 1.9 \times 10^{-8}\ \mathit{M}}$$

(e) Remove Ca^{2+} first because it is present in larger amounts.

16.122 $\text{pH} = \text{p}K_a + \log\frac{[\text{conjugate base}]}{[\text{acid}]}$

At pH = 1.0,

–COOH

$$1.0 = 2.3 + \log\frac{[-COO^-]}{[-COOH]}$$

$$\frac{[-COOH]}{[-COO^-]} = 20$$

$-NH_3^+$

$$1.0 = 9.6 + \log\frac{[-NH_2]}{[-NH_3^+]}$$

$$\frac{[-NH_3^+]}{[-NH_2]} = 4 \times 10^8$$

Therefore the **predominant species** is: $\mathbf{^+NH_3 - CH_2 - COOH}$

At pH = 7.0,

–COOH

$$7.0 = 2.3 + \log\frac{[-COO^-]}{[-COOH]}$$

$$\frac{[-COO^-]}{[-COOH]} = 5 \times 10^4$$

$-NH_3^+$

$$7.0 = 9.6 + \log\frac{[-NH_2]}{[-NH_3^+]}$$

$$\frac{[-NH_3^+]}{[-NH_2]} = 4 \times 10^2$$

Predominant species: $\mathbf{^+NH_3 - CH_2 - COO^-}$

At pH = 12.0,

–COOH $$12.0 = 2.3 + \log\frac{[\text{-COO}^-]}{[\text{-COOH}]}$$

$$\frac{[\text{-COO}^-]}{[\text{-COOH}]} = 5 \times 10^9$$

$-NH_3^+$ $$12.0 = 9.6 + \log\frac{[\text{-NH}_2]}{[\text{-NH}_3^+]}$$

$$\frac{[\text{-NH}_2]}{[\text{-NH}_3^+]} = 2.5 \times 10^2$$

Predominant species: $\mathbf{NH_2 - CH_2 - COO^-}$

16.124 (a) Before dilution:

$$\text{pH} = \text{p}K_\text{a} + \log\frac{[\text{CH}_3\text{COO}^-]}{[\text{CH}_3\text{COOH}]}$$

$$\text{pH} = 4.74 + \log\frac{[0.500]}{[0.500]} = 4.74$$

After a 10-fold dilution:

$$\text{pH} = 4.74 + \log\frac{[0.0500]}{[0.0500]} = 4.74$$

There is no change in the pH of a buffer upon dilution.

(b) Before dilution:

	$CH_3COOH(aq) + H_2O(l)$	$\rightleftharpoons$	$H_3O^+(aq)$	+	$CH_3COO^-(aq)$
Initial (M):	0.500		0		0
Change (M):	$-x$		$+x$		$+x$
Equilibrium (M):	$0.500 - x$		x		x

$$K_\text{a} = \frac{[\text{H}_3\text{O}^+][\text{CH}_3\text{COO}^-]}{[\text{CH}_3\text{COOH}]}$$

$$1.8 \times 10^{-5} = \frac{x^2}{0.500 - x} \approx \frac{x^2}{0.500}$$

$$x = 3.0 \times 10^{-3}\ M = [\text{H}_3\text{O}^+]$$

$$\text{pH} = -\log(3.0 \times 10^{-3}) = 2.52$$

After dilution:

$$1.8 \times 10^{-5} = \frac{x^2}{0.0500 - x} \approx \frac{x^2}{0.0500}$$

$$x = 9.5 \times 10^{-4}\ M = [H_3O^+]$$

$$pH = -\log(9.5 \times 10^{-4}) = 3.02$$

16.126 The reaction is:

$$NH_3 + HCl \rightarrow NH_4Cl$$

First, we calculate moles of HCl and NH_3.

$$n_{HCl} = \frac{PV}{RT} = \frac{\left(372\ \text{mmHg} \times \frac{1\ \text{atm}}{760\ \text{mmHg}}\right)(0.96\ \text{L})}{\left(0.0821\frac{\text{L}\cdot\text{atm}}{\text{mol}\cdot\text{K}}\right)(295\ \text{K})} = 0.0194\ \text{mol}$$

$$n_{NH_3} = \frac{0.57\ \text{mol NH}_3}{1\ \text{L soln}} \times 0.034\ \text{L} = 0.0194\ \text{mol}$$

The mole ratio between NH_3 and HCl is 1:1, so we have complete neutralization.

	NH_3	+	HCl	$\rightarrow$	NH_4Cl
Initial (mol):	0.0194		0.0194		0
Change (mol):	−0.0194		−0.0194		+0.0194
Final (mol):	0		0		0.0194

NH_4^+ is a weak acid. We set up the reaction representing the hydrolysis of NH_4^+.

	$NH_4^+(aq)$	+	$H_2O(l)$	$\rightleftharpoons$	$H_3O^+(aq)$	+	$NH_3(aq)$
Initial (M):	0.0194 mol/0.034 L				0		0
Change (M):	$-x$				$+x$		$+x$
Equilibrium (M):	$0.57 - x$				x		x

$$K_a = \frac{[H_3O^+][NH_3]}{[NH_4^+]}$$

$$5.6 \times 10^{-10} = \frac{x^2}{0.57 - x} \approx \frac{x^2}{0.57}$$

$$x = 1.79 \times 10^{-5}\ M = [H_3O^+]$$

$$\mathbf{pH} = -\log(1.79 \times 10^{-5}) = \mathbf{4.75}$$

16.128 The reaction is:

$$Ag_2CO_3(aq) + 2HCl(aq) \rightarrow 2AgCl(s) + CO_2(g) + H_2O(l)$$

The moles of $CO_2(g)$ produced will equal the moles of CO_3^{2-} in Ag_2CO_3 due to the stoichiometry of the reaction.

$$n_{CO_2} = \frac{PV}{RT} = \frac{\left(114\text{ mmHg} \times \frac{1\text{ atm}}{760\text{ mmHg}}\right)(0.019\text{ L})}{\left(0.0821\frac{\text{L}\cdot\text{atm}}{\text{mol}\cdot\text{K}}\right)(298\text{ K})} = 1.16 \times 10^{-4}\text{ mol}$$

Because the solution volume is 1.0 L, $[CO_3^{2-}] = 1.16 \times 10^{-4}\ M$. We set up a table representing the dissociation of Ag_2CO_3 to solve for K_{sp}.

	Ag_2CO_3	$\rightleftharpoons$	$2Ag^+$	+	CO_3^{2-}
Equilibrium (M):			$2(1.16 \times 10^{-4})$		1.16×10^{-4}

$$K_{sp} = [Ag^+]^2[CO_3^{2-}]$$

$$K_{sp} = (2.32 \times 10^{-4})^2(1.16 \times 10^{-4})$$

$$\boldsymbol{K_{sp} = 6.2 \times 10^{-12}} \text{ (at 5°C)}$$

CHAPTER 17
CHEMISTRY IN THE ATMOSPHERE

Almost all the types of problems in this chapter have been encountered in previous chapters. Therefore, we will not repeat the problem types here, but will refer you to problem types where appropriate.

SOLUTIONS TO SELECTED TEXT PROBLEMS

17.6 Using the information in Table 17.1 and Problem 17.5, 0.033 percent of the volume (and therefore the pressure) of dry air is due to CO_2. The partial pressure of CO_2 is:

$$P_{CO_2} = X_{CO_2}P_T = (3.3 \times 10^{-4})(754 \text{ mmHg}) \times \frac{1 \text{ atm}}{760 \text{ mmHg}} = \mathbf{3.3 \times 10^{-4} \text{ atm}}$$

17.8 From Problem 5.102, the total mass of air is 5.25×10^{18} kg. Table 17.1 lists the composition of air by volume. Under the same conditions of P and T, $V \propto n$ (Avogadro's law).

$$\text{Total moles of gases} = (5.25 \times 10^{21} \text{ g}) \times \frac{1 \text{ mol}}{29.0 \text{ g}} = 1.81 \times 10^{20} \text{ mol}$$

Mass of N_2 (78.03%):

$$(0.7803)(1.81 \times 10^{20} \text{ mol}) \times \frac{28.02 \text{ g}}{1 \text{ mol}} = 3.96 \times 10^{21} \text{ g} = \mathbf{3.96 \times 10^{18} \text{ kg}}$$

Mass of O_2 (20.99%):

$$(0.2099)(1.81 \times 10^{20} \text{ mol}) \times \frac{32.00 \text{ g}}{1 \text{ mol}} = 1.22 \times 10^{21} \text{ g} = \mathbf{1.22 \times 10^{18} \text{ kg}}$$

Mass of CO_2 (0.033%):

$$(3.3 \times 10^{-4})(1.81 \times 10^{20} \text{ mol}) \times \frac{44.01 \text{ g}}{1 \text{ mol}} = 2.63 \times 10^{18} \text{ g} = \mathbf{2.63 \times 10^{15} \text{ kg}}$$

17.12 See Problem Types 1 and 2, Chapter 7.

Strategy: We are given the wavelength of the emitted photon and asked to calculate its energy. Equation (7.2) of the text relates the energy and frequency of an electromagnetic wave.

$$E = h\nu$$

First, we calculate the frequency from the wavelength, then we can calculate the energy difference between the two levels.

Solution: Calculate the frequency from the wavelength.

$$\nu = \frac{c}{\lambda} = \frac{3.00 \times 10^{8} \text{ m/s}}{558 \times 10^{-9} \text{ m}} = 5.38 \times 10^{14} \text{ /s}$$

Now, we can calculate the energy difference from the frequency.

$$\Delta E = h\nu = (6.63 \times 10^{-34}\ \text{J}\cdot\text{s})(5.38 \times 10^{14}\ /\text{s})$$
$$\mathbf{\Delta E = 3.57 \times 10^{-19}\ J}$$

17.22 The quantity of ozone lost is:

$$(0.06)(3.2 \times 10^{12}\ \text{kg}) = 1.9 \times 10^{11}\ \text{kg of } O_3$$

Assuming no further deterioration, the kilograms of O_3 that would have to be manufactured on a daily basis are:

$$\frac{1.9 \times 10^{11}\ \text{kg } O_3}{100\ \text{yr}} \times \frac{1\ \text{yr}}{365\ \text{days}} = \mathbf{5.2 \times 10^6\ kg/day}$$

The standard enthalpy of formation (from Appendix 3 of the text) for ozone:

$$\tfrac{3}{2}O_2 \rightarrow O_3 \qquad \Delta H_f^\circ = 142.2\ \text{kJ/mol}$$

The *total* energy required is:

$$(1.9 \times 10^{14}\ \text{g of } O_3) \times \frac{1\ \text{mol } O_3}{48.00\ \text{g } O_3} \times \frac{142.2\ \text{kJ}}{1\ \text{mol } O_3} = \mathbf{5.6 \times 10^{14}\ kJ}$$

17.24 The energy of the photons of UV radiation in the troposphere is insufficient (that is, the wavelength is too long and the frequency is too small) to break the bonds in CFCs.

17.26 First, we need to calculate the energy needed to break one bond.

$$\frac{276 \times 10^3\ \text{J}}{1\ \text{mol}} \times \frac{1\ \text{mol}}{6.022 \times 10^{23}\ \text{molecules}} = 4.58 \times 10^{-19}\ \text{J/molecule}$$

The longest wavelength required to break this bond is:

$$\lambda = \frac{hc}{E} = \frac{(3.00 \times 10^8\ \text{m/s})(6.63 \times 10^{-34}\ \text{J}\cdot\text{s})}{4.58 \times 10^{-19}\ \text{J}} = 4.34 \times 10^{-7}\ \text{m} = \mathbf{434\ nm}$$

434 nm is in the visible region of the electromagnetic spectrum; therefore, CF_3Br will be decomposed in **both** the troposphere and stratosphere.

17.28 The Lewis structure of HCFC–123 is:

```
   F   H
   |   |
F—C—C—Cl
   |   |
   F   Cl
```

The Lewis structure for CF_3CFH_2 is:

```
   F   H
   |   |
F—C—C—H
   |   |
   F   F
```

Lone pairs on the outer atoms have been omitted.

17.40 See Problem Type 7, Chapter 3.

Strategy: Looking at the balanced equation, how do we compare the amounts of CaO and CO_2? We can compare them based on the mole ratio from the balanced equation.

Solution: Because the balanced equation is given in the problem, the mole ratio between CaO and CO_2 is known: 1 mole CaO $\simeq$ 1 mole CO_2. If we convert grams of CaO to moles of CaO, we can use this mole ratio to convert to moles of CO_2. Once moles of CO_2 are known, we can convert to grams CO_2.

$$\text{mass } CO_2 = (1.7 \times 10^{13} \text{ g CaO}) \times \frac{1 \text{ mol CaO}}{56.08 \text{ g CaO}} \times \frac{1 \text{ mol } CO_2}{1 \text{ mol CaO}} \times \frac{44.01 \text{ g}}{1 \text{ mol } CO_2}$$

$$= 1.3 \times 10^{13} \text{ g } CO_2 = \mathbf{1.3 \times 10^{10} \text{ kg } CO_2}$$

17.42 Ethane and propane are greenhouse gases. They would contribute to global warming.

17.50 Recall that ppm means the number of parts of substance per 1,000,000 parts. We can calculate the partial pressure of SO_2 in the troposphere.

$$P_{SO_2} = \frac{0.16 \text{ molecules of } SO_2}{10^6 \text{ parts of air}} \times 1 \text{ atm} = 1.6 \times 10^{-7} \text{ atm}$$

Next, we need to set up the equilibrium constant expression to calculate the concentration of H^+ in the rainwater. From the concentration of H^+, we can calculate the pH.

	SO_2	+	H_2O	$\rightleftharpoons$	H^+	+	HSO_3^-
Equilibrium:	1.6×10^{-7} atm				x		x

$$K = \frac{[H^+][HSO_3^-]}{P_{SO_2}} = 1.3 \times 10^{-2}$$

$$1.3 \times 10^{-2} = \frac{x^2}{1.6 \times 10^{-7}}$$

$$x^2 = 2.1 \times 10^{-9}$$

$$x = 4.6 \times 10^{-5}\ M = [H^+]$$

$$\mathbf{pH} = -\log(4.6 \times 10^{-5}) = \mathbf{4.34}$$

17.58 See Problem Type 2A, Chapter 5 and Problem Type 2, Chapter 13.

Strategy: This problem gives the volume, temperature, and pressure of PAN. Is the gas undergoing a change in any of its properties? What equation should we use to solve for moles of PAN? Once we have determined moles of PAN, we can convert to molarity and use the first-order rate law to solve for rate.

Solution: Because no changes in gas properties occur, we can use the ideal gas equation to calculate the moles of PAN. 0.55 ppm by volume means:

$$\frac{V_{PAN}}{V_T} = \frac{0.55 \text{ L}}{1 \times 10^6 \text{ L}}$$

Rearranging Equation (5.8) of the text, at STP, the number of moles of PAN in 1.0 L of air is:

$$n = \frac{PV}{RT} = \frac{(1\ \text{atm})\left(\frac{0.55\ \text{L}}{1 \times 10^6\ \text{L}} \times 1.0\ \text{L}\right)}{(0.0821\ \text{L} \cdot \text{atm/K} \cdot \text{mol})(273\ \text{K})} = 2.5 \times 10^{-8}\ \text{mol}$$

Since the decomposition follows first-order kinetics, we can write:

$$\text{rate} = k[\text{PAN}]$$

$$\textbf{rate} = (4.9 \times 10^{-4}\ /\text{s})\left(\frac{2.5 \times 10^{-8}\ \text{mol}}{1.0\ \text{L}}\right) = \mathbf{1.2 \times 10^{-11}}\ \boldsymbol{M/\text{s}}$$

17.60 The Gobi desert lacks the primary pollutants (nitric oxide, carbon monoxide, hydrocarbons) to have photochemical smog. The primary pollutants are present both in New York City and in Boston. However, the sunlight that is required for the conversion of the primary pollutants to the secondary pollutants associated with smog is more likely in a July afternoon than one in January. Therefore, answer **(b)** is correct.

17.66 See Problem Type 7, Chapter 3, and Problem Type 2A, Chapter 5.

Strategy: After writing a balanced equation, how do we compare the amounts of $CaCO_3$ and CO_2? We can compare them based on the mole ratio from the balanced equation. Once we have moles of CO_2, we can then calculate moles of air using the ideal gas equation. From the moles of CO_2 and the moles of air, we can calculate the percentage of CO_2 in the air.

Solution: First, we need to write a balanced equation.

$$CO_2 + Ca(OH)_2 \rightarrow CaCO_3 + H_2O$$

The mole ratio between $CaCO_3$ and CO_2 is: 1 mole $CaCO_3 \simeq$ 1 mole CO_2. If we convert grams of $CaCO_3$ to moles of $CaCO_3$, we can use this mole ratio to convert to moles of CO_2. Once moles of CO_2 are known, we can convert to grams CO_2.

Moles of CO_2 reacted:

$$0.026\ \text{g}\ CaCO_3 \times \frac{1\ \text{mol}\ CaCO_3}{100.1\ \text{g}\ CaCO_3} \times \frac{1\ \text{mol}\ CO_2}{1\ \text{mol}\ CaCO_3} = 2.6 \times 10^{-4}\ \text{mol}\ CO_2$$

The total number of moles of air can be calculated using the ideal gas equation.

$$n = \frac{PV}{RT} = \frac{\left(747\ \text{mmHg} \times \frac{1\ \text{atm}}{760\ \text{mmHg}}\right)(5.0\ \text{L})}{(0.0821\ \text{L} \cdot \text{atm/mol} \cdot \text{K})(291\ \text{K})} = 0.21\ \text{mol air}$$

The percentage by volume of CO_2 in air is:

$$\frac{V_{CO_2}}{V_{\text{air}}} \times 100\% = \frac{n_{CO_2}}{n_{\text{air}}} \times 100\% = \frac{2.6 \times 10^{-4}\ \text{mol}}{0.21\ \text{mol}} \times 100\% = \mathbf{0.12\%}$$

17.68 An increase in temperature has shifted the system to the right; the equilibrium constant has increased with an increase in temperature. If we think of heat as a reactant (endothermic)

$$\text{heat} + N_2 + O_2 \rightleftharpoons 2\,NO$$

based on Le Châtelier's principle, adding heat would indeed shift the system to the right. Therefore, the reaction is **endothermic**.

17.70 The concentration of O_2 could be monitored. Formation of CO_2 must deplete O_2.

17.72 In Problem 17.6, we determined the partial pressure of CO_2 in dry air to be 3.3×10^{-4} atm. Using Henry's law, we can calculate the concentration of CO_2 in water. See Problem Type 4, Chapter 12.

$$c = kP$$

$$[CO_2] = (0.032\ \text{mol/L}\cdot\cancel{\text{atm}})(3.3 \times 10^{-4}\ \cancel{\text{atm}}) = 1.06 \times 10^{-5}\ \text{mol/L}$$

We assume that all of the dissolved CO_2 is converted to H_2CO_3, thus giving us 1.06×10^{-5} mol/L of H_2CO_3. H_2CO_3 is a weak acid. Setup the equilibrium of this acid in water and solve for $[H^+]$.

The equilibrium expression is:

	H_2CO_3	$\rightleftharpoons$	H^+	+	HCO_3^-
Initial (M):	1.06×10^{-5}		0		0
Change (M):	$-x$		$+x$		$+x$
Equilibrium (M):	$(1.06 \times 10^{-5}) - x$		x		x

$$K\ (\text{from Table 15.5}) = 4.2 \times 10^{-7} = \frac{[H^+][HCO_3^-]}{[H_2CO_3]} = \frac{x^2}{(1.06 \times 10^{-5}) - x}$$

Solving the quadratic equation:

$$x = 1.9 \times 10^{-6}\ M = [H^+]$$

$$\mathbf{pH} = -\log(1.9 \times 10^{-6}) = \mathbf{5.72}$$

17.74 See Problem Types 4A and 5C, Chapter 6.

Strategy: From ΔH_f° values given in Appendix 3 of the text, we can calculate ΔH° for the reaction

$$NO_2 \rightarrow NO + O$$

Then, we can calculate ΔE° from ΔH°. The ΔE° calculated will have units of kJ/mol. If we can convert this energy to units of J/molecule, we can calculate the wavelength required to decompose NO_2.

Solution: We use the ΔH_f° values in Appendix 3 and Equation (6.18) of the text.

$$\Delta H_{\text{rxn}}^\circ = \sum n\Delta H_f^\circ(\text{products}) - \sum m\Delta H_f^\circ(\text{reactants})$$

Consider reaction (1):

$$\Delta H° = \Delta H_f°(NO) + \Delta H_f°(O) - \Delta H_f°(NO_2)$$

$$\Delta H° = (1)(90.4 \text{ kJ/mol}) + (1)(249.4 \text{ kJ/mol}) - (1)(33.85 \text{ kJ/mol})$$

$$\Delta H° = 306.0 \text{ kJ/mol}$$

From Equation (6.10) of the text, $\Delta E° = \Delta H° - RT\Delta n$

$$\Delta E° = (306.0 \times 10^3 \text{ J/mol}) - (8.314 \text{ J/mol·K})(298 \text{ K})(1)$$

$$\Delta E° = 304 \times 10^3 \text{ J/mol}$$

See Problem Type 2, Chapter 7. This is the energy needed to dissociate 1 mole of NO_2. We need the energy required to dissociate *one molecule* of NO_2.

$$\frac{304 \times 10^3 \text{ J}}{1 \text{ mol } NO_2} \times \frac{1 \text{ mol } NO_2}{6.022 \times 10^{23} \text{ molecules } NO_2} = 5.05 \times 10^{-19} \text{ J/molecule}$$

The longest wavelength that can dissociate NO_2 is:

$$\lambda = \frac{hc}{E} = \frac{(6.63 \times 10^{-34} \text{ J·s})(3.00 \times 10^8 \text{ m/s})}{5.05 \times 10^{-19} \text{ J}} = \mathbf{3.94 \times 10^{-7} \text{ m} = 394 \text{ nm}}$$

17.76 This reaction has a high activation energy.

17.78 The size of tree rings can be related to CO_2 content, where the number of rings indicates the age of the tree. The amount of CO_2 in ice can be directly measured from portions of polar ice in different layers obtained by drilling. The "age" of CO_2 can be determined by radiocarbon dating and other methods.

17.80 See Problem Type 7, Chapter 9.

$$Cl_2 + O_2 \rightarrow 2ClO$$

$$\Delta H° = \Sigma BE(\text{reactants}) - \Sigma BE(\text{products})$$

$$\Delta H° = (1)(242.7 \text{ kJ/mol}) + (1)(498.7 \text{ kJ/mol}) - (2)(206 \text{ kJ/mol})$$

$$\Delta H° = 329 \text{ kJ/mol}$$

See Problem Type 4A, Chapter 6.

$$\Delta H° = 2\Delta H_f°(ClO) - 2\Delta H_f°(Cl_2) - 2\Delta H_f°(O_2)$$

$$329 \text{ kJ/mol} = 2\Delta H_f°(ClO) - 0 - 0$$

$$\mathbf{\Delta H_f°(ClO)} = \frac{329 \text{ kJ/mol}}{2} = \mathbf{165 \text{ kJ/mol}}$$

17.82 In one second, the energy absorbed by CO_2 is 6.7 J. If we can calculate the energy of one photon of light with a wavelength of 14993 nm, we can then calculate the number of photons absorbed per second.

The energy of one photon with a wavelength of 14993 nm is:

$$E = \frac{hc}{\lambda} = \frac{(6.63 \times 10^{-34}\ \text{J}\cdot\text{s})(3.00 \times 10^{8}\ \text{m/s})}{14993 \times 10^{-9}\ \text{m}} = 1.3266 \times 10^{-20}\ \text{J}$$

The number of photons absorbed by CO_2 per second is:

$$6.7\ \text{J} \times \frac{1\ \text{photon}}{1.3266 \times 10^{-20}\ \text{J}} = \mathbf{5.1 \times 10^{20}\ photons}$$

17.84 **(a)** We use Equation (13.14) of the text.

$$\ln\frac{k_1}{k_2} = \frac{E_a}{R}\left(\frac{T_1 - T_2}{T_1T_2}\right)$$

$$\ln\frac{2.6 \times 10^{-7}\ \text{s}^{-1}}{3.0 \times 10^{-4}\ \text{s}^{-1}} = \frac{E_a}{8.314\ \text{J/mol}\cdot\text{K}}\left(\frac{233\ \text{K} - 298\ \text{K}}{(233\ \text{K})(298\ \text{K})}\right)$$

$$\boldsymbol{E_a} = 6.26 \times 10^{4}\ \text{J/mol} = \mathbf{62.6\ kJ/mol}$$

(b) The unit for the rate constant indicates that the reaction is first-order. The half-life is:

$$t_{\frac{1}{2}} = \frac{0.693}{k} = \frac{0.693}{3.0 \times 10^{-4}\ \text{s}^{-1}} = \mathbf{2.3 \times 10^{3}\ s} = \mathbf{38\ min}$$

17.86 In order to end up with the desired equation, we keep the second equation as written, but we must reverse the first equation and multiply by two.

$$2S(s) + 3O_2(g) \rightleftharpoons 2SO_3(g) \qquad K_2$$

$$2SO_2(g) \rightleftharpoons 2S(s) + 2O_2(g) \qquad K_1' = \frac{1}{(K_1)^2}$$

$$2SO_2(g) + O_2(g) \rightleftharpoons 2SO_3(g) \qquad K = K_2 \times \frac{1}{(K_1)^2}$$

$$K = K_2 \times \frac{1}{(K_1)^2} = (9.8 \times 10^{128}) \times \frac{1}{(4.2 \times 10^{52})^2}$$

$$\boldsymbol{K} = \mathbf{5.6 \times 10^{23}}$$

Thus, the reaction favors the formation of SO_3. But, this reaction has a high activation energy and requires a catalyst to promote it.

CHAPTER 18
ENTROPY, FREE ENERGY, AND EQUILIBRIUM

PROBLEM-SOLVING STRATEGIES AND TUTORIAL SOLUTIONS

TYPES OF PROBLEMS

Problem Type 1: Entropy.
- **(a)** Predicting entropy changes.
- **(b)** Calculating entropy changes of a system.

Problem Type 2: Free Energy.
- **(a)** Calculating standard free energy changes.
- **(b)** Calculating standard free energy change from enthalpy and entropy changes.
- **(c)** Entropy changes due to phase transitions.

Problem Type 3: Free Energy and Chemical Equilibrium.
- **(a)** Using the standard free energy change to calculate the equilibrium constant.
- **(b)** Using the free energy change to predict the direction of a reaction.

PROBLEM TYPE 1: ENTROPY

Entropy (*S*) is a direct measure of the randomness or disorder of a system. The greater the disorder of a system, the greater its entropy. Conversely, the more ordered a system, the lower its entropy.

A. Predicting entropy changes

For any substance, the particles in the solid are more ordered than those in the liquid state, which in turn are more ordered than those in the gaseous state. See Figure 18.3 of the text. Thus, for any substance, the entropy for the same molar amount always increases in the following order.

$$S_{\text{solid}} < S_{\text{liquid}} < S_{\text{gas}}$$

There are a number of other factors that you need to consider to predict entropy changes.

- Heating increases the entropy of a system because it increases the random motion of atoms and molecules.
- When a solid dissolves in water, the highly ordered structure of the solid and part of the order of the water are destroyed. Consequently, the solution possesses greater disorder than the pure solute and pure solvent.
- If a reaction produces more gas molecules than it consumes, the entropy of the system increases. If the total number of gas molecules diminishes during a reaction, the entropy of the system decreases. The dependence of the entropy of reaction on the number of gas molecules is due to the fact that gas molecules possess much greater entropy than either solid or liquid molecules.
- If there is no net change in the total number of gas molecules in a reaction, then the entropy of the system may increase or decrease, but it will be a relatively small change.

As discussed, the entropy is often described as a measure of disorder or randomness. While useful, these terms must be used with caution because they are subjective concepts. In general, it is preferable to view a change in entropy of a system in terms of the change in the number of microstates of the systems. If the number of microstates increases, entropy increases. Conversely, if the number of microstates decreases, the entropy decreases. See Section 18.3 of the text for a complete discussion.

EXAMPLE 18.1

Predict whether the entropy increases, decreases, or remains essentially unchanged for the following reactions.

(a) $H_2O_2(l) \longrightarrow H_2O(l) + \frac{1}{2}O_2(g)$

(b) $H^+(aq) + OH^-(aq) \longrightarrow H_2O(l)$

(c) $Ca(OH)_2(s) + CO_2(g) \longrightarrow CaCO_3(s) + H_2O(g)$

Strategy: To determine the entropy change in each case, we examine whether the number of microstates of the system increases or decreases. The sign of ΔS will be positive if there is an increase in the number of microstates and negative if the number of microstates decreases.

Solution:

(a) The number of moles of gaseous compounds in the products is greater than in the reactant. $\Delta S > 0$; increase in number of moles of gas (an increase in microstates).

(b) Two reactants combine into one product in this reaction. The number of microstates decreases and so entropy decreases ($\Delta S < 0$).

(c) The number of moles of gas phase products is the same as the reactants. The entropy remains essentially unchanged as the number of microstates is relatively constant.

PRACTICE EXERCISE

1. Predict whether the entropy increases, decreases, or remains essentially unchanged for the following reactions:

 (a) $CaO(s) + CO_2(g) \longrightarrow CaCO_3(s)$

 (b) $CuSO_4(s) \longrightarrow Cu^{2+}(aq) + SO_4^{2-}(aq)$

 (c) $2HCl(g) + Br_2(l) \longrightarrow 2HBr(g) + Cl_2(g)$

Text Problems: 18.10, 18.14

B. Calculating entropy changes of a system

The universe is made up of the system and the surroundings. The entropy change in the universe (ΔS_{univ}) for any process is the *sum* of the entropy changes in the system (ΔS_{sys}) and in the surroundings (ΔS_{surr}).

$$\Delta S_{univ} = \Delta S_{sys} + \Delta S_{surr}$$

For a spontaneous process, the entropy of the universe increases.

$$\Delta S_{univ} = \Delta S_{sys} + \Delta S_{surr} > 0$$

In chemistry, we typically focus on the entropy of the system. Let's suppose that the system is represented by the following reaction:

$$aA + bB \longrightarrow cC + dD$$

As is the case for the enthalpy of a reaction [see Equation (6.18) in the text], the standard entropy change $\Delta S°$ is given by:

$$\Delta S^\circ_{rxn} = [cS°(C) + dS°(D)] - [aS°(A) + bS°(B)]$$

or, using Σ to represent summation and m and n for the stoichiometric coefficients in the reaction,

$$\Delta S^\circ_{rxn} = \Sigma nS°(\text{products}) - \Sigma mS°(\text{reactants}) \qquad (18.7, \text{text})$$

Using the standard entropy values ($S°$) listed in Appendix 3 of the text, we can calculate ΔS°_{rxn}, which corresponds to ΔS_{sys}.

EXAMPLE 18.2

Use standard entropy values to calculate the standard entropy change (ΔS°_{rxn}) for the reaction

$$H_2(g) + \tfrac{1}{2}O_2(g) \longrightarrow H_2O(l)$$

Strategy: To calculate the standard entropy change of a reaction, we look up the standard entropies of reactants and products in Appendix 3 of the text and apply Equation (18.7) of the text. As in the calculation of enthalpy of reaction, the stoichiometric coefficients have no units, so ΔS°_{rxn} is expressed in units of J/K·mol.

Solution: The standard entropy change for a reaction can be calculated using the following equation.

$$\Delta S^\circ_{rxn} = \Sigma nS°(\text{products}) - \Sigma mS°(\text{reactants})$$

$$\Delta S^\circ_{rxn} = S°[H_2O(l)] - \{S°[H_2(g)] + \left(\tfrac{1}{2}\right)S°[O_2(g)]\}$$

$$\Delta S^\circ_{rxn} = (1)(69.9\ \text{J/K}\cdot\text{mol}) - [(1)(131.0\ \text{J/K}\cdot\text{mol}) + \left(\tfrac{1}{2}\right)(205.0\ \text{J/K}\cdot\text{mol})]$$

$$\Delta S^\circ_{rxn} = 69.9\ \text{J/K·mol} - 233.5\ \text{J/K·mol} = \mathbf{-163.6\ J/K·mol}$$

Tip: This reaction is known to be spontaneous, so you would think that ΔS should be positive. Remember, that the value of ΔS°_{rxn} applies only to the system; ΔS_{univ} will have a positive value.

PRACTICE EXERCISE

2. Hydrate lime or slaked lime, $Ca(OH)_2$, can be reformed into quicklime, CaO, by heating:

$$Ca(OH)_2(s) \xrightarrow{\text{heat}} CaO(s) + H_2O(g)$$

Use standard entropy values to calculate the standard entropy change (ΔS°_{rxn}) for this reaction.

Text Problem: 18.12

PROBLEM TYPE 2: FREE ENERGY

The second law of thermodynamics tells us that a spontaneous reaction increases the entropy of the universe; that is, $\Delta S_{univ} > 0$. To calculate ΔS_{univ}, we must calculate both ΔS_{surr} and ΔS_{sys}. However, it can be difficult to calculate ΔS_{surr}, and typically we are only concerned with what happens in a particular system.

Therefore, considering only the system, we use another thermodynamic function to determine if a reaction will occur spontaneously. This function, called **Gibbs free energy (*G*)**, or simply free energy is given by the following equation:

$$G = H - TS$$

where,

H is the enthalpy
S is the entropy
T is the temperature (in K)

The change in free energy (ΔG) of a system for a reaction at constant temperature is:

$$\Delta G = \Delta H - T\Delta S \qquad (18.10, text)$$

The sign of ΔG will allow us to predict whether a reaction is spontaneous. At constant temperature and pressure, we can summarize the following conditions in terms of ΔG.

- $\Delta G < 0$ A spontaneous process in the forward direction.
- $\Delta G > 0$ A nonspontaneous reaction as written. However, the reaction is spontaneous in the reverse direction.
- $\Delta G = 0$ The system is at equilibrium. There is no net change in the system.

A. Calculating standard free energy changes

Let's again suppose that the system is represented by the following reaction:

$$a\text{A} + b\text{B} \longrightarrow c\text{C} + d\text{D}$$

As is the case for the entropy of a reaction the standard free energy change $\Delta G°$ is given by:

$$\Delta G°_{rxn} = [c\,\Delta G°_f\,(\text{C}) + d\,\Delta G°_f\,(\text{D})] - [a\,\Delta G°_f\,(\text{A}) + b\,\Delta G°_f\,(\text{B})]$$

or,

$$\Delta G°_{rxn} = \Sigma n\Delta G°_f(\text{products}) - \Sigma m\Delta G°_f(\text{reactants}) \qquad (18.12, text)$$

where m and n are stoichiometric coefficients. The term $\Delta G°_f$ is the **standard free energy of formation** of a compound. It is the free energy change that occurs when 1 mole of the compound is synthesized from its elements in their standard states (see Table 18.2 of the text for conventions used for standard states). By definition, the standard free energy of formation of any element in its stable form is *zero*.

For example,

$$\Delta G°_f\,[\text{O}_2(g)] = 0 \quad \text{and} \quad \Delta G°_f\,[\text{Na}(s)] = 0$$

Other standard free energies of formation can be found in Appendix 3 of the text.

EXAMPLE 18.3

Calculate $\Delta G°_{rxn}$ at 25°C for the following reaction given that $\Delta G°_f\,(Fe_2O_3) = -741.0$ kJ/mol.

$$\mathbf{2Al(s) + Fe_2O_3(s) \longrightarrow Al_2O_3(s) + 2Fe(s)}$$

Strategy: To calculate the standard free-energy change of a reaction, we look up the standard free energies of formation of reactants and products in Appendix 3 of the text and apply Equation (18.12) of the text. Note that all the stoichiometric coefficients have no units so $\Delta G°_{rxn}$ is expressed in units of kJ/mol. The standard free energy of formation of any element in its stable allotropic form at 1 atm and 25°C is zero.

Solution: The standard free energy change for a reaction can be calculated using the following equation.

$$\Delta G^\circ_{rxn} = \Sigma n\Delta G^\circ_f(\text{products}) - \Sigma m\Delta G^\circ_f(\text{reactants})$$

$$\Delta G^\circ_{rxn} = \Delta G^\circ_f[Al_2O_3(s)] + 2\,\Delta G^\circ_f[Fe(s)] - \{2\,\Delta G^\circ_f[Al(s)] + \Delta G^\circ_f[Fe_2O_3(s)]\}$$

$$\Delta G^\circ_{rxn} = [(1)(-1576.41\text{ kJ/mol}) + (2)(0)] - [(2)(0) + (1)(-741.0\text{ kJ/mol})]$$

$$\boldsymbol{\Delta G^\circ_{rxn}} = -1576.41\text{ kJ/mol} + 741.0\text{ kJ/mol} = \mathbf{-835.4\ kJ/mol}$$

> **Tip:** Remember that the standard free energy of formation of any element in its stable form is *zero*. Therefore, the standard free energy of formation values for both Al(s) and Fe(s) are *zero*.

PRACTICE EXERCISE

3. Using Appendix 3 of the text, calculate ΔG°_{rxn} values for the following reactions:

(a) $3CaO(s) + 2Al(s) \longrightarrow 3Ca(s) + Al_2O_3(s)$

(b) $ZnO(s) \longrightarrow Zn(s) + \frac{1}{2}O_2(g)$

Text Problems: 18.18, 18.32

B. Calculating standard free energy change from enthalpy and entropy changes

We can determine the sign of ΔG if we know the signs of both ΔH and ΔS. A negative ΔH (an exothermic reaction) and a positive ΔS (increase in disorder), give a negative ΔG. In addition, temperature may influence the direction of a spontaneous reaction. There are four possible outcomes for the relationship

$$\Delta G = \Delta H - T\Delta S$$

- If both ΔH and ΔS are positive, ΔG will be negative only when $T\Delta S$ is greater in magnitude than ΔH. This condition is met when T is large (high temperature).
- If ΔH is positive and ΔS is negative, ΔG will always be positive, regardless of temperature. The reaction is nonspontaneous.
- If ΔH is negative and ΔS is positive, ΔG will always be negative, regardless of temperature. The reaction is spontaneous.
- If ΔH is negative and ΔS is negative, ΔG will be negative only when ΔH is greater in magnitude than $T\Delta S$. This condition is met when T is small (low temperature).

We can also calculate the value of ΔG if we know the values of ΔH, ΔS, and the temperature. Typically, ΔH° is calculated from standard enthalpies of formation.

$$\Delta H^\circ_{rxn} = \sum n\Delta H^\circ_f(\text{products}) - \sum m\Delta H^\circ_f(\text{reactants})$$

Also, ΔS° is calculated from standard entropy values (see Problem Type 1B).

$$\Delta S^\circ_{rxn} = \Sigma nS^\circ(\text{products}) - \Sigma mS^\circ(\text{reactants})$$

Then, for reactions carried out under standard-state conditions, we can substitute ΔH°, ΔS°, and the temperature into the following equation to calculate ΔG°.

$$\Delta G^\circ = \Delta H^\circ - T\Delta S^\circ \qquad (18.10,\ \text{text})$$

EXAMPLE 18.4

Calculate $\Delta G°$ for the following reaction at 298 K:

$$2H_2(g) + CO(g) \rightleftharpoons CH_3OH(g)$$

given that $\Delta H° = -90.7$ kJ/mol and $\Delta S° = -221.5$ J/K·mol for this process.

Strategy: The standard free energy change is given by, $\Delta G° = \Delta H° - T\Delta S°$ [Equation (18.10) of the text]. Make sure that the units of $\Delta H°$ and $\Delta S°$ are consistent. What unit of temperature should be used?

Solution: Let's convert $\Delta S°$ to units of kJ/K, so that we have consistent units.

$$-221.5\frac{\cancel{J}}{K \cdot mol} \times \frac{1\ kJ}{1000\cancel{J}} = -0.2215\ kJ/K \cdot mol$$

Substitute $\Delta H°$, $\Delta S°$, and the temperature (in K) into the Equation (18.10) to calculate $\Delta G°$.

$$\Delta G° = \Delta H° - T\Delta S°$$

$$\Delta G° = -90.7\ kJ/mol - [(298\ \cancel{K})(-0.2215\ kJ/\cancel{K}\cdot mol)]$$

$$\mathbf{\Delta G°} = -90.7\ kJ/mol + 66.0\ kJ/mol = \mathbf{-24.7\ kJ/mol}$$

PRACTICE EXERCISE

4. For the following reaction, $\Delta H° = -1204$ kJ/mol and $\Delta S° = -216.4$ J/K·mol.

$$2Mg(s) + O_2(g) \longrightarrow 2MgO(s)$$

At 25°C, are reactants or products favored at equilibrium under standard conditions?

Text Problem: 18.20

C. Entropy changes due to phase transitions

At the transition temperature, the melting or boiling point, a system is at equilibrium and $\Delta G = 0$. Thus, we can write,

$$\Delta G = \Delta H - T\Delta S$$

or

$$0 = \Delta H - T\Delta S$$

and

$$\Delta S = \frac{\Delta H}{T}$$

ΔS is the entropy change due to the phase transition.

EXAMPLE 18.5

The heat of fusion of water, ΔH_{fus}, at 0°C is 6.01 kJ/mol. What is ΔS_{fus} for 1 mole of H_2O at the melting point?

Strategy: At the melting point, solid and liquid phase water are at equilibrium, so $\Delta G = 0$. From equation (18.10) of the text, we have $\Delta G = 0 = \Delta H - T\Delta S$ or $\Delta S = \Delta H/T$. To calculate the entropy change for the solid water → liquid water transition, we write $\Delta S_{fus} = \Delta H_{fus}/T$. What temperature unit should we use?

Solution: The entropy change due to the phase transition (the melting of water), can be calculated using the following equation. Recall that the temperature must be in units of Kelvin (0°C = 273 K).

$$\Delta S = \frac{\Delta H}{T}$$

$$\Delta S = \frac{6.01 \text{ kJ/mol}}{273 \text{ K}} = \mathbf{+0.0220 \text{ kJ/mol}\cdot\text{K}} = \mathbf{+22.0 \text{ J/mol}\cdot\text{K}}$$

Check: The increase in entropy upon melting the solid corresponds to the increase in molecular disorder (increase in microstates) in the liquid state compared to the solid state.

PRACTICE EXERCISE

5. The enthalpy of vaporization of mercury is 58.5 kJ/mol and the normal boiling point is 630 K. What is the entropy of vaporization of mercury?

Text Problem: 18.60

PROBLEM TYPE 3: FREE ENERGY AND CHEMICAL EQUILIBRIUM

A. Using the standard free energy change to calculate the equilibrium constant

There is a relationship between the free energy change (ΔG) and the standard free energy change.

$$\Delta G = \Delta G° + RT \ln Q \qquad (18.13, \text{text})$$

where,

R is the gas constant (8.314 J/mol·K)
T is the absolute temperature of the reaction (in K)
Q is the reaction quotient (see Chapter 14)

At equilibrium, by definition, $\Delta G = 0$ and $Q = K$, where K is the equilibrium constant. Thus,

$$0 = \Delta G° + RT \ln K$$

or

$$\Delta G° = -RT \ln K \qquad (18.14, \text{text})$$

K can be either K_P, used for gases, or K_c used for reactions in solution. Note that the larger the value of K, the more negative the value of $\Delta G°$. This should make sense because a large value of K means that the equilibrium lies far to the right (toward products), and a negative value for $\Delta G°$ also means that products are favored at equilibrium. See Table 18.4 of the text for a summary of the relation between K and $\Delta G°$.

EXAMPLE 18.6

The standard free energy change for the reaction

$$\tfrac{1}{2}N_2(g) + \tfrac{3}{2}H_2(g) \rightleftharpoons NH_3(g)$$

is $\Delta G° = 26.9$ kJ/mol at 700 K. Calculate the equilibrium constant at this temperature.

Strategy: According to Equation (18.14) of the text, the equilibrium constant for the reaction is related to the standard free energy change; that is, $\Delta G° = -RT \ln K$. Since we are given the free energy change in the problem, we can solve for equilibrium constant. What temperature unit should be used?

Solution: The equilibrium constant is related to the standard free energy change by the following equation.

$$\Delta G° = -RT \ln K$$

R has units of J/mol·K, so we must convert $\Delta G°$ from units of kJ/mol to J/mol.

$$\frac{26.9\ \text{kJ}}{1\ \text{mol}} \times \frac{1000\ \text{J}}{1\ \text{kJ}} = 2.69 \times 10^4\ \text{J/mol}$$

Substitute $\Delta G°$, R, and T into Equation (18.14) to calculate the equilibrium constant, K_P. We are calculating K_P in this problem since this a gas-phase reaction.

$$\Delta G° = -RT \ln K_P$$

$$2.69 \times 10^4\ \text{J/mol} = -(8.314\ \text{J/mol}\cdot\text{K})(700\ \text{K}) \ln K_P$$

$$-4.62 = \ln K_P$$

Taking the antilog of both sides gives:

$$e^{-4.62} = K_P$$

$$\mathbf{K_P = 9.9 \times 10^{-3}}$$

Check: A positive $\Delta G°$ value indicates that reactants are favored at equilibrium. A small value for K also indicates that reactants are favored at equilibrium.

PRACTICE EXERCISE

6. In Chapter 14, we saw that for the reaction

$$H_2(g) + I_2(g) \longrightarrow 2HI(g)$$

the equilibrium constant at 400°C is $K_P = 64$. Calculate the value of $\Delta G°$ at this temperature.

Text Problems: 18.24, 18.26, 18.28, 18.30

B. Using the free energy change to predict the direction of a reaction

To predict the direction of a reaction, we can calculate ΔG from Equation (18.13) of the text.

$$\Delta G = \Delta G° + RT \ln Q \qquad (18.13, \text{text})$$

To calculate the free energy change (ΔG), you must be given or you must calculate $\Delta G°$, and you must calculate the reaction quotient Q. Substitute the values of $\Delta G°$ and Q into Equation (18.13) to solve for ΔG. Recall the meaning of the sign of ΔG:

- $\Delta G < 0$ A spontaneous process in the forward direction.
- $\Delta G > 0$ A nonspontaneous reaction as written. However, the reaction is spontaneous in the reverse direction.
- $\Delta G = 0$ The system is at equilibrium. There is no net change in the system.

EXAMPLE 18.7

Using the reaction and data given in Example 18.6, calculate ΔG at 700 K if the reaction mixture consists of 30.0 atm of H_2, 20.0 atm of N_2, and 0.500 atm of NH_3.

Strategy: From the information given we see that neither the reactants nor products are at their standard state of 1 atm. We use Equation (18.13) of the text to calculate the free-energy change under non-standard-state conditions. Note that the partial pressures are expressed as dimensionless quantities in the reaction quotient Q_P.

Solution: Under non-standard-state conditions, ΔG is related to the reaction quotient Q by the following equation.

$$\Delta G = \Delta G° + RT \ln Q_P$$

We are using Q_P in the equation because this is a gas-phase reaction.

$\Delta G°$ is given in Example 18.6. We must calculate Q_P.

$$\tfrac{1}{2}N_2(g) + \tfrac{3}{2}H_2(g) \rightleftharpoons NH_3(g)$$

$$Q_P = \frac{P_{NH_3}}{P_{N_2}^{\frac{1}{2}} \cdot P_{H_2}^{\frac{3}{2}}} = \frac{0.50}{(20.0)^{\frac{1}{2}}(30.0)^{\frac{3}{2}}} = 6.80 \times 10^{-4}$$

Substitute $\Delta G° = 2.69 \times 10^4$ J/mol and Q_P into Equation (18.13) to calculate ΔG.

$$\Delta G = \Delta G° + RT \ln Q_P$$

$$\Delta G = (2.69 \times 10^4 \text{ J/mol}) + (8.314 \text{ J/mol}\cdot\cancel{\text{K}})(700\,\cancel{\text{K}}) \ln(6.80 \times 10^{-4})$$

$$\mathbf{\Delta G} = (2.69 \times 10^4 \text{ J/mol}) - (4.24 \times 10^4 \text{ J/mol}) = \mathbf{-1.55 \times 10^4\ J/mol = -15.5\ kJ/mol}$$

Comment: By making the partial pressures of N_2 and H_2 high and that of NH_3 low, the reaction is spontaneous in the forward direction. This condition corresponds to $Q_P < K_P$, and so the reaction proceeds in the forward direction until $Q_P = K_P$ (see Chapter 14).

PRACTICE EXERCISE

7. Calculate ΔG for the following reaction at 25°C when the pressure of CO_2 is 0.0010 atm, given that $\Delta H° = 177.8$ kJ/mol and $\Delta S° = 160.5$ J/K·mol.

$$CaCO_3(s) \longrightarrow CaO(s) + CO_2(g)$$

Text Problem: 18.28

ANSWERS TO PRACTICE EXERCISES

1. **(a)** entropy decreases **(b)** entropy increases **(c)** entropy increases

2. $\Delta S°_{rxn} = +145.1$ J/K·mol

3. **(a)** $\Delta G°_{rxn} = 236$ kJ/mol **(b)** $\Delta G°_{rxn} = 318.2$ kJ/mol

4. $\Delta G° = -1.14 \times 10^3$ kJ/mol. Since $\Delta G° < 0$, products are favored at equilibrium.

5. $\Delta S_{vap} = 92.9$ J/mol·K

6. $\Delta G° = -23.2$ kJ/mol

7. $\Delta G = 113$ kJ/mol

SOLUTIONS TO SELECTED TEXT PROBLEMS

18.6 The probability (P) of finding all the molecules in the same flask becomes progressively smaller as the number of molecules increases. An equation that relates the probability to the number of molecules is given in the text.

$$P = \left(\frac{1}{2}\right)^N$$

where,

N is the total number of molecules present.

Using the above equation, we find:

(a) $\mathbf{P = 0.02}$ **(b)** $\mathbf{P = 9 \times 10^{-19}}$ **(c)** $\mathbf{P = 2 \times 10^{-181}}$

18.10 In order of increasing entropy per mole at 25°C:

(c) < (d) < (e) < (a) < (b)

(c) Na(s): ordered, crystalline material.
(d) NaCl(s): ordered crystalline material, but with more particles per mole than Na(s).
(e) H_2: a diatomic gas, hence of higher entropy than a solid.
(a) Ne(g): a monatomic gas of higher molar mass than H_2.
(b) $SO_2(g)$: a polyatomic gas of higher molar mass than Ne.

18.12 Calculating entropy changes of a system, Problem Type 1B.

Strategy: To calculate the standard entropy change of a reaction, we look up the standard entropies of reactants and products in Appendix 3 of the text and apply Equation (18.7). As in the calculation of enthalpy of reaction, the stoichiometric coefficients have no units, so ΔS°_{rxn} is expressed in units of J/K·mol.

Solution: The standard entropy change for a reaction can be calculated using the following equation.

$$\Delta S^\circ_{rxn} = \Sigma nS^\circ(\text{products}) - \Sigma mS^\circ(\text{reactants})$$

(a) $\Delta S^\circ_{rxn} = S^\circ(\text{Cu}) + S^\circ(\text{H}_2\text{O}) - [S^\circ(\text{H}_2) + S^\circ(\text{CuO})]$

$= (1)(33.3\text{ J/K·mol}) + (1)(188.7\text{ J/K·mol}) - [(1)(131.0\text{ J/K·mol}) + (1)(43.5\text{ J/K·mol})]$

$= \mathbf{47.5\ J/K·mol}$

(b) $\Delta S^\circ_{rxn} = S^\circ(\text{Al}_2\text{O}_3) + 3S^\circ(\text{Zn}) - [2S^\circ(\text{Al}) + 3S^\circ(\text{ZnO})]$

$= (1)(50.99\text{ J/K·mol}) + (3)(41.6\text{ J/K·mol}) - [(2)(28.3\text{ J/K·mol}) + (3)(43.9\text{ J/K·mol})]$

$= \mathbf{-12.5\ J/K·mol}$

(c) $\Delta S^\circ_{rxn} = S^\circ(\text{CO}_2) + 2S^\circ(\text{H}_2\text{O}) - [S^\circ(\text{CH}_4) + 2S^\circ(\text{O}_2)]$

$= (1)(213.6\text{ J/K·mol}) + (2)(69.9\text{ J/K·mol}) - [(1)(186.2\text{ J/K·mol}) + (2)(205.0\text{ J/K·mol})]$

$= \mathbf{-242.8\ J/K·mol}$

Why was the entropy value for water different in parts (a) and (c)?

18.14 **(a)** $\Delta S < 0$; gas reacting with a liquid to form a solid (decrease in number of moles of gas, hence a decrease in microstates).

(b) $\Delta S > 0$; solid decomposing to give a liquid and a gas (an increase in microstates).

(c) $\Delta S > 0$; increase in number of moles of gas (an increase in microstates).

(d) $\Delta S < 0$; gas reacting with a solid to form a solid (decrease in number of moles of gas, hence a decrease in microstates).

18.18 Calculating standard free energy changes, Problem Type 2A.

Strategy: To calculate the standard free-energy change of a reaction, we look up the standard free energies of formation of reactants and products in Appendix 3 of the text and apply Equation (18.12). Note that all the stoichiometric coefficients have no units so ΔG°_{rxn} is expressed in units of kJ/mol. The standard free energy of formation of any element in its stable allotropic form at 1 atm and 25°C is zero.

Solution: The standard free energy change for a reaction can be calculated using the following equation.

$$\Delta G^\circ_{rxn} = \Sigma n\Delta G^\circ_f(\text{products}) - \Sigma m\Delta G^\circ_f(\text{reactants})$$

(a) $$\Delta G^\circ_{rxn} = 2\Delta G^\circ_f(\text{MgO}) - [2\Delta G^\circ_f(\text{Mg}) + \Delta G^\circ_f(\text{O}_2)]$$

$$\boldsymbol{\Delta G^\circ_{rxn}} = (2)(-569.6\ \text{kJ/mol}) - [(2)(0) + (1)(0)] = \mathbf{-1139\ kJ/mol}$$

(b) $$\Delta G^\circ_{rxn} = 2\Delta G^\circ_f(\text{SO}_3) - [2\Delta G^\circ_f(\text{SO}_2) + \Delta G^\circ_f(\text{O}_2)]$$

$$\boldsymbol{\Delta G^\circ_{rxn}} = (2)(-370.4\ \text{kJ/mol}) - [(2)(-300.4\ \text{kJ/mol}) + (1)(0)] = \mathbf{-140.0\ kJ/mol}$$

(c) $$\Delta G^\circ_{rxn} = 4\Delta G^\circ_f[\text{CO}_2(g)] + 6\Delta G^\circ_f[\text{H}_2\text{O}(l)] - \{2\Delta G^\circ_f[\text{C}_2\text{H}_6(g)] + 7\Delta G^\circ_f[\text{O}_2(g)]\}$$

$$\boldsymbol{\Delta G^\circ_{rxn}} = (4)(-394.4\ \text{kJ/mol}) + (6)(-237.2\ \text{kJ/mol}) - [(2)(-32.89\ \text{kJ/mol}) + (7)(0)] = \mathbf{-2935.0\ kJ/mol}$$

18.20 **Reaction A:** Calculate ΔG from ΔH and ΔS.

$$\Delta G = \Delta H - T\Delta S = -126{,}000\ \text{J/mol} - (298\ \text{K})(84\ \text{J/K}\cdot\text{mol}) = -151{,}000\ \text{J/mol}$$

The free energy change is negative so the reaction is spontaneous at 298 K. Since ΔH is negative and ΔS is positive, **the reaction is spontaneous at all temperatures**.

Reaction B: Calculate ΔG.

$$\Delta G = \Delta H - T\Delta S = -11{,}700\ \text{J/mol} - (298\ \text{K})(-105\ \text{J/K}\cdot\text{mol}) = +19{,}600\ \text{J}$$

The free energy change is positive at 298 K which means the reaction is not spontaneous at that temperature. The positive sign of ΔG results from the large negative value of ΔS. At lower temperatures, the $-T\Delta S$ term will be smaller thus allowing the free energy change to be negative.

ΔG will equal zero when $\Delta H = T\Delta S$.

Rearranging,

$$\boldsymbol{T} = \frac{\Delta H}{\Delta S} = \frac{-11700\ \text{J/mol}}{-105\ \text{J/K}\cdot\text{mol}} = \mathbf{111\ K}$$

At temperatures **below 111 K**, ΔG will be negative and the reaction will be spontaneous.

18.24 Similar to using the standard free energy change to calculate the equilibrium constant, Problem Type 3A.

Strategy: According to Equation (18.14) of the text, the equilibrium constant for the reaction is related to the standard free energy change; that is, $\Delta G° = -RT\ln K$. Since we are given the equilibrium constant in the problem, we can solve for $\Delta G°$. What temperature unit should be used?

Solution: The equilibrium constant is related to the standard free energy change by the following equation.

$$\Delta G° = -RT\ln K$$

Substitute K_w, R, and T into the above equation to calculate the standard free energy change, $\Delta G°$. The temperature at which $K_w = 1.0 \times 10^{-14}$ is 25°C = 298 K.

$$\Delta G° = -RT\ln K_w$$

$$\mathbf{\Delta G°} = -(8.314\ \text{J/mol}\cdot\text{K})(298\ \text{K})\ln(1.0 \times 10^{-14}) = \mathbf{8.0 \times 10^4\ J/mol} = \mathbf{8.0 \times 10^1\ kJ/mol}$$

18.26 Use standard free energies of formation from Appendix 3 to find the standard free energy difference.

$$\Delta G°_{rxn} = 2\Delta G°_f[H_2(g)] + \Delta G°_f[O_2(g)] - 2\Delta G°_f[H_2O(g)]$$

$$\Delta G°_{rxn} = (2)(0) + (1)(0) - (2)(-228.6\ \text{kJ/mol})$$

$$\mathbf{\Delta G°_{rxn}} = \mathbf{457.2\ kJ/mol} = \mathbf{4.572 \times 10^5\ J/mol}$$

We can calculate K_P using the following equation. We carry additional significant figures in the calculation to minimize rounding errors when calculating K_P.

$$\Delta G° = -RT\ln K_P$$

$$4.572 \times 10^5\ \text{J/mol} = -(8.314\ \text{J/mol}\cdot\text{K})(298\ \text{K})\ln K_P$$

$$-184.54 = \ln K_P$$

Taking the antiln of both sides,

$$e^{-184.54} = K_P$$

$$\mathbf{K_P = 7.2 \times 10^{-81}}$$

18.28 **(a)** The equilibrium constant is related to the standard free energy change by the following equation.

$$\Delta G° = -RT\ln K$$

Substitute K_P, R, and T into the above equation to the standard free energy change, $\Delta G°$.

$$\Delta G° = -RT\ln K_P$$

$$\mathbf{\Delta G°} = -(8.314\ \text{J/mol}\cdot\text{K})(2000\ \text{K})\ln(4.40) = -2.464 \times 10^4\ \text{J/mol} = \mathbf{-24.6\ kJ/mol}$$

(b) Similar to using the free energy change to predict the direction of a reaction, Problem Type 3B.

Strategy: From the information given we see that neither the reactants nor products are at their standard state of 1 atm. We use Equation (18.13) of the text to calculate the free-energy change under non-standard-state conditions. Note that the partial pressures are expressed as dimensionless quantities in the reaction quotient Q_P.

Solution: Under non-standard-state conditions, ΔG is related to the reaction quotient Q by the following equation.

$$\Delta G = \Delta G^\circ + RT\ln Q_P$$

We are using Q_P in the equation because this is a gas-phase reaction.

Step 1: ΔG° was calculated in part (a). We must calculate Q_P. We carry additional significant figures in this calculation to minimize rounding errors.

$$Q_P = \frac{P_{H_2O} \cdot P_{CO}}{P_{H_2} \cdot P_{CO_2}} = \frac{(0.66)(1.20)}{(0.25)(0.78)} = 4.062$$

Step 2: Substitute $\Delta G^\circ = -2.46 \times 10^4$ J/mol and Q_P into the following equation to calculate ΔG.

$$\Delta G = \Delta G^\circ + RT\ln Q_P$$

$$\Delta G = -2.464 \times 10^4 \text{ J/mol} + (8.314 \text{ J/mol}\cdot\text{K})(2000 \text{ K})\ln(4.062)$$

$$\Delta G = (-2.464 \times 10^4 \text{ J/mol}) + (2.331 \times 10^4 \text{ J/mol})$$

$$\mathbf{\Delta G} = -1.33 \times 10^3 \text{ J/mol} = \mathbf{-1.33\ kJ/mol}$$

18.30 We use the given K_P to find the standard free energy change.

$$\Delta G^\circ = -RT\ln K$$

$$\Delta G^\circ = -(8.314 \text{ J/K}\cdot\text{mol})(298 \text{ K})\ln(5.62 \times 10^{35}) = 2.04 \times 10^5 \text{ J/mol} = -204 \text{ kJ/mol}$$

The standard free energy of formation of one mole of $COCl_2$ can now be found using the standard free energy of reaction calculated above and the standard free energies of formation of $CO(g)$ and $Cl_2(g)$.

$$\Delta G^\circ_{rxn} = \Sigma n\Delta G^\circ_f(\text{products}) - \Sigma m\Delta G^\circ_f(\text{reactants})$$

$$\Delta G^\circ_{rxn} = \Delta G^\circ_f[COCl_2(g)] - \{\Delta G^\circ_f[CO(g)] + \Delta G^\circ_f[Cl_2(g)]\}$$

$$-204 \text{ kJ/mol} = (1)\Delta G^\circ_f[COCl_2(g)] - [(1)(-137.3 \text{ kJ/mol}) + (1)(0)]$$

$$\mathbf{\Delta G^\circ_f[COCl_2(g)] = -341\ kJ/mol}$$

18.32 The standard free energy change is given by:

$$\Delta G^\circ_{rxn} = \Delta G^\circ_f(\text{graphite}) - \Delta G^\circ_f(\text{diamond})$$

You can look up the standard free energy of formation values in Appendix 3 of the text.

$$\mathbf{\Delta G^\circ_{rxn}} = (1)(0) - (1)(2.87 \text{ kJ/mol}) = \mathbf{-2.87\ kJ/mol}$$

Thus, the formation of graphite from diamond is **favored** under standard-state conditions at 25°C. However, the rate of the diamond to graphite conversion is very slow (due to a high activation energy) so that it will take millions of years before the process is complete.

18.36 The equation for the coupled reaction is:

glucose + ATP → glucose 6–phosphate + ADP

$$\Delta G° = 13.4 \text{ kJ/mol} - 31 \text{ kJ/mol} = -18 \text{ kJ/mol}$$

As an estimate:

$$\ln K = \frac{-\Delta G°}{RT}$$

$$\ln K = \frac{-(-18 \times 10^3 \text{ J/mol})}{(8.314 \text{ J/K}\cdot\text{mol})(298 \text{ K})} = 7.3$$

$$\boldsymbol{K = 1 \times 10^3}$$

18.38 In each part of this problem we can use the following equation to calculate ΔG.

$$\Delta G = \Delta G° + RT\ln Q$$

or,

$$\Delta G = \Delta G° + RT\ln [H^+][OH^-]$$

(a) In this case, the given concentrations are equilibrium concentrations at 25°C. Since the reaction is at equilibrium, $\boldsymbol{\Delta G = 0}$. This is advantageous, because it allows us to calculate $\Delta G°$. Also recall that at equilibrium, $Q = K$. We can write:

$$\Delta G° = -RT\ln K_w$$

$$\Delta G° = -(8.314 \text{ J/}\cancel{\text{K}}\cdot\text{mol})(298 \cancel{\text{K}})\ln(1.0 \times 10^{-14}) = 8.0 \times 10^4 \text{ J/mol}$$

(b) $\Delta G = \Delta G° + RT\ln Q = \Delta G° + RT\ln [H^+][OH^-]$

$$\boldsymbol{\Delta G} = (8.0 \times 10^4 \text{ J/mol}) + (8.314 \text{ J/}\cancel{\text{K}}\cdot\text{mol})(298 \cancel{\text{K}})\ln[(1.0 \times 10^{-3})(1.0 \times 10^{-4})] = \boldsymbol{4.0 \times 10^4 \textbf{ J/mol}}$$

(c) $\Delta G = \Delta G° + RT\ln Q = \Delta G° + RT\ln [H^+][OH^-]$

$$\boldsymbol{\Delta G} = (8.0 \times 10^4 \text{ J/mol}) + (8.314 \text{ J/}\cancel{\text{K}}\cdot\text{mol})(298 \cancel{\text{K}})\ln[(1.0 \times 10^{-12})(2.0 \times 10^{-8})] = \boldsymbol{-3.2 \times 10^4 \textbf{ J/mol}}$$

(d) $\Delta G = \Delta G° + RT\ln Q = \Delta G° + RT\ln [H^+][OH^-]$

$$\boldsymbol{\Delta G} = (8.0 \times 10^4 \text{ J/mol}) + (8.314 \text{ J/}\cancel{\text{K}}\cdot\text{mol})(298 \cancel{\text{K}})\ln[(3.5)(4.8 \times 10^{-4})] = \boldsymbol{6.4 \times 10^4 \textbf{ J/mol}}$$

18.40 One possible explanation is simply that no reaction is possible, namely that there is an unfavorable free energy difference between products and reactants ($\Delta G > 0$).

A second possibility is that the potential for spontaneous change is there ($\Delta G < 0$), but that the reaction is extremely slow (very large activation energy).

A remote third choice is that the student accidentally prepared a mixture in which the components were already at their equilibrium concentrations.

Which of the above situations would be altered by the addition of a catalyst?

18.42 For a solid to liquid phase transition (melting) the entropy always increases ($\Delta S > 0$) and the reaction is always endothermic ($\Delta H > 0$).

(a) Melting is always spontaneous above the melting point, so $\Delta G < 0$.

(b) At the melting point (−77.7°C), solid and liquid are in equilibrium, so $\Delta G = 0$.

(c) Melting is not spontaneous below the melting point, so $\Delta G > 0$.

18.44 If the process is *spontaneous* as well as *endothermic*, the signs of ΔG and ΔH must be negative and positive, respectively. Since $\Delta G = \Delta H - T\Delta S$, the sign of **$\Delta S$ must be positive ($\Delta S > 0$)** for ΔG to be negative.

18.46 **(a)** Using the relationship:

$$\frac{\Delta H_{vap}}{T_{b.p.}} = \Delta S_{vap} \approx 90 \text{ J/K} \cdot \text{mol}$$

benzene	ΔS_{vap} = 87.8 J/K·mol
hexane	ΔS_{vap} = 90.1 J/K·mol
mercury	ΔS_{vap} = 93.7 J/K·mol
toluene	ΔS_{vap} = 91.8 J/K·mol

Most liquids have ΔS_{vap} approximately equal to a constant value because the order of the molecules in the liquid state is similar. The order of most gases is totally random; thus, ΔS for liquid → vapor should be similar for most liquids.

(b) Using the data in Table 11.6 of the text, we find:

ethanol	ΔS_{vap} = 111.9 J/K·mol
water	ΔS_{vap} = 109.4 J/K·mol

Both water and ethanol have a larger ΔS_{vap} because the liquid molecules are more ordered due to hydrogen bonding (there are fewer microstates in these liquids).

18.48 **(a)** $2CO + 2NO \rightarrow 2CO_2 + N_2$

(b) The oxidizing agent is NO; the reducing agent is CO.

(c) $\Delta G° = 2\Delta G_f°(CO_2) + \Delta G_f°(N_2) - 2\Delta G_f°(CO) - 2\Delta G_f°(NO)$

$$\Delta G° = (2)(-394.4 \text{ kJ/mol}) + (0) - (2)(-137.3 \text{ kJ/mol}) - (2)(86.7 \text{ kJ/mol}) = -687.6 \text{ kJ/mol}$$

$$\Delta G° = -RT\ln K_P$$

$$\ln K_P = \frac{6.876 \times 10^5 \text{ J/mol}}{(8.314 \text{ J/K} \cdot \text{mol})(298 \text{ K})} = 277.5$$

$$\mathbf{K_P = 3 \times 10^{120}}$$

(d) $$Q_P = \frac{P_{N_2}P_{CO_2}^2}{P_{CO}^2 P_{NO}^2} = \frac{(0.80)(0.030)^2}{(5.0 \times 10^{-5})^2(5.0 \times 10^{-7})^2} = \mathbf{1.2 \times 10^{18}}$$

Since $Q_P << K_P$, the reaction will proceed from **left to right**.

(e) $\Delta H° = 2\Delta H_f°(CO_2) + \Delta H_f°(N_2) - 2\Delta H_f°(CO) - 2\Delta H_f°(NO)$

$\Delta H° = (2)(-393.5 \text{ kJ/mol}) + (0) - (2)(-110.5 \text{ kJ/mol}) - (2)(90.4 \text{ kJ/mol}) = -746.8 \text{ kJ/mol}$

Since $\Delta H°$ is negative, raising the temperature will decrease K_P, thereby increasing the amount of reactants and decreasing the amount of products. **No**, the formation of N_2 and CO_2 is not favored by raising the temperature.

18.50 The equilibrium reaction is:

$$AgCl(s) \rightleftharpoons Ag^+(aq) + Cl^-(aq)$$

$$K_{sp} = [Ag^+][Cl^-] = 1.6 \times 10^{-10}$$

We can calculate the standard enthalpy of reaction from the standard enthalpies of formation in Appendix 3 of the text.

$$\Delta H° = \Delta H_f°(Ag^+) + \Delta H_f°(Cl^-) - \Delta H_f°(AgCl)$$

$$\Delta H° = (1)(105.9 \text{ kJ/mol}) + (1)(-167.2 \text{ kJ/mol}) - (1)(-127.0 \text{ kJ/mol}) = 65.7 \text{ kJ/mol}$$

From Problem 18.49(a):

$$\ln\frac{K_2}{K_1} = \frac{\Delta H°}{R}\left(\frac{T_2 - T_1}{T_1 T_2}\right)$$

$K_1 = 1.6 \times 10^{-10}$ $\qquad$ $T_1 = 298 \text{ K}$

$K_2 = ?$ $\qquad$ $T_2 = 333 \text{ K}$

$$\ln\frac{K_2}{1.6 \times 10^{-10}} = \frac{6.57 \times 10^4 \text{ J}}{8.314 \text{ J/K}\cdot\text{mol}}\left(\frac{333 \text{ K} - 298 \text{ K}}{(333 \text{ K})(298 \text{ K})}\right)$$

$$\ln\frac{K_2}{1.6 \times 10^{-10}} = 2.79$$

$$\frac{K_2}{1.6 \times 10^{-10}} = e^{2.79}$$

$$\mathbf{K_2 = 2.6 \times 10^{-9}}$$

The increase in K indicates that the solubility increases with temperature.

18.52 Assuming that both $\Delta H°$ and $\Delta S°$ are temperature independent, we can calculate both $\Delta H°$ and $\Delta S°$.

$\Delta H° = \Delta H_f°(CO) + \Delta H_f°(H_2) - [\Delta H_f°(H_2O) + \Delta H_f°(C)]$

$\Delta H° = (1)(-110.5 \text{ kJ/mol}) + (1)(0)] - [(1)(-241.8 \text{ kJ/mol}) + (1)(0)]$

$\Delta H° = 131.3 \text{ kJ/mol}$

$\Delta S° = S°(CO) + S°(H_2) - [S°(H_2O) + S°(C)]$

$\Delta S° = [(1)(197.9 \text{ J/K}\cdot\text{mol}) + (1)(131.0 \text{ J/K}\cdot\text{mol})] - [(1)(188.7 \text{ J/K}\cdot\text{mol}) + (1)(5.69 \text{ J/K}\cdot\text{mol})]$

$\Delta S° = 134.5 \text{ J/K}\cdot\text{mol}$

It is obvious from the given conditions that the reaction must take place at a fairly high temperature (in order to have red–hot coke). Setting $\Delta G° = 0$

$$0 = \Delta H° - T\Delta S°$$

$$\boldsymbol{T} = \frac{\Delta H°}{\Delta S°} = \frac{131.3\ \text{kJ/mol} \times \frac{1000\ \text{J}}{1\ \text{kJ}}}{134.5\ \text{J/K} \cdot \text{mol}} = \mathbf{976\ K} = \mathbf{703°C}$$

The temperature must be greater than 703°C for the reaction to be spontaneous.

18.54 For a reaction to be spontaneous at constant temperature and pressure, $\Delta G < 0$. The process of crystallization proceeds with more order (less disorder), so $\boldsymbol{\Delta S < 0}$. We also know that

$$\Delta G = \Delta H - T\Delta S$$

Since ΔG must be negative, and since the entropy term will be positive ($-T\Delta S$, where ΔS is negative), then ΔH must be negative ($\boldsymbol{\Delta H < 0}$). The reaction will be exothermic.

18.56 For the reaction to be spontaneous, ΔG must be negative.

$$\Delta G = \Delta H - T\Delta S$$

Given that $\Delta H = 19$ kJ/mol = 19,000 J/mol, then

$$\Delta G = 19{,}000\ \text{J/mol} - (273\ \text{K} + 72\ \text{K})(\Delta S)$$

Solving the equation with the value of $\Delta G = 0$

$$0 = 19{,}000\ \text{J/mol} - (273\ \text{K} + 72\ \text{K})(\Delta S)$$

$$\boldsymbol{\Delta S} = \mathbf{55\ J/K{\cdot}mol}$$

This value of ΔS which we solved for is the value needed to produce a ΔG value of zero. The *minimum* value of ΔS that will produce a spontaneous reaction will be any value of entropy *greater than* 55 J/K·mol.

18.58 The second law states that the entropy of the universe must increase in a spontaneous process. But the entropy of the universe is the sum of two terms: the entropy of the system plus the entropy of the surroundings. One of the entropies can decrease, but not both. In this case, the decrease in system entropy is offset by an increase in the entropy of the surroundings. The reaction in question is exothermic, and the heat released raises the temperature (and the entropy) of the surroundings.

Could this process be spontaneous if the reaction were endothermic?

18.60 Entropy changes due to phase transitions, Problem Type 2C.

Strategy: At the boiling point, liquid and gas phase ethanol are at equilibrium, so $\Delta G = 0$. From Equation (18.10) of the text, we have $\Delta G = 0 = \Delta H - T\Delta S$ or $\Delta S = \Delta H/T$. To calculate the entropy change for the liquid ethanol $\rightarrow$ gas ethanol transition, we write $\Delta S_{vap} = \Delta H_{vap}/T$. What temperature unit should we use?

Solution: The entropy change due to the phase transition (the vaporization of ethanol), can be calculated using the following equation. Recall that the temperature must be in units of Kelvin (78.3°C = 351 K).

$$\Delta S_{vap} = \frac{\Delta H_{vap}}{T_{b.p.}}$$

$$\Delta S_{vap} = \frac{39.3\text{ kJ/mol}}{351\text{ K}} = 0.112\text{ kJ/mol}\cdot\text{K} = 112\text{ J/mol}\cdot\text{K}$$

The problem asks for the change in entropy for the vaporization of 0.50 moles of ethanol. The ΔS calculated above is for 1 mole of ethanol.

$$\mathbf{\Delta S\text{ for 0.50 mol}} = (112\text{ J/mol·K})(0.50\text{ mol}) = \mathbf{56\text{ J/K}}$$

18.62 For the given reaction we can calculate the standard free energy change from the standard free energies of formation (see Appendix 3 of the text). Then, we can calculate the equilibrium constant, K_P, from the standard free energy change.

$$\Delta G° = \Delta G_f^\circ[\text{Ni(CO)}_4] - [4\Delta G_f^\circ(\text{CO}) + \Delta G_f^\circ(\text{Ni})]$$

$$\Delta G° = (1)(-587.4\text{ kJ/mol}) - [(4)(-137.3\text{ kJ/mol}) + (1)(0)] = -38.2\text{ kJ/mol} = -3.82 \times 10^4\text{ J/mol}$$

Substitute $\Delta G°$, R, and T (in K) into the following equation to solve for K_P.

$$\Delta G° = -RT\ln K_P$$

$$\ln K_P = \frac{-\Delta G°}{RT} = \frac{-(-3.82 \times 10^4\text{ J/mol})}{(8.314\text{ J/K}\cdot\text{mol})(353\text{ K})}$$

$$\mathbf{K_P = 4.5 \times 10^5}$$

18.64 We carry additional significant figures throughout this calculation to minimize rounding errors. The equilibrium constant is related to the standard free energy change by the following equation:

$$\Delta G° = -RT\ln K_P$$

$$2.12 \times 10^5\text{ J/mol} = -(8.314\text{ J/mol·K})(298\text{ K})\ln K_P$$

$$K_P = 6.894 \times 10^{-38}$$

We can write the equilibrium constant expression for the reaction.

$$K_P = \sqrt{P_{O_2}}$$

$$P_{O_2} = (K_P)^2$$

$$\mathbf{P_{O_2}} = (6.894 \times 10^{-38})^2 = \mathbf{4.8 \times 10^{-75}\text{ atm}}$$

This pressure is far too small to measure.

18.66 Both (a) and (b) apply to a reaction with a negative $\Delta G°$ value. Statement (c) is not always true. An endothermic reaction that has a positive $\Delta S°$ (increase in entropy) will have a negative $\Delta G°$ value at high temperatures.

18.68 We write the two equations as follows. The standard free energy change for the overall reaction will be the sum of the two steps.

$$\text{CuO}(s) \rightleftharpoons \text{Cu}(s) + \tfrac{1}{2}\text{O}_2(g) \qquad \Delta G° = 127.2\text{ kJ/mol}$$

$$\text{C(graphite)} + \tfrac{1}{2}\text{O}_2(g) \rightleftharpoons \text{CO}(g) \qquad \Delta G° = -137.3\text{ kJ/mol}$$

$$\mathbf{CuO + C(graphite) \rightleftharpoons Cu(\mathit{s}) + CO(\mathit{g})} \qquad \Delta G° = -10.1\text{ kJ/mol}$$

We can now calculate the equilibrium constant from the standard free energy change, $\Delta G°$.

$$\ln K = \frac{-\Delta G°}{RT} = \frac{-(-10.1 \times 10^3 \text{ J/mol})}{(8.314 \text{ J/K}\cdot\text{mol})(673 \text{ K})}$$

$$\ln K = 1.81$$

$$\boldsymbol{K = 6.1}$$

18.70 As discussed in Chapter 18 of the text for the decomposition of calcium carbonate, a reaction favors the formation of products at equilibrium when

$$\Delta G° = \Delta H° - T\Delta S° < 0$$

If we can calculate $\Delta H°$ and $\Delta S°$, we can solve for the temperature at which decomposition begins to favor products. We use data in Appendix 3 of the text to solve for $\Delta H°$ and $\Delta S°$.

$$\Delta H° = \Delta H_f^\circ[\text{MgO}(s)] + \Delta H_f^\circ[\text{CO}_2(g)] - \Delta H_f^\circ[\text{MgCO}_3(s)]$$

$$\Delta H° = -601.8 \text{ kJ/mol} + (-393.5 \text{ kJ/mol}) - (-1112.9 \text{ kJ/mol}) = 117.6 \text{ kJ/mol}$$

$$\Delta S° = S°[\text{MgO}(s)] + S°[\text{CO}_2(g)] - S°[\text{MgCO}_3(s)]$$

$$\Delta S° = 26.78 \text{ J/K}\cdot\text{mol} + 213.6 \text{ J/K}\cdot\text{mol} - 65.69 \text{ J/K}\cdot\text{mol} = 174.7 \text{ J/K}\cdot\text{mol}$$

For the reaction to begin to favor products,

$$\Delta H° - T\Delta S° < 0$$

or

$$T > \frac{\Delta H°}{\Delta S°}$$

$$T > \frac{117.6 \times 10^3 \cancel{\text{J/mol}}}{174.7 \cancel{\text{J}}/\text{K}\cdot\cancel{\text{mol}}}$$

$$T > 673.2 \text{ K}$$

18.72 **(a)** $\Delta G° = \Delta G_f^\circ(\text{H}_2) + \Delta G_f^\circ(\text{Fe}^{2+}) - \Delta G_f^\circ(\text{Fe}) - 2\Delta G_f^\circ(\text{H}^+)]$

$$\Delta G° = (1)(0) + (1)(-84.9 \text{ kJ/mol}) - (1)(0) - (2)(0)$$

$$\Delta G° = -84.9 \text{ kJ/mol}$$

$$\Delta G° = -RT\ln K$$

$$-84.9 \times 10^3 \text{ J/mol} = -(8.314 \text{ J/mol}\cdot\text{K})(298 \text{ K})\ln K$$

$$\boldsymbol{K = 7.6 \times 10^{14}}$$

(b) $\Delta G° = \Delta G_f^\circ(\text{H}_2) + \Delta G_f^\circ(\text{Cu}^{2+}) - \Delta G_f^\circ(\text{Cu}) - 2\Delta G_f^\circ(\text{H}^+)]$

$$\Delta G° = 64.98 \text{ kJ/mol}$$

$$\Delta G° = -RT\ln K$$

$$64.98 \times 10^3 \text{ J/mol} = -(8.314 \text{ J/mol}\cdot\text{K})(298 \text{ K})\ln K$$

$$\boldsymbol{K = 4.1 \times 10^{-12}}$$

The activity series is correct. The very large value of K for reaction (a) indicates that *products* are highly favored; whereas, the very small value of K for reaction (b) indicates that *reactants* are highly favored.

18.74 **(a)** It is a "reverse" disproportionation redox reaction.

(b) $\Delta G° = (2)(-228.6\text{ kJ/mol}) - (2)(-33.0\text{ kJ/mol}) - (1)(-300.4\text{ kJ/mol})$

$\Delta G° = -90.8\text{ kJ/mol}$

$-90.8 \times 10^3\text{ J/mol} = -(8.314\text{ J/mol·K})(298\text{ K})\ln K$

$\mathbf{K = 8.2 \times 10^{15}}$

Because of the large value of K, this method is efficient for removing SO_2.

(c) $\Delta H° = (2)(-241.8\text{ kJ/mol}) + (3)(0) - (2)(-20.15\text{ kJ/mol}) - (1)(-296.1\text{ kJ/mol})$

$\Delta H° = -147.2\text{ kJ/mol}$

$\Delta S° = (2)(188.7\text{ J/K·mol}) + (3)(31.88\text{ J/K·mol}) - (2)(205.64\text{ J/K·mol}) - (1)(248.5\text{ J/K·mol})$

$\Delta S° = -186.7\text{ J/K·mol}$

$\Delta G° = \Delta H° - T\Delta S°$

Due to the negative entropy change, $\Delta S°$, the free energy change, $\Delta G°$, will become positive at higher temperatures. Therefore, the reaction will be **less effective** at high temperatures.

18.76 $2O_3 \rightleftharpoons 3O_2$

$\Delta G° = 3\Delta G_f°(O_2) - 2\Delta G_f°(O_3) = 0 - (2)(163.4\text{ kJ/mol})$

$\Delta G° = -326.8\text{ kJ/mol}$

$-326.8 \times 10^3\text{ J/mol} = -(8.314\text{ J/mol·K})(243\text{ K})\ln K_P$

$\mathbf{K_P = 1.8 \times 10^{70}}$

Due to the large magnitude of K, you would expect this reaction to be spontaneous in the forward direction. However, this reaction has a **large activation energy**, so the rate of reaction is extremely slow.

18.78 Heating the ore alone is not a feasible process. Looking at the coupled process:

$Cu_2S \rightarrow 2Cu + S$	$\Delta G° = 86.1\text{ kJ/mol}$
$S + O_2 \rightarrow SO_2$	$\Delta G° = -300.4\text{ kJ/mol}$
$Cu_2S + O_2 \rightarrow 2Cu + SO_2$	$\mathbf{\Delta G° = -214.3\text{ kJ/mol}}$

Since $\Delta G°$ is a large negative quantity, the coupled reaction is feasible for extracting sulfur.

18.80 First, we need to calculate $\Delta H°$ and $\Delta S°$ for the reaction in order to calculate $\Delta G°$.

$\Delta H° = -41.2\text{ kJ/mol}$ $\qquad\qquad$ $\Delta S° = -42.0\text{ J/K·mol}$

Next, we calculate $\Delta G°$ at 300°C or 573 K, assuming that $\Delta H°$ and $\Delta S°$ are temperature independent.

$$\Delta G^\circ = \Delta H^\circ - T\Delta S^\circ$$

$$\Delta G^\circ = -41.2 \times 10^3 \text{ J/mol} - (573 \text{ K})(-42.0 \text{ J/K·mol})$$

$$\Delta G^\circ = -1.71 \times 10^4 \text{ J/mol}$$

Having solved for ΔG°, we can calculate K_P.

$$\Delta G^\circ = -RT\ln K_P$$

$$-1.71 \times 10^4 \text{ J/mol} = -(8.314 \text{ J/K·mol})(573 \text{ K})\ln K_P$$

$$\ln K_P = 3.59$$

$$\boldsymbol{K_P = 36}$$

Due to the negative entropy change calculated above, we expect that ΔG° will become positive at some temperature higher than 300°C. We need to find the temperature at which ΔG° becomes zero. This is the temperature at which reactants and products are equally favored ($K_P = 1$).

$$\Delta G^\circ = \Delta H^\circ - T\Delta S^\circ$$

$$0 = \Delta H^\circ - T\Delta S^\circ$$

$$T = \frac{\Delta H^\circ}{\Delta S^\circ} = \frac{-41.2 \times 10^3 \text{ J/mol}}{-42.0 \text{ J/K} \cdot \text{mol}}$$

$$\boldsymbol{T = 981 \text{ K} = 708°\text{C}}$$

This calculation shows that at 708°C, $\Delta G^\circ = 0$ and the equilibrium constant $K_P = 1$. Above 708°C, ΔG° is positive and K_P will be smaller than 1, meaning that reactants will be favored over products. Note that the temperature 708°C is only an estimate, as we have assumed that both ΔH° and ΔS° are independent of temperature.

Using a more efficient catalyst will ***not*** increase K_P at a given temperature, because the catalyst will speed up both the forward and reverse reactions. The value of K_P will stay the same.

18.82 butane $\rightarrow$ isobutane

$$\Delta G^\circ = \Delta G_f^\circ(\text{isobutane}) - \Delta G_f^\circ(\text{butane})$$

$$\Delta G^\circ = (1)(-18.0 \text{ kJ/mol}) - (1)(-15.9 \text{ kJ/mol})$$

$$\Delta G^\circ = -2.1 \text{ kJ/mol}$$

For a mixture at equilibrium at 25°C:

$$\Delta G^\circ = -RT\ln K_P$$

$$-2.1 \times 10^3 \text{ J/mol} = -(8.314 \text{ J/mol·K})(298 \text{ K})\ln K_P$$

$$K_P = 2.3$$

$$K_P = \frac{P_{\text{isobutane}}}{P_{\text{butane}}} \propto \frac{\text{mol isobutane}}{\text{mol butane}}$$

$$2.3 = \frac{\text{mol isobutane}}{\text{mol butane}}$$

This shows that there are 2.3 times as many moles of isobutane as moles of butane. Or, we can say for every one mole of butane, there are 2.3 moles of isobutane.

$$\textbf{mol \% isobutane} = \frac{2.3 \text{ mol}}{2.3 \text{ mol} + 1.0 \text{ mol}} \times 100\% = \textbf{70\%}$$

By difference, the mole % of butane is **30%**.

Yes, this result supports the notion that straight-chain hydrocarbons like butane are less stable than branched-chain hydrocarbons like isobutane.

18.84 We can calculate K_P from $\Delta G°$.

$$\Delta G° = (1)(-394.4 \text{ kJ/mol}) + (0) - (1)(-137.3 \text{ kJ/mol}) - (1)(-255.2 \text{ kJ/mol})$$

$$\Delta G° = -1.9 \text{ kJ/mol}$$

$$-1.9 \times 10^3 \text{ J/mol} = -(8.314 \text{ J/mol·K})(1173 \text{ K})\ln K_P$$

$$K_P = 1.2$$

Now, from K_P, we can calculate the mole fractions of CO and CO_2.

$$K_P = \frac{P_{CO_2}}{P_{CO}} = 1.2 \qquad P_{CO_2} = 1.2P_{CO}$$

$$\boldsymbol{X_{CO}} = \frac{P_{CO}}{P_{CO} + P_{CO_2}} = \frac{P_{CO}}{P_{CO} + 1.2P_{CO}} = \frac{1}{2.2} = \mathbf{0.45}$$

$$\boldsymbol{X_{CO_2}} = 1 - 0.45 = \mathbf{0.55}$$

We assumed that $\Delta G°$ calculated from ΔG_f° values was temperature independent. The ΔG_f° values in Appendix 3 of the text are measured at 25°C, but the temperature of the reaction is 900°C.

18.86 For a phase transition, $\Delta G = 0$. We write:

$$\Delta G = \Delta H - T\Delta S$$

$$0 = \Delta H - T\Delta S$$

$$\Delta S_{sub} = \frac{\Delta H_{sub}}{T}$$

Substituting ΔH and the temperature, $(-78° + 273°)\text{K} = 195 \text{ K}$, gives

$$\Delta S_{sub} = \frac{\Delta H_{sub}}{T} = \frac{25.2 \times 10^3 \text{ J}}{195 \text{ K}} = 129 \text{ J/K} \cdot \text{mol}$$

This value of ΔS_{sub} is for the sublimation of 1 mole of CO_2. We convert to the ΔS value for the sublimation of 84.8 g of CO_2.

$$84.8\ \text{g CO}_2 \times \frac{1\ \text{mol CO}_2}{44.01\ \text{g CO}_2} \times \frac{129\ \text{J}}{\text{K} \cdot \text{mol}} = \mathbf{249\ J/K}$$

18.88 First, let's convert the age of the universe from units of years to units of seconds.

$$(13 \times 10^9\ \text{yr}) \times \frac{365\ \text{days}}{1\ \text{yr}} \times \frac{24\ \text{h}}{1\ \text{day}} \times \frac{3600\ \text{s}}{1\ \text{h}} = 4.1 \times 10^{17}\ \text{s}$$

The probability of finding all 100 molecules in the same flask is 8×10^{-31}. Multiplying by the number of seconds gives:

$$(8 \times 10^{-31})(4.1 \times 10^{17}\ \text{s}) = \mathbf{3 \times 10^{-13}\ s}$$

18.90 We can calculate ΔS_{sys} from standard entropy values in Appendix 3 of the text. We can calculate ΔS_{surr} from the ΔH_{sys} value given in the problem. Finally, we can calculate ΔS_{univ} from the ΔS_{sys} and ΔS_{surr} values.

$$\mathbf{\Delta S_{sys}} = (2)(69.9\ \text{J/K·mol}) - [(2)(131.0\ \text{J/K·mol}) + (1)(205.0\ \text{J/K·mol})] = \mathbf{-327\ J/K·mol}$$

$$\mathbf{\Delta S_{surr}} = \frac{-\Delta H_{sys}}{T} = \frac{-(-571.6 \times 10^3\ \text{J/mol})}{298\ \text{K}} = \mathbf{1918\ J/K \cdot mol}$$

$$\mathbf{\Delta S_{univ}} = \Delta S_{sys} + \Delta S_{surr} = (-327 + 1918)\ \text{J/K·mol} = \mathbf{1591\ J/K·mol}$$

18.92 q, and w are *not* state functions. Recall that state functions represent properties that are determined by the state of the system, regardless of how that condition is achieved. Heat and work are not state functions because they are not properties of the system. They manifest themselves only during a process (during a change). Thus their values depend on the path of the process and vary accordingly.

18.94 Since the adsorption is spontaneous, ΔG must be negative (**$\Delta G < 0$**). When hydrogen bonds to the surface of the catalyst, the system becomes more ordered (**$\Delta S < 0$**). Since there is a decrease in entropy, the adsorption must be exothermic for the process to be spontaneous (**$\Delta H < 0$**).

18.96 **(a)** Each CO molecule has two possible orientations in the crystal,

CO or OC

If there is no preferred orientation, then for one molecule there are two, or 2^1, choices of orientation. Two molecules have four or 2^2 choices, and for 1 mole of CO there are 2^{N_A} choices. From Equation (18.1) of the text:

$$S = k \ln W$$

$$S = (1.38 \times 10^{-23}\ \text{J/K}) \ln 2^{6.022 \times 10^{23}}$$

$$S = (1.38 \times 10^{-23}\ \text{J/K})(6.022 \times 10^{23}\ /\text{mol}) \ln 2$$

$$\mathbf{S = 5.76\ J/K·mol}$$

(b) The fact that the actual residual entropy is 4.2 J/K·mol means that the orientation is not totally random.

18.98 We use data in Appendix 3 of the text to calculate $\Delta H°$ and $\Delta S°$.

$$\Delta H° = \Delta H_{vap} = \Delta H_f°[C_6H_6(g)] - \Delta H_f°[C_6H_6(l)]$$

$$\mathbf{\Delta H°} = 82.93 \text{ kJ/mol} - 49.04 \text{ kJ/mol} = \mathbf{33.89 \text{ kJ/mol}}$$

$$\Delta S° = S°[C_6H_6(g)] - S°[C_6H_6(l)]$$

$$\mathbf{\Delta S°} = 269.2 \text{ J/K·mol} - 172.8 \text{ J/K·mol} = \mathbf{96.4 \text{ J/K·mol}}$$

We can now calculate $\Delta G°$ at 298 K.

$$\Delta G° = \Delta H° - T\Delta S°$$

$$\Delta G° = 33.89 \text{ kJ/mol} - (298 \text{ K})(96.4 \text{ J/K·mol}) \times \frac{1 \text{ kJ}}{1000 \text{ J}}$$

$$\mathbf{\Delta G° = 5.2 \text{ kJ/mol}}$$

$\Delta H°$ is positive because this is an endothermic process. We also expect $\Delta S°$ to be positive because this is a liquid $\rightarrow$ vapor phase change. $\Delta G°$ is positive because we are at a temperature that is below the boiling point of benzene (80.1°C).

18.100 We can calculate $\Delta G°$ at 872 K from the equilibrium constant, K_1.

$$\Delta G° = -RT \ln K$$

$$\Delta G° = -(8.314 \text{ J/mol·}\cancel{\text{K}})(872 \cancel{\text{K}}) \ln(1.80 \times 10^{-4})$$

$$\mathbf{\Delta G°} = 6.25 \times 10^4 \text{ J/mol} = \mathbf{62.5 \text{ kJ/mol}}$$

We use the equation derived in Problem 18.49 to calculate $\Delta H°$.

$$\ln\frac{K_2}{K_1} = \frac{\Delta H°}{R}\left(\frac{1}{T_1} - \frac{1}{T_2}\right)$$

$$\ln\frac{0.0480}{1.80 \times 10^{-4}} = \frac{\Delta H°}{8.314 \text{ J/mol·K}}\left(\frac{1}{872 \text{ K}} - \frac{1}{1173 \text{ K}}\right)$$

$$\mathbf{\Delta H° = 157.8 \text{ kJ/mol}}$$

Now that both $\Delta G°$ and $\Delta H°$ are known, we can calculate $\Delta S°$ at 872 K.

$$\Delta G° = \Delta H° - T\Delta S°$$

$$62.5 \times 10^3 \text{ J/mol} = (157.8 \times 10^3 \text{ J/mol}) - (872 \text{ K})\Delta S°$$

$$\mathbf{\Delta S° = 109 \text{ J/K·mol}}$$

CHAPTER 19
ELECTROCHEMISTRY

PROBLEM-SOLVING STRATEGIES AND TUTORIAL SOLUTIONS

TYPES OF PROBLEMS

Problem Type 1: Balancing Redox Equations.

Problem Type 2: Standard Reduction Potentials.
- **(a)** Comparing strengths of oxidizing agents.
- **(b)** Calculating the standard emf ($E°$) of a galvanic cell.

Problem Type 3: Spontaneity of Redox Reactions.
- **(a)** Predicting whether a redox reaction is spontaneous.
- **(b)** Calculating $\Delta G°$ and K from $E°$.

Problem Type 4: The Nernst Equation.
- **(a)** Using the Nernst equation to predict the spontaneity of a redox reaction.
- **(b)** Using the Nernst equation to calculate concentration.

Problem Type 5: Electrolysis.
- **(a)** Predicting the products of electrolysis.
- **(b)** Calculating the quantity of products in electrolysis.

PROBLEM TYPE 1: BALANCING REDOX EQUATIONS

We will use the **ion-electron method** to balance redox reactions. In this approach, the overall reaction is divided into two half-reactions, one for oxidation and one for reduction. The two equations are balanced separately and then added together to give the overall balanced equation. Example 19.1 demonstrates how to balance a redox reaction by the ion-electron method.

EXAMPLE 19.1

Balance the following redox reaction in an acidic solution.

$$\mathbf{Sn + NO_3^- \longrightarrow Sn^{2+} + NO}$$

Strategy: We follow the procedure for balancing redox reactions presented in Section 19.1 of the text.

Solution:

Step 1: Write the unbalanced equation for the reaction in ionic form.

The equation given in the problem is already in ionic form.

Step 2: Separate the equation into two half-reactions.

$$\overset{0}{Sn} \xrightarrow{\text{oxidation}} Sn^{2+}$$

$$\overset{+5}{N}O_3^- \xrightarrow{\text{reduction}} \overset{+2}{N}O$$

Step 3: We balance each half-reaction for number and type of atoms and charges.

The *oxidation half-reaction* is already balanced for Sn atoms. There are two net positive charges on the right, so we add two electrons to the same side to balance the charge.

$$Sn \longrightarrow Sn^{2+} + 2e^-$$

The *reduction half-reaction* is balanced for nitrogen atoms. To balance the O atoms, we add two water molecules on the right side.

$$NO_3^- \longrightarrow NO + 2H_2O$$

To balance the H atoms, we add 4 H^+ to the left-hand side.

$$4H^+ + NO_3^- \longrightarrow NO + 2H_2O$$

There are three net positive charges on the left, so we add three electrons to the same side to balance the charge.

$$4H^+ + NO_3^- + 3e^- \longrightarrow NO + 2H_2O$$

Step 4: We now add the oxidation and reduction half-reactions to give the overall reaction. In order to equalize the number of electrons, we need to multiply the oxidation half-reaction by 3 and the reduction half-reaction by 2.

$$3Sn \longrightarrow 3Sn^{2+} + 6e^-$$

$$8H^+ + 2NO_3^- + 6e^- \longrightarrow 2NO + 4H_2O$$

$$8H^+ + 2NO_3^- + 3Sn + 6e^- \longrightarrow 2NO + 3Sn^{2+} + 4H_2O + 6e^-$$

The electrons on both sides cancel, and we are left with the balanced net ionic equation:

$$\mathbf{8H^+ + 2NO_3^- + 3Sn \longrightarrow 2NO + 3Sn^{2+} + 4H_2O}$$

Step 5: Check to see that the equation is balanced by verifying that the equation has the same types and numbers of atoms and the same charges on both sides of the equation.

The equation is "atomically" balanced. There are 8 H, 2 N, 6 O, and 3 Sn atoms on each side of the equation. The equation is also "electrically" balanced. The net charge is +6 on each side of the equation.

PRACTICE EXERCISE

1. The redox reaction between permanganate ion and iron(II) ions in acidic solution can be used to analyze iron ore for its iron content. Balance this redox reaction.

$$MnO_4^-(aq) + Fe^{2+}(aq) \longrightarrow Fe^{3+}(aq) + Mn^{2+}(aq)$$

Text Problem: 19.2

PROBLEM TYPE 2: STANDARD REDUCTION POTENTIALS

A. Comparing strengths of oxidizing agents

For a reduction reaction at an electrode when all solutes are 1 *M* and all gases are at 1 atm, the voltage is called the **standard reduction potential**. Table 19.1 of the text lists the standard reduction potentials for a number of half-reactions. The more positive the value of $E°$, the greater the tendency for the substance to be reduced, and therefore the stronger its tendency to act as an oxidizing agent. Looking at Table 19.1, we see that F_2 is the strongest oxidizing agent and Li^+ the weakest.

EXAMPLE 19.2
Arrange the following species in order of increasing strength as oxidizing agents under standard-state conditions: Zn^{2+}, MnO_4^- (in acid solution), and Ag^+.

Strategy: The greater the tendency for the substance to be reduced, the stronger its tendency to act as an oxidizing agent. The species that has a stronger tendency to be reduced will have a larger reduction potential.

Solution: Consulting Table 19.1 of the text, we write the half-reactions in the order of decreasing standard reduction potentials, $E°$.

$$MnO_4^-(aq) + 8H^+(aq) + 5e^- \longrightarrow Mn^{2+}(aq) + 4H_2O(l) \qquad E° = +1.51\text{ V}$$
$$Ag^+(aq) + e^- \longrightarrow Ag(s) \qquad E° = +0.80\text{ V}$$
$$Zn^{2+}(aq) + 2e^- \longrightarrow Zn(s) \qquad E° = -0.76\text{ V}$$

Since the reduction of MnO_4^- has the highest reduction potential and Zn^{2+} has the lowest reduction potential, the order of increasing strength of oxidizing agents is:

$$Zn^{2+} < Ag^+ < MnO_4^-$$

The large positive reduction potential for MnO_4^- indicates the strong tendency for permanganate ion to be reduced, and therefore the stronger its tendency to act as an oxidizing agent.

PRACTICE EXERCISE

2. Arrange the following species in order of increasing strength as oxidizing agents: Ce^{4+}, O_2, H_2O_2, and SO_4^{2-}.

Text Problems: 19.14, **19.18**

B. Calculating the standard emf ($E°$) of a galvanic cell

In a galvanic cell, electrons flow from one electrode to the other. This indicates that there is a voltage difference between the two electrodes. This voltage difference is called the **electromotive force**, or **emf (*E*)**, and it can be measured by connecting a voltmeter to both electrodes. The electromotive force is also called the cell voltage or cell potential. It is usually measured in volts. The electrode at which reduction occurs is called the ***cathode***, and the electrode at which oxidation occurs is called the ***anode***.

If all solutes have a concentration of 1 *M* and all gases have a pressure of 1 atm (standard conditions), the voltage difference between the two electrodes of the cell is called the **standard emf ($E°_{cell}$)**. The ***standard emf*** which is composed of a contribution from the *anode* and a contribution from the *cathode*, is given by:

$$E°_{cell} = E°_{cathode} - E°_{anode}$$

where $E°_{cathode}$ and $E°_{anode}$ are the standard reduction potentials of the cathode and anode, respectively.

Under standard-state conditions, a positive $E°_{cell}$ means the redox reaction will favor the formation of products at equilibrium. Conversely, a negative standard cell emf, means that more reactants than products will be formed at equilibrium.

We can calculate the standard emf of the cell using a table of standard reduction potentials (Table 19.1 of the text). As an example, consider the galvanic cell represented by the following reaction:

$$Cu(s) + 2Ag^{+}(aq) \longrightarrow Cu^{2+}(aq) + 2Ag(s)$$

Let's break this reaction down into its two half-reactions.

$$Cu(s) \xrightarrow{\text{oxidation (anode)}} Cu^{2+}(aq) + 2e^{-}$$

$$2Ag^{+}(aq) + 2e^{-} \xrightarrow{\text{reduction (cathode)}} 2Ag(s)$$

We can calculate the standard cell emf by subtracting E°_{anode} from $E^{\circ}_{\text{cathode}}$.

$$E^{\circ}_{\text{cell}} = E^{\circ}_{\text{cathode}} - E^{\circ}_{\text{anode}} = E^{\circ}_{\text{Ag}^{+}/\text{Ag}} - E^{\circ}_{\text{Cu}^{2+}/\text{Cu}}$$

$$E^{\circ}_{\text{cell}} = 0.80\text{ V} - 0.34\text{ V} = 0.46\text{ V}$$

EXAMPLE 19.3

Calculate the standard cell potential for a galvanic cell in which the following reaction takes place:

$$\mathbf{Cl_2(g) + 2Br^{-}(aq) \longrightarrow Br_2(l) + 2Cl^{-}(aq)}$$

Strategy: Separate the reaction into half-reactions. Then look up reduction potentials in Table 19.1 of the text.

Solution:

$$Cl_2(g) + 2e^{-} \xrightarrow{\text{reduction (cathode)}} 2Cl^{-}(aq)$$

$$2Br^{-}(aq) \xrightarrow{\text{oxidation (anode)}} Br_2(l) + 2e^{-}$$

We can look up standard reduction potentials in Table 19.1 of the text.

$$E^{\circ}_{\text{cathode}} = E^{\circ}_{\text{Cl}_2/\text{Cl}^{-}} = +1.36\text{ V}$$

$$E^{\circ}_{\text{anode}} = E^{\circ}_{\text{Br}_2/\text{Br}^{-}} = +1.07\text{ V}$$

The standard cell emf is given by

$$E^{\circ}_{\text{cell}} = E^{\circ}_{\text{cathode}} - E^{\circ}_{\text{anode}} = E^{\circ}_{\text{Cl}_2/\text{Cl}^{-}} - E^{\circ}_{\text{Br}_2/\text{Br}^{-}}$$

$$\mathbf{E^{\circ}_{cell}} = +1.36\text{ V} - 1.07\text{ V} = \mathbf{+0.29V}$$

PRACTICE EXERCISE

3. Consider a uranium-bromine galvanic cell in which U is oxidized and Br_2 is reduced. The reduction half-reactions are:

$$U^{3+}(aq) + 3e^{-} \longrightarrow U(s) \qquad E^{\circ}_{\text{U}^{3+}/\text{U}} = ?$$

$$Br_2(l) + 2e^{-} \longrightarrow 2Br^{-}(aq) \qquad E^{\circ}_{\text{Br}_2/\text{Br}^{-}} = 1.07\text{ V}$$

If the standard cell emf is 2.91 V, what is the standard reduction potential for uranium?

Text Problem: 19.12

PROBLEM TYPE 3: SPONTANEITY OF REDOX REACTIONS

A. Predicting whether a redox reaction is spontaneous

There is a relationship between free energy change and cell emf.

$$\Delta G = -nFE_{\text{cell}} \qquad (19.2, \text{text})$$

where,

n is the number of moles of electrons transferred during the redox reaction
F is the Faraday constant, which is the electrical charge contained in 1 mole of electrons

$$1\ F = 96{,}500\ \text{C/mol} = 96{,}500\ \text{J/V·mol}$$

For a derivation of this relationship, see Section 19.4 of the text. Both n and F are positive quantities, and we know from Chapter 18 that ΔG is *negative* for a spontaneous process. Therefore, E_{cell} is *positive* for a spontaneous process.

For reactions in which reactants and products are in their standard states, Equation (19.2) of the text becomes:

$$\Delta G^\circ = -nFE^\circ_{\text{cell}} \qquad (19.3, \text{text})$$

A *negative* ΔG° and a *positive* E°_{cell} mean that formation of products is favored at equilibrium. A *positive* ΔG° and a *negative* E°_{cell} mean that formation of reactants is favored at equilibrium. Table 19.2 of the text summarizes the relationships between ΔG° and E°_{cell}.

EXAMPLE 19.4

Predict whether products or reactants are favored at equilibrium when the following reaction is run under standard-state conditions.

$$\mathbf{Fe^{2+} + Cr_2O_7^{2-} \longrightarrow Fe^{3+} + Cr^{3+}}$$

Strategy: A positive E°_{cell} means the reaction will favor the formation of products at equilibrium. Calculate the standard cell emf from the potentials for the two half-reactions.

$$E^\circ_{\text{cell}} = E^\circ_{\text{cathode}} - E^\circ_{\text{anode}}$$

Solution: Separate the reaction into half-reactions to calculate the standard cell emf.

$$Fe^{2+}(aq) \xrightarrow{\text{oxidation (anode)}} Fe^{3+}(aq) + e^- \qquad E^\circ_{\text{anode}} = +0.77\ \text{V}$$

At this point, we could balance the reduction half-reaction and then come up with the overall balanced equation. But, we are not asked to do that. We can save time by looking at Table 19.1 of the text and finding a reaction that contains both $Cr_2O_7^{2-}$ and Cr^{3+}.

$$Cr_2O_7^{2-}(aq) + 14H^+(aq) + 6e^- \xrightarrow{\text{reduction (cathode)}} 2Cr^{3+}(aq) + 7H_2O(l) \qquad E^\circ_{\text{cathode}} = +1.33\ \text{V}$$

We have not come up with a balanced equation, but we do not need a balanced equation to calculate E°_{cell}.

$$E^\circ_{\text{cell}} = E^\circ_{\text{cathode}} - E^\circ_{\text{anode}}$$

$$\mathbf{E^\circ_{cell}} = 1.33\ \text{V} - 0.77\ \text{V} = \mathbf{+0.56\ V}$$

Since E°_{cell} is positive, products are favored at equilibrium.

PRACTICE EXERCISE

4. Predict whether a spontaneous reaction will occur when the following reactants and products are in their standard states.

(a) $2Fe^{3+}(aq) + 2I^{-}(aq) \longrightarrow 2Fe^{2+}(aq) + I_2(s)$

(b) $Cu(s) + 2H^{+}(aq) \longrightarrow Cu^{2+}(aq) + H_2(g)$

Text Problem: 19.16

B. Calculating $\Delta G°$ and K from $E°$

Equation (19.3) of the text shows the relationship between standard free energy change and standard emf.

$$\Delta G° = -nFE°_{\text{cell}} \qquad (19.3, \text{text})$$

Also recall that in Section 18.6 of the text, we saw that the standard free energy change for a reaction is related to its equilibrium constant K by the following equation:

$$\Delta G° = -RT \ln K$$

Therefore, from the two equations we obtain:

$$-nFE°_{\text{cell}} = -RT \ln K$$

Solving for $E°_{\text{cell}}$, we obtain:

$$E°_{\text{cell}} = \frac{RT}{nF} \ln K \qquad (19.4, \text{text})$$

At 298 K, we can simplify Equation (19.4) of the text by substituting for R and F.

$$E°_{\text{cell}} = \frac{(8.314 \text{ J/mol} \cdot \text{K})(298 \text{ K})}{n(96{,}500 \text{ J/V} \cdot \text{mol})} \ln K$$

$$E°_{\text{cell}} = \frac{0.0257 \text{ V}}{n} \ln K \qquad (19.5, \text{text})$$

We now have relationships between $\Delta G°$ and $E°_{\text{cell}}$, between $\Delta G°$ and K, and between $E°_{\text{cell}}$ and K. Figure 19.5 of the text summarizes the relationships among $\Delta G°$, K, and $E°_{\text{cell}}$.

EXAMPLE 19.5

Calculate $\Delta G°$ and the equilibrium constant at 25°C for the reaction

$$2Br^{-}(aq) + I_2(s) \longrightarrow Br_2(l) + 2I^{-}(aq)$$

Strategy: The relationship between the standard free energy change and the standard emf of the cell is given by Equation (19.3) of the text: $\Delta G° = -nFE°_{\text{cell}}$. The relationship between the equilibrium constant, K, and the standard emf is given by Equation (19.5) of the text: $E°_{\text{cell}} = (0.0257 \text{ V}/n)\ln K$. Thus, if we can determine $E°_{\text{cell}}$, we can calculate $\Delta G°$ and K. We can determine the $E°_{\text{cell}}$ of a hypothetical galvanic cell made up of two couples (Br_2/Br^{-} and I_2/I^{-}) from the standard reduction potentials in Table 19.1 of the text.

Solution: The half-cell reactions are:

$$I_2(s) + 2e^- \xrightarrow{\text{reduction (cathode)}} 2I^-(aq) \qquad E^\circ_{\text{cathode}} = +0.53\text{ V}$$

$$2Br^-(aq) \xrightarrow{\text{oxidation (anode)}} Br_2(l) + 2e^- \qquad E^\circ_{\text{anode}} = +1.07\text{ V}$$

$$2Br^-(aq) + I_2(s) \longrightarrow Br_2(l) + 2I^-(aq) \qquad E^\circ_{\text{cell}} = E^\circ_{\text{cathode}} - E^\circ_{\text{anode}} = -0.54\text{ V}$$

Substitute E°_{cell} into Equation (19.3) of the text to calculate the standard free energy change. In the balanced equation above, 2 moles of electrons are transferred. Therefore, $n = 2$.

$$\Delta G^\circ = -nFE^\circ_{\text{cell}}$$

$$\Delta G^\circ = -(2)(96{,}500\text{ J/V}\cdot\text{mol})(-0.54\text{ V})$$

$$\boldsymbol{\Delta G^\circ = 1.04 \times 10^5\text{ J/mol} = 104\text{ kJ/mol}}$$

Both the positive value of ΔG° and the negative value for E°_{cell} indicate that the reactants are favored at equilibrium under standard-state conditions.

Rearrange Equation (19.5) of the text to solve for the equilibrium constant, K.

$$\ln K = \frac{nE^\circ}{0.0257\text{ V}}$$

$$\ln K = \frac{(2)(-0.54\text{ V})}{0.0257\text{ V}} = -42.0$$

Taking the anti-ln of both sides of the equation,

$$\boldsymbol{K} = e^{-42.0} = \boldsymbol{6 \times 10^{-19}}$$

You should notice that the magnitude of K relates to what we have learned from ΔG° and E°_{cell}. A very small value for the equilibrium constant indicates that reactants are highly favored at equilibrium. This agrees with the signs of ΔG° and E°_{cell}, which indicate that more reactants than products will be formed at equilibrium.

PRACTICE EXERCISE

5. Calculate the equilibrium constant for the following redox reaction at 25°C.

$$2Fe^{2+}(aq) + Ni^{2+}(aq) \longrightarrow 2Fe^{3+}(aq) + Ni(s)$$

Text Problems: 19.22, 19.24, **19.26**, 19.38

PROBLEM TYPE 4: THE NERNST EQUATION

Thus far, we have only focused on redox reactions in which the reactants and products are in their standard states. However, standard-state conditions are often difficult and sometimes impossible to maintain. Therefore, we need a relationship between cell emf and the concentrations of reactants and products under *nonstandard-state* conditions.

In Chapter 18, we encountered a relationship between the free energy change (ΔG) and the standard free energy change ($\Delta G°$).

$$\Delta G = \Delta G° + RT \ln Q$$

where,

Q is the reaction quotient.

We also know that:

$$\Delta G = -nFE_{\text{cell}}$$

and

$$\Delta G° = -nFE°_{\text{cell}}$$

Substituting for ΔG and $\Delta G°$ in the first equation, we find:

$$-nFE_{\text{cell}} = -nFE°_{\text{cell}} + RT \ln Q$$

Dividing both sides of the equation by $-nF$ and omitting "cell" for simplicity gives:

$$E = E° - \frac{RT}{nF} \ln Q \qquad (19.7,\ \text{text})$$

Equation (19.7) is known as the **Nernst equation**. At 298 K, Equation (19.7) of the text can be simplified by substituting R, T, and F into the equation.

$$E = E° - \frac{0.0257\ \text{V}}{n} \ln Q \qquad (19.8,\ \text{text})$$

At equilibrium, there is no net transfer of electrons, so $E = 0$ and $Q = K$, where K is the equilibrium constant of the redox reaction. Substituting into Equation (19.8) gives

$$E° = \frac{0.0257\ \text{V}}{n} \ln K \qquad (19.5,\ \text{text})$$

This is Equation (19.5) of the text derived in Problem Type 3B.

A. Using the Nernst equation to predict the spontaneity of a redox reaction

To predict the spontaneity of a redox reaction, we can use the Nernst equation to calculate the cell emf, E. If the cell emf is *positive*, the redox reaction is *spontaneous*. Conversely, if E is *negative*, the reaction is *not spontaneous* in the direction written.

EXAMPLE 19.6

Calculate the cell emf and predict whether the following reaction is spontaneous at 25°C.

$$\mathbf{Zn(s) + 2H^+(1 \times 10^{-4}\ M) \longrightarrow Zn^{2+}(1.5\ M) + H_2(1\ atm)}$$

Strategy: The standard emf ($E°$) can be calculated using the standard reduction potentials in Table 19.1 of the text. Because the reactions are not run under standard-state conditions (concentrations are not 1 M), we need Nernst's equation [Equation (19.8) of the text] to calculate the emf (E) of a hypothetical galvanic cell. Remember that solids do not appear in the reaction quotient (Q) term in the Nernst equation.

Solution: Calculate the standard cell emf, $E°$, from standard reduction potentials (Table 19.1). Separate the reaction into its half-reactions.

$$2H^+(aq) + 2e^- \xrightarrow{\text{reduction (cathode)}} H_2(g) \qquad E^\circ_{\text{cathode}} = 0.00\text{ V}$$

$$Zn(s) \xrightarrow{\text{oxidation (anode)}} Zn^{2+}(aq) + 2e^- \qquad E^\circ_{\text{anode}} = -0.76\text{ V}$$

$$Zn(s) + 2H^+(aq) \longrightarrow Zn^{2+}(aq) + H_2(g) \qquad E^\circ_{\text{cell}} = E^\circ_{\text{cathode}} - E^\circ_{\text{anode}} = +0.76\text{ V}$$

Calculate the cell emf, E, from the Nernst equation. Two moles of electrons were transferred during the reaction, so $n = 2$.

$$E = E° - \frac{0.0257\text{ V}}{n}\ln\frac{[Zn^{2+}]P_{H_2}}{[H^+]^2}$$

$$E = 0.76\text{ V} - \frac{0.0257\text{ V}}{2}\ln\frac{(1.5)(1)}{(1 \times 10^{-4})^2}$$

$$\boldsymbol{E} = 0.76\text{ V} - 0.24\text{ V} = \mathbf{0.52\text{ V}}$$

The reaction is spontaneous because the cell emf, E, is positive.

PRACTICE EXERCISE

6. Calculate the cell emf for the following reaction:

$$2Ag^+(0.10\ M) + H_2(1.0\text{ atm}) \longrightarrow 2Ag(s) + 2H^+\ (\text{pH} = 8.00)$$

Text Problems: 19.30, 19.32, 19.34

B. Using the Nernst equation to calculate concentration

Recall that the reaction quotient Q equals the concentrations of the products raised to the power of their stoichiometric coefficients divided by the concentrations of the reactants raised to the power of their stoichiometric coefficients.

$$Q = \frac{[\text{products}]^x}{[\text{reactants}]^y}$$

Since Q is in the Nernst equation, if the cell emf is measured and the standard cell emf is calculated, we can determine the concentration of one of the components if the concentrations of the other components are known. See Example 19.7 below.

EXAMPLE 19.7

A galvanic cell is constructed from a silver half-cell and a copper half-cell. The copper half-cell contains 0.10 *M* $Cu(NO_3)_2$ and the concentration of silver ions in the other half-cell is unknown. If the Ag electrode is the cathode and the cell emf is measured and found to be 0.10 V at 25°C, what is the Ag^+ ion concentration?

Strategy: We are given the cell emf, $E = 0.10$ V, in the problem. If we can write the reaction occurring in the cell, we can calculate the standard cell emf, $E°$. We are also given the Cu^{2+} concentration in the problem. Given all this information, we can calculate the Ag^+ ion concentration using the Nernst equation.

Solution: Write the half-reactions to calculate the standard cell emf, $E°$. We are told that the Ag electrode is the cathode (reduction occurs at the cathode). We can write:

$$Ag^+(aq) + e^- \xrightarrow{\text{reduction (cathode)}} Ag(s) \qquad E^\circ_{\text{cathode}} = +0.80\text{ V}$$

If Ag^+ is reduced, Cu must be oxidized.

$$Cu(s) \xrightarrow{\text{oxidation (anode)}} Cu^{2+}(aq) + 2e^- \qquad E^\circ_{anode} = +0.34\text{ V}$$

$$E^\circ_{cell} = E^\circ_{cathode} - E^\circ_{anode}$$

$$E^\circ_{cell} = 0.80\text{ V} - 0.34\text{ V} = \mathbf{+0.46\text{ V}}$$

We must multiply the Ag half-reaction by two so that the number of electrons gained equals the number of electrons lost. The balanced reaction is:

$$2Ag^+(aq) + Cu(s) \longrightarrow 2Ag(s) + Cu^{2+}(aq)$$

Substitute E, $E°$, and $[Cu^{2+}]$ into the Nernst equation to calculate the $[Ag^+]$. Also, $n = 2$ since two moles of electrons are transferred during the reaction.

$$E = E^\circ - \frac{0.0257\text{ V}}{n}\ln Q$$

$$E = E^\circ - \frac{0.0257\text{ V}}{n}\ln\frac{[Cu^{2+}]}{[Ag^+]^2}$$

$$0.10\text{ V} = 0.46\text{ V} - \frac{0.0257\text{ V}}{2}\ln\frac{[0.10]}{[Ag^+]^2}$$

$$\frac{2(0.10\text{ V} - 0.46\text{ V})}{-0.0257\text{ V}} = \ln\frac{0.10}{[Ag^+]^2}$$

$$28.0 = \ln\frac{0.10}{[Ag^+]^2}$$

Taking the anti-ln of both sides of the equation,

$$e^{28.0} = \frac{0.10}{[Ag^+]^2}$$

$$1.45 \times 10^{12} = \frac{0.10}{[Ag^+]^2}$$

$$[Ag^+] = \sqrt{\frac{0.10}{1.45 \times 10^{12}}}$$

$$\mathbf{[Ag^+] = 2.6 \times 10^{-7}\ \mathit{M}}$$

PRACTICE EXERCISE

7. When the concentration of Zn^{2+} is 0.15 *M*, the measured voltage of a Zn-Cu galvanic cell is 0.40 V. What is the Cu^{2+} ion concentration?

Text Problem: 19.110

PROBLEM TYPE 5: ELECTROLYSIS

Electrolysis is the process in which electrical energy is used to cause a nonspontaneous chemical reaction to occur. The same principles underlie electrolysis and the processes that take place in galvanic cells.

A. Predicting the products of electrolysis

EXAMPLE 19.8

Predict the products of the electrolysis of an aqueous $MgCl_2$ solution.

This aqueous solution contains several species that could be oxidized or reduced. Two species that could be reduced are the metal ion (Mg^{2+}) and H_2O. The reduction half-reactions that might occur at the cathode are

(1) $Mg^{2+}(aq) + 2e^- \xrightarrow{\text{reduction (cathode)}} Mg(s)$ $E^\circ_{\text{cathode}} = -2.37\text{ V}$

(2) $2H_2O(l) + 2e^- \xrightarrow{\text{reduction (cathode)}} H_2(g) + 2OH^-(aq)$ $E^\circ_{\text{cathode}} = -0.83\text{ V}$

(3) $2H^+(aq) + 2e^- \xrightarrow{\text{reduction (cathode)}} H_2(g)$ $E^\circ_{\text{cathode}} = 0.00\text{ V}$

We can rule out Reaction (1) immediately. The very negative standard reduction potential indicates that Mg^{2+} has essentially no tendency to undergo reduction. Reaction (3) is preferred over Reaction (2) under standard-state conditions. However, at a pH of 7 (as is the case for a $MgCl_2$ solution), they are equally probable. We generally use Reaction (2) to describe the cathode reaction because the concentration of H^+ ions is too low (about 1×10^{-7} *M*) to make (3) a reasonable choice.

The oxidation reactions that might occur at the anode are:

(4) $2Cl^-(aq) \xrightarrow{\text{oxidation (anode)}} Cl_2(g) + 2e^-$ $E^\circ_{\text{anode}} = +1.36\text{ V}$

(5) $2H_2O(l) \xrightarrow{\text{oxidation (anode)}} O_2(g) + 4H^+(aq) + 2e^-$ $E^\circ_{\text{anode}} = +1.23\text{ V}$

The standard reduction potentials of (4) and (5) are not very different, but the values do suggest that H_2O should be preferentially oxidized at the anode. However, we find by experiment that the gas liberated at the anode is Cl_2, not O_2. The large overvoltage for O_2 formation prevents its production when the Cl^- ion is there to compete (see Section 19.8 of the text).

The overall reaction is:

$$2H_2O(l) + 2e^- \longrightarrow H_2(g) + 2OH^-(aq)$$
$$2Cl^-(aq) \longrightarrow Cl_2(g) + 2e^-$$
$$\overline{2H_2O(l) + 2Cl^-(aq) \longrightarrow H_2(g) + 2OH^-(aq) + Cl_2(g)}$$

B. Calculating the quantity of products in electrolysis

During electrolysis, the mass of product formed (or reactant consumed) at an electrode is proportional to both the amount of electricity transferred at the electrode and the molar mass of the substance in question.

The steps involved in calculating the quantities of substances produced in electrolysis are shown below.

Current (amperes) and time	→	Charge in coulombs	→	Number of mol of electrons	→	Moles of substance reduced or oxidized	→	Grams of substance reduced or oxidized

To convert to coulombs, recall that

$$1\text{ C} = 1\text{ A} \times 1\text{ s}$$

To convert from coulombs to moles of electrons, use the conversion factor

$$1\text{ mol } e^- = 96{,}500\text{ C}$$

Finally, we need to consider the number of moles of electrons transferred during the reaction. For example, to reduce Al^{3+} ions to Al metal, 3 moles of electrons are needed to reduce 1 mole of Al^{3+} ions.

$$Al^{3+} + 3e^- \longrightarrow Al$$

EXAMPLE 19.9

How many moles of electrons are transferred in an electrolytic cell when a current of 12 amps flows for 16 hours?

Strategy: According to Figure 19.20 of the text, we can carry out the following conversion steps to calculate the moles of electrons transferred.

$$\text{current} \times \text{time} \rightarrow \text{coulombs} \rightarrow \text{mol } e^-$$

This is a large number of steps, so let's break it down into two parts. First, we calculate the coulombs of electricity that pass through the cell. Then, we will continue on to moles of electrons.

Solution: First, we can calculate the number of coulombs passing through the cell in 16 hours. Recall that

$$1\text{ C} = 1\text{ A·s}$$

Converting from amps to coulombs:

$$12\text{ A} \times \frac{1\text{ C}}{1\text{ A·s}} \times \frac{3600\text{ s}}{1\text{ h}} \times 16\text{ h} = 6.9 \times 10^5\text{ C}$$

Next, we convert from coulombs to moles of electrons using the following conversion factor.

$$1\text{ mol } e^- = 96{,}500\text{ C}$$

$$\textbf{? mol } \boldsymbol{e^-} = (6.9 \times 10^5\text{ C}) \times \frac{1\text{ mol e}^-}{96{,}500\text{ C}} = \textbf{7.2 mol } \boldsymbol{e^-}$$

EXAMPLE 19.10

How many grams of copper metal would be deposited from a solution of $CuSO_4$ by the passage of 3.0 A of electrical current through an electrolytic cell for 2.0 h?

Strategy: According to Figure 19.20 of the text, we can carry out the following conversion steps to calculate the quantity of Cu in grams.

$$\text{current} \times \text{time} \rightarrow \text{coulombs} \rightarrow \text{mol } e^- \rightarrow \text{mol Cu} \rightarrow \text{g Cu}$$

This is a large number of steps, so let's break it down into two parts. First, we calculate the coulombs of electricity that pass through the cell. Then, we will continue on to calculate grams of Cu.

Solution: Let's start by writing the half-reaction to determine the moles of electrons required to produce 1 mole of Cu. Start by writing the half-reaction for the reduction of Cu^{2+} to Cu metal.

$$Cu^{2+}(aq) + 2e^- \longrightarrow Cu(s)$$

This half-reaction tells us that 2 mol of e^- are required to produce 1 mol of Cu (*s*).

Next, we calculate the coulombs of electricity that pass through the cell.

$$3.0\,\cancel{A} \times \frac{1\text{ C}}{1\,\cancel{A}\cdot\cancel{s}} \times \frac{3600\,\cancel{s}}{1\,\cancel{h}} \times 2.0\,\cancel{h} = 2.2 \times 10^4\text{ C}$$

The grams of Cu produced at the cathode are:

$$\textbf{? g Cu} = (2.2 \times 10^4\,\cancel{C}) \times \frac{1\,\cancel{mol\,e^-}}{96{,}500\,\cancel{C}} \times \frac{1\,\cancel{mol}\text{ Cu}}{2\,\cancel{mol\,e^-}} \times \frac{63.55\text{ g Cu}}{1\,\cancel{mol}\text{ Cu}} = \mathbf{7.3\text{ g Cu}}$$

> **Note:** This predicted amount of Cu is based on a process that is 100 percent efficient. Any side reactions or any oxidation or reduction of impurities will cause the actual yield to be less than the theoretical yield.

EXAMPLE 19.11

How long will it take to electrodeposit (plate out) 1.0 g of Ni from a $NiSO_4$ solution using a current of 2.5 A?

Strategy: This calculation is opposite to Example 19.10. We start with grams of Ni and are asked to calculate the time required to electrodeposit 1.0 g. We follow the strategy:

$$\text{g Ni} \rightarrow \text{mol Ni} \rightarrow \text{mol } e^- \rightarrow \text{coulombs} \rightarrow \text{time}$$

Let's break the calculation down into two steps, first converting grams of Ni to mole of electrons and then converting from moles of electrons to time.

Solution: First, find the number of moles of electrons required to electrodeposit 1.0 g of Ni from solution. The half-reaction for the reduction of Ni^{2+} is:

$$Ni^{2+}(aq) + 2e^- \longrightarrow Ni(s)$$

2 moles of electrons are required to reduce 1 mol of Ni^{2+} ions to Ni metal. But, we are electrodepositing less than 1 mole of Ni(*s*). We need to complete the following conversions:

$$\text{g Ni} \rightarrow \text{mol Ni} \rightarrow \text{mol of } e^-$$

$$\text{? mol of } e^- = 1.0\,\cancel{g}\text{ Ni} \times \frac{1\,\cancel{mol}\text{ Ni}}{58.7\,\cancel{g}\text{ Ni}} \times \frac{2\text{ mol } e^-}{1\,\cancel{mol}\text{ Ni}} = 0.034\text{ mol } e^-$$

Determine how long it will take for 0.034 moles of electrons to flow through the cell when the current is 2.5 C/s. We need to complete the following conversions:

$$\text{mol of } e^- \rightarrow \text{coulombs} \rightarrow \text{seconds}$$

$$\textbf{? seconds} = 0.034\,\cancel{mol\,e^-} \times \frac{96{,}500\,\cancel{C}}{1\,\cancel{mol\,e^-}} \times \frac{1\text{ s}}{2.5\,\cancel{C}} = \mathbf{1.3 \times 10^3\text{ s}}\ (22\text{ min})$$

PRACTICE EXERCISES

8. How many moles of electrons are transferred in an electrolytic cell when a current of 2.0 amps flows for 6.0 hours?

9. How many grams of cobalt can be electroplated by passing a constant current of 5.2 A through a solution of $CoCl_3$ for 60.0 min?

10. How long will it take to produce 54 kg of Al metal by the reduction of Al^{3+} in an electrolytic cell using a current of 5.0×10^2 amps?

Text Problems: 19.46, 19.48, 19.52, 19.54, 19.56, 19.58, 19.60

ANSWERS TO PRACTICE EXERCISES

1. $5Fe^{2+}(aq) + MnO_4^-(aq) + 8H^+(aq) \longrightarrow 5Fe^{3+}(aq) + Mn^{2+}(aq) + 4H_2O(l)$

2. $SO_4^{2-} > O_2 > Ce^{4+} > H_2O_2$

3. $E^\circ_{U^{3+}/U} = -1.84$ V

4. **(a)** $E^\circ_{cell} = +0.24$ V. The positive value of the standard cell emf indicates that the reaction is spontaneous under standard-state conditions.

(b) $E^\circ_{cell} = -0.34$ V. A negative standard cell emf indicates that this reaction is not spontaneous under standard-state conditions. Copper will not dissolve in 1 *M* HCl.

5. $K = 3.0 \times 10^{-35}$

6. $E° = 1.21$ V

7. $[Cu^{2+}] = 3.1 \times 10^{-25}$ *M*

8. Number of mol e^- transferred = 0.45 mol e^-

9. Grams of Co electroplated = 3.81 g Co

10. Time = 322 h

SOLUTIONS TO SELECTED TEXT PROBLEMS

19.2 Balancing Redox Equations, Problem Type 1.

Strategy: We follow the procedure for balancing redox reactions presented in Section 19.1 of the text.

Solution:

(a)

Step 1: The unbalanced equation is given in the problem.

$$Mn^{2+} + H_2O_2 \longrightarrow MnO_2 + H_2O$$

Step 2: The two half-reactions are:

$$Mn^{2+} \xrightarrow{\text{oxidation}} MnO_2$$

$$H_2O_2 \xrightarrow{\text{reduction}} H_2O$$

Step 3: We balance each half-reaction for number and type of atoms and charges.

The *oxidation half-reaction* is already balanced for Mn atoms. To balance the O atoms, we add two water molecules on the left side.

$$Mn^{2+} + 2H_2O \longrightarrow MnO_2$$

To balance the H atoms, we add 4 H^+ to the right-hand side.

$$Mn^{2+} + 2H_2O \longrightarrow MnO_2 + 4H^+$$

There are four net positive charges on the right and two net positive charge on the left, we add two electrons to the right side to balance the charge.

$$Mn^{2+} + 2H_2O \longrightarrow MnO_2 + 4H^+ + 2e^-$$

Reduction half-reaction: we add one H_2O to the right-hand side of the equation to balance the O atoms.

$$H_2O_2 \longrightarrow 2H_2O$$

To balance the H atoms, we add $2H^+$ to the left-hand side.

$$H_2O_2 + 2H^+ \longrightarrow 2H_2O$$

There are two net positive charges on the left, so we add two electrons to the same side to balance the charge.

$$H_2O_2 + 2H^+ + 2e^- \longrightarrow 2H_2O$$

Step 4: We now add the oxidation and reduction half-reactions to give the overall reaction. Note that the number of electrons gained and lost is equal.

$$Mn^{2+} + 2H_2O \longrightarrow MnO_2 + 4H^+ + 2e^-$$

$$H_2O_2 + 2H^+ + 2e^- \longrightarrow 2H_2O$$

$$\overline{Mn^{2+} + H_2O_2 + 2e^- \longrightarrow MnO_2 + 2H^+ + 2e^-}$$

The electrons on both sides cancel, and we are left with the balanced net ionic equation in acidic medium.

$$Mn^{2+} + H_2O_2 \longrightarrow MnO_2 + 2H^+$$

Because the problem asks to balance the equation in basic medium, we add one OH^- to both sides for each H^+ and combine pairs of H^+ and OH^- on the same side of the arrow to form H_2O.

$$Mn^{2+} + H_2O_2 + 2OH^- \longrightarrow MnO_2 + 2H^+ + 2OH^-$$

Combining the H^+ and OH^- to form water we obtain:

$$\mathbf{Mn^{2+} + H_2O_2 + 2OH^- \longrightarrow MnO_2 + 2H_2O}$$

Step 5: Check to see that the equation is balanced by verifying that the equation has the same types and numbers of atoms and the same charges on both sides of the equation.

(b) This problem can be solved by the same methods used in part (a).

$$\mathbf{2Bi(OH)_3 + 3SnO_2^{2-} \longrightarrow 2Bi + 3H_2O + 3SnO_3^{2-}}$$

(c)

Step 1: The unbalanced equation is given in the problem.

$$Cr_2O_7^{2-} + C_2O_4^{2-} \longrightarrow Cr^{3+} + CO_2$$

Step 2: The two half-reactions are:

$$C_2O_4^{2-} \xrightarrow{\text{oxidation}} CO_2$$
$$Cr_2O_7^{2-} \xrightarrow{\text{reduction}} Cr^{3+}$$

Step 3: We balance each half-reaction for number and type of atoms and charges.

In the *oxidation half-reaction*, we first need to balance the C atoms.

$$C_2O_4^{2-} \longrightarrow 2CO_2$$

The O atoms are already balanced. There are two net negative charges on the left, so we add two electrons to the right to balance the charge.

$$C_2O_4^{2-} \longrightarrow 2CO_2 + 2e^-$$

In the *reduction half-reaction*, we first need to balance the Cr atoms.

$$Cr_2O_7^{2-} \longrightarrow 2Cr^{3+}$$

We add seven H_2O molecules on the right to balance the O atoms.

$$Cr_2O_7^{2-} \longrightarrow 2Cr^{3+} + 7H_2O$$

To balance the H atoms, we add $14H^+$ to the left-hand side.

$$Cr_2O_7^{2-} + 14H^+ \longrightarrow 2Cr^{3+} + 7H_2O$$

There are twelve net positive charges on the left and six net positive charges on the right. We add six electrons on the left to balance the charge.

$$Cr_2O_7^{2-} + 14H^+ + 6e^- \longrightarrow 2Cr^{3+} + 7H_2O$$

Step 4: We now add the oxidation and reduction half-reactions to give the overall reaction. In order to equalize the number of electrons, we need to multiply the oxidation half-reaction by 3.

$$3(C_2O_4^{2-} \longrightarrow 2CO_2 + 2e^-)$$
$$Cr_2O_7^{2-} + 14H^+ + 6e^- \longrightarrow 2Cr^{3+} + 7H_2O$$
$$\overline{3C_2O_4^{2-} + Cr_2O_7^{2-} + 14H^+ + 6e^- \longrightarrow 6CO_2 + 2Cr^{3+} + 7H_2O + 6e^-}$$

The electrons on both sides cancel, and we are left with the balanced net ionic equation in acidic medium.

$$\mathbf{3C_2O_4^{2-} + Cr_2O_7^{2-} + 14H^+ \longrightarrow 6CO_2 + 2Cr^{3+} + 7H_2O}$$

Step 5: Check to see that the equation is balanced by verifying that the equation has the same types and numbers of atoms and the same charges on both sides of the equation.

(d) This problem can be solved by the same methods used in part (c).

$$\mathbf{2Cl^- + 2ClO_3^- + 4H^+ \longrightarrow Cl_2 + 2ClO_2 + 2H_2O}$$

19.12 Calculating the standard emf of a galvanic cell, Problem Type 2B.

Strategy: At first, it may not be clear how to assign the electrodes in the galvanic cell. From Table 19.1 of the text, we write the standard reduction potentials of Al and Ag and apply the diagonal rule to determine which is the anode and which is the cathode.

Solution: The standard reduction potentials are:

$$Ag^+(1.0\,M) + e^- \rightarrow Ag(s) \qquad E° = 0.80\text{ V}$$
$$Al^{3+}(1.0\,M) + 3e^- \rightarrow Al(s) \qquad E° = -1.66\text{ V}$$

Applying the diagonal rule, we see that Ag^+ will oxidize Al.

Anode (oxidation):	$Al(s) \rightarrow Al^{3+}(1.0\,M) + 3e^-$
Cathode (reduction):	$3Ag^+(1.0\,M) + 3e^- \rightarrow 3Ag(s)$
Overall:	$\mathbf{Al(s) + 3Ag^+(1.0\,M) \rightarrow Al^{3+}(1.0\,M) + 3Ag(s)}$

Note that in order to balance the overall equation, we multiplied the reduction of Ag^+ by 3. We can do so because, as an intensive property, $E°$ is not affected by this procedure. We find the emf of the cell using Equation (19.1) and Table 19.1 of the text.

$$E°_{cell} = E°_{cathode} - E°_{anode} = E°_{Ag^+/Ag} - E°_{Al^{3+}/Al}$$

$$\mathbf{E°_{cell}} = 0.80\text{ V} - (-1.66\text{ V}) = \mathbf{+2.46\text{ V}}$$

Check: The positive value of $E°$ shows that the forward reaction is favored.

19.14 The half–reaction for oxidation is:

$$2H_2O(l) \xrightarrow{\text{oxidation (anode)}} O_2(g) + 4H^+(aq) + 4e^- \qquad E^\circ_{\text{anode}} = +1.23\text{ V}$$

The species that can oxidize water to molecular oxygen must have an E°_{red} more positive than +1.23 V. From Table 19.1 of the text we see that only $\mathbf{Cl_2}(g)$ and $\mathbf{MnO_4^-}(aq)$ in acid solution can oxidize water to oxygen.

19.16 Predicting whether a redox reaction is spontaneous, Problem Type 3A.

Strategy: E°_{cell} is *positive* for a spontaneous reaction. In each case, we can calculate the standard cell emf from the potentials for the two half-reactions.

$$E^\circ_{\text{cell}} = E^\circ_{\text{cathode}} - E^\circ_{\text{anode}}$$

Solution:

(a) $E° = -0.40\text{ V} - (-2.87\text{ V}) = \mathbf{2.47\ V}$. The reaction is spontaneous.

(b) $E° = -0.14\text{ V} - 1.07\text{ V} = \mathbf{-1.21\ V}$. The reaction is not spontaneous.

(c) $E° = -0.25\text{ V} - 0.80\text{ V} = \mathbf{-1.05\ V}$. The reaction is not spontaneous.

(d) $E° = 0.77\text{ V} - 0.15\text{ V} = \mathbf{0.62\ V}$. The reaction is spontaneous.

19.18 Similar to Problem Type 2A, Comparing the strengths of oxidizing agents.

Strategy: The greater the tendency for the substance to be oxidized, the stronger its tendency to act as a reducing agent. The species that has a stronger tendency to be oxidized will have a smaller reduction potential.

Solution: In each pair, look for the one with the smaller reduction potential. This indicates a greater tendency for the substance to be oxidized.

(a) Li **(b)** H_2 **(c)** Fe^{2+} **(d)** Br^-

19.22 Calculating $E°$ from K. Similar to Problem Type 3B.

Strategy: The relationship between the equilibrium constant, K, and the standard emf is given by Equation (19.5) of the text: $E^\circ_{\text{cell}} = (0.0257\text{ V}/n)\ln K$. Thus, knowing n (the moles of electrons transferred) and the equilibrium constant, we can determine E°_{cell}.

Solution: The equation that relates K and the standard cell emf is:

$$E^\circ_{\text{cell}} = \frac{0.0257\text{ V}}{n}\ln K$$

We see in the reaction that Mg goes to Mg^{2+} and Zn^{2+} goes to Zn. Therefore, two moles of electrons are transferred during the redox reaction. Substitute the equilibrium constant and the moles of e^- transferred ($n = 2$) into the above equation to calculate $E°$.

$$\boldsymbol{E°} = \frac{(0.0257\text{ V})\ln K}{n} = \frac{(0.0257\text{ V})\ln(2.69\times10^{12})}{2} = \mathbf{0.368\ V}$$

19.24 **(a)** We break the equation into two half–reactions:

$$Mg(s) \xrightarrow{\text{oxidation (anode)}} Mg^{2+}(aq) + 2e^- \qquad E^\circ_{\text{anode}} = -2.37\text{ V}$$

$$Pb^{2+}(aq) + 2e^- \xrightarrow{\text{reduction (cathode)}} Pb(s) \qquad E^\circ_{\text{cathode}} = -0.13\text{ V}$$

The standard emf is given by

$$E^\circ_{\text{cell}} = E^\circ_{\text{cathode}} - E^\circ_{\text{anode}} = -0.13\text{ V} - (-2.37\text{ V}) = 2.24\text{ V}$$

We can calculate $\Delta G°$ from the standard emf.

$$\Delta G° = -nFE^\circ_{\text{cell}}$$

$$\boldsymbol{\Delta G°} = -(2)(96500\text{ J/}\cancel{\text{V}}\cdot\text{mol})(2.24\ \cancel{\text{V}}) = \mathbf{-432\ kJ/mol}$$

Next, we can calculate K using Equation (19.5) of the text.

$$E^\circ_{\text{cell}} = \frac{0.0257\text{ V}}{n}\ln K$$

or

$$\ln K = \frac{nE^\circ_{\text{cell}}}{0.0257\text{ V}}$$

and

$$K = e^{\frac{nE°}{0.0257}}$$

$$\boldsymbol{K} = e^{\frac{(2)(2.24)}{0.0257}} = \mathbf{5\times10^{75}}$$

Tip: You could also calculate K_c from the standard free energy change, $\Delta G°$, using the equation: $\Delta G° = -RT\ln K_c$.

(b) We break the equation into two half–reactions:

$$Br_2(l) + 2e^- \xrightarrow{\text{reduction (cathode)}} 2Br^-(aq) \qquad E^\circ_{\text{cathode}} = 1.07\text{ V}$$

$$2I^-(aq) \xrightarrow{\text{oxidation (anode)}} I_2(s) + 2e^- \qquad E^\circ_{\text{anode}} = 0.53\text{ V}$$

The standard emf is

$$E^\circ_{\text{cell}} = E^\circ_{\text{cathode}} - E^\circ_{\text{anode}} = 1.07\text{ V} - 0.53\text{ V} = 0.54\text{ V}$$

We can calculate $\Delta G°$ from the standard emf.

$$\Delta G° = -nFE^\circ_{\text{cell}}$$

$$\boldsymbol{\Delta G°} = -(2)(96500\text{ J/}\cancel{\text{V}}\cdot\text{mol})(0.54\ \cancel{\text{V}}) = \mathbf{-104\ kJ/mol}$$

Next, we can calculate K using Equation (19.5) of the text.

$$K = e^{\frac{nE^\circ}{0.0257}}$$

$$\mathbf{K} = e^{\frac{(2)(0.54)}{0.0257}} = \mathbf{2 \times 10^{18}}$$

(c) This is worked in an analogous manner to parts (a) and (b).

$$E^\circ_{\text{cell}} = E^\circ_{\text{cathode}} - E^\circ_{\text{anode}} = 1.23\text{ V} - 0.77\text{ V} = 0.46\text{ V}$$

$$\Delta G^\circ = -nFE^\circ_{\text{cell}}$$

$$\mathbf{\Delta G^\circ} = -(4)(96500\text{ J/}\cancel{\text{V}}\cdot\text{mol})(0.46\,\cancel{\text{V}}) = \mathbf{-178\ kJ/mol}$$

$$K = e^{\frac{nE^\circ}{0.0257}}$$

$$\mathbf{K} = e^{\frac{(4)(0.46)}{0.0257}} = \mathbf{1 \times 10^{31}}$$

(d) This is worked in an analogous manner to parts (a), (b), and (c).

$$E^\circ_{\text{cell}} = E^\circ_{\text{cathode}} - E^\circ_{\text{anode}} = 0.53\text{ V} - (-1.66\text{ V}) = 2.19\text{ V}$$

$$\Delta G^\circ = -nFE^\circ_{\text{cell}}$$

$$\mathbf{\Delta G^\circ} = -(6)(96500\text{ J/}\cancel{\text{V}}\cdot\text{mol})(2.19\,\cancel{\text{V}}) = \mathbf{-1.27 \times 10^3\ kJ/mol}$$

$$K = e^{\frac{nE^\circ}{0.0257}}$$

$$\mathbf{K} = e^{\frac{(6)(2.19)}{0.0257}} = \mathbf{8 \times 10^{211}}$$

19.26 Calculating ΔG° and K from E°, Problem Type 3B.

Strategy: The relationship between the standard free energy change and the standard emf of the cell is given by Equation (19.3) of the text: $\Delta G^\circ = -nFE^\circ_{\text{cell}}$. The relationship between the equilibrium constant, K, and the standard emf is given by Equation (19.5) of the text: $E^\circ_{\text{cell}} = (0.0257\text{ V}/n)\ln K$. Thus, if we can determine E°_{cell}, we can calculate ΔG° and K. We can determine the E°_{cell} of a hypothetical galvanic cell made up of two couples (Cu^{2+}/Cu^+ and Cu^+/Cu) from the standard reduction potentials in Table 19.1 of the text.

Solution: The half-cell reactions are:

Anode (oxidation):	$Cu^+(1.0\ M) \rightarrow Cu^{2+}(1.0\ M) + e^-$
Cathode (reduction):	$Cu^+(1.0\ M) + e^- \rightarrow Cu(s)$
Overall:	$2Cu^+(1.0\ M) \rightarrow Cu^{2+}(1.0\ M) + Cu(s)$

$$E^\circ_{\text{cell}} = E^\circ_{\text{cathode}} - E^\circ_{\text{anode}} = E^\circ_{\text{Cu}^+/\text{Cu}} - E^\circ_{\text{Cu}^{2+}/\text{Cu}^+}$$

$$\boldsymbol{E^\circ_{\text{cell}}} = 0.52\text{ V} - 0.15\text{ V} = \mathbf{0.37\text{ V}}$$

Now, we use Equation (19.3) of the text. The overall reaction shows that $n = 1$.

$$\Delta G^\circ = -nFE^\circ_{\text{cell}}$$

$$\boldsymbol{\Delta G^\circ} = -(1)(96500\text{ J/}\cancel{\text{V}}\cdot\text{mol})(0.37\,\cancel{\text{V}}) = \mathbf{-36\text{ kJ/mol}}$$

Next, we can calculate K using Equation (19.5) of the text.

$$E^\circ_{\text{cell}} = \frac{0.0257\text{ V}}{n}\ln K$$

or

$$\ln K = \frac{nE^\circ_{\text{cell}}}{0.0257\text{ V}}$$

and

$$K = e^{\frac{nE^\circ}{0.0257}}$$

$$\boldsymbol{K} = e^{\frac{(1)(0.37)}{0.0257}} = e^{14.4} = \mathbf{2 \times 10^6}$$

Check: The negative value of ΔG° and the large positive value of K, both indicate that the reaction favors products at equilibrium. The result is consistent with the fact that E° for the galvanic cell is positive.

19.30 Using the Nernst equation, Problem Type 4.

Strategy: The standard emf (E°) can be calculated using the standard reduction potentials in Table 19.1 of the text. Because the reactions are not run under standard-state conditions (concentrations are not 1 *M*), we need Nernst's equation [Equation (19.8) of the text] to calculate the emf (E) of a hypothetical galvanic cell. Remember that solids do not appear in the reaction quotient (Q) term in the Nernst equation. We can calculate ΔG from E using Equation (19.2) of the text: $\Delta G = -nFE_{\text{cell}}$.

Solution:

(a) The half-cell reactions are:

Anode (oxidation):	$Mg(s) \rightarrow Mg^{2+}(1.0\ M) + 2e^-$
Cathode (reduction):	$Sn^{2+}(1.0\ M) + 2e^- \rightarrow Sn(s)$
Overall:	$Mg(s) + Sn^{2+}(1.0\ M) \rightarrow Mg^{2+}(1.0\ M) + Sn(s)$

$$E^\circ_{\text{cell}} = E^\circ_{\text{cathode}} - E^\circ_{\text{anode}} = E^\circ_{\text{Sn}^{2+}/\text{Sn}} - E^\circ_{\text{Mg}^{2+}/\text{Mg}}$$

$$\boldsymbol{E^\circ_{\text{cell}}} = -0.14\text{ V} - (-2.37\text{ V}) = \mathbf{2.23\text{ V}}$$

From Equation (19.8) of the text, we write:

$$E = E^\circ - \frac{0.0257\text{ V}}{n}\ln Q$$

$$E = E° - \frac{0.0257\text{ V}}{n}\ln\frac{[\text{Mg}^{2+}]}{[\text{Sn}^{2+}]}$$

$$\boldsymbol{E} = 2.23\text{ V} - \frac{0.0257\text{ V}}{2}\ln\frac{0.045}{0.035} = \mathbf{2.23\text{ V}}$$

We can now find the free energy change at the given concentrations using Equation (19.2) of the text. Note that in this reaction, $n = 2$.

$$\Delta G = -nFE_{\text{cell}}$$

$$\boldsymbol{\Delta G} = -(2)(96500\text{ J/V·mol})(2.23\text{ V}) = \mathbf{-430\text{ kJ/mol}}$$

(b) The half-cell reactions are:

Anode (oxidation):	$3[\text{Zn}(s) \rightarrow \text{Zn}^{2+}(1.0\ M) + 2e^-]$
Cathode (reduction):	$2[\text{Cr}^{3+}(1.0\ M) + 3e^- \rightarrow \text{Cr}(s)]$
Overall:	$3\text{Zn}(s) + 2\text{Cr}^{3+}(1.0\ M) \rightarrow 3\text{Zn}^{2+}(1.0\ M) + 2\text{Cr}(s)$

$$E°_{\text{cell}} = E°_{\text{cathode}} - E°_{\text{anode}} = E°_{\text{Cr}^{3+}/\text{Cr}} - E°_{\text{Zn}^{2+}/\text{Zn}}$$

$$\boldsymbol{E°_{\text{cell}}} = -0.74\text{ V} - (-0.76\text{ V}) = \mathbf{0.02\text{ V}}$$

From Equation (19.8) of the text, we write:

$$E = E° - \frac{0.0257\text{ V}}{n}\ln Q$$

$$E = E° - \frac{0.0257\text{ V}}{n}\ln\frac{[\text{Zn}^{2+}]^3}{[\text{Cr}^{3+}]^2}$$

$$\boldsymbol{E} = 0.02\text{ V} - \frac{0.0257\text{ V}}{6}\ln\frac{(0.0085)^3}{(0.010)^2} = \mathbf{0.04\text{ V}}$$

We can now find the free energy change at the given concentrations using Equation (19.2) of the text. Note that in this reaction, $n = 6$.

$$\Delta G = -nFE_{\text{cell}}$$

$$\boldsymbol{\Delta G} = -(6)(96500\text{ J/V·mol})(0.04\text{ V}) = \mathbf{-23\text{ kJ/mol}}$$

19.32 Let's write the two half-reactions to calculate the standard cell emf. (Oxidation occurs at the Pb electrode.)

$$\text{Pb}(s) \xrightarrow{\text{oxidation (anode)}} \text{Pb}^{2+}(aq) + 2e^- \qquad E°_{\text{anode}} = -0.13\text{ V}$$

$$2\text{H}^+(aq) + 2e^- \xrightarrow{\text{reduction (cathode)}} \text{H}_2(g) \qquad E°_{\text{cathode}} = 0.00\text{ V}$$

$$2\text{H}^+(aq) + \text{Pb}(s) \longrightarrow \text{H}_2(g) + \text{Pb}^{2+}(aq)$$

$$E°_{\text{cell}} = E°_{\text{cathode}} - E°_{\text{anode}} = 0.00\text{ V} - (-0.13\text{ V}) = 0.13\text{ V}$$

Using the Nernst equation, we can calculate the cell emf, E.

$$E = E^\circ - \frac{0.0257\text{ V}}{n}\ln\frac{[Pb^{2+}]P_{H_2}}{[H^+]^2}$$

$$\boldsymbol{E} = 0.13\text{ V} - \frac{0.0257\text{ V}}{2}\ln\frac{(0.10)(1.0)}{(0.050)^2} = \mathbf{0.083\text{ V}}$$

19.34 All concentration cells have the same standard emf: *zero* volts.

$$Mg^{2+}(aq) + 2e^- \xrightarrow{\text{reduction (cathode)}} Mg(s) \qquad E^\circ_{cathode} = -2.37\text{ V}$$

$$Mg(s) \xrightarrow{\text{oxidation (anode)}} Mg^{2+}(aq) + 2e^- \qquad E^\circ_{anode} = -2.37\text{ V}$$

$$E^\circ_{cell} = E^\circ_{cathode} - E^\circ_{anode} = -2.37\text{ V} - (-2.37\text{ V}) = 0.00\text{ V}$$

We use the Nernst equation to compute the emf. There are two moles of electrons transferred from the reducing agent to the oxidizing agent in this reaction, so $n = 2$.

$$E = E^\circ - \frac{0.0257\text{ V}}{n}\ln Q$$

$$E = E^\circ - \frac{0.0257\text{ V}}{n}\ln\frac{[Mg^{2+}]_{ox}}{[Mg^{2+}]_{red}}$$

$$\boldsymbol{E} = 0\text{ V} - \frac{0.0257\text{ V}}{2}\ln\frac{0.24}{0.53} = \mathbf{0.010\text{ V}}$$

What is the direction of spontaneous change in all concentration cells?

19.38 We can calculate the standard free energy change, ΔG°, from the standard free energies of formation, ΔG°_f using Equation (18.12) of the text. Then, we can calculate the standard cell emf, E°_{cell}, from ΔG°.

The overall reaction is:

$$C_3H_8(g) + 5O_2(g) \longrightarrow 3CO_2(g) + 4H_2O(l)$$

$$\Delta G^\circ_{rxn} = 3\Delta G^\circ_f[CO_2(g)] + 4\Delta G^\circ_f[H_2O(l)] - \{\Delta G^\circ_f[C_3H_8(g)] + 5\Delta G^\circ_f[O_2(g)]\}$$

$$\Delta G^\circ_{rxn} = (3)(-394.4\text{ kJ/mol}) + (4)(-237.2\text{ kJ/mol}) - [(1)(-23.5\text{ kJ/mol}) + (5)(0)] = -2108.5\text{ kJ/mol}$$

We can now calculate the standard emf using the following equation:

$$\Delta G^\circ = -nFE^\circ_{cell}$$

or

$$E^\circ_{cell} = \frac{-\Delta G^\circ}{nF}$$

Check the half-reactions on p. 843 of the text to determine that 20 moles of electrons are transferred during this redox reaction.

$$E^\circ_{\text{cell}} = \frac{-(-2108.5 \times 10^3 \text{ J/mol})}{(20)(96500 \text{ J/V}\cdot\text{mol})} = \mathbf{1.09\ V}$$

Does this suggest that, in theory, it should be possible to construct a galvanic cell (battery) based on any conceivable spontaneous reaction?

19.46 Calculating the quantity of products in electrolysis, Problem Type 5B.

(a) The only ions present in molten $BaCl_2$ are Ba^{2+} and Cl^-. The electrode reactions are:

$$\text{anode:} \quad 2Cl^-(aq) \longrightarrow Cl_2(g) + 2e^-$$

$$\text{cathode:} \quad Ba^{2+}(aq) + 2e^- \longrightarrow Ba(s)$$

This cathode half-reaction tells us that 2 moles of e^- are required to produce 1 mole of Ba(s).

(b)
Strategy: According to Figure 19.20 of the text, we can carry out the following conversion steps to calculate the quantity of Ba in grams.

$$\text{current} \times \text{time} \rightarrow \text{coulombs} \rightarrow \text{mol } e^- \rightarrow \text{mol Ba} \rightarrow \text{g Ba}$$

This is a large number of steps, so let's break it down into two parts. First, we calculate the coulombs of electricity that pass through the cell. Then, we will continue on to calculate grams of Ba.

Solution: First, we calculate the coulombs of electricity that pass through the cell.

$$0.50 \text{ A} \times \frac{1 \text{ C}}{1 \text{ A}\cdot\text{s}} \times \frac{60 \text{ s}}{1 \text{ min}} \times 30 \text{ min} = 9.0 \times 10^2 \text{ C}$$

We see that for every mole of Ba formed at the cathode, 2 moles of electrons are needed. The grams of Ba produced at the cathode are:

$$\textbf{? g Ba} = (9.0 \times 10^2 \text{ C}) \times \frac{1 \text{ mol } e^-}{96{,}500 \text{ C}} \times \frac{1 \text{ mol Ba}}{2 \text{ mol } e^-} \times \frac{137.3 \text{ g Ba}}{1 \text{ mol Ba}} = \mathbf{0.64\ g\ Ba}$$

19.48 The cost for producing various metals is determined by the moles of electrons needed to produce a given amount of metal. For each reduction, let's first calculate the number of tons of metal produced per 1 mole of electrons (1 ton = 9.072×10^5 g). The reductions are:

$$Mg^{2+} + 2e^- \longrightarrow Mg \qquad \frac{1 \text{ mol Mg}}{2 \text{ mol } e^-} \times \frac{24.31 \text{ g Mg}}{1 \text{ mol Mg}} \times \frac{1 \text{ ton}}{9.072 \times 10^5 \text{ g}} = 1.340 \times 10^{-5} \text{ ton Mg/mol } e^-$$

$$Al^{3+} + 3e^- \longrightarrow Al \qquad \frac{1 \text{ mol Al}}{3 \text{ mol } e^-} \times \frac{26.98 \text{ g Al}}{1 \text{ mol Al}} \times \frac{1 \text{ ton}}{9.072 \times 10^5 \text{ g}} = 9.913 \times 10^{-6} \text{ ton Al/mol } e^-$$

$$Na^+ + e^- \longrightarrow Na \qquad \frac{1 \text{ mol Na}}{1 \text{ mol } e^-} \times \frac{22.99 \text{ g Na}}{1 \text{ mol Na}} \times \frac{1 \text{ ton}}{9.072 \times 10^5 \text{ g}} = 2.534 \times 10^{-5} \text{ ton Na/mol } e^-$$

$$Ca^{2+} + 2e^- \longrightarrow Ca \qquad \frac{1 \text{ mol Ca}}{2 \text{ mol } e^-} \times \frac{40.08 \text{ g Ca}}{1 \text{ mol Ca}} \times \frac{1 \text{ ton}}{9.072 \times 10^5 \text{ g}} = 2.209 \times 10^{-5} \text{ ton Ca/mol } e^-$$

Now that we know the tons of each metal produced per mole of electrons, we can convert from \$155/ton Mg to the cost to produce the given amount of each metal.

(a) For aluminum :

$$\frac{\$155}{1\text{ ton Mg}} \times \frac{1.340 \times 10^{-5}\text{ ton Mg}}{1\text{ mol }e^-} \times \frac{1\text{ mol }e^-}{9.913 \times 10^{-6}\text{ ton Al}} \times 10.0\text{ tons Al} = \mathbf{\$2.10 \times 10^3}$$

(b) For sodium:

$$\frac{\$155}{1\text{ ton Mg}} \times \frac{1.340 \times 10^{-5}\text{ ton Mg}}{1\text{ mol }e^-} \times \frac{1\text{ mol }e^-}{2.534 \times 10^{-5}\text{ ton Na}} \times 30.0\text{ tons Na} = \mathbf{\$2.46 \times 10^3}$$

(c) For calcium:

$$\frac{\$155}{1\text{ ton Mg}} \times \frac{1.340 \times 10^{-5}\text{ ton Mg}}{1\text{ mol }e^-} \times \frac{1\text{ mol }e^-}{2.209 \times 10^{-5}\text{ ton Ca}} \times 50.0\text{ tons Ca} = \mathbf{\$4.70 \times 10^3}$$

19.50 **(a)** The half–reaction is:

$$2H_2O(l) \longrightarrow O_2(g) + 4H^+(aq) + 4e^-$$

First, we can calculate the number of moles of oxygen produced using the ideal gas equation.

$$n_{O_2} = \frac{PV}{RT}$$

$$n_{O_2} = \frac{(1.0\text{ atm})(0.84\text{ L})}{(0.0821\text{ L}\cdot\text{atm/mol}\cdot\text{K})(298\text{ K})} = 0.034\text{ mol }O_2$$

Since 4 moles of electrons are needed for every 1 mole of oxygen, we will need 4 F of electrical charge to produce 1 mole of oxygen.

$$\mathbf{?\ F} = 0.034\text{ mol }O_2 \times \frac{4\ F}{1\text{ mol }O_2} = \mathbf{0.14\ F}$$

(b) The half–reaction is:

$$2Cl^-(aq) \longrightarrow Cl_2(g) + 2e^-$$

The number of moles of chlorine produced is:

$$n_{Cl_2} = \frac{PV}{RT}$$

$$n_{Cl_2} = \frac{\left(750\text{ mmHg} \times \dfrac{1\text{ atm}}{760\text{ mmHg}}\right)(1.50\text{ L})}{(0.0821\text{ L}\cdot\text{atm/mol}\cdot\text{K})(298\text{ K})} = 0.0605\text{ mol }Cl_2$$

Since 2 moles of electrons are needed for every 1 mole of chlorine gas, we will need 2 F of electrical charge to produce 1 mole of chlorine gas.

$$? \, \mathbf{F} = 0.0605 \text{ mol } Cl_2 \times \frac{2 \, F}{1 \text{ mol } Cl_2} = \mathbf{0.121 \, F}$$

(c) The half–reaction is:

$$Sn^{2+}(aq) + 2e^- \longrightarrow Sn(s)$$

The number of moles of Sn(s) produced is

$$? \text{ mol Sn} = 6.0 \text{ g Sn} \times \frac{1 \text{ mol Sn}}{118.7 \text{ g Sn}} = 0.051 \text{ mol Sn}$$

Since 2 moles of electrons are needed for every 1 mole of Sn, we will need 2 F of electrical charge to reduce 1 mole of Sn^{2+} ions to Sn metal.

$$? \, \mathbf{F} = 0.051 \text{ mol Sn} \times \frac{2 \, F}{1 \text{ mol Sn}} = \mathbf{0.10 \, F}$$

19.52 **(a)** The half–reaction is:

$$\mathbf{Ag^+(aq) + e^- \longrightarrow Ag(s)}$$

(b) Since this reaction is taking place in an aqueous solution, the probable oxidation is the oxidation of water. (Neither Ag^+ nor NO_3^- can be further oxidized.)

$$\mathbf{2H_2O(l) \longrightarrow O_2(g) + 4H^+(aq) + 4e^-}$$

(c) The half-reaction tells us that 1 mole of electrons is needed to reduce 1 mol of Ag^+ to Ag metal. We can set up the following strategy to calculate the quantity of electricity (in C) needed to deposit 0.67 g of Ag.

$$\text{grams Ag} \rightarrow \text{mol Ag} \rightarrow \text{mol } e^- \rightarrow \text{coulombs}$$

$$0.67 \text{ g Ag} \times \frac{1 \text{ mol Ag}}{107.9 \text{ g Ag}} \times \frac{1 \text{ mol } e^-}{1 \text{ mol Ag}} \times \frac{96500 \text{ C}}{1 \text{ mol } e^-} = \mathbf{6.0 \times 10^2 \, C}$$

19.54 **(a)** First find the amount of charge needed to produce 2.00 g of silver according to the half–reaction:

$$Ag^+(aq) + e^- \longrightarrow Ag(s)$$

$$2.00 \text{ g Ag} \times \frac{1 \text{ mol Ag}}{107.9 \text{ g Ag}} \times \frac{1 \text{ mol } e^-}{1 \text{ mol Ag}} \times \frac{96500 \text{ C}}{1 \text{ mol } e^-} = 1.79 \times 10^3 \text{ C}$$

The half–reaction for the reduction of copper(II) is:

$$Cu^{2+}(aq) + 2e^- \longrightarrow Cu(s)$$

From the amount of charge calculated above, we can calculate the mass of copper deposited in the second cell.

$$(1.79 \times 10^3 \cancel{C}) \times \frac{1 \cancel{mol\ e^-}}{96500 \cancel{C}} \times \frac{1 \cancel{mol\ Cu}}{2 \cancel{mol\ e^-}} \times \frac{63.55\text{ g Cu}}{1 \cancel{mol\ Cu}} = \mathbf{0.589\ g\ Cu}$$

(b) We can calculate the current flowing through the cells using the following strategy.

Coulombs → Coulombs/hour → Coulombs/second

Recall that 1 C = 1 A·s

The current flowing through the cells is:

$$(1.79 \times 10^3\text{ A}\cdot\cancel{s}) \times \frac{1 \cancel{h}}{3600 \cancel{s}} \times \frac{1}{3.75 \cancel{h}} = \mathbf{0.133\ A}$$

19.56 *Step 1:* Balance the half–reaction.

$$Cr_2O_7^{2-}(aq) + 14H^+(aq) + 12e^- \longrightarrow 2Cr(s) + 7H_2O(l)$$

Step 2: Calculate the quantity of chromium metal by calculating the volume and converting this to mass using the given density.

Volume Cr = thickness × surface area

$$\text{Volume Cr} = (1.0 \times 10^{-2} \cancel{mm}) \times \frac{1\text{ m}}{1000 \cancel{mm}} \times 0.25\text{ m}^2 = 2.5 \times 10^{-6}\text{ m}^3$$

Converting to cm^3,

$$(2.5 \times 10^{-6}\text{ m}^3) \times \left(\frac{1\text{ cm}}{0.01\text{ m}}\right)^3 = 2.5\text{ cm}^3$$

Next, calculate the mass of Cr.

Mass = density × volume

$$\text{Mass Cr} = 2.5 \cancel{cm^3} \times \frac{7.19\text{ g}}{1 \cancel{cm^3}} = 18\text{ g Cr}$$

Step 3: Find the number of moles of electrons required to electrodeposit 18 g of Cr from solution. The half-reaction is:

$$Cr_2O_7^{2-}(aq) + 14H^+(aq) + 12e^- \longrightarrow 2Cr(s) + 7H_2O(l)$$

Six moles of electrons are required to reduce 1 mol of Cr metal. But, we are electrodepositing less than 1 mole of Cr(s). We need to complete the following conversions:

g Cr → mol Cr → mol e^-

$$?\text{ faradays} = 18 \cancel{g}\text{ Cr} \times \frac{1 \cancel{mol\ Cr}}{52.00 \cancel{g}\text{ Cr}} \times \frac{6\text{ mol }e^-}{1 \cancel{mol\ Cr}} = 2.1\text{ mol }e^-$$

Step 4: Determine how long it will take for 2.1 moles of electrons to flow through the cell when the current is 25.0 C/s. We need to complete the following conversions:

$$\text{mol } e^- \rightarrow \text{coulombs} \rightarrow \text{seconds} \rightarrow \text{hours}$$

$$\textbf{? h} = 2.1 \text{ mol } e^- \times \frac{96{,}500 \text{ C}}{1 \text{ mol } e^-} \times \frac{1 \text{ s}}{25.0 \text{ C}} \times \frac{1 \text{ h}}{3600 \text{ s}} = \textbf{2.3 h}$$

Would any time be saved by connecting several bumpers together in a series?

19.58 Based on the half-reaction, we know that one faraday will produce half a mole of copper.

$$Cu^{2+}(aq) + 2e^- \longrightarrow Cu(s)$$

First, let's calculate the charge (in C) needed to deposit 0.300 g of Cu.

$$(3.00 \text{ A})(304 \text{ s}) \times \frac{1 \text{ C}}{1 \text{ A}\cdot\text{s}} = 912 \text{ C}$$

We know that one faraday will produce half a mole of copper, but we don't have a half a mole of copper. We have:

$$0.300 \text{ g Cu} \times \frac{1 \text{ mol Cu}}{63.55 \text{ g Cu}} = 4.72 \times 10^{-3} \text{ mol}$$

We calculated the number of coulombs (912 C) needed to produce 4.72×10^{-3} mol of Cu. How many coulombs will it take to produce 0.500 moles of Cu? This will be Faraday's constant.

$$\frac{912 \text{ C}}{4.72 \times 10^{-3} \text{ mol Cu}} \times 0.500 \text{ mol Cu} = \mathbf{9.66 \times 10^4 \text{ C}} = \mathbf{1}\ \boldsymbol{F}$$

19.60 First we can calculate the number of moles of hydrogen produced using the ideal gas equation.

$$n_{H_2} = \frac{PV}{RT}$$

$$n_{H_2} = \frac{\left(782 \text{ mmHg} \times \dfrac{1 \text{ atm}}{760 \text{ mmHg}}\right)(0.845 \text{ L})}{(0.0821 \text{ L}\cdot\text{atm/K}\cdot\text{mol})(298 \text{ K})} = 0.0355 \text{ mol}$$

The number of faradays passed through the solution is:

$$0.0355 \text{ mol } H_2 \times \frac{2\ F}{1 \text{ mol } H_2} = \mathbf{0.0710}\ \boldsymbol{F}$$

19.62 If you have difficulty balancing redox equations, see Problem Type 1. The balanced equation is:

$$\mathbf{Cr_2O_7^{2-} + 6\,Fe^{2+} + 14H^+ \longrightarrow 2Cr^{3+} + 6Fe^{3+} + 7H_2O}$$

The remainder of this problem is a solution stoichiometry problem.

The number of moles of potassium dichromate in 26.0 mL of the solution is:

$$26.0 \text{ mL} \times \frac{0.0250 \text{ mol}}{1000 \text{ mL soln}} = 6.50 \times 10^{-4} \text{ mol } K_2Cr_2O_7$$

From the balanced equation it can be seen that 1 mole of dichromate is stoichiometrically equivalent to 6 moles of iron(II). The number of moles of iron(II) oxidized is therefore

$$(6.50 \times 10^{-4} \text{ mol } Cr_2O_7^{2-}) \times \frac{6 \text{ mol } Fe^{2+}}{1 \text{ mol } Cr_2O_7^{2-}} = 3.90 \times 10^{-3} \text{ mol } Fe^{2+}$$

Finally, the molar concentration of Fe^{2+} is:

$$\frac{3.90 \times 10^{-3} \text{ mol}}{25.0 \times 10^{-3} \text{ L}} = 0.156 \text{ mol/L} = \mathbf{0.156\ \mathit{M}\ Fe^{2+}}$$

19.64 The balanced equation is:

$$MnO_4^- + 5Fe^{2+} + 8H^+ \longrightarrow Mn^{2+} + 5Fe^{3+} + 4H_2O$$

First, let's calculate the number of moles of potassium permanganate in 23.30 mL of solution.

$$23.30 \text{ mL} \times \frac{0.0194 \text{ mol}}{1000 \text{ mL soln}} = 4.52 \times 10^{-4} \text{ mol } KMnO_4$$

From the balanced equation it can be seen that 1 mole of permanganate is stoichiometrically equivalent to 5 moles of iron(II). The number of moles of iron(II) oxidized is therefore

$$(4.52 \times 10^{-4} \text{ mol } MnO_4^-) \times \frac{5 \text{ mol } Fe^{2+}}{1 \text{ mol } MnO_4^-} = 2.26 \times 10^{-3} \text{ mol } Fe^{2+}$$

The mass of Fe^{2+} oxidized is:

$$\text{mass } Fe^{2+} = (2.26 \times 10^{-3} \text{ mol } Fe^{2+}) \times \frac{55.85 \text{ g } Fe^{2+}}{1 \text{ mol } Fe^{2+}} = 0.126 \text{ g } Fe^{2+}$$

Finally, the mass percent of iron in the ore can be calculated.

$$\text{mass \% Fe} = \frac{\text{mass of iron}}{\text{total mass of sample}} \times 100\%$$

$$\mathbf{\%Fe} = \frac{0.126 \text{ g}}{0.2792 \text{ g}} \times 100\% = \mathbf{45.1\%}$$

19.66 **(a)** The half–reactions are:

(i) $MnO_4^-(aq) + 8H^+(aq) + 5e^- \longrightarrow Mn^{2+}(aq) + 4H_2O(l)$

(ii) $C_2O_4^{2-}(aq) \longrightarrow 2CO_2(g) + 2e^-$

We combine the half-reactions to cancel electrons, that is, [2 × equation (i)] + [5 × equation (ii)]

$$2MnO_4^-(aq) + 16H^+(aq) + 5C_2O_4^{2-}(aq) \longrightarrow 2Mn^{2+}(aq) + 10CO_2(g) + 8H_2O(l)$$

(b) We can calculate the moles of $KMnO_4$ from the molarity and volume of solution.

$$24.0\ \text{mL}\ KMnO_4 \times \frac{0.0100\ \text{mol}\ KMnO_4}{1000\ \text{mL soln}} = 2.40 \times 10^{-4}\ \text{mol}\ KMnO_4$$

We can calculate the mass of oxalic acid from the stoichiometry of the balanced equation. The mole ratio between oxalate ion and permanganate ion is 5:2.

$$(2.40 \times 10^{-4}\ \text{mol}\ KMnO_4) \times \frac{5\ \text{mol}\ H_2C_2O_4}{2\ \text{mol}\ KMnO_4} \times \frac{90.04\ \text{g}\ H_2C_2O_4}{1\ \text{mol}\ H_2C_2O_4} = 0.0540\ \text{g}\ H_2C_2O_4$$

Finally, the percent by mass of oxalic acid in the sample is:

$$\textbf{\% oxalic acid} = \frac{0.0540\ \text{g}}{1.00\ \text{g}} \times 100\% = \mathbf{5.40\%}$$

19.68 The balanced equation is:

$$2MnO_4^- + 5C_2O_4^{2-} + 16H^+ \longrightarrow 2Mn^{2+} + 10CO_2 + 8H_2O$$

Therefore, 2 mol MnO_4^- reacts with 5 mol $C_2O_4^{2-}$

$$\text{Moles of } MnO_4^- \text{ reacted} = 24.2\ \text{mL} \times \frac{9.56 \times 10^{-4}\ \text{mol}\ MnO_4^-}{1000\ \text{mL soln}} = 2.31 \times 10^{-5}\ \text{mol}\ MnO_4^-$$

Recognize that the mole ratio of Ca^{2+} to $C_2O_4^{2-}$ is 1:1 in CaC_2O_4. The mass of Ca^{2+} in 10.0 mL is:

$$(2.31 \times 10^{-5}\ \text{mol}\ MnO_4^-) \times \frac{5\ \text{mol}\ Ca^{2+}}{2\ \text{mol}\ MnO_4^-} \times \frac{40.08\ \text{g}\ Ca^{2+}}{1\ \text{mol}\ Ca^{2+}} = 2.31 \times 10^{-3}\ \text{g}\ Ca^{2+}$$

Finally, converting to mg/mL, we have:

$$\frac{2.31 \times 10^{-3}\ \text{g}\ Ca^{2+}}{10.0\ \text{mL}} \times \frac{1000\ \text{mg}}{1\ \text{g}} = \mathbf{0.231\ mg\ Ca^{2+}/mL\ blood}$$

19.70 **(a)** The half–reactions are:

$$2H^+(aq) + 2e^- \longrightarrow H_2(g) \qquad E^\circ_{\text{anode}} = 0.00\ \text{V}$$

$$Ag^+(aq) + e^- \longrightarrow Ag(s) \qquad E^\circ_{\text{cathode}} = 0.80\ \text{V}$$

$$\boldsymbol{E^\circ_{\text{cell}}} = E^\circ_{\text{cathode}} - E^\circ_{\text{anode}} = 0.80\ \text{V} - 0.00\ \text{V} = \mathbf{0.80\ V}$$

(b) The spontaneous cell reaction under standard-state conditions is:

$$\mathbf{2Ag^+(\textit{aq}) + H_2(\textit{g}) \longrightarrow 2Ag(\textit{s}) + 2H^+(\textit{aq})}$$

(c) Using the Nernst equation we can calculate the cell potential under nonstandard-state conditions.

$$E = E^\circ - \frac{0.0257\ \text{V}}{n} \ln \frac{[H^+]^2}{[Ag^+]^2 P_{H_2}}$$

(i) The potential is:

$$E = 0.80\text{ V} - \frac{0.0257\text{ V}}{2}\ln\frac{(1.0\times10^{-2})^2}{(1.0)^2(1.0)} = \mathbf{0.92\ V}$$

(ii) The potential is:

$$E = 0.80\text{ V} - \frac{0.0257\text{ V}}{2}\ln\frac{(1.0\times10^{-5})^2}{(1.0)^2(1.0)} = \mathbf{1.10\ V}$$

(d) From the results in part (c), we deduce that this cell is a pH meter; its potential is a sensitive function of the hydrogen ion concentration. Each 1 unit increase in pH causes a voltage increase of 0.060 V.

19.72 The overvoltage of oxygen is not large enough to prevent its formation at the anode. Applying the diagonal rule, we see that water is oxidized before fluoride ion.

$$F_2(g) + 2e^- \longrightarrow 2F^-(aq) \qquad E° = 2.87\text{ V}$$

$$O_2(g) + 4H^+(aq) + 4e^- \longrightarrow 2H_2O(l) \qquad E° = 1.23\text{ V}$$

The very positive standard reduction potential indicates that F^- has essentially no tendency to undergo oxidation. The oxidation potential of chloride ion is much smaller (−1.36 V), and hence $Cl_2(g)$ can be prepared by electrolyzing a solution of NaCl.

This fact was one of the major obstacles preventing the discovery of fluorine for many years. HF was usually chosen as the substance for electrolysis, but two problems interfered with the experiment. First, any water in the HF was oxidized before the fluoride ion. Second, pure HF without any water in it is a nonconductor of electricity (HF is a weak acid!). The problem was finally solved by dissolving KF in liquid HF to give a conducting solution.

19.74 We can calculate the amount of charge that 4.0 g of MnO_2 can produce.

$$4.0\text{ g }MnO_2 \times \frac{1\text{ mol}}{86.94\text{ g}} \times \frac{2\text{ mol }e^-}{2\text{ mol }MnO_2} \times \frac{96500\text{ C}}{1\text{ mol }e^-} = 4.44\times10^3\text{ C}$$

Since a current of one ampere represents a flow of one coulomb per second, we can find the time it takes for this amount of charge to pass.

$$0.0050\text{ A} = 0.0050\text{ C/s}$$

$$(4.44\times10^3\text{ C})\times\frac{1\text{ s}}{0.0050\text{ C}}\times\frac{1\text{ h}}{3600\text{ s}} = \mathbf{2.5\times10^2\ h}$$

19.76 Since this is a concentration cell, the standard emf is zero. (Why?) Using the Nernst equation, we can write equations to calculate the cell voltage for the two cells.

$$(1) \quad E_{\text{cell}} = -\frac{RT}{nF}\ln Q = -\frac{RT}{2F}\ln\frac{[Hg_2^{2+}]\text{soln A}}{[Hg_2^{2+}]\text{soln B}}$$

$$(2) \quad E_{\text{cell}} = -\frac{RT}{nF}\ln Q = -\frac{RT}{1F}\ln\frac{[Hg^{+}]\text{soln A}}{[Hg^{+}]\text{soln B}}$$

In the first case, two electrons are transferred per mercury ion ($n = 2$), while in the second only one is transferred ($n = 1$). Note that the concentration ratio will be 1:10 in both cases. The voltages calculated at 18°C are:

$$(1) \quad E_{cell} = \frac{-(8.314 \text{ J/K}\cdot\text{mol})(291 \text{ K})}{2(96500 \text{ J}\cdot\text{V}^{-1}\text{mol}^{-1})}\ln 10^{-1} = 0.0289 \text{ V}$$

$$(2) \quad E_{cell} = \frac{-(8.314 \text{ J/K}\cdot\text{mol})(291 \text{ K})}{1(96500 \text{ J}\cdot\text{V}^{-1}\text{mol}^{-1})}\ln 10^{-1} = 0.0577 \text{ V}$$

Since the calculated cell potential for cell (1) agrees with the measured cell emf, we conclude that the mercury(I) ion exists as $\mathbf{Hg_2^{2+}}$ in solution.

19.78 We begin by treating this like an ordinary stoichiometry problem (see Chapter 3).

Step 1: Calculate the number of moles of Mg and Ag^+.

The number of moles of magnesium is:

$$1.56 \text{ g Mg} \times \frac{1 \text{ mol Mg}}{24.31 \text{ g Mg}} = 0.0642 \text{ mol Mg}$$

The number of moles of silver ion in the solution is:

$$\frac{0.100 \text{ mol Ag}^+}{1 \text{ L}} \times 0.1000 \text{ L} = 0.0100 \text{ mol Ag}^+$$

Step 2: Calculate the mass of Mg remaining by determining how much Mg reacts with Ag^+.

The balanced equation for the reaction is:

$$2\text{Ag}^+(aq) + \text{Mg}(s) \longrightarrow 2\text{Ag}(s) + \text{Mg}^{2+}(aq)$$

Since you need twice as much Ag^+ compared to Mg for complete reaction, Ag^+ is the limiting reagent. The amount of Mg consumed is:

$$0.0100 \text{ mol Ag}^+ \times \frac{1 \text{ mol Mg}}{2 \text{ mol Ag}^+} = 0.00500 \text{ mol Mg}$$

The amount of magnesium remaining is:

$$(0.0642 - 0.00500) \text{ mol Mg} \times \frac{24.31 \text{ g Mg}}{1 \text{ mol Mg}} = \mathbf{1.44 \text{ g Mg}}$$

Step 3: Assuming complete reaction, calculate the concentration of Mg^{2+} ions produced.

Since the mole ratio between Mg and Mg^{2+} is 1:1, the mol of Mg^{2+} formed will equal the mol of Mg reacted. The concentration of Mg^{2+} is:

$$[\text{Mg}^{2+}]_0 = \frac{0.00500 \text{ mol}}{0.100 \text{ L}} = 0.0500\ M$$

Step 4: We can calculate the equilibrium constant for the reaction from the standard cell emf.

$$E^\circ_{\text{cell}} = E^\circ_{\text{cathode}} - E^\circ_{\text{anode}} = 0.80\text{ V} - (-2.37\text{ V}) = 3.17\text{ V}$$

We can then compute the equilibrium constant.

$$K = e^{\frac{nE^\circ_{\text{cell}}}{0.0257}}$$

$$K = e^{\frac{(2)(3.17)}{0.0257}} = 1 \times 10^{107}$$

Step 5: To find equilibrium concentrations of Mg^{2+} and Ag^+, we have to solve an equilibrium problem.

Let x be the small amount of Mg^{2+} that reacts to achieve equilibrium. The concentration of Ag^+ will be $2x$ at equilibrium. Assume that essentially all Ag^+ has been reduced so that the initial concentration of Ag^+ is zero.

	$2Ag^+(aq)$	+	$Mg(s)$	$\rightleftharpoons$	$2Ag(s)$	+	$Mg^{2+}(aq)$
Initial (M):	0.0000						0.0500
Change (M):	$+2x$						$-x$
Equilibrium (M):	$2x$						$(0.0500 - x)$

$$K = \frac{[Mg^{2+}]}{[Ag^+]^2}$$

$$1 \times 10^{107} = \frac{(0.0500 - x)}{(2x)^2}$$

We can assume $0.0500 - x \approx 0.0500$.

$$1 \times 10^{107} \approx \frac{0.0500}{(2x)^2}$$

$$(2x)^2 = \frac{0.0500}{1 \times 10^{107}} = 0.0500 \times 10^{-107}$$

$$(2x)^2 = 5.00 \times 10^{-109} = 50.0 \times 10^{-110}$$

$$2x = 7 \times 10^{-55}\ M$$

$$\mathbf{[Ag^+] = 2x = 7 \times 10^{-55}\ M}$$

$$[Mg^{2+}] = 0.0500 - x = \mathbf{0.0500\ M}$$

19.80 **(a)** Since this is an acidic solution, the gas must be hydrogen gas from the reduction of hydrogen ion. The two electrode reactions and the overall cell reaction are:

anode: $Cu(s) \longrightarrow Cu^{2+}(aq) + 2e^-$

cathode: $2H^+(aq) + 2e^- \longrightarrow H_2(g)$

$$Cu(s) + 2H^+(aq) \longrightarrow Cu^{2+}(aq) + H_2(g)$$

Since 0.584 g of copper was consumed, the amount of hydrogen gas produced is:

$$0.584 \text{ g Cu} \times \frac{1 \text{ mol Cu}}{63.55 \text{ g Cu}} \times \frac{1 \text{ mol } H_2}{1 \text{ mol Cu}} = 9.20 \times 10^{-3} \text{ mol } H_2$$

At STP, 1 mole of an ideal gas occupies a volume of 22.41 L. Thus, the volume of H_2 at STP is:

$$V_{H_2} = (9.20 \times 10^{-3} \text{ mol } H_2) \times \frac{22.41 \text{ L}}{1 \text{ mol}} = \mathbf{0.206 \text{ L}}$$

(b) From the current and the time, we can calculate the amount of charge:

$$1.18 \text{ A} \times \frac{1 \text{ C}}{1 \text{ A} \cdot \text{s}} \times (1.52 \times 10^3 \text{ s}) = 1.79 \times 10^3 \text{ C}$$

Since we know the charge of an electron, we can compute the number of electrons.

$$(1.79 \times 10^3 \text{ C}) \times \frac{1\, e^-}{1.6022 \times 10^{-19} \text{ C}} = 1.12 \times 10^{22}\, e^-$$

Using the amount of copper consumed in the reaction and the fact that 2 mol of e^- are produced for every 1 mole of copper consumed, we can calculate Avogadro's number.

$$\frac{1.12 \times 10^{22} \text{ e}^-}{9.20 \times 10^{-3} \text{ mol Cu}} \times \frac{1 \text{ mol Cu}}{2 \text{ mol } e^-} = \mathbf{6.09 \times 10^{23} \text{ /mol } e^-}$$

In practice, Avogadro's number can be determined by electrochemical experiments like this. The charge of the electron can be found independently by Millikan's experiment.

19.82 **(a)** We can calculate $\Delta G°$ from standard free energies of formation.

$$\Delta G° = 2\Delta G_f°(N_2) + 6\Delta G_f°(H_2O) - [4\Delta G_f°(NH_3) + 3\Delta G_f°(O_2)]$$

$$\Delta G = 0 + (6)(-237.2 \text{ kJ/mol}) - [(4)(-16.6 \text{ kJ/mol}) + 0]$$

$$\mathbf{\Delta G = -1356.8 \text{ kJ/mol}}$$

(b) The half-reactions are:

$$4NH_3(g) \longrightarrow 2N_2(g) + 12H^+(aq) + 12e^-$$

$$3O_2(g) + 12H^+(aq) + 12e^- \longrightarrow 6H_2O(l)$$

The overall reaction is a 12-electron process. We can calculate the standard cell emf from the standard free energy change, $\Delta G°$.

$$\Delta G° = -nFE°_{\text{cell}}$$

$$E°_{\text{cell}} = \frac{-\Delta G°}{nF} = \frac{-\left(\frac{-1356.8 \text{ kJ}}{1 \text{ mol}} \times \frac{1000 \text{ J}}{1 \text{ kJ}}\right)}{(12)(96500 \text{ J/V} \cdot \text{mol})} = \mathbf{1.17 \text{ V}}$$

19.84 The reduction of Ag^+ to Ag metal is:

$$Ag^+(aq) + e^- \longrightarrow Ag$$

We can calculate both the moles of Ag deposited and the moles of Au deposited.

$$? \text{ mol Ag} = 2.64 \text{ g Ag} \times \frac{1 \text{ mol Ag}}{107.9 \text{ g Ag}} = 2.45 \times 10^{-2} \text{ mol Ag}$$

$$? \text{ mol Au} = 1.61 \text{ g Au} \times \frac{1 \text{ mol Au}}{197.0 \text{ g Au}} = 8.17 \times 10^{-3} \text{ mol Au}$$

We do not know the oxidation state of Au ions, so we will represent the ions as Au^{n+}. If we divide the mol of Ag by the mol of Au, we can determine the ratio of Ag^+ reduced compared to Au^{n+} reduced.

$$\frac{2.45 \times 10^{-2} \text{ mol Ag}}{8.17 \times 10^{-3} \text{ mol Au}} = 3$$

That is, the same number of electrons that reduced the Ag^+ ions to Ag reduced only one-third the number of moles of the Au^{n+} ions to Au. Thus, each Au^{n+} required three electrons per ion for every one electron for Ag^+. The oxidation state for the gold ion is **+3**; the ion is Au^{3+}.

$$Au^{3+}(aq) + 3e^- \longrightarrow Au$$

19.86 We reverse the first half–reaction and add it to the second to come up with the overall balanced equation

$$Hg_2^{2+} \longrightarrow 2Hg^{2+} + 2e^- \qquad E^\circ_{\text{anode}} = +0.92 \text{ V}$$

$$Hg_2^{2+} + 2e^- \longrightarrow 2Hg \qquad E^\circ_{\text{cathode}} = +0.85 \text{ V}$$

$$2Hg_2^{2+} \longrightarrow 2Hg^{2+} + 2Hg \qquad E^\circ_{\text{cell}} = 0.85 \text{ V} - 0.92 \text{ V} = -0.07 \text{ V}$$

Since the standard cell potential is an intensive property,

$$Hg_2^{2+}(aq) \longrightarrow Hg^{2+}(aq) + Hg(l) \qquad E^\circ_{\text{cell}} = -0.07 \text{ V}$$

We calculate ΔG° from E°.

$$\mathbf{\Delta G^\circ} = -nFE^\circ = -(1)(96500 \text{ J/V·mol})(-0.07 \text{ V}) = \mathbf{6.8\ kJ/mol}$$

The corresponding equilibrium constant is:

$$K = \frac{[Hg^{2+}]}{[Hg_2^{2+}]}$$

We calculate K from ΔG°.

$$\Delta G^\circ = -RT\ln K$$

$$\ln K = \frac{-6.8 \times 10^3 \text{ J/mol}}{(8.314 \text{ J/K}\cdot\text{mol})(298 \text{ K})}$$

$$\mathbf{K = 0.064}$$

19.88 The reactions for the electrolysis of NaCl(aq) are:

Anode:	$2Cl^-(aq) \longrightarrow Cl_2(g) + 2e^-$
Cathode:	$2H_2O(l) + 2e^- \longrightarrow H_2(g) + 2OH^-(aq)$
Overall:	$2H_2O(l) + 2Cl^-(aq) \longrightarrow H_2(g) + Cl_2(g) + 2OH^-(aq)$

From the pH of the solution, we can calculate the OH^- concentration. From the $[OH^-]$, we can calculate the moles of OH^- produced. Then, from the moles of OH^- we can calculate the average current used.

$$\text{pH} = 12.24$$

$$\text{pOH} = 14.00 - 12.24 = 1.76$$

$$[OH^-] = 1.74 \times 10^{-2}\ M$$

The moles of OH^- produced are:

$$\frac{1.74 \times 10^{-2}\text{ mol}}{1\text{ L}} \times 0.300\text{ L} = 5.22 \times 10^{-3}\text{ mol OH}^-$$

From the balanced equation, it takes 1 mole of e^- to produce 1 mole of OH^- ions.

$$(5.22 \times 10^{-3}\text{ mol OH}^-) \times \frac{1\text{ mol }e^-}{1\text{ mol OH}^-} \times \frac{96500\text{ C}}{1\text{ mol }e^-} = 504\text{ C}$$

Recall that 1 C = 1 A·s

$$504\text{ C} \times \frac{1\text{ A}\cdot\text{s}}{1\text{ C}} \times \frac{1\text{ min}}{60\text{ s}} \times \frac{1}{6.00\text{ min}} = \mathbf{1.4\text{ A}}$$

19.90 The reaction is:

$$Pt^{n+} + ne^- \longrightarrow Pt$$

Thus, we can calculate the charge of the platinum ions by realizing that n mol of e^- are required per mol of Pt formed.

The moles of Pt formed are:

$$9.09\text{ g Pt} \times \frac{1\text{ mol Pt}}{195.1\text{ g Pt}} = 0.0466\text{ mol Pt}$$

Next, calculate the charge passed in C.

$$C = 2.00\text{ h} \times \frac{3600\text{ s}}{1\text{ h}} \times \frac{2.50\text{ C}}{1\text{ s}} = 1.80 \times 10^4\text{ C}$$

Convert to moles of electrons.

$$?\text{ mol e}^- = (1.80 \times 10^4\text{ C}) \times \frac{1\text{ mol }e^-}{96500\text{ C}} = 0.187\text{ mol }e^-$$

We now know the number of moles of electrons (0.187 mol e^-) needed to produce 0.0466 mol of Pt metal. We can calculate the number of moles of electrons needed to produce 1 mole of Pt metal.

$$\frac{0.187 \text{ mol } e^-}{0.0466 \text{ mol Pt}} = 4.01 \text{ mol } e^-/\text{mol Pt}$$

Since we need 4 moles of electrons to reduce 1 mole of Pt ions, the charge on the Pt ions must be **+4**.

19.92 The half–reaction for the oxidation of water to oxygen is:

$$2H_2O(l) \xrightarrow{\text{oxidation (anode)}} O_2(g) + 4H^+(aq) + 4e^-$$

Knowing that one mole of any gas at STP occupies a volume of 22.41 L, we find the number of moles of oxygen.

$$4.26 \text{ L } O_2 \times \frac{1 \text{ mol}}{22.41 \text{ L}} = 0.190 \text{ mol } O_2$$

Since four electrons are required to form one oxygen molecule, the number of electrons must be:

$$0.190 \text{ mol } O_2 \times \frac{4 \text{ mol e}^-}{1 \text{ mol } O_2} \times \frac{6.022 \times 10^{23} e^-}{1 \text{ mol}} = 4.58 \times 10^{23} e^-$$

The amount of charge passing through the solution is:

$$6.00 \text{ A} \times \frac{1 \text{ C}}{1 \text{ A} \cdot \text{s}} \times \frac{3600 \text{ s}}{1 \text{ h}} \times 3.40 \text{ h} = 7.34 \times 10^4 \text{ C}$$

We find the electron charge by dividing the amount of charge by the number of electrons.

$$\frac{7.34 \times 10^4 \text{ C}}{4.58 \times 10^{23} e^-} = \mathbf{1.60 \times 10^{-19} \text{ C}/e^-}$$

In actual fact, this sort of calculation can be used to find Avogadro's number, not the electron charge. The latter can be measured independently, and one can use this charge together with electrolytic data like the above to calculate the number of objects in one mole. See also Problem 19.80.

19.94 Cells of higher voltage require very reactive oxidizing and reducing agents, which are difficult to handle. (From Table 19.1 of the text, we see that 5.92 V is the theoretical limit of a cell made up of Li^+/Li and F_2/F^- electrodes under standard-state conditions.) Batteries made up of several cells in series are easier to use.

19.96 The half-reactions are:

$$Zn(s) + 4OH^-(aq) \rightarrow Zn(OH)_4^{2-}(aq) + 2e^- \qquad E^\circ_{\text{anode}} = -1.36 \text{ V}$$

$$Zn^{2+}(aq) + 2e^- \rightarrow Zn(s) \qquad E^\circ_{\text{cathode}} = -0.76 \text{ V}$$

$$Zn^{2+}(aq) + 4OH^-(aq) \rightarrow Zn(OH)_4^{2-}(aq) \qquad E^\circ_{\text{cell}} = -0.76 \text{ V} - (-1.36 \text{ V}) = 0.60 \text{ V}$$

$$E^\circ_{\text{cell}} = \frac{0.0257 \text{ V}}{n} \ln K_f$$

$$K_f = e^{\frac{nE°}{0.0257}} = e^{\frac{(2)(0.60)}{0.0257}} = \mathbf{2 \times 10^{20}}$$

19.98 **(a)** Since electrons flow from X to SHE, $E°$ for X must be negative. Thus $E°$ for Y must be positive.

(b)

$Y^{2+} + 2e^- \rightarrow Y$	$E°_{cathode} = 0.34\text{ V}$
$X \rightarrow X^{2+} + 2e^-$	$E°_{anode} = -0.25\text{ V}$
$X + Y^{2+} \rightarrow X^{2+} + Y$	$\mathbf{E°_{cell}} = 0.34\text{ V} - (-0.25\text{ V}) = \mathbf{0.59\text{ V}}$

19.100 **(a)** Gold does not tarnish in air because the reduction potential for oxygen is insufficient to result in the oxidation of gold.

$$O_2 + 4H^+ + 4e^- \rightarrow 2H_2O \qquad E°_{cathode} = 1.23\text{ V}$$

That is, $E°_{cell} = E°_{cathode} - E°_{anode} < 0$, for either oxidation by O_2 to Au^+ or Au^{3+}.

$$E°_{cell} = 1.23\text{ V} - 1.50\text{ V} < 0$$

or

$$E°_{cell} = 1.23\text{ V} - 1.69\text{ V} < 0$$

(b)

$3(Au^+ + e^- \rightarrow Au)$	$E°_{cathode} = 1.69\text{ V}$
$Au \rightarrow Au^{3+} + 3e^-$	$E°_{anode} = 1.50\text{ V}$
$3Au^+ \rightarrow 2Au + Au^{3+}$	$E°_{cell} = 1.69\text{ V} - 1.50\text{ V} = 0.19\text{ V}$

Calculating ΔG,

$$\Delta G° = -nFE° = -(3)(96{,}500\text{ J/V·mol})(0.19\text{ V}) = -55.0\text{ kJ/mol}$$

For spontaneous electrochemical equations, $\Delta G°$ must be negative. Thus, **the disproportionation occurs spontaneously**.

(c) Since the most stable oxidation state for gold is Au^{3+}, the predicted reaction is:

$$\mathbf{2Au + 3F_2 \rightarrow 2AuF_3}$$

19.102 The balanced equation is: $5Fe^{2+} + MnO_4^- + 8H^+ \longrightarrow Mn^{2+} + 5Fe^{3+} + 4\,H_2O$

Calculate the amount of iron(II) in the original solution using the mole ratio from the balanced equation.

$$23.0\text{ mL} \times \frac{0.0200\text{ mol KMnO}_4}{1000\text{ mL soln}} \times \frac{5\text{ mol Fe}^{2+}}{1\text{ mol KMnO}_4} = 0.00230\text{ mol Fe}^{2+}$$

The concentration of iron(II) must be:

$$\mathbf{[Fe^{2+}]} = \frac{0.00230\text{ mol}}{0.0250\text{ L}} = \mathbf{0.0920\ \mathit{M}}$$

The total iron concentration can be found by simple proportion because the same sample volume (25.0 mL) and the same $KMnO_4$ solution were used.

$$[\text{Fe}]_{\text{total}} = \frac{40.0 \text{ mL KMnO}_4}{23.0 \text{ mL KMnO}_4} \times 0.0920\ M = 0.160\ M$$

$$[\text{Fe}^{3+}] = [\text{Fe}]_{\text{total}} - [\text{Fe}^{2+}] = \mathbf{0.0680\ M}$$

Why are the two titrations with permanganate necessary in this problem?

19.104 From Table 19.1 of the text.

$$H_2O_2(aq) + 2H^+(aq) + 2e^- \rightarrow 2H_2O(l) \qquad E^\circ_{\text{cathode}} = 1.77 \text{ V}$$

$$H_2O_2(aq) \rightarrow O_2(g) + 2H^+(aq) + 2e^- \qquad E^\circ_{\text{anode}} = 0.68 \text{ V}$$

$$2H_2O_2(aq) \rightarrow 2H_2O(l) + O_2(g) \qquad E^\circ_{\text{cell}} = E^\circ_{\text{cathode}} - E^\circ_{\text{anode}} = 1.77 \text{ V} - (0.68 \text{ V}) = 1.09 \text{ V}$$

Because $E°$ is positive, the decomposition is **spontaneous**.

19.106 **(a)** unchanged **(b)** unchanged **(c)** squared **(d)** doubled **(e)** doubled

19.108 $F_2(g) + 2H^+(aq) + 2e^- \rightarrow 2HF(g)$

$$E = E^\circ - \frac{RT}{2F}\ln\frac{P_{HF}^2}{P_{F_2}[H^+]^2}$$

With increasing $[H^+]$, E will be larger. F_2 will become a **stronger oxidizing agent**.

19.110

$$Pb \rightarrow Pb^{2+} + 2e^- \qquad E^\circ_{\text{anode}} = -0.13 \text{ V}$$

$$2H^+ + 2e^- \rightarrow H_2 \qquad E^\circ_{\text{cathode}} = 0.00 \text{ V}$$

$$Pb + 2H^+ \rightarrow Pb^{2+} + H_2 \qquad E^\circ_{\text{cell}} = 0.00 \text{ V} - (-0.13 \text{ V}) = 0.13 \text{ V}$$

$$\text{pH} = 1.60$$

$$[H^+] = 10^{-1.60} = 0.025\ M$$

$$E = E^\circ - \frac{RT}{nF}\ln\frac{[Pb^{2+}]P_{H_2}}{[H^+]^2}$$

$$0 = 0.13 - \frac{0.0257 \text{ V}}{2}\ln\frac{(0.035)P_{H_2}}{0.025^2}$$

$$\frac{0.26}{0.0257} = \ln\frac{(0.035)P_{H_2}}{0.025^2}$$

$$\boldsymbol{P_{H_2} = 4.4 \times 10^2 \text{ atm}}$$

19.112 **(a)** The half-reactions are:

Anode: $Zn \rightarrow Zn^{2+} + 2e^-$

Cathode: $\frac{1}{2}O_2 + 2e^- \rightarrow O^{2-}$

Overall: $Zn + \frac{1}{2}O_2 \rightarrow ZnO$

To calculate the standard emf, we first need to calculate $\Delta G°$ for the reaction. From Appendix 3 of the text we write:

$$\Delta G° = \Delta G_f°(ZnO) - [\Delta G_f°(Zn) + \tfrac{1}{2}\Delta G_f°(O_2)]$$
$$\Delta G° = -318.2 \text{ kJ/mol} - [0 + 0]$$
$$\Delta G° = -318.2 \text{ kJ/mol}$$

$$\Delta G° = -nFE°$$
$$-318.2 \times 10^3 \text{ J/mol} = -(2)(96{,}500 \text{ J/V·mol})E°$$
$$\mathbf{E° = 1.65\ V}$$

(b) We use the following equation:

$$E = E° - \frac{RT}{nF}\ln Q$$

$$E = 1.65 \text{ V} - \frac{0.0257 \text{ V}}{2}\ln\frac{1}{P_{O_2}}$$

$$E = 1.65 \text{ V} - \frac{0.0257 \text{ V}}{2}\ln\frac{1}{0.21}$$

$$E = 1.65 \text{ V} - 0.020 \text{ V}$$

$$\mathbf{E = 1.63\ V}$$

(c) Since the free energy change represents the maximum work that can be extracted from the overall reaction, the maximum amount of energy that can be obtained from this reaction is the free energy change. To calculate the energy density, we multiply the free energy change by the number of moles of Zn present in 1 kg of Zn.

$$\textbf{energy density} = \frac{318.2 \text{ kJ}}{1 \text{ mol Zn}} \times \frac{1 \text{ mol Zn}}{65.39 \text{ g Zn}} \times \frac{1000 \text{ g Zn}}{1 \text{ kg Zn}} = \mathbf{4.87 \times 10^3\ kJ/kg\ Zn}$$

(d) One ampere is 1 C/s. The charge drawn every second is given by nF.

$$\text{charge} = nF$$

$$2.1 \times 10^5 \text{ C} = n(96{,}500 \text{ C/mol } e^-)$$

$$n = 2.2 \text{ mol } e^-$$

From the overall balanced reaction, we see that 4 moles of electrons will reduce 1 mole of O_2; therefore, the number of moles of O_2 reduced by 2.2 moles of electrons is:

$$\text{mol } O_2 = 2.2 \text{ mol } e^- \times \frac{1 \text{ mol } O_2}{4 \text{ mol } e^-} = 0.55 \text{ mol } O_2$$

The volume of oxygen at 1.0 atm partial pressure can be obtained by using the ideal gas equation.

$$V_{O_2} = \frac{nRT}{P} = \frac{(0.55 \text{ mol})(0.0821 \text{ L·atm/mol·K})(298 \text{ K})}{(1.0 \text{ atm})} = 13 \text{ L}$$

Since air is 21 percent oxygen by volume, the volume of air required every second is:

$$V_{\textbf{air}} = 13 \text{ L } O_2 \times \frac{100\% \text{ air}}{21\% \ O_2} = \mathbf{62\ L\ of\ air}$$

19.114 We can calculate ΔG°_{rxn} using the following equation.

$$\Delta G^\circ_{rxn} = \Sigma n\Delta G^\circ_f(\text{products}) - \Sigma m\Delta G^\circ_f(\text{reactants})$$

$$\Delta G^\circ_{rxn} = 0 + 0 - [(1)(-293.8 \text{ kJ/mol}) + 0] = 293.8 \text{ kJ/mol}$$

Next, we can calculate E° using the equation

$$\Delta G^\circ = -nFE^\circ$$

We use a more accurate value for Faraday's constant.

$$293.8 \times 10^3 \text{ J/mol} = -(1)(96485.3 \text{ J/V·mol})E^\circ$$

$$\mathbf{E^\circ = -3.05 \text{ V}}$$

19.116 First, we need to calculate E°_{cell}, then we can calculate K from the cell potential.

$$H_2(g) \rightarrow 2H^+(aq) + 2e^- \qquad E^\circ_{anode} = 0.00 \text{ V}$$

$$2H_2O(l) + 2e^- \rightarrow H_2(g) + 2OH^- \qquad E^\circ_{cathode} = -0.83 \text{ V}$$

$$2H_2O(l) \rightarrow 2H^+(aq) + 2OH^-(aq) \qquad E^\circ_{cell} = -0.83 \text{ V} - 0.00 \text{ V} = -0.83 \text{ V}$$

We want to calculate K for the reaction: $H_2O(l) \rightarrow H^+(aq) + OH^-(aq)$. The cell potential for this reaction will be the same as the above reaction, but the moles of electrons transferred, n, will equal one.

$$E^\circ_{cell} = \frac{0.0257 \text{ V}}{n}\ln K_w$$

$$\ln K_w = \frac{nE^\circ_{cell}}{0.0257 \text{ V}}$$

$$K_w = e^{\frac{nE^\circ}{0.0257}}$$

$$\mathbf{K_w} = e^{\frac{(1)(-0.83)}{0.0257}} = e^{-32} = \mathbf{1 \times 10^{-14}}$$

19.118 **(a)** $1\text{A·h} = 1\text{A} \times 3600\text{s} = 3600 \text{ C}$

(b) Anode: $Pb + SO_4^{2-} \rightarrow PbSO_4 + 2e^-$

Two moles of electrons are produced by 1 mole of Pb. Recall that the charge of 1 mol e^- is 96,500 C. We can set up the following conversions to calculate the capacity of the battery.

mol Pb $\rightarrow$ mol e^- $\rightarrow$ coulombs $\rightarrow$ ampere hour

$$406 \text{ g Pb} \times \frac{1 \text{ mol Pb}}{207.2 \text{ g Pb}} \times \frac{2 \text{ mol } e^-}{1 \text{ mol Pb}} \times \frac{96500 \text{ C}}{1 \text{ mol } e^-} = (3.74 \times 10^5 \text{ C}) \times \frac{1 \text{ h}}{3600 \text{ s}} = \mathbf{104 \text{ A·h}}$$

This ampere·hour cannot be fully realized because the concentration of H_2SO_4 keeps decreasing.

(c) $E^\circ_{\text{cell}} = 1.70\text{ V} - (-0.31\text{ V}) = \mathbf{2.01\text{ V}}$ (From Table 19.1 of the text)

$$\Delta G^\circ = -nFE^\circ$$

$$\Delta G^\circ = -(2)(96500\text{ J/V}\cdot\text{mol})(2.01\text{ V}) = \mathbf{-3.88 \times 10^5\text{ J/mol}}$$

Spontaneous as expected.

19.120 The surface area of an open cylinder is $2\pi rh$. The surface area of the culvert is

$$2\pi(0.900\text{ m})(40.0\text{ m}) \times 2 \text{ (for both sides of the iron sheet)} = 452\text{ m}^2$$

Converting to units of cm^2,

$$452\text{ m}^2 \times \left(\frac{100\text{ cm}}{1\text{ m}}\right)^2 = 4.52 \times 10^6\text{ cm}^2$$

The volume of the Zn layer is

$$0.200\text{ mm} \times \frac{1\text{ cm}}{10\text{ mm}} \times (4.52 \times 10^6\text{ cm}^2) = 9.04 \times 10^4\text{ cm}^3$$

The mass of Zn needed is

$$(9.04 \times 10^4\text{ cm}^3) \times \frac{7.14\text{ g}}{1\text{ cm}^3} = 6.45 \times 10^5\text{ g Zn}$$

$$Zn^{2+} + 2e^- \rightarrow Zn$$

$$Q = (6.45 \times 10^5\text{ g Zn}) \times \frac{1\text{ mol Zn}}{65.39\text{ g Zn}} \times \frac{2\text{ mol }e^-}{1\text{ mol Zn}} \times \frac{96500\text{ C}}{1\text{ mol }e^-} = 1.90 \times 10^9\text{ C}$$

$$1\text{ J} = 1\text{ C} \times 1\text{ V}$$

$$\text{Total energy} = \frac{(1.90 \times 10^9\text{ C})(3.26\text{ V})}{0.95 \leftarrow \text{(efficiency)}} = 6.52 \times 10^9\text{ J}$$

$$\text{Cost} = (6.52 \times 10^9\text{ J}) \times \frac{1\text{ kw}}{1000\frac{\text{J}}{\text{s}}} \times \frac{1\text{ h}}{3600\text{ s}} \times \frac{\$0.12}{1\text{ kwh}} = \mathbf{\$217}$$

19.122 It might appear that because the sum of the first two half-reactions gives Equation (3), E°_3 is given by $E^\circ_1 + E^\circ_2 = 0.33$ V. This is not the case, however, because emf is not an extensive property. We cannot set $E^\circ_3 = E^\circ_1 + E^\circ_2$. On the other hand, the Gibbs energy is an extensive property, so we can add the separate Gibbs energy changes to obtain the overall Gibbs energy change.

$$\Delta G^\circ_3 = \Delta G^\circ_1 + \Delta G^\circ_2$$

Substituting the relationship $\Delta G° = -nFE°$, we obtain

$$n_3FE_3^\circ = n_1FE_1^\circ + n_2FE_2^\circ$$

$$E_3^\circ = \frac{n_1E_1^\circ + n_2E_2^\circ}{n_3}$$

$n_1 = 2$, $n_2 = 1$, and $n_3 = 3$.

$$\boldsymbol{E_3^\circ} = \frac{(2)(-0.44\text{ V}) + (1)(0.77\text{ V})}{3} = \mathbf{-0.037\text{ V}}$$

19.124 First, calculate the standard emf of the cell from the standard reduction potentials in Table 19.1 of the text. Then, calculate the equilibrium constant from the standard emf using Equation (19.5) of the text.

$$E_{\text{cell}}^\circ = E_{\text{cathode}}^\circ - E_{\text{anode}}^\circ = 0.34\text{ V} - (-0.76\text{ V}) = 1.10\text{ V}$$

$$\ln K = \frac{nE_{\text{cell}}^\circ}{0.0257\text{ V}}$$

$$K = e^{\frac{nE_{\text{cell}}^\circ}{0.0257\text{ V}}} = e^{\frac{(2)(1.10\text{ V})}{0.0257\text{ V}}}$$

$$\boldsymbol{K} = \mathbf{2 \times 10^{37}}$$

The very large equilibrium constant means that the oxidation of Zn by Cu^{2+} is virtually complete.

CHAPTER 20
METALLURGY AND THE CHEMISTRY OF METALS

This chapter is of a very descriptive nature. As such, there are essentially no defined problem types. Please read Chapter 20 of the text carefully before answering the end-of-chapter problems.

SOLUTIONS TO SELECTED TEXT PROBLEMS

20.12 The cathode reaction is: $Cu^{2+}(aq) + 2e^- \longrightarrow Cu(s)$

First, let's calculate the number of moles of electrons needed to reduce 5.0 kg of Cu.

$$5.00 \text{ kg Cu} \times \frac{1000 \text{ g}}{1 \text{ kg}} \times \frac{1 \text{ mol Cu}}{63.55 \text{ g Cu}} \times \frac{2 \text{ mol e}^-}{1 \text{ mol Cu}} = 1.57 \times 10^2 \text{ mol e}^-$$

Next, let's determine how long it will take for 1.57×10^2 moles of electrons to flow through the cell when the current is 37.8 C/s.

$$(1.57 \times 10^2 \text{ mol e}^-) \times \frac{96{,}500 \text{ C}}{1 \text{ mol e}^-} \times \frac{1 \text{ s}}{37.8 \text{ C}} \times \frac{1 \text{ h}}{3600 \text{ s}} = \mathbf{111\ h}$$

20.14 The sulfide ore is first roasted in air:

$$2ZnS(s) + 3O_2(g) \longrightarrow 2ZnO(s) + 2SO_2(g)$$

The zinc oxide is then mixed with coke and limestone in a blast furnace where the following reductions occur:

$$ZnO(s) + C(s) \longrightarrow Zn(g) + CO(g)$$

$$ZnO(s) + CO(g) \longrightarrow Zn(g) + CO_2(g)$$

The zinc vapor formed distills from the furnace into an appropriate receiver.

20.16 **(a)** We first find the mass of ore containing 2.0×10^8 kg of copper.

$$(2.0 \times 10^8 \text{ kg Cu}) \times \frac{100\% \text{ ore}}{0.80\% \text{ Cu}} = 2.5 \times 10^{10} \text{ kg ore}$$

We can then compute the volume from the density of the ore.

$$(2.5 \times 10^{10} \text{ kg}) \times \frac{1000 \text{ g}}{1 \text{ kg}} \times \frac{1 \text{ cm}^3}{2.8 \text{ g}} = \mathbf{8.9 \times 10^{12}\ cm^3}$$

(b) From the formula of chalcopyrite it is clear that two moles of sulfur dioxide will be formed per mole of copper. The mass of sulfur dioxide formed will be:

$$(2.0 \times 10^8 \text{ kg Cu}) \times \frac{1 \text{ mol Cu}}{0.06355 \text{ kg Cu}} \times \frac{2 \text{ mol SO}_2}{1 \text{ mol Cu}} \times \frac{0.06407 \text{ kg SO}_2}{1 \text{ mol SO}_2} = \mathbf{4.0 \times 10^8\ kg\ SO_2}$$

20.18 Iron can be produced by reduction with coke in a blast furnace; whereas, aluminum is usually produced electrolytically, which is a much more expensive process.

20.28 **(a)** $2Na(s) + 2H_2O(l) \longrightarrow 2NaOH(aq) + H_2(g)$

(b) $2NaOH(aq) + CO_2(g) \longrightarrow Na_2CO_3(aq) + H_2O(l)$

(c) $Na_2CO_3(s) + 2HCl(aq) \longrightarrow 2NaCl(aq) + CO_2(g) + H_2O(l)$

(d) $NaHCO_3(aq) + HCl(aq) \longrightarrow NaCl(aq) + CO_2(g) + H_2O(l)$

(e) $2NaHCO_3(s) \longrightarrow Na_2CO_3(s) + CO_2(g) + H_2O(g)$

(f) $Na_2CO_3(s) \longrightarrow$ no reaction. Unlike $CaCO_3(s)$, $Na_2CO_3(s)$ is not decomposed by moderate heating.

20.30 The balanced equation is: $Na_2CO_3(s) + 2HCl(aq) \longrightarrow 2NaCl(aq) + CO_2(g) + H_2O(l)$

$$\text{mol } CO_2 \text{ produced} = 25.0\text{ g }Na_2CO_3 \times \frac{1\text{ mol }Na_2CO_3}{106.0\text{ g }Na_2CO_3} \times \frac{1\text{ mol }CO_2}{1\text{ mol }Na_2CO_3} = 0.236\text{ mol }CO_2$$

$$V_{CO_2} = \frac{nRT}{P} = \frac{(0.236\text{ mol})(0.0821\text{ L}\cdot\text{atm/K}\cdot\text{mol})(283\text{ K})}{\left(746\text{ mmHg} \times \dfrac{1\text{ atm}}{760\text{ mmHg}}\right)} = \mathbf{5.59\text{ L}}$$

20.34 First magnesium is treated with concentrated nitric acid (redox reaction) to obtain magnesium nitrate.

$$3Mg(s) + 8HNO_3(aq) \longrightarrow 3Mg(NO_3)_2(aq) + 4H_2O(l) + 2NO(g)$$

The magnesium nitrate is recovered from solution by evaporation, dried, and heated in air to obtain magnesium oxide:

$$2Mg(NO_3)_2(s) \longrightarrow 2MgO(s) + 4NO_2(g) + O_2(g)$$

20.36 The electron configuration of magnesium is $[Ne]3s^2$. The $3s$ electrons are outside the neon core (shielded), so they have relatively low ionization energies. Removing the third electron means separating an electron from the neon (closed shell) core, which requires a great deal more energy.

20.38 Even though helium and the Group 2A metals have ns^2 outer electron configurations, helium has a closed shell noble gas configuration and the Group 2A metals do not. The electrons in He are much closer to and more strongly attracted by the nucleus. Hence, the electrons in He are not easily removed. Helium is inert.

20.40 **(a)** quicklime: $CaO(s)$ **(b)** slaked lime: $Ca(OH)_2(s)$

(c) limewater: an aqueous suspension of $Ca(OH)_2$

20.44 The reduction reaction is: $Al^{3+}(aq) + 3e^- \rightarrow Al(s)$

First, we can calculate the amount of charge needed to deposit 664 g of Al.

$$664\text{ g Al} \times \frac{1\text{ mol Al}}{26.98\text{ g Al}} \times \frac{3\text{ mol }e^-}{1\text{ mol Al}} \times \frac{96{,}500\text{ C}}{1\text{ mol }e^-} = 7.12 \times 10^6\text{ C}$$

Since a current of one ampere represents a flow of one coulomb per second, we can find the time it takes to pass this amount of charge.

$32.6\text{ A} = 32.6\text{ C/s}$

$$(7.12 \times 10^6\ \text{C}) \times \frac{1\ \text{s}}{32.6\ \text{C}} \times \frac{1\ \text{h}}{3600\ \text{s}} = \mathbf{60.7\ h}$$

20.46 **(a)** The relationship between cell voltage and free energy difference is:

$$\Delta G = -nFE$$

In the given reaction $n = 6$. We write:

$$E = \frac{-\Delta G}{nF} = \frac{-594 \times 10^3\ \text{J/mol}}{(6)(96500\ \text{J/V}\cdot\text{mol})} = -1.03\ \text{V}$$

The balanced equation shows *two* moles of aluminum. Is this the voltage required to produce *one* mole of aluminum? If we divide everything in the equation by two, we obtain:

$$\tfrac{1}{2}Al_2O_3(s) + \tfrac{3}{2}C(s) \rightarrow Al(l) + \tfrac{3}{2}CO(g)$$

For the new equation $n = 3$ and ΔG is $\left(\frac{1}{2}\right)(594\ \text{kJ/mol}) = 297\ \text{kJ/mol}$. We write:

$$E = \frac{-\Delta G}{nF} = \frac{-297 \times 10^3\ \text{J/mol}}{(3)(96500\ \text{J/V}\cdot\text{mol})} = -1.03\ \text{V}$$

The minimum voltage that must be applied is **1.03 V** (a negative sign in the answers above means that 1.03 V is required to produce the Al). The voltage required to produce one mole or one thousand moles of aluminum is the same; the amount of *current* will be different in each case.

(b) First we convert 1.00 kg (1000 g) of Al to moles.

$$(1.00 \times 10^3\ \text{g Al}) \times \frac{1\ \text{mol Al}}{26.98\ \text{g Al}} = 37.1\ \text{mol Al}$$

The reaction in part (a) shows us that three moles of electrons are required to produce one mole of aluminum. The voltage is three times the minimum calculated above (namely, –3.09 V or –3.09 J/C). We can find the electrical energy by using the same equation with the other voltage.

$$\mathbf{\Delta G} = -nFE = -(37.1)\left(\frac{3\ \text{mol e}^-}{1\ \text{mol Al}} \times \frac{96500\ \text{C}}{1\ \text{mol e}^-}\right)\left(\frac{-3.09\ \text{J}}{1\ \text{C}}\right) = \mathbf{3.32 \times 10^7\ J/mol} = \mathbf{3.32 \times 10^4\ kJ/mol}$$

This equation can be used because electrical work can be calculated by multiplying the voltage by the amount of charge transported through the circuit (joules = volts × coulombs). The nF term in Equation (19.2) of the text used above represents the amount of charge.

What is the significance of the positive sign of the free energy change? Would the manufacturing of aluminum be a different process if the free energy difference were negative?

20.48 $4Al(NO_3)_3(s) \longrightarrow 2Al_2O_3(s) + 12NO_2(g) + 3O_2(g)$

20.50 The "bridge" bonds in Al_2Cl_6 break at high temperature: $Al_2Cl_6(g) \rightleftharpoons 2AlCl_3(g)$.

This increases the number of molecules in the gas phase and causes the pressure to be higher than expected for pure Al_2Cl_6.

If you know the equilibrium constants for the above reaction at higher temperatures, could you calculate the expected pressure of the $AlCl_3$–Al_2Cl_6 mixture?

20.52 In Al_2Cl_6, each aluminum atom is surrounded by 4 bonding pairs of electrons (AB_4–type molecule), and therefore each aluminum atom is ***sp*^{3} hybridized**. VSEPR analysis shows $AlCl_3$ to be an AB_3–type molecule (no lone pairs on the central atom). The geometry should be trigonal planar, and the aluminum atom should therefore be ***sp*^{2} hybridized**.

20.54 The formulas of the metal oxide and sulfide are MO and MS (why?). The balanced equation must therefore be:

$$2MS(s) + 3O_2(g) \rightarrow 2MO(s) + 2SO_2(g)$$

The number of moles of MO and MS are equal. We let x be the molar mass of metal. The number of moles of metal oxide is:

$$0.972\text{ g} \times \frac{1\text{ mol}}{(x + 16.00)\text{g}}$$

The number of moles of metal sulfide is:

$$1.164\text{ g} \times \frac{1\text{ mol}}{(x + 32.07)\text{g}}$$

The moles of metal oxide equal the moles of metal sulfide.

$$\frac{0.972}{(x + 16.00)} = \frac{1.164}{(x + 32.07)}$$

We solve for x.

$$0.972(x + 32.07) = 1.164(x + 16.00)$$

$$\mathbf{x = 65.4\text{ g/mol}}$$

20.56 Copper(II) ion is more easily reduced than either water or hydrogen ion (How can you tell? See Section 19.3 of the text.) Copper metal is more easily oxidized than water. Water should not be affected by the copper purification process.

20.58 Using Equation (18.12) from the text:

(a) $\Delta G^\circ_{rxn} = 4\Delta G^\circ_f(Fe) + 3\Delta G^\circ_f(O_2) - 2\Delta G^\circ_f(Fe_2O_3)$

$\Delta G^\circ_{rxn} = (4)(0) + (3)(0) - (2)(-741.0\text{ kJ/mol}) = \mathbf{1482\text{ kJ/mol}}$

(b) $\Delta G^\circ_{rxn} = 4\Delta G^\circ_f(Al) + 3\Delta G^\circ_f(O_2) - 2\Delta G^\circ_f(Al_2O_3)$

$\Delta G^\circ_{rxn} = (4)(0) + (3)(0) - (2)(-1576.4\text{ kJ/mol}) = \mathbf{3152.8\text{ kJ/mol}}$

20.60 At high temperature, magnesium metal reacts with nitrogen gas to form magnesium nitride.

$$3Mg(s) + N_2(g) \longrightarrow Mg_3N_2(s)$$

Can you think of any gas other than a noble gas that could provide an inert atmosphere for processes involving magnesium at high temperature?

20.62 **(a)** In water the aluminum(III) ion causes an increase in the concentration of hydrogen ion (lower pH). This results from the effect of the small diameter and high charge (3+) of the aluminum ion on surrounding water molecules. The aluminum ion draws electrons in the O–H bonds to itself, thus allowing easy formation of H^+ ions.

(b) $Al(OH)_3$ is an amphoteric hydroxide. It will dissolve in strong base with the formation of a complex ion.

$$Al(OH)_3(s) + OH^-(aq) \longrightarrow Al(OH)_4^-(aq)$$

The concentration of OH^- in aqueous ammonia is too low for this reaction to occur.

20.64 Calcium oxide is a base. The reaction is a neutralization.

$$CaO(s) + 2HCl(aq) \longrightarrow CaCl_2(aq) + H_2O(l)$$

20.66 Metals have closely spaced energy levels and (referring to Figure 20.10 of the text) a very small energy gap between filled and empty levels. Consequently, many electronic transitions can take place with absorption and subsequent emission continually occurring. Some of these transitions fall in the visible region of the spectrum and give rise to the flickering appearance.

20.68 NaF is used in toothpaste to fight tooth decay.

Li_2CO_3 is used to treat mental illness.

$Mg(OH)_2$ is an antacid.

$CaCO_3$ is an antacid.

$BaSO_4$ is used to enhance X ray images of the digestive system.

$Al(OH)_2NaCO_3$ is an antacid.

20.70 Both Li and Mg form oxides (Li_2O and MgO). Other Group 1A metals (Na, K, etc.) also form peroxides and superoxides. In Group 1A, only Li forms nitride (Li_3N), like Mg (Mg_3N_2).

Li resembles Mg in that its carbonate, fluoride, and phosphate have low solubilities.

20.72 You might know that Ag, Cu, Au, and Pt are found as free elements in nature, which leaves **Zn** by process of elimination. You could also look at Table 19.1 of the text to find the metal that is easily oxidized. Looking at the table, the standard oxidation potential of Zn is +0.76 V. The positive value indicates that Zn is easily oxidized to Zn^{2+} and will not exist as a free element in nature.

20.74 Because only B and C react with 0.5 *M* HCl, they are more electropositive than A and D. The fact that when B is added to a solution containing the ions of the other metals, metallic A, C, and D are formed indicates that B is the most electropositive metal. Because A reacts with 6 *M* HNO_3, A is more electropositive than D. The metals arranged in increasing order as reducing agents are:

$$\mathbf{D < A < C < B}$$

Examples are: D = Au, A = Cu, C = Zn, B = Mg

20.76 First, we calculate the density of O_2 in KO_2 using the mass percentage of O_2 in the compound.

$$\frac{32.00\text{ g O}_2}{71.10\text{ g KO}_2} \times \frac{2.15\text{ g}}{1\text{ cm}^3} = 0.968\text{ g/cm}^3$$

Now, we can use Equation (5.11) of the text to calculate the pressure of oxygen gas that would have the same density as that provided by KO_2.

$$\frac{0.968\text{ g}}{1\text{ cm}^3} \times \frac{1000\text{ cm}^3}{1\text{ L}} = 968\text{ g/L}$$

$$d = \frac{P\mathcal{M}}{RT}$$

or

$$\boldsymbol{P} = \frac{dRT}{\mathcal{M}} = \frac{\left(\frac{968\text{ g}}{1\text{ L}}\right)\left(0.0821\frac{\text{L}\cdot\text{atm}}{\text{mol}\cdot\text{K}}\right)(293\text{ K})}{\left(\frac{32.00\text{ g}}{1\text{ mol}}\right)} = \mathbf{727\text{ atm}}$$

Obviously, using O_2 instead of KO_2 is not practical.

CHAPTER 21
NONMETALLIC ELEMENTS AND THEIR COMPOUNDS

This chapter is of a very descriptive nature. As such, there are essentially no defined problem types. Please read Chapter 21 of the text carefully before answering the end-of-chapter problems.

SOLUTIONS TO SELECTED TEXT PROBLEMS

21.12 **(a)** Hydrogen reacts with alkali metals to form ionic hydrides:

$$2Na(s) + H_2(g) \rightarrow 2NaH(s)$$

The oxidation number of hydrogen drops from 0 to −1 in this reaction.

(b) Hydrogen reacts with oxygen (combustion) to form water:

$$2H_2(g) + O_2(g) \rightarrow 2H_2O(l)$$

The oxidation number of hydrogen increases from 0 to +1 in this reaction.

21.14 Hydrogen forms an interstitial hydride with palladium, which behaves almost like a solution of hydrogen atoms in the metal. At elevated temperatures hydrogen atoms can pass through solid palladium; other substances cannot.

21.16 The number of moles of deuterium gas is:

$$n = \frac{PV}{RT} = \frac{(0.90\ \text{atm})(2.0\ \text{L})}{(0.0821\ \text{L}\cdot\text{atm/K}\cdot\text{mol})(298\ \text{K})} = 0.074\ \text{mol}$$

If the abundance of deuterium is 0.015 percent, the number of moles of water must be:

$$0.074\ \text{mol D}_2 \times \frac{100\%\ H_2O}{0.015\%\ D_2} = 4.9\times10^2\ \text{mol}\ H_2O$$

At a recovery of 80 percent the amount of water needed is:

$$\frac{4.9\times10^2\ \text{mol}\ H_2O}{0.80} \times \frac{0.01802\ \text{kg}\ H_2O}{1.0\ \text{mol}\ H_2O} = \mathbf{11\ kg\ H_2O}$$

21.18 **(a)** $H_2 + Cl_2 \rightarrow 2HCl$

(b) $3H_2 + N_2 \rightarrow 2NH_3$

(c) $2Li + H_2 \rightarrow 2LiH$

$LiH + H_2O \rightarrow LiOH + H_2$

21.26 The Lewis structure is:

$$[:C\equiv C:]^{2-}$$

21.28 **(a)** The reaction is: $2NaHCO_3(s) \rightarrow Na_2CO_3(s) + H_2O(g) + CO_2(g)$

Is this an endo- or an exothermic process?

(b) The hint is generous. The reaction is:

$$Ca(OH)_2(aq) + CO_2(g) \rightarrow CaCO_3(s) + H_2O(l)$$

The visual proof is the formation of a white precipitate of $CaCO_3$. Why would a water solution of NaOH be unsuitable to qualitatively test for carbon dioxide?

21.30 Heat causes bicarbonates to decompose according to the reaction:

$$2HCO_3^- \rightarrow CO_3^{2-} + H_2O + CO_2$$

Generation of carbonate ion causes precipitation of the insoluble $MgCO_3$.

Do you think there is much chance of finding natural mineral deposits of calcium or magnesium bicarbonates?

21.32 The wet sodium hydroxide is first converted to sodium carbonate:

$$2NaOH(aq) + CO_2(g) \rightarrow Na_2CO_3(aq) + H_2O(l)$$

and then to sodium hydrogen carbonate: $Na_2CO_3(aq) + H_2O(l) + CO_2(g) \rightarrow 2NaHCO_3(aq)$

Eventually, the sodium hydrogen carbonate precipitates (the water solvent evaporates since $NaHCO_3$ is not hygroscopic). Thus, most of the white solid is $NaHCO_3$ plus some Na_2CO_3.

21.34 Carbon monoxide and molecular nitrogen are isoelectronic. Both have 14 electrons. What other diatomic molecules discussed in these problems are isoelectronic with CO?

21.40 **(a)** $2NaNO_3(s) \rightarrow 2NaNO_2(s) + O_2(g)$

(b) $NaNO_3(s) + C(s) \rightarrow NaNO_2(s) + CO(g)$

21.42 The balanced equation is: $2NH_3(g) + CO_2(g) \rightarrow (NH_2)_2CO(s) + H_2O(l)$

If pressure increases, the position of equilibrium will shift in the direction with the smallest number of molecules in the gas phase, that is, to the right. Therefore, the reaction is best run at high pressure.

Write the expression for Q_p for this reaction. Does increasing pressure cause Q_p to increase or decrease? Is this consistent with the above prediction?

21.44 The density of a gas depends on temperature, pressure, and the molar mass of the substance. When two gases are at the same pressure and temperature, the ratio of their densities should be the same as the ratio of their molar masses. The molar mass of ammonium chloride is 53.5 g/mol, and the ratio of this to the molar mass of molecular hydrogen (2.02 g/mol) is 26.8. The experimental value of 14.5 is roughly half this amount. Such results usually indicate breakup or dissociation into smaller molecules in the gas phase (note the temperature). The measured molar mass is the average of all the molecules in equilibrium.

$$NH_4Cl(g) \rightleftharpoons NH_3(g) + HCl(g)$$

Knowing that ammonium chloride is a stable substance at 298 K, is the above reaction exo- or endothermic?

21.46 The highest oxidation state possible for a Group 5A element is +5. This is the oxidation state of nitrogen in nitric acid (HNO_3).

21.48 Nitric acid is a strong oxidizing agent in addition to being a strong acid (see Table 19.1 of the text, $E^\circ_{red} = +0.96V$). The primary action of a good reducing agent like zinc is reduction of nitrate ion to ammonium ion.

$$4Zn(s) + NO_3^-(aq) + 10H^+(aq) \rightarrow 4Zn^{2+}(aq) + NH_4^+(aq) + 3H_2O(l)$$

21.50 One of the best Lewis structures for nitrous oxide is:

$$^{-}\ddot{\underset{\cdot\cdot}{N}}=\overset{+}{N}=\ddot{\underset{\cdot\cdot}{O}}$$

There are no lone pairs on the central nitrogen, making this an AB_2 VSEPR case. All such molecules are linear. Other resonance forms are:

$$:N\equiv\overset{+}{N}-\ddot{\underset{\cdot\cdot}{O}}:^{-} \qquad\qquad ^{2-}:\ddot{\underset{\cdot\cdot}{N}}-\overset{+}{N}\equiv\overset{+}{O}:$$

Are all the resonance forms consistent with a linear geometry?

21.52 $\Delta H^\circ = 4\Delta H^\circ_f[NO(g)] + 6\Delta H^\circ_f[H_2O(l)] - \{4\Delta H^\circ_f[NH_3(g)] + 5\Delta H^\circ_f[O_2(g)]\}$

$\boldsymbol{\Delta H^\circ} = (4)(90.4 \text{ kJ/mol}) + (6)(-285.8 \text{ kJ/mol}) - [(4)(-46.3 \text{ kJ/mol}) + (5)(0)] = \mathbf{-1168\ kJ/mol}$

21.54 $\Delta T_b = K_b m = 0.409°C$

$$\text{molality} = \frac{0.409°C}{2.34°C/m} = 0.175\ m$$

The number of grams of white phosphorus in 1 kg of solvent is:

$$\frac{1.645 \text{ g phosphorus}}{75.5 \text{ g } CS_2} \times \frac{1000 \text{ g}}{1 \text{ kg}} = 21.8 \text{ g phosphorus/kg } CS_2$$

The molar mass of white phosphorus is:

$$\frac{21.8 \text{ g phosphorus/kg } CS_2}{0.175 \text{ mol phosphorus/kg } CS_2} = \mathbf{125\ g/mol}$$

Let the molecular formula of white phosphorus be P_n so that:

$$n \times 30.97 \text{ g/mol} = 125 \text{ g/mol}$$

$$n = 4$$

The molecular formula of white phosphorus is $\mathbf{P_4}$.

21.56 The balanced equation is:

$$\mathbf{P_4O_{10}(s) + 4HNO_3(aq) \rightarrow 2N_2O_5(g) + 4HPO_3(l)}$$

The theoretical yield of N_2O_5 is :

$$79.4\ \text{g P}_4\text{O}_{10} \times \frac{1\ \text{mol P}_4\text{O}_{10}}{283.9\ \text{g P}_4\text{O}_{10}} \times \frac{2\ \text{mol N}_2\text{O}_5}{1\ \text{mol P}_4\text{O}_{10}} \times \frac{108.0\ \text{g N}_2\text{O}_5}{1\ \text{mol N}_2\text{O}_5} = \mathbf{60.4\ g\ N_2O_5}$$

21.58 PH_4^+ is similar to NH_4^+. The hybridization of phosphorus in PH_4^+ is $\boldsymbol{sp^3}$.

21.66 $\Delta G° = \Delta G_f°(NO_2) + \Delta G_f°(O_2) - [\Delta G_f°(NO) + \Delta G_f°(O_3)]$

$\boldsymbol{\Delta G°} = (1)(51.8\ \text{kJ/mol}) + (0) - [(1)(86.7\ \text{kJ/mol}) + (1)(163.4\ \text{kJ/mol})] = \mathbf{-198.3\ kJ/mol}$

$$\Delta G° = -RT \ln K_p$$

$$\ln K_p = \frac{-\Delta G°}{RT} = \frac{198.3 \times 10^3\ \text{J/mol}}{(8.314\ \text{J/K}\cdot\text{mol})(298\ \text{K})}$$

$$\boldsymbol{K_p = 6 \times 10^{34}}$$

Since there is no change in the number of moles of gases, $\boldsymbol{K_c}$ is *equal* to $\boldsymbol{K_p}$.

21.68 Following the rules given in Section 4.4 of the text, we assign hydrogen an oxidation number of +1 and **fluorine** an oxidation number of **−1**. Since HFO is a neutral molecule, the oxidation number of **oxygen** is **zero**. Can you think of other compounds in which oxygen has this oxidation number?

21.70 First, let's calculate the moles of sulfur in 48 million tons of sulfuric acid.

$$(48 \times 10^6\ \text{tons H}_2\text{SO}_4) \times \frac{2000\ \text{lb}}{1\ \text{ton}} \times \frac{453.6\ \text{g}}{1\ \text{lb}} \times \frac{1\ \text{mol H}_2\text{SO}_4}{98.09\ \text{g H}_2\text{SO}_4} \times \frac{1\ \text{mol S}}{1\ \text{mol H}_2\text{SO}_4} = \mathbf{4.4 \times 10^{11}\ mol\ S}$$

Converting to grams of sulfur:

$$(4.4 \times 10^{11}\ \text{mol S}) \times \frac{32.07\ \text{g S}}{1\ \text{mol S}} = \mathbf{1.4 \times 10^{13}\ g\ S}$$

21.72 There are actually several steps involved in removing sulfur dioxide from industrial emissions with calcium carbonate. First calcium carbonate is heated to form carbon dioxide and calcium oxide.

$$CaCO_3(s) \rightleftharpoons CaO(s) + CO_2(g)$$

The CaO combines with sulfur dioxide to form calcium sulfite.

$$CaO(s) + SO_2(g) \rightarrow CaSO_3(s)$$

Alternatively, calcium sulfate forms if enough oxygen is present.

$$2CaSO_3(s) + O_2(g) \rightarrow 2CaSO_4(s)$$

The amount of calcium carbonate (limestone) needed in this problem is:

$$50.6\ \text{g SO}_2 \times \frac{1\ \text{mol SO}_2}{64.07\ \text{g SO}_2} \times \frac{1\ \text{mol CaCO}_3}{1\ \text{mol SO}_2} \times \frac{100.1\ \text{g CaCO}_3}{1\ \text{mol CaCO}_3} = \mathbf{79.1\ g\ CaCO_3}$$

The calcium oxide–sulfur dioxide reaction is an example of a Lewis acid-base reaction (see Section 15.12 of the text) between oxide ion and sulfur dioxide. Can you draw Lewis structures showing this process? Which substance is the Lewis acid and which is the Lewis base?

21.74 The usual explanation for the fact that no chemist has yet succeeded in making SCl_6, SBr_6 or SI_6 is based on the idea of excessive crowding of the six chlorine, bromine, or iodine atoms around the sulfur. Others suggest that sulfur in the +6 oxidation state would oxidize chlorine, bromine, or iodine in the −1 oxidation state to the free elements. In any case, none of these substances has been made as of the date of this writing.

It is of interest to point out that thirty years ago all textbooks confidently stated that compounds like ClF_5 could not be prepared.

Note that PCl_6^- is a known species. How different are the sizes of S and P?

21.76 First we convert gallons of water to grams of water.

$$(2.0 \times 10^2\ \text{gal}) \times \frac{3.785\ \text{L}}{1\ \text{gal}} \times \frac{1000\ \text{mL}}{1\ \text{L}} \times \frac{1.00\ \text{g H}_2\text{O}}{1\ \text{mL}} = 7.6 \times 10^5\ \text{g H}_2\text{O}$$

An H_2S concentration of 22 ppm indicates that in 1 million grams of water, there will be 22 g of H_2S. First, let's calculate the number of moles of H_2S in 7.6×10^5 g of H_2O:

$$(7.6 \times 10^5\ \text{g H}_2\text{O}) \times \frac{22\ \text{g H}_2\text{S}}{1.0 \times 10^6\ \text{g H}_2\text{O}} \times \frac{1\ \text{mol H}_2\text{S}}{34.09\ \text{g H}_2\text{S}} = 0.49\ \text{mol H}_2\text{S}$$

The mass of chlorine required to react with 0.49 mol of H_2S is:

$$0.49\ \text{mol H}_2\text{S} \times \frac{1\ \text{mol Cl}_2}{1\ \text{mol H}_2\text{S}} \times \frac{70.90\ \text{g Cl}_2}{1\ \text{mol Cl}_2} = \mathbf{35\ g\ Cl_2}$$

21.78 A check of Table 19.1 of the text shows that sodium ion cannot be reduced by any of the substances mentioned in this problem; it is a "spectator ion". We focus on the substances that are actually undergoing oxidation or reduction and write half-reactions for each.

$$2\text{I}^-(aq) \rightarrow \text{I}_2(s)$$

$$\text{H}_2\text{SO}_4(aq) \rightarrow \text{H}_2\text{S}(g)$$

Balancing the oxygen, hydrogen, and charge gives:

$$2\text{I}^-(aq) \rightarrow \text{I}_2(s) + 2e^-$$

$$\text{H}_2\text{SO}_4(aq) + 8\text{H}^+(aq) + 8e^- \rightarrow \text{H}_2\text{S}(g) + 4\text{H}_2\text{O}(l)$$

Multiplying the iodine half-reaction by four and combining gives the balanced redox equation.

$$\text{H}_2\text{SO}_4(aq) + 8\text{I}^-(aq) + 8\text{H}^+(aq) \rightarrow \text{H}_2\text{S}(g) + 4\text{I}_2(s) + 4\text{H}_2\text{O}(l)$$

The hydrogen ions come from extra sulfuric acid. We add one sodium ion for each iodide ion to obtain the final equation.

$$\mathbf{9H_2SO_4(\mathit{aq}) + 8NaI(\mathit{aq}) \rightarrow H_2S(\mathit{g}) + 4I_2(\mathit{s}) + 4H_2O(\mathit{l}) + 8NaHSO_4(\mathit{aq})}$$

21.82 Sulfuric acid is added to solid sodium chloride, not aqueous sodium chloride. Hydrogen chloride is a gas at room temperature and can escape from the reacting mixture.

$$\mathbf{H_2SO_4(\mathit{l}) + NaCl(\mathit{s}) \rightarrow HCl(\mathit{g}) + NaHSO_4(\mathit{s})}$$

The reaction is driven to the right by the continuous loss of $HCl(g)$ (Le Châtelier's principle).

What happens when sulfuric acid is added to a water solution of NaCl? Could you tell the difference between this solution and the one formed by adding hydrochloric acid to aqueous sodium sulfate?

21.84 The reaction is: $2Br^-(aq) + Cl_2(g) \rightarrow 2Cl^-(aq) + Br_2(l)$

The number of moles of chlorine needed is:

$$167 \text{ g Br}^- \times \frac{1 \text{ mol Br}^-}{79.90 \text{ g Br}^-} \times \frac{1 \text{ mol Cl}_2}{2 \text{ mol Br}^-} = 1.05 \text{ mol Cl}_2(g)$$

Use the ideal gas equation to calculate the volume of Cl_2 needed.

$$V_{\text{Cl}_2} = \frac{nRT}{P} = \frac{(1.05 \text{ mol})(0.0821 \text{ L}\cdot\text{atm/K}\cdot\text{mol})(293 \text{ K})}{(1 \text{ atm})} = \mathbf{25.3\ L}$$

21.86 As with iodide salts, a redox reaction occurs between sulfuric acid and sodium bromide.

$$2H_2SO_4(aq) + 2NaBr\ (aq) \rightarrow SO_2(g) + Br_2(l) + 2H_2O(l) + Na_2SO_4(aq)$$

21.88 The balanced equation is:

$$Cl_2(g) + 2Br^-(aq) \rightarrow 2Cl^-(aq) + Br_2(g)$$

The number of moles of bromine is the same as the number of moles of chlorine, so this problem is essentially a gas law exercise in which P and T are changed for some given amount of gas.

$$V_2 = \frac{P_1V_1}{T_1} \times \frac{T_2}{P_2} = \frac{(760 \text{ mmHg})(2.00 \text{ L})}{288 \text{ K}} \times \frac{373 \text{ K}}{700 \text{ mmHg}} = \mathbf{2.81\ L}$$

21.90 The balanced equation is:

$$\mathbf{I_2O_5(\mathit{s}) + 5CO(\mathit{g}) \rightarrow I_2(\mathit{s}) + 5CO_2(\mathit{g})}$$

The oxidation number of iodine changes from +5 to 0 and the oxidation number of carbon changes from +2 to +4. **Iodine** is **reduced**; **carbon** is **oxidized**.

21.92 **(a)** $SiCl_4$ **(b)** F^- **(c)** F **(d)** CO_2

21.94 There is no change in oxidation number; it is zero for both compounds.

21.96 **(a)** $2Na + 2D_2O \rightarrow 2NaOD + D_2$

(b) $2D_2O \xrightarrow{\text{electrolysis}} 2D_2 + O_2$

$D_2 + Cl_2 \rightarrow 2DCl$

(c) $Mg_3N_2 + 6D_2O \rightarrow 3Mg(OD)_2 + 2ND_3$

(d) $CaC_2 + 2D_2O \rightarrow C_2D_2 + Ca(OD)_2$

(e) $Be_2C + 4D_2O \rightarrow 2Be(OD)_2 + CD_4$

(f) $SO_3 + D_2O \rightarrow D_2SO_4$

21.98 **(a)** At elevated pressures, water boils above 100°C.

(b) Water is sent down the outermost pipe so that it is able to melt a larger area of sulfur.

(c) Sulfur deposits are structurally weak. There will be a danger of the sulfur mine collapsing.

21.100 The oxidation is probably initiated by breaking a C–H bond (the rate-determining step). The C–D bond breaks at a slower rate than the C–H bond; therefore, replacing H by D decreases the rate of oxidation.

21.102 Organisms need a source of energy to sustain the processes of life. Respiration creates that energy. Molecular oxygen is a powerful oxidizing agent, reacting with substances such as glucose to release energy for growth and function. Molecular nitrogen (containing the nitrogen-to-nitrogen triple bond) is too unreactive at room temperature to be of any practical use.

21.104 We know that $\Delta G° = -RT\ln K$ and $\Delta G° = \Delta H° - T\Delta S°$. We can first calculate $\Delta H°$ and $\Delta S°$ using data in Appendix 3 of the text. Then, we can calculate $\Delta G°$ and lastly K.

$$\Delta H° = 2\Delta H_f^\circ[CO(g)] - \{\Delta H_f^\circ[C(s)] + \Delta H_f^\circ[CO_2(g)]\}$$

$$\Delta H° = (2)(-110.5\text{ kJ/mol}) - (0 + -393.5\text{ kJ/mol}) = 172.5\text{ kJ/mol}$$

$$\Delta S° = 2S°[CO(g)] - \{S°[C(s)] + S°[CO_2(g)]\}$$

$$\Delta S° = (2)(197.9\text{ J/K·mol}) - (5.69\text{ J/K·mol} + 213.6\text{ J/K·mol}) = 176.5\text{ J/K·mol}$$

At 298 K (25°C),

$$\Delta G° = \Delta H° - T\Delta S° = (172.5 \times 10^3\text{ J/mol}) - (298\cancel{K})(176.5\text{ J/}\cancel{K}\cdot\text{mol}) = 1.199 \times 10^5\text{ J/mol}$$

$$\Delta G° = -RT\ln K$$

$$\boldsymbol{K} = e^{\frac{-\Delta G°}{RT}} = e^{\frac{-(1.199 \times 10^5\text{ J/mol})}{(8.314\text{ J/K·mol})(298\text{ K})}} = \mathbf{9.61 \times 10^{-22}}$$

At 1273 K (1000°C),

$$\Delta G° = (172.5 \times 10^3\text{ J/mol}) - (1273\cancel{K})(176.5\text{ J/}\cancel{K}\cdot\text{mol}) = -5.218 \times 10^4\text{ J/mol}$$

$$\boldsymbol{K} = e^{\frac{-\Delta G°}{RT}} = e^{\frac{-(-5.218 \times 10^4\text{ J/mol})}{(8.314\text{ J/K·mol})(1273\text{ K})}} = \mathbf{138}$$

The much larger value of K at the higher temperature indicates the formation of CO is favored at higher temperatures (achieved by using a blast furnace).

21.106 The reactions are:

$$P_4(s) + 5O_2(g) \rightarrow P_4O_{10}(s)$$
$$P_4O_{10}(s) + 6H_2O(l) \rightarrow 4H_3PO_4(aq)$$

First, we calculate the moles of H_3PO_4 produced. Next, we can calculate the molarity of the phosphoric acid solution. Finally, we can determine the pH of the H_3PO_4 solution (a weak acid).

$$10.0 \text{ g } P_4 \times \frac{1 \text{ mol } P_4}{123.9 \text{ g } P_4} \times \frac{1 \text{ mol } P_4O_{10}}{1 \text{ mol } P_4} \times \frac{4 \text{ mol } H_3PO_4}{1 \text{ mol } P_4O_{10}} = 0.323 \text{ mol } H_3PO_4$$

$$\text{Molarity} = \frac{0.323 \text{ mol}}{0.500 \text{ L}} = 0.646\ M$$

We set up the ionization of the weak acid, H_3PO_4. The K_a value for H_3PO_4 can be found in Table 15.5 of the text.

	$H_3PO_4(aq) + H_2O$	$\rightleftharpoons$ $H_3O^+(aq)$	+ $H_2PO_4^-(aq)$
Initial (M):	0.646	0	0
Change (M):	$-x$	$+x$	$+x$
Equilibrium (M):	$0.646 - x$	x	x

$$K_a = \frac{[H_3O^+][H_2PO_4^-]}{[H_3PO_4]}$$

$$7.5 \times 10^{-3} = \frac{(x)(x)}{(0.646 - x)}$$

$$x^2 + 7.5 \times 10^{-3}x - 4.85 \times 10^{-3} = 0$$

Solving the quadratic equation,

$$x = 0.066\ M = [H_3O^+]$$

Following the procedure in Problem 15.118 and the discussion in Section 15.8 of the text, we can neglect the contribution to the hydronium ion concentration from the second and third ionization steps. Thus,

$$\textbf{pH} = -\log(0.066) = \textbf{1.18}$$

CHAPTER 22
TRANSITION METAL CHEMISTRY AND COORDINATION COMPOUNDS

PROBLEM-SOLVING STRATEGIES AND TUTORIAL SOLUTIONS

TYPES OF PROBLEMS

Problem Type 1: Assigning Oxidation Numbers to the Metal Atom in Coordination Compounds.

Problem Type 2: Naming Coordination Compounds.

Problem Type 3: Writing Formulas for Coordination Compounds.

Problem Type 4: Predicting the Number of Unpaired Spins in a Coordination Compound.

PROBLEM TYPE 1: ASSIGNING OXIDATION NUMBERS TO THE METAL ATOM IN COORDINATION COMPOUNDS

Transition metals exhibit variable oxidation states in their compounds. The charge on the central metal atom and its surrounding ligands sum to zero in a neutral coordination compound. In a complex ion, the charges on the central metal atom and the surrounding ligands sum to the net charge of the ion.

EXAMPLE 22.1

Specify the oxidation number of the central metal atom in each of the following compounds:
(a) $[Co(NH_3)_6]Cl_3$ and (b) $Co(CN)_6^{3-}$

Strategy: The oxidation number of the metal atom is equal to its charge. First we look for known charges in the species. Recall that alkali metals are +1 and alkaline earth metals are +2. Also determine if the ligand is a charged or neutral species. From the known charges, we can deduce the net charge of the metal and hence its oxidation number.

Solution:

(a) NH_3 is a neutral species. Since each chloride ion carries a −1 charge, and there are three Cl^- ions, the oxidation number of Co must be +3.

(b) Each cyanide ion has a charge of −1. The sum of the oxidation number of Co and the −6 charge for the six cyanide ions is −3. Therefore, the oxidation number of Co must be +3.

PRACTICE EXERCISE

1. Specify the oxidation number of the central metal atom in each of the following compounds:
   (a) $[Pt(NH_3)_3Cl_3]NO_3$ and (b) $Ni(CO)_4$.

Text Problems: 22.12, **22.14**

PROBLEM TYPE 2: NAMING COORDINATION COMPOUNDS

The complete set of rules for naming coordination compounds is given in Section 22.3 of the text. Presented below is an abridged version of those rules.

1. The cation is named before the anion, as is the case for other ionic compounds.
2. Within a complex ion the ligands are named first, in alphabetical order, and the metal ion is named last.
3. The names of anionic ligands end with the letter *o*, whereas a neutral ligand is usually called by the name of the molecule. Exceptions are listed in Table 22.4 of the text.
4. Greek prefixes are used to indicate the number of ligands of a particular kind present. If the ligand itself contains a Greek prefix, the prefixes *bis* (2), *tris* (3), and *tetrakis* (4) are used to indicate the number of ligands present.
5. The oxidation number of the metal is written in Roman numerals following the name of the metal.
6. If the complex is an anion, its name ends in "–ate".

EXAMPLE 22.2

Name the following coordination compounds and complex ions: (a) $[Ni(NH_3)_6]^{2+}$, (b) $K_2[Cu(CN)_4]$, (c) $[Pt(NH_3)_4Cl_2]Cl_2$.

Strategy: We follow the procedure for naming coordination compounds outlined in Section 22.3 of the text and refer to Tables 22.4 and 22.5 of the text for names of ligands and anions containing metal atoms.

Solution:

(a) NH_3 is a neutral species; therefore, the Ni must have a +2 charge. Ammonia as a ligand in a coordination compound is called ammine. The complex ion is called **hexaamminenickel(II) ion**.

(b) Potassium is an alkali metal; it always has a +1 charge in ionic compounds. Since the two potassium ions have a total charge of +2, the charge on the complex ion $[Cu(CN)_4]$ must be −2. Cyanide ion has a −1 charge; therefore, Cu must have a +2 charge. Potassium is the cation and is named first. The compound is called **potassium tetracyanocuprate(II)**. We use an "–ate" ending because the complex is an anion.

(c) The complex ion has a +2 charge balanced by the total charge of −2 for the two chloride ions. Focusing on the complex ion, NH_3 is a neutral species and the two chloride ligands each have a −1 charge. Since the charge of the complex ion is +2, platinum must have a +4 charge. This compound is called **tetraamminedichloroplatinum(IV) chloride**.

PRACTICE EXERCISE

2. Name the following coordination compounds and complex ions:

(a) $[Cr(OH)_4]^-$ (b) $[Pt(NH_3)_3Cl_3]NO_3$.

Text Problem: 22.16

PROBLEM TYPE 3: WRITING FORMULAS FOR COORDINATION COMPOUNDS

Follow the nomenclature rules given in Problem Type 2 above. Remember that the cation is named before the anion.

EXAMPLE 22.3

Write the formulas for the following compounds or complex ions:
(a) tetrahydroxoaluminate(III) ion, (b) potassium hexachloropalladate(IV), (c) diaquodicyanocopper(II)

Strategy: We follow the procedure in Section 22.3 of the text and refer to Tables 22.4 and 22.5 of the text for names of ligands and anions containing metal atoms.

Solution:

(a) Tetrahydroxo refers to four hydroxide ligands. The "–ate" of aluminate indicates that the complex is an anion. The Roman numeral (III) indicates a +3 charge on aluminum. Since each hydroxide ligand has a −1 charge, the charge on the ion is −1. The formula for the complex ion is **$[Al(OH_4)]^-$**.

(b) The complex anion, hexachloropalladate(IV), has a −2 charge. Each of the six chloride ligands has a −1 charge (total of −6) and palladium has a +4 charge. To balance the −2 charge of the anion, there must be two potassium +1 ions. The formula for the compound is **$K_2[PdCl_6]$**.

(c) Diaquo refers to two water ligands, and dicyano refers to two cyanide ligands. Each cyanide ligand has a −1 charge (total of −2), which balances the +2 charge on copper. The compound is electrically neutral. The formula for the compound is **$[Cu(H_2O)_2(CN)_2]$**.

PRACTICE EXERCISE

3. Write the formulas for the following compounds:

(a) tris(ethylenediamine)nickel(II) sulfate
(b) tetraamminediaquocobalt(III) chloride
(c) potassium hexacyanoferrate(II)

Text Problem: 22.18

PROBLEM TYPE 4: PREDICTING THE NUMBER OF UNPAIRED SPINS IN A COORDINATION COMPOUND

First, we must write the electron configuration for the transition metal of the coordination compound. Remember that the *ns* shell fills before the $(n - 1)d$ shell. When the number of *d* electrons in the transition metal is known, we must decide how to place them in the *d* orbitals.

Crystal Field Theory tells us that in an octahedral complex, the five *d* orbitals are *not* equivalent in energy. The d_{xy}, d_{xz}, and d_{yz} are degenerate and at a lower energy than the degenerate set, d_{z^2} and $d_{x^2-y^2}$. The energy difference between these two sets of d orbitals is called the **crystal field splitting (Δ)**.

d_{z^2} $d_{x^2-y^2}$

Δ

d_{xy} d_{xz} d_{yz}

Due to the crystal field splitting, for metals with electron configurations of d^4, d^5, d^6, or d^7, there are two ways to place the electrons in the five *d* orbitals. Let's consider a d^4 case. According to Hund's rule (see Section 7.8 of the text), maximum stability is reached if the electrons are placed in four separate orbitals with parallel spin. But, this arrangement can be achieved only at a cost; one of the four electrons must be energetically promoted to the higher energy d_{z^2} or $d_{x^2-y^2}$ orbital. This energy investment is not needed if all four electrons enter the d_{xy}, d_{xz}, and d_{yz} orbitals. However, in this electron arrangement, we must pair up two of the electrons. This pairing also takes energy. The two possible electron configurations are shown below.

high spin complex

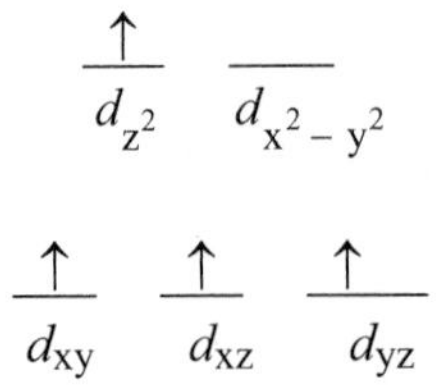

$\uparrow$ d_{z^2} ___ $d_{x^2-y^2}$

$\uparrow$ d_{xy} $\uparrow$ d_{xz} $\uparrow$ d_{yz}

---

low spin complex

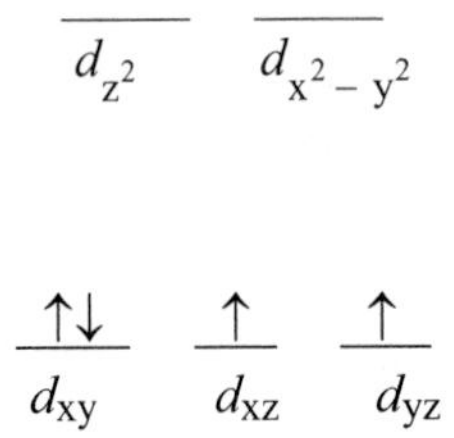

___ d_{z^2} ___ $d_{x^2-y^2}$

$\uparrow\downarrow$ d_{xy} $\uparrow$ d_{xz} $\uparrow$ d_{yz}

Whether a complex is high spin or low spin depends on the ligands that are bonded to the metal center. If a ligand is a weak-field ligand, the crystal field splitting (Δ) is small. Therefore, only a small energy expenditure is needed to keep all spins parallel, resulting in *high spin complexes*. On the other hand, the crystal field splitting is larger when a strong-field ligand is bound to the metal center. In these cases, it is energetically more favorable to pair up the electrons, resulting in *low spin complexes*.

In summary, the actual arrangement of the *d*-electrons is determined by the amount of stability gained by having maximum parallel spins versus the investment in energy required to promote electrons to higher energy *d* orbitals.

EXAMPLE 22.4

Predict the number of unpaired spins in the $[Co(CN)_6]^{3-}$ ion.

Strategy: The electron configuration of Co^{3+} is $[Ar]3d^6$. Since CN^- is a strong-field ligand, we expect $[Co(CN)_6]^{3-}$ to be a low spin complex.

Solution: All six electrons will be placed in the lower energy *d* orbitals (d_{xy}, d_{xz}, and d_{yz}), and there will be no unpaired spins.

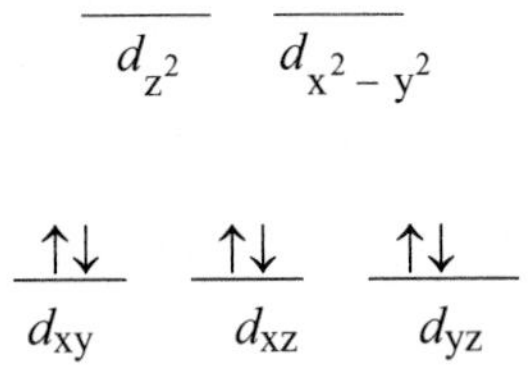

___ d_{z^2} ___ $d_{x^2-y^2}$

$\uparrow\downarrow$ d_{xy} $\uparrow\downarrow$ d_{xz} $\uparrow\downarrow$ d_{yz}

PRACTICE EXERCISE

4. Predict the number of unpaired spins in the $[Fe(H_2O)_6]^{3+}$ ion. Water is a weak-field ligand.

Text Problem: 22.60

ANSWERS TO PRACTICE EXERCISES

1. **(a)** Pt, +4 **(b)** Ni, 0
2. **(a)** tetrahydroxochromate(III) ion **(b)** triamminetrichloroplatinum(IV) nitrate
3. **(a)** $[Ni(NH_2CH_2CH_2NH_2)_3]SO_4$ **(b)** $[Co(NH_3)_4(H_2O)_2]Cl_3$ **(c)** $K_4[Fe(CN)_6]$
4. five unpaired spins

SOLUTIONS TO SELECTED TEXT PROBLEMS

22.12 **(a)** The oxidation number of Cr is **+3**.

(b) The coordination number of Cr is **6**.

(c) **Oxalate ion** ($C_2O_4^{2-}$) is a bidentate ligand.

22.14 Assigning Oxidation Numbers to the Metal Atom in Coordination Compounds, Problem Type 1.

Strategy: The oxidation number of the metal atom is equal to its charge. First we look for known charges in the species. Recall that alkali metals are +1 and alkaline earth metals are +2. Also determine if the ligand is a charged or neutral species. From the known charges, we can deduce the net charge of the metal and hence its oxidation number.

Solution:

(a) Since **sodium** is always **+1** and the oxygens are −2, **Mo** must have an oxidation number of **+6**.

(b) **Magnesium** is **+2** and oxygen −2; therefore **W** is **+6**.

(c) CO ligands are neutral species, so the iron atom bears no net charge. The oxidation number of **Fe** is **0**.

22.16 Naming Coordination Compounds, Problem Type 2.

Strategy: We follow the procedure for naming coordination compounds outlined in Section 22.3 of the text and refer to Tables 22.4 and 22.5 of the text for names of ligands and anions containing metal atoms.

Solution:

(a) Ethylenediamine is a neutral ligand, and each chloride has a −1 charge. Therefore, cobalt has a oxidation number of +3. The correct name for the ion is ***cis*–dichlorobis(ethylenediammine)cobalt(III)**. The prefix *bis* means two; we use this instead of *di* because *di* already appears in the name ethylenediamine.

(b) There are four chlorides each with a −1 charge; therefore, Pt has a +4 charge. The correct name for the compound is **pentaamminechloroplatinum(IV) chloride**.

(c) There are three chlorides each with a −1 charge; therefore, Co has a +3 charge. The correct name for the compound is **pentaamminechlorocobalt(III) chloride**.

22.18 Writing Formulas for Coordination Compounds, Problem Type 3.

Strategy: We follow the procedure in Section 22.3 of the text and refer to Tables 22.4 and 22.5 of the text for names of ligands and anions containing metal atoms.

Solution:

(a) There are two ethylenediamine ligands and two chloride ligands. The correct formula is $\mathbf{[Cr(en)_2Cl_2]^+}$.

(b) There are five carbonyl (CO) ligands. The correct formula is $\mathbf{Fe(CO)_5}$.

(c) There are four cyanide ligands each with a −1 charge. Therefore, the complex ion has a −2 charge, and two K^+ ions are needed to balance the −2 charge of the anion. The correct formula is $\mathbf{K_2[Cu(CN)_4]}$.

(d) There are four NH_3 ligands and two H_2O ligands. Two chloride ions are needed to balance the +2 charge of the complex ion. The correct formula is $\mathbf{[Co(NH_3)_4(H_2O)Cl]Cl_2}$.

22.24 **(a)** In general for any MA_2B_4 octahedral molecule, only **two** geometric isomers are possible. The only real distinction is whether the two A–ligands are *cis* or *trans*. In Figure 22.11 of the text, (a) and (c) are the same compound (Cl atoms *cis* in both), and (b) and (d) are identical (Cl atoms *trans* in both).

(b) A model or a careful drawing is very helpful to understand the MA_3B_3 octahedral structure. There are only **two** possible geometric isomers. The first has all A's (and all B's) *cis*; this is called the facial isomer. The second has two A's (and two B's) at opposite ends of the molecule (*trans*). Try to make or draw other possibilities. What happens?

22.26 **(a)** There are *cis* and *trans* geometric isomers (See Problem 22.24). No optical isomers.

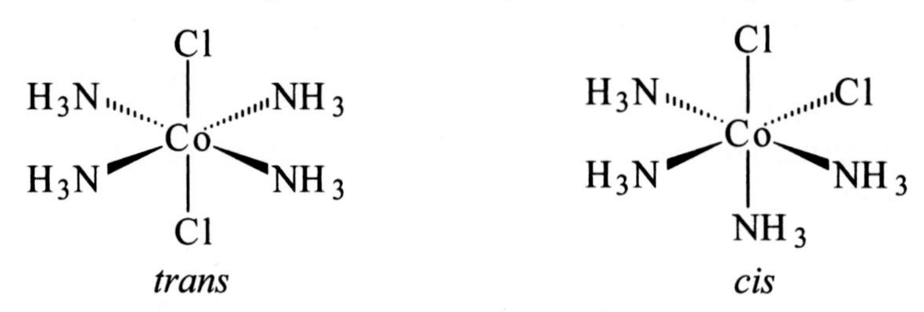

(b) There are two optical isomers. See Figure 22.7 of the text. The three bidentate en ligands are represented by the curved lines.

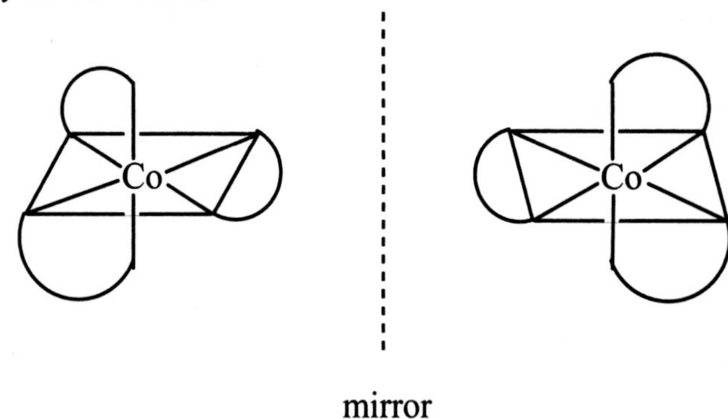

22.34 When a substance appears to be yellow, it is absorbing light from the blue-violet, high energy end of the visible spectrum. Often this absorption is just the tail of a strong absorption in the ultraviolet. Substances that appear green or blue to the eye are absorbing light from the lower energy red or orange part of the spectrum.

Cyanide ion is a very strong field ligand. It causes a larger crystal field splitting than water, resulting in the absorption of higher energy (shorter wavelength) radiation when a *d* electron is excited to a higher energy *d* orbital.

22.36 **(a)** Wavelengths of 470 nm fall between blue and blue-green, corresponding to an observed color in the **orange** part of the spectrum.

(b) We convert wavelength to photon energy using the Planck relationship.

$$\Delta E = \frac{hc}{\lambda} = \frac{(6.63 \times 10^{-34}\ \text{J}\cdot\cancel{\text{s}})(3.00 \times 10^{8}\ \cancel{\text{m}}/\cancel{\text{s}})}{470 \times 10^{-9}\ \cancel{\text{m}}} = 4.23 \times 10^{-19}\ \text{J}$$

$$\frac{4.23 \times 10^{-19}\ \cancel{\text{J}}}{1\ \cancel{\text{photon}}} \times \frac{6.022 \times 10^{23}\ \cancel{\text{photons}}}{1\ \text{mol}} \times \frac{1\ \text{kJ}}{1000\ \cancel{\text{J}}} = \mathbf{255\ kJ/mol}$$

22.38 *Step 1:* The equation for freezing-point depression is

$$\Delta T_f = K_f m$$

Solve this equation algebraically for molality (m), then substitute ΔT_f and K_f into the equation to calculate the molality.

$$m = \frac{\Delta T_f}{K_f} = \frac{0.56°\text{C}}{1.86°\text{C}/m} = 0.30\ m$$

Step 2: Multiplying the molality by the mass of solvent (in kg) gives moles of unknown solute. Then, dividing the mass of solute (in g) by the moles of solute, gives the molar mass of the unknown solute.

$$\text{? mol of unknown solute} = \frac{0.30 \text{ mol solute}}{1 \text{ kg water}} \times 0.0250 \text{ kg water} = 0.0075 \text{ mol solute}$$

$$\text{molar mass of unknown} = \frac{0.875 \text{ g}}{0.0075 \text{ mol}} = 117 \text{ g/mol}$$

The molar mass of $Co(NH_3)_4Cl_3$ is 233.4 g/mol, which is twice the computed molar mass. This implies dissociation into two ions in solution; hence, there are **two moles** of ions produced per one mole of $Co(NH_3)_4Cl_3$. The formula must be:

$[Co(NH_3)_4Cl_2]Cl$

which contains the complex ion $[Co(NH_3)_4Cl_2]^+$ and a chloride ion, Cl^-. Refer to Problem 22.26 (a) for a diagram of the structure of the complex ion.

22.42 Use a radioactive label such as $^{14}CN^-$ (in NaCN). Add NaCN to a solution of $K_3Fe(CN)_6$. Isolate some of the $K_3Fe(CN)_6$ and check its radioactivity. If the complex shows radioactivity, then it must mean that the CN^- ion has participated in the exchange reaction.

22.44 The white precipitate is copper(II) cyanide.

$$Cu^{2+}(aq) + 2CN^-(aq) \rightarrow Cu(CN)_2(s)$$

This forms a soluble complex with excess cyanide.

$$Cu(CN)_2(s) + 2CN^-(aq) \rightarrow Cu(CN)_4^{2-}(aq)$$

Copper(II) sulfide is normally a very insoluble substance. In the presence of excess cyanide ion, the concentration of the copper(II) ion is so low that CuS precipitation cannot occur. In other words, the cyanide complex of copper has a very large formation constant.

22.46 The formation constant expression is:

$$K_f = \frac{[Fe(H_2O)_5NCS^{2+}]}{[Fe(H_2O)_6^{3+}][SCN^-]}$$

Notice that the original volumes of the Fe(III) and SCN^- solutions were both 1.0 mL and that the final volume is 10.0 mL. This represents a tenfold dilution, and the concentrations of Fe(III) and SCN^- become 0.020 M and 1.0×10^{-4} M, respectively. We make a table.

	$Fe(H_2O)_6^{3+}$ +	SCN^-	$\rightleftharpoons$	$Fe(H_2O)_5NCS^{2+}$ + H_2O
Initial (*M*):	0.020	1.0×10^{-4}		0
Change (*M*):	-7.3×10^{-5}	-7.3×10^{-5}		$+7.3 \times 10^{-5}$
Equilibrium (*M*):	0.020	2.7×10^{-5}		7.3×10^{-5}

$$K_f = \frac{7.3 \times 10^{-5}}{(0.020)(2.7 \times 10^{-5})} = \mathbf{1.4 \times 10^2}$$

22.48 Mn^{3+} is $3d^4$ and Cr^{3+} is $3d^5$. Therefore, $\mathbf{Mn^{3+}}$ has a greater tendency to accept an electron and is a stronger oxidizing agent. The $3d^5$ electron configuration of Cr^{3+} is a stable configuration.

22.50 Ti is +3 and Fe is +3.

22.52 A 100.00 g sample of hemoglobin contains 0.34 g of iron. In moles this is:

$$0.34 \text{ g Fe} \times \frac{1 \text{ mol}}{55.85 \text{ g}} = 6.1 \times 10^{-3} \text{ mol Fe}$$

The amount of hemoglobin that contains one mole of iron must be:

$$\frac{100.00 \text{ g hemoglobin}}{6.1 \times 10^{-3} \text{ mol Fe}} = \mathbf{1.6 \times 10^4 \text{ g hemoglobin/mol Fe}}$$

We compare this to the actual molar mass of hemoglobin:

$$\frac{6.5 \times 10^4 \text{ g hemoglobin}}{1 \text{ mol hemoglobin}} \times \frac{1 \text{ mol Fe}}{1.6 \times 10^4 \text{ g hemoglobin}} = 4 \text{ mol Fe/1 mol hemoglobin}$$

The discrepancy between our minimum value and the actual value can be explained by realizing that one hemoglobin molecule contains **four** iron atoms.

22.54 **(a)** $[Cr(H_2O)_6]Cl_3$, **(b)** $[Cr(H_2O)_5Cl]Cl_2 \cdot H_2O$, **(c)** $[Cr(H_2O)_4Cl_2]Cl \cdot 2H_2O$

The compounds can be identified by a conductance experiment. Compare the conductances of equal molar solutions of the three compounds with equal molar solutions of NaCl, $MgCl_2$, and $FeCl_3$. The solution that has similar conductance to the NaCl solution contains (c); the solution with the conductance similar to $MgCl_2$ contains (b); and the solution with conductance similar to $FeCl_3$ contains (a).

22.56

$$Zn(s) \rightarrow Zn^{2+}(aq) + 2e^- \qquad E^\circ_{anode} = -0.76 \text{ V}$$

$$2[Cu^{2+}(aq) + e^- \rightarrow Cu^+(aq)] \qquad E^\circ_{cathode} = 0.15 \text{ V}$$

$$Zn(s) + 2Cu^{2+}(aq) \rightarrow Zn^{2+}(aq) + 2Cu^+(aq) \qquad E^\circ_{cell} = E^\circ_{cathode} - E^\circ_{anode} = 0.15 \text{ V} - (-0.76 \text{ V}) = 0.91 \text{ V}$$

We carry additional significant figures throughout the remainder of this calculation to minimize rounding errors.

$$\Delta G^\circ = -nFE^\circ = -(2)(96500 \text{ J/V·mol})(0.91 \text{ V}) = -1.756 \times 10^5 \text{ J/mol} = \mathbf{-1.8 \times 10^2 \text{ kJ/mol}}$$

$$\Delta G° = -RT\ln K$$

$$\ln K = \frac{-\Delta G°}{RT} = \frac{-(-1.756 \times 10^5 \text{ J/mol})}{(8.314 \text{ J/K}\cdot\text{mol})(298 \text{ K})}$$

$$\ln K = 70.88$$

$$\boldsymbol{K} = e^{70.88} = \mathbf{6 \times 10^{30}}$$

22.58 Iron is much more abundant than cobalt.

22.60 Oxyhemoglobin absorbs higher energy light than deoxyhemoglobin. Oxyhemoglobin is diamagnetic (low spin), while deoxyhemoglobin is paramagnetic (high spin). These differences occur because oxygen (O_2) is a strong–field ligand. The crystal field splitting diagrams are:

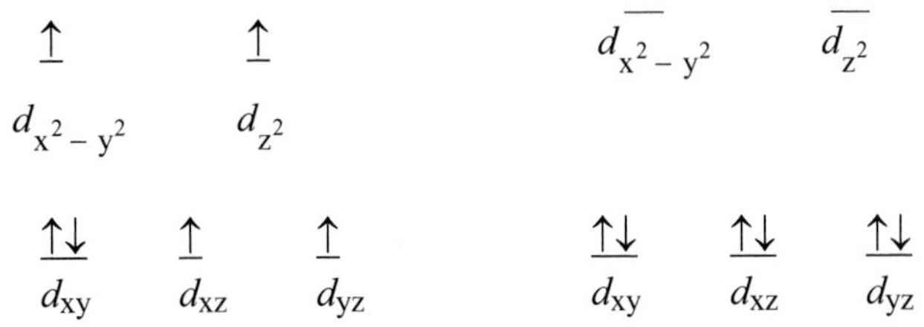

deoxyhemoglobin:

$\uparrow$ $d_{x^2-y^2}$ $\uparrow$ d_{z^2}

$\uparrow\downarrow$ d_{xy} $\uparrow$ d_{xz} $\uparrow$ d_{yz}

oxyhemoglobin:

— $d_{x^2-y^2}$ — d_{z^2}

$\uparrow\downarrow$ d_{xy} $\uparrow\downarrow$ d_{xz} $\uparrow\downarrow$ d_{yz}

22.62 Complexes are expected to be colored when the highest occupied orbitals have between one and nine *d* electrons. Such complexes can therefore have $d \rightarrow d$ transitions (that are usually in the visible part of the electromagnetic radiation spectrum). The ions V^{5+}, Ca^{2+}, and Sc^{3+} have d^0 electron configurations and Cu^+, Zn^{2+}, and Pb^{2+} have d^{10} electron configurations: these complexes are colorless. The other complexes have outer electron configurations of d^1 to d^9 and are therefore colored.

22.64 Dipole moment measurement. Only the *cis* isomer has a dipole moment.

22.66 EDTA sequesters metal ions (like Ca^{2+} and Mg^{2+}) which are essential for growth and function, thereby depriving the bacteria to grow and multiply.

22.68 The square planar complex shown in the problem has **3** geometric isomers. They are:

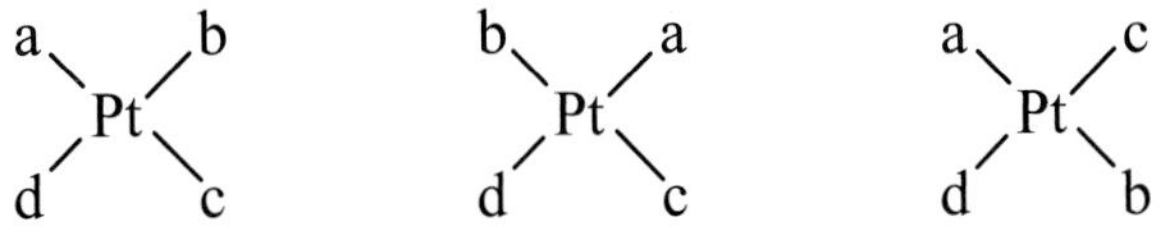

Structure 1: a (upper left), b (upper right), d (lower left), c (lower right) around Pt
Structure 2: b (upper left), a (upper right), d (lower left), c (lower right) around Pt
Structure 3: a (upper left), c (upper right), d (lower left), b (lower right) around Pt

Note that in the first structure a is *trans* to c, in the second a is *trans* to d, and in the third a is *trans* to b. Make sure you realize that if we switch the positions of b and d in structure 1, we do not obtain another geometric isomer. A 180° rotation about the a–Pt–c axis gives structure 1.

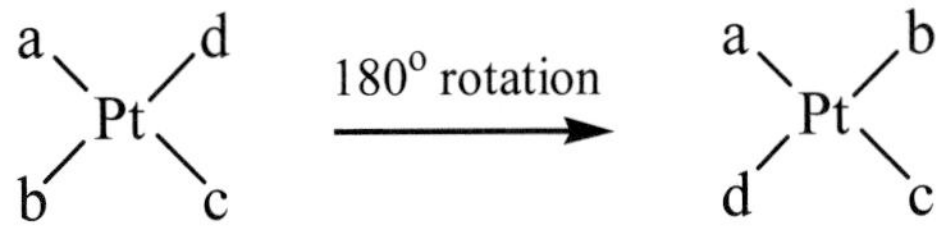

a (upper left), d (upper right), b (lower left), c (lower right) around Pt —180° rotation→ a (upper left), b (upper right), d (lower left), c (lower right) around Pt

22.70 The reaction is: $Ag^+(aq) + 2CN^-(aq) \rightleftharpoons Ag(CN)_2^-(aq)$

$$K_f = 1.0 \times 10^{21} = \frac{[Ag(CN)_2^-]}{[Ag^+][CN^-]^2}$$

First, we calculate the initial concentrations of Ag^+ and CN^-. Then, because K_f is so large, we assume that the reaction goes to completion. This assumption will allow us to solve for the concentration of Ag^+ at equilibrium. The initial concentrations of Ag^+ and CN^- are:

$$[CN^-] = \frac{\frac{5.0\text{ mol}}{1\text{ L}} \times 9.0\text{ L}}{99.0\text{ L}} = 0.455\ M$$

$$[Ag^+] = \frac{\frac{0.20\text{ mol}}{1\text{ L}} \times 90.0\text{ L}}{99.0\text{ L}} = 0.182\ M$$

We set up a table to determine the concentrations after complete reaction.

	$Ag^+(aq)$	+ $2CN^-(aq)$	$\rightleftharpoons Ag(CN)_2^-(aq)$
Initial (M):	0.182	0.455	0
Change (M):	−0.182	−(2)(0.182)	+0.182
Final (M):	0	0.0910	0.182

$$K_f = \frac{[Ag(CN)_2^-]}{[Ag^+][CN^-]^2}$$

$$1.0 \times 10^{21} = \frac{0.182\ M}{[Ag^+](0.0910\ M)^2}$$

$$\mathbf{[Ag^+] = 2.2 \times 10^{-20}\ M}$$

22.72 **(a)** The equilibrium constant can be calculated from $\Delta G°$. We can calculate $\Delta G°$ from the cell potential.

From Table 19.1 of the text,

$Cu^{2+} + 2e^- \rightarrow Cu$ $\quad E° = 0.34\text{ V}$ and $\Delta G° = -(2)(96500\text{ J/V·mol})(0.34\text{ V}) = -6.562 \times 10^4\text{ J/mol}$

$Cu^{2+} + e^- \rightarrow Cu^+$ $\quad E° = 0.15\text{ V}$ and $\Delta G° = -(1)(96500\text{ J/V·mol})(0.15\text{ V}) = -1.448 \times 10^4\text{ J/mol}$

These two equations need to be arranged to give the disproportionation reaction in the problem. We keep the first equation as written and reverse the second equation and multiply by two.

$Cu^{2+} + 2e^- \rightarrow Cu$ $\quad \Delta G° = -6.562 \times 10^4\text{ J/mol}$

$2Cu^+ \rightarrow 2Cu^{2+} + 2e^-$ $\quad \Delta G° = +(2)(1.448 \times 10^4\text{ J/mol})$

$2Cu^+ \rightarrow Cu^{2+} + Cu$ $\quad \Delta G° = -6.562 \times 10^4\text{ J/mol} + 2.896 \times 10^4\text{ J/mol} = -3.666 \times 10^4\text{ J/mol}$

We use Equation (18.14) of the text to calculate the equilibrium constant.

$$\Delta G° = -RT \ln K$$

$$K = e^{-\Delta G°/RT}$$

$$K = e^{-(-3.666 \times 10^4 \text{ J/mol})/(8.314 \text{ J/mol·K})(298 \text{ K})}$$

$$\boldsymbol{K = 2.7 \times 10^6}$$

(b) Free Cu^+ ions are unstable in solution [as shown in part (a)]. Therefore, the only stable compounds containing Cu^+ ions are insoluble.

22.74 **(a)** Cu^{3+} would not be stable in solution because it can be easily reduced to the more stable Cu^{2+}. From Figure 22.3 of the text, we see that the 3rd ionization energy is quite high, about 3500 kJ/mol. Therefore, Cu^{3+} has a great tendency to accept an electron.

(b) **Potassium hexafluorocuprate(III)**. Cu^{3+} is $3d^8$. CuF_6^{3-} has an octahedral geometry. According to Figure 22.17 of the text, CuF_6^{3-} should be paramagnetic, containing two unpaired electrons. (Because it is $3d^8$, it does not matter whether the ligand is a strong or weak-field ligand. See Figure 22.22 of the text.)

(c) We refer to Figure 22.24 of the text. The splitting pattern is such that all the square-planar complexes of Cu^{3+} should be diamagnetic.

CHAPTER 23
NUCLEAR CHEMISTRY

PROBLEM-SOLVING STRATEGIES AND TUTORIAL SOLUTIONS

TYPES OF PROBLEMS

Problem Type 1: Balancing Nuclear Equations.

Problem Type 2: Nuclear Stability

Problem Type 3: Calculating Nuclear Binding Energy.

Problem Type 4: Kinetics of Radioactive Decay.

Problem Type 5: Balancing Nuclear Transmutation Equations.

PROBLEM TYPE 1: BALANCING NUCLEAR EQUATIONS

Writing a nuclear equation differs slightly from writing equations for chemical reactions. In addition to writing the symbols for various chemical elements, we must also explicitly indicate protons, neutrons, and electrons. In fact, we must show the numbers of protons and neutrons present in *every* species in the reaction.

You must know the symbols for elementary particles in order to balance a nuclear equation.

$^{1}_{1}\text{p}$ or $^{1}_{1}\text{H}$	$^{1}_{0}\text{n}$	$^{0}_{-1}e$ or $^{0}_{-1}\beta$	$^{0}_{+1}e$ or $^{0}_{+1}\beta$	$^{4}_{2}\text{He}$ or $^{4}_{2}\alpha$
proton	neutron	electron	positron	α particle

In balancing any nuclear equation, we observe the following rules:

- The total number of protons plus neutrons in the products and in the reactants must be the same (conservation of mass number).
- The total number of nuclear charges in the products and in the reactants must be the same (conservation of atomic number).

If the atomic numbers and mass numbers of all the species but one in a nuclear equation are known, the unknown species can be identified by applying the above rules.

EXAMPLE 23.1
Identify X in the following nuclear reactions:

$$\text{(a)}\quad {}^{14}_{7}\text{N} + {}^{1}_{0}\text{n} \longrightarrow {}^{14}_{6}\text{C} + \text{X}$$

$$\text{(b)}\quad {}^{226}\text{Ra} \longrightarrow {}^{4}_{2}\alpha + \text{X}$$

Strategy: In balancing nuclear equations, note that the sum of atomic numbers and that of mass numbers must match on both sides of the equation.

Solution:

(a) Because the sum of the mass numbers must be conserved, the unknown product will have a mass number of 1. Because the sum of the atomic numbers must be conserved, the nuclear charge of the unknown product must be 1.

The particle is a proton.

$$^{14}_{7}N + ^{1}_{0}n \longrightarrow ^{14}_{6}C + \mathbf{^{1}_{1}p}$$

(b) Note that the atomic number of radium is missing. Look at a periodic table to find that Ra is element number 88. Balancing the mass numbers first, we find that the unknown product must have a mass of 222. Balancing the nuclear charges, we find that the nuclear charge of the unknown must be 86. Element number 86 is radon (Rn).

$$^{226}_{88}Ra \longrightarrow ^{4}_{2}\alpha + \mathbf{^{222}_{86}Rn}$$

PRACTICE EXERCISE

1. Identify X in the following nuclear reactions:

(a) $^{239}_{94}Pu \longrightarrow ^{235}_{92}U + X$

(b) $^{90}_{38}Sr \longrightarrow ^{90}_{37}Rb + X$

Text Problems: 23.6, 23.28

PROBLEM TYPE 2: NUCLEAR STABILITY

The principal factor for determining the stability of a nucleus is the *neutron-to-proton ratio* (n/p). For stable elements of low atomic number, the n/p ratio is close to 1. As the atomic number increases, the n/p ratios of stable nuclei become greater than 1. The deviation in the n/p ratio at higher atomic numbers arises because a larger number of neutrons is needed to stabilize the nucleus by counteracting the strong repulsion among the large number of protons. Figure 23.1 of the text shows a plot of the number of neutrons versus the number of protons in various isotopes. The stable nuclei are located in an area of the graph called the *belt of stability*. Most radioactive nuclei lie outside the belt.

The following rules are useful in predicting nuclear stability.

- Nuclei that contain 2, 8, 20, 50, 82, or 126 protons or neutrons are generally more stable than nuclei that do not possess these numbers. These numbers are called *magic numbers*.
- Nuclei with even numbers of both protons and neutrons are generally more stable than those with odd numbers of these particles (see Table 23.2 of the text).
- All isotopes of the elements starting with polonium (Po, Z = 84) are radioactive. All isotopes of technetium (Tc, Z = 43) and promethium (Pm, Z = 61) are also radioactive.

EXAMPLE 23.2

Rank the following nuclei in order of increasing nuclear stability:

$\mathbf{^{40}_{20}Ca}$ $\quad$ $\mathbf{^{39}_{20}Ca}$ $\quad$ $\mathbf{^{11}_{5}B}$

Strategy: The principal factor for determining the stability of a nucleus is the *neutron-to-proton ratio* (n/p). For stable elements of low atomic number, the n/p ratio is close to 1. As the atomic number increases, the n/p ratios of stable nuclei become greater than 1. The following rules are useful in predicting nuclear stability.

1) Nuclei that contain 2, 8, 20, 50, 82, or 126 protons or neutrons are generally more stable than nuclei that do not possess these numbers. These numbers are called *magic numbers*.
2) Nuclei with even numbers of both protons and neutrons are generally more stable than those with odd numbers of these particles (see Table 23.2 of the text).

Solution:

Boron-11 has both an odd number of protons and an odd number of neutrons; therefore, it should be the least stable of the three nuclei. Calcium-39 has a magic number of protons (20), but an odd number of neutrons. Calcium-40 has both a magic number of protons and neutrons (20), so it should be more stable than calcium-39. The order of increasing nuclear stability is

$$^{11}_{5}\text{B} \quad < \quad ^{39}_{20}\text{Ca} \quad < \quad ^{40}_{20}\text{Ca}$$

PRACTICE EXERCISE

2. Rank the following isotopes in order of increasing nuclear stability:

$^{39}_{20}\text{Ca}$ $\quad$ $^{222}_{86}\text{Rn}$ $\quad$ $^{98}_{43}\text{Tc}$

Text Problems: 23.14, 23.16

PROBLEM TYPE 3: CALCULATING NUCLEAR BINDING ENERGY

A quantitative measure of nuclear stability is the **nuclear binding energy**, which is the energy required to break up a nucleus into its component protons and neutrons. The concept of nuclear binding energy evolved from studies showing that the masses of nuclei are always less than the sum of the masses of the **nucleons** (the protons and neutrons in a nucleus). The difference between the mass of an atom and the sum of the masses of its protons, neutrons, and electrons is called the **mass defect**. According to Einstein's mass-energy equivalence relationship

$$E = mc^2$$

where,

E is energy
m is mass
c is the velocity of light

the loss in mass shows up as energy (heat) given off to the surroundings. We can calculate the amount of energy released by writing:

$$\Delta E = (\Delta m)c^2$$

where,

ΔE = energy of products – energy of reactants

Δm = mass of products – mass of reactants

See Example 23.3 below for a detailed calculation.

EXAMPLE 23.3

Calculate the nuclear binding energy of the light isotope of helium, ^{3}He. The atomic mass of $^{3}_{2}\text{He}$ is 3.01603 amu.

Strategy: To calculate the nuclear binding energy, we first determine the difference between the mass of the nucleus and the mass of all the protons and neutrons, which gives us the mass defect. Next, we apply Einstein's mass-energy relationship [$\Delta E = (\Delta m)c^2$].

Solution: The binding energy is the energy required for the process

$$^{3}_{2}\text{He} \longrightarrow 2\,^{1}_{1}\text{p} + ^{1}_{0}\text{n}$$

There are 2 protons and 1 neutron in the helium nucleus. The mass of 2 protons is

$$(2)(1.007825\text{ amu}) = 2.015650\text{ amu}$$

and the mass of a neutron is 1.008665 amu

Therefore, the predicted mass of ${}^{3}_{2}He$ is 2.015650 + 1.008665 = 3.024315 amu, and the mass defect is

$$\Delta m = 3.024315 \text{ amu} - 3.01603 \text{ amu} = 0.00829 \text{ amu}$$

The energy change (ΔE) for the process is

$$\Delta E = (\Delta m)c^2$$

$$\Delta E = (0.00829 \text{ amu})(3.00 \times 10^8 \text{ m/s})^2$$

$$\Delta E = 7.46 \times 10^{14} \frac{\text{amu} \cdot \text{m}^2}{\text{s}^2}$$

Let's convert to more familiar energy units (J/He atom).

$$7.46 \times 10^{14} \frac{\text{amu} \cdot \text{m}^2}{\text{s}^2} \times \frac{1.00 \text{ g}}{6.022 \times 10^{23} \text{ amu}} \times \frac{1 \text{ kg}}{1000 \text{ g}} \times \frac{1 \text{ J}}{\frac{1 \text{ kg} \cdot \text{m}^2}{\text{s}^2}} = \mathbf{1.24 \times 10^{-12} \text{ J/atom}}$$

This is the nuclear binding energy. It's the energy required to break up one helium-3 nucleus into 2 protons and 1 neutron.

When comparing the stability of any two nuclei we must account for the fact that they have different numbers of nucleons. For this reason, it is more meaningful to use the *nuclear binding energy per nucleon*, defined as

$$\text{nuclear binding energy per nucleon} = \frac{\text{nuclear binding energy}}{\text{number of nucleons}}$$

For the helium-3 nucleus,

$$\text{nuclear binding energy per nucleon} = \frac{1.24 \times 10^{-12} \text{ J/He atom}}{3 \text{ nucleons/He atom}} = \mathbf{4.13 \times 10^{-13} \text{ J/nucleon}}$$

PRACTICE EXERCISE

3. (a) Calculate the binding energy and the binding energy per nucleon of ${}^{27}_{13}Al$. The atomic mass of ${}^{27}_{13}Al$ is 26.98154 amu.

(b) Compare the result from part (a) to the binding energy of ${}^{28}_{14}Si$, which has an even number of protons and neutrons. The atomic mass of ${}^{28}_{14}Si$ is 27.976928 amu.

Text Problem: 23.20

PROBLEM TYPE 4: KINETICS OF RADIOACTIVE DECAY

All radioactive decays obey first-order kinetics. To review first-order reactions, see Sections 13.2 and 13.3 of the text. The decay rate at any time *t* is given by

$$\text{rate of decay at time } t = \lambda N$$

where,

λ is the first-order rate constant
N is the number of radioactive nuclei present at time t

The relationship between the number of radioactive nuclei present at time zero (N_0) and the number of nuclei remaining at a later time t (N_t) is given by:

$$\ln\frac{N_t}{N_0} = -\lambda t$$

The corresponding half-life for a first-order reaction is given by:

$$t_{\frac{1}{2}} = \frac{0.693}{\lambda}$$

Unlike ordinary chemical reactions, the rate constants for nuclear reactions are unaffected by changes in environmental conditions such as temperature and pressure.

EXAMPLE 23.4

Cobalt-60 is used in radiation therapy. It has a half-life of 5.26 years.

(a) Calculate the rate constant for radioactive decay.

(b) What fraction of a certain sample will remain after 12 years?

Strategy: (a) According to Equation (13.6) of the text, the half-life of first-order decay is:

$$t_{\frac{1}{2}} = \frac{0.693}{\lambda}$$

We can solve for the first-order rate constant (λ) given the half-life.

(b) According to Equation (13.3) of the text, the number of radioactive nuclei at time zero (N_0) and time t (N_t) is:

$$\ln\frac{N_t}{N_0} = -\lambda t$$

We want to calculate the fraction of the sample remaining, which will equal $\frac{N_t}{N_0}$.

Solution:

(a) The rate constant for the decay can be calculated from the half-life.

$$t_{\frac{1}{2}} = \frac{0.693}{\lambda}$$

Rearrange the equation to solve for the rate constant (λ).

$$\lambda = \frac{0.693}{t_{\frac{1}{2}}} = \frac{0.693}{5.26\text{ yr}} = \mathbf{0.132\ yr^{-1}}$$

(b) The fraction of a certain sample that will remain after 12 years is

$$\frac{N_t}{N_0}$$

where $t = 12$ yr.

Rearrange the equation, $\ln\frac{N_t}{N_0} = -\lambda t$, to solve for $\frac{N_t}{N_0}$.

$$\ln \frac{N_t}{N_0} = -\lambda t$$

$$\frac{N_t}{N_0} = e^{-\lambda t}$$

$$\frac{N_t}{N_0} = e^{-(0.132\,\cancel{\text{yr}^{-1}})(12\,\cancel{\text{yr}})} = \mathbf{0.205}$$

In other words, 20.5 percent of the original sample will remain after 12 years.

PRACTICE EXERCISE

4. Estimate the age of a bottle of wine that has a tritium $\left({}^{3}_{1}\text{H}\right)$ content 75.0 percent of the tritium content of environmental water obtained from the area where the grapes were grown. $t_{\frac{1}{2}} = 12.3$ yr.

Text Problems: 23.24, 23.26, 23.30

PROBLEM TYPE 5: BALANCING NUCLEAR TRANSMUTATION EQUATIONS

Nuclei can undergo change as a result of bombardment by neutrons, protons, or other nuclei. This process is called **nuclear transmutation**. Unlike radioactive decay, nuclear transmutation is *not* a spontaneous process; consequently, nuclear transmutation reactions have more than one reactant. As an example, consider the synthesis of neptunium (Np), which was the first transuranium element to be synthesized by scientists.

First, uranium-238 is bombarded with neutrons to produce uranium-239.

$${}^{238}_{92}\text{U} + {}^{1}_{0}\text{n} \longrightarrow {}^{239}_{92}\text{U}$$

This is a nuclear transmutation reaction. Uranium-239 is unstable and decays spontaneously to neptunium-239 by emitting a β particle.

$${}^{239}_{92}\text{U} \longrightarrow {}^{239}_{93}\text{Np} + {}^{0}_{-1}\beta$$

To balance a nuclear transmutation reaction, follow the same rules used to balance nuclear equations, Problem Type 1.

- The total number of protons plus neutrons in the products and in the reactants must be the same (conservation of mass number).
- The total number of nuclear charges in the products and in the reactants must be the same (conservation of atomic number).

EXAMPLE 23.5
Write and balance the following reactions. (a) When aluminum-27 is bombarded with α particles, phosphorus-30 and one other particle are produced. (b) Phosphorus-30 has a low *n*/*p* ratio and decays spontaneously by positron emission.

Strategy: (a) The first reaction is a nuclear transmutation. You are given both reactants and one of the two products in the problem. To balance the equation, remember that both mass number and atomic number must be conserved. (b) The second reaction is spontaneous so there is only one reactant, phosphorus-30. The problem indicates that phosphorus-30 decays by positron emission.

Solution:

(a) Let's start by writing what is given in the problem.

$$^{27}_{13}Al + ^{4}_{2}\alpha \longrightarrow ^{30}_{15}P + X$$

To balance the mass number, the missing particle (X) must have a mass number of 1. To balance the atomic number, X must have an atomic number of 0. X must be a neutron.

$$^{27}_{13}Al + ^{4}_{2}\alpha \longrightarrow ^{30}_{15}P + ^{1}_{0}n$$

(b) The second reaction is spontaneous so there is only one reactant, phosphorus-30. The problem indicates that phosphorus-30 decays by positron emission. Let's write down what we know so far.

$$^{30}_{15}P \longrightarrow X + ^{0}_{+1}\beta$$

To balance the mass number, the missing element (X) must have a mass number of 30. To balance the atomic number, X must have an atomic number of 14. X must be silicon-30.

$$^{30}_{15}P \longrightarrow ^{30}_{14}Si + ^{0}_{+1}\beta$$

PRACTICE EXERCISE

5. Write and balance the following reactions. When chlorine-37 is bombarded with neutrons, only one product is produced. The product is unstable and spontaneously decays by beta emission.

Text Problems: 23.34, 23.36

ANSWERS TO PRACTICE EXERCISES

1. **(a)** $X = ^{4}_{2}He$

(b) $X = ^{0}_{+1}\beta$

2. Technetium-98 should be unstable because it has an odd number of both protons and neutrons. In fact, all isotopes of technetium are radioactive. Radon-222 will perhaps be slightly more stable than technetium-98, because it has an even number of both protons and neutrons. Remember, however, that all isotopes of the elements starting with polonium (Po, Z = 84) are radioactive. Calcium-39 has 20 protons (a "magic number") and should be the most stable of the three isotopes. The correct order of increasing nuclear stability is

$$^{98}_{43}Tc \quad < \quad ^{222}_{86}Rn \quad < \quad ^{39}_{20}Ca$$

3. **(a)** Binding energy $= 3.6 \times 10^{-11}$ J/Al atom $= 1.3 \times 10^{-12}$ J/nucleon

(b) Binding energy $= 3.8 \times 10^{-11}$ J/Si atom $= 1.4 \times 10^{-12}$ J/nucleon

When comparing the stability of any two nuclei, it is best to compare the binding energy in units of J/nucleon. Silicon has the greater binding energy per nucleon and hence is more stable than Al. You should have expected this result, because Si has an even number of protons and neutrons; whereas, Al has an odd number of protons and an even number of neutrons. Nuclei with even numbers of both protons and neutrons are generally more stable than those with odd numbers of these particles.

4. You can calculate the rate constant (λ) from the half-life.

$$\lambda = 0.0563 \text{ yr}^{-1}$$

Next, $\frac{N_t}{N_0} = 0.750$. The age (t) of the bottle of wine can then be calculated from the equation,

$$\ln\frac{N_t}{N_0} = -\lambda t$$

$$\mathbf{t = 5.07\ yr}$$

5. ${}^{1}_{0}\text{n} + {}^{37}_{17}\text{Cl} \longrightarrow {}^{38}_{17}\text{Cl}$

${}^{38}_{17}\text{Cl} \longrightarrow {}^{38}_{18}\text{Ar} + {}^{0}_{-1}\beta$

SOLUTIONS TO SELECTED TEXT PROBLEMS

23.6 Balancing Nuclear Equations, Problem Type 1.

Strategy: In balancing nuclear equations, note that the sum of atomic numbers and that of mass numbers must match on both sides of the equation.

Solution:

(a) The sum of the mass numbers must be conserved. Thus, the unknown product will have a mass number of 0. The atomic number must be conserved. Thus, the nuclear charge of the unknown product must be −1. The particle is a β particle.

$$^{135}_{53}\mathrm{I} \longrightarrow ^{135}_{54}\mathrm{Xe} + ^{\mathbf{0}}_{\mathbf{-1}}\boldsymbol{\beta}$$

(b) Balancing the mass numbers first, we find that the unknown product must have a mass of 40. Balancing the nuclear charges, we find that the atomic number of the unknown must be 20. Element number 20 is calcium (Ca).

$$^{40}_{19}\mathrm{K} \longrightarrow ^{0}_{-1}\beta + ^{\mathbf{40}}_{\mathbf{20}}\mathbf{Ca}$$

(c) Balancing the mass numbers, we find that the unknown product must have a mass of 4. Balancing the nuclear charges, we find that the nuclear charge of the unknown must be 2. The unknown particle is an alpha (α) particle.

$$^{59}_{27}\mathrm{Co} + ^{1}_{0}\mathrm{n} \longrightarrow ^{56}_{25}\mathrm{Mn} + ^{\mathbf{4}}_{\mathbf{2}}\boldsymbol{\alpha}$$

(d) Balancing the mass numbers, we find that the unknown products must have a combined mass of 2. Balancing the nuclear charges, we find that the combined nuclear charge of the two unknown particles must be 0. The unknown particles are neutrons.

$$^{235}_{92}\mathrm{U} + ^{1}_{0}\mathrm{n} \longrightarrow ^{99}_{40}\mathrm{Zr} + ^{135}_{52}\mathrm{Te} + 2\,^{\mathbf{1}}_{\mathbf{0}}\mathbf{n}$$

23.14 Nuclear Stability, Problem Type 2.

Strategy: The principal factor for determining the stability of a nucleus is the *neutron-to-proton ratio* (*n*/*p*). For stable elements of low atomic number, the *n*/*p* ratio is close to 1. As the atomic number increases, the *n*/*p* ratios of stable nuclei become greater than 1. The following rules are useful in predicting nuclear stability.

2) Nuclei that contain 2, 8, 20, 50, 82, or 126 protons or neutrons are generally more stable than nuclei that do not possess these numbers. These numbers are called *magic numbers*.

2) Nuclei with even numbers of both protons and neutrons are generally more stable than those with odd numbers of these particles (see Table 23.2 of the text).

Solution:

(a) **Lithium-9** should be less stable. The neutron-to-proton ratio is too high. For small atoms, the *n*/*p* ratio will be close to 1:1.

(b) **Sodium-25** is less stable. Its neutron-to-proton ratio is probably too high.

(c) **Scandium-48** is less stable because of odd numbers of protons and neutrons. We would not expect calcium-48 to be stable even though it has a magic number of protons. Its *n*/*p* ratio is too high.

23.16 (a) **Neon-17** should be radioactive. It falls below the belt of stability (low *n*/*p* ratio).

(b) **Calcium-45** should be radioactive. It falls above the belt of stability (high *n*/*p* ratio).

(c) All **technetium** isotopes are radioactive.

(d) **Mercury-195** should be radioactive. Mercury-196 has an even number of both neutrons and protons.

(e) All **curium** isotopes are unstable.

23.18 We can use the equation, $\Delta E = \Delta mc^2$, to solve the problem. Recall the following conversion factor:

$$1\ \text{J} = \frac{1\ \text{kg}\cdot\text{m}^2}{\text{s}^2}$$

The energy loss in one second is:

$$\Delta m = \frac{\Delta E}{c^2} = \frac{\dfrac{5\times10^{26}\ \text{kg}\cdot\text{m}^2}{1\ \text{s}^2}}{\left(3.00\times10^8\ \dfrac{\text{m}}{\text{s}}\right)^2} = 6\times10^9\ \text{kg}$$

Therefore the rate of mass loss is **6 × 10⁹ kg/s**.

23.20 Calculating Nuclear Binding Energy, Problem Type 3.

Strategy: To calculate the nuclear binding energy, we first determine the difference between the mass of the nucleus and the mass of all the protons and neutrons, which gives us the mass defect. Next, we apply Einstein's mass-energy relationship [$\Delta E = (\Delta m)c^2$].

Solution:

(a) The binding energy is the energy required for the process

$$^{4}_{2}\text{He} \rightarrow 2\ ^{1}_{1}\text{p} + 2\ ^{1}_{0}\text{n}$$

There are 2 protons and 2 neutrons in the helium nucleus. The mass of 2 protons is

$$(2)(1.007825\ \text{amu}) = 2.015650\ \text{amu}$$

and the mass of 2 neutrons is

$$(2)(1.008665\ \text{amu}) = 2.017330\ \text{amu}$$

Therefore, the predicted mass of $^{4}_{2}\text{He}$ is 2.015650 + 2.017330 = 4.032980 amu, and the mass defect is

$$\Delta m = 4.032980\ \text{amu} - 4.0026\ \text{amu} = 0.0304\ \text{amu}$$

The energy change (ΔE) for the process is

$$\Delta E = (\Delta m)c^2$$

$$= (0.0304\ \text{amu})(3.00\times10^8\ \text{m/s})^2$$

$$= 2.74\times10^{15}\ \frac{\text{amu}\cdot\text{m}^2}{\text{s}^2}$$

Let's convert to more familiar energy units (J/He atom).

$$\frac{2.74 \times 10^{15}\ \cancel{\text{amu}} \cdot \cancel{\text{m}^2}}{1\ \cancel{\text{s}^2}} \times \frac{1.00\ \cancel{\text{g}}}{6.022 \times 10^{23}\ \cancel{\text{amu}}} \times \frac{1\ \cancel{\text{kg}}}{1000\ \cancel{\text{g}}} \times \frac{1\ \text{J}}{\frac{1\ \cancel{\text{kg}} \cdot \cancel{\text{m}^2}}{\cancel{\text{s}^2}}} = \mathbf{4.55 \times 10^{-12}\ J}$$

This is the nuclear binding energy. It's the energy required to break up one helium-4 nucleus into 2 protons and 2 neutrons.

When comparing the stability of any two nuclei we must account for the fact that they have different numbers of nucleons. For this reason, it is more meaningful to use the *nuclear binding energy per nucleon*, defined as

$$\text{nuclear binding energy per nucleon} = \frac{\text{nuclear binding energy}}{\text{number of nucleons}}$$

For the helium-4 nucleus,

$$\text{nuclear binding energy per nucleon} = \frac{4.55 \times 10^{-12}\ \text{J/He}\ \cancel{\text{atom}}}{4\ \text{nucleons/He}\ \cancel{\text{atom}}} = \mathbf{1.14 \times 10^{-12}\ J/nucleon}$$

(b) The binding energy is the energy required for the process

$${}^{184}_{74}\text{W} \rightarrow 74\ {}^{1}_{1}\text{p} + 110\ {}^{1}_{0}\text{n}$$

There are 74 protons and 110 neutrons in the tungsten nucleus. The mass of 74 protons is

$$(74)(1.007825\ \text{amu}) = 74.57905\ \text{amu}$$

and the mass of 110 neutrons is

$$(110)(1.008665\ \text{amu}) = 110.9532\ \text{amu}$$

Therefore, the predicted mass of ${}^{184}_{74}\text{W}$ is 74.57905 + 110.9532 = 185.5323 amu, and the mass defect is

$$\Delta m = 185.5323\ \text{amu} - 183.9510\ \text{amu} = 1.5813\ \text{amu}$$

The energy change (ΔE) for the process is

$$\Delta E = (\Delta m)c^2$$

$$= (1.5813\ \text{amu})(3.00 \times 10^8\ \text{m/s})^2$$

$$= 1.42 \times 10^{17}\ \frac{\text{amu} \cdot \text{m}^2}{\text{s}^2}$$

Let's convert to more familiar energy units (J/W atom).

$$\frac{1.42 \times 10^{17}\ \cancel{\text{amu}} \cdot \cancel{\text{m}^2}}{1\ \cancel{\text{s}^2}} \times \frac{1.00\ \cancel{\text{g}}}{6.022 \times 10^{23}\ \cancel{\text{amu}}} \times \frac{1\ \cancel{\text{kg}}}{1000\ \cancel{\text{g}}} \times \frac{1\ \text{J}}{\frac{1\ \cancel{\text{kg}} \cdot \cancel{\text{m}^2}}{\cancel{\text{s}^2}}} = \mathbf{2.36 \times 10^{-10}\ J}$$

This is the nuclear binding energy. It's the energy required to break up one tungsten-184 nucleus into 74 protons and 110 neutrons.

When comparing the stability of any two nuclei we must account for the fact that they have different numbers of nucleons. For this reason, it is more meaningful to use the *nuclear binding energy per nucleon*, defined as

$$\text{nuclear binding energy per nucleon} = \frac{\text{nuclear binding energy}}{\text{number of nucleons}}$$

For the tungsten-184 nucleus,

$$\text{nuclear binding energy per nucleon} = \frac{2.36 \times 10^{-10}\ \text{J/W}\ \cancel{\text{atom}}}{184\ \text{nucleons/W}\ \cancel{\text{atom}}} = \mathbf{1.28 \times 10^{-12}\ J/nucleon}$$

23.24 Kinetics of Radioactive Decay, Problem Type 4.

Strategy: According to Equation (13.3) of the text, the number of radioactive nuclei at time zero (N_0) and time t (N_t) is

$$\ln\frac{N_t}{N_0} = -\lambda t$$

and the corresponding half-life of the reaction is given by Equation (13.6) of the text:

$$t_{\frac{1}{2}} = \frac{0.693}{\lambda}$$

Using the information given in the problem and the first equation above, we can calculate the rate constant, λ. Then, the half-life can be calculated from the rate constant.

Solution: We can use the following equation to calculate the rate constant λ for each point.

$$\ln\frac{N_t}{N_0} = -\lambda t$$

From day 0 to day 1, we have

$$\ln\frac{389}{500} = -\lambda(1\ \text{d})$$

$$\lambda = 0.251\ \text{d}^{-1}$$

Following the same procedure for the other days,

t (d)	mass (g)	λ (d^{-1})
0	500	
1	389	0.251
2	303	0.250
3	236	0.250
4	184	0.250
5	143	0.250
6	112	0.249

The average value of λ is **0.250 d^{-1}**.

We use the average value of λ to calculate the half-life.

$$t_{\frac{1}{2}} = \frac{0.693}{\lambda} = \frac{0.693}{0.250\ \text{d}^{-1}} = \mathbf{2.77\ d}$$

23.26 Since all radioactive decay processes have first–order rate laws, the decay rate is proportional to the amount of radioisotope at any time. The half-life is given by the following equation:

$$t_{\frac{1}{2}} = \frac{0.693}{\lambda} \qquad (1)$$

There is also an equation that relates the number of nuclei at time zero (N_0) and time t (N_t).

$$\ln\frac{N_t}{N_0} = -\lambda t$$

We can use this equation to solve for the rate constant, λ. Then, we can substitute λ into Equation (1) to calculate the half-life.

The time interval is:

(2:15 p.m., 12/17/92) – (1:00 p.m., 12/3/92) = 14 d + 1 hr + 15 min = 20,235 min

$$\ln\left(\frac{2.6\times 10^{4}\ \text{dis/min}}{9.8\times 10^{5}\ \text{dis/min}}\right) = -\lambda(20{,}235\ \text{min})$$

$$\lambda = 1.8\times 10^{-4}\ \text{min}^{-1}$$

Substitute λ into equation (1) to calculate the half-life.

$$t_{\frac{1}{2}} = \frac{0.693}{\lambda} = \frac{0.693}{1.8\times 10^{-4}\ \text{min}^{-1}} = \mathbf{3.9\times 10^{3}\ min\ or\ 2.7\ d}$$

23.28 The equation for the overall process is:

$$^{232}_{90}\text{Th} \longrightarrow 6\,^{4}_{2}\text{He} + 4\,^{0}_{-1}\beta + \text{X}$$

The final product isotope must be $\mathbf{^{208}_{82}Pb}$.

23.30 Let's consider the decay of A first.

$$\lambda = \frac{0.693}{t_{\frac{1}{2}}} = \frac{0.693}{4.50\ \text{s}} = 0.154\ \text{s}^{-1}$$

Let's convert λ to units of day^{-1}.

$$0.154\frac{1}{\cancel{\text{s}}}\times\frac{3600\ \cancel{\text{s}}}{1\ \cancel{\text{h}}}\times\frac{24\ \cancel{\text{h}}}{1\ \text{d}} = 1.33\times 10^{4}\ \text{d}^{-1}$$

Next, use the first-order rate equation to calculate the amount of A left after 30 days.

$$\ln\frac{[A]_t}{[A]_0} = -\lambda t$$

Let x be the amount of A left after 30 days.

$$\ln\frac{x}{100} = -(1.33 \times 10^4\ d^{-1})(30\ d) = -3.99 \times 10^5$$

$$\frac{x}{100} = e^{(-3.99 \times 10^5)}$$

$$x \approx 0$$

Thus, **no A remains**.

For B: As calculated above, all of A is converted to B in less than 30 days. In fact, essentially all of A is gone in less than 1 day! This means that at the beginning of the 30 day period, there is 1.00 mol of B present. The half life of B is 15 days, so that after two half-lives (30 days), there should be **0.25 mole of B** left.

For C: As in the case of A, the half-life of C is also very short. Therefore, at the end of the 30–day period, **no C is left**.

For D: D is not radioactive. 0.75 mol of B reacted in 30 days; therefore, due to a 1:1 mole ratio between B and D, there should be **0.75 mole of D** present after 30 days.

23.34 **(a)** $^{80}_{34}\text{Se} + ^{2}_{1}\text{H} \longrightarrow \mathbf{^{81}_{34}Se} + ^{1}_{1}\text{p}$

(b) $\mathbf{^{9}_{4}Be} + ^{2}_{1}\text{H} \longrightarrow ^{9}_{3}\text{Li} + 2\,^{1}_{1}\text{p}$

(c) $^{10}_{5}\text{B} + ^{1}_{0}\text{n} \longrightarrow \mathbf{^{7}_{3}Li} + ^{4}_{2}\alpha$

23.36 Upon bombardment with neutrons, mercury–198 is first converted to mercury–199, which then emits a proton. The reaction is:

$$^{198}_{80}\text{Hg} + ^{1}_{0}\text{n} \longrightarrow ^{199}_{80}\text{Hg} \longrightarrow ^{198}_{79}\text{Au} + ^{1}_{1}\text{p}$$

23.48 The fact that the radioisotope appears only in the I_2 shows that the IO_3^- is formed only from the IO_4^-. Does this result rule out the possibility that I_2 could be formed from IO_4^- as well? Can you suggest an experiment to answer the question?

23.50 Add iron-59 to the person's diet, and allow a few days for the iron–59 isotope to be incorporated into the person's body. Isolate red blood cells from a blood sample and monitor radioactivity from the hemoglobin molecules present in the red blood cells.

23.52 Apparently there is a sort of Pauli exclusion principle for nucleons as well as for electrons. When neutrons pair with neutrons and when protons pair with protons, their spins cancel. Even–even nuclei are the only ones with no net spin.

23.54 **(a)** One millicurie represents 3.70×10^7 disintegrations/s. The rate of decay of the isotope is given by the rate law: rate = λN, where N is the number of atoms in the sample. We find the value of λ in units of s^{-1}:

$$\lambda = \frac{0.693}{t_{\frac{1}{2}}} = \frac{0.693}{2.20 \times 10^6 \text{ yr}} \times \frac{1 \text{ yr}}{365 \text{ d}} \times \frac{1 \text{ d}}{24 \text{ h}} \times \frac{1 \text{ h}}{3600 \text{ s}} = 9.99 \times 10^{-15} \text{ s}^{-1}$$

The number of atoms (N) in a 0.500 g sample of neptunium–237 is:

$$0.500 \text{ g} \times \frac{1 \text{ mol}}{237.0 \text{ g}} \times \frac{6.022 \times 10^{23} \text{ atoms}}{1 \text{ mol}} = 1.27 \times 10^{21} \text{ atoms}$$

$$\text{rate of decay} = \lambda N$$

$$= (9.99 \times 10^{-15} \text{ s}^{-1})(1.27 \times 10^{21} \text{ atoms}) = 1.27 \times 10^7 \text{ atoms/s}$$

We can also say that:

$$\text{rate of decay} = 1.27 \times 10^7 \text{ disintegrations/s}$$

The activity in millicuries is:

$$(1.27 \times 10^7 \text{ disintegrations/s}) \times \frac{1 \text{ millicurie}}{3.70 \times 10^7 \text{ disintegrations/s}} = \mathbf{0.343\ millicuries}$$

(b) The decay equation is:

$$^{237}_{93}\text{Np} \longrightarrow {}^{4}_{2}\alpha + {}^{233}_{91}\text{Pa}$$

23.56 We use the same procedure as in Problem 23.20.

	Isotope	Atomic Mass (amu)	Nuclear Binding Energy (J/nucleon)
(a)	^{10}B	10.0129	$\mathbf{1.040 \times 10^{-12}}$
(b)	^{11}B	11.00931	$\mathbf{1.111 \times 10^{-12}}$
(c)	^{14}N	14.00307	$\mathbf{1.199 \times 10^{-12}}$
(d)	^{56}Fe	55.9349	$\mathbf{1.410 \times 10^{-12}}$

23.58 When an isotope is above the belt of stability, the neutron/proton ratio is too high. The only mechanism to correct this situation is beta emission; the process turns a neutron into a proton. Direct neutron emission does not occur.

$$\mathbf{^{18}_{7}N \longrightarrow {}^{18}_{8}O + {}^{0}_{-1}\beta}$$

Oxygen–18 is a stable isotope.

23.60 The age of the fossil can be determined by radioactively dating the age of the deposit that contains the fossil.

23.62 **(a)** $^{209}_{83}\text{Bi} + {}^{4}_{2}\alpha \longrightarrow {}^{211}_{85}\text{At} + 2\,{}^{1}_{0}\text{n}$

(b) $^{209}_{83}\text{Bi}(\alpha, 2\text{n}){}^{211}_{85}\text{At}$

23.64 Because of the relative masses, the force of gravity on the sun is much greater than it is on Earth. Thus the nuclear particles on the sun are already held much closer together than the equivalent nuclear particles on the earth. Less energy (lower temperature) is required on the sun to force fusion collisions between the nuclear particles.

23.66 *Step 1:* The half-life of carbon-14 is 5730 years. From the half-life, we can calculate the rate constant, λ.

$$\lambda = \frac{0.693}{t_{\frac{1}{2}}} = \frac{0.693}{5730\ \text{yr}} = 1.21 \times 10^{-4}\ \text{yr}^{-1}$$

Step 2: The age of the object can now be calculated using the following equation.

$$\ln\frac{N_t}{N_0} = -\lambda t$$

N = the number of radioactive nuclei. In the problem, we are given disintegrations per second per gram. The number of disintegrations is directly proportional to the number of radioactive nuclei. We can write,

$$\ln\frac{\text{decay rate of old sample}}{\text{decay rate of fresh sample}} = -\lambda t$$

$$\ln\frac{0.186\ \text{dps/g C}}{0.260\ \text{dps/g C}} = -(1.21 \times 10^{-4}\ \text{yr}^{-1})t$$

$$\mathbf{t = 2.77 \times 10^3\ yr}$$

23.68 **(a)** The balanced equation is:

$$^{40}_{19}\text{K} \longrightarrow {}^{40}_{18}\text{Ar} + {}^{0}_{+1}\beta$$

(b) First, calculate the rate constant λ.

$$\lambda = \frac{0.693}{t_{\frac{1}{2}}} = \frac{0.693}{1.2 \times 10^9\ \text{yr}} = 5.8 \times 10^{-10}\ \text{yr}^{-1}$$

Then, calculate the age of the rock by substituting λ into the following equation. ($N_t = 0.18N_0$)

$$\ln\frac{N_t}{N_0} = -\lambda t$$

$$\ln\frac{0.18}{1.00} = -(5.8 \times 10^{-10}\ \text{yr}^{-1})t$$

$$\mathbf{t = 3.0 \times 10^9\ yr}$$

23.70 **(a)** In the ^{90}Sr decay, the mass defect is:

$$\Delta m = (\text{mass } ^{90}\text{Y} + \text{mass e}^-) - \text{mass } ^{90}\text{Sr}$$

$$= [(89.907152\ \text{amu} + 5.4857 \times 10^{-4}\ \text{amu}) - 89.907738\ \text{amu}] = -3.743 \times 10^{-5}\ \text{amu}$$

$$= (-3.743 \times 10^{-5}\ \text{amu}) \times \frac{1\ \text{g}}{6.022 \times 10^{23}\ \text{amu}} = -6.216 \times 10^{-29}\ \text{g} = -6.216 \times 10^{-32}\ \text{kg}$$

The energy change is given by:

$$\Delta E = (\Delta m)c^2$$

$$= (-6.126 \times 10^{-32} \text{ kg})(3.00 \times 10^8 \text{ m/s})^2$$

$$= -5.59 \times 10^{-15} \text{ kg m}^2/\text{s}^2 = -5.59 \times 10^{-15} \text{ J}$$

Similarly, for the ^{90}Y decay, we have

$$\Delta m = (\text{mass } ^{90}\text{Zr} + \text{mass e}^-) - \text{mass } ^{90}\text{Y}$$

$$= [(89.904703 \text{ amu} + 5.4857 \times 10^{-4} \text{ amu}) - 89.907152 \text{ amu}] = -1.900 \times 10^{-3} \text{ amu}$$

$$= (-1.900 \times 10^{-3} \text{ amu}) \times \frac{1 \text{ g}}{6.022 \times 10^{23} \text{ amu}} = -3.156 \times 10^{-27} \text{ g} = -3.156 \times 10^{-30} \text{ kg}$$

and the energy change is:

$$\Delta E = (-3.156 \times 10^{-30} \text{ kg})(3.00 \times 10^8 \text{ m/s})^2 = -2.84 \times 10^{-13} \text{ J}$$

The energy released in the above two decays is **5.59×10^{-15} J** and **2.84×10^{-13} J**. The total amount of energy released is:

$$(5.59 \times 10^{-15} \text{ J}) + (2.84 \times 10^{-13} \text{ J}) = 2.90 \times 10^{-13} \text{ J.}$$

(b) This calculation requires that we know the rate constant for the decay. From the half-life, we can calculate λ.

$$\lambda = \frac{0.693}{t_{\frac{1}{2}}} = \frac{0.693}{28.1 \text{ yr}} = 0.0247 \text{ yr}^{-1}$$

To calculate the number of moles of ^{90}Sr decaying in a year, we apply the following equation:

$$\ln\frac{N_t}{N_0} = -\lambda t$$

$$\ln\frac{x}{1.00} = -(0.0247 \text{ yr}^{-1})(1.00 \text{ yr})$$

where x is the number of moles of ^{90}Sr nuclei left over. Solving, we obtain:

$$x = 0.9756 \text{ mol } ^{90}\text{Sr}$$

Thus the number of moles of nuclei which decay in a year is

$$(1.00 - 0.9756) \text{ mol} = 0.0244 \text{ mol} = \mathbf{0.024 \text{ mol}}$$

This is a reasonable number since it takes 28.1 years for 0.5 mole of ^{90}Sr to decay.

(c) Since the half–life of ^{90}Y is much shorter than that of ^{90}Sr, we can safely assume that *all* the ^{90}Y formed from ^{90}Sr will be converted to ^{90}Zr. The energy changes calculated in part (a) refer to the decay of individual nuclei. In 0.024 mole, the number of nuclei that have decayed is:

$$0.0244 \text{ mol} \times \frac{6.022 \times 10^{23} \text{ nuclei}}{1 \text{ mol}} = 1.47 \times 10^{22} \text{ nuclei}$$

Realize that there are two decay processes occurring, so we need to add the energy released for each process calculated in part (a). Thus, the heat released from 1 mole of ^{90}Sr waste in a year is given by:

$$\textbf{heat released} = (1.47 \times 10^{22}\ \text{nuclei}) \times \frac{2.90 \times 10^{-13}\ \text{J}}{1\ \text{nucleus}} = \mathbf{4.26 \times 10^{9}\ J} = \mathbf{4.26 \times 10^{6}\ kJ}$$

This amount is roughly equivalent to the heat generated by burning 50 tons of coal! Although the heat is released slowly during the course of a year, effective ways must be devised to prevent heat damage to the storage containers and subsequent leakage of radioactive material to the surroundings.

23.72 First, let's calculate the number of disintegrations/s to which 7.4 mC corresponds.

$$7.4\ \text{mC} \times \frac{1\ \text{Ci}}{1000\ \text{mC}} \times \frac{3.7 \times 10^{10}\ \text{disintegrations/s}}{1\ \text{Ci}} = 2.7 \times 10^{8}\ \text{disintegrations/s}$$

This is the rate of decay. We can now calculate the number of iodine-131 atoms to which this radioactivity corresponds. First, we calculate the half-life in seconds:

$$t_{\frac{1}{2}} = 8.1\ \text{d} \times \frac{24\ \text{h}}{1\ \text{d}} \times \frac{3600\ \text{s}}{1\ \text{h}} = 7.0 \times 10^{5}\ \text{s}$$

$$\lambda = \frac{0.693}{t_{\frac{1}{2}}} \qquad \text{Therefore, } \lambda = \frac{0.693}{7.0 \times 10^{5}\ \text{s}} = 9.9 \times 10^{-7}\ \text{s}^{-1}$$

$$\text{rate} = \lambda N$$

$$2.7 \times 10^{8}\ \text{disintegrations/s} = (9.9 \times 10^{-7}\ \text{s}^{-1})N$$

$$N = \mathbf{2.7 \times 10^{14}\ iodine\text{-}131\ atoms}$$

23.74 One curie represents 3.70×10^{10} disintegrations/s. The rate of decay of the isotope is given by the rate law: rate = λN, where N is the number of atoms in the sample and λ is the first-order rate constant. We find the value of λ in units of s^{-1}:

$$\lambda = \frac{0.693}{t_{\frac{1}{2}}} = \frac{0.693}{1.6 \times 10^{3}\ \text{yr}} = 4.3 \times 10^{-4}\ \text{yr}^{-1}$$

$$\frac{4.3 \times 10^{-4}}{1\ \text{yr}} \times \frac{1\ \text{yr}}{365\ \text{d}} \times \frac{1\ \text{d}}{24\ \text{h}} \times \frac{1\ \text{h}}{3600\ \text{s}} = 1.4 \times 10^{-11}\ \text{s}^{-1}$$

Now, we can calculate N, the number of Ra atoms in the sample.

$$\text{rate} = \lambda N$$

$$3.7 \times 10^{10}\ \text{disintegrations/s} = (1.4 \times 10^{-11}\ \text{s}^{-1})N$$

$$N = 2.6 \times 10^{21}\ \text{Ra atoms}$$

By definition, 1 curie corresponds to exactly 3.7×10^{10} nuclear disintegrations per second which is the decay rate equivalent to that of *1 g of radium*. Thus, the mass of 2.6×10^{21} Ra atoms is 1 g.

$$\frac{2.6 \times 10^{21}\ \text{Ra atoms}}{1.0\ \text{g Ra}} \times \frac{226.03\ \text{g Ra}}{1\ \text{mol Ra}} = \mathbf{5.9 \times 10^{23}\ atoms/mol} = \boldsymbol{N_A}$$

23.76 All except gravitational have a nuclear origin.

23.78 U–238, $t_{\frac{1}{2}} = 4.5 \times 10^9$ yr and Th–232, $t_{\frac{1}{2}} = 1.4 \times 10^{10}$ yr.

They are still present because of their long half lives.

23.80 $E = \dfrac{hc}{\lambda}$

$$\lambda = \frac{hc}{E} = \frac{(3.00 \times 10^8 \text{ m/s})(6.63 \times 10^{-34} \text{ J}\cdot\text{s})}{2.4 \times 10^{-13} \text{ J}} = 8.3 \times 10^{-13} \text{ m} = \mathbf{8.3 \times 10^{-4}\ nm}$$

This wavelength is clearly in the γ-ray region of the electromagnetic spectrum.

23.82 Only ^{3}**H** has a suitable half-life. The other half-lives are either too long or too short to accurately determine the time span of 6 years.

23.84 Obviously, a small scale chain reaction took place. Copper played the crucial role of reflecting neutrons from the splitting uranium-235 atoms back into the uranium sphere to trigger the chain reaction. Note that a sphere has the most appropriate geometry for such a chain reaction. In fact, during the implosion process prior to an atomic explosion, fragments of uranium-235 are pressed roughly into a sphere for the chain reaction to occur (see Section 23.5 of the text).

23.86 In this problem, we are asked to calculate the molar mass of a radioactive isotope. Grams of sample are given in the problem, so if we can find moles of sample we can calculate the molar mass. The rate constant can be calculated from the half-life. Then, from the rate of decay and the rate constant, the number of radioactive nuclei can be calculated. The number of radioactive nuclei can be converted to moles.

First, we convert the half-life to units of minutes because the rate is given in dpm (disintegrations per minute). Then, we calculate the rate constant from the half-life.

$$(1.3 \times 10^9 \text{ yr}) \times \frac{365 \text{ days}}{1 \text{ yr}} \times \frac{24 \text{ h}}{1 \text{ day}} \times \frac{60 \text{ min}}{1 \text{ h}} = 6.8 \times 10^{14} \text{ min}$$

$$\lambda = \frac{0.693}{t_{\frac{1}{2}}} = \frac{0.693}{6.8 \times 10^{14} \text{ min}} = 1.0 \times 10^{-15} \text{ min}^{-1}$$

Next, we calculate the number of radioactive nuclei from the rate and the rate constant.

$$\text{rate} = \lambda N$$

$$2.9 \times 10^4 \text{ dpm} = (1.0 \times 10^{-15} \text{ min}^{-1})N$$

$$N = 2.9 \times 10^{19} \text{ nuclei}$$

Convert to moles of nuclei, and then determine the molar mass.

$$(2.9 \times 10^{19} \text{ nuclei}) \times \frac{1 \text{ mol}}{6.022 \times 10^{23} \text{ nuclei}} = 4.8 \times 10^{-5} \text{ mol}$$

$$\textbf{molar mass} = \frac{\text{g of substance}}{\text{mol of substance}} = \frac{0.0100 \text{ g}}{4.8 \times 10^{-5} \text{ mol}} = \mathbf{2.1 \times 10^2\ g/mol}$$

23.88 **(a)** $^{238}_{94}\text{Pu} \rightarrow\, ^{4}_{2}\text{He} +\, ^{234}_{92}\text{U}$

(b) At $t = 0$, the number of ^{238}Pu atoms is

$$(1.0 \times 10^{-3}\ \text{g}) \times \frac{1\ \text{mol}}{238\ \text{g}} \times \frac{6.022 \times 10^{23}\ \text{atoms}}{1\ \text{mol}} = 2.53 \times 10^{18}\ \text{atoms}$$

The decay rate constant, λ, is

$$\lambda = \frac{0.693}{t_{\frac{1}{2}}} = \frac{0.693}{86\ \text{yr}} = 0.00806\frac{1}{\text{yr}} \times \frac{1\ \text{yr}}{365\ \text{d}} \times \frac{1\ \text{d}}{24\ \text{h}} \times \frac{1\ \text{h}}{3600\ \text{s}} = 2.56 \times 10^{-10}\ \text{s}^{-1}$$

$$\text{rate} = \lambda N_0 = (2.56 \times 10^{-10}\ \text{s}^{-1})(2.53 \times 10^{18}\ \text{atoms}) = 6.48 \times 10^{8}\ \text{decays/s}$$

$$\text{Power} = (\text{decays/s}) \times (\text{energy/decay})$$

$$\text{Power} = (6.48 \times 10^{8}\ \text{decays/s})(9.0 \times 10^{-13}\ \text{J/decay}) = 5.8 \times 10^{-4}\ \text{J/s} = \mathbf{5.8 \times 10^{-4}\ W} = \mathbf{0.58\ mW}$$

At $t = 10$ yr,

$$\text{Power} = (0.58\ \text{mW})(0.92) = \mathbf{0.53\ mW}$$

23.90 1 Ci = 3.7×10^{10} decays/s

Let R_0 be the activity of the injected 20.0 mCi ^{99m}Tc.

$$R_0 = (20.0 \times 10^{-3}\ \text{Ci}) \times \frac{3.70 \times 10^{10}\ \text{decays/s}}{1\ \text{Ci}} = 7.4 \times 10^{8}\ \text{decays/s}$$

$R_0 = \lambda N_0$, where N_0 = number of ^{99m}Tc nuclei present.

$$\lambda = \frac{0.693}{t_{\frac{1}{2}}} = \frac{0.693}{6.0\ \text{h}} = 0.1155\frac{1}{\text{h}} \times \frac{1\ \text{h}}{3600\ \text{s}} = 3.208 \times 10^{-5}\ \text{s}^{-1}$$

$$N_0 = \frac{R_0}{\lambda} = \frac{7.4 \times 10^{8}\ \text{decays/s}}{3.208 \times 10^{-5}\ /\text{s}} = 2.307 \times 10^{13}\ \text{decays} = 2.307 \times 10^{13}\ \text{nuclei}$$

Each of the nuclei emits a photon of energy 2.29×10^{-14} J. The total energy absorbed by the patient is

$$\boldsymbol{E} = \frac{2}{3}(2.307 \times 10^{13}\ \text{nuclei}) \times \left(\frac{2.29 \times 10^{-14}\ \text{J}}{1\ \text{nuclei}}\right) = \mathbf{0.352\ J}$$

The rad is:

$$\frac{0.352\ \text{J}/10^{-2}\ \text{J}}{70} = \mathbf{0.503\ rad}$$

Given that RBE = 0.98, the rem is:

$$(0.503)(0.98) = \mathbf{0.49\ rem}$$

23.92 The ignition of a fission bomb requires an ample supply of neutrons. In addition to the normal neutron source placed in the bomb, the high temperature attained during the chain reaction causes a small scale nuclear fusion between deuterium and tritium.

$$^{2}_{1}H + ^{3}_{1}H \rightarrow ^{4}_{2}He + ^{1}_{0}n$$

The additional neutrons produced will enhance the efficiency of the chain reaction and result in a more powerful bomb.

CHAPTER 24
ORGANIC CHEMISTRY

PROBLEM-SOLVING STRATEGIES AND TUTORIAL SOLUTIONS

TYPES OF PROBLEMS

Problem Type 1: Determining the Number of Structural Isomers.

Problem Type 2: Organic Nomenclature.
- **(a)** Alkanes.
- **(b)** Alkenes.
- **(c)** Alkynes.
- **(d)** Aromatic Compounds.

Problem Type 3: Addition Reactions.

Problem Type 4: Distinguishing between Structural and Geometric Isomers.

Problem Type 5: Functional Groups.

Problem Type 6: Chirality.

PROBLEM TYPE 1: DETERMINING THE NUMBER OF STRUCTURAL ISOMERS

Structural isomers are molecules that have the same molecular formula but a different order of linking the atoms. For small hydrocarbon molecules (eight or fewer carbon atoms), it is relatively easy to determine the number of structural isomers by trial and error.

EXAMPLE 24.1

How many structural isomers can be identified for hexane, C_6H_{14}?

Strategy: For small hydrocarbon molecules (eight or fewer carbons), it is relatively easy to determine the number of structural isomers by trial and error.

Solution: Start by writing the straight-chain structure.

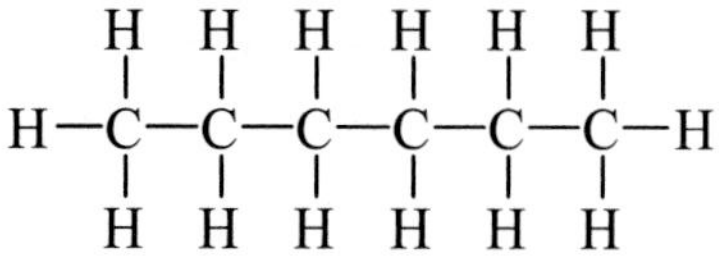

```
    H  H  H  H  H  H
    |  |  |  |  |  |
H—C—C—C—C—C—C—H
    |  |  |  |  |  |
    H  H  H  H  H  H
```

n-hexane

By necessity, the other structures must have branched chains. First, try single methyl substituents.

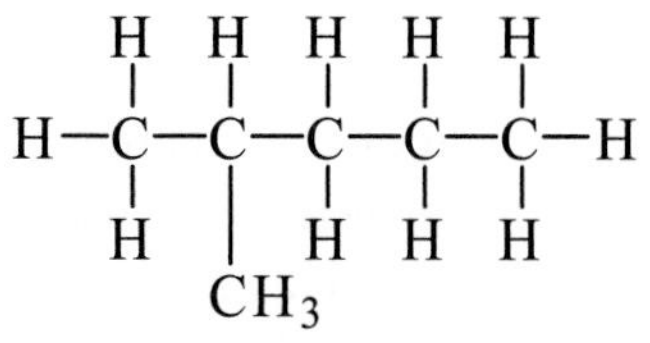

```
    H  H  H  H  H
    |  |  |  |  |
H—C—C—C—C—C—H
    |  |  |  |  |
    H  |  H  H  H
      CH3
```

2-methylpentane

```
    H  H  H  H  H
    |  |  |  |  |
H—C—C—C—C—C—H
    |  |  |  |  |
    H  H  |  H  H
         CH3
```

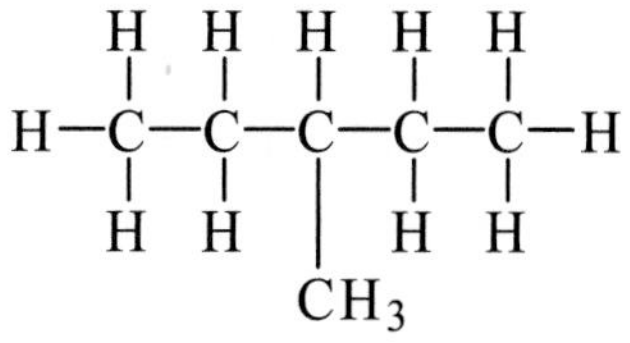

3-methylpentane

Then, try structures that have two methyl groups.

```
        CH3
  H      |   H   H
  |      |   |   |
H—C——C——C——C—H
  |      |   |   |
  H      |   H   H
        CH3
2,2-dimethylbutane
```

```
              CH3
  H   H        |   H
  |   |        |   |
H—C——C——C——C—H
  |   |        |   |
  H   |        H   H
     CH3
2,3-dimethylbutane
```

Hexane has five structural isomers, in which the numbers of carbon and hydrogen atoms remain unchanged despite the differences in structure.

PRACTICE EXERCISE

1. How many structural isomers are there of $C_4H_{10}O$? **Hint:** Consider both alcohols and ethers.

Text Problems: 24.12, 24.14

PROBLEM TYPE 2: ORGANIC NOMENCLATURE

The first step in learning the nomenclature of hydrocarbons is to know the names of the first ten straight-chain alkanes. Except for the first four members, the number of carbon atoms in each alkane is identified by the Greek prefix. See Table 24.1 below.

TABLE 24.1
The First Ten Straight-Chain Alkanes

Name of hydrocarbon	Molecular formula
Methane	CH_4
Ethane	$CH_3 - CH_3$
Propane	$CH_3 - CH_2 - CH_3$
Butane	$CH_3 - (CH_2)_2 - CH_3$
Pentane	$CH_3 - (CH_2)_3 - CH_3$
Hexane	$CH_3 - (CH_2)_4 - CH_3$
Heptane	$CH_3 - (CH_2)_5 - CH_3$
Octane	$CH_3 - (CH_2)_6 - CH_3$
Nonane	$CH_3 - (CH_2)_7 - CH_3$
Decane	$CH_3 - (CH_2)_8 - CH_3$

Next, you need to learn the names of substituents (other than hydrogen) attached to hydrocarbon chains or aromatic compounds. For example, when a hydrogen atom is removed from methane, a $-CH_3$ fragment is left, which is called a *methyl* group. Similarly, removing a hydrogen atom from an ethane molecule gives an *ethyl* group, $-CH_2CH_3$.

These groups or substituents that are derived from alkanes are called *alkyl* groups. See Table 24.2 of the text for other common alkyl groups.

A. Alkanes

Alkanes have the general formula C_nH_{2n+2}, where $n = 1, 2, \ldots$. Alkanes contain only single covalent bonds. The bonds are said to be saturated because no more hydrogen atoms can be added to the carbon atoms. Thus, alkanes are also called *saturated hydrocarbons*.

There are other rules to follow when naming hydrocarbons.

1. Find the longest carbon chain in the molecule. The parent name of the compound is based on the longest carbon chain. For example, if the longest carbon chain contains five C atoms, the parent name of the compound is *pentane*. See Table 24.1.
2. Next, you must specify the name and location of any groups attached to the longest carbon chain. See Table 24.2 of the text for naming *alkyl* groups. To specify the location of the group or groups, start numbering the longest chain from the end that is closer to the carbon atom bearing the substituent group.
3. If there is more than one of a particular group attached to the longest carbon chain, you must specify the number of groups with a prefix. The prefixes are di– (2 groups), tri– (3 groups), tetra– (4 groups), and so on.
4. There can be many different types of substituents other than alkyl groups. Table 24.3 of the text lists the names of some common functional groups.

EXAMPLE 24.2
Give the correct name for the following structure:

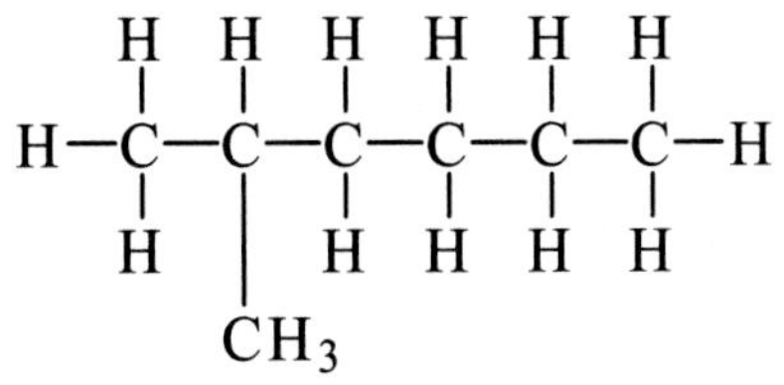

```
    H   H   H   H   H   H
    |   |   |   |   |   |
H — C — C — C — C — C — C — H
    |   |   |   |   |   |
    H   |   H   H   H   H
        |
       CH3
```

Strategy: We follow the IUPAC rules and use the information in Table 24.2 of the text to name the compound. How many C atoms are there in the longest chain?

Solution: Find the longest carbon chain. The longest chain has *six* carbons. Therefore, the parent name of the compound is *hexane*.

Next, specify the name and location of any groups attached to the longest carbon chain. You should see that there is a methyl group attached to the second carbon from the left of the chain. You should number the carbon chain from the end that is closer to the carbon atom bearing the substituent group. If you start numbering from the left, the methyl group is on the second carbon in the chain. However, if you start numbering from the right, the methyl group is on the fifth carbon in the chain. Numbering from the left is correct.

The correct name for this compound is **2-methylhexane**.

> **Tip:** You should always put a dash (–) between a number and a "word" when naming an organic compound.

EXAMPLE 24.3

Give the correct name for the following structure:

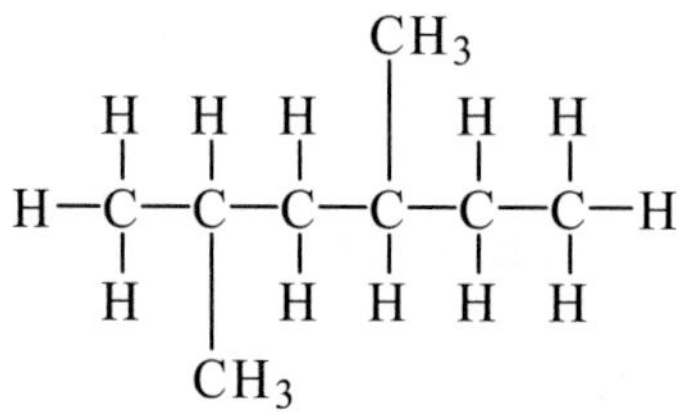

```
                CH3
   H   H   H    |    H   H
   |   |   |    |    |   |
H—C—C—C—C—C—C—H
   |   |   |    |    |   |
   H   |   H    H    H   H
       |
      CH3
```

Strategy: We follow the IUPAC rules and use the information in Table 24.2 of the text to name the compound. How many C atoms are there in the longest chain?

Solution: Find the longest carbon chain. You should find that the longest carbon chain has six carbons. The parent name is *hexane.*

Numbering from the left, there are methyl groups on the second and fourth carbons. Since there are two methyl groups, we must specify this by using the prefix di–.

The correct name for the compound is **2,4-dimethylhexane**.

Comment: Let's examine the name of the compound if we had numbered the longest chain from right to left. The methyl groups would be on the third and fifth carbons. Hence, the name would be 3,5-dimethylhexane, which is incorrect. The correct name should always have the lowest numbering scheme as possible.

> **Tip:** Always use commas to separate numbers when naming organic compounds.

EXAMPLE 24.4

Give the correct name for each of the following structures:

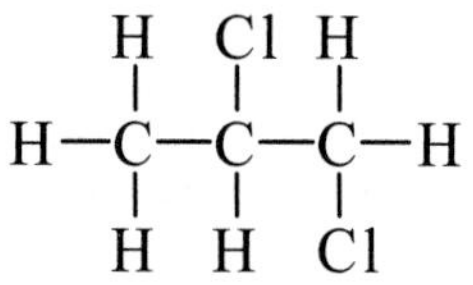

```
(a)                          (b)
   H   Cl  H                      H
   |   |   |                      |
H—C—C—C—H                  H—C—NO2
   |   |   |                      |
   H   H   Cl                     H
```

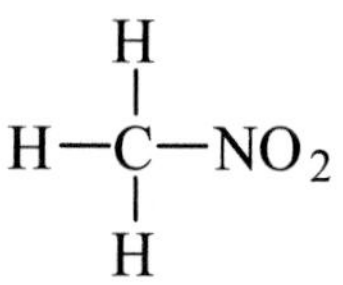

Strategy: We follow the IUPAC rules and use the information in Table 24.2 of the text to name the compound. How many C atoms are there in the longest chain? See Table 24.3 of the text for the names of common substituent groups.

Solution:

(a) You should find that the longest carbon chain in the molecule is *three* carbons. The parent name is *propane.*

Referring to Table 24.3 of the text, you should find that a $-Cl$ group is called a chloro group. There are chloro groups on the 1 and 2 carbons, numbering from right to left. (Why not number from left to right?)

Hence, the correct name for the compound is **1,2-dichloropropane**.

(b) There is only one carbon in the molecule, so the parent name is *methane.*

Referring to Table 24.3 of the text, you should find that a $-NO_2$ group is called a nitro group. The correct name for the compound is **nitromethane**.

Why isn't the correct name 1-nitromethane?

PRACTICE EXERCISE

2. Give the systematic name for each of the following structural formulas:

(a) $CH_3(CH_2)_7CH_3$

```
                CH3·CHCH3
                    |
(b) CH3CH2CH2CHCHCH2CH2CH3
                      |
                     CH2CH3
```

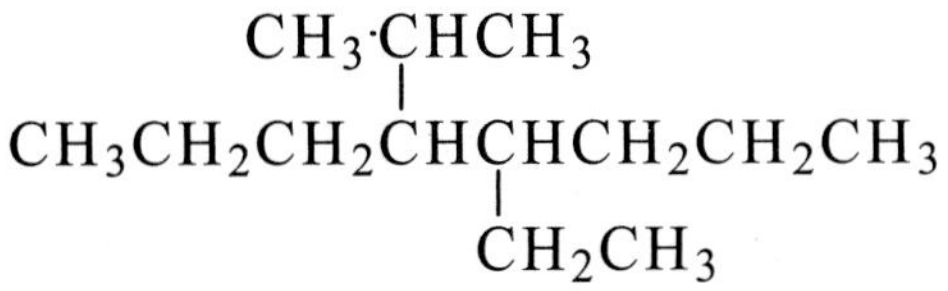

(c) $BrCH_2CH_2CHBrCH_2Br$

Text Problem: 24.26

B. Alkenes

Alkenes are molecules that contain at least one carbon-carbon double bond. Alkenes have the general formula C_nH_{2n}, where $n = 2, 3, \ldots$.

To name alkenes, you follow the same rules as outlined for alkanes; except, you must specify the position(s) of the carbon-carbon double bond(s), and the name of alkenes ends with an "–ene" suffix.

EXAMPLE 24.5

Give the correct name for the following structure:

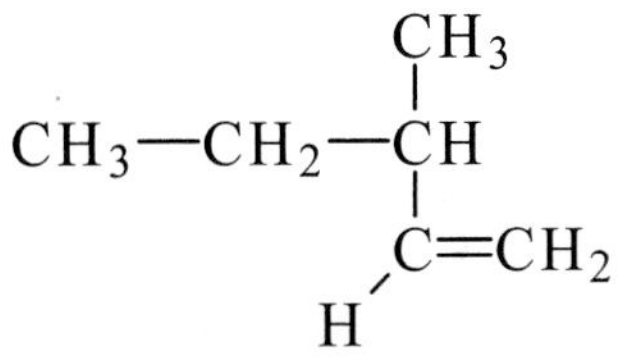

```
                 CH3
                 |
CH3—CH2—CH
                 |
                 C=CH2
               /
              H
```

Strategy: We follow the IUPAC rules and use the information in Table 24.2 of the text to name the compound. How many C atoms are there in the longest chain? For alkenes, the suffix is "–ene".

Solution: The longest carbon chain in the molecule is five carbons. The parent name is pent*ene*, because the molecule has a double bond.

For this molecule, we number from right to left, because we want the double bond to be at the lowest number possible. In this case, the double bond starts on the first carbon if we number from right to left. There is also a methyl substituent on the third carbon.

The correct name for this compound is **3-methyl-1-pentene**. The number 1 before pentene specifies that the double bond starts on the first carbon.

PRACTICE EXERCISE

3. Draw the structural formula for 4-methyl-2-hexene.

Text Problems: 24.26, 24.28

C. Alkynes

Alkynes contain at least one carbon-carbon triple bond. They have the general formula C_nH_{2n-2}, where $n = 2, 3, \ldots$.

To name alkynes, you follow the same rules as outlined above for alkanes; except, you must specify the position(s) of the carbon-carbon triple bond(s), and the name of alkynes ends with an "–yne" suffix.

EXAMPLE 24.6
Give the correct name for the following structure:

$$\begin{array}{l} CH_3{-}CH_2{-}C{\equiv}C{-}CH_2{-}\underset{\displaystyle \begin{array}{c}|\\ CH_2\\ |\\ CH_2\\ |\\ CH_3\end{array}}{CH}{-}CH_3 \end{array}$$

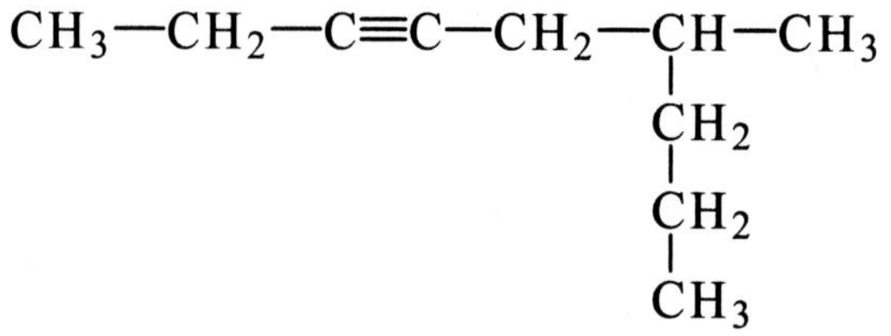

Strategy: We follow the IUPAC rules and use the information in Table 24.2 of the text to name the compound. How many C atoms are there in the longest chain? For alkynes the suffix is "–yne".

Solution: The longest carbon chain in the molecule contains nine carbons. The parent name is non*yne* because there is a triple bond.

You should number the carbon chain from left to right, placing the triple bond on the third carbon. If you numbered from right to left, the triple bond would be on the sixth carbon. You should also notice that there is a methyl substituent on the sixth carbon.

The correct name for this molecule is **6-methyl-3-nonyne.**

PRACTICE EXERCISE

4. Give an acceptable name for each of the following structures:

   (a) $(CH_3)_3CC{\equiv}CCH_2CH_3$

   (b) 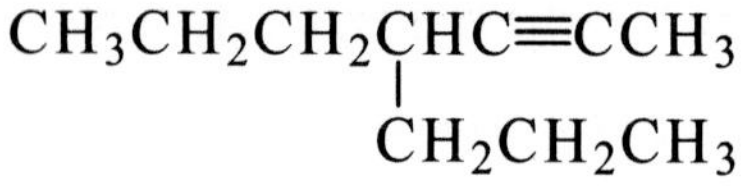 $CH_3CH_2CH_2\underset{\displaystyle \begin{array}{c}|\\ CH_2CH_2CH_3\end{array}}{C}HC{\equiv}CCH_3$

Text Problems: 24.26, 24.28

D. Aromatic Compounds

Benzene is the parent compound of this large family of organic substances. In naming aromatic compounds, we will consider both mono- and di-substituted benzene rings.

The naming of mono-substituted benzene rings (benzenes in which one H atom has been replaced by another atom or groups of atoms) is straightforward. You simply name the substituent followed by the name "benzene".

EXAMPLE 24.7

Give the correct name for each of the following structures:

Strategy: If a benzene ring contains only one substituent, simply name the substituent followed by "benzene".

Solution:

(a) The substituent is a methyl group, so the correct name is **methylbenzene**. However, there are older names that are still in common use for many compounds. Methylbenzene is usually called **toluene**.

(b) The substituent is a bromo group, so the correct name is **bromobenzene**.

For di-substituted benzenes, we must indicate the location of the second group relative to the first. For example, three different dichlorobenzenes are possible.

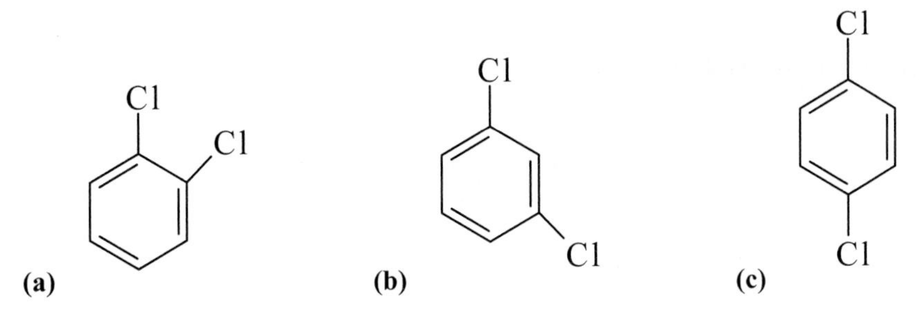

The systematic way to name these molecules is to number the carbon atoms of the benzene ring as follows:

1
6 2
5 3
4

Thus, (a) would be named 1,2-dichlorobenzene, (b) 1,3-dichlorobenzene, and (c) 1,4-dichlorobenzene. However, the prefixes *o*- (*ortho*-), *m*- (*meta*), and *p*- (*para*-) are used more often to denote the relative positions of the two substituted groups. *Ortho*- designates 1,2 substituents, *meta*- designates 1,3 substituents, and *para*- designates 1,4 substituents. Thus, (a) is named *o*-dichlorobenzene, (b) *m*-dichlorobenzene, and (c) *p*-dichlorobenzene.

In compounds with two different substituted groups, the positions of the substituents can be specified with numbers or with the *o*-, *m*-, or *p*- prefixes.

PRACTICE EXERCISE

5. Draw structural formulas for:

(a) ethylbenzene (b) *m*-bromochlorobenzene (c) *p*-nitrotoluene

Text Problems: 24.26, **24.32**

Finally in some molecules, benzene is named as a substituent. A benzene molecule minus a hydrogen atom (C_6H_5) is called a *phenyl* group.

EXAMPLE 24.8
Give the correct name for the following structure:

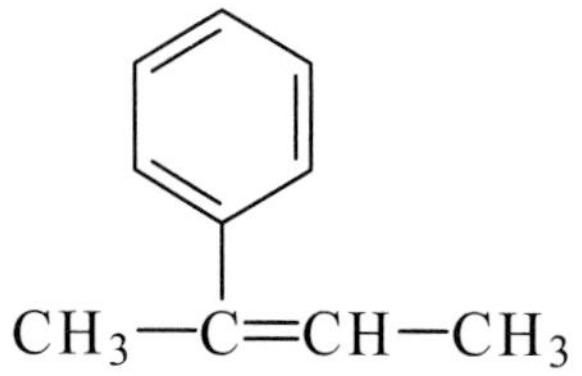

$$CH_3—\underset{}{C}(C_6H_5)=CH—CH_3$$

Strategy: For this molecule, it would be difficult to name the carbon chain as a substituent on the benzene ring. Therefore, we name the benzene ring as a substituent on the carbon chain.

Solution: The longest carbon chain contains four carbons. The parent name is but*ene* because the molecule contains a double bond.

You should number the carbon chain from left to right so that both the double bond and the phenyl group are on the second carbon.

The correct name for the molecule is **2-phenyl-2-butene**.

PROBLEM TYPE 3: ADDITION REACTIONS

Alkenes, alkynes, and aromatic compounds are called *unsaturated hydrocarbons*, compounds with double or triple carbon-carbon bonds. Unsaturated hydrocarbons commonly undergo **addition reactions** in which one molecule adds to another to form a single product. An example of an addition reaction is *hydrogenation*, which is the addition of molecular hydrogen to compounds containing C=C and C≡C bonds.

$$CH_2{=}CH_2 + H_2 \longrightarrow CH_3—CH_3$$

$$CH{\equiv}CH + 2\,H_2 \longrightarrow CH_3—CH_3$$

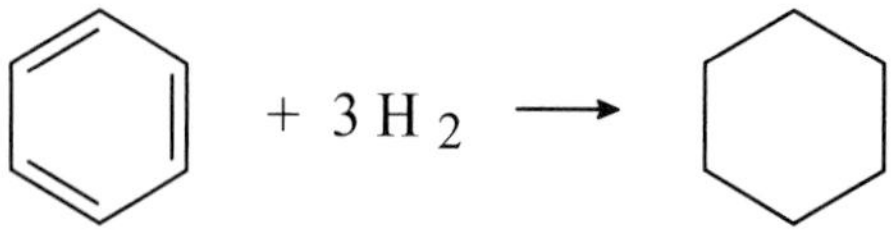

$$C_6H_6 \text{ (benzene)} + 3\,H_2 \longrightarrow C_6H_{12} \text{ (cyclohexane)}$$

Alkanes are called *saturated hydrocarbons* because no more hydrogen can be added to the carbon atoms. The carbon atoms in a saturated hydrocarbon are already bonded to the maximum number of H atoms.

$$CH_3—CH_3 + H_2 \longrightarrow \text{no reaction}$$

Other addition reactions involve an addition of HX or X_2 to the multiple bond, where X represents a halogen (Cl, Br, or I). Examples of these addition reactions are:

$$CH_2{=}CH_2 + HCl \longrightarrow CH_3—CH_2Cl$$

$$CH_2{=}CH_2 + Cl_2 \longrightarrow CH_2Cl—CH_2Cl$$

PROBLEM TYPE 4: DISTINGUISHING BETWEEN STRUCTURAL AND GEOMETRIC ISOMERS

Recall that **structural isomers** are molecules that have the same molecular formula but a different order of linking the atoms. **Geometric isomers** are molecules with the same type, number, and order of attachment of atoms and the same chemical bonds, but different spatial arrangements of the atoms. For example, 1,2-dichloroethene exists as two geometric isomers.

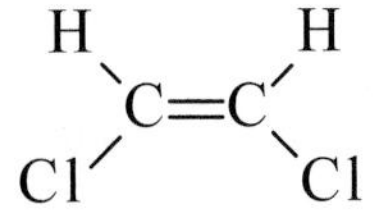

```
 H       H              Cl       H
  \     /                 \     /
   C=C                     C=C
  /     \                 /     \
Cl       Cl              H       Cl
```

cis-1,2-dichloroethene 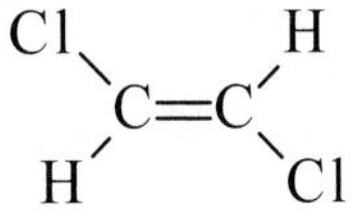   *trans*-1,2-dichloroethene

Note that in the *cis* isomer, the two Cl atoms (and the two H atoms) are adjacent to each other, whereas in the *trans* isomer, the Cl atoms are on the opposite side of the C=C bond.

EXAMPLE 24.9

Draw Lewis structures for all compounds, not including cyclic compounds, with the molecular formula C_5H_{10} and determine which are geometric isomers.

Strategy: Alkenes have the general formula C_nH_{2n}. Thus the structures with molecular formula C_5H_{10} should be alkenes.

Solution: To draw the correct Lewis structures, follow the procedure outlined in Chapter 9. We can draw six Lewis structures as follows:

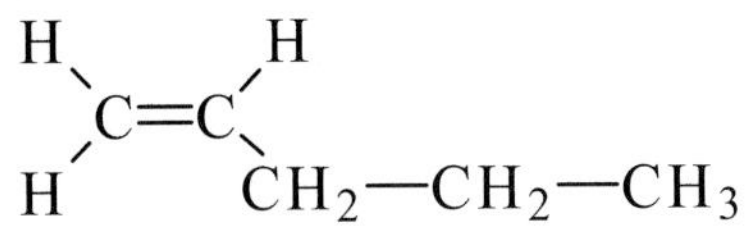

```
H       H                 H       H               CH3      H
 \     /                   \     /                   \     /
  C=C                       C=C                       C=C
 /     \                   /     \                   /     \
H       CH2—CH2—CH3     CH3       CH2—CH3           H       CH2—CH3
```

1-pentene 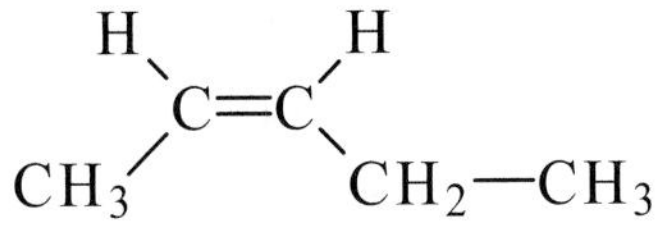   *cis*-2-pentene 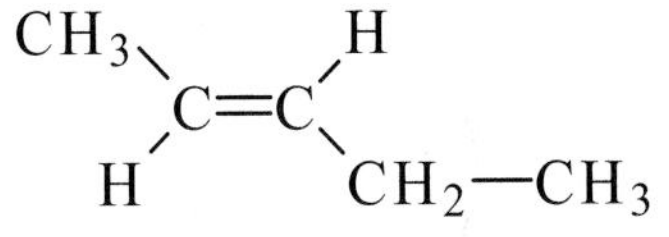   *trans* -2-pentene

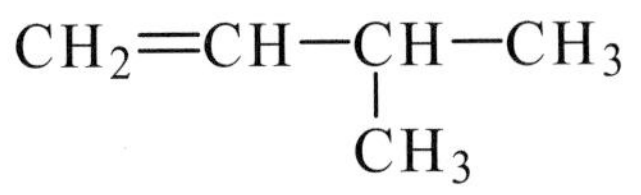

```
CH2=CH—CH—CH3        CH2=C—CH2—CH3        CH3—C=CH—CH3
       |                  |                   |
       CH3                CH3                 CH3
```

3-methyl-1-butene 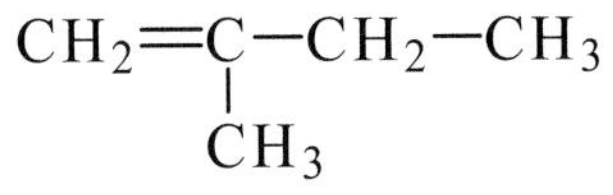   2-methyl-1-butene 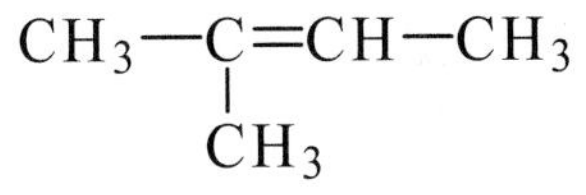   2-methyl-2-butene

The geometric isomers are *cis*- and *trans*-2-pentene. 1-pentene, 3-methyl-1-butene, 2-methyl-1-butene, and 2-methyl-2-butene are structural isomers to the geometric isomers because the atoms are attached in a different order.

Why don't 1-pentene, 3-methyl-1-butene, 2-methyl-1-butene, and 2-methyl-2-butene have geometric isomers?

PRACTICE EXERCISE

6. Draw the structural formula for 3-bromo-2,5-dimethyl-*trans*-3-hexene.

Text Problem: 24.24

PROBLEM TYPE 5: FUNCTIONAL GROUPS

We will discuss the most common organic functional groups, with emphasis on classes of compounds in which the functional groups include oxygen or nitrogen.

1. **Alcohols.** All alcohols contain the hydroxyl group, –OH. Figure 24.8 of the text shows some common alcohols. Alcohols typically undergo esterification (formation of an ester) reactions with carboxylic acids. Oxidation to aldehydes, ketones, and carboxylic acids are also common reactions.

2. **Ethers.** Ethers contain the R–O–R' linkage, where R and R' are either an alkyl group or a group derived from an aromatic hydrocarbon.

3. **Aldehydes and Ketones.** These compounds contain the carbonyl functional group.

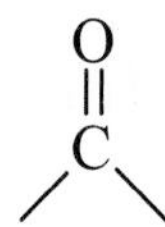

The difference between aldehydes and ketones is that in aldehydes at least one hydrogen atom is bonded to the carbon atom of the carbonyl group. In ketones, no hydrogen atoms are bonded to the carbonyl carbon atom. Common reactions include reduction to yield alcohols. Oxidation of aldehydes yields carboxylic acids.

4. **Carboxylic Acids.** These compounds contain the carboxyl group, –COOH.

O
||
C
OH

Common reactions include esterification with alcohols and reaction with phosphorus pentachloride to yield acid chlorides.

5. **Esters.** Esters have the general formula R'COOR.

O
||
C
R' OR

R' can be H, an alkyl, or an aromatic hydrocarbon group, and R is an alkyl or an aromatic hydrocarbon group. A common reaction is hydrolysis to yield acids and alcohols.

6. **Amines.** Amines are organic bases. They have the general formula R_3N, where one of the R groups must be an alkyl group or an aromatic hydrocarbon group. A common reaction is formation of ammonium salts with acids.

Table 24.4 of the text summarizes important functional groups. For more detailed discussions of reactions, see Section 24.4 of the text.

EXAMPLE 24.10

Identify the functional groups in the following molecules: (a) $C_5H_{11}OH$, (b) CH_3CHO, (c) $C_3H_7OCH_3$, (d) $CH_3COC_2H_5$, (e) CH_3COOCH_3.

Strategy: Learning to recognize functional groups requires memorization of their structural formulas. Table 24.4 of the text shows a number of the important functional groups.

Solution:

(a) $C_5H_{11}OH$ contains a *hydroxyl* group. It is an *alcohol.*

(b) CH_3CHO is a way to represent

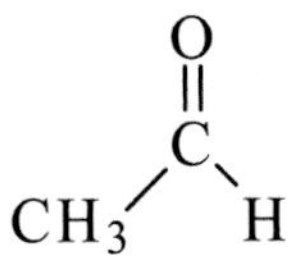

$$\begin{array}{c} O \\ \| \\ CH_3{-}C{-}H \end{array}$$

on a single line of type. C=O is a carbonyl group. Since there is a hydrogen atom bonded to the carbonyl carbon, CH_3CHO is an *aldehyde.*

(c) $C_3H_7OC_2H_5$ contains a C–O–C group and is therefore an *ether.*

(d) $CH_3COC_2H_5$ contains a carbonyl group. Since a hydrogen atom is not bonded to the carbonyl carbon, this molecule is a *ketone.*

(e) CH_3COOCH_3 is a condensed structural formula for an *ester.*

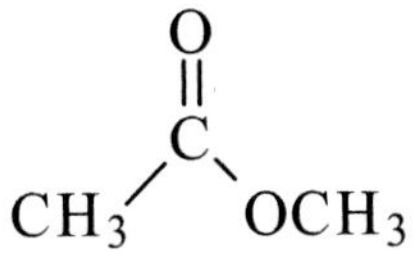

$$\begin{array}{c} O \\ \| \\ CH_3{-}C{-}OCH_3 \end{array}$$

PRACTICE EXERCISE

7. Indicate the functional groups by name that are in the following molecules.

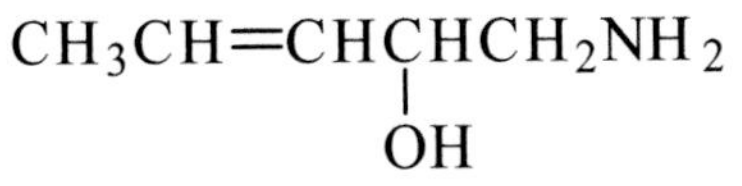

(a)

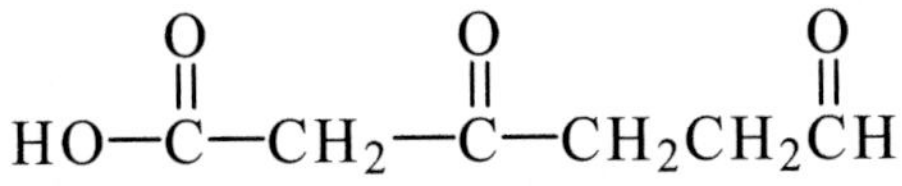

$$CH_3CH{=}CH\underset{\displaystyle OH}{\underset{|}{C}}HCH_2NH_2$$

(b)

$$HO{-}\overset{\displaystyle O}{\overset{\|}{C}}{-}CH_2{-}\overset{\displaystyle O}{\overset{\|}{C}}{-}CH_2CH_2\overset{\displaystyle O}{\overset{\|}{C}}H$$

Text Problems: 24.36, 24.38, 24.40, 24.42

PROBLEM TYPE 6: CHIRALITY

Compounds that come as mirror-image pairs can be compared with left-handed and right-handed gloves and are thus referred to as **chiral**, or handed, molecules. While every molecule can have a mirror image, the *chiral* mirror-image pairs are *nonsuperimposable*. Conversely, achiral (nonchiral) pairs are superimposable. See Figure 24.3 of the text for an illustration of superimposable compared to nonsuperimposable.

Observations show that most simple chiral molecules contain at least one *asymmetric* carbon atom; that is, a carbon atom bonded to four different atoms or groups of atoms.

> **Tip:** Consider your hands when thinking about chirality. If you view your left hand in a mirror, the mirror image of your left hand is a right hand. However, your left and right hands are nonsuperimposable. To verify this, try putting a "left-handed" glove on your right hand.

EXAMPLE 24.11

Classify the following objects as chiral or achiral:

(a) shoe **(b) screw** **(c) fork** **(d) coffee cup**

A shoe and a screw are chiral. The mirror image of a right shoe is a left shoe. A left shoe is not superimposable on a right shoe. A mirror image of a screw with clockwise threads will have counterclockwise threads. The screws would not be superimposable.

A fork and a coffee cup are achiral. Their mirror images are superimposable.

PRACTICE EXERCISE

8. Are the following molecules chiral?

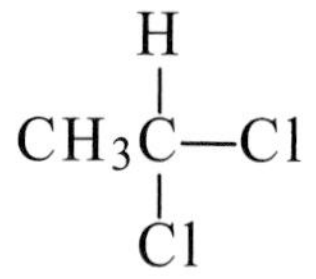

```
      H                    H
      |                    |
  CH3C—Cl              CH3C—Cl
      |                    |
      Cl                   Br
```

(a) 1,1-dichloroethane 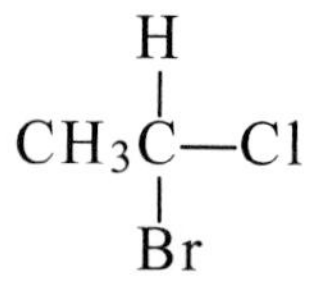 (b) 1-bromo-1-chloroethane

Text Problem: 24.56

ANSWERS TO PRACTICE EXERCISES

1. There are seven structural isomers—four alcohols and three ethers. The structures are shown below.

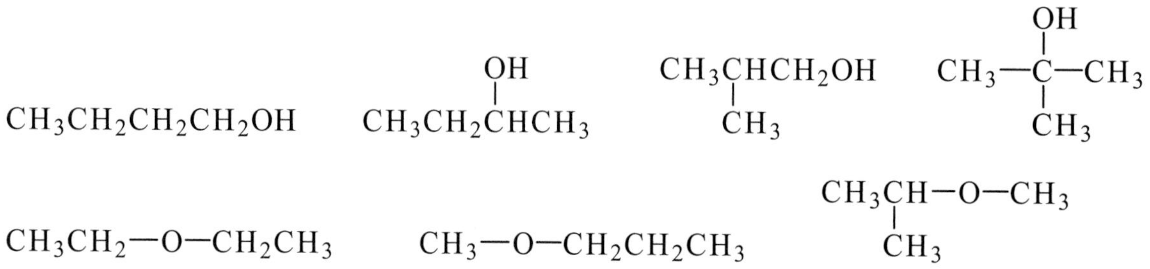

```
                                                                   OH
                             OH                                    |
                             |          CH3CHCH2OH          CH3—C—CH3
CH3CH2CH2CH2OH       CH3CH2CHCH3            |                      |
                                            CH3                    CH3

                                                      CH3CH—O—CH3
                                                          |
CH3CH2—O—CH2CH3       CH3—O—CH2CH2CH3                     CH3
```

2. **(a)** nonane **(b)** 4-ethyl-5-isopropyloctane **(c)** 1,2,4-tribromobutane

3. $$CH_3CH{=}CH\overset{\displaystyle CH_3}{\overset{|}{C}}HCH_2CH_3$$

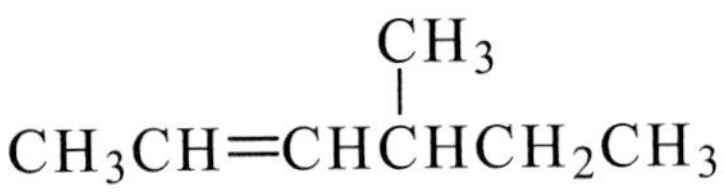

4. **(a)** 2,2-dimethyl-3-hexyne **(b)** 4-propyl-2-heptyne

5. The structures are:

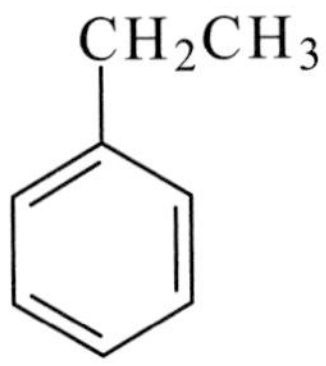

(a) ethylbenzene (benzene ring bearing CH_2CH_3)

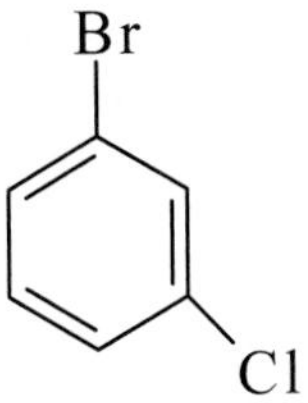

(b) benzene ring bearing Br and Cl (meta)

(c) benzene ring bearing CH_3 and NO_2 (para)

6. The structure of 3-bromo-2,5-dimethyl-*trans*-3-hexene is:

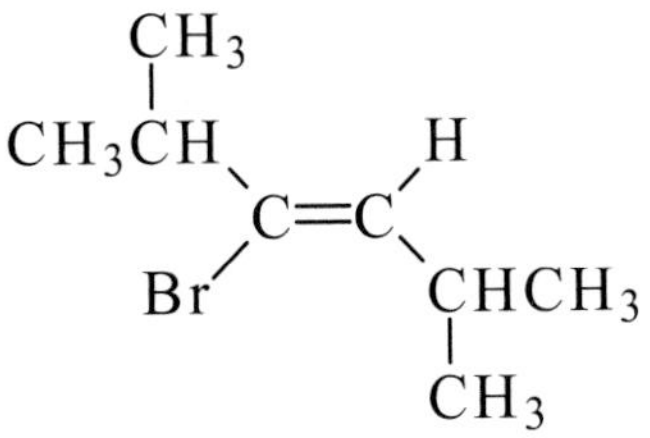

$$\begin{array}{ccc} CH_3 & & \\ | & & \\ CH_3CH & & H \\ & \diagdown\ \ \diagup & \\ & C{=}C & \\ & \diagup\ \ \diagdown & \\ Br & & CHCH_3 \\ & & | \\ & & CH_3 \end{array}$$

7. **(a)** carbon-carbon double bond, hydroxyl (–OH), and amine ($R{-}NH_2$).

(b) carboxyl (–COOH), carbonyl (ketone), and carbonyl (aldehyde).

8. **(a)** There are only three different groups bonded to the second carbon atom. This molecule is *achiral.*

(b) Replacing one of the chloro groups with a bromo group places four different groups on the second carbon atom. This molecule is *chiral.*

SOLUTIONS TO SELECTED TEXT PROBLEMS

24.12 Determining the Number of Structural Isomers, Problem Type 1.

Strategy: For small hydrocarbon molecules (eight or fewer carbons), it is relatively easy to determine the number of structural isomers by trial and error.

Solution: We are starting with *n*-pentane, so we do not need to worry about any branched chain structures. In the chlorination reaction, a Cl atom replaces one H atom. There are three different carbons on which the Cl atom can be placed. Hence, *three* structural isomers of chloropentane can be derived from *n*–pentane:

$CH_3CH_2CH_2CH_2CH_2Cl$ $CH_3CH_2CH_2CHClCH_3$ $CH_3CH_2CHClCH_2CH_3$

24.14 Both alkenes and cycloalkanes have the general formula C_nH_{2n}. Let's start with C_3H_6. It could be an alkene or a cycloalkane.

```
                              H   H
                               \ /
                                C
H        CH3              H    / \    H
 \      /                  \  /   \  /
  C==C                      C-----C
 /      \                  /       \
H        H                H         H
```

Now, let's replace one H with a Br atom to form C_3H_5Br. *Four* isomers are possible.

```
                                                                      H   H
                                                                       \ /
                                                                        C
H        CH3      Br        CH3      H        CH2Br             H     / \    H
 \      /           \      /          \      /                   \   /   \  /
  C==C               C==C              C==C                        C-----C
 /      \           /      \          /      \                    /       \
Br       H         H        H        H        H                  H         Br
```

There is only one isomer for the cycloalkane. Note that all three carbons are equivalent in this structure.

24.16 **(a)** This compound could be an **alkene** or a **cycloalkane**; both have the general formula, C_nH_{2n}.

(b) This could be an **alkyne** with general formula, C_nH_{2n-2}. It could also be a hydrocarbon with two double bonds (a diene). It could be a cyclic hydrocarbon with one double bond (a cycloalkene).

(c) This must be an **alkane**; the formula is of the C_nH_{2n+2} type.

(d) This compound could be an **alkene** or a **cycloalkane**; both have the general formula, C_nH_{2n}.

(e) This compound could be an **alkyne** with one triple bond, or it could be a cyclic alkene (unlikely because of ring strain).

24.18 If cyclobutadiene were square or rectangular, the C–C–C angles must be 90°. If the molecule is diamond-shaped, two of the C–C–C angles must be less than 90°. Both of these situations result in a great deal of distortion and strain in the molecule. Cyclobutadiene is very unstable for these and other reasons.

24.20 One compound is an alkane; the other is an alkene. Alkenes characteristically undergo addition reactions with hydrogen, with halogens (Cl_2, Br_2, I_2) and with hydrogen halides (HCl, HBr, HI). Alkanes do not react with these substances under ordinary conditions.

24.22 In this problem you are asked to calculate the standard enthalpy of reaction. This type of problem was covered in Chapter 6.

$$\Delta H^\circ_{\text{rxn}} = \Sigma n\Delta H^\circ_{\text{f}}(\text{products}) - \Sigma m\Delta H^\circ_{\text{f}}(\text{reactants})$$

$$\Delta H^\circ_{\text{rxn}} = \Delta H^\circ_{\text{f}}(C_6H_6) - 3\Delta H^\circ_{\text{f}}(C_2H_2)$$

You can look up $\Delta H^\circ_{\text{f}}$ values in Appendix 3 of your textbook.

$$\boldsymbol{\Delta H^\circ_{\text{rxn}}} = (1)(49.04\text{ kJ/mol}) - (3)(226.6\text{ kJ/mol}) = \mathbf{-630.8\text{ kJ/mol}}$$

24.24 In this problem you must distinguish between *cis* and *trans* isomers. Recall that *cis* means that two particular atoms (or groups of atoms) are adjacent to each other, and *trans* means that the atoms (or groups of atoms) are on opposite sides in the structural formula.

In (a), the Cl atoms are adjacent to each other. This is the *cis* isomer. In (b), the Cl atoms are on opposite sides of the structure. This is the *trans* isomer.

The names are: **(a) *cis*-1,2-dichlorocyclopropane**; and **(b) *trans*-1,2-dichlorocyclopropane**.

Are any other dichlorocyclopropane isomers possible?

24.26 **(a)** This is a branched hydrocarbon. The name is based on the longest carbon chain. The name is **2–methylpentane**.

(b) This is also a branched hydrocarbon. The longest chain includes the C_2H_5 group; the name is based on hexane, not pentane. This is an old trick. Carbon chains are flexible and don't have to lie in a straight line. The name is **2,3,4–trimethylhexane**. Why not 3,4,5–trimethylhexane?

(c) How many carbons in the longest chain? It doesn't have to be straight! The name is **3–ethylhexane**.

(d) An alkene with two double bonds is called a diene. The name is **3–methyl–1,4–pentadiene**.

(e) The name is **2–pentyne**.

(f) The name is **3–phenyl–1–pentene**.

24.28 The hydrogen atoms have been omitted from the skeletal structure for simplicity.

(a) H_3C, H / C=C / H, CH_2CH_3

(b) H, CH_2CH_3 / C=C / H, CH_2CH_3

(c) H_3C, H / C=C / H, HC($CH_2CH_2CH_3$)(CH_2CH_3)

(d) HC≡C–C(H)(–CH_3)–C_6H_5

24.32 Organic Nomenclature, Problem Type 2.

Strategy: We follow the IUPAC rules and use the information in Table 24.2 of the text. When a benzene ring has more than *two* substituents, you must specify the location of the substituents with numbers. Remember to number the ring so that you end up with the lowest numbering scheme as possible, giving preference to alphabetical order.

Solution:

(a) Since a chloro group comes alphabetically before a methyl group, let's start by numbering the top carbon of the ring as 1. If we number clockwise, this places the second chloro group on carbon 3 and a methyl group on carbon 4.

This compound is **1,3–dichloro–4–methylbenzene**.

(b) If we start numbering counterclockwise from the bottom carbon of the ring, the name is 2–ethyl–1,4–dinitrobenzene. Numbering clockwise from the top carbon gives 3–ethyl–1,4–dinitrobenzene.

Numbering as low as possible, the correct name is **2–ethyl–1,4–dinitrobenzene**.

(c) Again, keeping the numbers as low as possible, the correct name for this compound is **1,2,4,5–tetramethylbenzene**. You should number clockwise from the top carbon of the ring.

24.36 Functional Groups, Problem Type 5.

Strategy: Learning to recognize functional groups requires memorization of their structural formulas. Table 24.4 of the text shows a number of the important functional groups.

Solution:

(a) $H_3C–O–CH_2–CH_3$ contains a C–O–C group and is therefore an **ether**.

(b) This molecule contains an RNH_2 group and is therefore an **amine**.

(c) This molecule is an **aldehyde**. It contains a carbonyl group in which one of the atoms bonded to the carbonyl carbon is a hydrogen atom.

(d) This molecule also contains a carbonyl group. However, in this case there are no hydrogen atoms bonded to the carbonyl carbon. This molecule is a **ketone**.

(e) This molecule contains a carboxyl group. It is a **carboxylic acid**.

(f) This molecule contains a hydroxyl group (–OH). It is an **alcohol**.

(g) This molecule has both an RNH_2 group and a carboxyl group. It is therefore both an *amine* and a *carboxylic acid*, commonly called an **amino acid**.

24.38 Alcohols react with carboxylic acids to form esters. The reaction is:

$$HCOOH + CH_3OH \longrightarrow HCOOCH_3 + H_2O$$

The structure of the product is:

$$H-\overset{\overset{\displaystyle O}{\|}}{C}-O-CH_3 \qquad \text{(methyl formate)}$$

24.40 The fact that the compound does not react with sodium metal eliminates the possibility that the substance is an alcohol. The only other possibility is the ether functional group. There are three ethers possible with this molecular formula:

$CH_3–CH_2–O–CH_2–CH_3$ $\quad$ $CH_3–CH_2–CH_2–O–CH_3$ $\quad$ $(CH_3)_2CH–O–CH_3$

Light–induced reaction with chlorine results in substitution of a chlorine atom for a hydrogen atom (the other product is HCl). For the first ether there are only two possible chloro derivatives:

$ClCH_2–CH_2–O–CH_2–CH_3$ $\quad$ $CH_3–CHCl–O–CH_2–CH_3$

For the second there are four possible chloro derivatives. Three are shown below. Can you draw the fourth?

$CH_3–CHCl–CH_2–O–CH_3$ $\quad$ $CH_3–CH_2–CHCl–O–CH_3$ $\quad$ $CH_2Cl–CH_2–CH_2–O–CH_3$

For the third there are three possible chloro derivatives:

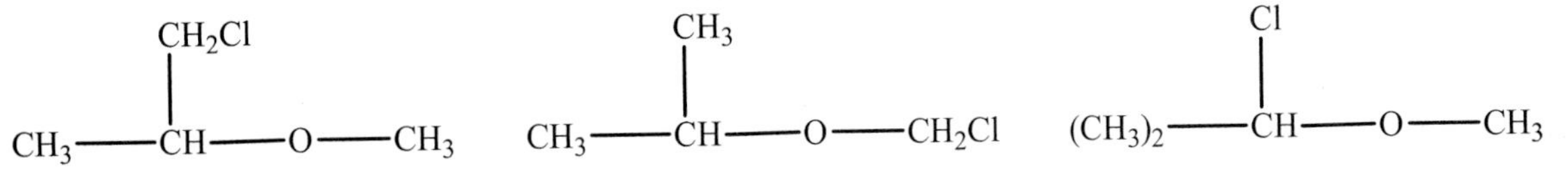

The **(CH₃)₂CH–O–CH₃** choice is the original compound.

24.42 **(a)** ketone **(b)** ester **(c)** ether

24.44 This is a Hess's Law problem. See Chapter 6.

If we rearrange the equations given and multiply times the necessary factors, we have:

$$2CO_2(g) + 2H_2O(l) \longrightarrow C_2H_4(g) + 3O_2(g) \qquad \Delta H^\circ = 1411 \text{ kJ/mol}$$
$$C_2H_2(g) + \tfrac{5}{2}O_2(g) \longrightarrow 2CO_2(g) + H_2O(l) \qquad \Delta H^\circ = -1299.5 \text{ kJ/mol}$$
$$H_2(g) + \tfrac{1}{2}O_2(g) \longrightarrow H_2O(l) \qquad \Delta H^\circ = -285.8 \text{ kJ/mol}$$
$$C_2H_2(g) + H_2(g) \longrightarrow C_2H_4(g) \qquad \boldsymbol{\Delta H^\circ = -174 \text{ kJ/mol}}$$

The heat of hydrogenation for acetylene is **−174 kJ/mol.**

24.46 To form a hydrogen bond *with water* a molecule must have at least one H–F, H–O, or H–N bond, *or* must contain an O, N, or F atom. The following can form hydrogen bonds with water:

(a) carboxylic acids **(c)** ethers **(d)** aldehydes **(f)** amines

24.48 **(a)** rubbing alcohol **(b)** vinegar **(c)** moth balls **(d)** organic synthesis
(e) organic synthesis **(f)** antifreeze **(g)** fuel (natural gas) **(h)** synthetic polymers

24.50 **(a)** 2–butyne has **three** C–C sigma bonds.

(b) Anthracene is:

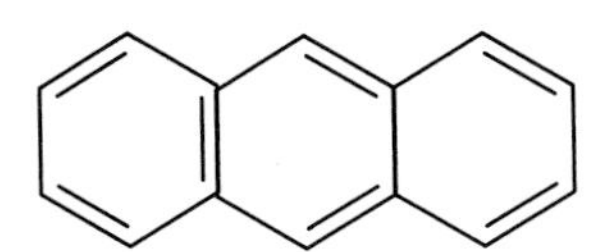

There are **sixteen** C–C sigma bonds.

(c)

```
    |   |
C—C—C—C—C
```

There are **six** C–C sigma bonds.

24.52 **(a)** The easiest way to calculate the mg of C in CO_2 is by mass ratio. There are 12.01 g of C in 44.01 g CO_2 or 12.01 mg C in 44.01 mg CO_2.

$$\textbf{? mg C} = 57.94 \text{ mg } CO_2 \times \frac{12.01 \text{ mg C}}{44.01 \text{ mg } CO_2} = \textbf{15.81 mg C}$$

Similarly,

$$\textbf{? mg H} = 11.85 \text{ mg } H_2O \times \frac{2.016 \text{ mg H}}{18.02 \text{ mg } H_2O} = \textbf{1.326 mg H}$$

The mg of oxygen can be found by difference.

$$\textbf{? mg O} = 20.63 \text{ mg Y} - 15.81 \text{ mg C} - 1.326 \text{ mg H} = \textbf{3.49 mg O}$$

(b) *Step 1:* Calculate the number of moles of each element present in the sample. Use molar mass as a conversion factor.

$$\text{? mol C} = (15.81 \times 10^{-3} \text{ g C}) \times \frac{1 \text{ mol C}}{12.01 \text{ g C}} = 1.316 \times 10^{-3} \text{ mol C}$$

Similarly,

$$\text{? mol H} = (1.326 \times 10^{-3} \text{ g H}) \times \frac{1 \text{ mol H}}{1.008 \text{ g H}} = 1.315 \times 10^{-3} \text{ mol H}$$

$$\text{? mol O} = (3.49 \times 10^{-3} \text{ g O}) \times \frac{1 \text{ mol O}}{16.00 \text{ g O}} = 2.18 \times 10^{-4} \text{ mol O}$$

Thus, we arrive at the formula $C_{1.316 \times 10^{-3}}H_{1.315 \times 10^{-3}}O_{2.18 \times 10^{-4}}$, which gives the identity and the ratios of atoms present. However, chemical formulas are written with whole numbers.

Step 2: Try to convert to whole numbers by dividing all the subscripts by the smallest subscript.

$$\textbf{C:} \ \frac{1.316 \times 10^{-3}}{2.18 \times 10^{-4}} = 6.04 \approx 6 \qquad \textbf{H:} \ \frac{1.315 \times 10^{-3}}{2.18 \times 10^{-4}} = 6.03 \approx 6 \qquad \textbf{O:} \ \frac{2.18 \times 10^{-4}}{2.18 \times 10^{-4}} = 1.00$$

This gives us the empirical formula, $\mathbf{C_6H_6O}$.

(c) The presence of six carbons and a corresponding number of hydrogens suggests a benzene derivative. A plausible structure is shown below.

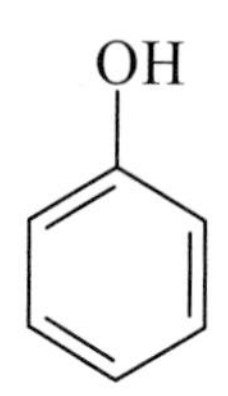

24.54 First, calculate the moles of each element.

$$\text{C:} \quad (9.708 \times 10^{-3}\ \text{g CO}_2) \times \frac{1\ \text{mol CO}_2}{44.01\ \text{g CO}_2} \times \frac{1\ \text{mol C}}{1\ \text{mol CO}_2} = 2.206 \times 10^{-4}\ \text{mol C}$$

$$\text{H:} \quad (3.969 \times 10^{-3}\ \text{g H}_2\text{O}) \times \frac{1\ \text{mol H}_2\text{O}}{18.02\ \text{g H}_2\text{O}} \times \frac{2\ \text{mol H}}{1\ \text{mol H}_2\text{O}} = 4.405 \times 10^{-4}\ \text{mol H}$$

The mass of oxygen is found by difference:

$$3.795\ \text{mg compound} - (2.649\ \text{mg C} + 0.445\ \text{mg H}) = 0.701\ \text{mg O}$$

$$\text{O:} \quad (0.701 \times 10^{-3}\ \text{g O}) \times \frac{1\ \text{mol O}}{16.00\ \text{g O}} = 4.38 \times 10^{-5}\ \text{mol O}$$

This gives the formula is $C_{2.206 \times 10^{-4}}H_{4.405 \times 10^{-4}}O_{4.38 \times 10^{-5}}$. Dividing by the smallest number of moles gives the empirical formula, $\mathbf{C_5H_{10}O}$.

We calculate moles using the ideal gas equation, and then calculate the molar mass.

$$n = \frac{PV}{RT} = \frac{(1.00\ \text{atm})(0.0898\ \text{L})}{(0.0821\ \text{L}\cdot\text{atm/K}\cdot\text{mol})(473\ \text{K})} = 0.00231\ \text{mol}$$

$$\textbf{molar mass} = \frac{\text{g of substance}}{\text{mol of substance}} = \frac{0.205\ \text{g}}{0.00231\ \text{mol}} = \mathbf{88.7\ g/mol}$$

The formula mass of $C_5H_{10}O$ is 86.13 g, so this is also the molecular formula. Three possible structures are:

[Structure 1: six-membered ring of CH2, CH2, CH2, CH2, CH2 (H2C) and O — tetrahydropyran]

[Structure 2: four CH2/CH ring with O: H2C—CH2, H2C, CH (with CH3), O]

$H_2C{=}CH{-}CH_2{-}O{-}CH_2{-}CH_3$

24.56 A carbon atom is asymmetric if it is bonded to four different atoms or groups. In the given structures the asymmetric carbons are marked with an asterisk (*).

(a) $CH_3{-}CH_2{-}\overset{*}{C}H(CH_3){-}\overset{*}{C}H(NH_2){-}C({=}O){-}NH_2$

(b) [Three-membered ring structure: carbon with H, H (bottom) and two carbons marked *, one bearing H and Br inside the ring, the other bearing H and Br]

24.58 Acetone is a ketone with the formula, CH_3COCH_3. We must write the structure of an aldehyde that has the same number and types of atoms (C_3H_6O). Removing the aldehyde functional group (–CHO) from the formula leaves C_2H_5. This is the formula of an ethyl group. The aldehyde that is a structural isomer of acetone is:

$$CH_3CH_2\overset{O}{\overset{\|}{C}}{-}H$$

24.60 **(a)** alcohol **(b)** ether **(c)** aldehyde **(d)** carboxylic acid **(e)** amine

24.62 In Chapter 11, we found that salts with their electrostatic intermolecular attractions had low vapor pressures and thus high boiling points. Ammonia and its derivatives (amines) are molecules with dipole–dipole attractions. If the nitrogen has one direct N–H bond, the molecule will have hydrogen bonding. Even so, these molecules will have much weaker intermolecular attractions than ionic species and hence higher vapor pressures. Thus, if we could convert the neutral ammonia–type molecules into salts, their vapor pressures, and thus associated odors, would decrease. Lemon juice contains acids which can react with ammonia–type (amine) molecules to form ammonium salts.

$$NH_3 + H^+ \longrightarrow NH_4^+ \qquad\qquad RNH_2 + H^+ \longrightarrow RNH_3^+$$

24.64 Marsh gas (methane, CH_4); grain alcohol (ethanol, C_2H_5OH); wood alcohol (methanol, CH_3OH); rubbing alcohol (isopropyl alcohol, $(CH_3)_2CHOH$); antifreeze (ethylene glycol, CH_2OHCH_2OH); mothballs (naphthalene, $C_{10}H_8$); vinegar (acetic acid, CH_3COOH).

24.66 The asymmetric carbons are shown by asterisks:

(a)
$$\begin{array}{ccccccc} & & H & & H & & H \\ & & | & & {}^{*}| & & | \\ H & - & C & - & C & - & C & - Cl \\ & & | & & | & & | \\ & & H & & Cl & & H \end{array}$$

(b)
$$\begin{array}{ccccccc} & & OH & & CH_3 & & \\ & & {}^{*}| & & {}^{*}| & & \\ CH_3 & - & C & - & C & - & CH_2OH \\ & & | & & | & & \\ & & H & & H & & \end{array}$$

(c) All of the carbon atoms in the ring are asymmetric. Therefore there are **five** asymmetric carbon atoms.

24.68 The red bromine vapor absorbs photons of blue light and dissociates to form bromine atoms.

$$Br_2 \rightarrow 2Br\bullet$$

The bromine atoms collide with methane molecules and abstract hydrogen atoms.

$$Br\bullet + CH_4 \rightarrow HBr + \bullet CH_3$$

The methyl radical then reacts with Br_2, giving the observed product and regenerating a bromine atom to start the process over again:

$$\bullet CH_3 + Br_2 \rightarrow CH_3Br + Br\bullet$$

$$Br\bullet + CH_4 \rightarrow HBr + \bullet CH_3 \qquad \text{and so on...}$$

24.70 2–butanone is

$$\begin{array}{ccccccc} & & O & & & & \\ & & \| & & & & \\ H_3C & - & C & - & CH_2 & - & CH_3 \end{array}$$

Reduction with $LiAlH_4$ produces 2-butanol.

$$\begin{array}{ccccccc} & & OH & & & & \\ & & | & & & & \\ H_3C & - & C & - & CH_2 & - & CH_3 \\ & & | & & & & \\ & & H & & & & \end{array}$$

This molecule possesses an asymmetric carbon atom and should be chiral. However, the reduction produces an equimolar *d* and *l* isomers; that is, a racemic mixture (see Section 22.4 of the text). Therefore, the optical rotation as measured in a polarimeter is zero.

24.72 To help determine the molecular formula of the alcohol, we can calculate the molar mass of the carboxylic acid, and then determine the molar mass of the alcohol from the molar mass of the acid. Grams of carboxylic acid are given (4.46 g), so we need to determine the moles of acid to calculate its molar mass.

The number of moles in 50.0 mL of 2.27 *M* NaOH is

$$\frac{2.27 \text{ mol NaOH}}{1000 \text{ mL soln}} \times 50.0 \text{ mL} = 0.1135 \text{ mol NaOH}$$

The number of moles in 28.7 mL of 1.86 *M* HCl is

$$\frac{1.86 \text{ mol HCl}}{1000 \text{ mL soln}} \times 28.7 \text{ mL} = 0.05338 \text{ mol HCl}$$

The difference between the above two numbers is the number of moles of NaOH reacted with the carboxylic acid.

$$0.1135 \text{ mol} - 0.05338 \text{ mol} = 0.06012 \text{ mol}$$

This is the number of moles present in 4.46 g of the carboxylic acid. The molar mass is

$$\mathcal{M} = \frac{4.46 \text{ g}}{0.06012 \text{ mol}} = 74.18 \text{ g/mol}$$

A carboxylic acid contains a –COOH group and an alcohol has an –OH group. When the alcohol is oxidized to a carboxylic acid, the change is from $-CH_2OH$ to –COOH. Therefore, the molar mass of the alcohol is

$$74.18 \text{ g} - 16 \text{ g} + (2)(1.008 \text{ g}) = 60.2 \text{ g/mol}$$

With a molar mass of 60.2 g/mol for the alcohol, there can only be 1 oxygen atom and 3 carbon atoms in the molecule, so the formula must be C_3H_8O. The alcohol has one of the following two molecular formulas.

$CH_3CH_2CH_2OH$ $\qquad$ $H_3C{-}CH(OH){-}CH_3$

24.74 **(a)** Reaction between glycerol and carboxylic acid (formation of an ester).

(b)

$$\begin{array}{l} CH_2{-}O{-}\overset{O}{\overset{\|}{C}}{-}R \\ | \\ CH{-}O{-}\overset{O}{\overset{\|}{C}}{-}R' \\ | \\ CH_2{-}O{-}\overset{O}{\overset{\|}{C}}{-}R'' \end{array} \xrightarrow[H_2O]{NaOH} \begin{array}{l} CH_2{-}OH \\ | \\ CH{-}OH \\ | \\ CH_2{-}OH \end{array} + 3\ R{-}\overset{O}{\overset{\|}{C}}{-}O^-\ Na^+$$

A fat or oil $\qquad$ Glycerol $\qquad$ Fatty acid salts (soap)

(c) Molecules having more C=C bonds are harder to pack tightly together. Consequently, the compound has a lower melting point.

(d) H_2 gas with either a heterogeneous or homogeneous catalyst would be used. See Section 13.6 of the text.

(e) Number of moles of $Na_2S_2O_3$ reacted is:

$$20.6\text{ mL} \times \frac{1\text{ L}}{1000\text{ mL}} \times \frac{0.142\text{ mol Na}_2\text{S}_2\text{O}_3}{1\text{ L}} = 2.93 \times 10^{-3}\text{ mol Na}_2\text{S}_2\text{O}_3$$

The mole ratio between I_2 and $Na_2S_2O_3$ is 1:2. The number of grams of I_2 left over is:

$$(2.93 \times 10^{-3}\text{ mol Na}_2\text{S}_2\text{O}_3) \times \frac{1\text{ mol I}_2}{2\text{ mol Na}_2\text{S}_2\text{O}_3} \times \frac{253.8\text{ g I}_2}{1\text{ mol I}_2} = 0.372\text{ g I}_2$$

Number of grams of I_2 reacted is: $(43.8 - 0.372)\gamma = 43.4\text{ g I}_2$

The *iodine number* is the number of grams of iodine that react with 100 g of corn oil.

$$\textit{\textbf{iodine number}} = \frac{43.4\text{ g I}_2}{35.3\text{ g corn oil}} \times 100\text{ g corn oil} = \mathbf{123}$$

CHAPTER 25
SYNTHETIC AND NATURAL ORGANIC POLYMERS

PROBLEM-SOLVING STRATEGIES AND TUTORIAL SOLUTIONS

TYPES OF PROBLEMS

Problem Type 1: Synthetic Organic Polymers.
- **(a)** Addition reactions.
- **(b)** Condensation reactions.

Problem Type 2: Proteins.

Problem Type 3: Nucleic Acids.

PROBLEM TYPE 1: SYNTHETIC ORGANIC POLYMERS

The word **polymer** means "many parts". A **polymer** is a compound with an unusually high molecular mass, consisting of a large number of molecular units linked together. The small unit that is repeated many times is called a **monomer**. A typical polymer contains chains of monomers several thousand units long.

A. Addition reactions

Addition polymers are made by adding monomer to monomer until a long chain is produced. Ethylene (ethene) and its derivatives are excellent monomers for forming addition polymers. In Chapter 24, we saw that addition reactions occur with unsaturated compounds containing C=C and C≡C bonds. In an addition reaction, the polymerization process is initiated by a radical or an ion. When ethylene is heated to 250°C under high pressure (1000–3000 atm) in the presence of a little oxygen or benzoyl peroxide (the initiator), addition polymers with masses of about 30,000 amu are obtained. This reaction is represented by

$$n\ \mathrm{H_2C{=}CH_2} \longrightarrow \mathrm{-CH_2-CH_2-CH_2-CH_2-CH_2-CH_2-CH_2-CH_2-}$$

ethylene — a segment of polyethylene

The general equation for addition polymerization is:

$$n\ \mathrm{H_2C{=}CH_2} \longrightarrow \left[\mathrm{-CH_2-CH_2-}\right]_n$$

monomer — repeating unit

Substitution for one or more hydrogen atoms in ethylene with Cl atoms, acetate, CN, or F provides a wide selection of monomers from which to make addition polymers with various properties. For instance, substitution of a Cl atom for a H atom in ethylene gives the monomer called vinyl chloride, $CH_2{=}CHCl$. Polymerization of vinyl chloride yields the polymer, polyvinyl chloride (PVC).

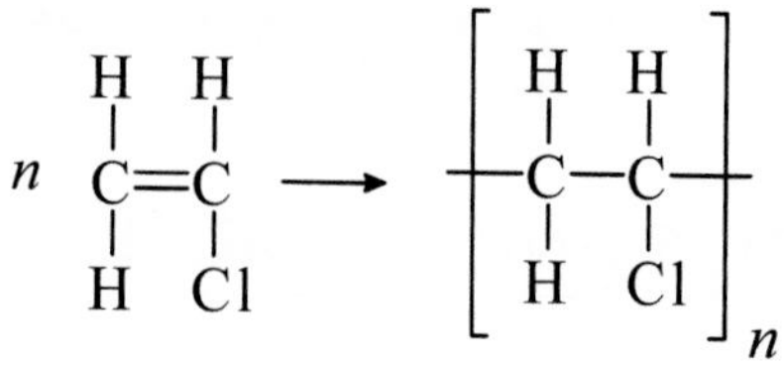

$$n\ \begin{matrix} H & & H \\ | & & | \\ C & = & C \\ | & & | \\ H & & Cl \end{matrix} \longrightarrow \left[\begin{matrix} H & & H \\ | & & | \\ -C & - & C- \\ | & & | \\ H & & Cl \end{matrix} \right]_n$$

vinyl chloride monomer — polyvinyl chloride repeating unit

Polymers that are made from one type of monomer such as polyvinyl chloride are called **homopolymers**. Table 25.1 of the text gives the names, structures, and uses of a number of monomers and addition polymers.

For more detailed information on the reaction mechanism of addition polymerization, see Section 25.2 of the text.

EXAMPLE 25.1

Write the formulas of the monomers used to prepare the following polymers: (a) Teflon, (b) Polystyrene, and (c) PVC.

Refer to Table 25.1 of the text.

(a) Teflon is an addition polymer with the formula

$$\left(CF_2 - CF_2 \right)_n$$

It is prepared from the monomer tetrafluoroethylene ($CF_2{=}CF_2$).

(b) The monomer used to prepare polystyrene is styrene.

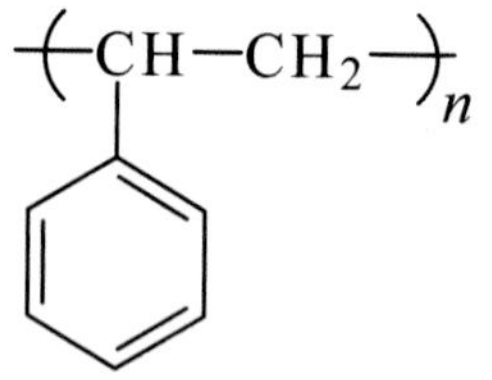

$$\left(\underset{\displaystyle C_6H_5}{\underset{|}{CH}} - CH_2 \right)_n \qquad \underset{\displaystyle C_6H_5}{\underset{|}{CH}} = CH_2$$

polystyrene 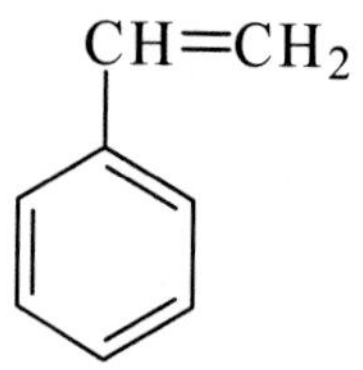 — styrene

(c) PVC (polyvinylchloride) is prepared by the successive addition of vinyl chloride molecules ($CH_2{=}CHCl$).

PRACTICE EXERCISE

1. Draw the structures of the monomers from which the following polymers are formed:

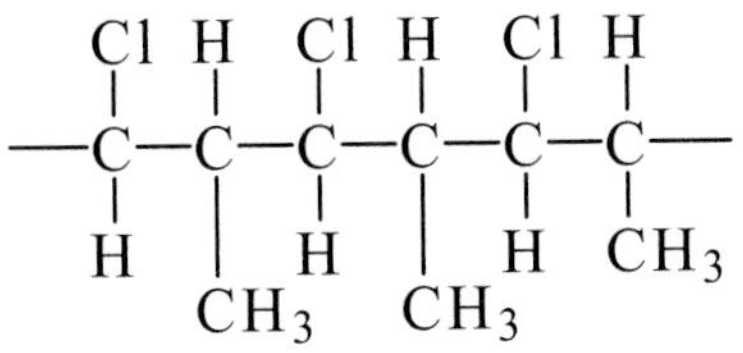

(a)

```
 Cl H  Cl H  Cl H
 |  |  |  |  |  |
—C—C—C—C—C—C—
 |  |  |  |  |  |
 H  |  H  |  H  CH3
   CH3   CH3
```

(b) $—CH_2—CCl_2—CH_2—CCl_2—CH_2—CCl_2—$

Text Problems: 25.8, 25.10, 25.12a

B. Condensation reactions

Copolymers are polymers that contain two or more different types of monomers. Polyesters, such as the well-known Dacron, are copolymers. In Dacron, one monomer is an alcohol and the other is a carboxylic acid, which can be joined by an *esterification* reaction. The alcohol and the acid both must contain two functional groups. The monomers of Dacron are the dicarboxylic acid called terephthalic acid and the dialcohol called ethylene glycol.

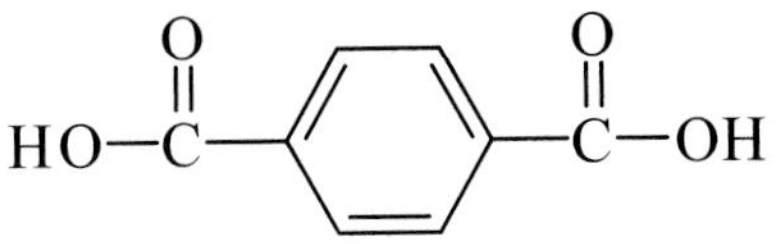

$HO-C(=O)-C_6H_4-C(=O)-OH$

terephthalic acid

$HOCH_2-CH_2OH$ (H—C(OH)(H)—C(OH)(H)—H)

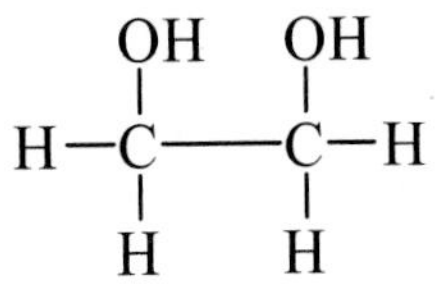

ethylene glycol

Condensation reactions differ from addition reactions in that the former always result in the formation of a small molecule such as water. When terephthalic acid and ethylene glycol react to form an ester, the first products are:

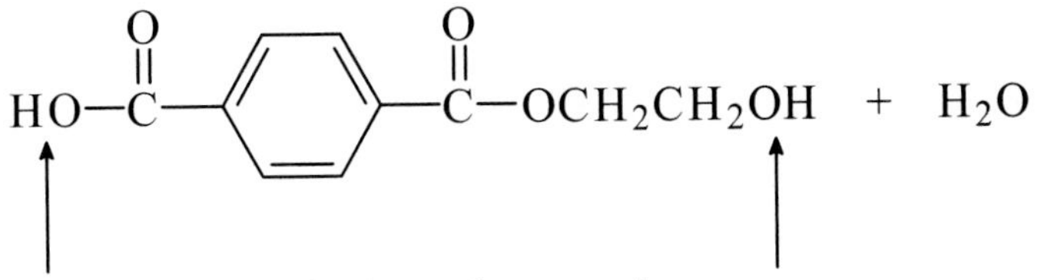

$$HO-C(=O)-C_6H_4-C(=O)-OCH_2CH_2OH + H_2O$$

sites for further condensation reactions

When this product reacts with another molecule of the diacid, the polymer chain grows longer.

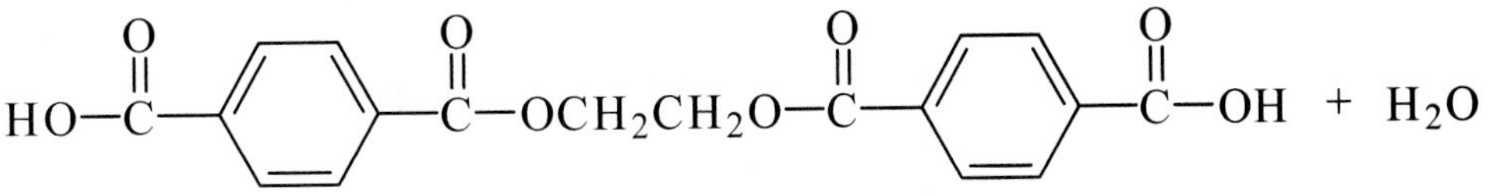

$$HO-C(=O)-C_6H_4-C(=O)-OCH_2CH_2O-C(=O)-C_6H_4-C(=O)-OH + H_2O$$

segment of a condensation polymer chain

The general formula for the polyester, Dacron, is:

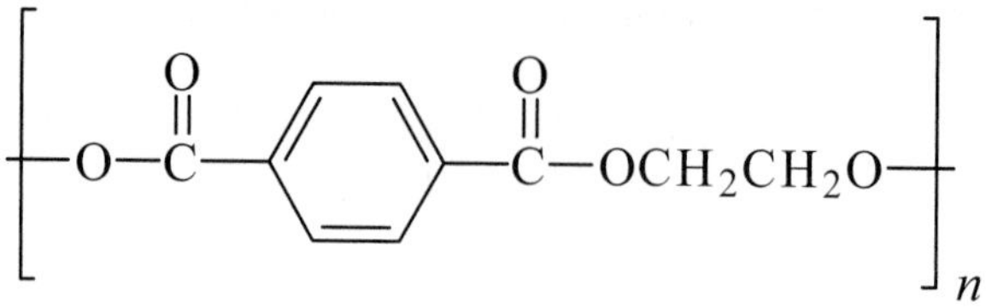

$$\left[-O-\overset{\overset{O}{\|}}{C}-C_6H_4-\overset{\overset{O}{\|}}{C}-OCH_2CH_2O- \right]_n$$

EXAMPLE 25.2
Draw structures for the monomers used to make the following polyester:

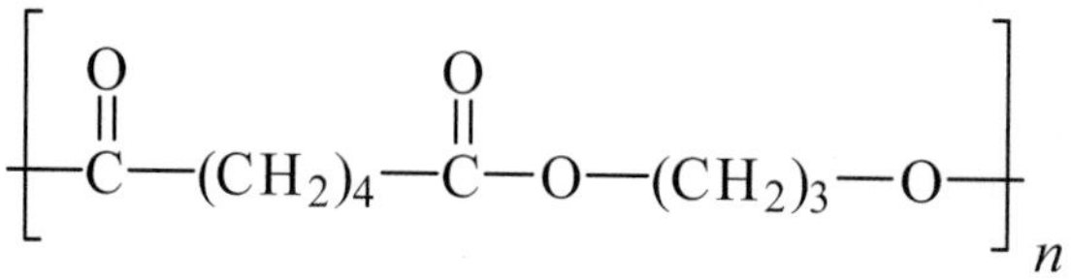

$$\left[-\overset{\overset{O}{\|}}{C}-(CH_2)_4-\overset{\overset{O}{\|}}{C}-O-(CH_2)_3-O- \right]_n$$

Answer

$$HO-\overset{\overset{O}{\|}}{C}-(CH_2)_4-\overset{\overset{O}{\|}}{C}-OH \qquad HO-(CH_2)_3-OH$$

PRACTICE EXERCISE

2. List two examples of condensation polymers discussed in the textbook.

Text Problem: 25.12b

PROBLEM TYPE 2: PROTEINS

Proteins are *polymers of amino acids*. Proteins are truly giant molecules having molecular masses that range from about 10,000 to several million amu. Proteins play many roles in living organisms, where they function as catalysts (enzymes), transport molecules (hemoglobin), contractile fibers (muscle), protective agents (blood clots), hormones (chemical messengers), and structural members (feathers, horns, and nails). The word *protein* comes from the Greek work *proteios*, meaning "first". From the partial list of protein functions above, it is easy to see why proteins occupy "first place" among biomolecules in their importance to life.

Even though each protein is unique, all proteins are built from a set of only **20 amino acids**. An amino acid consists of an amino group, a carboxylic acid group, a hydrogen atom, and a distinctive R group, all bonded to the same carbon atom.

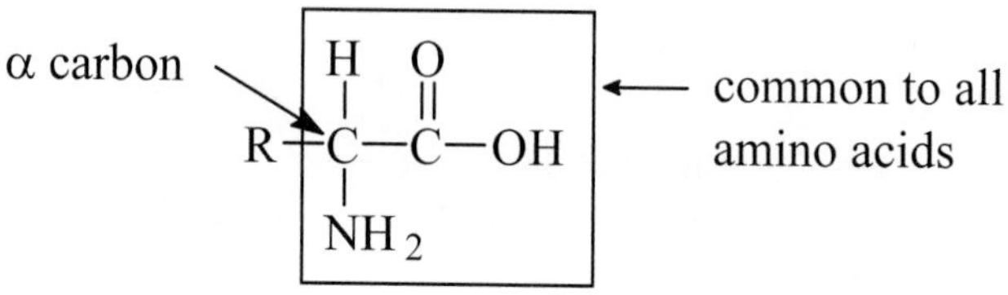

$$R-\overset{\overset{H}{|}}{\underset{\underset{NH_2}{|}}{C}}-\overset{\overset{O}{\|}}{C}-OH$$

α carbon (arrow points to the central C); boxed portion ← common to all amino acids

All amino acids in proteins have a common structural feature that is the attachment of the amino group ($-NH_2$) to the carbon atom adjacent to the carboxylic acid group. This carbon is called the α carbon. The difference in amino acids is due to different R groups. Twenty different R groups are found in proteins from natural sources. In fact, all proteins in all species, from bacteria to humans, are constructed from the same set of 20 amino acids. The structural formulas of the 20 amino acids essential to living organisms are shown in Table 25.2 of the text.

In proteins, the amino acid units are joined together to form a *polypeptide* chain. The carboxyl group of one amino acid is joined to the amino group of another amino acid by the formation of a peptide bond.

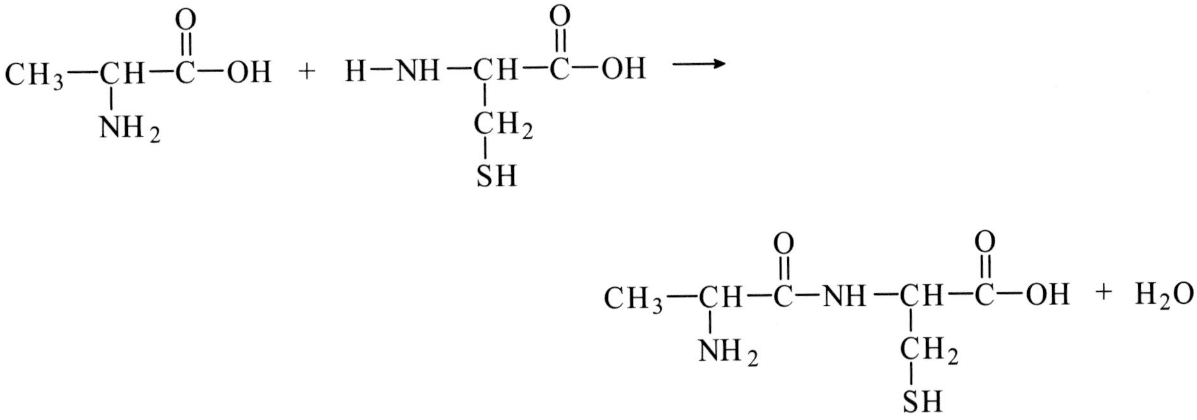

$$CH_3-\underset{\displaystyle NH_2}{\underset{|}{CH}}-\overset{\displaystyle O}{\overset{\|}{C}}-OH \;+\; H-NH-\underset{\displaystyle \underset{SH}{\underset{|}{CH_2}}}{\underset{|}{CH}}-\overset{\displaystyle O}{\overset{\|}{C}}-OH \longrightarrow$$

$$CH_3-\underset{\displaystyle NH_2}{\underset{|}{CH}}-\overset{\displaystyle O}{\overset{\|}{C}}-NH-\underset{\displaystyle \underset{SH}{\underset{|}{CH_2}}}{\underset{|}{CH}}-\overset{\displaystyle O}{\overset{\|}{C}}-OH \;+\; H_2O$$

This type of reaction is another example of a condensation reaction. The new C–N covalent bond is called a *peptide* or an *amide* bond. The amide functional group present in all proteins is:

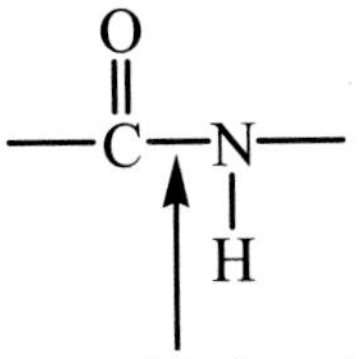

$$-\overset{\displaystyle O}{\overset{\|}{C}}-\underset{\displaystyle H}{\underset{|}{N}}-$$

peptide bond

The molecule above, in which two amino acids are joined, is called a *dipeptide*. Peptides are structures intermediate in size between amino acids and proteins. The term *polypeptide* refers to long molecular chains containing many amino acid units. An amino acid unit in a polypeptide chain is called a *residue*.

Proteins are so complex that four levels of structural features have been identified. The structure of proteins is extremely important in determining how efficiently and effectively a protein will function. We will discuss the first two levels of protein structure.

1. **The Primary Structure**. Each protein has a unique amino acid sequence of its polypeptide chain. It is the amino acid sequence that distinguishes one protein from another. Proteins also differ in the numbers and types of amino acids, but the main difference is the sequence of amino acid residues.

2. **The Secondary Structure**. This refers to the spatial relationship of amino acid units that are close to one another in sequence. The configuration that appears in many proteins is the α helix, shown in Figure 25.11 of the textbook. In this configuration, the polypeptide is coiled much like the arrangement of stairs in a spiral staircase. Figure 25.11 also shows that the α helix is stabilized by the presence of intermolecular hydrogen bonds (dashed lines). The C=O group of each amino acid is hydrogen bonded to the NH group of the amino acid that is located four amino acids ahead in the sequence. The α helix is the main structural feature of the oxygen-transport protein, hemoglobin, and of many other proteins. The structure of hemoglobin is shown in Figure 25.13 of the text.

   The β-pleated sheet is another common secondary structure. In this structure, a polypeptide chain interacts strongly with adjacent chains by forming many hydrogen bonds. See Figure 25.12 of the text.

EXAMPLE 25.3

What are the five chemical elements found in proteins?

Proteins are made up of the same elements as amino acids. Therefore, proteins contain C, H, O, and N. Two amino acids, methionine and cysteine, also contain sulfur, S.

PRACTICE EXERCISE

3. Sketch a portion of the polypeptide chain consisting of the amino acids glycine, valine, and alanine, in that order.

Text Problems: 25.20, 25.22

PROBLEM TYPE 3: NUCLEIC ACIDS

The chemical composition of the cell nucleus was first studied in the 1860s by Friedrich Miescher. He found the major components to be protein and a new material not previously isolated. This material was found to be acidic, and so it was referred to as **nucleic acid**.

Nucleic acids are polymers that store genetic information and control protein synthesis. There are two types of nucleic acids, deoxyribonucleic acid (DNA) and ribonucleic acid (RNA). ***DNA*** carries all the genetic information necessary to carry on reproduction and to sustain an organism throughout its lifetime. ***RNA*** transcribes these instructions and controls the synthesis of proteins needed to implement them.

DNA molecules are among the largest known with molar masses up to 10 billion g/mol. Hydrolysis of nucleic acids show that they are composed of only four types of building blocks: a phosphate group, a sugar, purines, and pyrimidines.

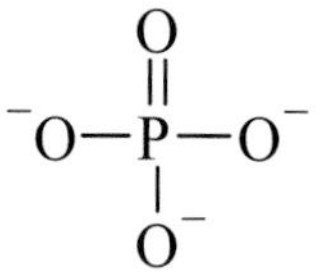

Phosphate

One of two sugars is present, ribose in RNA and deoxyribose in DNA.

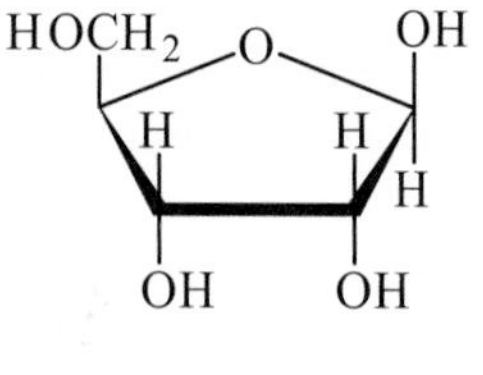

Ribose

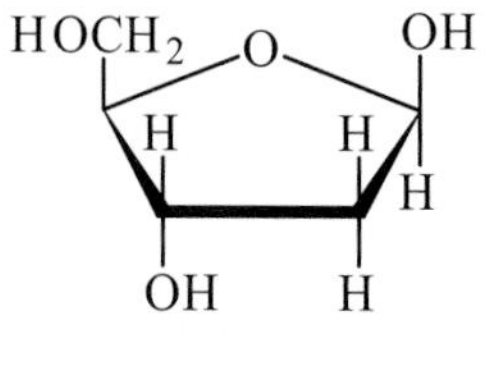

Deoxyribose

Two purines, adenine and guanine, are found in both DNA and RNA.

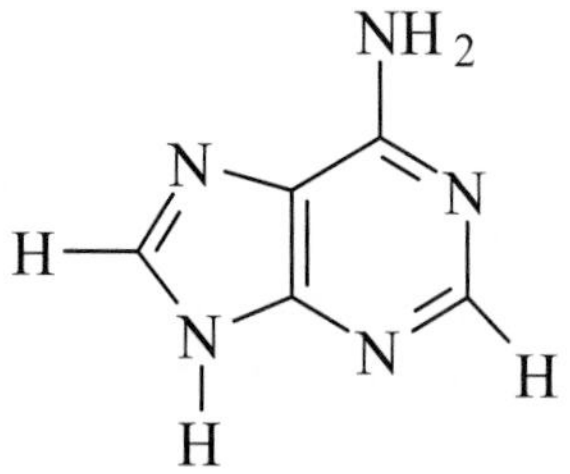

Adenine

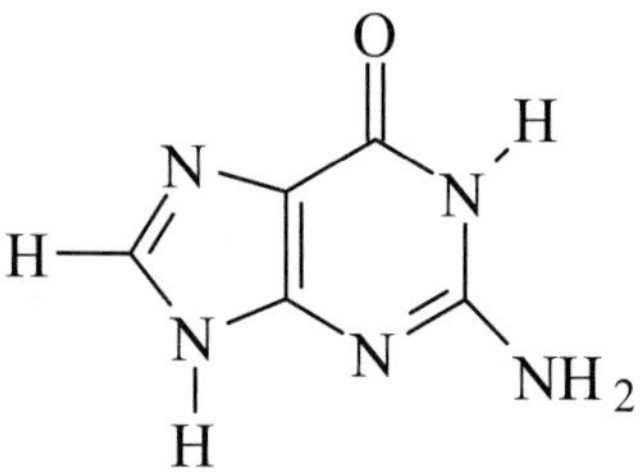

Guanine

DNA also contains the pyrimidines thymine and cytosine. RNA also contains thymine and another pyrimidine, uracil.

Thymine Cytosine Uracil

The purines and pyrimidines are collectively called ***bases***. The table below summarizes the building blocks of DNA and RNA.

Building Blocks of DNA and RNA

	DNA	**RNA**
	Phosphate (P)	Phosphate (P)
Sugar	Deoxyribose (D)	Ribose (R)
Purines	Adenine (A)	Adenine (A)
	Guanine (G)	Guanine (G)
Pyrimidines	Thymine (T)	Thymine (T)
	Cytosine (C)	Uracil (U)

In the 1940s, Edwin Chargaff studied the composition of DNA. His analysis of the base composition of DNA showed that the amount of adenine always equaled that of thymine and that the amount of guanine equaled that of cytosine.

In 1953, Watson and Crick proposed a double-helical structure for DNA. The repeating unit in each strand is called a **nucleotide**, which consists of a base-deoxyribose-phosphate linkage (see Figure 25.18 of the text). The two strands are *not* identical, rather they are complementary. Adenine in one strand is always paired with thymine in the other strand. Guanine is always paired with cytosine. The base pairing is consistent with Chargaff's rules. Base pairing and the resulting association of the two strands are the result of *hydrogen bonding*. Hydrogen atoms in a base in one strand are attracted to unshared electron pairs on oxygen and nitrogen atoms of the base attached to the other strand (see Figure 25.19 of the text).

RNA, on the other hand, does not follow the base-pairing rules. X ray data and other evidence ruled out a double-helical structure for RNA. RNA is single-stranded.

EXAMPLE 25.4

What types of forces cause base pairing in the double-stranded helical DNA molecule?

The purine bases (adenine and guanine) in one strand of DNA form hydrogen bonds to the pyrimidine bases (thymine and cytosine) in the other DNA strand. Hydrogen atoms covalently bonded to nitrogen carry a partial positive charge. These H atoms are attracted to lone-electron pairs on oxygen and nitrogen atoms of another base. Since two complementary bases are attached to different strands, the hydrogen bonds hold the strands together.

PRACTICE EXERCISE

4. If the base sequence in one strand of DNA is A, T, G, C, T, what is the base sequence in the complementary strand?

Text Problem: 25.28

ANSWERS TO PRACTICE EXERCISES

1. **(a)**

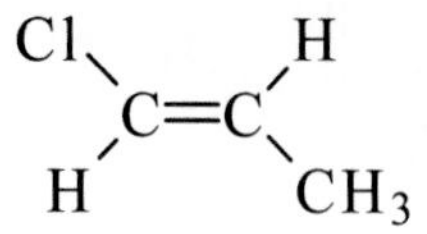

$$\begin{matrix} Cl & & H \\ & C{=}C & \\ H & & CH_3 \end{matrix}$$

(b)

$$\begin{matrix} H & & Cl \\ & C{=}C & \\ H & & Cl \end{matrix}$$

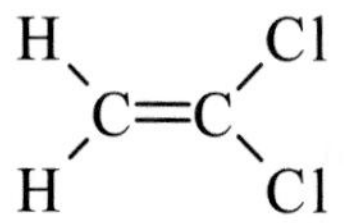

2. Nylon 66 (first prepared in 1931) and polyester.

3.

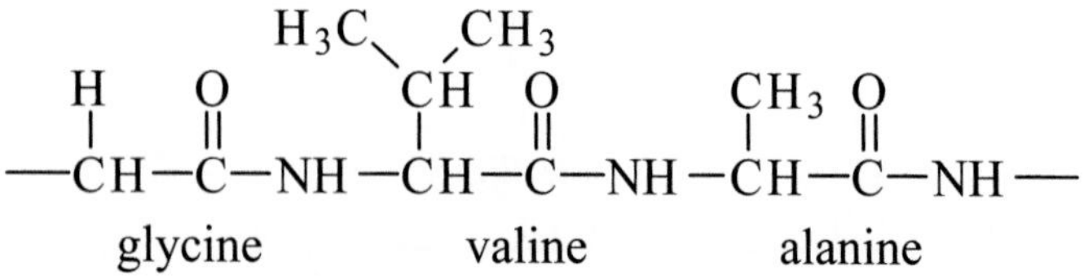

$$-\underset{\text{glycine}}{\overset{\overset{\displaystyle H}{|}}{CH}-\overset{\overset{\displaystyle O}{||}}{C}}-NH-\underset{\text{valine}}{\overset{\overset{\displaystyle H_3C\diagdown\diagup CH_3}{\overset{\displaystyle CH}{|}}}{CH}-\overset{\overset{\displaystyle O}{||}}{C}}-NH-\underset{\text{alanine}}{\overset{\overset{\displaystyle CH_3}{|}}{CH}-\overset{\overset{\displaystyle O}{||}}{C}}-NH-$$

4. Recall that according to Chargaff's rules, adenine (A) is always paired with thymine (T) and guanine (G) is always paired with cytosine (C). The complementary base sequence is T, A, C, G, A.

SOLUTIONS TO SELECTED TEXT PROBLEMS

25.8 The repeating structural unit of the polymer is:

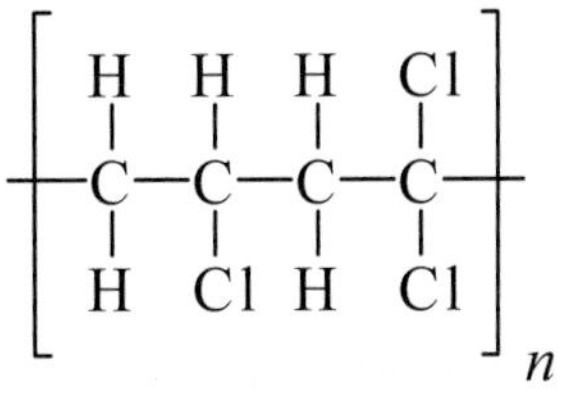

$$\left[\begin{array}{cccc} H & H & H & Cl \\ | & | & | & | \\ -C- & C- & C- & C- \\ | & | & | & | \\ H & Cl & H & Cl \end{array}\right]_n$$

Does each carbon atom still obey the octet rule?

25.10 Polystyrene is formed by an addition polymerization reaction with the monomer, styrene, which is a phenyl–substituted ethylene. The structures of styrene and polystyrene are shown in Table 25.1 of your text.

25.12 The structures are shown.

(a) $H_2C{=}CH{-}CH{=}CH_2$

(b) $HO{-}C({=}O){-}CH_2{-}CH_2{-}CH_2{-}CH_2{-}CH_2{-}CH_2{-}NH_2$

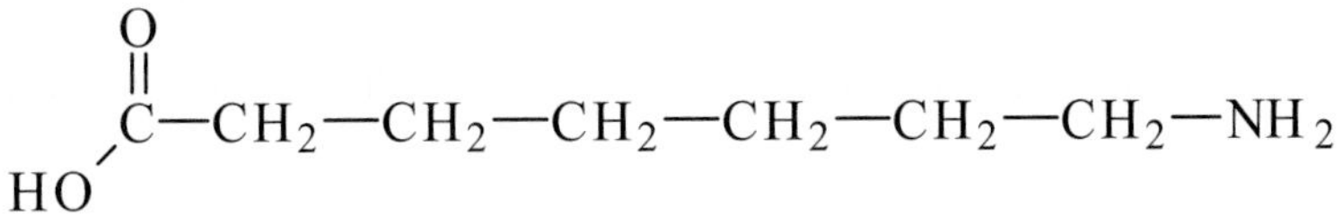

25.20 The main backbone of a polypeptide chain is made up of the α carbon atoms and the amide group repeating alternately along the chain.

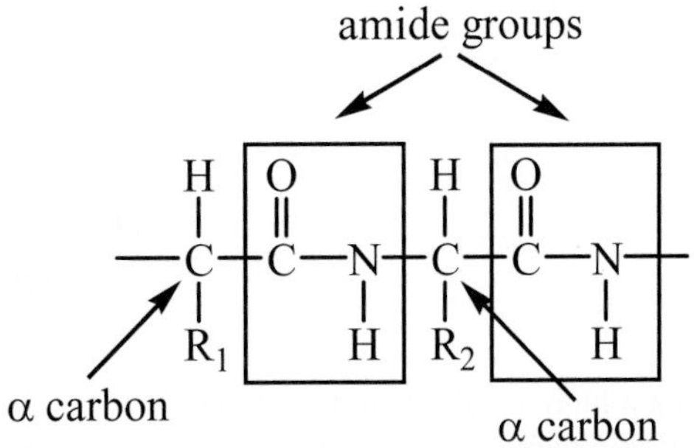

amide groups

$$-\underset{R_1}{\overset{H}{C}}-\overset{O}{\overset{\|}{C}}-\underset{H}{N}-\underset{R_2}{\overset{H}{C}}-\overset{O}{\overset{\|}{C}}-\underset{H}{N}-$$

α carbon α carbon

For each R group shown above, substitute the distinctive side groups of the two amino acids. Their are two possible dipeptides depending on how the two amino acids are connected, either glycine–lysine or lysine–glycine. The structures of the dipeptides are:

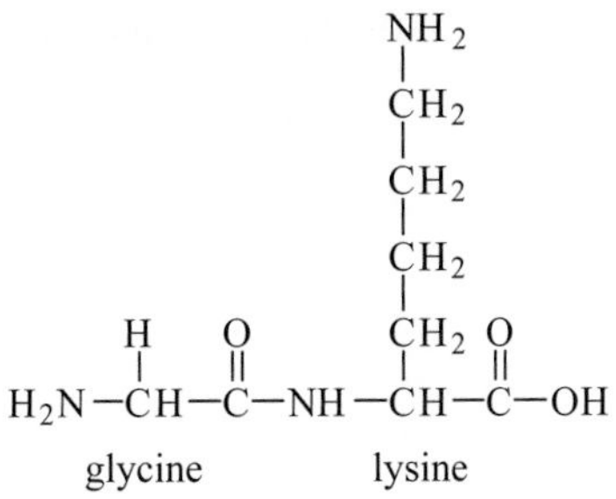

$$H_2N-\underset{\text{glycine}}{\overset{H}{CH}}-\overset{O}{\overset{\|}{C}}-NH-\underset{\text{lysine}}{\overset{CH_2CH_2CH_2CH_2NH_2}{CH}}-\overset{O}{\overset{\|}{C}}-OH$$

and

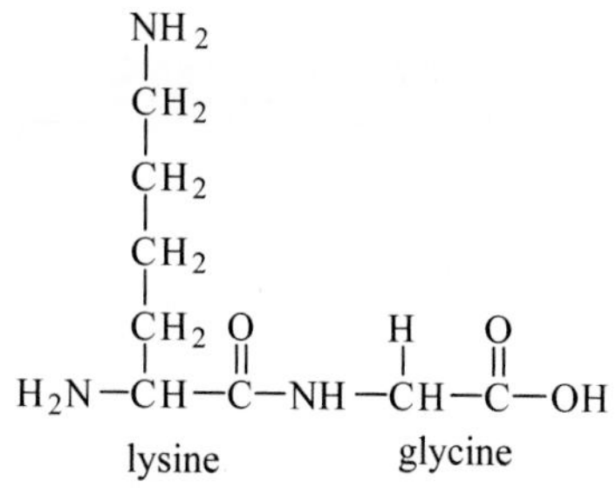

$$\mathrm{H_2N{-}\underset{\underset{\underset{\underset{\underset{NH_2}{|}}{CH_2}}{|}}{CH_2}}{CH}{-}\overset{O}{\overset{\|}{C}}{-}NH{-}\overset{H}{\overset{|}{C}H}{-}\overset{O}{\overset{\|}{C}}{-}OH}$$

(side chain on the lysine α-carbon: $-CH_2-CH_2-CH_2-CH_2-NH_2$)

lysine glycine

25.22 The rate increases in an expected manner from 10°C to 30°C and then drops rapidly. The probable reason for this is the loss of catalytic activity of the enzyme because of denaturation at high temperature.

25.28 Nucleic acids play an essential role in protein synthesis. Compared to proteins, which are made of up to 20 different amino acids, the composition of nucleic acids is considerably simpler. A DNA or RNA molecule contains only four types of building blocks: purines, pyrimidines, furanose sugars, and phosphate groups. Nucleic acids have simpler, uniform structures because they are primarily used for protein synthesis, whereas proteins have many uses.

25.30 The sample that has the higher percentage of C–G base pairs has a higher melting point because C–G base pairs are held together by three hydrogen bonds. The A–T base pair interaction is relatively weaker because it has only two hydrogen bonds. Hydrogen bonds are represented by dashed lines in the structures below.

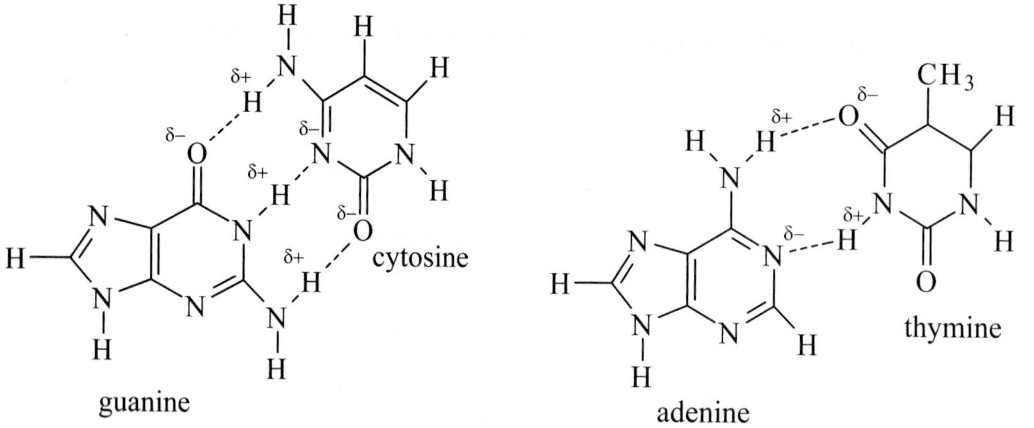

guanine cytosine

adenine thymine

25.32 Leg muscles are active having a high metabolism, which requires a high concentration of myoglobin. The high iron content from myoglobin makes the meat look dark after decomposition due to heating. The breast meat is "white" because of a low myoglobin content.

25.34 Insects have blood that contains no hemoglobin. Thus, they rely on simple diffusion to supply oxygen. It is unlikely that a human-sized insect could obtain sufficient oxygen by diffusion alone to sustain its metabolic requirements.

25.36 From the mass % Fe in hemoglobin, we can determine the mass of hemoglobin.

$$\%\ \text{Fe} = \frac{\text{mass of Fe}}{\text{mass of compound (hemoglobin)}} \times 100\%$$

$$0.34\% = \frac{55.85\ \text{g}}{\text{mass of hemoglobin}} \times 100\%$$

$$\textbf{minimum mass of hemoglobin} = \mathbf{1.6 \times 10^4\ g}$$

Hemoglobin must contain **four Fe atoms per molecule** for the actual molar mass to be four times the minimum value calculated.

25.38 The type of intermolecular attractions that occur are mostly attractions between nonpolar groups. This type of intermolecular attraction is called a **dispersion force**.

25.40 This is as much a puzzle as it is a chemistry problem. The puzzle involves breaking up a nine-link chain in various ways and trying to deduce the original chain sequence from the various pieces. Examine the pieces and look for patterns. Remember that depending on how the chain is cut, the same link (amino acid) can show up in more than one fragment.

Since there are only seven different amino acids represented in the fragments, at least one must appear more than once. The nonapeptide is:

Gly–Ala–Phe–Glu–His–Gly–Ala–Leu–Val

Do you see where all the pieces come from?

25.42 No, the milk would *not* be fit to drink. Enzymes only act on one of two optical isomers of a compound.

25.44 We assume $\Delta G = 0$, so that

$$\Delta G = \Delta H - T\Delta S$$

$$0 = \Delta H - T\Delta S$$

$$\boldsymbol{T} = \frac{\Delta H}{\Delta S} = \frac{125 \times 10^3\ \text{J/mol}}{397\ \text{J/K}\cdot\text{mol}} = \mathbf{315\ K} = \mathbf{42°C}$$

25.46

$$\left[-\overset{\overset{\displaystyle O}{\|}}{C}-O-\langle\bigcirc\rangle-\underset{\underset{\displaystyle CH_3}{|}}{\overset{\overset{\displaystyle CH_3}{|}}{C}}-\langle\bigcirc\rangle-O- \right]$$

25.48 **(a)** The –COOH group is more acidic because it has a smaller $\text{p}K_\text{a}$.

(b) We use the Henderson-Hasselbalch equation, Equation (16.4) of the text.

$$\text{pH} = \text{p}K_\text{a} + \log\frac{[\text{conjugate base}]}{[\text{acid}]}$$

At pH = 1.0,

–COOH $$1.0 = 2.32 + \log\frac{[\text{-COO}^-]}{[\text{-COOH}]}$$

$$\frac{[\text{-COOH}]}{[\text{-COO}^-]} = 21$$

$-NH_3^+$ $$1.0 = 9.62 + \log\frac{[\text{-NH}_2]}{[\text{-}\overset{+}{\text{N}}\text{H}_3]}$$

$$\frac{[\text{-}\overset{+}{\text{N}}\text{H}_3]}{[\text{-NH}_2]} = 4.2 \times 10^8$$

Therefore the **predominant species** is: $\mathbf{CH(CH_3)_2-CH(\overset{+}{N}H_3)-COOH}$

At pH = 7.0,

–COOH $$7.0 = 2.32 + \log\frac{[\text{-COO}^-]}{[\text{-COOH}]}$$

$$\frac{[\text{-COO}^-]}{[\text{-COOH}]} = 4.8 \times 10^4$$

$-NH_3^+$ $$7.0 = 9.62 + \log\frac{[\text{-NH}_2]}{[\text{-}\overset{+}{\text{N}}\text{H}_3]}$$

$$\frac{[\text{-}\overset{+}{\text{N}}\text{H}_3]}{[\text{-NH}_2]} = 4.2 \times 10^2$$

Predominant species: $\mathbf{CH(CH_3)_2-CH(\overset{+}{N}H_3)-COO^-}$

At pH = 12.0,

–COOH $$12.0 = 2.32 + \log\frac{[\text{-COO}^-]}{[\text{-COOH}]}$$

$$\frac{[\text{-COO}^-]}{[\text{-COOH}]} = 4.8 \times 10^9$$

$-NH_3^+$ $$12.0 = 9.62 + \log\frac{[\text{-NH}_2]}{[\text{-}\overset{+}{\text{N}}\text{H}_3]}$$

$$\frac{[\text{-NH}_2]}{[\text{-}\overset{+}{\text{N}}\text{H}_3]} = 2.4 \times 10^2$$

Predominant species: $\mathbf{CH(CH_3)_2-CH(NH_2)-COO^-}$

(c) $$\text{pI} = \frac{pK_{a_1} + pK_{a_2}}{2} = \frac{2.32 + 9.62}{2} = \mathbf{5.97}$$

25.50 **(a)** All the sulfur atoms will have an octet of electrons and be sp^3 hybridized.

(b) cysteine

(c) Denaturation will lead to more disorder (more microstates). ΔS is positive. To break a bond, energy must be supplied (endothermic). ΔH is positive. Consider the equation $\Delta G = \Delta H - T\Delta S$. This type of process with a positive ΔH and a positive ΔS is favored as the temperature is raised. The $T\Delta S$ term will become a larger negative number as the temperature is raised eventually leading to a negative ΔG (spontaneous).

(d) If we assume that the probability of forming a disulfide bond between any two cysteine residues is the same, then, statistically, the total number of structurally different isomers formed from eight cysteine residues is given by $7 \times 5 \times 3 = 105$. Note that the first cysteine residue has seven choices in forming an S–S bond, the next cysteine residue has only five choices, and so on. This relationship can be generalized to $(N-1)(N-3)(N-5) \cdots 1$, where N is the total (even) number of cysteine residues present. The observed activity of the mixture—the "scrambled protein"—is less than 1% of that of the native enzyme ($1/105 < 0.01$). This finding is consistent with the fact that only one out of every 105 possible structures corresponds to the original state.

(e) Oxidation causes sulfur atoms in two molecules to link, similar to the cross-linking described in the problem. The new compound formed has less odor compared to the compound secreted by the skunk.

Crit
"How to Write a
By Gen

"This how-to book is full of good advice for college-bound students. Thirty essays with brief introductory notes give teens concrete examples of what works. Twelve essays that 'bombed' are also analyzed, pointing out pitfalls to avoid. Examples of interview questions and answers are also included. An excellent guide for all students who are hoping to continue their education."

–*School Library Journal*

"A wealth of tips, tricks, techniques, advice and useful strategies…absolute 'must-read' for anyone interested in competing for scholarship funds."

–*Bookwatch, Midwest Book Review*

"Sound advice for the college bound and useful for counselors as well as for libraries."

–*Paula Rohrlick, KLIATT*

Praise for Other Books by Gen and Kelly Tanabe

Authors of *Get into Any College, Get Free Cash for College* and *Accepted! 50 Successful College Admission Essays*

"Upbeat, well-organized and engaging, this comprehensive tool is an exceptional investment for the college-bound."

–*Publishers Weekly*

"Upbeat tone and clear, practical advice."

–*Book News*

"What's even better than all the top-notch tips is that the book is written in a cool, conversational way."

–*College Bound Magazine*

"Invaluable information ranging from the elimination of admission myths to successfully tapping into scholarship funds."

–Leonard Banks, *The Journal Press*

"A present for anxious parents."

–Mary Kaye Ritz, *The Honolulu Advertiser*

"When you consider the costs of a four-year college or university education nowadays, think about forking out (the price) for this little gem written and produced by two who know."

–Don Denevi, *Palo Alto Daily News*

"The Tanabes literally wrote the book on the topic."

–*Bull & Bear Financial Report*

"Filled with student-tested strategies."

–Pam Costa, *Santa Clara Vision*

"The first book to feature the strategies and stories of real students."

–*New Jersey Spectator Leader*

HOW TO WRITE A WINNING SCHOLARSHIP ESSAY

INCLUDING 30 ESSAYS THAT WON OVER $3 MILLION IN SCHOLARSHIPS

You can win or lose a scholarship with your essay. Learn how to write an essay that wins.

Step-by-step instructions on how to craft a winning scholarship essay and ace the interview.

Get valuable advice from actual scholarship judges and winners.

This book is a 'must have' for high school and college students who want to learn how to win scholarships.

Gen and Kelly Tanabe

Harvard graduates and award-winning authors of

The Ultimate Scholarship Book ▪ *Get Free Cash for College*

1001 Ways to Pay for College

Special contributions by Gregory James Yee

How to Write a Winning Scholarship Essay
By Gen and Kelly Tanabe

Published by SuperCollege, LLC
2713 Newlands Avenue
Belmont, CA 94002
650-618-2221
www.supercollege.com

Credits: Cover design by TLC Graphics, www.TLCGraphics.com. Design: Monica Thomas. Cover photograph © iStockphoto.com/Jerry Moorman. All essays in this book are used with the permission of their authors.

ISBN13: 978-1-61760-007-4

Manufactured in the United States of America
10 9 8 7 6 5 4 3 2 1

Library of Congress Cataloging-in-Publication Data

Tanabe, Gen S.
How to write a winning scholarship essay / by Gen and Kelly Tanabe.
p. cm.
ISBN 978-1-61760-007-4 (alk. paper)
1. Scholarships. 2. College applications--United States. 3. Universities and colleges--United States--Admission. 4. Exposition (Rhetoric) I. Tanabe, Kelly Y. II. Title.
LB2338.T363 2012
808'.066378--dc22

2012026470

CONTENTS AT A GLANCE

TABLE OF CONTENTS

Chapter 7. Judges' Roundtable: The Scholarship Essay / 151

Chapter 8. Winning Interview Strategies / 159

Chapter 9. Real Interview Questions & Answers / 175

Chapter 10. Judges' Roundtable: The Interview / 203

Chapter 11. Final Thoughts / 211

Scholarship Directory / 215

A directory of 125 scholarships including information on how to apply.

Index / 251

About the Authors / 256

SPECIAL FEATURES

Stories & Advice from Winners & Judges

These stories of success and failure from students and advice from judges and experts are both entertaining and enlightening.

Judges' Roundtable: Chapters 3 - 7 - 10

We sat down with judges and experts from around the country to find out what it takes to win a scholarship. Read their frank advice on what students have done right and wrong.

This book would not have been possible without the selfless contributions of scholarship winners, judges and experts. They gave their time and shared their knowledge for your benefit.

We would like to recognize the special contributions made by Gregory James Yee.

We dedicate this book to all of the people who helped to make it possible. And to you, our dear reader, we hope you will use these lessons to create your own winning scholarship essays.

CHAPTER
ONE

THE SECRET TO WINNING A SCHOLARSHIP

In this chapter:

- **Why you are 60 seconds away from the trash bin**
- **How your essay or interview will make or break your chances of winning a scholarship**
- **Why you can win even without straight A's**
- **Who we are and why we know so much about winning scholarships**

You Are 60 Seconds Away from the Trash Bin

If you witnessed the judging of a scholarship competition, you'd be surprised at how quickly decisions are made. It's not unusual for scholarship judges to decide in less than 60 seconds whether your application advances to the next round or gets tossed into the rejection pile.

With so little time, how do you capture the attention of these discriminating judges and improve your chances of winning? For most competitions the secret is in your **essay** or **interview**. These give the judges the insight they need to separate the winners from the runners-up.

If you are like the majority of scholarship applicants, you probably aren't sure what makes an essay or interview a "winner." And how would you know? It's unlikely that you have access to dozens of past scholarship winners who can share their experiences and show you their winning essays and interview answers. And it is even more unlikely that you've had the opportunity to speak with actual scholarship judges who can tell you exactly what qualities cause a candidate to stand out as a winner.

Obviously, anyone with this kind of access and knowledge has a tremendous advantage for winning a scholarship.

Now you can have this advantage.

In this book you will learn about scholarship winners, read their essays and interviews and learn the strategies that they have used to win more than $3 million in awards. (That amount is not a misprint. In fact, one incredible student received over $250,000 in scholarship offers!) You will also have the opportunity to hear from scholarship judges who will take you inside the selection process to reveal what they look for when choosing winners.

Unlike any other book on scholarships, by reading *How to Write a Winning Scholarship Essay* you will learn how to:

> **Find the scholarships you are most likely to win.** We identify the best places to find awards and strategies for selecting those that you have the best chances of winning.

Craft a winning scholarship essay. The best way to learn how to win scholarships is from the experiences of others. You will read 30 actual scholarship essays that were used by students like you to win free money for college. Our complete essay-writing workshop will also guide you step-by-step through selecting a topic, writing about it and editing your own winning scholarship essay.

Avoid costly essay mistakes. Failure can be a great teacher, too. Through 12 essays that were less than exemplary (in other words they bombed), you will see exactly how to avoid the mistakes that have doomed others.

Interview confidently and skillfully. With comprehensive interview strategies and examples of over 20 questions you are likely to be asked along with successful answers, you will have everything you need to ace the scholarship interview.

Discover what scholarship judges want. Three special chapters called *The Judges' Roundtables* reveal what scholarship judges seek in selecting winners. This knowledgeable group of experts has seen thousands of applications and decided the fate of many applicants. Their lessons and advice are indispensable.

While the thought of making a winning impression in 60 seconds is certainly daunting, it is also an opportunity. By investing the time to learn the skills presented in this book, you will give yourself a tremendous advantage over other applicants. While most of them will be quickly eliminated within the first 60 seconds, your application will steadily progress through each round and ultimately help you to emerge from the competition a winner.

The Scholarship Strategies You'll Learn

This book is jam packed with secrets, tips and strategies. It's also filled with plenty of examples to show you how these strategies work in the real word. The following is a brief summary of what you will learn in each chapter:

Chapter 2: Where to Find Great Scholarships. Before you can win a scholarship you need to find them. This chapter shows you the best places to find scholarships and also how to select the awards that you have the best chance of winning.

Chapter 3: Judges' Roundtable: Inside the Selection Process. Meet actual judges and understand from their perspective what it takes to win scholarships.

Chapter 4: Essay Writing Workshop. In this information-packed chapter we give you everything you need to craft a powerful essay. We guide you step-by-step from selecting a topic to using effective writing techniques to avoiding common mistakes.

Chapter 5: 30 Winning Scholarship Essays. See how strategy is put into action in these essays that won $3 million in scholarships. See how your essay compares and be inspired by these successes.

Chapter 6: 12 Essays That Bombed. The best lessons often come from failure–and preferably the failure of someone else! These disaster essays illustrate important lessons of what not to do in the essay.

Chapter 7: Judges' Roundtable: The Scholarship Essay. Many of the scholarship judges have read hundreds if not thousands of essays and know what works and what doesn't. See what the judges have to say about the making of a winning essay.

Chapter 8: Winning Interview Strategies. Learn how to deliver a knockout interview. Discover what every good interviewee knows and how to overcome interview nervousness.

Chapter 9: Real Interview Questions & Answers. Preview typical interview questions and review example responses to learn what makes a powerful answer.

Not Just Looking for Straight A Students

Discover Card Tribute Award Scholarship

You might think that scholarships seek only students with perfect SAT scores and flawless GPAs. This is not true. Many scholarships including those awarded by Discover® Card are looking beyond grades and test scores.

For the Discover Card Tribute Award® Scholarship, the judges look at many factors besides grades, and test scores are not even requested in the application.

"We believe that success in life is more than a GPA. It's also what you can accomplish in your personal life and in your community," says Shirley Kennedy Keller.

Keller encourages all students to apply, even those with less than perfect academic records or test scores. "This scholarship applies to virtually every high school junior who has a B minus to C plus grade point average who wants to continue their education or training beyond high school. That hopefully provides some encouragement to students who say that scholarships don't apply to me," she says.

The message is clear: If you don't have perfect grades or test scores don't let that prevent you from applying for awards.

The Discover Card Tribute Award Scholarship is no longer offered, but this profile still illustrates how scholarship organizations consider more than grades.

Chapter 10: Judges' Roundtable: The Interview. Find out what the judges are really listening for when they interview you.

Chapter 11: Final Thoughts. A few final tips before you embark on the journey to winning free cash for college.

Why We Know So Much about Scholarships

You may be wondering who we are and what we know about winning scholarships. As the authors of several books on college admission and scholarships including *The Ultimate Scholarship Book, Get Free Cash for College, 1001 Ways to Pay for College* and *Get into Any College*, we have had the unique opportunity to meet thousands of students and scholarship judges. In writing this book we conducted extensive research and interviewed dozens of scholarship judges and scholarship-winning students. We have distilled all of this research into the easy-to-read pages of this book.

But just as important as the research that went into this book is the fact that we've been where you are today. Both before and during college we were fanatical about applying for scholarships. Using the strategies that we have incorporated into this book, we won more than $100,000 in merit-based scholarships. This money was instrumental in allowing us to graduate from Harvard University debt-free.

In addition, we have served as scholarship judges for numerous competitions. One scholarship that we judge is the SuperCollege.com Scholarship, which receives over 10,000 applications each year. Reading these application essays has given us valuable experience, and we have seen what works as well as what mistakes students make over and over. We are also expert interviewers, having conducted both scholarship interviews and admission interviews for Harvard.

The sum total of this experience, research and know-how is contained within these pages. If you follow these strategies, you too can become a scholarship winner. Always remember that someone is going to win every scholarship that is out there, and there is no reason for that someone not to be you.

CHAPTER
TWO

WHERE TO FIND GREAT SCHOLARSHIPS

In this chapter:

- **Scholarship goldmines**
- **How to decide which scholarships to apply to and which to avoid**
- **How to uncover the mission of any scholarship**

Exploring Scholarship Goldmines

When we were looking for scholarships, we found them in nearly every place imaginable. We discovered some in the dusty collection of books at our library, others by serendipitous newspaper announcements of past winners. We even found an award advertised on a supermarket shopping bag.

Having personally spent hundreds of hours scouring the planet for scholarships and meeting hundreds of other successful scholarship winners, we have learned where most scholarships are hidden. To help make your scholarship hunt more efficient, we present what we believe are the best places to look for scholarships.

As you search, keep in mind that not every scholarship you find is one you should apply for. As the list of possibilities grows, evaluate each award to determine if it is right for you. Don't worry–we'll show you how to do this later in the chapter. But knowing which awards to pass on is vital since it lets you focus your time and energy on those awards that you have the best chance of winning.

The first two obvious places to find scholarships are:

Books. There are a number of good scholarship books from which to choose. When looking for a guide, seek one that offers detailed descriptions of the awards. Most importantly, make sure that the book has an easy-to-use index. You don't have time to read through every scholarship, so an index will help narrow your choices quickly. For example, our scholarship directory *The Ultimate Scholarship Book* not only contains thousands of awards but also has indexes based on criteria like field of study, ethnicity, athletics, hobbies, talents and much more to help you pick awards that match your talents and abilities.

Internet Websites. A great way to find scholarships is through the Internet. One of the benefits of online scholarship databases is that they can be updated often. Check out the free scholarship search on our website at www.supercollege.com. By creating a personal profile you can let our database do the work of finding awards that match you. Here are some websites we recommend to help you get started:

- SuperCollege (www.supercollege.com)
- CollegeAnswer (www.collegeanswer.com)
- MoolahSPOT (www.moolahspot.com)
- The College Board (www.collegeboard.com)
- Scholarships.com (www.scholarships.com)
- AdventuresinEducation (www.adventuresineducation.org)
- CollegeNet (www.collegenet.com)
- Mario Einaudi Center for International Studies (www.einaudi.cornell.edu/funding)

Regardless of which websites you use, always make sure that they are free, (i.e. there are no fees for using the service). Never pay to do an online scholarship search.

While scholarship books and online databases are easy ways to find scholarships, you also need to do your own detective work. With literally millions of scholarships available it is impossible for any one book or website to list them all.

We believe that one of the best places to find scholarships is right in your own backyard–your community. Start with the following:

Counselor or financial aid officer. Do this right now. Call your counselor or financial aid officer and make an appointment. Before the meeting, determine how much money you will need for college and prepare a resume or list of your activities and awards. During the meeting explain your situation and ask if there are any scholarships that your counselor or advisor can recommend. These counselors and financial aid officers probably know more about the awards available in the community and on campus than anyone else. But, it is up to you to take the initiative to meet with them and give them enough information so they can recommend appropriate awards.

It's important whenever you speak to a counselor (either in high school or college) that you inquire about any scholarships that require a nomination. With these competitions, the applicant pool is almost always smaller. The most difficult hurdle is that you need to get nominated. You have nothing to lose by asking, and if nothing else, it shows how serious you are about financing your education.

Ambassadors Wanted

Rotary International Ambassadorial Scholarships

Designed to promote international understanding, the Ambassadorial Scholarships program has assisted more than 30,000 men and women from 100 nations since the program began in 1947.

"We're looking for the people who want to give something back to the community, people who we hope will make a difference in the world," says Russ Hobbs, a district scholarship chairman in the San Francisco Bay Area. "We're not looking for someone who will go back to their community or company and not give anything back."

Previous recipients in Hobbs' district have included an aspiring politician who speaks five languages, a former teacher who has interned with the United Nations and the holder of a patent for a shallow water pump.

One of the reasons why recipients of the Rotary award must be committed to making a difference is that they are ambassadors representing their home country abroad. As a part of the program, scholars make presentations about their home country in the countries in which they study.

Brent Drage of Rotary International says that the organization seeks people who will "represent our culture and appreciate the culture they are visiting." These are also qualities that are valued by members of the Rotary club.

And the organization hopes that scholars who return stay involved with the program, speaking to future applicants and even becoming Rotarians themselves.

Activities. Many clubs and organizations on campus offer awards for their members. Meet with the officers or advisors to see what is available. Also check with the national parent organization, if the group has one, since it may also provide scholarship funds.

Professional associations. One or more professional associations exist for practically every career field. These groups often offer awards for students in their field. For example, the American Dental Association and American Medical Association provide scholarships for students who want to become future dentists and doctors. If you have a strong idea of what you want to do after college, these professional associations can be a real gold mine of scholarships.

Community organizations. You don't have to belong to an organization to win a scholarship. In fact, many community groups raise money with the intent of giving it away to members of their community who are prospective students. Local Rotary clubs, American Legions and Lions clubs often offer scholarships for outstanding students in the community. These groups view their scholarship programs as part of their service to the community. Open a phone book or go online and call the 10 largest organizations in your area.

Hometown professional sports team. Is your city the home of a professional sports team? If so contact the front office to see if they offer scholarships. Many teams offer scholarships that have nothing to do with athletic ability. You can also visit the official website of your hometown professional teams and look for a "community," "foundation" or "player's foundation" link.

Employer. If you have a full- or part-time job, check with your employer for awards. Many companies offer educational support as an employee benefit. If your employer doesn't offer a scholarship, suggest that they start one.

Parents' employer. Companies often award scholarships to the children of their employees. Ask your parents to speak with their human resources department about scholarships and other educational programs offered to employees and their families.

Parents' union. Many unions also sponsor scholarships for the children of their members. Again, have your parents speak with the union officers about union-sponsored scholarships and other educational programs.

Church or religious organizations. Religious organizations may provide scholarships for members. Inquire both locally at your house of worship as well as with the national organization, if any.

Local government. Often, local city council members and state representatives have a scholarship fund for the students who live in their districts. Even if you didn't vote for them, call their offices and ask if they offer any scholarships.

Local businesses. Local businesses often provide awards to students in the community. Start inquiring at your local Chamber of Commerce or similar business organization.

Local newspaper. Most community newspapers make announcements about local students who win scholarships. Keep a record of the scholarship announcements or go to the library and look at back issues of the newspaper. Check last year's spring issues and you'll probably find announcements of scholarship recipients. Contact the sponsoring organizations to see if you're eligible to enter the next competition.

After you have exhausted the resources in your own community, you can then expand your search to your entire state or even the nation. Many large corporations offer scholarships (think: Coca-Cola, Microsoft, Intel, Best Buy, etc.). Fortunately, most of these larger state-wide and nation-wide awards will be listed in scholarship books and on websites. The downside is that there will be a lot of competition for these awards.

Work Experience & Financial Aid

Knight Ridder Minority Scholars Program

Students who become Knight Ridder Minority Scholars not only receive a $40,000 scholarship, they also receive four summer internships and a full-time job after they graduate from college.

Nominated by one of the 31 local newspapers owned by Knight Ridder, scholars are selected in their senior year of high school. Even before they start college, the students work at a 12-week internship during the summer between their senior year of high school and freshman year of college. For the next three summers, the students intern for Knight Ridder and then work for the company for one year after they graduate.

"It's building a relationship between students and the company early on at a time when students want to learn more about the newspaper business," says Jacqui Love Marshall, vice president of human resources, diversity and development.

In addition to the internships and job opportunity, the students also meet annually at a Scholars Retreat. The retreat allows the students to meet executives of the company, interact with each other and gain additional training. A recent retreat brought the students to the conference of the Asian American Journalists Association and offered the students additional workshops on subjects such as workplace political skills, business social skills and strategic planning in the newspaper industry.

"For us it's a great opportunity to find new talent. For students it's a wonderful opportunity for them to get a diverse experience before graduating," Marshall says.

The Knight Ridder Minority Scholars Program is no longer offered, but this profile still provides insight into why companies sponsor scholarships.

Which Scholarship Is Right for You?

Recently, a student wrote to us with a problem. He had conducted a search for scholarships on the Internet. Anticipating that he would find a handful of awards to apply for, he was shocked to find not handfuls but bucketfuls–more than 100 potential scholarships. Unless he made applying for scholarships his full-time job, there was no way that he could apply for all of them.

You will likely find yourself in a similar situation. Since there are so many scholarships available, the problem may not be finding awards but deciding which ones you have the best chance of winning. Although

When & Where to Find Scholarships

CollegeAnswer.com

The best advice that Michael Darne, director of business development for Sallie Mae's CollegeAnswer.com, has for students is to start early and use both high and low tech search tools.

"It really helps to start early. Students should begin thinking about scholarships in their sophomore year," recommends Darne. Even if you are not able to apply for every award that you find, keeping a running list now will give you a huge advantage once you become a senior.

As for where to find scholarships, Darne recommends two distinctly different methods. "You can use a scholarship search engine like the CollegeAnswer search engine," says Darne. But besides the Internet, Darne also advises to not overlook the people around you.

"Meet with your guidance counselor. They're going to know about most of the good local scholarships. Get the word out to everyone you know that you're going to college and you need the money. Some of these scholarships may not be big money, but $500 here or $1,000 there can add up quickly," he says.

there is no way to predict if you will win a scholarship, there are some techniques you can use to select those that fit you best and therefore offer you the best chance of winning. Naturally, these are the ones for which you should apply.

The key is to realize that almost every scholarship organization has a mission or goal for giving away its money. Few groups give away free money for no reason. For example, a nature group might sponsor a scholarship with the goal of promoting conservation and encouraging students to be environmentally conscious. To this end the group will reward students who have demonstrated a concern for the environment and have some plan to contribute to this cause in the future.

Understanding the mission of the scholarship is important because it will clue you into the kind of student the organization is interested in finding. If you have the background, interests and accomplishments that match this mission, then it is a scholarship you have a good chance of winning.

In our example of the nature group, if you are passionate about conservation, are active in an organization like the Sierra Club and know that you could write a compelling essay about your interest in global warming, then you would be a great candidate. If, on the other hand, you can't remember the last time you spent more than an hour outdoors, this award is not for you and you would be wasting your time by applying.

By understanding the mission of the scholarship, you can determine if you are the kind of student the organization wants to reward.

Make Learning Their Mission Your Mission

There is no mystery to figuring out why organizations give away money. In most cases, the organizations come right out and tell you what they are trying to achieve with the award.

Start by carefully reading the award description. Oftentimes organizations spell out what they are looking for in the description of who is eligible for the award. Sometimes they provide the criteria that they use for judging the competition. Criteria can include qualities such as academic achievement, community involvement, leadership, specific career goals and character.

These requirements are valuable clues. Is there a minimum GPA? If there is and it's high, then academic achievement is probably important. Does the application provide a half page to list your activities? If so, then your involvement in organizations and projects outside of school is probably a fairly significant part of the selection criteria. Do you need to submit an essay on a specific topic or a project to demonstrate your proficiency in a field of study? All of these requirements are clues about what the scholarship committee thinks is important. Visualize yourself filling out their application. Would you have enough information to fill all of the blanks and answer all of the questions? If not, then you may want to consider passing on this award to focus on one that you are more qualified to win.

After reading the application, research the awarding organization. What is the group's mission? Who are its members? What do they hope to accomplish? You can probably guess what kind of student will impress a group of physicists versus poets. All things being equal, most clubs and organizations want to reward students who are most similar to their membership. If you don't know much about the organization, contact them to find out more. Check out their website. Read their brochures or publications. The more you know about why the organization is giving the award, the better you'll be able to understand how you may or may not fit with their expectations.

For a local scholarship you may actually know the previous winner. If you do, definitely contact him or her and learn as much as you can about the selection process. Ask for advice. Don't forget to ask winners why they think they won. Often, their familiarity with the contest and experience of having gone through the competition will give them an impression of why they were selected.

Making the Match

As your list of scholarships grows, you need to start prioritizing. Create an ordered list with the scholarships that fit you best written at the top. As you find new scholarships, you can decide where in the list they should go. When you start to complete applications, just start at the top of the list and work your way down. (Don't forget to list the deadlines since this may affect the order in which you list the awards.) The goal is not to get through the entire list but to get through as many as possible while still allowing enough time to create a quality application for each competition.

Prioritizing not only gives you an easy way to approach each competition but it also forces you to really think about what the scholarship committee is looking for. You will want to ask critical questions such as these:

- Do you have the background to fit with the expectations of the committee?
- Are your talents synonymous with those necessary to win?
- Have you accomplished tasks or won awards that show achievement in the area of the award?

You also avoid wasting time on awards that at first glance sound good (perhaps they have huge prizes) but you really have no chance of winning.

Finding awards that match your background, achievements and interests is extremely important. All of the essay and interview strategies presented in the following chapters are most effective if you have spent time selecting awards that are a match to your qualifications.

CHAPTER
THREE

JUDGES' ROUNDTABLE: INSIDE THE SELECTION PROCESS

In this chapter:

- **Get the inside scoop from real scholarship judges and experts**
- **Discover how the selection process actually works**
- **See who judges the competitions**
- **Understand what makes the difference between being a winner and a runner-up**

Meet the Scholarship Judges

This is the first of three Judges' Roundtables in this book. We present these candid conversations with actual scholarship judges and experts so that you can hear in their own words what works and what doesn't.

Imagine that you were trying to win $1,000. One way to do this would be to buy a lottery ticket. Another way would be to apply for a scholarship. The difference between the two is that you cannot affect your chances of winning the lottery, but you can affect your chances of winning a scholarship.

When applying for scholarship competitions, one of the most important things to understand is why the organization is giving the award. Every organization has a reason for providing funds to students. These reasons can be quite varied. The organization may want to contribute to the communities in which it does business, increase the status of members of a minority or underrepresented group or build morale among its employees.

By understanding what the organization hopes to achieve through its scholarship program, you can develop essays and interview answers that best show how you fulfill the mission of the award. Here are some examples of purposes of scholarship programs. Try to imagine that you are applying to these awards and think about how knowing the purpose of each scholarship would help you to create a better application.

What is the purpose of your award? Why are you giving away free money?

Georgina Salguero
Hispanic Heritage Awards Foundation
"Our mission is to promote and reward Hispanic excellence and to provide a greater understanding of the contribution of Hispanic Americans in the U.S. One way we do this is through identifying and rewarding outstanding youth who in turn will serve as the role models for other youth.

"In addition to a scholarship award, we also provide each winner with a $1,000 check to donate to an established nonprofit organization of the student's choice. This is a way for the student to give right back to the community, to never forget who they are and where they came from. After all, how many 17-year-olds do you know who have a way to thank the community that has helped them get to where they are?"

Cathy Edwards
Discover Financial Services Inc.
"We wanted to structure a program that would not only demonstrate our commitment to education but also invest in our children's futures."

Ellen Frishberg
Johns Hopkins University
"The Hodson award goes to students who present the best credentials in our incoming class academically and in leadership. We're looking not just for local leadership but leadership on a regional or national level."

Brent Drage
Rotary International Ambassadorial Scholarship Program
"Our scholarship program was started to further international understanding by providing money for students to study abroad. That's why we call our winners ambassadors. Recipients are expected to be ambassadors of good will."

Tracey Wong Briggs
All-USA Academic Teams
"Our program is really an editorial project for *USA Today.* What we are trying to do is tell the stories about what outstanding students can do. We are looking to reward students who have a good story to tell about something they have accomplished either academically or extracurricularly."

Marie M. Ishida
California Interscholastic Federation Scholar-Athlete of the Year
"We want to acknowledge and recognize outstanding scholars who are also good athletes."

Jacqui Love Marshall
Knight Ridder Minority Scholars Program
"We're looking to find people early in their career planning who have an interest in newspapering, either journalism or working on the business side. We also want to raise interest and awareness about careers in newspapers among students."

Corisa Moreno
The Music Center Spotlight Awards
"Our award is meant to encourage high school students in the visual and performing arts to continue to follow their passion by providing support."

Wanda Carroll
National Association of Secondary School Principals
"We administer several awards that have different goals. For example, the Prudential Spirit of Community Award seeks to identify and honor middle level high school students on the basis of their volunteer work. For Wendy's High School Heisman Award we are looking to honor and promote both citizenship and athletic ability. For the Principal's Leadership Award we want to recognize an outstanding student leader."

Bob Murray
USA Funds
"Our scholarship program is aimed toward lower-income students in an attempt to narrow the gap in college attendance rates between lower-income and higher-income families."

Who typically judges the scholarship competition?

When applying for scholarships, it's important to keep in mind who the judges are. If you know that the judges will be local leaders, you may focus on how you have contributed to the community. If you can, ask who will be on the selection committee. This may help you select which essay subject and interview topics of conversation are the most appropriate.

Georgina Salguero
Hispanic Heritage Awards Foundation
"Within the 12 cities of our regional competition we try to get local leaders to be the judges. For example, we work with the head of the chamber of commerce, people at various universities, clergy, civic leaders and other leaders within the community. We feel that these people know best how to evaluate the students in their communities."

Tracey Wong Briggs
All-USA Academic Teams
"For the high school competition, the judges include representatives from the National Education Association and National Association of Secondary School Principals. For the college competition, the judges include representatives from the American Council on Education, National Association of Independent Colleges and Universities, American Association of Colleges for Teacher Education and a former winner."

Laura DiFiore
FreSch! Free Scholarship Search
"I try to get a diverse group of people. Recent judges have included an Episcopalian minister, nurse, former librarian, substitute teacher, accountant, retired homemaker, computer programmer and high school junior."

Jacqui Love Marshall
Knight Ridder Minority Scholars Program
"We form our judging committee with two or three people at the corporate office level, sometimes a person from our local newspaper (*The San Jose Mercury News* or a Monterey newspaper)

and then I usually ask someone from the community or local industry. Each year I aim to create a diverse committee."

Russ Hobbs
Rotary International Ambassadorial Scholarship Program
"We have a six-member judging panel. Three are former Ambassadorial Scholars. The other three are Rotarians who have a love of the scholarship program."

How does your scholarship selection process work?

Understanding how the selection process works as well as realizing that not all competitions are the same will help you see where to focus your energies. If you can, learn about the scholarship program's selection process, what happens at each stage and how the applicants are evaluated.

Kimberly Hall
United Negro College Fund
"Students submit their applications directly to the UNCF. We perform a preliminary screening and send the strongest applications that we feel match the goals of specific donors on to those donors. Typically, the donor will then select the winner."

Trisha Bazemore
Coca-Cola Scholars Foundation
"The foundation first selects semifinalists based on a quantitative analysis of students' initial applications. Then, a 27-member Program Review Committee reviews the semifinalists' applications and essays. From this judging, 250 students are advanced as finalists. These students are then asked to travel to Atlanta at the expense of the Scholars Foundation to attend the Scholars Weekend, where they meet the other finalists and are interviewed by members of the National Selection Committee. Based on their applications and this interview, finalists are designated either as one of 50 National Scholars, receiving a four-year, $20,000 scholarship or as one of 200 Regional Scholars, receiving a four-year, $4,000 scholarship."

Corisa Moreno
The Music Center Spotlight Awards
"We start with our preliminary audition and there we screen for the basics. For example, let's say you're a singer. We would look for basic skills like tone and pitch. For dance we would look at the quality of your dance skills and musicality. As the applicants advance, the competition gets more intense because you've got a handful of very qualified dancers and musicians. At the final level everyone is extremely talented and what often makes the winner stand out is that extra level of professional persona."

Tracey Wong Briggs
All-USA Academic Teams
"We have two steps. The preliminary judges score each applicant against a score sheet. That's why it's real important for people to read the nomination form carefully since every item on the nomination form is what we are judging for on the score sheet. During the finals, the judges meet and they try to build an academic team. They read all of the finalist applications and pick the first, second and third teams."

Wanda Carroll
National Association of Secondary School Principals
"When you are dealing with thousands of applications, you have to find a way to narrow the field. All of the applications are screened using an algorithm. We determine a set number of criteria that they have to meet and a certain number of points that they need to have to be considered for the finals. We narrow down our applicants to the top 1,000 or 2,000 applications. Each application is then read by a team of readers. A team will select a state and read all of the applications from that state. The number of awards for a specific state is based on the population of that state."

Laura DiFiore
FreSch! Free Scholarship Search
"There are two stages for the judging. I personally pull out the no go's. These are the applicants that on first glance are clearly not qualified to move forward in the competition. Those that pass this stage are sent to judging. During the judging stage you have to impress all 10 judges. We use a point scale, and the

winning applicant and essay has received the highest marks from all of the judges. We usually pick out the winner in about three hours. The second and third prizes tend to take us two days. Out of all the entries, there will be about 50 stories and 70 poems going to final judging."

What qualities do you look for when selecting the winner?

If you know what qualities the organization is seeking in the winner, you can highlight those achievements that best showcase these qualities. Carefully read the organization's literature, website and publications to figure out what the judges seek. Or contact the organization directly to ask. Imagine how much stronger your application will be if you apply for these awards and know what the judges themselves think is most important.

Bob Murray
USA Funds
"In addition to the income requirement, we consider past academic performance and future potential, leadership, participation in school and community activities, work experience and career and educational aspirations and goals. Each of these additional criteria carry approximately equal weight."

Trisha Bazemore
Coca-Cola Scholars Foundation
"We are primarily looking at overall achievements in leadership, not only in school activities but also within their communities."

Georgina Salguero
Hispanic Heritage Awards Foundation
"Because the students select from seven career or talent categories in which to enter, the selection committee asks, 'Do we see this student excelling in this category 10 years from now?' With any selection, the cream rises to the top. We look at their academics, community service, leadership, as well as their fit and potential within their category."

Jacqui Love Marshall
Knight Ridder Minority Scholars Program
"We're looking for primarily academic strength. This is a person who is likely to do well in college so they won't be struggling. We also want a person who has a genuine interest in either journalism or the newspaper business."

Cathy Edwards
Discover Financial Services Inc.
"We're asking students to not only list community service efforts but to describe why they've been meaningful. It makes them stop and reflect. It helps them demonstrate their well-roundedness. It's not the students who have the highest GPA. We're trying to help the students who might not have scholarship dollars available to them."

Tracey Wong Briggs
All-USA Academic Teams
"The judges are looking for how you use your intellectual skills outside of the classroom, how you take academic excellence beyond getting an 'A' in class. Judges ask things like how are you using the knowledge that you're gaining in the classroom? Do you have a rigorous curriculum and are you challenging yourself?"

Corisa Moreno
The Music Center Spotlight Awards
"We look at the end result. Our past finalists are at a very professional level. A lot of them go on to major dance companies and institutions such as Juilliard. They are at that professional level where you could stick them on a stage anywhere and they would be a crowd pleaser."

Laura DiFiore
FreSch! Free Scholarship Search
"We're looking for originality and creativity, what we call the goosebump effect. When you get to the last sentence of the essay, you lean back and say 'wow.'"

Wanda Carroll
National Association of Secondary School Principals
"The National Honor Society Scholarships are based on the students' character, service and leadership as exemplified in their application and essay. As the name suggests, winners for the Principal's Leadership Award are selected based on their leadership. Our selection committee seeks a student who is class president, who is involved in athletics being captain or co-captain or who demonstrates community service as the head of their youth group. In short we are looking for someone who is obviously showing leadership qualities. The Prudential Spirit of Community Award, on the other hand, is based entirely on community service. One recent winner started a *Suitcases for Kids* program to provide suitcases for children in homeless shelters to move their belongings."

Russ Hobbs
Rotary International Ambassadorial Scholarship Program
"We're not looking for that person who has a 4.0 GPA and spends all his time in the library. We're looking for the people who want to give something back to the community, people who we hope will make a difference in the world and who believe in the values that we believe in Rotary. It's our hope that at some point in the future our winners will become Rotarians."

What sets the winner apart from the runner-up?

One of the most frequent observations from scholarship judges is that there are many more students qualified to win than there are scholarships available. This means that they must look for that little something extra to separate the winners from the almost winners. Here is some guidance to help you understand what can set you apart from the other applicants.

Shirley Kennedy Keller
American Association of School Administrators
"It's about what the students have accomplished in more than just one area of their lives. We believe that success in life is

more than a GPA. It's also what you can accomplish in your personal life, in your community. We're looking for the best all-around applicants. We're looking for students who are going to be successful in their lives."

Corisa Moreno
The Music Center Spotlight Awards
"They really have to enjoy what they're doing. That comes off quite a bit in their performances. You can see when they have a deep love for their art form."

Brent Drage
Rotary International Ambassadorial Scholarship Program
"Students who have a little more concrete plan tend to edge out others who don't. We understand that at that age you need to be flexible about your future, but we also want to see that you have some grasp on where you're headed."

Q What advice do you have for future applicants?

Some of the best advice you can get is from those who will be judging your scholarship applications. Since many have seen hundreds if not thousands of applications, they see firsthand how you can find awards, create a powerful application and essay and avoid the mistakes that other students make.

Kimberly Hall
United Negro College Fund
"Start looking for scholarships and applying early. I encourage students to start in their sophomore year of high school. Of course, they can start even earlier than that too."

Tracey Wong Briggs
All-USA Academic Teams
"Pursue the interests you love. One of the things we see is students who just care so much about what they're doing. If you do that, the awards will come."

Mario A. De Anda
Hispanic Scholarship Fund
"Read the application before you start filling it out. Spend time on the personal statement. It's our first impression of you so it needs to be good."

Wanda Carroll
National Association of Secondary School Principals
"There is nothing that disqualifies a student quicker than not following instructions. Make sure that all the signatures are there. Make sure you have every 'i' dotted and every 't' crossed. We have hundreds each year who don't make the first cutoff. It's not that they are not qualified but they just haven't followed the instructions."

Russ Hobbs
Rotary International Ambassadorial Scholarship Program
"Sell us on the things that you have done outside of the academic area. You need to let us know why we should want to fund you over 30 or 40 other candidates in our district."

Ellen Frishberg
Johns Hopkins University
"Focus and do well. We're looking for the academic program not just grades. We also want to see that students do well in activities and that they don't just join lots of organizations for the sake of listing them on their application. Focus on leadership since this award tends to go to students who invented, chaired, captained or did something that shows leadership."

Laura DiFiore
FreSch! Free Scholarship Search
"The bottom line is that we want to give you the money. It's up to you to give us a reason to say 'yes.' Please, please, please give us reasons to say 'yes.'"

Participating Judges & Experts

Trisha Bazemore, Program Assistant, Coca-Cola Scholars Foundation

Tracey Wong Briggs, Coordinator, *USA Today* All-USA Academic and Teacher Teams

Wanda Carroll, Program Manager, National Association of Secondary School Principals

Mario A. De Anda, Director of Scholarship Programs, Hispanic Scholarship Fund

Laura DiFiore, Founder, FreSch! Free Scholarship Search Let's Get Creative Short Story and Poetry Scholarship Contest

Brent Drage, Resource Development Assistant, Rotary International Ambassadorial Scholarship Program

Cathy Edwards, Manager, Public Relations and Charitable Sponsorships, Discover Financial Services Inc.

Ellen Frishberg, Director of Student Financial Services, Johns Hopkins University

Kimberly Hall, Peer Program Manager, United Negro College Fund

Russ Hobbs, District Scholarship Chairman, Rotary International Ambassadorial Scholarship Program

Marie M. Ishida, Executive Director, California Interscholastic Federation (CIF) Scholar-Athlete of the Year

Shirley Kennedy Keller, Program Director, American Association of School Administrators

Jacqui Love Marshall, Vice President of Human Resources, Diversity and Development, Knight Ridder Minority Scholars Program

Corisa Moreno, Project Coordinator, The Music Center Spotlight Awards

Bob Murray, Manager of Corporate Communications, USA Funds

Georgina Salguero, Senior Manager, Programs and Events, Hispanic Heritage Youth Awards

CHAPTER
FOUR

ESSAY WRITING WORKSHOP

In this chapter:

- **Why the essay is critical**
- **What judges look for in the scholarship essay**
- **The three common features of all money-winning essays**
- **How to find the perfect topic**
- **The keys to crafting a powerful essay**
- **Why sob stories don't work**
- **Hard to believe but true essay mistakes**
- **How to recycle your essay**

Why the Essay Is Critical to Winning

Let's imagine for a moment that you are a scholarship judge. You have an enormous pile of applications in front of you. From the application forms, you can get basic information about each applicant such as grades, test scores and brief descriptions of their activities.

But without being able to meet each candidate, how do you get a sense of who they are so that you can determine if they are the most deserving of your money? One of the best (and sometimes only) ways to get to know the applicants beyond their cut and dry statistics is through their essays.

This is why for many scholarship competitions the essay is the most important part of the application and where you should spend the most time. Scholarship judges view the essay as their window into who you are, your passions and your potential. It is their way of getting to know you without actually meeting you. And it is where you can make the strongest and most meaningful impression.

There are some interesting implications depending on the type of student that you are. If you are a straight "A" student with excellent test scores and a flawless academic record, you may be tempted to rely on these achievements to carry your application all the way to the final round. However, if you neglect the essay, your achievements (no matter how impressive) may advance you beyond the preliminary round, but you won't win the big prize.

On the other hand, if you are an average student and know that other applicants will have better academic achievements, you can use the opportunity that the essay provides to make yourself stand out. In many cases you will actually be able to beat applicants who have higher GPAs and test scores.

Regardless of your accomplishments and academic achievements, you need to write a powerful essay if you want to win a scholarship.

The Making of a Powerful Essay

Every summer there is a blockbuster adrenaline-laden action movie complete with pumped-up action hero, oversized guns and unbelievable

Focus on Solutions

United Negro College Fund

Some students seem to think that the more tears they can get the selection committee to shed, the better their chances will be to win. It's true that scholarship judges will feel sympathy for students who have gone through difficult times. But they will reward those who have done so and succeeded or who have a plan for succeeding despite these obstacles.

In the over 400 scholarship programs that the United Negro College Fund administers, essays play an important role. "It gives you a sense of who the student is and what they want to do with the money if they win. It gives you more of a picture of the student as a whole as opposed to just a name," says Kimberly Hall, peer program manager.

While many students write about serious issues or hardships, Hall advises students to take a positive approach when writing their essays. Instead of focusing only on their problems, students should explain what they have faced and then describe their plan of action for the future. The best essays have a sense of "purpose and direction," she says.

car chases. In the midst of exploding buildings and the hero tearing away in a red sports car, it would be out of place for the background music to be a polka. The sights and sounds need to fit together to create the desired atmosphere. If one element is out of place (like a polka during the climax of an action scene) it destroys the effect of the entire movie.

Similarly, what makes a good essay is that it fits within the context of the overall application. In other words, the essay and all other elements in your application package–such as your list of activities and teacher recommendations (if required)–must fit together to create the effect you want.

Let's say that you are applying for an award based on community service. In the application you list all of the community service groups that you belong to and service project awards that you've won. But in the essay you vent about your disgust for the homeless and how they should find jobs instead of blocking your passage on sidewalks. Your essay may be brilliantly conceived and written, but if its message is not in line with the rest of your application, it will create a conflicting message and keep you out of the winners' bracket.

Even if we reverse this example, the result is the same. Imagine that you wrote a brilliant essay about community service but had no related activities to back up the commitment you profess in the essay. The essay, no matter how well written, will not make up for a lack of actual involvement in community service work.

When you think about the essay, consider it within the context of the entire application. You want to present a cohesive message with the essay as the centerpiece. Each piece of the application should add to this unified message.

At this point many students ask, "How do I know what the message or theme of my essay and application should be?" The answer is actually quite simple and goes back to why you decided to apply for the scholarship in the first place.

The theme of your essay and application is almost always determined by the goal of the award or why the organization is giving away the money.

For example, a minority advocacy organization may provide an award to help members of an under-represented ethnic group to pursue higher education. A private foundation may give an award to preserve the memory of a late benefactor who supported students entering teaching. A professional organization may award money to encourage students to enter their profession.

As you learned in Chapter 2, it's important to research and uncover the purpose of each award. Then you can use this information to guide the essay and application.

Once you know the goal of the organization, use that knowledge to choose which aspect of your life to highlight as the general theme of the essay. If you are applying for the award for under-represented students, you may want to focus on your potential and how you will be a role model for others in the future. To apply for the educator or other professional awards, you'd want to highlight your future in education or the field of the awarding organization. In other words, use the goal of the award as a guide for the essay.

Four Common Features of All Winning Essays

Let's imagine that you have done research on the scholarship organization and have a sense of what they hope to gain by giving away their money. You have even thought of a few themes that you could write about in the essay. No matter what topic you ultimately choose, there are qualities that are shared by all successful scholarship essays. It doesn't matter what type of student wrote the essay or what it is about, to write a winning essay you need to keep these points in mind.

#1 Originality

For your essay to be a winner, it needs to be original. Remember that your essay will be among thousands of other essays that are being judged. If your essay does not stand out, it will be forgotten along with your chances of winning.

There are two ways to be original. The first is to find a unique topic. Think about what makes you...well...*you.* What point of view or life experience can you share that is unique? One judge we know uses the "thumb test." Place your thumb over your name at the top of the essay, and ask yourself if any of your classmates could have written this essay. If the answer is "yes" then it fails the thumb test and is probably not original.

Unfortunately, finding a unique topic is very difficult, and that leads us to the second way that you can be original. Instead of racking your brains to come up with a 100% original topic, take an ordinary topic and approach it in an original way. For example, if you were writing about how your mother is a role model you would not want to approach it in the same way that everyone else will. Many applicants will write about how their mothers taught them the importance of education or showed them how to persevere in the face of adversity. If your essay is going to have any chance of winning, it needs to be different from those written by other competitors. So spend some time thinking–not

Hard to Believe But True Essay Mistakes

FreSch! Let's Get Creative Scholarship Contest

Laura DiFiore, the founder of FreSch! Free Scholarship Search, has seen many mistakes. Some have prompted her to separate essays into three piles: the good, the bad and the ugly. Those essays designated as the ugly are put into a box with a tombstone drawn on it.—Gen and Kelly

Bathroom humor. One applicant wrote an entire essay about excrement. "It met the requirements of originality but not creativity. It was gross."

Spelling mistakes. "While some mistakes are tolerated, if you can't spell your own major, you're not getting a scholarship."

Inferior, illegible printing. DiFiore received an essay in which the middle section of the applicant's printing ribbon ran out. This meant that she could only see the top third and bottom third of each letter. The entire middle of all of the letters was totally illegible.

Plagiarism. During a recent competition, DiFiore received essays from applicants who took last year's winning essay which was posted on her website, modified it and submitted it as their own.

Copycats. DiFiore received essays from three sisters who all wrote the same basic story.

Threats. One essay writer threatened the selection committee, "If you don't give me money I'm going to hunt you down."

Anonymous applicants. DiFiore has received applications with no name or address on them. It's difficult to award a scholarship to an anonymous applicant.

writing–about your mother. What is it specifically that she has done or said that has been so influential? Can you cite a concrete example? Maybe your mother has a secret recipe for meatloaf that she has shared with no one except you. Perhaps the moment that she revealed to you her treasured secret recipe was a milestone in your relationship. Focusing on this event and examining and analyzing it may yield a very powerful and certainly original essay. The truth is that we all have experiences and people that make us unique, and the key is to zero in on these and use them in your scholarship essays.

When you read the example essays in the next chapter, look at how each is unique. Pay special attention to how the authors present their topics. Notice how they often bring in points of view that help to make their essays original.

One sure way to ensure that your essay is original is to avoid common topics or approaches of other essay writers. In fact, these mistakes are so common that we have an entire chapter devoted to them. Be sure to carefully study the essays in Chapter 6, *12 Essays That Bombed*, to make sure that your topic or approach does not resemble any of these failures.

#2 Answer the Underlying Question

Have you ever been asked one question but felt like there was an underlying question that was really being asked? Maybe a parent has asked you something like, "Tell me about your new friend Karen." But what your parent is really asking is, "Tell me about your new friend Karen. Are her 12 earrings and tattoo-laden arms a sign that you shouldn't be spending so much time with her?"

In most cases the essay question is just a springboard for you to answer the real question the scholarship judges want addressed. An organization giving an award for students who plan to study business might ask, "Why do you want to study business?" But the underlying question they are asking is, "Why do you want to study business, and why are you the best future business person we should gift with our hard earned money?"

For every scholarship you will be competing with students who share similar backgrounds and goals. If you are applying to an award that supports students who want to become doctors, you can bet that 99% of the students applying also want to become doctors. Therefore, the goal of every scholarship judge is to determine the *best* applicant out of a pool of applicants who at first glance look very similar.

So let's distill the underlying question that the scholarship judges really want answered; that is, Why do you deserve to win? (Your answer should not be, "Because I need the money!")

Think about these two hypothetical essay topics: The Farmers Association asks about the future of farming. The Historical Society wants an analysis of the importance of history. While at first these two questions seem unrelated, they are both driving at the same thing: Tell us why you deserve to win.

In addressing either of these topics, you would need to recognize the underlying question. When writing the Farmers Association essay, you could discuss the general condition of farms and farmers, but you'd better be sure to include how you fit into the future of farming. Similarly when answering the Historical Society's question, you could write about history in any way that you please; but you should also include if not focus on your own past and future contributions to the field of historical research or preservation. Use the essay question as a way to prove to the scholarship committee that you are the worthiest applicant for the award.

#3 Share a Slice of Life

As you are explaining why you deserve to win, it is important that you also reveal something about yourself. Obviously, in the short space of 500 to 1,000 words you can't cover everything about you. This is why one of the most effective techniques is to share just a "slice of your life." In other words, don't try to explain everything. Just focus on one aspect of your life.

If you are writing about your involvement in an activity, it may be tempting to summarize your involvement over the years and list numerous accomplishments. However, this would sound more like a resume and it would not tell the judges something that they could not learn by reading your resume. However, if you focus on just one aspect or one day of an experience, you could spend some time below the surface and share something about who you are. In other words, you would be sharing a slice of your life.

Since many students write about activities in which they are involved, here are a few topics that you might want to consider. These will help you focus the essay and force you to share a slice of your life:

- What motivated you to get involved with this activity?
- How do you personally benefit from participating?
- How do you stay motivated during challenging times?
- Is there a person that you've met through this activity that has inspired you? How?
- What one accomplishment are you most proud of? Why?
- Have you ever considered quitting this activity? Why didn't you?
- What is one thing you learned from being involved?

These types of questions make you examine yourself and find a specific incident, moment or thought to share. Even if the subject of the essay is an activity that you enjoy, it is important that the judges who read it come away knowing more about you.

#4 Passion

As a student you have written a lot of essays. And let's be honest–most were probably on topics you didn't care much about. You might be tempted to approach the scholarship essay in the same way that you did when writing about the Roman Aqueducts, but this would be a tragic mistake. The last common feature of all winning essays is that they are written on subjects about which the author is truly passionate.

It is very difficult to fake passion for a subject. (Just try to be excited throughout your Uncle Larry's hourlong slideshow of his tonsil operation.) But when you are genuinely enthusiastic about something it does not take much effort for that energy to naturally show through in your writing. Therefore, when you are choosing a topic, be sure it is something you truly care about and are interested in. Without even trying, you will find that your sentences convey an excitement that the reader can almost feel.

When you read the examples in the next chapter, you will quickly see that the writers all cared deeply for their topics.

How to Find the Perfect Topic

When we were taught to color inside the lines, our artwork may have been neater but it was at the expense of creativity. The best time-tested method to develop creative ideas that lead to a great topic is through brainstorming. By thinking without restrictions, creativity flourishes. We have found that the best way to do this is to keep a notebook with you and write down ideas for topics whenever they pop into your head. Also set aside some time for a dedicated brainstorm session where you force yourself to generate new ideas.

When brainstorming topics don't be critical of the ideas you write down. Let your imagination roam. Also, ask your parents and friends for suggestions.

The one shortcoming of brainstorming is that sometimes a good idea does not make for a good essay. A thought may be too complex to write about within the limitations of the essay requirements.

The only way to really tell if an idea is good is to start writing. So from your list of ideas pick several that are the most promising and start composing an essay. Again, don't pay attention to the quality of the writing just yet. You are basically testing the topic to see if it has the potential to become a great essay.

If you get stuck and think the topic may not work then set it aside and try another. We have found that most students will try and then abandon two or three ideas for every good one they find. That means that you need a long list of ideas and must be willing to cut your losses and ditch a topic that does not pan out.

Putting Words onto Paper

At some point you can't escape the need to start writing. The best way to begin is the same as removing a bandage–just do it, and do it quickly. To help get you going, here are some strategies for writing:

Go beyond the Superficial

Knight Ridder Minority Scholars Program

To apply for the Knight Ridder Minority Scholars Program, students must write a personal statement. Many turn to the Knight Ridder website to get background information on the company. For some, this is a mistake.

"You realize they went to the site and cut and pasted material from it into their essay. But it's clear that they have little idea what these facts mean," says Jacqui Love Marshall, vice president of human resources, diversity and development.

The students cite statistics about the number of newspapers owned by Knight Ridder or the number of Pulitzer Prizes the media company has won. She adds, "It's almost the difference between writing a book report by having read the CliffsNotes versus having read the book."

What's more important than regurgitating statistics found on an awarding organization's website is finding a personal connection to the organization.

"When you're looking at dozens of these essays in the middle of the night you begin to differentiate between someone who put their heart and soul in it, that there is a level of commitment there versus essays that have all the requisite information but not a personal involvement," says Marshall.

The Knight Ridder Minority Scholars Program is no longer offered, but this profile gives insight into the importance of learning about the sponsoring organization.

Remember to Focus on Originality. While it is not always possible to come up with an original topic–especially if the question is the same for everyone–make sure that the essay contains originality or that the topic is approached in a novel way.

If you are writing about involvement in a sport, don't use common topics like how sports taught you the value of teamwork or how you scored the winning touchdown, goal or point. These are repetitive topics. Using them risks having your essay lost among the hundreds of others that sound similar to yours. It's perfectly fine to write about common topics like sports, but think of a different angle. Maybe you had a unique experience or can focus on an aspect of athletics that is often overlooked.

Be Specific. A common mistake in essay writing is to use general statements instead of specific ones. Don't write, "Education is the key to success." Instead, give the judges a slice of your life. Show them how education has impacted your life in a single experience or realization.

If you are writing about your desire to become an astronaut you might explain how this began when your father bought you a model rocket for Christmas. Focusing on a specific example of your life will help readers relate to your experiences and ensure that your essay is memorable and (as a bonus) original.

Share Something Personal. While some questions ask about a national or international problem or event, the scholarship committee still would like to know something about you. After all, they are considering giving their money to you.

Some of the better essays written about serious issues like drug abuse or nuclear proliferation have also found ways to incorporate information about the author. One student who wrote about the U.S. arms policy spoke about his personal involvement in a club at school that hosts an annual peace conference. He was able to tie in the large international policy issues with the more personal aspect of what he was doing on an individual level. It was a great policy essay, which also revealed something about the author.

Have a Thesis. It sounds obvious, but many students' essays don't have a clear point. Whether you are describing the influence of your father or the effect of World War II on race relations, you must have a central idea to communicate to the reader.

To see if your essay has a central thesis, try this simple exercise. Ask yourself, "What is the point of my essay in a single sentence?" Here are some answers that would satisfy the question for essays on independence and drug addition, respectively:

"Growing up in the country taught me to be independent."

"Treatment of addiction is the only way to win the war on drugs."

If you cannot condense the point of your essay into a single sentence, then the main point may not be clear enough. Or worse, your essay may not have a thesis.

Expand on Your Accomplishments. Winning a scholarship is about impressing the judges and showing them why you are the best candidate for a monetary award. Your accomplishments, activities, talents and awards all help to prove that you are the best fit. Since you will probably list your activities on the application form, use the essay to expand on one or two of the most important ones.

However, don't just parrot back what is on the application. Use the opportunity to focus on a specific accomplishment, putting it into the proper context. Share details. Listing on the application that you were a stage manager for a play does not explain that you also had to design and build all of the sets in a week. The essay allows you to expand on an achievement to demonstrate its significance.

Beware of Meaningless Facts. Some students approach the essay like a research paper, cramming it with statistics and survey results. You might think that the facts and figures "wow" judges. While this does display research skills, facts and figures alone hardly make a good essay. In particular, if you are trying to impress a corporation with your knowledge of their sales and global markets, don't just repeat facts from their website. You may use facts about the sponsoring organization, but be sure that they are essential to the essay. Don't repeat statistics without a reason, and don't think that the more you have the better.

Use Examples & Illustrate

Mark R. Eadie, Coca-Cola Scholars Regional Winner

"Make sure to use examples to illustrate points. Instead of saying 'I was active in high school,' describe your high school activities. Also, focus on one or two activities that had special meaning to you.

"There's a fine line between bragging and too much humility. Be honest about yourself and what you've done, and the scholarship committee will recognize this. Your essay is going to be read by real people who are intelligent and wise so don't make things up. Trying to trick them is like trying to trick parents; it just doesn't work.

"Also, the essay readers may have to read hundreds of essays, so give them something to remember you by."

Avoid Clichés. We are all guilty of using a cliché in our writing. "Don't cry over spilled milk." "Good things come to those who wait." "Try and try and you will succeed." These are all common clichés. It's important to avoid using them in the essay. Why? First of all, the use of clichés is just lazy writing. You are using a common phase instead of taking the time to come up with your own words. Second it's not your words and therefore it's not original. When you use a cliché you are penalized for being both lazy and unoriginal. It's just not worth it. If you find yourself writing a cliché, stop, and rewrite the idea in your own words.

Don't Write a Sob Story. Tear-jerking stories may be popular subjects for television specials and song lyrics, but they rarely, if ever, win scholarships. A common theme students write about is why they need the scholarship money to continue their education. While this is a perfectly legitimate topic, it is often answered with an essay filled with family tragedies and hardships–a sob story. Again, there is nothing wrong with writing about this topic, but don't expect to win if the intent of your essay is to evoke pity.

If your main point (remember our test) is this: "I deserve money because of the suffering I've been through," you have a problem. Scholarship committees are not as interested in problems as they are in solutions. What have you accomplished despite these hardships? How have you succeeded despite the challenges you've faced? This is more significant and memorable than merely cataloging your misfortunes.

Plus, don't forget that to win you have to be an original. The sob story is one of the more common types of essays, and it is hard to compete when you are telling the same story that literally hundreds of other students are also writing. Remember that every applicant has faced difficulties. What's different and individual to you is how you've *overcome* those difficulties.

Show Positive Energy. Mom has probably said: "If you don't have anything nice to say, don't say anything at all." Everyone likes an uplifting story. Especially, since you have your entire future ahead of you, scholarship judges want to feel your enthusiasm. In fact, one reason adults love to volunteer to be scholarship judges is to meet positive and enthusiastic young adults who do not have the cynicism or closed minds of adults.

Try to stay away from essays that are overly pessimistic, antagonistic or critical. This doesn't mean that you have to put a happy spin on every word or that you can't write about a serious problem. But it does mean that you should not concentrate only on the negative. If you are writing about a problem try to present some solutions.

Your optimism is what makes organizations excited about giving you money to pursue your passion for changing the world. Don't shy away from this fact.

The Importance of Editors

There is an old writer's saying: "Behind every good writer is an even better editor." If you want to create a masterpiece, you need the help of others. You don't need a professional editor or even someone who is good at writing. You just need people who can read your work and provide useful and constructive feedback.

Roommates, friends, family members, teachers, professors or advisors all make great editors. When others read your essay, they will find errors that you missed and help make the essay clearer to someone who is not familiar with the topic.

You will find that some editors catch grammar and spelling mistakes but will not comment on the overall quality of the essay. Others will miss the technical mistakes but give you great advice on making the substance of your essay better. It's essential to find both types of editors.

As you find others to help improve your essay, be careful that they do not alter your work so much that your voice is lost. Editing is essential but your writing should always be your own.

Recycle & Reuse

Recycling in the context of this discussion has no relation to aluminum cans or newspapers. What we mean is that you should reuse essays that you have written for college applications, classes or even other scholarships. Writing a good essay takes a lot of time and effort. When you have a good essay you'll want to edit it and reuse it as much as possible.

Sometimes, to recycle an essay, you must change the introduction. Try experimenting with this. You may find that while you might have to write a few new paragraphs you can still use the body of the original essay.

One word of caution: Don't try to recycle an essay when it just doesn't fit. The essay must answer the question given by the scholarship organization. It's better to spend the extra time to write an appropriate essay than to submit one that doesn't match the scholarship requirements.

How to Write a Great Introduction or Conclusion

Great novels have two things in common–a gripping introduction and a conclusion that leaves the reader with something to think about. Great essays share similar traits.

The first impression that the judges get is from the introduction. If it does not catch their attention and make them want to read further then you will lose even before you have had a chance. Here are some strategies for beginning any essay:

Create action or movement. Use an example or short story to create action right at the beginning. Have you noticed how most movies begin with a striking scene that quickly draws you in? Do the same with your introduction.

Pose a question. Questions draw attention as the readers think about their answers and are curious to see how you answer them in the essay. You can also use an interesting or surprising fact in place of a question.

Use descriptions. If you can create a vivid image for readers, they will be more likely to want to read on. Just be sure to do so succinctly since you don't want the introduction to be filled with detail that does not move the plot forward.

Conclusions are just as important as introductions since they are the last impression you will leave with the reader (the scholarship judge). Here are a few tips for the closing remarks.

Be thoughtful. The conclusion should end with something insightful. You may even decide to withhold a thought from the essay so that you have something for the conclusion.

Essays Get Better with Each Revision

Kristin N. Javaras, Rhodes Scholar

"I highly recommend showing your essay to people who have won fellowships themselves or who have read successful fellowship application essays before (and the more people the better). I feel that the revision process was crucial for my essay: I went through about seven or eight drafts of my personal statement before I was satisfied!"

Don't just summarize. Since the reader has just finished the essay, there is no need for a restatement of the points that you made. It's okay to wrap up your thoughts in one sentence, but try to add to the conclusion as a whole by making an extra point.

Don't be too quick to end. Too many students tack on a meaningless conclusion or even worse, don't have one at all. Have a decent conclusion that connects with the rest of the essay and that doesn't consist of two words, "The End."

As you look at the essay ask yourself: Will they think about what I have said after they have finished reading? If the answer is yes, then you have written a conclusion that you can be proud of.

Stay Motivated

Writing scholarship essays may not be the ideal way to spend a Friday night or Sunday afternoon. But remember that these essays can win you hundreds, if not thousands, of dollars for college. Try to keep this in mind when you feel burned out. If you really get down on writing take a break. Go outside. Watch some meaningless television. Then when you are refreshed get back to your essay.

In the next chapter are the actual essays that won the writers thousands of dollars in scholarships. At some point each of these writers got tired or disgusted and contemplated quitting. But each persevered and didn't give up. They pushed ahead and finished their essays. If they had given up they would never have won the money that they did and that all important college diploma would have been a far more expensive (and for some impossible) accomplishment.

CHAPTER
FIVE

30 WINNING SCHOLARSHIP ESSAYS

In this chapter:

- **30 real essays about challenges, family, issues, community service, career plans, leadership, academics, athletics and artistic talents**
- **Learn from and be inspired by these successful essays**

The Money-Winning Essays

You sit down at the computer, eyes focused on the monitor and fingers poised above the keyboard. You are ready to start writing your money-winning scholarship essay. But something is missing. Aha! What you need is inspiration.

In this chapter, we want to give you this inspiration. One of the best ways to learn how to write a successful essay is to read actual essays that won. While there is no single way to write a winning essay, most successful ones share traits such as originality, demonstrating why the author deserves to win and passion.

As you read these essays imagine that you are a scholarship judge. What image of the writer does the essay create? How do the essays make you feel? Would you give away your money to these writers?

Remember, unlike a creative writing assignment, the goal of a scholarship essay is to show the scholarship committee why you deserve to win. Keep in mind that these essays are meant to be examples of what worked for these particular students. Naturally, your essays will be individual to you. While your essays will surely differ in style, tone, language and subject matter, they should convey the same powerful impressions.

Ultimately, we want you to use these successful essays as inspiration to write your own masterpiece.

Experiences & Challenges

Brian C. Babcock, Marshall & Truman Scholarship Winner

The path to becoming a Marshall Scholar and Truman Scholar is a long one. Brian's journey began at Bowie High School in Bowie, Maryland, when he was elected the president of the Russian Club. Since that time, he has studied at the U.S. Military Academy at West Point and worked as a Russian linguist.

Brian is one of 40 students in the nation to win the Marshall Scholarship and one of 80 students to win the Truman Scholarship. With the Marshall Scholarship, he will study at Oxford after graduating. Brian plans to use the $30,000 Truman Award to support his future graduate studies in foreign service and history and would eventually like to become the Defense Attache to Russia, working with the governments of the former Soviet republics to assist them in dismantling their nuclear, biological and chemical weapons. In this essay for the Truman Scholarship, he describes how at age 17 he embarked on a hike of the Appalachian Trail from Maine to Georgia. The solo hike lasted six months.

Lessons from the Outdoors

The outdoors has always played a large role in my life, whether in Boy Scouts, on my own or with the military thus far. However, there is one outdoor experience of mine that did not involve my being in a club. I also did not get any awards for this experience, yet it has had a more profound impact on who I am than any other single event in my life, my "thru-hike" of the Appalachian Trail from Maine to Georgia.

I started my thru-hike when I was 17 years old, three weeks after I graduated from high school. It took me just over six months to complete. In those six months, I learned more about myself than in the previous 17 years or in the five years since. There is nothing with which it can compare.

I financed the hike with money that I saved during my last semester of high school, working 40 hours per week on top of my full-time student schedule. I was determined to reach Maine and hike south to Georgia. This was the first real goal that I had ever made for myself, and I reached it alone on a cold January morning.

The lessons from the trail are ones that have affected me in everything I have done since. Because of those six months, I see the world differently, in a way that is sometimes impossible to explain to someone else, though I might try.

Brian's contributions reflect his own opinions, not those of the U.S. military.

My life was not difficult growing up, but I found a need to put myself through the difficulties of trail life. From this time, I gained an appreciation for the little things, like clean water to drink and a dry place to sleep (both of which were sometimes lacking). I met people from all walks of life, as they crossed paths with my walk in life. From that experience I am better able to deal with those whose backgrounds do not resemble mine, a skill I have used often in the military.

Now I have turned my life 180 degrees. I no longer have hair to the middle of my back or a beard. I have traded my Birkenstock sandals for combat boots. Yet, somehow, everything I did on the trail applies to what I have done since. Whether it's suffering in a foxhole during field training, or sleeping in a cold, dank lean-to on my hike, the lessons are not all that different.

Though my journey in life has wandered back onto the beaten path, I know that if the nation needs me to lead soldiers into the brush or assist in providing humanitarian aid, I have my previous experience to draw from. Because I have been there, I have a common bond of suffering with millions throughout the world and another bond to all of my soldiers. I am still amazed at how my former life as a free-spirited wanderer has better prepared me for life as a disciplined soldier.

Daniel Heras, Scantron Scholarship Winner

Daniel dreams of becoming a teacher to inspire students to learn in the same way that he has been inspired by his teachers. In this essay, he describes one of the most meaningful experiences he has had in high school through the Environmental Science Club, which took him to real life locales to learn about science not from textbooks but from seeing, touching and experiencing science first hand. Student body president and captain of the baseball team at Woodrow Wilson High School in Los Angeles, Daniel won more than $17,000 in scholarships to attend U.C. Berkeley.

Inspired to Teach

In the ninth grade, I was introduced to the Environmental Science Club and to Mr. Quezada, my science teacher and advisor. Outside of the classroom and through the club, I saw an entirely different side to education. The science club took me to far and exotic destinations, such as the Islands of Hawaii, the underwater wonderlands of the Cayman Islands, the temperate climates of the Florida Everglades, the deep blue waters of the Mexican Riviera and the High Sierras of Northern California.

We learned that one cannot experience these things in class behind a small cramped desk made for 10-year-olds. I was able to hold, smell and sometimes taste, foreign artifacts. I have seen the migration patterns of the Humpback whale, have become a certified scuba diver, learned to surf, rock climb, snowboard and trail the mountains of the world, all while learning about science. Our trips have also given me the life skills of communication, learning to intermingle with people of the world.

It only took a year to see that teaching was my future. Why would someone not want to get paid for helping his or her community, to enlighten the future generation and best of all, do the things that bring joy to one's life all while on the job? I was given experiences I would not have received anywhere else, and I want to do the same for the next generation to come. There is a world that one can hold, smell and sometimes taste. I want to show people that there really is a world out there beyond the pictures in textbooks.

Mark R. Eadie, Coca-Cola Scholars Regional Winner

When Mark visits his 90-year-old grandmother, the two turn the volume on the television up. Though they are separated by almost 70 years in age, they share a similar problem: hearing impairment, Mark's grandmother because of age and Mark because of a childhood injury that left him partially deaf. This injury has not stopped Mark. If anything, it has sparked a passion to be a role model for others.

While a student at Columbia High School, Mark trained for hours as the lead of the school musical, perfecting his singing without the benefit of stereo hearing. The performance garnered rave reviews. From Rensselaer, New York, Mark received over $50,000 in scholarships to attend the University of Michigan, where he is a member of the national champion solar car team and is studying aerospace and mechanical engineering. He hopes that through his research in engineering he can develop solutions for others like him.

Invisible Handicap

Who would think a game of catch would change my life? At age 10 I lost hearing in one ear and had to struggle with the challenges resulting from this "invisible handicap." Through this I have become more sensitive to people's problems and handicaps, learned the value of my support community, refined career goals and challenged myself in new and difficult situations to help others.

My catching skills were not what my older brother thought, and his fast ball missed my glove and hit my cheek bone. After a severe concussion and cochlear surgery, I was totally deaf in the right ear. I had lost all stereophonic hearing and musicality. My voice started to become monotone. I could not tell from where sounds were coming, hear notes I was singing or distinguish voices in a noisy room. The hardest part was exhaustion from having to focus on everything going on. School became far harder. Conventional hearing aids don't work with total deafness, so I tried a microphone and receiver system in class. However, it was more frustrating than helpful. After that, my teachers were dazzled with my attentiveness, not realizing I was reading their lips. All this has been very tough emotionally.

A Scholarship Support Network

Emanuel Pleitez, Recipient of $30,000 in Scholarships

Emanuel Pleitez remembers falling asleep while writing his scholarship essays. In fact, he worked so late, that he went to the post office at the Los Angeles International airport to mail his applications because that was the only post office open until midnight. Fortunately, he had a classmate who joined him on these late night drives.

"You have to surround yourself with friends who are motivated like you, who want to go to college and apply for scholarships," he says.

Emanuel and his friend developed an informal support network for each other when applying for awards. He used his friend's computer because he didn't have one and the two helped edit each other's work. In fact, Emanuel encouraged his friend to apply for an award that he found. His friend ultimately won the award.

His classmate wasn't the only person that Emanuel relied on for help. He received encouragement from his coach, who was also his senior class advisor and school's dean. His counselor mentored him after Emanuel approached him during his freshman year of high school to explain his ambition to attend a selective college. He gained interviewing experience at a program he attended to prepare students for internships and essay-writing help from the Quest summer school program he attended at Stanford.

"I was really lucky to be surrounded by all these good people. At every stage there was always a couple of people who I could turn to for help," Emanuel says.

With my parents' help I learned not only to cope but to grow. Most people never know I have a severe hearing problem. I turn my head or move so they're on my "good ear" side. I ask people to clarify when statements are unclear. I still play sports, especially lacrosse, though my coach nearly goes hoarse yelling to me.

I challenged myself, joining symphonic band and chamber singers and taking a lead in "The Fantasticks" musical. Enormous hours were spent pinging on the piano, trying to match my voice to notes. The support of friends and teachers was wonderful, and we received rave reviews for the performances.

The struggle has brought me closer to my 90-year-old grandmother, who is losing her hearing. We visit daily and watch PBS together on weekends, the volume blasting. We empathize with each other, laughing and crying over the frustrations of deafness. When volunteering in the hospital cardiac care unit, I comfort older patients by comparing hearing aids. They laugh and do not feel quite so old.

Spring three years ago brought an incredible gift. A doctor developed a trans-cranial hearing aid. It transmits sounds powerfully from the deaf ear, through the skull, to the nerves in the "good" ear. Now, I hear some stereophonic sound and tonality. This cutting edge solution has helped me decide to study engineering, to help others as I have been helped. Engineering is a noble profession; its goal is to alleviate the human condition. I seek to examine and solve problems by creating new visions that combine innovation with technological development.

My invisible handicap makes communication difficult, but I wanted to help other youth grow and develop life skills, faith and values. So, I pushed myself and took increased leadership in Boy Scouts and in my church. These positions require good communication, making me work extremely hard. But the results have been worth the effort. As the leader of the Presbyterian Youth Connection Council for the Synod of the Northeast, I have worked with youth and adults from eight states. We hold training events to improve youth leadership. I went to Colorado to help the Synod of the Rocky Mountains establish a youth council. Twice we have planned conferences for nearly 200 youth. I have learned to work until a task is completed.

Though unable to say I'm glad it happened, I have benefited from my hearing loss. I have learned to use my limitations to help others and to never give up. My no longer monotone voice now reflects the non-monotone life I have developed.

Nhia TongChai Lee, Knight Ridder Minority Scholar

Nhia comes from a Hmong family where tradition is important. If his parents had their way, he would never date or even have friends of the opposite sex. While he respects his parents and his family's values, he feels that it's important that Hmong of his generation take steps toward independence and leadership. As he says, "Just make sure you take big steps and not little baby steps." And through writing he wants to be someone who influences those of his generation.

It's because of this desire to inspire others that Nhia got involved in his newspaper at Lansing Everett High School in Lansing, Michigan, and is now majoring in journalism at Michigan State University. His passion to affect others through his writing has been recognized. He has won more than $60,000 in scholarships including the Michigan State University Distinguished Freshman Achievement Award and *Detroit Free Press* Journalism Award. When asked his planned career field, he says that he will become the editor of *Rolling Stone.*

Only the Strong Survive

Our lives are not predetermined but rather a journey that each individual must decide for himself. Events that transpire along the way do not just disrupt the journey but sometimes occur to benefit it. During the Vietnam War, my family was forced out of their homeland Laos and into Thailand, where they sought refuge for five years. All was left behind to take a stab at giving my siblings and me a possible future. The only life they knew had been wiped clear of existence. The familiar air breathed, land cultivated and faces seen all seemed like a lost dream.

Relocating in Thailand did not manifest into the Promised Land everyone had heard about. Instead of the beautiful lands and abundance of food, what they found were crowded camps and no food. Hunger spread like wildfire and people died by the handful. What many thought was a safe haven was in actuality a waiting deathbed.

Only the strong survived the refugee camps. My family members were just more faces in the crowd of thousands in the same situation. It was there that I was born into a life deprived of the simple good things in life. Finding food was always a problem and just trying to survive to the next day was a top priority. My parents knew that in order to survive we had to leave the refugee camps. If you were lucky, you were sponsored to move to America. Along with thousands of others, we had nothing to do but wait. Wait for a reply to our pleas to leave.

In 1985 my family finally received word that our prayers out of Thailand were answered. A church in Michigan sponsored our family and that was our ticket out. We immigrated to the United States to start anew. We had to adopt a new language, a new culture and a whole new way of life. Through it all, we continued to practice our culture and customs. That was something my parents wanted to keep and pass on to generations to come. It was the only thing about the past that remained with us. My parents wanted us to grow up to be traditional Hmong boys and girls.

I knew what I wanted in life, but knowing that traditionally Hmong children married at a young age, it was hard to break out of that mold. By choosing journalism as a career path, I hope to set an example: following the traditional rules is not the only option, even though that's the only life we know. I want to complete school and have my writing reach a vast audience. I hope to make a difference with writing and show the youth of my culture that we can balance both worlds at the same time. We can still have respect for our parents and compassion for our culture while changing along with modern society. There is a lot more out there for us, a world beyond marriage and children, a world that can show a whole different perspective on life.

I want to show that growing up impoverished can still lead to being published in a national newspaper or writing a Pulitzer Prize-winning article. I want to be that role model for Hmong kids who sometimes feel trapped within the walls that are built around them.

I believe that if I can live my life the way I want it and not how my parents want it, then others can follow. Instead of marrying into a burden-filled life, I can become the anchor for that change. I want to take the path that my parents never spoke of. I know that in the end that will be the difference between what is and what could have been. Hopefully young people, not just Hmong kids, but anyone who feels lost can look at what I have achieved and find their own path.

Jennifer Chiu, Telluride Association Summer Program Scholarship Winner

Jennifer had the opportunity to experience college life while she was still in high school. As a junior at Hunter College High School in New York City, she won a full scholarship to the Telluride Association Summer Program on constitutional law at Cornell University. The program exposes students from a variety of backgrounds to college life and courses. Jennifer used her experience to make the transition to Yale University. In addition to this award, she also won the *New York Times* College Scholarship, National Merit Scholarship and Yale Club of New York City Scholarship. She gives the following advice about applying for awards, "Don't stress too much over sending in the perfect application. Behind every piece of paper is a person, not a robot."

Lessons from a Pitbull

Every time I walked down 52nd Avenue on my way home from the library, I passed a mean pitbull that always barked at and tried to attack strangers. For some reason, he seemed to hate me especially. I suppose that dogs instinctively protect their territory against all intruders and that I qualified. Yet, I was a very poor intruder at best. Whenever I saw him, I cowered next to the hedges, but he would always smell my fear and start his tirade. Perhaps it was my fault for not crossing over to the opposite sidewalk. I didn't want to admit to myself that I was scared.

One afternoon, after having had an especially bad day, I passed him once again. When he started to yap as usual, something snapped inside of me, and I growled back. I think that I could have been heard all the way into the next street. When the dog's owner came out to see what was going on, I ran away.

After that, I avoided the house.

On the surface, the conflict was simple: a struggle for territory. That dog simply did not want me around, while I insisted on it. But deeper down, the problem was my refusal to admit that I was scared of him. My foolish courage rested on the notion that I had a fear of being afraid. I refused to believe that every day is a struggle for survival, since humans have supposedly evolved beyond this. Obviously this is untrue, and now I realize that I, like any other creature, experience terror.

As I battled the dog, I felt conflict with myself at a deeper level. I realized I had a superiority complex, since I was better off in some ways than other people. That would boost my motivation to succeed, but it came at the cost of being alienated and eternally conscious of my weaknesses. I always watched my back, even when it was not necessary. I was intimidated by other people just like I was intimidated even by the dog. I paid the price of needless self-torture and confusion.

This barking episode was one decisive moment in my life. Though it is embarrassing, after all, I proved myself worthy against a dog, not all lessons can be picture-perfect. I'm glad I learned it the hard way than never at all. I realized that I am allowed to admit that I am afraid sometimes, as long as I am willing to work to mediate the anxiety.

Last week, I walked past the house again. It was abandoned and a "For Sale" sign adorned the front yard. I turned on my heels and left.

Essays about Family

Rodolfo Valadez, Cohen Foundation Scholarship Winner

Going to Thomas Jefferson High School in Los Angeles, California, Rodolfo discovered his passion: filmmaking. While making movies including his critically acclaimed documentary "los angeles," which was screened at the Sundance Film Festival, is his passion, Rodolfo's inspiration is his mother. Rodolfo freely admits that his mother is "my support, my help, my guidance, my friend, my hero." With this essay, Rodolfo won a $6,000 scholarship from the Cohen Foundation and $7,000 scholarship from the Hispanic Heritage Youth Awards to attend the University of California at Los Angeles.

A Mother's Sacrifice

My mother sat in between the dry grass growing out of the puddle of dirt water under a bridge in the hills separating Mexico from San Diego. In that exact moment I sat aboard a plane with strangers and in possession of a name other than mine in order to be granted admission into the United States. It has now been 14 years, and my mother still sacrifices her own comfort for mine. She works nine hours a day, six days a week in a machine-like position, humped over a sewing machine, altering clothes for strangers. At the same time I sit in class, socializing, enjoying a productive school day.

However, my mother is fulfilled knowing that her children take advantage of the vast opportunities this country has to offer despite the hardships she has to endure for a weekly paycheck. A paycheck she has vowed to invest into my college education at UCLA, an institution requiring almost $16,000 a year. At her hourly salary and after taxes she would have to work 2,560 hours in order to pay a year's tuition, which equals working 10,240 hours over the four years of college. If asked to, she would be more than willing to undergo the task of paying for my education knowing fully the standards and strife she would burden herself with.

In my sophomore year, I was among 24 honor students sequestered into a film course. I became one of the first students to attend the annual Telluride Film Festival in Colorado. In my second year in the course I attended the Sundance Film Festival where I was able to display my own work. Again, in my final year, we were invited to Sundance to show more films from our Academy of Film and Theatre Arts.

The film I showcased in the student forum was a documentary film commemorating my mother's struggle and sacrifice ever since her departure from a small oasis named Los Angeles in Durango, Mexico, only to move into the cold, industrial city of Los Angeles, California. The film depicts her struggle and reason for doing so. It also illuminates the fact that more like her exist all around us.

The film course at my high school has been my passion. For three years the course taught me to acquire a more perceptive and critical view of the world. Throughout the course we studied Aristotle's philosophies and the evolution of cinematography, and we analyzed films and wrote essays comparing the motifs they translate into a sequence of shots. I have become more creative and just recently started working on a new 16 mm film entitled "Love Story." As clichéd as the title suggests, the story is a satire of what the title represents. Such projects motivate me to work even when class is over. In the last three years I have found myself in class Saturday mornings and afternoons, editing and brainstorming ideas with classmates.

I have my mother to thank for being able to pursue my passion in filmmaking. I realize she gave up everything for me, and I will do what I can to make it well worth it.

Jessica Haskins, SuperCollege.com Scholarship Winner

Jessica's dream is to write fantasy novels and short stories. Throughout her time at Saratoga Springs High School in Saratoga Springs, New York, Jessica took challenging classes and focused on obtaining a wide breadth of knowledge that would be useful in her future career. She is studying creative writing at Bard College. Outside of classes, she keeps up her writing with day and dream journals. When writing her essay, Jessica had a difficult time with the length requirement. "Editing is terrible," she admits, but after much cutting she was able to pare down her essay to meet the requirements. Although Jessica was worried that her essay had lost much of its power, her editors assured her that it had not. Obviously, the judges concurred.

Thank You, Dr. Seuss

(With Special Recognition for the Trenton, Georgia, School System)

More than anything else I can think up as a reason, my mother is why I'm going to college. Because of her, there could be no other decision. Not that I'm being forced or anything, but she has heavily influenced me and my decision. In a good way.

She always regretted that she could never go to college. Her parents, her teachers and her school counselors somehow, even though I still have trouble understanding it, simply never arranged it for her. College has always been a foregone conclusion for me, so this seems bizarre. To this day, she's never been able to explain it to my satisfaction. Nor to her own.

My mother is a very intelligent woman and was one of the best students in her small-town, athletics-minded Georgia high school (3rd or so in her class, where rank was unweighted and the valedictorian did as little as she could to get 100's in easy classes). She's certain that she could have accomplished a lot in life if she'd only been able to get a college education. She swears her education actually stopped at 9th grade, when she moved from Illinois to Georgia. I've seen the white sticker she uses as a bookmark in her gigantic Random House Dictionary of the English Language. It says "I [heart] Georgia," with the heart crossed out in black marker. She does love things like warm weather, big flowers and Southern cooking but despises their educational system.

I grew up knowing how keenly she lamented her missed opportunity, and she passed on her appreciation for the value of education to me. From the start she raised me to be an intellectual. I could read by age 3. I have a vivid memory of lying on the couch with her in the living room of our old house right on Route 9, where the cars would streak past day and night. My mother was reading Dr. Seuss to me—"We run for fun in the hot, hot sun," from *One Fish, Two Fish, Red Fish, Blue Fish*—and afternoon sunlight was pouring in through the windows, warming the whole room. As she read, I followed along. That was my first memory of ever actually reading the words, rather than just being read to. Whenever I think of my educational success, I attribute it first and foremost to learning to read at an early age, and the very next thing I think of is that reading lesson in the sun.

I have another memory. Me, at about the same age, confronting my father in the bathroom and asking, "Daddy, when can I go to school?" I just couldn't wait. I went to two years of preschool, where I did very well, except that I wasn't very generous. Recently I was poking through my old school files and found a couple of reports from one of my preschools. In fact, let me go get them so that I can quote it exactly—ah, here it is. "Jessica needs prompting to share." I found that very amusing. Now that I think about it, I was pretty attached to that Viewmaster.

When I was finally old enough I went to kindergarten, but nothing there was a challenge for me, and my parents and I all wanted me to skip a grade. It took a little battling with the school, which was reluctant to move me ahead, and some extensive testing, but they finally agreed to have me skip first grade. I'm glad that I did. Even though I was younger than the other kids in my grade, I took advanced classes whenever there was an opportunity. Because of the importance my mother always placed on education, I was always ready to take on harder material.

As I said before, it was always a given that I would go to college. My mother wanted it for me, and I wanted it for myself. And not just any college—my aim was never to just get a degree and a good job, but to continue to enrich myself. The "good job" isn't even guaranteed. After all, I want to be a writer, and there's no ticket to success in that field without a good bit of luck. So my standards for college are slightly different. Basically what I want is the most liberal of liberal arts. I want to continue the educational path that started way back when I was lying on the living room couch reading *One Fish, Two Fish, Red Fish, Blue Fish* with my mother, who knew I'd someday get the college experience she never had. So thank you, Dr. Seuss. And thank you, Mom.

The Intangible Benefits of Applying

U.C. Berkeley Scholarship Connection

You have something to gain by applying for scholarships even if you don't win says Leah Carroll, coordinator of U.C. Berkeley's Haas Scholars program and former program coordinator of the university's Scholarship Connection. In her roles Carroll has assisted Berkeley students with applying for awards, especially for the highly competitive scholarships, including the Rhodes, Marshall and Truman.

While Carroll gives students feedback on their essays and practices interviewing them, she reminds them that there is more than scholarship dollars at stake.

"I also emphasize the fringe benefits. For starters you get to know your professors better than before since you need to speak with them," she says. Carroll adds, "You also get practice presenting yourself in interviews and on paper." This is helpful for students who will soon be applying for jobs or for graduate school. She says that the essays can even serve as rough drafts of graduate school admission essays.

And looking at the big picture, Carroll says that applying for one of these awards "forces you to analyze your own life." She says that one of the things she enjoys most about her job is helping students clarify their purpose in life through the process of applying for scholarships.

Donald H. Matsuda, Jr., Truman Scholarship Winner

Working as an intern for a health clinic, Donald read an article in the *New York Times.* The headline was, "Forty-four Million Americans without Health Insurance." When he learned through the article that over one-third of these Americans were children, he decided to take action. With the help of the clinic's director, he secured the funding for and developed a series of insurance drives for Asian immigrant children. In addition to his work with the clinic, Donald is the founder of the San Mateo Children's Health Insurance Program, the national director of United Students for Veterans' Health and the founder of the Nepal Pediatric Clinical Internship.

A Stanford University student from Sacramento, California, where he attended Jesuit High School, Donald plans to use the $30,000 Truman Scholarship to obtain a medical degree and master's degree in public administration and would eventually like to be the medical director of a nonprofit clinic to aid underserved populations and the uninsured.

When Drinking Water

When drinking water, my grandmother would often proclaim, "Never forget its source." For some reason, I always enjoyed hearing her repeat these words of wisdom from her book of ancient Asian proverbs. Perhaps it was because I had grown to fully appreciate its true meaning—that one must always remember and treasure their ancestry and elders, who are viewed as the ultimate source of life. Or, perhaps it was because I felt this proverb effectively expressed my own sentiments about my life.

Growing up as an only child, I developed a very close relationship with my entire family and I greatly valued the time I was able to spend in the company of my elders, especially my grandmother. As a survivor of the Japanese American internment camps, she maintained an unbridled idealism, an impeccable work ethic and a genuine compassion for those in need. Moreover, she was intent on instilling these values in me when I was a young boy. I often looked to her as my true source of strength, for she always infused me with energy, passion and ideals.

Two years ago, I received a call from my parents urging me to return home. When I got there, I saw my mother was on the verge of tears as she told me what was wrong: "Grandma passed away today. She had a massive stroke and the doctors did everything they could, but..." I embraced my mother and we cried for what seemed like an eternity. I soon realized that I had lost not only my grandmother, but also a precious source of inspiration and strength.

Since that tragic day, I have become a much stronger person. I have internalized grandma's work ethic, idealism and compassion so that my source of strength now comes from within. It is this new motivation that fuels my convictions and drives my passion for a life dedicated to public service.

Every day, when I pass by the elegantly sculpted water fountains on my way to class, I pause as cherished memories of my grandmother fill my mind, and I know in my heart that I will never forget my true source.

Essay Advice from the Winners

Scholarship Winners

Here are some essay tips from scholarship winners. Having survived various competitions these winners have a unique understanding of what goes into crafting a winning essay.—Gen & Kelly

Sara Bei
Stanford University student and scholarship winner
"Sometimes the ones you end up winning are the ones you almost didn't apply for. Even if it takes a long time to fill out applications and write essays, think of it as being paid $500 an hour if you win."

Jason Morimoto
U.C. Berkeley student and scholarship winner
"Use your essay to craft a story showing why you are a unique candidate. Include personal experiences, lessons learned and how you are trying to improve yourself."

Chheng Sok, Chicago Scholars Foundation Winner

Every time Chheng announced good news to her parents, they gave her their special smile. Her parents grew up in Cambodia, where her father's education ended in grade school and her mother did not receive a formal education. So it held special meaning when Chheng was accepted to the University of Chicago. She received her parents' special smile. And when she won more than $35,000 in scholarships, their smile broadened even wider.

Graduating from Lane Technical School in Chicago, Chheng was president of the Chinese Club and involved in public service. She encourages others to apply for scholarships. As she says, "I'm not exactly the best student, but I still got scholarships." Majoring in East Asian Language and Civilization and Economics at the University of Chicago, she plans a career in education or international business.

My Family's Hope

My family and I immigrated to the United States from Cambodia to flee the ravages of the Khmer Rouge when I was only a year old. We did not have a single penny when we came to the United States. I remember seeing my father diligently collect soda cans on the streets to trade in to the local recycling center for a penny each. I remember watching my family silently endure the rudeness of waiters and salespeople because we did not speak grammatically correct English and realizing at the age of five how much illiteracy paralyzes a person.

I am the youngest out of my parent's nine children, yet I possess the greatest amount of education. My father can barely read English. My mother is totally illiterate. Due to my family's financial situation, none of my eight siblings have completed college. Throughout my elementary and high school years, I oftentimes had difficulty with my schoolwork. I remember staying up late at night, sometimes until two in the morning, just so that I could figure out the answer to a homework problem. My parents and older siblings, as much as they wanted to help me on my assignments, were unable to because they simply did not understand the material. They would quietly sit by me and bring me refreshments from time to time and offer me encouragement. My siblings make me realize how priceless knowledge is and to make the most out of one's education.

My parents look at me as my family's hope for the future. They dream that I will some day graduate from an American university. They want to be able to send back letters to our relatives in China and in Cambodia, telling them about how one person in the family has gained an American diploma. I want to be the realization of their dream and my dream. I dream of graduating from one of the finest colleges in America, the University of Chicago. I hope that someday I will be able to repay my parents for all the years they have lovingly supported me. I want to be able to financially and intellectually provide for my family so that we no longer have to endure the discrimination toward illiteracy.

In addition, I strive to succeed in school because I want to be a role model for my nieces and nephews. As a student, I personally know how tough it can be to excel in school. I want to be there to help them if they need help on a class assignment, to guide them through their first multiplication table and to be their mentor when they start the college selection process.

With knowledge, one need not fear being cheated by a salesman or being looked down upon by an egotistical snob. Education is the door that opens the path to knowledge. With knowledge, I am in control of my life and my destiny.

Essay Advice from the Winners

Scholarship Winners

Here are some essay tips from scholarship winners. Having survived various competitions these winners have a unique understanding of what goes into crafting a winning essay.—Gen & Kelly

Donald H. Matsuda, Jr.
Stanford University student and Truman Scholar
"If readers can connect with you, feel your emotions and feel they know you, that you're such a dynamic person, that comes across in your writing. That really is a plus. You can espouse all the pros and cons of think tanks in the American political system but that doesn't really help the committee learn who you are."

Dalia Alcázar
U.C. Berkeley student and scholarship winner
"Your essays are the most time consuming part of applying. Some of the questions are very similar. You might have a couple essays already written that you can modify. For many of the scholarships I won I used the same essay with slight modifications."

Jessica Haskins
Bard College student and scholarship winner
"The topic doesn't have to be profound. You don't have to write about the time you saved someone's life, or describe an earth-shattering experience—I personally think that a simple, thoughtful and honest reflection carries more weight than an elevated epic of love, loss and life's lessons in 500 words or less."

Alex Dao, Gates Millennium Scholarship

A student at Stanford University, Alex won several scholarships including the Maria Hart Becker Scholarship Fund, Sam Walton Scholarship and Robert C. Byrd Scholarship. He says he didn't expect to win all of the scholarships that he did and advises students, "If you don't apply, you'll never win. Let the scholarship judges know who you really are."

Childhood

Every time I open up our photo albums during Christmas and family gatherings I feel a sense of nostalgia. With each turn of the page, each resonance of laughter, each event and each year—precisely remembered—all the problems of today vanish: my parents never divorced, my father never lost his job and my family never moved. Instead, life is filled with memories of happy and exuberant times. Although those days are now rooted far in the past, the memories of life as a child stay vivid and clear.

Life had always been carefree and pleasant. I had cousins who loved me, parents who disciplined me and girls who teased me. I felt all the warmth and comfort any child could want; however, it was more than just this that made my childhood "perfect." I had always been close to my brothers, and the most memorable moments of my childhood embrace the love and affection my brothers and I shared. We spent countless summer days playing and dreaming on the front lawn. We wrestled and fought, imitating those we saw on television. Yet, with our short attention spans, it wasn't long before we sat down together and started talking about our hopes for the future, our ambitions and goals, our future wives and children. Innocently, I had always thought becoming a superhero was a realistic goal. I talked about all the superpowers I would somehow acquire and how people would tell stories of my accomplishments for generations to come. My brothers, although younger, laughed and made fun; after all, they had more realistic ambitions, hoping to become doctors or lawyers. Then the debates began. We went on and on for hours, talking about how each of us would be better than the others. Although I did not always claim first place, I looked forward to the next day when we'd come back out and start our discussions anew. As simple as it may seem, their presence was more than enough to make me happy. These experiences understandably may not seem like much to an outside observer, but for me they are among the best days of my life. To this day I can still think of no better way I could have spent my summer days than just sitting in the front yard, enjoying the company of my siblings. Nothing even comes close.

Although it's been many years since then, I have always longed to return to this past: every day was an experience in its own and filled with nothing but excitement and joy. As I look back on my childhood, I contemplate the things that made it so enjoyable—the simplicity of life as a child. I was devoid of responsibilities, satisfied with life and hopeful for the future. My childhood was instrumental in shaping who I have become: someone driven to succeed but optimistic even in the face of failure. Through years past, I have realized that life will never be the same, but then again, when does life ever stay the same? Each day presents a new set of problems. I can no longer just sit in the front yard with my brothers, dreaming the day away. Instead, I must confront these challenges and do my best to resolve them. Although times have changed and obstacles have arisen, I still view the future with the same optimism and anticipation I have always viewed it with. My experiences have hardly been "perfect," but life continues to amaze and excite me at every turn. The problems of "yesterday" should not affect the futures of "tomorrow."

National or International Issues

Elizabeth Ashlea Wood, Optimist International Essay Contest Winner

After having witnessed a nuclear disaster, Elizabeth knew that her life was changed. She says, "It really opened my eyes. It got me away from thinking that I'm young and can live forever." Touched, she wrote about the experience for this award to share her fear and realizations with others. A graduate of the Classen School of Advanced Studies for Performing and Visual Arts in Oklahoma City, she is studying literature, writing and the arts at Eugene Lang College in New York City.

The Tokai Nuclear Disaster

Last year on an October evening in Japan, I enjoyed the rain, walking slowly to my host family's farm for the night. After I yelled the customary "Tadaima" and removed my shoes my host mother pulled out a heavy English dictionary. She searched for a word and then pointed excitedly. Above her finger I read "radiation." The Tokai nuclear power plant two miles away was experiencing a severe accident. Soon trucks driving by screamed warnings in Japanese to prepare for nuclear disaster. My body was numb.

I had been to Hiroshima the week before. All I could imagine were the grotesque pictures of goiters and dripping flesh. Photographs of the burnt remains of an ancient city flashed in my head. I remembered seeing the "Daisy Girl" commercial from LBJ's presidential campaign in government class. It slowly played in my mind, a blonde child holding a daisy, framed by a green-gray mushroom cloud. My imagination forced me to expect the worst.

The air terrified me. I thought I was suffocating. In that moment I could not understand how my life had led to this crucial moment. I had left my home for a beautiful opportunity to live in Japan and experience the culture. I had joined an exchange program in a small village by the ocean. This succession of serendipitous events led me to the only place in the entire world where a severe nuclear disaster was occurring. My choices had exposed me to the ultimate weapon of our time; I was waiting for radiation to subside. The rice paper windows and layers of silk robes provided little comfort. There was nothing I could do to protect myself from the danger. In that moment I could only learn.

The world's issues no longer can disappear as I close a schoolbook. On that autumn evening I was suddenly a part of one of the nemeses of the twentieth century: nuclear energy. Ironically, my frightening experience was only an accident. When I decided to embrace a three-month adventure I never expected to trade in theater and friends for a serious nuclear disaster. My eyes were pried open to make me realize that the world's issues are not separate from my American life. I realized that I had been educated about the world to understand cause and effects, the cycles of history and of the future, but I had not metacognitively incorporated them into a worldview.

That evening reached into my mind and opened a door to the realities of this world. The Tokai disaster threatened my life, but it also demonstrated the capabilities for any person to experience the same shocking circumstances. In Japan, quarantined for days on a Buddhist farm, I could see no separation of myself from other cultures. I realized that I could no longer segregate America from other countries, my race from other races, Oklahoma from Japan. Three days after the Tokai nuclear disaster I stepped out of the farmhouse into the fresh sunshine of a glorious oriental garden. Over a cup of green tea I determined to be committed to my new perception of the world as an entirety.

Elisa Tatiana Juárez, Target All-Around Scholarship Winner

Based on her research in osteoporosis and gerontology, Elisa has placed first and best in show in a number of science competitions including the Intel International Science and Engineering Fair and the South Florida Science and Engineering Fair. But each time she entered a science competition, she noticed that economically disadvantaged students were underrepresented. She did something to change this. Working with the Miami Museum of Science and Big Brothers Big Sisters of Greater Miami, she founded the Students and Teachers Advocating Research Science (STARS) program to provide assistance to economically-disadvantaged middle school children.

In addition to STARS, Elisa has been recognized by the United States Air Force for her research in gerontology and was selected to present her work to the Florida state legislature for her research on osteoporosis. Her commitment to the sciences has paid off. She has won scholarships including the National Hispanic Heritage Youth Award for Science and Technology and the Science Silver Knight Award. A graduate of Coral Reef Senior High School in Miami, she is a student at Brown University.

STARS

I developed the STARS (Students and Teachers Advocating Research Science) Project to provide information, materials and, most importantly, mentors to help middle school students from at-risk environments to complete and present successful science projects. STARS now helps the Miami Museum of Science and Big Brothers Big Sisters of Greater Miami to "support and empower single-parent families to actively engage in their children's science education."

STARS grew as an extension of my own involvement with science research projects that have been successful at local, regional, state and international levels. I had a lot of support from family, teachers and mentors and wanted to find a way to offer similar support and opportunities for students who didn't have all that. American students are losing ground internationally in science and math, and I wanted to find a way to share my passion in tangible ways.

Why science? I feel that it is important for every kid to be involved in science. The Third International Mathematics and Science Study (TIMSS) showed that U.S. 12th graders outperformed only two (Cyprus and South Africa) of the 21 participating countries in math and science. All kids need to have access to better opportunities in science, and this project has allowed kids who had never even thought about science research to discover that they can do anything they want to do.

I think the most memorable part of this project was to see the kids' eyes sparkle when they talked about their ideas for projects and then began to see the results of actually doing the research. I remember at the end of a workshop, one of the girls came up to me and said that she really enjoyed the day. I asked her what she meant, and she said with a smile, "Well, I don't really do science. I'm more into English and literature. When I heard about this, I didn't really want to come, but I am so glad I did. I had so much fun learning that I could do a good project. I think my teacher will be proud."

I am now designing a STARS science curriculum that correlates with the National Science Education Standards. I hope that Big Brothers Big Sisters chapters around the country will be able to use the curriculum to encourage their "littles" and "bigs" to participate more actively in science.

Based on this experience so far, I would tell other young people that there is nothing more rewarding than realizing that you can make a difference. That you can identify a need, develop a solution, find people to help you accomplish your goals. Of course, I would also tell them that science is everywhere and is exciting and can help you learn about the universe and about yourself.

Elisa Tatiana Juárez, Presbyterian Church USA Scholarship Winner

The Power to Change the World

When people ask me what I want to do when I grow up, I answer them quite simply and firmly, "I am going to change the world." I am 17 years old, but I have known for a very long time that I would, in some way, be responsible for shaping the world of the future. Crazy? Maybe. Impossible? Definitely not.

Unfortunately, in my experience, it has been kids my age who tell me that I am just a dreamer and that there is no way I could possibly make a difference in the world. "Come on Elisa," they tell me, "You're just a kid. No one in his or her right mind is going to listen to some high school girl. Don't bother; no one cares anyway. Someone else will do it."

I think that the greatest opportunity facing youth today is the power to better the world around us by using new tools, new technologies and a new understanding of the global community. By the same token I believe that the most urgent problem facing youth today is indifference. The general attitude about everything and anything is "Who cares? I am not that important, there is nothing I can do about it." I find this incredibly sad and distressing. God gives us the intelligence to build the tools; we only need to use them with the guidance of His Spirit guided by His love.

My generation is very cynical when it comes to helping out. They claim that what they have to say couldn't possibly be important enough to be heard by others. What they don't understand are two very important concepts. First of all, the majority of the youth today don't realize that there are plenty of problems in their own community. Making a difference doesn't always mean moving to Somalia to end hunger. It could mean something along the lines of helping a migrant family learn the basics of the English language. Second of all, youth today don't realize how something very simple can change someone's entire world. By teaching that family English, for example, they will feel more comfortable in this country.

Growing up I heard a story about an old man who goes down to the sea one morning. He notices that a young girl is reaching down and throwing starfish into the water. Curious, he walks over to the girl and asks her what she is doing. She replies, "Well, the tide is awfully low, and if I don't throw the starfish into the water the sun will dry them out." The old man looked at her and laughed. There were miles of shore with thousands of starfish. The little girl couldn't possibly throw all the starfish back in the sea. He told her she wouldn't be able to make a difference. The little girl bent down scooping up yet another starfish. She turned it over in her hand processing what the man had told her. Then, looking at the old man, she placed the starfish in his hands and helped him throw it back into the sea and moved on to the next starfish. She looked over her shoulder and said, "Well, to that starfish, I made a world of difference."

This is where the story traditionally ends. I have added on to it. The man, realizing the power this little girl had over the lives of the starfish, called up his grandchildren. Together they worked at saving the stranded starfish. That day, maybe not all the starfish were saved, but those that were, I'm sure, were very grateful. They continued living because of the determination of a little girl who knew that she could make a difference and could find ways to get others involved.

We must each find our starfish. If we throw our stars wisely and well, the world will be blessed. I constantly am praying for the strength to carry on and for the courage to help others find the power within them to help shape the world of today.

Shashank Bengali, Scripps-Howard College Journalism Scholarship Winner

Shashank knew that his parents wouldn't be able to foot the entire bill for his education at a private college, especially since his younger brother would soon follow him to college. Rather than look for a less-expensive school, Shashank decided to take action by applying for scholarships. His advisers at Whitney High School in Cerritos, California, suggested that he apply for the awards for which he would be the best candidate. Journalism awards were a natural fit.

Since the age of 13, Shashank has worked for his school newspaper. As the editor, he won a national competition for the Knight Ridder Minority Journalism Scholarship, which allowed him to intern at four newspapers across the country while in school. In addition, he won a full-tuition scholarship from the University of Southern California and Scripps-Howard College Journalism Scholarship, awarded to 10 college journalists in the country. He recently graduated with a degree in broadcast journalism, political science and French and is working as a Missouri state correspondent for *The Kansas City Star*.

Media Misunderstandings

On the day of the New Hampshire primary this year, the online magazine Slate.com posted early exit-poll results on its site before voting had closed for the day, inciting an enormous outcry in the traditional media. The major newspapers and television networks, bound by contract to honor an embargo on those results, said Slate violated the law—and a journalistic trust with the people. Slate disagreed, and went ahead publishing exit-poll results on the days of several other primaries this season, before being threatened with a lawsuit.

On the question of law, at least, Slate never agreed to any embargo. The other question, of the people's trust, is murkier, because it's unclear whether knowing preliminary election results actually deters voters from going to the polls. Tempers ran high on both sides here, with Slate columnist Jack Schafer going so far as to write in a *Wall Street Journal* op-ed piece that the Big Three networks' coverage of primary results is "all an act."

I bring up this story to illustrate a point: the relationship between "old" and "new" media is in real disrepair. Finger-pointing and mutual misunderstanding rule the day, when diverse news organizations hardly seem, most times, to understand each other's roles. The fact is, there are now three ways to disseminate information in the world—in print, in broadcast and online. All three have merit, and all three, in some form or another, appear to be here to stay.

What's needed, then, is greater cooperation and less co-optation. If, as many believe, the best news organizations of the near future will be those that are diversified and bring elements of print, broadcast and online media to bear together on their coverage, those journalists who are trained in each form will be the most valuable. I subscribe to this view and I will be one of those "hybrid" journalists, helping guide a shift to more complete, convergent news coverage that encompasses the three forms because I know the power of each.

I consider myself lucky to be around for the old media-new media debate because it's a vital one. I believe that much of the disdain on both sides is a product of misconceptions—and a lack of experience. My generation's position is unique: I've grown up with 24-hour news channels and the Internet, so I know their immediacy and reach in a visceral way. But the reason I got into journalism in the first place was that I love the written word and the prudence of print media. I know that people will always need their daily newspaper and their weekly newsmagazine, no matter how quickly TV and websites can give them their fix.

So I've explored all three realms of journalism during my time in college, and I've discovered that I'm a journalist first and foremost, gladly unfettered by any other labels. Internships at newspapers have reinforced my newsgathering and writing skills. My time as executive producer of USC's nightly newscast has taught me that "print values" can be tuned for broadcast—for visual impact and swift, assured responses to breaking-news situations. And each time a distant friend or relative e-mails me to say they saw our webcast and have feedback, or (better yet) to offer story ideas from their part of the world, I learn the power of the Internet as a medium of news.

Good journalism can be done with these three modes working in concert, but the best journalism can only come with responsible and well-trained leaders at the top, who know how to direct multifaceted coverage because they believe in it and have done it before. That's what I want to do and am learning how to do—help lead the new wave.

I would encourage the sort of synergy that Tribune is pursuing—and that reporters at CNN practice daily when they submit their stories for that network's TV, radio and online products. There is no reason consumers shouldn't have the benefit of the fullest possible picture in a news story; after all, it's what we should strive to give them each time around. Already, on a small scale, I am working to achieve that convergence on our campus. Next year, our newscast and the student newspaper—two entirely distinct organizations—will team up for one in-depth story. We will each pursue the angles to that story that are best suited to our particular medium—visual stories for TV, for example, and longer analysis pieces for print—and we'll use our websites to complete the coverage, including any long documents or transcripts of interviews that can't fit newshole or airtime.

A news organization should be dedicated to the kind of public service in journalism that may uproot the company's tradition for the sake of its work. It makes good business sense, because through responsible convergence you can reach more people with greater speed than your competition. But most important, it improves the product—and that is good journalism.

Of course, I don't know exactly how I'm going to get to a position where I can help implement this vision I share. One thing that's certain about this changing market is that nothing's certain. For the time being, after graduating college, I plan to write for a newspaper—because that's my first love, and still the traditional journalist's ground zero. That's a personal bias, I admit, but my experience in the other realms will probably make me a better print reporter. From that first job to wherever I end up, it's a yet-to-be-paved road. But I am confident in the future of journalism in this new era, and I'll remain dedicated to it.

Lindsay Hyde, National React Take Action Award & Toyota Scholar Winner

When her grandmother received a corneal transplant that saved her vision, it motivated Lindsay to ensure that others were educated about the benefits of organ donation. From this single experience the Organ Donor Project was born.

Over three years, the Miami student secured corporate sponsorships to produce an educational curriculum and informational video for other students to view and create their own organ donation awareness programs. The project expanded from Lindsay's own high school, Southwest Miami High School, to 12 schools across the nation and five in Malaysia, Australia, Costa Rica and the United Kingdom.

The Importance of Getting Editors

Donald H. Matsuda, Jr., Truman Scholar

The Truman Scholarship competition requires that each applicant write a detailed policy statement. This is a rigorous academic paper about a topic of national or international importance. However, there is nothing that says you cannot seek help. In fact, we recommend that regardless of your essay topic—whether a policy statement or personal narrative—you find others to read the essay and provide constructive feedback.

Donald maximized the knowledge of the people around him when writing his essay. He says, "I obtained advice from at least 10 different people, professors, experts in health care policy, the director of fellowships, my honors research advisor, a number of other Stanford students who had won Trumans in the past, friends and parents."

In fact, Donald credits his win to the many people who helped. He says, "It's a long process and requires quite a bit of emotional reflection. I really am indebted to them. I don't think I could have been successful without their help."

As a result of her efforts, Lindsay won a number of scholarships including being a National Coca-Cola Scholar, National Toyota Scholar and National React Take Action Award winner. At Harvard University, she is studying sociology and women's studies as well as continuing her volunteer work as the founder of Strong Women, Strong Girls, designed to assist at-risk elementary school girls.

Organ Donor Project

Imagine a stadium filled with 100,000 football fans. The stadium would overflow with people, the sound of cheering would be deafening. Now, imagine the game has finished, but the fans are unable to leave. Instead they sit waiting, waiting for...the unknown. By next year, experts estimate that 100,000 people will be waiting for life-saving organ transplants.

For several months, my grandmother was a fan in that stadium, waiting for a sight-saving corneal transplant. My grandmother was fortunate to receive her transplant, thanks to the generosity of individuals who made the decision to become organ donors.

As a result of my grandmother's experience, I realized the importance of organ donation and the need for accurate organ donation education. To meet this need, I developed a community service project during my sophomore year to provide teenagers with accurate organ donation information. The Organ Donor Project was introduced with a three-day awareness event that included a pep rally, speakers day and fundraising paintball tournament. Over 700 teachers and students participated in these awareness activities, 400 of which made the decision to become organ donors. As project founder, I procured over $5,000 in prize giveaway and in-kind donations, secured community support and coordinated the three-day event. That same year, the Organ Donor Project was recognized at the International Community Problem Solving Forum for its outstanding service to the Dade County community with a first place international award.

The second year of the Organ Donor Project brought tremendous growth. As a part of my efforts to increase teen awareness of the myths and misconceptions surrounding organ donation, I scripted an informational "Fact or Fiction" video. To make this video a reality, I secured the $10,000 in production costs from Burger King Corporation. The video received a

Special Achievement Award from the Miami Children's Film Festival for its outstanding educational value.

Also in the second year of the Organ Donor Project, I authored a step-by-step workbook designed to guide teenagers through the process of creating an organ donation awareness project. Recognizing the value of the student workbook, Hoffman-La Roche Laboratories underwrote its publication. This workbook is currently being utilized by the Transplant Foundation of South Florida in Dade, Broward, Palm Beach and Monroe counties as a part of its educational outreach efforts.

In response to the outpouring of enthusiasm demonstrated for the Project by educators at my school, I realized that classroom curriculum would serve as a means of educating young people of the need for organ donation. Using the Sunshine State Standards for Education as a guide, I wrote interdisciplinary organ donation curriculum for grades 3 to 12. This curriculum was utilized by teachers at Southwest High, Riviera Middle and Cypress Elementary schools. The curriculum was recognized in a national contest by Co-NECT, Inc. for its outstanding community value.

Now in its third year, the Organ Donor Project has expanded into 17 schools nationally and internationally, through a partnership with the Interact Service Club. As project founder, I am coordinating the efforts of students at Southwest High School serving as "e-bassadors" to schools in the United States, Malaysia, Australia, Ireland, Costa Rica and Panama. Utilizing Internet resources and communication, the students are exchanging information and ideas for organ donation education activities. The schools involved are utilizing the Organ Donor Project Student Workbook, "Fact or Fiction" video and interdisciplinary curriculum to increase awareness of the need for organ donation in their communities. The final product of this collaboration will be the creation of an International Organ Donation Information Exchange that will explore the cultural, legal and ethical implications of organ donation internationally. Through the efforts of the Organ Donor Project, thousands of teens have become aware of the desperate need for organ donation.

Community Service & Volunteerism

Vanessa Deanne Perplies, Target All-Around Scholarship Winner

As a volunteer for the Los Angeles Police Department Explorer Scout Program, Vanessa has assisted with crime prevention surveillance, evidence searches and police ride alongs. She volunteered nearly 400 hours with the program designed for students who are interested in law enforcement and community service. Her work was one of the reasons she won the Target All-Around Scholarship, which is based on community service. In addition to volunteering with the Explorer Scout Program, Vanessa also raised funds and walked with the North Hollywood High School Zoo Magnet AIDS team and volunteered for Project Chicken Soup preparing and delivering food to AIDS patients. A student at U.C. Santa Barbara, she is majoring in sociology and plans to become either a sociologist or journalist.

Serving & Protecting

One Sunday morning, bright and early, approximately 25 Girl Scouts arrived for a tour of the Los Angeles Police Department Foothill Police Station.

As a Los Angeles Police Department Explorer Scout, it was my task to help these girls. Experiences like this one have shaped and solidified my career goals, in addition to benefiting the children of Los Angeles.

The girls, ranging in ages from 4 to 11, had not only an enjoyable adventure but also learned about important issues such as 911, acting in emergencies and overall safety. I helped the girls try on riot gear, turn on the lights and sirens of police cars and use the police radios. I escorted them through empty jail cells, reminding them of the dangers of the world and teaching them to stay safe.

For young girls who are rarely taught self-reliance, this experience taught them how to take care of themselves. I was keeping people safe as well as helping them grow up to be stronger, wiser women. I could see their delight and curiosity at the unfamiliar environment of a police station, and I was happy to demystify law enforcement in such a positive manner.

The gratitude of the little girls showed me the simple appreciation in a child's smile was a priceless feeling. The Girl Scouts were not the only ones who learned; I realized that the things that make me happiest also make others happy. I have been inspired and challenged to learn and do more, and especially to reach out and share the knowledge I can, changing the lives of others for the better.

More Essay Advice from the Winners

Scholarship Winners

Here are some more tips on crafting your essay from scholarship winners.—Gen & Kelly

Jason Morimoto
U.C. Berkeley student and scholarship winner
"The way to shine is by crafting a story in your essay that brings out your strengths. I like to give a lot of personal examples as to why I am involved in certain activities. I try to avoid the generic responses like 'it was a good learning experience' or 'I wanted to try something new.' For example, I often use the example of how I walked on to a national champion rugby team as a mere 5'6" player with no prior experience. The coach took one look at me and wanted to laugh. I told him that all I wanted was a tryout. With a lot of determination and hard work, I proved myself capable of playing with world class athletes."

Kristin N. Javaras
Rhodes Scholar
"I highly recommend showing it to people who have won fellowships themselves or who have read successful fellowship application essays before (and the more people the better). Also, I feel that the revision process was crucial for my essay: I went through about seven or eight drafts of my personal statement!"

Svati Singla, Discover Card Tribute Award Scholarship Winner

Svati says that she has never let society's perception of age stop her. This was one of the factors that led her to publish an abstract in the American Journal of Hypertension after years of research–at the age of 11. Throughout junior high and high school, she continued her research at East Carolina University on fetal alcohol syndrome, won accolades from the U.S. Navy and Army for her research and spent three years shadowing surgeons at East Carolina University Health Systems.

After graduating from J.H. Rose High School in Greenville, North Carolina, Svati is studying biology with a concentration in genetics at Duke University. She has won an extraordinary $1 million in scholarships including the Discover Card Gold Tribute Award, Benjamin N. Duke Leadership Scholarship, Boy Scouts of America National Scholarship and National Merit Scholarship. After graduating, she plans to attend medical school.

Giving Back to My Community

I dedicate many hours of my time to significant community service activities. Through my participation in such service projects and activities, I have learned many valuable lessons about the significance of each individual in the community.

As a literacy volunteer, I am given the opportunity to see the glow on a mother's face when she realizes that her son will finally be given the gift of the ability to read. I am given the satisfaction of knowing that my time is positively contributing to another's life.

Another community service activity that has significantly influenced the community is my involvement in Teen Court. Teen Court is an alternative program to the court system that provides graduated penalties for juvenile offenders. It is an innovative program that benefits teens on both sides of the court system. The teens who are brought before the Teen Court learn to accept the consequences of their actions, without having a flaw in their permanent record. On the other hand, the teens that comprise the court system are educated about the justice system while they work together for awareness and compliance with the law. As a member of the Teen Court program, I am able to provide far-reaching benefits for all members of the community by keeping the youth well disciplined and well educated.

Recognizing the need for volunteers at a facility for mentally retarded children, I immediately seized the opportunity to make a difference in the lives of these children. As I read and play with them, I realize how simple pleasures bring so much satisfaction and joy to their hearts. I take great pride, knowing that I am spreading a feeling of warmth and happiness with my actions. At the local Boy's and Girl's Club, I have initiated a program, called "Bookworm", which encourages young children to read. As I go and read to these children, I realize that I am not only increasing their interest in reading, but that I am also serving as a role model to them. The children are all motivated to learn and make great strides in their reading. Their interest in education creates a positive attitude towards learning that is beneficial to the community.

As I volunteer for other organizations such as the Salvation Army, American Cancer Society, Knights of Columbus, Greenville Community Shelter, East Carolina University Health Systems and more, I realize how my actions can be compared to a pebble in a pond. Despite the size of the pebble, once it is thrown into the pond, the entire pond feels the pebble's impact through the ripples. Similarly, though I am just one individual in a large community, I am able to make a difference. Dedicated to community service, I am the pebble in the pond.

Donald H. Matsuda, Jr., Truman Scholarship Winner

Camp ReCreation

Working with mentally and physically disabled children over the past four summers has been one of the most amazing and rewarding experiences of my undergraduate career. Before volunteering with Camp ReCreation for Disabled Kids, I shied away from any interaction with the disabled community and remained distant from this group of people whose lives and problems seemed so very different from my own. Nevertheless, I felt compelled to bridge this gap, and I decided to board the bus for my first experience at Camp ReCreation.

During this first summer, I took care of a deaf boy named Michael. At first, I was quite frustrated because I was unable to establish any means of communication with him. However, I did not see this problem as an insurmountable obstacle; instead, I viewed it as a challenge that could be overcome with some dedication and perseverance on my part. Over the next week, I voraciously read all books I could find on sign language, and I devoted most of my nights to mastering this very complex form of communication. My tireless efforts paid off, as Michael began to recognize my signs and responded with frequent smiles, indicating his understanding and acceptance. As Michael began to open up and even sign back, I realized that we had developed a special and meaningful relationship—one that provided him with happiness and one that solidified my genuine love of service.

I eagerly returned to camp for three more summers. One summer I worked with Nick, a mildly autistic teenager to improve his communication skills. Last year, I helped Brittney, a young girl with a neurological disorder, in developing better motor coordination. Despite the differing needs of each camper, I still maintained the camp's mission: to provide a positive and healthy summer experience for disabled youth. In return, I gained the love and friendship of disabled kids and learned that these children have needs that are not unlike my own. I truly value my Camp ReCreation experience because it has fueled my passion to protect and promote the rights of children nationwide.

Emily Kendall, Association for Women in Science Scholarship Winner

To tutor a struggling math student, Emily drew on the patience that her own teachers had shown her when she was younger. This is the topic of Emily's essay, which she wrote to win a number of scholarships and to apply to Harvard, Duke, MIT, Washington University in St. Louis, Caltech, Vanderbilt and the University of Chicago. In fact, she not only gained admission to all but also received a number of offers of full scholarships. In addition to her volunteer work, she has been named a national semifinalist in both the Intel and Siemens-Westinghouse national science research competitions, led her high school academic team to two state championships and been one of two delegates from her state to participate in the U.S. Senate Youth Program. A graduate of North High School in Evansville, Indiana, she is now studying physics at Harvard University.

A Lesson for Both of Us

June pursed her lips and furrowed her brow as I plunged into yet another problem demonstrating least common denominators. Recognizing June's confusion and exasperation, I wracked my brain for a simpler approach, but as I spoke, the blank expression on June's face foretold my impending failure. I waited hopefully as she puzzled over the final step, but she dropped her pencil and sighed, "Negative numbers just don't make sense!"

Clearly, June was frustrated. So was I. All my life, I had ceaselessly soaked up knowledge from books and asked questions about everything around me. I explored new ideas; I pushed myself to achieve and to learn far more than my classes required; and when it came to math, I would gladly spend days pondering a challenging problem. Why couldn't this girl be more like I am? Or at least more like my other pupils? My other math students studied faithfully; my freshman debaters shared insightful new arguments with me daily; even little Hernando, whom I tutored in inner-city Chicago, had been thrilled to have a friend who would study with him.

Part of me wanted to give up on June, but then I realized that some of my own difficult experiences, which had shaped me greatly, could help me to help June. I recalled my painful rejection from the first grade "select choir."

That experience, although it had taught me to accept failures with grace and learn from them, had still hurt terribly, so I resolved never to let anyone label June a failure at math. I also recalled my determination to make the eighth grade volleyball team and the countless hours I spent lifting weights and repeatedly serving the volleyball alone in my backyard before I earned a school uniform. Recognizing the persistence necessary to achieve something difficult, I determined to work my hardest until June mastered her algebra. Finally, I recalled my enthusiastic middle school teacher whose coaching helped and inspired me to win local MATHCOUNTS competitions, qualify for the state team and advance with my teammates to place second nationally. Seeing the value of a committed, motivated, enthusiastic teacher who puts more faith in you than you put in yourself, I promised myself that, despite my frustration, I would not give up on my student.

As I move on to college and beyond, I intend to excel, but more importantly, I want my endeavors to have a positive influence on others. As I work toward this end, I am thankful for my talents and my successes; however, I recognize that some of the greatest gifts I can offer are the perseverance, humility, compassion or strength arising from apparent defeat. Drawing on my natural abilities and life's lessons, I can continue to help others as I helped June, whose blank looks turned to expressions of understanding, whose signs of frustration became promising smiles and who, though I still have much to teach her, finally believes she can succeed.

Career Plans or Field of Study

Danny Fortson, Rotary International Ambassadorial Scholarship Winner

Danny applied for the Rotary scholarship as a way of not only studying journalism but also doing so abroad. He will use the award to gain formal training in journalism and to continue his studies in Spanish. A graduate of U.C. Santa Barbara, Danny's experiences include writing for two San Francisco-based publications, studying abroad in Costa Rica and interning for the Center for Strategic and International Studies, a foreign policy think tank. He thinks he won the award because of his persistence. An unsuccessful applicant the previous year, he used the time in between to increase his qualifications with more journalism experience which is clearly evident in his essay.

International Journalism

Several months ago, I walked into the local bicycle shop, picked out one of the few two-wheelers that fit my lanky frame, strapped a Styrofoam helmet to my head, and set about riding my bike to work every day through the hectic downtown San Francisco traffic. I did not do this out of a strange compulsion to tempt fate but because the day prior I had signed up for the California AIDSRide, a 575-mile, week-long sojourn from San Francisco to Los Angeles to raise money for AIDS treatment and awareness. When I first heard about the event, I had trepidation: "Why not just drive your car? There is no way I'm going to ride my bike more than 500 miles in one week!"

However, I was drawn to the event not only by the desire to challenge myself but to serve a worthy cause. It is these guiding principles, to always push myself and to impact people beyond my own circle, that I look to for direction in all the things I do and something that I feel journalism fulfills. That is why I have chosen to pursue it as a profession, because to be able to write news that is relevant to people's lives is a way to tangibly serve society.

My primary job is as a reporter at *The Daily Deal*, a financial newspaper aimed at investment bankers, corporate lawyers and company executives. Knowing nothing of the complexities of the financial world, I was thrown into the job initially as an editorial assistant, which required me to write authoritatively for a very sophisticated audience. After about three months of intensive learning, I moved up from my post as an editorial assistant to reporter. It was and still is a daunting task, especially with the daily four-hour deadline, but I love the challenge.

My work as a freelance reporter at The Independent Newspaper Group, on the other hand, provides a whole new set of challenges and resplendent rewards. For The Independent I cover stories with a human aspect, issues that are relevant to the people of the local community. Whether I am covering the community service of a local congregation or the phenomenon of the ever-popular scooter, I enjoy the work because I can engage with people in the community and talk about issues important to them.

Working for two drastically different publications, I have learned that so much is determined by larger economic and political factors, but that it all ultimately trickles down to people on the local level. Making that connection with the community is incredibly rewarding, and to be able to extend my reach to a wider audience but retain the local interest is what I am ultimately aiming to do. A year abroad would provide the crucial stepping-stone toward that end.

Of course, the issues affecting local communities are most tangible when you experience them first hand through interactions with its less fortunate, something I have always sought.

In Costa Rica, I found myself at the other end of this teacher-student spectrum. The seven months I spent living and studying at The University of Costa Rica were my first experience as a foreigner in a foreign land. Staying with a local family and attending university classes, I had no choice but to learn Spanish, and fast, at a level that six years of classroom instruction did not afford. In addition, I was in the middle of a crash course in Central American culture and the pertinent issues to the people there, and it was that experience of fully delving into Costa Rican life that imbued me with a passion for the region.

I took that passion to Washington D.C. where I worked as an intern for The Center for Strategic and International Studies, a foreign policy think tank. While there, I worked for The Mexico Project, a program that brought together dignitaries from both sides of the border to foster a unified binational policy. My time in Guatemala also is what fueled my desire to work closely with the Americas Program director to coauthor an election study of that country's presidential race that was disseminated throughout Capitol Hill. Writing that piece for a crowd so far removed from the issues highlighted what I have found to be a recurring dichotomy between the local concerns of one community and the disconnect with a potential audience so many miles away. That is the gap I want to be able to bridge.

It has been three years since I used Spanish in my daily life in Costa Rica, and I desperately want to reach the level of fluency necessary to use in a public forum. In seeking to go abroad once more I intend to do so, as well as get the formal journalistic education to buttress my practical training. Moreover, I could absorb the sense of a foreign society, history and culture that is only attainable through living the daily experience of another country. I would then be equipped to communicate more effectively and to a much wider audience, taking into account the sensibilities of another people and culture and balancing that with the knowledge I have from my own.

Cecilia A. Oleck, Knight Ridder Minority Scholar

Cecilia has competed athletically on the court, field and track. But it's her sense of competition in the newsroom that makes her want to pursue a career in journalism. While a student at West Catholic High School in Grand Rapids, Michigan, she wrote for the newspaper, tackling topics such as the double standards for athletes and gender stereotypes. As a result of her journalism experience, she won over $80,000 in awards including the Knight Ridder Minority Journalism Scholarship and a scholarship from the *Detroit Free Press*, where she has worked as an intern. A student at Saint Mary's College in South Bend, Indiana, she is majoring in communications, preparing for a career in journalism.

The Right Fit

For more people, the future is uncertain; the direction their life will take is not spelled out for them. Each person is responsible for the choices she will make that will determine the course of her life. One of the choices that has an incredible impact on her life is what she will choose for her career.

This can be a difficult decision to make, as it will affect almost every aspect of a person's life to some degree. Most people are also not fortunate enough to receive a startling revelation directing them on the right course for their lives. Instead, the most powerful way that a person is able to determine her direction is not through an earthshaking revelation but through the quiet confidence that this is what she is called to do and to be.

In the same way, I have never received any startling revelations that I should pursue a career in journalism, but as I look back on my life, I am able to see that there were many little steps along the way that have led me to this choice. I have often heard my parents tell the story of how, on my first day of kindergarten, I came home crying because I had not been taught how to read. I have always loved to read and from that has developed a great respect and fascination for the written word.

Joseph Pulitzer stated the purpose of a journalist should be to "Put it before them briefly so they will read it, clearly so they will appreciate it, picturesquely so they will remember it, and above all, accurately so they will be guided by its light." His words serve as a reminder for me of the many different dimensions of writing. Journalism encompasses both creativity and technicality. It is a way of expressing individuality and of communicating with others.

Many of my personal qualities convince me that a career in journalism is my calling. I find that I am a person who responds well to challenges. Perhaps it is because of my competitive nature that challenges motivate me, and I discover that my biggest competitor is usually myself. I think this is why I enjoy trying to combine both the creative and technical aspects of writing. Each time I begin to write, I am presented with a fresh challenge.

Every day that I live, I realize more the impact other people have on me and the impact I have on them. Writing is a very personal way of reaching out to people I may never meet, but who I am still connected to because of our identities as human beings. Journalism allows the opportunity for the sharing of information, thoughts, feelings and ideas between people from all walks of lives. It enables people to see things through another's eyes and gain new perspectives on the world around them.

It is because of the many dimensions of journalism that I desire to pursue a career in this field. I feel that as a journalist, I will be able to use the talents that I already have, as well as learn new ones. I believe it is a chance for me to be an instrument to bring people together.

Chris Kennedy, National Merit Semi-Finalist

From Leawood, Kansas, Chris has always been around animals. He has volunteered at a wildlife rehabilitation center, studied butterflies and moths and worked for his family's canine rescue group. In the ninth grade, Chris and his parents decided that he should be home-schooled so he could better pursue his interests. In addition to studying at home, he has remained active in local science groups and taken courses from a charter high school and college. With this essay on his plans to become an avian veterinarian, he was recognized as a National Merit Semi-Finalist. He gives this advice on applying for scholarships: "Don't put it off. The closer it gets to the deadline, the more terrifying it becomes. So start now."

Avian Veterinarian

I am the only person I know who dreams of becoming an avian veterinarian. That's a bird doctor if your Latin is rusty. I got my first bird—a cockatiel named Sunny for his cheerful disposition—slightly over a year ago. One look into those sweet, intelligent eyes, and I was hooked.

I first gained experience caring for birds of prey, songbirds and waterfowl volunteering with Operation Wildlife, a wildlife rehabilitation and education center that serves the eastern half of Kansas. One particularly memorable day at the center had me tube-feeding a hummingbird. I also participate in the Idalia Society, a group of lepidopterists that studies butterflies, moths and their environments. The magic of nature has always fascinated me, and I am lucky to have found a passion that will let me explore the world of birds and nature in my eventual career.

After seventh grade I chose unorthodox schooling that allowed me to explore my avian interests in more depth while still covering all academic subjects at an advanced level. I have thus pursued a blend of home-school, public school and college coursework. Each year's educational program has been different. In ninth grade I chose a year of home-schooling in literature and world history complemented with work at the college level in geology and biology at Johnson County Community College and volunteering with Operation Wildlife and Farley's Angels, my family's private canine-rescue effort.

At the end of that year I decided I wanted to earn a Kansas Regent's high school diploma but also wanted to continue working at my own pace, exploring topics with greater depth than might be possible in a high school classroom. Those goals came together at Basehor-Linwood Virtual Charter School. I am completing the four-year curriculum in three years, substituting college courses in the sciences. Thanks to the Basehor-Linwood program I have had the latitude to pursue my interest in one facet of my heritage by substituting three years of intensive immersion education in Norwegian for more traditional language programming. At home, I'm currently completing Cornell Ornithology Lab's Bird Biology distance learning course, the University of Missouri-Columbia honors physics course and continuing my volunteer work.

My hard work earned me prizes in physics in the Kansas City Science Fair, in science and social studies in the State of Kansas Scholarship Contest and recognition at the national level by Duke University's Talent Identification (TIP) Program. My unusual curriculum allowed me to compete with students from traditional academic backgrounds while serving my interests as no public school could.

Boy Scouts gave me the chance to enjoy the outdoors while learning about myself and others and gaining valuable leadership skills. I applied those leadership lessons last summer in the American Legion Boys' State of Kansas, a once-in-a-lifetime opportunity to learn about state government. Together these programs taught me about the importance of individual participation and good leadership in groups.

At the most basic level my goals are like those of many others. I want to find the best college environment, study my chosen field of biology or pre-veterinary medicine, go to graduate school and have a fulfilling career. Of course, my particular career goal—to start my own avian veterinary clinic—is unusual, but it serves the same human need. I hope to be a good husband and father, to learn from mistakes, help others and do the right thing while remaining young at heart. With my preparation thus far and a lot more hard work, I will achieve my goals.

Andrea Setters, Dow Scholarship Winner

While she had always known that her future was in the sciences, it took an inspiring teacher for Andrea to discover that her passion was for chemistry. Her teacher at Fairborn High School in Fairborn, Ohio, taught her about the laws of thermodynamics, periodic trends, redox reactions and moles, and in doing so also became a mentor and friend. Andrea says about her teacher, "She brought out the best in me."

Andrea, majoring in chemistry, won scholarships including the Dow Chemistry Scholarship and Furman Founder's Scholarship to attend Furman University in South Carolina. She hopes to eventually earn a Ph.D. and work in pharmaceutical research.

Scientific Inspiration

I often sit in front of blank pieces of notebook paper and half-finished applications wondering why in the world I am still doing all this, wishing it were all over a little sooner. After all, basket weavers make a nice living: they create wonderful pieces with both aesthetic and functional purposes and I bet they didn't have to fill out 20-page questionnaires about what they have done with every waking hour of their lives for the last four years. Then again, I don't have any desire to be a basket weaver. Honestly, I would not have the slightest clue how to begin a basket and I hate splinters. I do, however, have a love of chemistry and it is this love that pushes me to continue with all the forms and essays with a little more enthusiasm than I may have started them with.

When I began to consider future careers, I set two basic criteria: I had to be decent at it, and it had to be something I never seemed to grow tired of. Science became the obvious answer. The not-so-easy question became what I wanted to do with science and what specific discipline. Then, during my sophomore and junior years, two experiences helped narrow down my field: my chemistry classes and being a teaching aide.

When I signed up for honors chemistry, I had absolutely no idea what I was getting myself into. I soon discovered, however, that I had nothing to worry about. The guidance gods smiled on me when they were assigning classes and I was placed in Mrs. Roshto's class. Mrs. Roshto's teaching ability and knowledge of chemistry continue to amaze me to this day. She had the knack to get to know each of her students personally and then be able to offer individual direction. I progressed through the class in a constant state of awe with the new world that was being opened to my eyes each day. I knew chemistry was the field I wanted to devote the rest of my life to.

My junior year I took AP Chemistry, which was also taught by Mrs. Roshto, and during my study hall I aided her Chemistry I class. The positive influence Mrs. Roshto had over so many students was inspiring and I fell in love with the idea that I could introduce students to this world of chemistry that they never could have imagined before. It was during that year that I decided teaching would become one of my career aspirations.

I have solved part of my original dilemma by finding that chemistry is the scientific field I want to work with most. I have yet to discover that perfect career but I believe with classroom experience and guidance I can find it and be able to devote myself to it wholeheartedly.

Essays about Leadership

Swati Deshmukh, Discover Card Tribute Award Scholarship Winner

Swati has helped collect 500 pairs of shoes for hurricane victims in India, raised funds for flood victims in Venezuela and spearheaded a bottled water sale fundraiser to aid flood victims in Mozambique. And these are just some of her efforts abroad.

At home at East Lyme High School in Connecticut, Swati has committed herself to public service, crocheting blankets for premature babies, tutoring students who are refugees from Burma and organizing a book drive for needy libraries.

It is her service and leadership that helped her become one of the nine national winners of the Discover Card Tribute Award. In addition to volunteering, Swati has won numerous awards for her writing, including first place in her state for the National History Day competition. Academically, she has a passion for research, studying organic synthesis of piezoelectric molecules as a participant in NASA's Sharp Program. She would like to eventually attend medical school.

A Fight against Discrimination

One good example of my continuing leadership is my efforts to diminish prejudice and spread feelings of well-being throughout the school. Nearly all of the students who attend my school are white, and I am in a very small minority. For this reason, I feel almost obliged or rather chosen to carry the torch and lead the warriors of unbiased acceptance in an endless war against discrimination.

At the end of my freshman year, I wrote a proposal for a Multi-Cultural Club to recognize and celebrate the minorities at our school. However, the school felt that this subject was covered by other clubs. Disagreeing with the administration, a faculty member at my school invited me to visit Westbrook High School where the Anti-Defamation League was running its program "Names Can Really Hurt Us."

Amazed at what I saw, I labored to bring this program to my school to combat our problems. I convinced my principal to let the Anti-Defamation League come to our school, and I raised the $4,500 that was necessary. With faculty members and other students, I formed the Diversity Team to help run the program. We selected 30 students to form the team, which would be trained by the ADL to help lead the program.

Now, every other week, I run meetings for the Diversity Team in which we prepare for the program. I have set up an e-mail system to contact them, and I have organized a special retreat for us. I also initiated a paper chain project in which every student in the school was given a slip of paper to decorate. The slips of paper will be linked together to form a chain.

My efforts to combat prejudice in the school have turned me into a leader of my peers. In guiding and directing others, I have discovered that I have the ability to lead others and motivate them to achieve great things. I plan to continue my leadership and maintain diversity programs at our school.

Academic Accomplishments

Jonathan Bloom, National Merit Scholarship Winner

When applying for the National Merit Scholarship, Jonathan chose to write about the subject for which he has the most passion–mathematics. While still a student at West Bloomfield High School in Michigan, he took college level math courses and conducted research through an internship with General Motors. He advises other students applying for awards to write about their interests. He says, "They have many students to choose from so you can't be too modest or you won't stand out. Be excited, tell them your passions and write with a goal in mind." Jonathan is studying mathematics at Harvard University and plans to pursue a career in the field. In addition to math, he enjoys tae kwon do, volunteering at a student-run homeless shelter and juggling knives, torches and balls.

Cryptography & Encryption

I would like to pursue an academic life in the field of mathematics. This interest in mathematics developed not only through my coursework but more importantly through my independent research. After successfully completing AP Calculus BC and AP Physics as a sophomore, I felt the need for a greater challenge than that derived from "spoon-fed" instruction. Therefore, the following summer, I attended the Ross Young Scholars Program at Ohio State University. While fully immersed in number theory for eight weeks, I developed a burning hunger for the in-depth study of mathematics. I learned how to think scientifically and perform research independently. The founder of the program honored me with an invitation to return as a junior counselor.

During my junior year, I used a graduate text to guide my research in one of the most active fields of applied number theory, cryptography. My investigation into both public and private key cryptosystems led to an award-winning science project through the development of software, which demonstrated RSA and other encryption algorithms. I completed my directed study in the area of cryptography for 500-level credit at Wayne State University.

Last summer, instead of returning to Ohio State, I completed an internship in the Operations Research Department of the General Motors Truck Group. I conducted a study that identified the most efficient method for accurately approximating the vehicle weight distribution for their product lines. Ten weeks later, I had completed the project, documented the results and given five presentations to increasing levels of management. General Motors is now looking at ways to quickly implement my findings, which will result in the company saving millions of dollars. The vice president has requested that I return next summer.

Currently, in addition to taking 400- and 500-level math courses, I am conducting an independent study at the University of Michigan-Dearborn. I am constructing the first small-scale prototype of "The Weizmann Institute Key Locating Engine" (TWINKLE). As described by Adi Shamir in 1998, this device utilizes optoelectronics to factor large integers presently considered not factorable, effectively threatening the security of 512-bit RSA encryption. The project requires extensive knowledge in mathematics, computer science and electrical engineering. My research has the full support of my faculty advisor, a Ph.D. in mathematics and professor in the CIS department. Professors at three universities have expressed an interest in my progress. Even if I don't reach my goal of a working model for the science fair this spring, I will still have amassed a tremendous amount of knowledge.

As chairman of our school's Science Research Committee and president of the National Honor Society, I have the opportunity to personally encourage classmates to do independent research and to facilitate their entry into the Science and Engineering Fair of Metropolitan Detroit. I hope that I am able to instill into some of my classmates even a fraction of the enthusiasm and motivation that I gained from the Ross Young Scholars Program. With the desire in place, the opportunities are endless.

Note: Jonathan makes one clarification about his essay. He was able to construct only part of the TWINKLE device. However he says, it wasn't "such a disappointment. It's just that my original goal was a little too ambitious."

Svati Singla, Discover Card Tribute Award Scholarship Winner

11-Year-Old Scientist

As an active and innovative student, I am always seeking unique opportunities that will broaden my realm of experience. At a very early age, I became involved in the field of scientific research as a very unique endeavor.

Though I was only 11 years old, I was determined not to let my age hinder my extreme ambition and interest in higher level research. Thus, I independently contacted the head of the Nuclear Cardiology Department at the local university and requested the opportunity to conduct research in his laboratory. Recognizing my genuine interest and scientific aptitude, he immediately introduced me to the lab methods and I began a detailed study, which demanded many hours of my time.

Since I was extremely young, I found the research concepts to be very difficult in the beginning; however, with determination and a positive mental outlook, I was able to comprehend all the research methods. The findings of my research were very significant and were published in the *American Journal of Hypertension*.

After the conclusion of this study, I continued my interest in research by initiating another experimental study in the Department of Biochemistry. This study dealt with drug abuse during pregnancy and fetal alcohol syndrome. The research was presented at local science competitions and was awarded top honors by the U.S. Army and Navy.

Another challenging activity, which I initiated, was to coach a young Odyssey of the Mind team. Odyssey of the Mind is a program that encourages creative thinking, problem solving, and teamwork; I have been involved in this program for over six years and have found it to be a rewarding experience. Thus, when a group of interested first and second graders needed a coach for their team, I readily stepped up and volunteered to accept this massive responsibility. It is very rare for students to coach Odyssey of the Mind teams. Thus, this was a very creative and unique endeavor, which I initiated for the benefit of the young team. Though it was a strenuous time commitment, I obtained a priceless feeling of satisfaction knowing that this had been a positive experience for everyone who was involved.

Both of these unique endeavors have taught me that age should never be a hindrance in the way of learning or sharing knowledge with others.

Seeking Genuineness

Coca-Cola Scholars Foundation

You may think that with the right mix of perseverance and success you can create a winning essay. The truth is that there is no single winning formula for creating a masterpiece. As you'll see in comments from the Coca-Cola Scholars Foundation, the best thing that you can do in your essay is be yourself. —Gen and Kelly

Beyond the essay topic provided, students participating at the Semifinalist level are given no instruction as to how to write their essays for the Coca-Cola Scholars Foundation's four-year Scholars Program.

"We don't provide instruction because we see the essay as an opportunity for each student to sincerely express themselves," says Trisha Bazemore, program assistant.

The 27 members of the Program Review Committee, comprised of college admission officers and high school guidance counselors, are chosen for their expertise in evaluating students' writing.

So what is the committee seeking as it reviews the 2,000 Semifinalist applications? In a word, genuineness.

"You can tell when you read an essay if it's 'real,' expressing an individual's heartfelt experience, or if it's an essay derived more from an awareness of presentation," says Bazemore. She says that it's important that students not try to write what they think the review committee wants to read.

"Be yourself," she says.

Emily Heikamp, Angier B. Duke Memorial Scholarship Winner

When exploring colleges in high school, it took Emily and her mother 14 hours to drive from Metairie, Louisiana, to the North Carolina campus of Duke University. But it was time well spent. After her visit, Emily fell in love with the college. She later wrote this essay to gain admission to and earn a full-tuition scholarship from Duke. In all, the self-described "science nerd" earned over $250,000 in scholarships including full-tuition awards from Texas A&M and Tulane University, which she declined to attend her dream school. A graduate of Archbishop Chapelle High School, she is majoring in biology and mathematics and plans to earn an M.D.Ph.D. in immunology or oncology.

Science Nerd

AGTCCGGAATT is the genetic code for Tumor Necrosis Factor (TNF), a human cytokine that may have deleterious, even fatal effects if produced in excess or inadequate quantities. For the past two years, I have performed research to study the effects of alcohol and glucocorticoids on the TNF response in murine macrophage cells. One may ask why I am interested in such an obscure topic. Well, I am a science nerd.

Scientific research fascinates me, as experiments raise many questions and always provide new challenges. Research also supplies knowledge of the most intimate interactions of the human body, giving a glimpse of processes that are invisible to the naked eye. My research provides me with this knowledge and the ability to share it with others, and it has given my life direction and purpose.

I discovered my passion for research when I was 15 years old. The summer after my sophomore year, I decided to trade in cherished lazy afternoons with tennis buddies, waking up at 1 p.m. and two months of dormancy for my tired brain. I became an employee of the physiology department at Louisiana State University Medical Center. My buddies became lab technicians, I woke up at 7 a.m. instead of sleeping in and my tired brain was forced into overdrive as I learned about Tumor Necrosis Factor and Lipopolysaccharides. And I loved my job.

I worked for Dr. Gregory Bagby, a professor and researcher of the Alcohol Research Center at LSUMC. His lab studies the effects of Simian Immunodeficiency Virus and Ethanol on Rhesus Macaques. In other words, how SIV-infected drunken monkeys can get really sick. Nonetheless, his research fascinated me, and I had so many questions. Perhaps what fascinated me even more was Dr. Bagby spent time explaining his world of Lipopolysaccharides (LPS) and Tumor Necrosis Factor (TNF) to me, a lowly high school sophomore! I began to perform experiments and assays for his lab, while also doing secretarial work for his research grants. Eventually, I began my own research on alcohol and stress hormones. My research taught me so much and gave my life new direction. I finally knew what career I would pursue. But more importantly, I learned what kind of researcher I want to be. Dr. Bagby had shown me that being a great scientist is more than Nobels, prestige, and grant money. It is about sharing what you have learned with others, even lowly high school sophomores!

Being able to work with others who share my passion and enthusiasm has helped me to shape my dream. I plan to earn an M.D.Ph.D., specializing in immunology. As a physician scientist, I will see patients while also performing research to find new medications or a cure for their illnesses. I feel blessed to be a healthy young woman, and so I want to serve those who are not as fortunate by doing what I love most—research.

Nancy Pan, National Merit Scholar

Nancy is from Covina, California, and was attending South Hills High School when she wrote this essay for the National Merit Scholarship. She has won more than $10,000 in academic scholarships and is currently attending Stanford University. Her advice to students writing personal scholarship essays is to "dig deep within yourself to find something that uniquely represents you. From there show how your actions or achievements illustrate that particular aspect of you. The key is to have both the internal passion and the evidence that supports it."

Superpan

My name is Nancy Pan. Growing up, my parents would always tease me, "You're not Superpan, you know."

But in my mind I was. Not only did I boast a red cape tied across my shoulders, I was also always pushing my limits. At age four, I would secretly practice on the courts for hours with ambitions of beating my six-foot tall dad in basketball. In third grade, I dedicated my entire summer at the library to writing my first 62-page novel, complete with hand illustrations. By the time I entered middle school, I had managed to skip a total of four years in mathematics while remaining #1 in a class with high school juniors.

I am obviously not a superhero, but my life has been characterized by the dual roles which typify one–doing what others expect of me and doing what I expect of myself. It is with my choice to establish a profound difference between the two that I have optimized my high school experience.

These last three years, my academic life has been fueled by my passions for writing and mathematics. In writing, I am fascinated by its polar nature. At school, I've enrolled in Advanced Placement writing courses to understand the objective aspect of writing, dissecting written works based on both the content and presentation of the author's message. I achieved a perfect score on my SAT Verbal and AP Literature exams, but I did not stop at being a good student. Rather, as an individual, I wanted to express myself in a way that was uniquely my own and yet still capable of moving others. I saw the development of my analytical abilities as a means of advancing my true passion, creative writing. Although such writing is more liberated and subjective, it too is built on a similar ability to dissect, analyze and understand plot and theme construction. I exploited what I learned in class, and in my own time, wrote volumes of poetry and short stories. In doing so, I won several city-wide writing contests, a poetry competition with Barnes & Nobles, a local publication and the luxury of putting my soul to words.

Perhaps in a way completely antonymous to my attraction for creative writing, I am fascinated by the objective purity of mathematics. However to me, math is not solely an abstract science but also a way to practically understand the world with numbers. Prior to high school, I extended my knowledge of mathematics outside of class, so that by the time I was a freshman, I had completed the AP Calculus curriculum. My school did not offer an official AP Calculus BC class, so I independently prepared for the exam and received a 5. Outside of class, I am enrolled in community college math courses, active in the Science Bowl with a focus on Mathematics and am additionally, the school representative for the Mathematics Olympiad. Although there are limitations in the math coursework provided by my school, my knowledge and passion for the subject has continued to thrive through my search for and involvement in outside opportunities.

Writing and mathematics are only two examples of areas in which I have recognized my potential to achieve and acted accordingly. However, I am an individual with many working passions. You will find in my application that I am additionally the Captain and All-League Finalist of my Varsity Tennis Team, a Valedictorian candidate, a winner of various scholarships, an active executive/officer in several extracurricular clubs, an avid volunteer, an employed instructor at a learning center and many other positions, each listed neatly but constrainedly upon the allotted line. I am all of these things, but they themselves are simply manifestations of my desire to reach my peak as an individual.

Over the years, my parents adapted their mocking tone and started calling me Superpan with affection. As for me, it's been years since I've put on that red cape again, but my mentality has not waived. I will continue to push my limits only to someday realize that there are none.

Essays about Athletics

Sara Bei, CIF Scholar-Athlete of the Year

From Montgomery High School in Santa Rosa, California, Sara was still excited about her team's underdog victory at the state cross country meet when she wrote this essay. One of the most profound lessons that she learned was how to motivate her team, a topic she uses as the centerpiece of this essay. Along with a half-tuition scholarship to Stanford, Sara also won over $6,000 in scholarships. She encourages all students to be relentless about applying for scholarships since in her words, "sometimes the ones you end up winning are the one you almost didn't apply to."

Inspiring Greatness

As a three-time state cross country champion entering my senior year, I hadn't expected this season to be much different than the others. I planned on working hard to achieve my goal of winning state, and I looked forward to having fun with my teammates in the process. In previous years, our girls' team hadn't been very motivated, leaving me to take it upon myself to make it to state as an individual. Little did I know that a completely different challenge lay ahead of me for my senior year.

At the beginning of the year, I was pleasantly surprised to find two newcomers fresh out of junior high, who had decided to come out for the team to give us the fourth and fifth runners that we so desperately needed. Immediately I began to ponder what our team's potential was, and as always, I shot high. I organized a team sleep-over and, while beading necklaces and watching movies, tried to instill in them the goal of winning the state championship. Most of them were doubtful, even shocked, that I thought a team who failed to even make it to the state meet the previous year could have a shot at winning it. However, I was prepared to help them to not only realize their potential and believe in themselves, but to work together as a group and strengthen one another in the process.

Throughout the road to the state meet, I was busy trying to find ways to motivate the girls to train harder. I gave them little weekly gifts and notes, made breakfast as incentive for morning runs before school, organized team bonding activities outside of practice and even made a "State Champion Challenge Chart." I tried everything possible to get them to do the necessary preparation to be the best, as well as have fun and come together as a team. In the process, I found myself devoting so much time to the team that I was hardly channeling any energy into my own training. Although this concerned my coach, I reassured her that by working with the team, we were helping each other and improving together.

Finally, the day arrived in a flutter of nerves, anxiety and excitement. After giving them a pep talk, we toed the line together and I thought back on all the months we spent training, planning and dreaming for this moment. True, I was out to become the first person to win four state titles, but as I chanted our cheer with each of the girls, I realized that my real drive to win was coming from our team's need for every point we could get. That day, we upset the first- and second-ranked teams, with each girl running the race of her lifetime to become the Division II State Champions! Seeing the smiles and tears of pure joy on the faces of my teammates, I realized that beyond the medals and championships, there lies a treasure of value that far surpasses any other individual award in inspiring greatness in others.

Artistic Talents

Andrew Koehler, Fulbright Grant Winner

Andrew has been a serious student of music since he was 5 years old, when he first began to play the violin. Originally from Oreland, Pennsylvania, he has performed as a violinist with the Philadelphia Youth Orchestra, the Yale Symphony Orchestra and numerous festival orchestras. While at Yale, from which he graduated with honors as a double German Studies and Music major, he began to conduct seriously and is now pursuing a career in conducting. After spending a summer as a conducting student at the Aspen Music Festival and School, he is continuing his studies at the University for Music and Art in Vienna on a Fulbright Grant.

Turningpoint

My parents both spent the first part of their childhood overseas. As Ukrainian immigrants in America, they faced both the immediate difficulty of learning a new language and the eventual difficulty of acting as translators for their families, who never were able to learn English adequately. Though I, too, was raised speaking Ukrainian, my parents ultimately wanted to give me a means of communication that would transcend all others, a language of international recognition; this, they decided, was to be music.

The choice they gave me, for whatever reason, consisted of only two instruments: violin and piano. Such was the decision I was to make at the age of 5, when I might otherwise have been happier to continue playing uninterruptedly with my action figures. I arbitrarily chose violin, blissfully unaware of the impending consequences. I cannot pretend that something in my soul stirred the first time I held the instrument. I played dutifully enough, though I, like any other child, did not really enjoy practicing. I went through a long series of mediocre teachers who failed to generate any real excitement in me. They gave me enough encouragement, however, to realize that I had at least some talent, and with this ray of hope, my parents continued to push me.

Much later, during the first rehearsal of a youth orchestra I had recently joined, something astonishing occurred. Sitting in the back of the section, I was intimidated in part by the music on the stand in front of me but mostly by the conductor on the podium, an extraordinarily temperamental man who could be blisteringly honest in passing judgment on one's abilities as a musician. We began the rehearsal by sight-reading the first piece in our program. We muddled through admirably enough, but he was clearly unsatisfied. He leapt from his stool and spat, "You play like you're older than I am," with each strong syllable of his phrase strengthened by ferocious baton swats on his stand. He paused and composed himself. "Do you know what this music is about?" No one dared breathe a word. He then spoke of love and death and of the profound tragedy that links the two with an eloquence unbefitting his audience. He stopped suddenly and disgustedly offered, "but you don't understand; you're too young," and then looked up at us questioningly, as if to ask whether or not he was right. He might well have been, but we desperately wanted to understand for that man. And so we played again, and this time, I began to hear music in an utterly different way, as did many, I suspect. His passion for this art was simply infectious. Where I had never previously known music to be more than just pleasant, I began to understand what defined greatness, both in interpretation and in composition. Poor performances could make me cringe, but the most sublime moments in great performances made my hair stand on end. I began to approach music with a penetrating enthusiasm which, save for the charisma of a great teacher, might never have been.

Years later, and as my focus shifts from violin to conducting, this enthusiasm continues only to strengthen. Knowing the difficulties of a life in music, I have tried seriously to pursue other interests, but no matter how engaging I find them, none generate the same passion in me that music does. I am resigned to my fate; I wish to be a musician. It is a perilous fate that neither my parents nor certainly I could have imagined when this endeavor began, but it is a fate nevertheless tinged with joy, for I may count myself among the few who have found something they sincerely love.

CHAPTER
SIX

12 ESSAYS THAT BOMBED

In this chapter:

- **12 disastrous essays**
- **"Where's the Point"**
- **"Attempted Tearjerker"**
- **"Miss America"**
- **"Behold! My Statistics"**
- **"My Life as Seen on TV"**

Learning from Failure

History has shown that some of our greatest successes have been inspired by failure. Akio Morita's automatic rice cooker was a huge failure and burned the rice it was supposed to boil. In desperation, Morita and a partner turned to building cheap tape recorders. From this single product came Sony Corp. Across the ocean a high school coach cut a young varsity football player. That athlete's name was none other than Michael Jordan. The founder of the automobile industry, Henry Ford twice filed for bankruptcy before he finally stumbled onto the product that would launch his company, the Model T.

In keeping with history's tradition we bring you 12 essays that failed. These essays, however, provide an extremely important lesson: they help you learn what not to do. As you read each essay along with our comments you will understand why they fell short of the mark. While you are writing essays, keep the lessons from this chapter in mind. These essay writers lost the various competitions in which they entered, but at least in doing so they are helping you to avoid the same fate.

Where's the Point?

Reading an essay without a point is like getting on an airplane without knowing where it's going. Yet many students turn in essays without any clear message. Consider the following essay and see if you can locate the message the author is trying to convey:

Where Has Time Gone?

As I sit at the lunch table, it suddenly hits me. Where has all the time gone? I am a senior in high school who is about to graduate in a matter of months and I have just realized that I might never see my friends after we receive our diplomas.

Surely we'll see each other at reunions, but what will become of the great moments that we have shared?

I will never forget when one of my friends and I were given the responsibility of putting together a class beach party. My friend wasn't a very creative or outgoing person, however, he was nearly twice as strong as me. So we came up with a plan. I would do all of the promotion for the party as well as decide the theme and menu. He would be responsible for making sure all of the food, sound equipment and decorations were transported to the beach. It was the perfect plan. Most of our classmates told us afterward that it was the best activity that they'd ever been to.

I sometimes wish I could stay in my school forever. I have learned so much in the last four years. Before I came to high school I didn't even know what I was capable of intellectually. My teachers have been some of the most inspirational people in my life.

I know that college will bring with it many new memories and experiences and I am looking forward to it. However, I will never forget the friends who stayed by my side and the teachers who cared throughout the good times and the bad.

Why This Essay Bombed

In this essay, the author simply has no meaning in his writing. The essay covers a range of feelings and experiences. By the end of the essay we wonder what we just read. Is the point that the author will miss his friends? Is it that he is able to solve the problem of working with a friend who is not creative or outgoing? Or is it that his teachers have been the most influential people in his life? There is no connection between the disjointed ideas. We are left confused and unimpressed.

How to Avoid This Mistake

As you are writing, think about what you are trying to convey. Ask the question, "What's the significance of this essay?" If you can't answer this question in a single sentence, then it probably means you need to make your message stronger and more clearly defined.

Since there is often limited space for the essay, it is better to stick to a single topic. Select one and develop it throughout the essay. Don't confuse the scholarship committee by writing about a number of things that have little or no connection to each other.

The Attempted Tearjerker

There were few dry eyes at the end of the movie "Titanic," and the director wanted it that way. Movies about tragedy are intended to evoke emotion from viewers. Some students do the same thing with their scholarship essay, attempting to win the reader over with dejected accounts of loss, desperation and hopelessness.

Unfortunately, these essays do not appeal to scholarship judges. They do not want to read about how difficult your past has been except within the context of how you've faced the challenges or your plans for improving the situation. They want to be inspired by what you have done and see that you are working to make your life better.

My So-Called Life

Someone once said, "Life is like a bowl of cherries--sometimes it's the pits." There could be no more accurate saying to describe my life thus far.

Even before I was born there was trouble. When my mother was pregnant she got into a car accident and nearly lost her baby--me! While I don't recall this event it was clearly an omen of things to come.

Throughout my childhood my parents were never rich. I remember one Christmas how jealous I was when I went back to school and my friends had the newest clothes and toys. Sure, I got gifts but not the kind of expensive presents that my friends had received.

When I was 15 years old I returned home one day and noticed that something was different. Half of the stuff that we owned in our apartment was gone. We had been robbed. The burglar had taken most of the good stuff that we owned. That year my brothers and I had to share a single 21-inch television.

As if things could not get any worse, the next year I learned that I had diabetes. While not life threatening it was enough to send me into a depression that took months to get out from.

Now that I am about to graduate I feel lucky to even be here given the hardships of my past 17 years. Going to college has been a life-long dream. This scholarship would help me pay for college and build a better life.

Why This Essay Bombed

While it is hard to not feel sympathy for an applicant who has suffered misfortunes and hardships, there are almost no scholarships that give money based on how much you have suffered. Rather scholarship judges want to see how you have excelled despite the obstacles in your life. The focus should be on what you have accomplished or what you plan to accomplish in spite of setbacks.

How to Avoid This Mistake

If the past has been rough, you can certainly write about it. But don't expect the hardships themselves to make your essay a winner. Make sure to include what you have achieved or what you have learned from these challenges. Write about how the hardships will influence your choices or affect the future. While scholarship judges know that many students have had to endure difficulties, what they want to see is someone who has survived and thrived.

Miss America Essay

We've all seen the Miss America Pageants. And we've all heard (and made fun of) the speeches contestants make. "I want to cure the world of hunger," "I want to save and give back to mother nature" and "I want to make sure that every person on the planet has a place they can happily call home." These ideals are just too lofty to take seriously. It is amazing how many scholarship applicants write about these very ideals that, despite their good intentions, are just too idealistic to be considered seriously.

My Dedication to the World

Through five years of community service, I've learned many things. I've seen the empty hearts in the children without parents and the broken hearts of seniors who get no visitors. Because of these experiences, I've learned that only through service I can be a fulfilled person.

Therefore, I have decided to work to end the suffering of all people who face the perils of being without food, clothing or shelter. This is now my life goal.

After college I plan to start a shelter for orphans. This orphanage will take care of children who have been abandoned and will attempt to create as normal a family life as possible. Once my first orphanage is established I will branch out to other areas and countries. My dream is to build a global network that would once and for all end the suffering of children.

Once I have accomplished this I plan on running for public office so that I can affect change on an even broader scale. As senator or president I will make laws and convince other countries to do what they can to protect each and every human. For it is only by committing ourselves to ending human suffering that progress can be made.

As humans we are here to make the world a better place, and if each person does his or her part, like I plan to do, the world *would* be a much better place.

Why This Essay Bombed

The applicant's heart is in the right place but the ideas are just too farfetched to be taken seriously. This just sounds too much like a Miss America answer and does not show that the applicant has any basis in reality.

How to Avoid This Mistake

This type of essay should be avoided altogether. There's no doubt that if each of us were given the chance, we would end worldwide hunger or save Mother Nature, but let's face it, this isn't realistic. Focusing on a few issues and describing what you have done can make a great essay. Keep a positive attitude and enthusiasm but ground your ideas in reality, and focus on what you have done instead of what you would do in a limitless world.

The Life-Changing Voyage

Whether backpacking across Europe or climbing Mt. McKinley, there are those students who have traveled the world. A part of their experience is the wealth of memories they brought back home. Thus, travel is a common topic when it comes to essays. However, essays about travel too often make sweeping generalizations, depict the superficial aspects of the trip or cover the events of two weeks in two pages. Here is an example of a travel essay gone awry:

My Trip to Europe

Two years ago I had the privilege of traveling to six European countries. There I met many interesting people and saw many interesting sights. In England I got to stand next to the guard who cannot be disturbed from his upright, staring position. In France I got to look out to the horizon from the famous Eiffel Tower. In Belgium I ate frites, which are essentially Belgian french fries. In Germany I saw where the Berlin Wall stood not too many years ago. In Italy I saw the Colosseum, where the Gladiators fought. And finally, in Switzerland I saw the Alps and ate fondue.

Besides having a great time seeing new places and meeting new faces, I also learned a great deal about the cultures of different European countries. I learned that people from different countries are, well, different. They have different mindsets about certain aspects of life and different ways of thinking. However, I also learned that people are, in a way, all the same. All the different ideas and concepts centered on the same areas of thinking and are therefore merely different interpretations of the same thing.

My visit to Europe has definitely changed my view of the world. I hope that someday everyone will have a chance to visit Europe or another foreign land and learn how diverse and similar our world really is.

Why This Essay Bombed

This essay is too much of a diary of sights seen, activities done and food eaten. Virtually *any* student who visited Europe could have written this–and many will. The essay also makes a general observation of travelers–that while people from different countries have differences we are all essentially the same. The result is that this essay hardly stands out from any other essay about travel.

How to Avoid This Mistake

Whether your travels have taken you to the Museé du Louvre in Paris or your grandmother's house in Tulsa, you probably have numerous experiences that could become good essays. However, when you're considering the possibilities, try to separate those events that could happen to many travelers from those that were truly unique to your visit. Focus on a specific event and elaborate on what it has taught you or how it has affected your life. Instead of writing about all seven days of travel, narrow it to one day or even one hour. Also avoid sweeping generalizations about the people of a country or humanity at large.

Convoluted Vocabulary

How many times have you read a passage in a standardized test or in an advanced work of literature and found that each word made you more confused than the last? If used properly, word choice can convey sophistication and demonstrate a writer's command of the language. However, when used incorrectly or only to impress, the results are convoluted, conceited or just plain incorrect. Here is an example of an essay that was intended to awe. See if you are impressed enough with this writer to hand over *your* money.

Educationality

That education is my utmost priorative focus is verified in my multitude of academic, extracurricular and intercurricular activities. I insinuate myself in learning and acquiring a plethora of knowledge. I am a person that doesn't approbate no for an answer when it comes to enhancing the prominence of my mind.

This pontifical accolade is an integral part of my scholarization, and without it, my temperament would fall short of instructured. My transcendent achievements speak for themselves and deserve accolades.

Why This Essay Bombed

It appears as if this essay had a head-on collision with a thesaurus. Using SAT words is fine as long as you use them correctly. Scholarship judges are not interested in how complicated a sentence you can construct, but rather how meaningful you can make it. Plus, some of the words this applicant uses were made up!

How to Avoid This Mistake

Don't venture into areas of the English language where you are a stranger. It is okay to use multisyllabic words when you see fit, but to use big words just for the sake of using them is a mistake.

Behold! My Statistics

Have you ever read the back of a baseball card? It is filled with statistics reflecting the player's performance during the season. It may show that a player is one of the fastest men on the field or that he performs well in the playoffs, but the statistics say nothing about a player as a person. Keep this in mind when writing scholarship application essays. If you just list statistics such as GPA, classes and activities, the judges will never get meaningful insight into who you really are.

My Name Is Brooke

Hello, my name is Brooke. I will be a senior at Central High School in Topeka, Kansas.

I was born on October 29. I have interests in writing and mathematics. My schedule junior year was as follows: AP English Language, Honors Physics, AP Calculus AB, AP United States History, Honors III Spanish and P.E. My extracurricular activities are varsity cheer and Key Club.

Here are my standardized test scores: 2170 SAT, 620 SAT Literature Subject Test, 610 SAT US History Subject Test, 680 SAT Math Level 2 Subject Test, 4 on AP English, 3 on AP US History and 3 on AP Calculus.

I have worked hard throughout my four years in high school to maintain a 3.7 GPA. I plan to graduate with honors next year. From there I will go to college.

I plan to major in either communications or business in college. The reason I will major in either communications or business is because I love to work with people and I am seriously interested in getting into the entertainment industry.

My favorite subjects in school this past year were AP English and AP Calculus. My hobbies include sewing, playing piano, singing and writing short stories. I currently have a job at a local restaurant.

Why This Essay Bombed

This essay gives a great deal of information about the writer but it says almost nothing about her motivation, dreams or beliefs. Qualities that show your character are the ones in which scholarship judges are most interested. They want a sense of who you are. This isn't conveyed through a list of statistics. Also, the application often asks for most of this information. So why repeat a list of activities, classes and GPA when the scholarship judges already have your application and transcript?

How to Avoid This Mistake

It's easy to write an essay in which you rattle off your status in life and a list of accomplishments. What's more difficult is putting your place in life and achievements into perspective and making sense of them. Focus on a few of the more important achievements and expand on those. Since the application form has a place to list activities, grades and test scores, don't repeat that information in the essay. Use the essay to go beyond your statistics and provide context for their significance.

The Most Influential Person in the World

A common essay topic is the person who has had the most influence on your life. You can imagine the countless essays that students write about parents, grandparents, siblings, friends and idols. The challenge is to write about this influential person in a way that is different from what other students write and that reveals something about you as well. This essay falls short on both counts.

I Love My Family

There is no one person who has had the most influence in my life; instead, it is a group. That group consists of the most important people in my life: my family. My family is made up of four people: my mom, my dad, my younger sister and me. My parents are my role models; they provide the home in which I live, the food that I eat and the money to buy essential items. They have set for me a good example of what kind of life I should lead. They have always been there for me, through the good times and the bad, to support, love and cherish me.

My mom, in particular, has always been very supportive of me. She has been the one to tell me bedtime stories when I go to sleep. She has been my unacknowledged chauffeur, taking me places such as the occasional baseball game or regular piano lesson even if she had more important things to do.

My father, on the other hand, has always been the advice giver. To me, he is all knowing, for he always has a good answer to the questions I have. My father, who is an engineer, helps me with my math and science homework; I wouldn't be as successful at math if it weren't for him. I love both my parents because they have both contributed to my life so much.

My sister has also played an important role in my life. She has always helped me whenever I was in need; when I couldn't solve a problem, when I couldn't think of a good design for my visual aid or just when I needed someone to talk to. I love my sister, and even though I'd be embarrassed to tell it to her face, she's my best friend. And even though we might fight every so often about issues we shouldn't even care about, our friendship is a strong one; those fights are just testaments to how real it really is.

Why This Essay Bombed

The problem with this essay is that it is too ordinary. Many applicants will write about what their parents do for them and how they hope to pattern their lives after them. This essay just does not stand out.

How to Avoid This Mistake

It may be difficult to choose just one person who stands above the rest. And the person you select may be mom or dad. These essays will work if, and *only* if, a unique angle is taken. A generic description about your father, such as, "My father is always there for me, through the good times and the bad," will go nowhere. It's important to be as specific as possible in describing exactly what it is that you admire about your father. Fortunately, there are unique things about all of our siblings, parents, friends and idols that make them special. Focus on those aspects.

Creativity Overload

We've all heard the various analogies for life–"life is like a box of chocolates," "life is a sport" and so on. Wouldn't it be clever if, somehow, you could create your *own* analogy of life? Surely it would show how deep of a thinker you are and how well you can write, right? Being clever is usually good. But sometimes it can go too far, wearing out a novel idea.

The Highway of Life

Life is like a highway with cars going in all directions. People are constantly coming and going from all sorts of places. Sometimes, when too many people want to go to the same place, traffic jams form, just as when too many people apply for a single position at a company.

Be True

Elisa Juárez & Emanuel Pleitez

Being truthful in the essay is not just the right thing to do but it also makes for a much better essay. Many essay disasters are created when students decide to write about something they don't know much or care about.—Gen & Kelly

Elisa Tatiana Juárez
Brown University student and scholarship winner
"Be honest. I have a lot of friends who said you could lie. You could but in the end people will really know who you are."

Emanuel Pleitez
Stanford University student and scholarship winner
"Be true. Don't try to fake anyone out. It's not going to work. You don't have to be the greatest writer. I write what I really feel. If you really believe in what you're writing, then you should be well off."

The Highway Patrol is akin to my parents because whenever I feel like breaking the rules, as any driver would, the presence of my parents always prevents me from doing so.

Throughout my young life I have been on a highway full of cars passing interesting exits. As I pass each exit--the doctor exit, the lawyer exit, the CEO exit--I realize that my highway of life is full of so many possibilities.

However, none of these intriguing possibilities can be reached without the integral element of the automobile: gas. In my mind, my education is the gas that will run my car that will take me to these places.

Why This Essay Bombed

The student might think that this concept of comparing life to a highway is quite inspired. In fact, she might have even talked about it with her friends and they might have been impressed. However, somewhere along the line she must have missed adult contact. This essay really just makes the applicant appear silly. She starts with an original idea but takes it too far in an overly simplified way. Being creative is good. But don't go overboard and end up with a laughable essay.

How to Avoid This Mistake

Have several other people read your essay. If you feel that maybe it is too creative or may border on being trite, ask what they think. Sometimes we just get too caught up in our own writing to make good judgments.

The Future Me

Scholarship organizations will often ask applicants where they see themselves in 10 years (or some other time in the not so distant future). Now we know what you're thinking–steady job, happily married, living in a nice house, two adorable children and dog named Spike. Stop right there, because this is your worst enemy. This idea is exactly what the 10,000 other applicants are planning to write about.

But wait a minute, this idea is *also* your best friend. You now have the perfect idea of what *not* to write about, and this information can prove quite useful when it comes to writing an original essay. Here is an example of an essay where the writer falls for this trap:

Ten Years from Now

Ten years from now, I see myself as a college graduate from a local private university. I will also have a steady job at a company for which I love to work. Hopefully I will be married, and, if I am, I'll probably have one or two kids. I plan to spend my free time with my family and maybe indulge in a few sporting events.

My job will probably be as an accountant, and I will give it my best. I'll like my job, because I've always liked to work with numbers. The only drawback will be that an accountant's salary will not be as much as I would desire to make. So I plan to achieve a state of wealth by investing in the stock market. By doing so, I will enable myself to retire at an early age.

In retirement I will continue to invest intelligently in the stock market so I can pay for my children's educations. I will travel the world with my family during the summers and donate to various charitable organizations.

My future will be a bright one provided that I get my own education. Ten years from now, the brightness will just be beginning to unveil, and I will be stepping into the happiest phase of my life.

Why This Essay Bombed

While this may be how the applicant truly feels it makes for one boring essay. It reflects the goals that almost every person has of being successful and happy, making the essay ordinary and unoriginal.

How to Avoid This Mistake

When asked where you see yourself in 10 years, focus on a single desire. You might write about an unfulfilled dream or a specific contribution that you hope to make. To make the essay more interesting, don't approach your future in the same way that nearly every other student will view his or her life. Make the essay interesting by including your motivations, challenges, inspirations, rationale and expectations for the future.

My Life as Seen on TV

The average American spends dozens of hours watching television each week. It's not surprising then that entertainment from television, film and music finds its way into scholarship essays. While entertainment is an influence, it is a mistake to draw parallels to something with which there really is little or no connection. Whatever topic you choose to write about should have a relationship to your life.

Liza

A movie I recently saw struck a heartfelt tone in my mind. I realized that the main character, Liza, was forced to struggle through circumstances similar to mine.

After losing her father in World War II, Liza's mother raised her as a single parent. Through the hardships, Liza's mother grew physically and mentally stronger because she knew she had to make sure that her daughter would be all right. Liza, however, did not appreciate her mother's efforts. Liza was an aspiring actress. Her favorite acting part during her senior year was playing Dorothy in *The Wizard of Oz*, the final play of her high school career.

On the night of the final performance, Liza's mother had a commitment that she could not afford to miss. Liza begged her mother to come in time to hear the last song--a reasonable request--and her mother lovingly promised that she would be there. Liza gave the greatest performance of her life that night and waited patiently for her mother to arrive, until the final song was about to be cued. At that point, Liza began to lose hope, and by the time the song was over she completely hated her mother. Liza waited around after the play until a police officer suddenly pulled up to the school to tell her that her mother had died at the hands of a drunk driver while on her way to the play. Devastated, Liza finally realized how much her mother really loved her.

The movie was especially touching because Liza lived her entire childhood trying to get away from her mother, but lived the remainder of her life appreciating how much her mother actually meant to her. My life story is similar to Liza's because I too am aspiring to be someone, except I aspire to be a psychologist. Also, I have been fortunate enough not to lose either of my parents, although my father has a strenuous work schedule, and I only see him in the mornings.

I have been inspired by this movie to appreciate the love, support and encouragement that my parents have provided for me in the past and which they continue to provide.

Why This Essay Bombed

Touching story, the movie that is. Not only is the writer's life nothing like that of Liza's, the inspiration the writer has received tells us nothing about the writer herself. This essay would work a lot better for someone who *has* suffered such tragedies and hardships.

How to Avoid This Mistake

While choosing a topic, it is fine to select one that shares a touching story, but this story should relate to your life. A good story is entertaining, but a better one gives insight into who you are.

Excuses, Excuses, Excuses

Nobody is perfect. We all make mistakes. Maybe you skipped a few too many 8 a.m. classes in freshman year. Perhaps you didn't put all of your effort into a science project. All of this is normal. However, what is not normal or useful is using the essay to explain past mistakes. The essay should be used to highlight your strengths, not call the judges' attention to or make excuses for your shortcomings.

It's Not My Fault

I received horrible grades all throughout high school, but hardly any of them were due to my own actions. Let me explain: you'll notice that during freshman year I earned a 2.5 GPA. The reason for this is that I had just gone through a difficult move to a new city, the first time our family had relocated. It was hard for me to adjust to the new environment of living in a big city and I made few friends. Because of this, I had a very difficult time in all my classes, which, by the way, were chosen all by my overprotective mother.

Then, during sophomore year, I finally started to make some pretty good friends, but one day in the middle of October, my dog died. That devastated me. I took the SAT I that month and my results definitely reflect this loss. Throughout the year I couldn't recover from such a loss, because I had my dog since I was 5 years old.

Then came junior year. Emotionally, this was my worst year. I went through a terrible breakup with my girlfriend of five months in the middle of winter break. Since then, school has been in the way of my recovery, and I have performed poorly as a result. I tried taking the SAT I again, and my emotional weakness once again reflects my scores. APs and SAT IIs were no different. I joined the basketball team in the beginning of the year, but I was almost immediately cut.

Senior year I tried to make a comeback, but my GPA remained the same. This discouraged me because none of my friends believed I was intelligent.

I tried to join clubs in the hopes that community involvement would cure my woes, but in fact the impersonality of clubs altogether discouraged me further so I haven't stayed in any.

Meanwhile, my family is running out of money. My dad has been laid off from his job for about six months now. I feel like the world is against me now, and I could really use this scholarship to help my college career, if I have one.

Why This Essay Bombed

What this essay does is call attention to the writer's deficiencies instead of his strengths. We all have things about ourselves that we are not proud of so why put faults on display when the object of the scholarship essay is to impress the judges? Compounding this effect is that the student tries to avoid taking responsibility for his shortcomings by blaming everything and everyone else. If you are going to admit to a mistake then at least take responsibility for your actions.

How to Avoid This Mistake

Showcase your strengths in the essay. If you do need to reveal a weakness or shortcoming, explain how you have grown from the experience. We all make mistakes but what is important is that we learn from each of them. Whatever you do, don't avoid responsibility for your actions.

Complex Problem, Simple Solution

Some scholarship essays are about an issue of national or international importance. Scholarship committees often choose topics that ask difficult questions regarding complex issues so that they can discover what is important to you. The judges want to weigh your thoughts and to check your understanding of complicated interests and viewpoints. It would seem fairly obvious that the biggest mistake in writing this type of essay is to know nothing about the problem or to present a wholly unrealistic solution. Observe.

Nuclear Nightmare

Imagine a nuclear nightmare. Bombs exploding. Millions of people vaporized. Entire cities destroyed. This is the reality we face even after the Cold War has been won.

The major problem is that nuclear weapons are all around us. Nearly every country has them even if they don't openly admit it. Worse yet is the fact that anyone can easily build a bomb from plans posted on various Internet websites. All you need is a small quantity of uranium which is supposedly easily available at many hospitals and pharmacies.

The danger posed by nuclear weapons is all too real and something must be done to combat this threat to the world. I propose the following solution.

First, we must collect and destroy all nuclear weapons and sources of radiation. We need someone like the UN to collect all of the missiles and bombs and destroy them once and for all.

Then, we must erase all knowledge about building nuclear bombs. Since man cannot be trusted with this knowledge we need to destroy all plans and instructions on building bombs. While science is important this is one area of knowledge that it can do without. Once all plans and documents relating to nuclear bombs are destroyed it will take centuries for man to relearn how to build them.

These two simple but decisive steps could rid the world of the threat of nuclear destruction. Once and for all we could all sleep at night without the fear of a nuclear nightmare.

Why This Essay Bombed

The problem with this essay is that the writer clearly does not have any knowledge about nuclear proliferation. Not only do some of the facts cited seem to have come from urban legends but there is no mention of the international aspect or diplomatic dimension of the issue. The second weak point of the essay is the solution–it is entirely unrealistic.

How do we go about collecting weapons? Is it even possible to remove knowledge once it is known? The writer assumes these are easy solutions when in fact they are extremely difficult, if not impossible.

How to Avoid This Mistake

Don't write about an issue without understanding it. You don't have to be an expert, but you should read about it and even discuss it with some teachers or professors. By speaking about the issues with others, you will gain a better understanding of a problem and this will also help you to generate innovative solutions. Your suggestions do not have to be easy or even doable, but they do need to show some thought and understanding of the difficulty involved in solving any large national or international problem.

If possible, draw a connection between the issue and your own life. Have you taken action even on a personal level? Remember that the scholarship judges don't want a lecture on the issue as much as they want to learn about you and your ability to analyze a complex problem.

CHAPTER
SEVEN

JUDGES' ROUNDTABLE: THE SCHOLARSHIP ESSAY

In this chapter:

- **Why the scholarship essay is so important**
- **Common qualities of money-winning essays**
- **How to avoid essay mistakes**

Meet the Scholarship Judges

In this second roundtable, judges and experts provide insight into the importance of scholarship essays and the qualities of those that win.

How important is the essay to winning a scholarship?

For many scholarship competitions the essay plays a vital role. It allows the selection committee to get to know you beyond a list of courses and achievements. In some competitions, the essay alone is the deciding factor that separates those who receive awards from those who don't.

Kimberly Hall
United Negro College Fund
"The essay is a very important piece of the application because it is often what the donor, who makes the final decision, will use to see the student's aspirations. It gives us a sense of who the student is and what they want to do with the money. It gives us a more complete picture of the student as a whole person as opposed to just a name."

Jacqui Love Marshall
Knight Ridder Minority Scholars Program
"In the end if we had to look at a student who had a little lower score and a little lower GPA but wrote an outstanding essay from the heart and had some experience or testimonials to back up their strengths, we'd be inclined to award the student with the strong essay. What really differentiates one applicant from another is a genuinely written essay."

Q What qualities make an essay powerful?

One of the keys to writing a powerful scholarship essay is to be honest and to write from the heart. The scholarship judges and experts have stressed that they can see through an essay that is not honest and that the best essays are about something for which the applicants are truly passionate.

Georgina Salguero
Hispanic Heritage Youth Awards
"We look at how you develop your thesis statement. What are you going to talk about and do you stick to those points? Do your paragraph structures make sense? We can tell how passionate you are for your subject by how you write. Don't just write that you want to attend college because you know you have to go to school. Instead, tell us why. Why do you want to go to college? What drives you? What gives you the strength to keep going?"

Shirley Kennedy Keller
American Association of School Administrators
"It's very important for them to be specific, to give specific examples of their leadership, special talents, obstacles or community service. The more specific they can be and the more they can back up their statements, the better they're going to fare in the judging process."

Kimberly Hall
United Negro College Fund
"Essays should be well developed in terms of the paragraph structure. The essay should have a definite purpose and direction. By the end of the essay we should know where you have started from and where you are heading. We want to see what you have dealt with and what your plan of action is. We also want to see where you see yourself in the future."

Trisha Bazemore
Coca-Cola Scholars Foundation
"You can tell when you read an essay if it's a real expression of something the student really cares about or if it was written just to impress. We intentionally don't provide students with instructions for the essay. We want to give each student the opportunity to be genuine."

Jacqui Love Marshall
Knight Ridder Minority Scholars Program
"Your essay needs to fit the scholarship. Sometimes when reading an essay you get the feeling that the essay was written generically for 60 different scholarships and the author just substituted newspapering for engineering. If you want to win our scholarship your essay needs to tie into your involvement to the things we care about like the newspaper, photography or the sales or marketing side of the business."

Wanda Carroll
National Association of Secondary School Principals
"Go back to your basic English lessons and remember all that your English teacher taught. Make sure your essay is concise and there's a point to what you're writing."

Laura DiFiore
FreSch! Free Scholarship Search
"If you can make the reader laugh, cry or get angry, even when you're just writing about yourself, you've already won half the game. That's the bottom line. Get an emotional response out of the reader."

What common mistakes do students make on the essay?

As you're writing the essay, it's important to know what works. But sometimes it can be even more helpful to know what doesn't work. By knowing what mistakes kill an otherwise good essay you can avoid them in your own writing. After having read hundreds or even thousands of essays, our panel has encountered many common mistakes that students make in their scholarship essays.

Kimberly Hall
United Negro College Fund
"Some students write their essays about the difficulties that they have faced but do so in a negative way and don't explain how they've overcome the difficulties. I would recommend that students present a positive light. Here are some of the challenges that I've had to overcome and here is how I did it. Stay positive."

Tracey Wong Briggs
All-USA Academic Teams
"There are some students who have outstanding biographical information, but when you read the essay all they've done is recount the facts that are in their application form. We want you to use the essay to go deeper and beyond what is listed in your application. We lose very good nominees that way. They just don't give you a clear idea of who they are beyond the basic facts."

Mario A. De Anda
Hispanic Scholarship Fund
"One of the common mistakes is that they use the same personal statement for many scholarships. They even forget to change the name of the scholarship they are applying to. We encourage students to make sure they write a personal statement specifically for the program."

Laura DiFiore
FreSch! Free Scholarship Search
"A huge mistake is what I call the crush, when we're getting 40 to 50 percent of our applications in the last three days. I think a lot of students would be better off if they didn't apply in the last two weeks before the deadline. The ones rushing to get in by the deadline would probably be better off spending more time on their essays and applying next year."

Wanda Carroll
National Association of Secondary School Principals
"Spelling. You should use your computer's spell check. We wouldn't disqualify an applicant solely on spelling, but the committee does see the mistakes and it does distract from the quality of the essay. If they had a choice between two equally well-written essays, they would choose the essay without spelling errors."

Leah Carroll
U.C. Berkeley Haas Scholars Program
"The most common error I run into is people who are trying to say what the foundation wants to hear. It ends up sounding inauthentic. I tell students to write as if they are trying to explain something to a friend. Just write from the heart. They seem to always come out better that way. Another mistake is that students, at Berkeley in particular, often sell themselves short. You should not be afraid to call attention to all of your achievements."

Michael Darne
CollegeAnswer
"When approaching the essay a lot of students are eager to dump a huge laundry list of achievements–a list of everything that they've done. But what scholarship providers are looking for is to get an understanding of who this person is and where they're going in life. They don't just want a list of accomplishments. If you can paint some picture of yourself, where you're going and how you're going to get there, you're going to be in a much better situation."

Georgina Salguero
Hispanic Heritage Youth Awards
"We're not giving the award to the best sob story. We're not looking for someone who can write the best woe is me story. Please don't give us this kind of essay."

Participating Judges & Experts

Trisha Bazemore, Program Assistant, Coca-Cola Scholars Foundation

Tracey Wong Briggs, Coordinator, *USA Today* All-USA Academic and Teacher Teams

Leah Carroll, Coordinator, U.C. Berkeley Haas Scholars Program and former program coordinator, U.C. Berkeley Scholarship Connection

Wanda Carroll, Program Manager, National Association of Secondary School Principals

Michael Darne, Director of Business Development, CollegeAnswer.com, the website of Sallie Mae

Mario A. De Anda, Director of Scholarship Programs, Hispanic Scholarship Fund

Laura DiFiore, Founder, FreSch! Free Scholarship Search Let's Get Creative Short Story and Poetry Scholarship Contest

Kimberly Hall, Peer Program Manager, United Negro College Fund

Shirley Kennedy Keller, Program Director, American Association of School Administrators

Jacqui Love Marshall, Vice President of Human Resources, Diversity and Development, Knight Ridder Minority Scholars Program

Georgina Salguero, Senior Manager, Programs and Events, Hispanic Heritage Youth Awards

CHAPTER
EIGHT

WINNING INTERVIEW STRATEGIES

In this chapter:

- **The two types of interviews**
- **How to ace the interview**
- **Who are the interviewers**
- **How to dress and act**
- **What to do if you have a disaster interview**

Face-to-Face with the Interview

Let's start with some good news. If you are asked to do an interview for a scholarship competition it means that you are a serious contender. Most competitions only interview a small number of finalists who make it through the initial round based on their application and essay. The bad news is that you will now undergo the nerve-wracking scrutiny of an interview with one or more scholarship judges. If the thought of this makes your palms moisten or you get a sinking feeling in your stomach, you are not alone.

The best way to overcome a fear of the interview is to know exactly what to expect and to be prepared for the questions you might be asked. In this chapter, we discuss what scholarship interviewers are looking for in your answers, and we will share some strategies to help you prepare.

Many students wonder why they have to do an interview in the first place. While some scholarships are awarded solely based on the written application, many scholarship committees like to perform face-to-face interviews to make the final decision. Particularly, if the scholarship is for a significant amount of money, the selection committee wants to be sure to give it to the most deserving student.

Having sat on both sides of the interview table, we can attest to the fact that an interview can shed significant insight on an applicant. Before we discuss how to make the most of the interview, let's cover the two situations you may face.

Friendly & Hostile Interviews

There are basically two types of interviews: 1) friendly and 2) less than friendly or even hostile. The friendly interview is fairly straight forward with the scholarship judges asking easy to answer questions that will help them get to know you better. While most interviews fall into the friendly camp, others especially for highly competitive and prestigious awards such as the Rhodes or Truman are far less pleasant. In these interviews the scholarship judges want to test you to see how you react to stressful and difficult questions. A hostile interview creates an environment for the judges to be able to evaluate how you react to pressure.

Whenever you encounter hostile judges or interview situations, keep in mind that they are not trying to personally attack you or diminish your accomplishments. Rather, they are observing how you respond to the situation. It is really a test of your ability to deal with difficult questions. Also, keep in mind that they will act the same toward all applicants.

How to Ace the Interview

Regardless of the type of interview, the keys to success are the same.

> **First, remember that scholarship interviewers are real people.** This is especially true for hostile situations in which you may have to fight feelings of anger or frustration with the interviewer. Your goal is to create as engaging a conversation as possible. This means you can't give short, one-sentence answers and you certainly should not be afraid to ask questions. Most interviewers enjoy conversations over interrogations.
>
> **The second key to the interview is to practice.** The more you practice interviewing, the easier and more natural your answers will be. Practice can take the form of asking and answering your own questions out loud or finding someone to conduct a mock interview. Consider taping your mock interview so you can review your technique. Having someone simulate a hostile interview is very good practice and will give you a tremendous edge over applicants who have not experienced this yet.

Transform Any Interview from an Interrogation into a Conversation

The reason most people volunteer to be scholarship judges is because they are passionate about the organization or award they support. Being an interviewer is hard work. In most cases, interviewers have a few questions to begin with but then hope the interviewee can help carry the conversation and direct it into other interesting areas. In fact, it is very difficult to interview an applicant who quickly and succinctly answers the questions but offers nothing else to move the conversation forward.

As the interviewee you are an essential part of determining where the conversation goes and whether or not it is easy or difficult for the interviewer. Your job is to supply the interviewer not only with complete answers but also with information that leads to other interesting topics of conversation.

It helps to know something about the interviewers. One thing you know is that they care about their organization. They may be members of the organization or long-time supporters. The more you learn about the organization and its membership the better idea you'll have about the interviewers and what interests them.

This knowledge is useful in choosing how to answer questions that require you to highlight a specific area of your life or achievements. It will also give you a feel for topics to avoid and questions you should ask.

Before every interview, do homework on the award and the awarding organization, which includes knowing the following:

Purpose of the scholarship. What is the organization hoping to accomplish by awarding the scholarship? Whether it's promoting students to enter a certain career, encouraging a hobby or interest or rewarding students for leadership, every scholarship has a mission. By understanding why the organization is giving away the money, you can share with the interviewers how you meet their priorities.

Criteria for selecting the winner. Use the scholarship materials to get a reasonable idea of what the selection committee is looking for when choosing the winner. From the kinds of information they request in the application to the topic of the essay question, each piece is a clue about what is important to the scholarship committee.

Background of the awarding organization. Do a little digging on the organization itself. Check out its website or publications. Attend a meeting or speak with a member. From this detective work, you will get a better idea of who the organization's members are and what they are trying to achieve. Knowing something about the organization will also prevent you from making obvious blunders during the interview.

Advice from a Rhodes Scholar

Kristin N. Javaras, Oxford University

The interview is one of last hurdles to becoming a prestigious Rhodes Scholar. Kristin, who is working on a doctorate in statistics at Oxford University, says about interviews, "The best advice I can offer is to be yourself, as trite as that may sound."

But what happens when you are stumped for an answer? "If you just don't know the answer to a question, don't be afraid to admit it," advises Kristin.

Regarding the type of questions that she was asked, Kristin recalls, "Almost every question was at least tangentially and often directly related to topics and experiences mentioned in my personal statement or included in my list of activities and jobs."

Once you've done the detective work, think about how the information can help you. Let's take a look at an example piece of information. Imagine you discovered that the organization offering the scholarship values leadership. In addition, you discover from reading the organization's website that all of its members are invited to join only if they have led large companies. Knowing this you could guess that the interviewers will probably be business leaders and will be most impressed if you highlight leadership and entrepreneurial activities. If asked about your greatest achievement you can insightfully highlight being president of your school's business club over anything else.

Knowing something about the interviewers beforehand will also help you think of appropriate and engaging questions. Most interviewers allow time to ask a few questions toward the end of the interview. By asking intelligent questions (i.e., not the ones that can be answered by simply reading the group's website), you will hopefully be able to touch upon something the interviewer really cares about that will lead to further conversation.

Going back to the example, you might ask a question such as, "As the president of the business club one of my greatest challenges has been to get funding from businesses for new projects and ideas. What advice do you have for young business people to secure seed money from established businesses?"

This question not only demonstrates that you know the background of the interviewers but also poses a question that they can answer with their expertise, and it could start a new conversation about how to fund a business idea.

You Are Not the Center of the Universe

Despite what you think, you are not the center of the universe–at least not yet! Therefore, in the interview you need to keep it interactive by not just focusing on yourself.

This can be accomplished by asking questions and engaging in two-way conversation. If you don't ask any questions, it will appear that you are not attentive or that you haven't put much thought into the interview. Beforehand, develop a list of questions you may want to ask. Of course you don't have to ask all of the questions, but be prepared to ask a few.

To get you started, we've developed some suggestions. Adapt these questions to the specific scholarship you are applying for, and personalize them.

- How did you get involved with this organization?
- How did you enter this field? What was your motivation for entering this field?
- Who were your mentors? Heroes?
- What do you think are the most exciting things about this field?
- What professional advice do you have?
- What do you see as the greatest challenges?
- What do you think will be the greatest advancements in the next 10 years?
- What effect do you think technology will have on this field?

The Group Interview

Key Strategies

So it's you on one side of the table and a panel of six on the other side. It's certainly not the most natural way to have a conversation. How do you stay calm when you are interviewed by a council of judges? Here's how:

Think of the group as individuals. Instead of thinking it's you versus the team, think of each of the interviewers as an individual. Try to connect with each separately.

Try to get everyone's name if you can. Have a piece of paper to jot down everyone's name and role so that you can refer to them in the conversation and be able to target your answers to appeal to each of the constituents. For example, if you are interviewing with a panel of employees from a company and you know that Sue works in accounting while Joe works in human resources, you can speak about your analytical skills to appeal to Sue and your people skills to appeal to Joe.

Make eye contact. Look into the eyes of each of the panelists. Don't stare, but show them that you are confident. Be careful not to focus on only one or two panelists.

Respect the hierarchy. You may find that there is a leader in the group like the scholarship chair or the CEO of the company. Pay a little more attention to stroke the ego of the head. A little kissing up never hurt anyone.

Try to include everyone. In any group situation, there are usually one or two more vocal members who take the lead. Don't focus all of your attention only on the loud ones. Spread your attention as evenly as possible.

Ultimately, the more interaction you have and the more you engage the interviewers the better their impression of you. You want to leave them with the feeling that you are a polite and intelligent person who is as interested in what they have to say as in what you do yourself.

Dress & Act the Part

Studies have shown than in speeches, the audience remembers what you look like and how you sound more than what you actually say. While it may seem unimportant, presentation style and presence are probably more significant than you think.

Think about the delivery of your answers and keep the following points in mind:

Sit up straight. During interviews, don't slouch. Sitting up straight with your shoulders back conveys confidence, strength and intelligence. It communicates that you are interested in the conversation.

Speak in a positive tone of voice. One thing that keeps interviewers engaged is your tone. Make sure to speak with positive inflection in your words. Convey confidence in your answers by speaking loudly enough for the judges to hear you clearly. This will not only maintain your interviewers' interest but will also suggest that you have an optimistic outlook toward life.

Don't be monotonous. Speaking at the same rate and tone of voice without variation is a good way to give the interviewers very heavy eyelids. Tape record yourself and pay attention to your tone of voice. There should be a natural variation in your timbre.

Speak at a natural pace. If you're like most people, the more nervous you are the faster you speak. Combat this by speaking on the slower side of your natural pace. During the interview you might think that you are speaking too slowly, but in reality you are probably speaking at just the right pace.

Make natural gestures. Let your hands and face convey action and emotions. Use them as tools to illustrate anecdotes and punctuate important points.

Make eye contact. Eye contact engages interviewers and conveys self-assurance and honesty. If it is a group interview, make eye contact with all of the interviewers–don't just focus on one. Maintaining good eye contact can be difficult, but just imagine little dollar signs in your interviewers' eyes and you shouldn't have any trouble. Ka-ching!

What It's Like to Be an Interviewer

Rotary International Ambassadorial Scholarships

Selection committees are typically composed of volunteers who sign up for a long day of interviews. By understanding their role, you can see the importance of interacting with them, keeping their attention and giving them a reason to want to listen to what you have to say.

Each year the selection committee for Rotary International in the San Francisco Bay Area interviews about 15 applicants. The six or seven selection committee members, Rotarians and previous scholarship winners start the day at 8:30 a.m. and end at 6 p.m., with 45 minutes for lunch and a couple of stretch breaks. They spend about 30 minutes with each applicant.

"After the interview, we score and go to the next one," says Russ Hobbs, district scholarship chairman.

Surprisingly, Hobbs says there is no advantage to interviewing earlier in the day than later. Still, to be fair, they schedule the interviews randomly instead of alphabetically. Despite the long day Rotary has little trouble finding volunteers, he says, "Because the applicants are such phenomenally interesting people."

Smile. There's nothing more depressing than having a conversation with someone who never smiles. Don't smile nonstop, but show some teeth at least once in a while.

Dress appropriately. This means business attire. No-no's include: caps, bare midriffs, short skirts or shorts, open-toe shoes and wrinkles. Think about covering obtrusive tattoos or removing extra ear/nose/tongue/eyebrow rings. Don't dress so formally that you feel uncomfortable, but dress nicely. It may not seem fair, but your dress will affect the impression you make and influence the judgment of the committee.

By using these tips, you will have a flawless look and sound to match what you're saying. All of these attributes together create a powerful portrait of who you are. Remember that not all of these things come naturally, so you'll need to practice before they become unconscious actions.

How to Make Practice into Perfect

The best way to prepare for an interview is to do a dress rehearsal before the real thing. This allows you to run through answering questions you might be asked, practice honing your demeanor and feel more comfortable when it comes time for the actual interview. Force yourself to set aside some time to run through a practice session at least once. Here's how:

Find mock interviewers. Bribe or coerce a friend or family member to be a mock interviewer. Parents, teachers or professors make great interviewers.

Prep your mock interviewers. Give them questions (such as those in the next chapter) and also ask them to think of some of their own. Share with them what areas of your presentation you are trying to improve so that they can pay attention and give constructive feedback. For example, if you know that you fidget during the interview ask your interviewers to pay special attention to your posture and movements during the practice.

Capture yourself on tape. If you have a tape recorder or camcorder, set it up to tape yourself so that you can review the mock interview. Position the camera behind your interviewer so you can observe how you appear from the right perspective.

Get feedback. After you are finished the practice interview, get constructive criticism from your mock interviewer. Find out what you did well and what you need to work on. What were the best parts of the interview? Which of your answers were strong, and which were weak? When did you capture or lose your interviewer's attention? Was your conversation one-way or two-way?

Review the tape. If you can, watch or listen to the tape with your mock interviewer for additional feedback. Listen carefully to how you answer questions to improve on them. Pay attention to your tone of voice. Watch your body language to see what you communicate.

Do it again. If you have the time and your mock interviewer has the energy or you can find another person willing to help, do a second interview. If you can't find anyone, do it solo. Practice your answers, and focus on making some of the weaker ones more interesting.

The bottom line is this: the more you practice, the better you'll do.

The Long-Distance Interview

Interviewing over the telephone is a real challenge. While most interviews are held in person, sometimes you just can't meet face-to-face. When this happens, the telephone is the only option. The most difficult aspect of a long-distance interview is that you can't judge the reactions of the interviewers. You have no idea if what you are saying is making them smile or frown. While there is no way to overcome the inherent disadvantage of a phone interview, here are some tips that should help to bridge the distance:

Find a quiet place. Do the interview in a place where you won't be interrupted. You need to be able to pay full attention to the conversation.

The Hostile Interview

U.C. Berkeley & Truman Scholarship Winners

In some scholarship competitions, particularly ones for prestigious awards like the Rhodes, Marshall or Truman, the interviews are designed to challenge you. To do well you need to prepare and have the right mindset for these provocative interviews. —Gen and Kelly

During his interview to become a Truman Scholar, one of the eight panelists asked Brian C. Babcock to name a good funny novel he had recently read. Brian hadn't read a humorous novel recently, but he did have a children's book that he thought was funny.

He started to say, "It's not a novel, but..." and before he finished the interviewer interrupted him to say, "No, I want a novel."

The sentiment in most scholarship interviews is friendly and cordial. But for some competitions, particularly the prestigious ones with fierce national competition, the setting is often challenging and even adversarial.

"There's a kind of devil's advocate interviewing style for these competitions. The phrasing and tone is more antagonistic," says Leah Carroll, coordinator of U.C. Berkeley's Haas Scholars Program and former program coordinator of the university's Scholarship Connection office, which assists students who are applying for awards.

She coaches students to view these kinds of interviews as "intellectual sparring," and advises them to "practice interviews with friends and to tell their friends to be mean."

Donald H. Matsuda, Jr. experienced this intellectual sparring first hand. A student at Stanford University, Matsuda is also a Truman Scholar. The panel challenged his policy plan on health care for children asking why they should "continue to waste millions of our federal budget to help this situation that has no clear cut solution."

Brian's contributions reflect his own opinions, not those of the U.S. military.

Donald was also asked to define music. The panel gave him the option of defining it or singing a definition. He chose to define it. He said, "I see music as the ultimate way a person can express himself. I chose not to sing, which is why I think I won the Truman."

A student at the U.S. Military Academy, Brian applied for the award to receive a master's in foreign service and history and certificate in Russian area studies. In his interview, the selection committee asked Brian questions about gays in the Boy Scouts, an example of bad leadership and why he wanted to work in public service instead of make millions of dollars.

They challenged his choice of topic for his essay, asking why he chose as an example of leadership when he led one other person instead of when he led many. He answered, "If you can't lead one person how can you expect to lead a group?"

And, they questioned his grades, which weren't perfect but still high. He says, "I explained that to me it was more important to get the breadth of knowledge and take the classes while I have the time and it's free. I take as much as I can handle. If that means that my grades slip from a 3.9 to a 3.75 so be it."

Besides preparing for the interview, what may have helped Brian was his frame of mind. He says, "I didn't have an interview. I had a talk with eight people around the table."

Know who's on the other end of the line. You may interview with a panel of people. Write down each of their names and positions when they first introduce themselves to you. They will be impressed when you are able to respond to them individually and thank each of them by name.

Use notes from the practice interviews. One of the advantages of doing an interview over the telephone is that you can refer to notes. Take advantage of this.

Look and sound like you would in person. Pretend the interviewers are in the room with you, and use the same gestures and facial expressions that you would if you were meeting in person. It may sound strange, but the interviewers will actually be able to hear through your voice when you are smiling, when you are paying attention and when you are enthusiastic about what you're saying. Don't do the interview lying down in bed or slouched back in a recliner.

Don't use a speaker phone, cordless phone or cell phone. Speaker phones often echo and pick up distracting noise. Cordless and cell phones can generate static, and the battery can die at the worst possible moment.

Turn off call waiting. Nothing is more annoying than hearing the call waiting beep while you are trying to focus and deliver an important thought. (And, this may sound obvious, but don't click over to take a second call during the interview.)

The Disaster Interview

Even after doing interview homework and diligently practicing mock interviews, you may still find that you and the interviewer just don't connect or that you just don't seem to have the right answers. If you spend some time preparing, this is very unlikely. Interviewers are not trying to trick you or make you feel bad. They are simply trying to find out more about you and your fit with the award. Still, if you think that you've bombed, here are some things to keep in mind:

Avoid "should have," "would have," "could have." Don't replay the interview in your head again and again, thinking of all the things you "should have" said. It's too easy to look back and have the best answers. Instead, use what you've learned to avoid making the same mistakes in the next interview.

There are no right answers. Remember that in reality there are no right answers. Your answers may have not been perfect, but that doesn't mean they were wrong. There are countless ways to answer the same question.

The toughest judge is you. Realize that you are your own greatest critic. While you may think that you completely bombed an interview, the interviewer will most likely not have as harsh an opinion.

Post-Interview

After you complete the interviews, follow up with a thank you note. Remember that interviewers are typically volunteers and have made the time to meet with you. If you feel that there is very important information that you forgot to share in the interview, mention it briefly in a thank you note. If not, a simple thank you will suffice.

Make Sure You Make Your Point

Jason Morimoto, State Farm Exceptional Student

"I have been involved in practically every type of interview whether it be a single interviewer, a panel or a phone interview.

"The toughest by far are the phone interviews because the scholarship committee cannot physically see who you are and your facial expressions.

"I personally prefer the panel interviews because it gives you a chance to make a strong impression on multiple people. I have found great success with panels.

"However, no matter what the format of the interview, the most important thing is to make sure that you get across your main strengths. If they do not ask you directly, try to weave it in with a related story or tie it in as a closing statement. You always want to give the most information possible to the interviewers so that they can understand your uniqueness as a person."

CHAPTER
NINE

REAL INTERVIEW QUESTIONS & ANSWERS

In this chapter:

- **See what makes a great answer**
- **Questions you'll likely face in the scholarship interview**
- **Interview tips from winners**

Giving the Right Answers

Imagine that your professor gave you the questions to an exam before you took it. Your score would certainly be higher. In this chapter we give you precisely this advantage by sharing the questions you're likely to be asked in scholarship interviews. Plus, we show examples of how to answer. You will be a fly on the wall, observing a typical scholarship interview.

Before the preview, we have a couple of caveats. Remember that these questions and answers are meant to be examples. Each of these is not the only acceptable way to answer a question. In fact, there are innumerable ways to answer each question successfully.

Also, since your background and achievements are different, your answers will inevitably be different too. Don't focus on the specific details of the answers. Instead, look at the overall message and impression that each answer conveys.

The comments that follow each question are based on interviews with actual scholarship judges as well as our own experience in competing for scholarships. To get the most out of this chapter, we suggest that you read a question first. Pause to think about how you would answer it. Then read the response and comments. Keeping the comments in mind, analyze how the judges might react to your response. Be tough on yourself and think of ways to strengthen your answers. If your parents or friends are helping you practice interviewing, ask them to read through a couple of the questions and answers to get a better idea of what kind of questions to ask and what to look for in responses.

Achievements & Leadership Questions

Q: What achievement are you the most proud of?

> **A:** This may not seem like an achievement to many people, but it is for me. Last year I learned how to swim. Ever since I can remember I've had a grave fear of the water. Anything above knee level was a frightening experience. Last year my little brother fell into a pool and had to be rescued by the life-

guard. As I stood by not able to help, I realized that I needed to learn to swim. Twice a week I went to swimming lessons. It was kind of embarrassing to be in a swimming class with elementary school students but I was determined to learn. It took an entire class for me to feel comfortable walking in the water up to my neck, but after eight weeks of lessons, I could actually swim several laps. I never thought that I could learn. I'm proud of this accomplishment not because it was difficult to learn but because of the huge fear I had to face and overcome to learn it.

A: Unlike other schools, ours never had a debate team. Because I plan to be an attorney, I wanted to get practice in debating so I decided to form a team. None of the teachers at my school had the time to be the faculty adviser so I contacted local attorneys in the yellow pages. I finally found one who despite her busy schedule volunteered to help us. I recruited 12 other students to join and became the team captain. In our first year we made it to the district competition and won several rounds. For me this was my biggest accomplishment especially since now we have a core group of debaters who will continue the team after I graduate.

Comments

This is a challenging question because in addition to selecting and describing an accomplishment, you need to put it into context and explain its significance. The first answer vividly illustrates how this student overcomes his fears to learn how to swim. Everyone has something that he or she is deathly afraid of, and it is likely that the judges can easily relate to this accomplishment. Notice how the answer reveals why the student decided to face his fear of swimming and gives enough detail to create a mental picture. There is also a nice element of humor in the story that makes you smile.

The second answer is an excellent example of how to highlight an impressive achievement. While the scholarship judges may notice that this student is the founder of her school's debate team in the application, this answer underscores just how difficult it was to start the team. It also reveals the student's desire to be an attorney. The applicant ends nicely by emphasizing how her achievement has affected others and will continue to make a positive impact on the lives of her fellow students.

Q: How have you been a leader or displayed leadership?

A: I am the chair of my dorm committee, which consists of six officers. My job is to oversee the committee as well as 500 student residents. My responsibilities include planning the orientation for new students, organizing social activities and directing our dorm's annual charity event. It's a challenge to get students motivated for a special event because there are so many other ways that they can spend their time. I am most proud of the way that I have been able to mobilize the students in our dorm to support our annual charity event, the bowl-a-thon for lung cancer research. To make this event a success I knew that I needed the help of others in our dorm. I recruited and trained hall representatives to personally contact all 500 students in our dorm and encourage them to participate in the bowl-a-thon. In the end over 50 percent of the students participated. We had a higher participation rate and donation level than any other dorm at our school.

A: This year I organized an event to collect toys for underprivileged children. I started by writing an article for the school newspaper to raise awareness and to get students to donate toys. I had volunteers who also went to local businesses and asked for donations as well as a group that decided which families in our town would receive the toys. The toy drive was a huge success. We were able to provide toys to over 200 families and we solicited donations from over 50 local businesses.

Comments

What's notable about both of these answers is that the applicants don't just list off a bunch of titles and positions. Instead the students focus on one specific leadership position or activity and give enough detail to show the depth of their commitment. Citing concrete accomplishments like getting half of the dorm to participate or giving toys to more than 200 families also helps judges to better gauge the significance of each achievement. The second answer illustrates that you don't have to hold an official title or elected position in order to show leadership. This applicant, who does not hold an elected position, is still able to

Tip #1 from a Scholarship Winner

Elisa Tatiana Juárez, Brown University

"Be proud of what you've done. Don't be falsely modest, but also make sure that you don't give the impression of being egotistical.The ability to talk positively about my accomplishments took me a long time to learn. I was afraid showing people what I've done would make me sound too conceited. Always remember that the judges want to know why they should pick you. Show them."

answer this question impressively by describing how she organized an event. You can certainly be a leader and motivator even if you don't have an official title.

Personal Questions

Q: What is your greatest strength and weakness?

A: One of my strengths is my ability to lead. For example, at my school we didn't have a recycling program. The janitors wouldn't pick up the paper for recycling because it wasn't in their contracts. I met with our principal to discuss the problem, but he said the school didn't have the budget to pay for a recycling program. So I started a program myself. I got donations to buy bins to put in each of the classrooms and went to each class to make a speech to get volunteers to collect the papers for recycling. Every month, I gathered the volunteers for a meeting to discuss any changes or problems.

One of my weaknesses is impatience. I get frustrated when I see a problem but nothing is happening to fix it. I like to see people working toward solutions. I got very frustrated when I first found out that there was no recycling program at my school and especially when the janitors said they wouldn't pick up the recycling even though I thought it didn't require that much extra work. But I guess it was this frustration that led me to do something about it.

A: My strength is in math. Ever since elementary school I have been talented in math. In school, when everyone else was struggling with algebra and geometry, I didn't have any trouble. I just imagined the problems in my head, visualizing the pyramids, spheres and cones. My math teacher even asked me to grade homework assignments. And I've represented our school each year for the county math competition.

My weakness is creative writing. I think because of the way that my mind works, it can be difficult to write creative essays. This is one of the reasons that I took a creative writing class last summer, and it really helped. The instructor had us pretend to be another person in the class and write from the other person's perspective. We also went outside and imagined being the grass, trees and sun. I never thought like that before and it's really opened up my mind to some new possibilities. This is one area that I know I need to work to improve.

Comments

It's easy to say that your strength is that you work hard. But what will really prove this to the judges is an example. Use an example to illustrate your strength so that the judges can see what you mean. It's not enough to say that your strength is leadership. How have you led? What kind of results have come from your leadership? Why do you do it? Both of these strength answers are good in giving complete examples. But more importantly they also help to contrast and balance the weakness. In the second response where the applicant admits to not being a skilled creative writer, this not only reveals an honest flaw but also gives him an opportunity to show what action he has taken to improve. There is nothing wrong with acknowledging a weakness, but it is very impressive to see that you are also taking steps to transform that weakness into a strength.

Misinterpreting the Judges

Silver Knight Scholarship Winner

It's tempting to try to read the judges during the scholarship interview. You might think that the longer the interview or the more involved the selection committee gets, the better your performance. Unfortunately, it's very hard to accurately interpret the thoughts of others. When she interviewed for the Silver Knight scholarship, Elisa Tatiana Juárez was scared. After all, the Silver Knight award is a highly competitive program and is given to only 14 of the top students in Miami-Dade County.

Elisa had been on the other side of the interview table before as an interviewer so she thought she had a good sense of how judges react when they are really interested in a candidate.

"I thought that you could tell what the judges think of you by their responses," she says.

In the interview, Elisa described the STARS (Students and Teachers Advocating Research Science) program that she started at the Miami Museum of Science to provide opportunities for minority and economically disadvantaged students in the sciences. During her interview, she was disappointed that the judges seemed too enthusiastic about her work to the point that she thought they were faking their level of interest.

"In my interview, they were too encouraging," she says. "My image was they thought, `Good luck, but try again next year.'"

Thankfully, her interpretation of their reactions was completely opposite from reality and she won the award.

Q: Who is a role model for you?

A: Oprah Winfrey. I admire Oprah not because of her wealth but because through sheer determination and hard work she has built one of the largest media companies in the country. I am inspired by people like Oprah who didn't inherit wealth or fame but who built it on their own by setting goals and working hard to achieve them. She is also motivating because she has chosen to do what she wanted, not necessarily what was seen as the most popular thing to do. For example, instead of making her talk show about sex and violence, she's taken a different route to make it about the positive things in life. I also want to live my life by what I think is right, even if that is at odds with what the majority feel I should do.

A: My father is my role model. He has taught me to endure difficult times with resolve. I always knew that my father didn't make a lot of money at his job and that our family's finances were stretched. But I only recently learned how stretched our finances were. There were points where my parents weren't sure how they were going to pay the bills. But looking at my father, you would never detect the stress that he was under. He always made sure that we kids had everything that we needed. We didn't have the brightest or newest things, but we were always cared for. I remember one Christmas when I was 12. My friends received the newest "in" toys for Christmas while I received a wooden car and airplane that my father had made. I still have those toys and plan to give them to my kids someday. That is what I admire about my father and hope that in the face of adversity I too can be as calm and innovative as he is.

Comments

When judges ask this question, their intent is to learn something about you through your choice of whom you admire. If you just say that your role model is golfing superstar Tiger Woods but offer no explanation why, you aren't sharing much about yourself. The judges won't know if Tiger is your role model because he's a good golfer, a Stanford graduate or something else. Both of these answers give specific reasons why the applicants idolize these people, and both support their choice with concrete and memorable examples. No matter whom you choose as a hero, be sure to know enough about him or her to explain what specific quality you want to emulate. Also, know their shortcomings since you may be asked about that as a follow-up question.

Q: What is your favorite book?

A: *Les Miserables* by Victor Hugo. When we first were assigned this book to read, it was pretty daunting. But as I started reading, I couldn't put it down. I became consumed with the characters, feeling their emotions. The book took me through the low points of Jean Valjean's arrests and the high point of his final release. Throughout, the book made me think about the line between right and wrong and whether or not someone who was wrong in the past could make up for his or her mistakes to experience true freedom. It really made me think about the ethics I live by and about the mistakes I've made in the past. It also inspired me to be more forgiving of people with whom I've had disagreements.

A: *The Day Lincoln Was Shot* by Jim Bishop. This book chronicles the last 24 hours of Lincoln's life from the perspective of the assassins and the government officials who were the target of the assassination plot. I normally don't get drawn into history books, but this one was an exception because the detail allowed me to imagine everything that was happening and I felt like I was actually there. After reading the book, I took a drive to Washington, D.C., to see the Ford Theater and the boarding house where Lincoln died. I could almost see Booth approaching the President from behind and the doctors working on the President in vain. This book brought history alive for me, and I have a whole new interest in the American presidency.

Comments

Neither of these answers are book reports–which is good since the judges are not asking for a summary of the book. What the judges want to learn is who you are through your selection of a book and why you say reading the book is important. Both of these answers show

how the book affected the reader. When thinking about which book to choose, ask yourself if your selection made you think differently or compelled you to take action. Ask yourself what specifically made you relate to a character. Also, don't feel that you have to select a classic. It's fine to say that your favorite book is *Charlotte's Web* or *Green Eggs and Ham.* What's important is not your selection of the book but why it is meaningful to you.

Why You Deserve to Win Questions

Q: Why do you think you deserve to win this scholarship?

A: I believe that by giving this award you are trying to help students who show academic promise and who will contribute to the community. Since my first day of school, my parents have instilled in me a commitment to academics, and I have a nearly perfect academic record. I am on track to graduating with highest honors. I have also been contributing to my community for many years. I started a program to provide books for a local elementary school's library. By using funds from book fairs, I increased the number of books at the elementary school from 500 to 2,500. My commitment to learning and public service are two things that I believe in very strongly and I will continue to do so throughout my life.

A: This award is meant to assist students who are interested in business. I have been an entrepreneur since I was a kid and convinced my parents that I could organize a neighborhood-wide garage sale. We raised several hundred dollars that way. In school, I started a tutor-matching business. Students let me know what kind of help they needed, and I matched them to an appropriate tutor. Through ventures like these I've learned the value of marketing, building relationships and having a business plan. I'm planning on majoring in business and have a business internship lined up this summer. Ultimately I would like to be a professor at a business school so that I can continue to learn and pass on to others the knowledge and skills that will make them successful in business.

Comments

Both of these applicants do a good job of focusing on the purpose of the award to clearly explain how their background and achievements fulfill this purpose. It's important to address how you meet the mission of the award or the awarding organization. Be as specific as possible. Don't just say that you should win the scholarship because you are a good student. Give details and examples to support what you say.

Q: What would winning this scholarship mean to you?

A: For me, winning this scholarship could mean the difference between going to college or working full-time. Without this award, I will need to work for a couple years to save up enough money to go to college. I've been accepted to the college that I want to attend, but I simply don't have the money to pay for it. My parents didn't go to college, and I'll be the first in my family. And I will go. The question is whether it will be now or in a couple years.

A: My parents have spent the last 17 years taking care of me. Now I have a chance to do something to help them by winning scholarships. I feel that I owe it to my parents to try as best I can to help pay for my education. Winning this award would help to reduce their burden and help me to fulfill my goal of repaying my parents for all that they have done for me.

A: While I plan to work during the school year to earn money, winning this scholarship would mean that I could work fewer hours. Instead of working 20 hours a week, I could work only 10 hours and spend my extra time on my studies. It's been tough to balance working with my studies, and winning this award would help immensely.

Comments

Impact is important. Scholarship committees are trying to get the maximum benefit from their award. If the award will make the difference between your being able to attend a college or not, say so. The judges will understand that this award is more meaningful to you than to a student who already has a way to pay for his or her education. But be careful not to unload all of the challenges you face in the form of a sob story. Remember that many of the other applicants also have financial needs.

Education & College Questions

Q: Why is education important to you?

A: I want a job that makes me personally satisfied and my dream is to work in the medical field. An important part of this is being able to help people on a daily basis. I also know that I need to be challenged to be happy. So the medical profession is a perfect match since it allows me to contribute to society while working in an intellectually challenging environment. This past summer I volunteered at our county hospital and worked closely with a neurologist. It intrigued me that he was able to look at a set of symptoms and test results and figure out what was wrong with someone. It was like being a detective except that solving the mystery meant helping a person get better. But I also know that my ability as a doctor will depend on how well I am educated. I know that for some students going to college is about getting grades and a general education. But for me it's not only about learning because in the future someone's life may depend on how much I learned.

A: To me, education represents a limitless future. At this point in my life, I can be anything. I can be a doctor, teacher, computer programmer or artist. There are a thousand different directions I could go, but the only way to get anywhere is through education. I am studying English with a minor in music. In my classes, I have done a variety of things from writing a research paper on the role of women writers to composing an original piece performed by my school's orchestra. This is what education is all about: exploring interests and discovering what really excites you. So I guess what education really means to me is the chance to find out who I am by being able to try, succeed and sometimes even fail.

Tip #2 from a Scholarship Winner

Brian C. Babcock, U.S. Military Academy

"One of the best things that I did was do a couple of practice interviews. The way we did it was that all of the Truman applicants at my school would meet over lunch with an expert on a topic and talk about it. For example, we'd have someone come in and we'd discuss the 'don't ask don't tell policy.' We were able to get the former drug czar to sit down and talk to us for an hour and a half about the drug policy. Not a bad person to speak with about the drug policy.

"My advice is to go into the interview thinking that you're just going to have a very fun discussion. I didn't have an interview. I had a talk with eight people around the table. They asked me difficult questions but it wasn't hostile."

Brian's contributions reflect his own opinions, not those of the U.S. military.

Comments

Both of these students make personal what could otherwise be a very general answer. Instead of recounting the history of education or statistics from the latest national survey on education, these students reveal how education has personally affected them. When you are answering this question ask yourself: What have you gained personally from education or what do you hope to gain? What benefits have you received from the educational system? Try to be specific. It's not enough to say that you value education. Who doesn't? Try to get to the root of why education is important. Give specific examples so that the judges will understand your personal reason for pursuing a degree.

Q: What has influenced you to get a college education?

A: I am the first person from my family to attend college. My parents immigrated to the U.S. when I was a child. Without a college education my parents turned to what they knew, which was running a restaurant. I have also had to work in the family restaurant since I was a child so I know how difficult the work is. They said that their dream was for me to go to college so that I would have a wide choice of careers. They don't regret their decision coming to the U.S. because they expect me to go to college and succeed. Attending college means that I will have opportunities that my parents never had and that I will reach not only my goals but the goals of my parents as well. They have sacrificed their lives for me to have a better one. I don't plan on letting them down.

A: Since most of the students from my high school go to college, this seems like a strange question since it was almost assumed by my parents, teachers and friends that college was the next step after graduating. But, I look at going to college as my chance to pursue what I love–which is to design and build robots. I don't know what the job market is for "robot builders" but I intend to find out, and the first step is to get a solid education. I am choosing which schools to apply to based on whether or not they offer classes in robotics. College represents the first step in my ultimate goal of merging what I love to do with a career.

Comments

Both of these answers go beyond what's expected. Almost everyone can say that they want to go to college because they think education is important. What makes these answers strong is that they are specific to the individual. Try to personalize your answer by explaining why you have been inspired to get a college degree. What specific incident or person motivated you? What do you hope to gain? Be as specific as possible to give the selection committee insight into what inspires you and to avoid relying on overused generalizations. Also, be sure to stay away from saying that you are going to college just to earn more moolah. On a practical level, earning a degree will enable you to earn more money, but you should focus on less-materialistic factors.

Q: Why did you choose your college?

A: When I was researching colleges, I figured out that I had three priorities. First, I wanted a college with a strong program in biology and opportunities for doing hands-on research as an undergraduate. This was important because I plan to become a researcher after graduating and want to get useful experience during my college years. Second, I wanted to attend a school in which classes were taught by professors and not graduate students. I learn best when I am inspired, and I knew that I would be best inspired by learning directly from professors who are shaping the field of biology. My third priority was to attend a school with diversity. I think that college is a place not only for book learning but for personal learning as well. It's my chance to meet people with different ideas and from different backgrounds.

Comments

This is an excellent example of how to reveal something about yourself through your answer. You don't want to be a tour guide, describing the well-known assets of the college. Explain why the college's features are important to you. Instead of saying that you chose the college because of its research facilities, explain how you plan to make use of the facilities. The more details and specifics you can give the better. If appropriate, walk the judges through the thought process you went through when selecting the college. This will help them understand what is important to you and also show them how seriously you consider a college education.

Academic Questions

Q: What is your favorite subject in school and why?

A: I enjoy studying English because I like writing. When I write, I feel like I can be myself or I can be a totally different person. I can step into the shoes of someone in the past or be someone who I'm not–like an explorer in the Sahara. Writing lets me see through another person's eyes and forces me to experience what their life must be like. The most difficult thing I ever wrote was a short story that won an award from my school's literary magazine. Because the story was about a woman, the hardest part of writing it was that as a male, I had to completely reevaluate how I viewed my world through the eyes of a woman. I can tell you that I have a whole new understanding of what life might be like for the opposite sex.

A: My favorite subject is civics. Most people don't understand the way that our government works and why so many checks and balances have been put into place. It's a system that doesn't always produce the results that I'd like to see, but I am fascinated by our attempts to make a system that is as close to perfect as possible. I can see myself working in government in the future.

Comments

Both of these answers clearly explain the applicants' choices. It would be easy just to name a favorite subject and leave it at that. But the judges are trying to understand why you like what you like. When answering a question like this, give reasons or examples for your selection. Don't state the obvious. If you are asked why English is your favorite subject, give more than "Because I like it" or "Because I'm good at it." You can also use a question like this as an opportunity to talk about an achievement or award. If you say that your favorite subject is English, you can speak about a writing competition that you won or the reading marathon that you started. This is a good springboard question which you can expand to bring your impressive achievements into the conversation.

Q: Why did you select your major?

A: I'm majoring in history. History is intriguing because there are so many ways to describe the same event. A good example is the Second World War. There are many different viewpoints depending on which country the writer is from, whether he was in the military or a civilian or at what level in the leadership chain he was. It's the historian's job to present the information in the most objective way possible while still understanding that there are a lot of subjective elements to history. To me it's like unraveling a mystery, except that the mystery is real.

Tip #3 from a Scholarship Winner

Emanuel Pleitez, Stanford University

"Be confident. Confidence is going to help you out in immeasurable ways. Some people think the interviewer is there just to ask questions and make you falter. Usually the interviewers are really nice and they want to get to know you. As long as you let the interviewer know what you're really about, you'll be fine. Smile and let the interviewer get to know you. To do this all you really need is confidence."

A: My major is sociology. I hadn't planned to study sociology, but in my first year I took a class on women and the law. We covered the history and effect of laws on women including laws covering maternity leave, pornography and employment. It was an eye-opening class in which we got to interact with women who had been personally affected by these laws. I was hooked. One of my most recent research papers is based on a survey of working-class women. I really care about this field and can relate to it in a personal way.

Comments

Both answers share the applicants' inspiration for selecting their majors. Try to bring the judges into your mind so that they understand why you are passionate about your field. Examples also help the judges, who may have no idea what your major is about, understand why you chose the field. Bring up some interesting facts about the major or hot issues in the field. You might try to also think about how the degree will help you after graduation. What effect on your future might your choice have? What are your plans for using the degree in the future?

Q: Which educator has had the most influence on you?

A: Without a doubt my economics professor. In his lectures he made the theories come to life by showing real life examples of how they worked. He also took the time to meet with each of us individually to get our feedback and to see if we were interested in majoring in economics. My meeting with him lasted for two hours. I explained that I was thinking about majoring in economics but I wasn't sure if I had the mathematical ability. He convinced me that I could work on my math skills as long as I had a passion for learning. I met with him several other times, and he agreed to be my thesis advisor. Of all my professors, he's the one who has made the most effort to make sure that I was learning and excited about the field.

A: I had a professor last year who taught design. He assigned us projects that I never imagined I'd be doing. At the start of class, he had a handful of toothpicks. He asked us, "What is this?" We all said matter-of-factly, "Toothpicks."

"No," he said. "It's a bridge."

So our assignment was to build a bridge that could support the weight of a bowling ball out of toothpicks. This professor taught me to look at everyday things in a different way, to notice the shape of a gate, the color of the sky after it rained or the shadow of a building on the ground. I learned to pause and appreciate all of the efforts that went into creating what's around me.

Comments

As much as possible try to illustrate the specific influence of a teacher or professor. Give concrete examples of what he or she has done to help you learn. This will give the judges insight into your learning style and what motivates you.

Be sure that you don't just select an educator who was cool, friendly or popular. If you had a teacher or professor with whom you shared a love for baseball but not for the subject matter, this is not a great choice.

The judges want to learn about an educator who has inspired you to learn, not one who was a buddy.

Also, don't criticize other educators. In describing an influential teacher or professor, it is tempting to point out the negative traits of the others. Try not to do this. In many cases, the judges will be educators themselves or will be well-connected with educators. It would be a mistake to insult the profession. Focus on the positive aspects of the educator you choose.

Q: Can you tell me about an academic class, project or other experience that was meaningful for you?

> **A:** One of the most meaningful projects I did was in my English class. We were assigned to develop a plan for the future of our local community. The problem was that the local agriculture industry was deteriorating and within the next five years would leave hundreds in our small community jobless and farmland without a use. We were assigned to create a five-year development plan. We planned the growth of our town, training programs for the displaced workers and redevelopment plans for the land. We each published a report and built a scale model of what the town would look like. Mine was one of the handful selected to be presented at a meeting of our city council.

> **A:** For one of my sociology classes, I had to write a paper based on primary interviews. I knew that in order to be motivated to do the paper, I needed to write about something close to me. So I chose to write about the ethnic identities of first- and second-generation Asian American teenagers. Speaking with the teens, I learned how those in the first generation still had close ties to their home country while those in the second generation strived to just fit in with being American. It made me think about my own identity as a third-generation Asian American, and I started to ask questions of my own family. I could see that my family too went through a similar experience. What began as a class assignment actually helped me learn more about my family history.

Comments

A question of this type is a great opportunity to show off an impressive project. Be sure to give a lot of detail and demonstrate why the project or class was so meaningful. If appropriate, select a subject or project that relates to the scholarship since it will help demonstrate why you deserve to win the award.

Your Career & Future Questions

Q: Why do you want to enter this career?

A: I want to be a journalist because I want people to react to my writing. Whether I am uncovering an injustice or celebrating a hero, I want to invoke readers to respond. I recently wrote an article on ethnic barriers on our college's campus–how students tend to socialize with others of the same ethnic background. That series of articles sparked a huge controversy on campus. The minority clubs asked me to be on a panel discussion on the topic. Over 300 students came. There were a lot of tense moments, but I think they were necessary. My article made people think about a tough subject.

A: It might sound like a cliché to say that I want to become a doctor because of a television show, but for me it's true. I became inspired after watching "ER." I know that the show is a fictionalized drama with the purpose of entertaining and that it is as much about the personal lives of the doctors as the medicine that they practice. But what inspires me is seeing the characters act selflessly for the good of their patients. It's heartening to see that even though people see medical care as impersonal and bureaucratic, these characters give a human and humane face to the field. That's a trait that I think is important and that I want to carry with me when I become a doctor.

Tip #4 from a Scholarship Winner

Dalia Alcázar, U.C. Berkeley

"Sometimes you fill out an application, send it off and it was something you did at 3 a.m. Then you get called in for an interview and you have completely forgotten what you had written.

"Before every interview I made sure I knew about the organization, what I had written on my application and what I had written about for the essay. I needed to know what information they had about me. I also took in resumes and I always carried a portfolio with letters of recommendation, a personal profile, certificates and some examples of my writing."

Comments

When judges ask this kind of question, what they really want to know is what inspires you. They want to see that you have a rationale for entering a profession. Be sure to give a reason even if it is something as simple as being influenced by a TV show. Help the judges understand your inspiration by using lots of examples. They will not only comprehend why you want to work in the industry but also what motivates you in general.

Q: What are your career plans?

A: Eventually I would like to be the managing editor of a major newspaper. I know that I will need to start out working as a journalist for a small circulation paper and slowly move up to higher circulation newspapers. During an internship with a newspaper last summer, I had the opportunity to meet with the managing editor. She described a day in her life. What really struck me was the number of serious decisions she had to make and that every day was different. I don't want to be in a job where I do the same thing every day. I want a job that constantly makes me think and interact with a variety of people. I know it will take years of hard work and perseverance, but I think that I have the decision-making and management ability to do this kind of job well.

A: My goal is to start my own nonprofit organization to provide programs for inner-city youth. In high school and in college I have volunteered with groups to help underprivileged kids. It's been great to see how much of a difference a few hours a week can make in the life of a child. I'm a big sister to a sixth grade student now, and we've developed such a strong relationship that she asked me to go to her sixth grade graduation. I'd begin by working for a nonprofit organization. Eventually I'd like to start my own nonprofit group. I think I can help the most children this way.

Comments

Both of these applicants are aiming high, which is a very good thing. It's important to show the scholarship committee that you have high ambitions, will hold a leadership role in the future and that you are striving to make significant achievements in the career field. Explain how you would like your career to progress and what you would like to achieve. Remember that organizations awarding scholarships have limited funds and want their dollars to have the largest impact possible. This makes it important for scholarship committees to provide their awards to students who will make contributions to the field, who will be role models for others and who may directly participate in their organization in the future. Share with the selection committee what kind of influence you intend to make moving forward. Of course, you don't need to have your entire future planned out. But the scholarship judges do expect you to have a general idea of what you perceive is ahead of you.

Q Where do you see yourself 10 years from now?

A: I've asked myself that very question before and while nothing is set in stone, I do have a general idea of what I want to do and where I want to be in a decade. First, I plan to graduate from college with a degree in marketing that focuses on marketing communications. After college I'd like to work for a consumer products company to gain practical experience.

In 10 years, I'd like to take my big company experience and be a marketing manager at a smaller company, perhaps a start-up. I see myself working in marketing communications because I enjoy writing and I like the challenge of communicating complicated ideas to potential buyers. However, I think ultimately I would enjoy working in a smaller, more intimate environment, which is why I see myself at a smaller company after getting some experience. I also hope to have a family. My career will be important but not as important as my family.

A: I am majoring in political science. In 10 years, I'd like to run a nonprofit organization to help women gain equality in the workplace. Even though there have been gains for women with more females serving on executive boards and with increasing equity in pay, there is still a long way to go. Through the nonprofit organization, I'd like to provide training and recruitment programs to help women advance in business. I would also like to survey the track records of the representation of women at the executive level and lobby politicians to support women's issues.

A: In 10 years, I hope to be working for a clinic as a pediatrician. I believe that all children have the right to quality medical care whether their parents can afford it or not. Those who have lower incomes do not deserve second-class medical care. I'd like to be a part of eradicating that situation, working at a clinic and helping the children who need it the most. I would get more personal satisfaction from this than anything else.

Comments

The key to answering this question is acknowledging that the judges want to understand your motivation, not just the fact that you want to be CEO of a company. These students' responses demonstrate their inspiration. Judges also like to see the passion of students who still have their entire future ahead of them. Giving this answer with energy and enthusiasm is essential. Of course, it is possible to go too far and sound naïve. Ideally your answers should be a mix of a healthy dose of youthful idealism with a touch of adult reality.

Activity Questions

Q: What activities are you involved in?

For an award for student-athletes:

A: The main activities that I'm involved in are soccer, the academic decathlon and student government. I am the captain of the soccer team, which has won the county championship for the past three years. I plan to continue to play that sport in college. I'm also the co-captain of the academic decathlon team. For the first time in our school's history we've made it to the state level competition. While it was a team effort, my co-captain and I recruited a teacher to coach us and organized extra study sessions that I think made the difference. My participation in student government includes serving as the vice president of our school. During my tenure I have directed our school's international festival, canned food drive and election process.

For an award for writing:

A: My most important activity is writing for my school's literary magazine. One piece that I wrote received an award in a writing contest. After traveling to Italy, I decided to write a creative piece about how my life would have been different had my great-grandparents not immigrated to America. I explained what kind of relationship I would have had with my family, my education and my vocation. In my piece, I incorporated memories that my grandparents had of their home country.

Comments

Don't give a laundry list of activities. Instead of telling all 12 clubs that you are a member of, select a handful in which you've made significant contributions. This will be more meaningful to the selection committee and will better capture their attention. Be sure to also highlight activities that match the goal of the awards. If you are applying for a writing award, speak about your writing experience. If you are applying for an award in medicine, speak about your medical-related experience, studies or volunteer work. Make the activities relevant to the selection committee.

Q: How have you contributed to your community?

A: One of the ways that I have contributed to my community is volunteering over 200 hours at our local library for the children's reading program. Three times a week I go to the library after school to read stories to the children and lead them in arts and crafts activities related to the books. I do this because I think it's important to get kids excited about reading and to expose them to new ideas. The artwork gets them to interact with the material and to be creative. I know that the volunteer work that I'm doing is making a difference because parents tell me that their children have learned to enjoy reading more because of my efforts.

A: I have contributed to my community by being a voice for teenagers. Last year there was a series of articles about how teens felt like they were second-class consumers. When we go to a store, we are frequently followed around so that we don't shoplift or we are treated poorly because of our age. I thought that this kind of treatment was unnecessary and wanted to send a message to the stores that were the main culprits. I organized a protest in front of four of these stores, getting teens to carry picket signs. We got media coverage, and there was another article to follow up on the changes the stores made. We were able to get the owners of all four of the stores to sign pledges to treat teens fairly.

Comments

Show the judges how significant your contribution has been by describing the effects. How many people were affected? In what way? Have you been honored for your contributions? Contributing to your community can go beyond volunteering. Remember that there are other ways to play a role in your community such as being an advocate for a cause. Your efforts do not have to be part of a formal organization or club.

Opinion Questions

Q: What is the most important issue to you?

A: I have been personally affected by underage drinking. A friend of mine was killed in an accident caused by a drunk driver. The accident was devastating not only to her family but also to our entire school. Since that happened, I began volunteering for a group that provides rides for people who have been drinking. Even though the people I drive home are in terrible shape and I ask myself how they could let themselves go so far, I treat them well knowing that at least they had the sense to get a ride home instead of driving themselves. This issue is important to me because I know that my friend's death could have been prevented.

A: It's important to me that children are exposed to the arts. In our school district, funding was cut for the music program at the elementary schools. I volunteered after school to teach students how to play the flute and had 10 students I taught regularly. I believe that the arts encourage students to think creatively, to recognize that there are different ways of communicating and to appreciate the beauty of music. I was exposed to music when I was a child and I think it's important that I help pass on that experience to other children.

Comments

When you are identifying a problem, try to also suggest some solutions. It is even more significant if you have tried to be a part of the solution. Of course, be careful not to sound like a Miss America contestant. Don't proclaim that you are going to single-handedly end the world's problems. Be realistic about your role in affecting the issue.

Q: Is there anything else you want to add?

A: We spoke about the activities that I'm involved in, but there's an important one that I forgot to mention. I've been volunteering at the local art museum as a docent, and the

Tip #5 from a Scholarship Winner

Donald H. Matsuda, Jr., Stanford University

"I think the idea with the Truman Scholarship interview is that it's supposed to be somewhat controversial. You go into a room and there is a panel of eight people. They are all distinguished public servants who are trying to test your commitment to your views and get a sense of who you are beyond what you put on paper. You need to be psychologically and emotionally ready for this type of interview so that you won't be surprised."

experience has been great. In classes, I've studied modern art. Working at the museum, however, has given me the opportunity to share my appreciation and knowledge with tour groups every week. More than anything else this experience has solidified my desire to become an art curator in the future.

A: I would like to emphasize how committed I am to obtaining a degree and becoming a teacher. Given the purpose of your award, I think that my background including my work with the after-school program and the awards that I have won for working with children shows that I have fulfilled many of the goals of the award. After graduating, I plan to work as a teacher in my school district. I think it's important that everyone recognizes where they got their start so that they can help others in the same way. That's the only way to make improvements.

Comments

Don't be shy about bringing up something important that the judges didn't ask you about. If you've forgotten to speak about something or a topic never came up during the conversation, now is the time to say so. Use this question to bring up a strong point or two that wasn't discussed. The last impression you leave is often the strongest. If you think you've already left a strong impression, then you don't have to say anything. But if you think you need to reemphasize an important point, this is the time to make a final statement. Use this opportunity to make sure that you have made it clear why you deserve to win the award.

CHAPTER
TEN

JUDGES' ROUNDTABLE: THE INTERVIEW

In this chapter:

- **Get the inside story from real scholarship judges**
- **See what the judges think are the keys to a great interview**
- **Learn what mistakes students make in the interview**

Meet the Scholarship Judges

This is the last of three roundtables in this book. In this roundtable, scholarship judges and experts provide insight into the importance of interviews and what you can do to ace them.

If applying for scholarships is like running a race, then the interview is the last lap, the last step toward winning or losing and oftentimes the most important. It is also what worries students the most. Unlike the essay which you can write in the safety of your bedroom, for the interview you actually have to sit across the table from live human beings who are watching and evaluating your every word.

The key to acing the interview is to know what to expect and then to practice. While you may never shed the butterflies in your stomach, you can calm them down. In this roundtable we ask scholarship judges and experts what they are looking for in the interview.

How important is the interview in determining who wins the award?

Brent Drage
Rotary International Ambassadorial Scholarship Program
"I would say that the most important part of our process is the interviews. It's important because it allows the Rotarians to speak with the applicants, see what they're like and get an idea of how they would act if they go abroad."

Jacqui Love Marshall
Knight Ridder Minority Scholars Program
"The interview can make a pretty big difference when evaluating two candidates. The interview plays the largest role in the final selection since we use it to narrow down our eight semifinalists to the three finalists."

What are some typical questions that you ask?

Because scholarship committees have similar goals–to get to know you beyond the written application and to determine your fit with the award–they ask similar questions. Here are some actual questions you are likely to be asked.

Brent Drage
Rotary International Ambassadorial Scholarship Program
"We typically ask questions related to the autobiographical essay that the student has submitted. We ask applicants to expand on their thoughts and ideas in their essays. We also ask them how they plan on contributing to the world when they graduate. How are they planning on making an impact?"

Jacqui Love Marshall
Knight Ridder Minority Scholars Program
"Typical questions include: Tell me about what you know about the newspaper business. Why do you want to enter into this career? How do your parents feel about your decision to pursue a career in newspapering?"

From your experience what are qualities of a good interview?

While there is no single correct way to answer an interview question, there are certain qualities that help to make a good interview stand out in the minds of scholarship judges. It can be difficult to pinpoint what the qualities are, but our panel offers some guidance. Keep in mind that there are a limitless number of ways to have great interviews and these points are meant to provide guidance and a starting point. A good interview does not necessarily need to embody every one of these qualities.

Jacqui Love Marshall
Knight Ridder Minority Scholars Program
"I think being very honest and straightforward is important. We can tell when a student is genuinely interested and passionate about the newspaper business. It is very hard to fake this level of enthusiasm."

What common mistakes do students make in interviews?

Scholarship interviews are one place where it may seem like there's nowhere to hide and every misspoken word is magnified. In most cases, you are probably your harshest critic, noticing your errors more than the interviewers. But we found that there are some mistakes that judges notice more than others. Fortunately, these mistakes are all avoidable.

Russ Hobbs
Rotary International Ambassadorial Scholarship Program
"Not doing your homework on our organization is an easy mistake to avoid. We had one applicant who attended an Ivy League college and flew across the country for his interview. When the selection committee asked him what he knew about Rotary he didn't have a clue. He relied completely on the fact that he was an Ivy League graduate. It was like applying for a job at IBM without knowing what IBM does. This applicant figured all he had to do was show up and sign for the check."

Leah Carroll
U.C. Berkeley Haas Scholars Program
"When we advise Berkeley students who are about to go into a difficult interview we remind them that they have to see this as intellectual sparring. You need to be prepared for an interviewer to challenge your ideas. You need to be able to defend your views and even poke back. We tell our students to go in with the attitude that it's challenging but also fun. We have found that a lot of judges are most impressed when students

are willing to defend who they are and feel good about their beliefs. Not being ready to do this or not practicing for this kind of interview is a huge mistake."

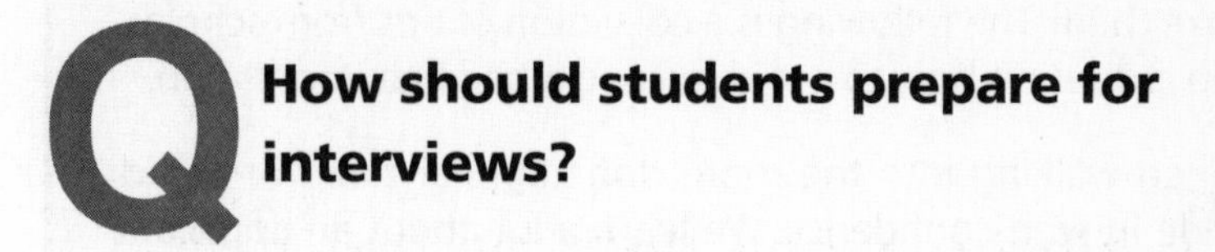

How should students prepare for interviews?

While interviews can be stress-inducing, there is something you can do to combat the tension–prepare. Here is some guidance for getting ready for the main event.

Russ Hobbs
Rotary International Ambassadorial Scholarship Program
"There are certain things that you can do to prepare for the interview. The no-brainer is knowing a little bit about Rotary. The interview is not designed to be a test like do you know all the capitals in the U.S. We presuppose that everyone is smart. What we're more interested in is the applicants themselves. What do you believe in? What do you stand for? Those aren't things that you can bone up on. They're either part of who you are or not."

Georgina Salguero
Hispanic Heritage Youth Awards
"The interview is your 15 minutes of glory. This is not the time to be modest. You have bragging rights. Use this opportunity. Prepare for it and know what you want to say."

Poor Ways to Begin an Interview

Various Scholarship Judges

First impressions are priceless since you only get one shot to make them. The following is a collection of tips from scholarship judges on how to avoid making a bad first impression.

"When walking into the room don't appear timid or afraid. Stride in with confidence. We learn a lot about an applicant from the way he or she crosses the distance from the door to the chair."

"Drink some water right before the interview. When nervous your throat naturally dries out and you don't want the first words the judges hear to sound unnaturally hoarse."

"Don't forget to look at all members of the judging panel when you speak. Some applicants look only at the members sitting directly in front of them but forget to turn to address those on the sides."

"Don't get our names wrong. Since we interview with a panel of five judges we don't expect applicants to remember our names. It's better not to use our names in conversation if you can't remember them than to call everyone by the wrong name."

"Sit up straight and still. Don't slouch or fidget. It can be very distracting."

"Don't hold pens or paper in your hands. It's too easy to unconsciously play with these objects while talking."

"Smile when you first walk in even if you feel nervous."

Participating Judges & Experts

Leah Carroll, Coordinator, U.C. Berkeley Haas Scholars Program and former Program Coordinator, U.C. Berkeley Scholarship Connection

Brent Drage, Resource Development Assistant, Rotary International Ambassadorial Scholarship Program

Russ Hobbs, District Scholarship Chairman, Rotary International Ambassadorial Scholarship Program

Jacqui Love Marshall, Vice President of Human Resources, Diversity and Development, Knight Ridder Minority Scholars Program

Georgina Salguero, Senior Manager, Programs and Events, Hispanic Heritage Youth Awards

CHAPTER
ELEVEN

FINAL THOUGHTS

In this chapter:

- **A personal ending**
- **A special request**

A Personal Ending

When you are just beginning the scholarship search it may seem like a daunting–if not downright impossible–task. But you need to keep in mind that the rewards of finding and applying for scholarships are substantial. Every student we interviewed recalled how that when starting it seemed like winning was a long shot. However, each student did apply and ultimately won.

In this book you have met students who have won tens of thousands of dollars in scholarship money. You no doubt have noticed they all have different backgrounds, achievements and aspirations. When applying, take the time to highlight your strengths. Show the scholarship judges why you deserve to win.

You can do it. And the fact that you have made your way to the end of this book shows not only your commitment to winning but also gives you a tremendous advantage. Now you know what it takes to write a powerful essay and deliver a knockout interview. You have been witness to success and failure and learned from both.

We wrote this book because we wish that we had known what we do now back when we were applying. Although we were successful, we also learned some hard lessons that we want you to avoid.

We would like to end with a personal story. When I (Kelly) was applying for scholarships I found one offered by my father's company. I was a junior in high school and didn't have any idea what it took to win a scholarship.

I thought that scholarships were based entirely on grades and test scores. Since I had good grades and high PSAT scores I thought I would win. I quickly filled out the application and wrote an essay. When it came time to interview I didn't even practice. I went in cold and "winged it." The whole time I assumed that I would win based on my academic achievements. In fact, I was so confident that I actually spent the rest of the summer waiting for the check to arrive.

But the check never came.

When I found out who won I was shocked. The student had lower grades and test scores than I did! Why did he win? How unfair!

That's when I realized that the scholarship committee was looking for more than good grades and test scores. The following year I spent time on my essay. I also practiced for the interview with a friend.

While I had spent much more time and effort this time I was rewarded when one day, out of the blue, an envelope arrived with a $2,500 check enclosed.

You can win a scholarship through your essay and interview. Even though you might be tempted like I was to bang out an essay and run into an interview cold, don't. You'll spend less time but you won't win.

It takes time and effort to craft a powerful essay and hone your interview skills. But there are a lot of awards out there and someone has to win. Let's make sure that it's you.

Special Request

Before you embark on your own quest for scholarships, we have a special request. We would love to hear about your experiences with scholarships. We want to know what works and what doesn't and how this book has helped you. Please send us a note after you've finished your own winning scholarship essays and interviews. You can reach us at:

Gen and Kelly Tanabe
c/o SuperCollege
2713 Newlands Avenue
Belmont, CA 94002

gen@supercollege.com
kelly@supercollege.com

$1,000 Moolahspot Scholarship

MoolahSPOT
2713 Newlands Avenue
Belmont CA 94002
http://www.moolahspot.com/index.cfm?scholarship=1

Purpose: To help students pay for college or graduate school.

Eligibility: Students must be at least 16 years or older and plan to attend or currently attend college or graduate school. Applicants may study any major or plan to enter any career field at any accredited college or graduate school. A short essay is required.

Amount: $1,000.

Number of Awards: Varies.

Deadline: August 31.

How to Apply: Applications are available only online.

$1,500 College JumpStart Scholarship

4546 B10 El Camino Real
No. 325
Los Altos CA 94022
http://www.jumpstart-scholarship.net

Purpose: To recognize students who are committed to using education to better their life and that of their family and/or community.

Eligibility: Applicants must be 10th, 11th or 12th grade high school, college, or adult students. Applicants may study any major and attend any college in the U.S. Applicants must be legal residents of the U.S. and complete the online application form including the required personal statement. The award may be used for tuition, room and board, books or any related educational expense.

Amount: Up to $1,500.

Number of Awards: 3.

Deadline: October 17.

How to Apply: Applications are available online.

$1,500 Scholarship Detective Launch Scholarship

Scholarship Detective
http://www.scholarshipdetective.com/scholarship/index.cfm

Purpose: To help college and adult students pay for college or graduate school.

Eligibility: Applicants must be high school, college or graduate students (including adult students) who are U.S. citizens or permanent residents. Students may study any major. The funds may be used to attend an accredited U.S. institution for undergraduate or graduate education.

Amount: $1,500.

Number of Awards: 2.

Deadline: May 31.

How to Apply: Applications are available online.

AAUW Educational Foundation Career Development Grants

Dept. 60
301 ACT Drive
Iowa City IA 52243-4030
Phone: 319-337-1716 x60
Email: aauw@act.org
http://www.aauw.org

Purpose: To support college-educated women who need additional training to advance their careers, re-enter the workforce or change careers.

Eligibility: Applicants must be U.S. citizens, hold a bachelor's degree and enroll in courses at a regionally-accredited program related to their professional development, including two- and four-year colleges, technical schools and distance learning programs. Special preference is given to women of color, AAUW members and women pursuing their first advanced degree or credentials in a nontraditional field.

Amount: $2,000-$12,000.

Number of Awards: Varies.

Deadline: December 15.
How to Apply: Applications are available online from August 1-December 15.

Academic Competitiveness Grant

Federal Student Aid
U.S. Department of Education
400 Maryland Avenue, SW
Washington DC 20202
Phone: 800-433-3243
http://studentaid.ed.gov
Purpose: To help students who have finished a rigorous secondary school program of study.
Eligibility: Applicants must be full-time students who are Federal Pell Grant recipients and have enrolled or been accepted by a two- or four-year degree-granting institution of higher education. The grants are available to students for the first and second years of college with up to $750 for the first year and up to $1,300 for the second year. Second year students must also have a minimum 3.0 GPA.
Amount: Up to $1,300.
Number of Awards: Varies.
Scholarship may be renewable.
Deadline: Varies.
How to Apply: Applicants must complete the Free Application for Federal Student Aid (FAFSA).

Adult Students in Scholastic Transition (ASIST)

Executive Women International (EWI)
515 South 700 East
Suite 2A
Salt Lake City UT 84102
Phone: 801-355-2800
Email: ewi@ewiconnect.com
http://www.executivewomen.org
Purpose: To assist adult students who face major life transitions.
Eligibility: Applicants may be single parents, individuals just entering the workforce or displaced workers.
Amount: $2,000-$10,000.
Number of Awards: Varies.
Deadline: Varies by Chapter.
How to Apply: Contact your local EWI chapter.

Advancing Aspirations Global Scholarship

Womenetics
99 West Paces Ferry Road NW
Suite 200
Atlanta GA 30305
Phone: 404-816-7224
Email: info@womenetics.com
http://www.womenetics.com
Purpose: To award outstanding college student essayists.
Eligibility: Applicants must be U.S. citizens or legal residents. They must be undergraduate students at an accredited postsecondary institution and write a 2,500-word essay on a sponsor-determined topic. Selection is based on essay presentation, thoroughness and quality of research.
Amount: Up to $5,000.
Number of Awards: 10.
Deadline: Unknown.
How to Apply: Application instructions are available online.

AFSA National Essay Contest

American Foreign Service Association (AFSA)
2101 East Street NW
Washington DC 20037
Phone: 202-944-5504
Email: dec@afsa.org
http://www.afsa.org
Purpose: To support students interested in writing an essay on foreign service.
Eligibility: Students must attend a public, private, parochial school, home school or participate in a high school correspondence program in any of the 50 states, the District of Columbia or U.S. territories or must be U.S. citizens attending schools overseas. Students must be the dependent of a US

government Foreign Service employee (active, retired with pension, deceased or separated). The current award is $2,500 to the student, $500 to his/her school and all expenses paid trip to Washington, DC, for the winner and parents.
Amount: $2,500.
Number of Awards: 1.
Deadline: April 15.
How to Apply: The registration form is available online.

Akash Kuruvilla Memorial Scholarship

P.O. Box 140900
Gainesville FL 32614
Email: info@akmsf.com
http://www.akmscholarship.com
Purpose: To continue the legacy of Akash Jacob Kuruvilla.
Eligibility: Applicants must be entering or current full-time college students at an accredited U.S. four-year college or university. They must demonstrate academic achievement, leadership, integrity and excellence in diversity. Selection is based on character, financial need and the applicant's potential to impact his or her community.
Amount: $1,000.
Number of Awards: 2.
Deadline: June 1.
How to Apply: Applications are available online.

American Fire Sprinkler Association Scholarship Program

12750 Merit Drive
Suite 350
Dallas TX 75251
Phone: 214-349-5965
Email: acampbell@firesprinkler.org
http://www.afsascholarship.org
Purpose: To provide financial aid to high school seniors and introduce them to the fire sprinkler industry.
Eligibility: Applicants must be high school seniors who plan to attend a U.S. college, university or certified trade school. Students must read the "Fire Sprinkler Essay" available online and then take an online quiz. Applicants receive one entry in the scholarship drawing for each question answered correctly.
Amount: $2,000.
Number of Awards: 10.
Scholarship may be renewable.
Deadline: April 5.
How to Apply: Applications are available online.

Americorps Vista

1201 New York Avenue NW
Washington DC 20525
Phone: 202-606-5000
Email: questions@americorps.org
http://www.americorps.gov
Purpose: To provide education assistance in exchange for community service.
Eligibility: Applicants must be United States citizens who are at least 17 years of age. They must be available to serve full-time for one year at a nonprofit organization or local government agency with an objective that may include to fight illiteracy, improve health services, create businesses or strengthen community groups.
Amount: Up to $5,550.
Number of Awards: Varies.
Deadline: Varies.
How to Apply: Applications are available online.

Anthem Essay Contest

Department W
P.O. Box 57044
Irvine CA 92619-7044
Phone: 949-222-6550
Email: essay@aynrand.org
http://www.aynrand.org
Purpose: To honor high school students who distinguish themselves in their

understanding of Ayn Rand's novel *Anthem.*
Eligibility: Applicants must be high school freshmen or sophomores who submit a 600-1200 word essay that will be judged on both style and content, with an emphasis on writing that is clear, articulate and logically organized. Winning essays must demonstrate an outstanding grasp of the philosophic meaning of *Anthem.*
Amount: $30-$2,000.
Number of Awards: 236.
Deadline: March 20.
How to Apply: Application request information is available online.

Art Awards

Scholastic
557 Broadway
New York NY 10012
Phone: 212-343-6100
Email: a&wgeneralinfo@scholastic.com
http://www.artandwriting.org
Purpose: To reward America's best student artists.
Eligibility: Applicants must be in grades 7 through 12 in American or Canadian schools and must submit artwork in one of the following categories: art portfolio, animation, ceramics and glass, computer art, design, digital imagery, drawing, mixed media, painting, photography, photography portfolio, printmaking, sculpture or video and film. There are regional and national levels.
Amount: Varies.
Number of Awards: Varies.
Deadline: Varies.
How to Apply: Applications are available online.

Atlas Shrugged Essay Contest

Department W
P.O. Box 57044
Irvine CA 92619-7044
Phone: 949-222-6550
Email: essay@aynrand.org
http://www.aynrand.org
Purpose: To honor college students who distinguish themselves in their understanding of Ayn Rand's novel *Atlas Shrugged.*
Eligibility: Applicants must be college students and high school students entering college in the fall who submit an 800-1,600 word essay which will be judged on both style and content with an emphasis on writing that is clear, articulate and logically organized. Winning essays must demonstrate an outstanding grasp of the philosophic meaning of *Atlas Shrugged.*
Amount: Up to $10,000.
Number of Awards: 84.
Deadline: September 17.
How to Apply: Application request information is available online.

AXA Achievement Community Scholarship

c/o Scholarship America
One Scholarship Way
St. Peter MN 56082
Phone: 800-537-4180
Email: axaachievement@scholarshipamerica.org
http://www.axa-equitable.com/axa-foundation/community-scholarships.html
Purpose: To aid outstanding college-bound high school seniors.
Eligibility: Applicants must be U.S. citizens or legal residents and be high school seniors in the U.S. or Puerto Rico. They must be planning to attend an accredited two- or four-year postsecondary institution no later than the fall following high school graduation. They must be ambitious, goal-oriented, respectful of others and able to succeed in college. Selection is based on the overall strength of the application.
Amount: $2,000.
Number of Awards: Varies.

Deadline: February 1.
How to Apply: Applications are available online.

AXA Achievement Scholarships

c/o Scholarship America
One Scholarship Way
St. Peter MN 56082
Phone: 800-537-4180
Email: axaachievement@scholarshipamerica.org
http://www.axa-equitable.com/axa-foundation/community-scholarships.html
Purpose: To provide financial assistance to ambitious students.
Eligibility: Applicants must be U.S. citizens or legal residents who are current high school seniors and are planning to enroll full-time in an accredited college or university in the fall following their graduation. They must show ambition and drive evidenced by outstanding achievement in school, community or workplace activities. A recommendation from an unrelated adult who can vouch for the student's achievement is required.
Amount: $10,000-$25,000.
Number of Awards: 52.
Deadline: December 1.
How to Apply: Applications are available online.

Best Buy @15 Scholarship Program

Best Buy Children's Foundation
7601 Penn Avenue S.
Richfield MN 55423
Phone: 612-292-6397
Email: bestbuy@scholarshipamerica.org
http://www.bestbuy-communityrelations.com/scholarship.htm
Purpose: To assist students in grades 9-12 who will be attending a college, university or vocational school immediately following high school graduation.
Eligibility: Applicants must be currently in grades 9-12 living in the United States or Puerto Rico, have a minimum cumulative GPA of 2.5 (on a 4.0 scale), and demonstrate commitment to and involvement in community volunteer service or work experience. Selection is based on academic performance and exemplary community volunteer service or work experience. Consideration may also be given to participation and leadership in school activities. Financial need is not considered. Best Buy employees and relatives of employees who meet the requirements are eligible to apply.
Amount: $1,000.
Number of Awards: Up to 1200.
Deadline: February 15.
How to Apply: Applications are only available online.

Big Dig Scholarship

Antique Trader
4216 Pacific Coast Highway
#302
Torrance CA 90505
Phone: 310-294-9981
Email: henryk@antiquetrader.tv
http://www.antiquetrader.tv
Purpose: To aid the student who has written the most interesting essay on a specific antiques-related topic.
Eligibility: Applicants must be current undergraduate freshmen or sophomores or must be graduating high school seniors who are planning to enroll in college during the year following the application due date year. They must write an essay on a sponsor-defined question that relates to antiques. Selection is based on the content, grammatical correctness and originality of the essay.
Amount: $3,000.
Number of Awards: 1.
Deadline: June 1.

How to Apply: Application instructions are available online.

Burger King Scholars Program

5505 Blue Lagoon Drive
Miami FL 33126
Phone: 305-378-3000
Email: bdorado@whopper.com
http://www.haveityourwayfoundation.org/burger_king_scholars_program.html

Purpose: To provide financial assistance for high school seniors who have part-time jobs.

Eligibility: Applicants may apply from public, private, vocational, technical, parochial and alternative high schools in the United States, Canada and Puerto Rico and must be U.S. or Canadian residents. Students must also have a minimum 2.5 GPA, work part-time an average of 15 hours per week unless there are extenuating circumstances, participate in community service or other activities, demonstrate financial need and plan to enroll in an accredited two- or four-year college, university or vocational/technical school by the fall term of the graduating year. Applicants do NOT need to work at Burger King, but Burger King employees are eligible.

Amount: $1,000-$25,000.

Number of Awards: Varies.

Deadline: January 10.

How to Apply: Applications are available online and may only be completed online.

C.I.P. Scholarship

College Is Power
1025 Alameda de las Pulgas
No. 215
Belmont CA 94002
http://www.collegeispower.com/scholarship.cfm

Purpose: To assist adult students age 18 and over with college expenses.

Eligibility: Applicants must be adult students currently attending or planning to attend a two-year or four-year college or university within the next 12 months. Students must be 18 years or older and U.S. citizens or permanent residents. The award may be used for full- or part-time study at either on-campus or online schools.

Amount: $1,500.

Number of Awards: Varies.

Deadline: May 31.

How to Apply: Applications are available online.

Carson Scholars

305 W Chesapeake Avenue
Suite L-020
Towson MD 21204
Phone: 877-773-7236
Email: caitlin@carsonscholars.org
http://www.carsonscholars.org

Purpose: To recognize students who demonstrate academic excellence and commitment to the community.

Eligibility: Applicants must be nominated by their school. They must be in grades 4 through 11 and have a GPA of 3.75 or higher in English, reading, language arts, math, science, social studies and foreign language. They must have participated in some form of voluntary community service beyond what is required by their school. Scholarship recipients must attend a four-year college or university upon graduation to receive funds.

Amount: $1,000.

Number of Awards: Varies.

Deadline: January 13.

How to Apply: Applications are available from the schools of those nominated. Only one student per school may be nominated.

Castle Ink's Green Scholarship

37 Wyckoff Street
Greenlawn NY 11740
Phone: 800-399-5193
Email: scholarships@castleink.com
http://www.castleink.com

Purpose: To aid environmentally-minded students.
Eligibility: Applicants must be U.S. citizens or permanent residents who are college-bound high school seniors or current college students at an accredited institution. They must have a GPA of 2.5 or higher and submit an essay describing their own efforts in recycling and conservation. Selection is based on the strength of the essay.
Amount: $2,500.
Number of Awards: 1.
Deadline: June 30.
How to Apply: Application instructions are available online.

Charles Shafae' Scholarship

Papercheck
Phone: 866-693-3348
Email: scholarships@papercheck.com
http://www.papercheck.com
Purpose: To reward the students who have written the most compelling essays.
Eligibility: Applicants must be legal residents of the U.S. or must hold a valid student visa and be enrolled as full-time undergraduates at an accredited U.S. four-year postsecondary institution. They must have a GPA of 3.2 or higher, must be in good academic standing and must write an essay that addresses a question that has been predetermined by the scholarship sponsor.
Amount: $500.
Number of Awards: 2.
Deadline: May 1.
How to Apply: Application instructions are available online.

Church Hill Classics "Frame My Future" Scholarship

594 Pepper Street
Monroe CT 06468
Phone: 800-477-9005
Email: info@diplomaframe.com
http://www.framemyfuture.com
Purpose: To help success-driven students attain their higher education goals.
Eligibility: Applicants must be high school seniors or otherwise eligible for graduation in the school year of application or current college students. They must plan to enroll in college full-time the following academic year. Applicants must be residents of the United States, including APO/FPO addresses but excluding Puerto Rico. A photograph, essay, painting or other creative entry is required to demonstrate what the applicants want to achieve in their personal and professional life after college.
Amount: $1,000.
Number of Awards: 5.
Deadline: March 6.
How to Apply: Applications are available online.

Coca-Cola All-State Community College Academic Team

P.O. Box 442
Atlanta GA 30301
Phone: 800-306-2653
Email: questions@coca-colascholars.org
http://www.coca-colascholars.org
Purpose: To assist community college students with college expenses.
Eligibility: Applicants must be enrolled in community college, have a minimum GPA of 3.5 on a four-point scale and be on track to earn an associate's or bachelor's degree. Students attending community college in the U.S. do NOT need to be members of Phi Theta Kappa. One student from each state will win a $2,000 scholarship. Fifty students will win a $1,500 scholarship, fifty students will win a $1,250 scholarship and fifty students will win a $1,000 scholarship.
Amount: $1,000-$2,000.
Number of Awards: 200.
Deadline: December 1.
How to Apply: Applications are available online. Nomination from the

designated nominator at your school is required. A list of nominators is available at http://www.ptk.org/schol/allusacontacts/

Coca-Cola Scholars Program

P.O. Box 442
Atlanta GA 30301
Phone: 800-306-2653
Email: questions@coca-colascholars.org
http://www.coca-colascholars.org

Purpose: Begun in 1986 to celebrate the Coca-Cola Centennial, the program is designed to contribute to the nation's future and to assist a wide range of students.

Eligibility: Applicants must be high school seniors in the U.S. and must use the awards at an accredited U.S. college or university. Selection is based on character, personal merit and commitment. Merit is shown through leadership, academic achievement and motivation to serve and succeed.

Amount: $10,000-$20,000.

Number of Awards: 250.

Scholarship may be renewable.

Deadline: October 31.

How to Apply: Applications are available online.

College Answer $1,000 Scholarship

Sallie Mae
300 Continental Drive
Newark, DE 19713
http://www.collegeanswer.com

Purpose: To help students pay for college.

Eligibility: Applicants may be high school, undergraduate or graduate students and must register on the CollegeAnswer website. Each month one registered user is selected in a random drawing to receive the scholarship. When you are registered for the website or the Sallie Mae Scholarship Search, you are automatically entered into the scholarship drawing.

Amount: $1,000.

Number of Awards: 1 per month.

Deadline: Monthly.

How to Apply: Enter the scholarship by registering on the website.

Create Real Impact Contest

Impact Teen Drivers
Attn.: Create Real Impact Contest
P.O. Box 161209
Sacramento CA 95816
Phone: 916-733-7432
Email: info@impactteendrivers.org
http://www.createrealimpact.com

Purpose: To raise awareness of the dangers of distracted driving and poor decision making.

Eligibility: Applicants must be legal U.S. residents who are between the ages of 15 and 22. They must be enrolled full-time at an accredited secondary or post-secondary school. They must submit an original, videotaped creative project that offers a solution to the problem of distracted driving. Selection is based on project concept, message effectiveness and creativity.

Amount: $500.

Number of Awards: Up to 20.

Deadline: June 30.

How to Apply: Contest entry instructions are available online. A videotaped creative project is required.

CrossLites Scholarship Contest

1000 Holt Avenue 1178
Winter Park FL 32789
Phone: 407-833-3886
Email: crosslites@gmail.com
http://www.crosslites.com

Purpose: To encourage students to learn about Dr. Charles Parker.

Eligibility: Applicants must be high school, undergraduate or graduate students. There are no minimum GPA, SAT, ACT, GMAT, GRE or any other test score requirements. Students must write a reflective essay of 400 to 600 words based on one of Dr. Charles

Parker's quotes or messages, which are listed on the website. There are winners for the high school, undergraduate and graduate school levels. Selection is based on the judges' score (10 percent) and votes from website visitors (90 percent).
Amount: Up to $2,000.
Number of Awards: 33.
Deadline: December 15.
How to Apply: Applications are available online. An essay, contact information and transcript are required.

Davidson Fellows Award

Davidson Institute for Talent Development
9665 Gateway Drive, Suite B
Reno NV 89521
Phone: 775-852-3483
Email: davidsonfellows@ditd.org
http://www.davidson-institute.org
Purpose: To award young people for their works in mathematics, science, technology, music, literature, philosophy or "outside the box."
Eligibility: Applicants must be under the age of 18 and be able to attend the awards reception in Washington, DC. In addition to the monetary award, the institute will pay for travel and lodging expenses. Three nominator forms, three copies of a 15-minute DVD or VHS videotape and additional materials are required.
Amount: $10,000-$50,000.
Number of Awards: Varies.
Deadline: February 1.
How to Apply: Applications are available online.

Davis-Putter Scholarship Fund

P.O. Box 7307
New York NY 10116
Email: information@davisputter.org
http://www.davisputter.org
Purpose: To assist students who are both academically capable and who aid the progressive movement for peace and justice both on campus and in their communities.
Eligibility: Applicants must be undergraduate or graduate students who participate in the progressive movement, acting in the interests of issues such as expansion of civil rights and international solidarity, among others. Applicants must also have demonstrated financial need as well as a solid academic record.
Amount: Up to $10,000.
Number of Awards: Varies.
Deadline: April 1.
How to Apply: Applications are available online.

Delete Cyberbullying Scholarship Award

2261 Market Street #291
San Francisco, CA 94114
Email: applications@deletecyberbullying.org
http://www.deletecyberbullying.org/scholarship/
Purpose: To get students committed to the cause of deleting cyberbullying.
Eligibility: Applicants must be a U.S. citizen or permanent resident and attending or planning to attend an accredited U.S. college or university for undergraduate or graduate studies. Applicants must also be a high school, college or graduate student or a student planning to enter college.
Amount: $1,500.
Number of Awards: 2.
Deadline: June 30.
How to Apply: Applications are available online.

Dell Scholars Program

Michael and Susan Dell Foundation
P.O. Box 163867
Austin TX 78716
Phone: 512-329-0799
Email: apply@dellscholars.org
http://www.dellscholars.org

Purpose: To support underprivileged high school seniors.
Eligibility: Students must be participants in an approved college readiness program, and they must have at least a 2.4 GPA. Applicants must be pursuing a bachelor's degree in the fall directly after graduation. Students must also be U.S. citizens or permanent residents and demonstrate financial need. Selection is based on "individual determination to succeed," future goals, hardships that have been overcome, self motivation and financial need.
Amount: Varies.
Number of Awards: Varies. Scholarship may be renewable.
Deadline: January 15.
How to Apply: Applications are available online. An online application is required.

Directron.com College Scholarship

10402 Harwin Drive
Houston TX 77036
Phone: 713-773-3636 x1500
Email: customer_service@directron.us
http://www.directron.com
Purpose: To aid U.S. college students.
Eligibility: Applicants must be high school seniors or current college students and submit an essay on the provided topic related to computers. Essays are judged on academic merit (50 percent) and originality and creativity (50 percent). Photos are not required but recommended.
Amount: $300-$1,000.
Number of Awards: 6.
Deadline: May 1.
How to Apply: No application is required. Applicants should send their contact information and essays to information@directron.com.

Discus Awards College Scholarships

7101 Wisconsin Avenue, Suite 750
Bethesda MD 20814
Email: info@discusawards.com
http://www.discusawards.com
Purpose: To aid well-rounded, college-bound high school students.
Eligibility: Applicants must be U.S.-based high school students who have involvement or achievements in at least three of the following areas: academics, arts, athletics, community service, faith, government, green, technology, work or other achievements. Students may be in grades 9 to 12. Selection is based on merit.
Amount: $2,000.
Number of Awards: 10.
Deadline: Monthly.
How to Apply: Applications are available online. An application form and supporting materials are required.
Distinguished Young Women Scholarship Program

Distinguished Young Women

P.O. Box 2786
Mobile AL 36652
Phone: 251-438-3621
Email: lynne@ajm.org
http://www.ajm.org
Purpose: To provide scholarship opportunities and encourage personal development for high school girls through a competitive pageant stressing academics and talent as well as self-expression and fitness.
Eligibility: Teen girls are selected from state competitions to participate in a national pageant. Contestants are judged on a combination of scholastics, personal interview, talent, fitness and self-expression. Applicants should be a high school student at least in their sophomore year. Students must be U.S. citizens, have never been married and have never been pregnant.
Amount: Varies.

Number of Awards: Varies.
Deadline: Varies.
How to Apply: Applications are available online.

Do Something Awards

24-32 Union Square East
4th Floor
New York NY 10003
Phone: 212-254-2390
Email: tacklehunger@dosomething.org
http://www.dosomething.org
Purpose: To award scholarships and community grants to young social entrepreneurs who make a measurable difference in their communities.
Eligibility: Young community leaders up to age 25 may apply. Emphasis is on those who take a leadership role in creating a positive, lasting impact on the community. Focus areas include health, environment and community building.
Amount: Varies.
Number of Awards: Varies.
Deadline: March 1.
How to Apply: Applications are available online.

Doing Good Campaign Scholarship Program

11710 Plaza America Drive
Suite 530
Reston VA 20190
Phone: 703-476-1000
Email: info@applyists.com
http://www.doinggood.com
Purpose: To aid students who seek to help the underserved.
Eligibility: Applicants must be U.S. citizens, legal permanent residents or nationals. They must be high school graduates or GED recipients. High school graduates must have a cumulative GPA of 3.0 or higher. They must show a commitment to helping others in the community. Selection is based on academic achievement, community involvement, essay and financial need.
Amount: $5,000-$25,000.
Number of Awards: 5.
Deadline: February 15.
How to Apply: Applications are available online. An application form, essay and other supporting materials are required.

Dollars for Scholars Scholarship

Citizens' Scholarship Foundation of America
One Scholarship Way
P.O. Box 297
St. Peter MN 56082
Phone: 800-537-4180
http://www.scholarshipamerica.org
Purpose: To encourage students to aim for and achieve loftier educational goals.
Eligibility: Applicants must be members of a local Dollars for Scholars chapter. There are more than 1,200 Dollars for Scholars chapters that award more than $29 million in awards each year.
Amount: Varies.
Number of Awards: Varies.
Deadline: Varies.
How to Apply: Contact your local Dollars for Scholars chapter for more information. A list of chapters is available online.

Dr. Arnita Young Boswell Scholarship

National Hook-Up of Black Women Inc.
1809 East 71st Street, Suite 205
Chicago IL 60649
Phone: 773-667-7061
Email: nhbwdir@aol.com
http://www.nhbwinc.com
Purpose: To reward adult students for their academic achievement.
Eligibility: Applicants must be undergraduate or graduate continuing education students. Selection is based on academic accomplishments as well as involvement in school and community activities and an essay.

Amount: $1,000.
Number of Awards: Varies.
Scholarship may be renewable.
Deadline: March 1.
How to Apply: Applications are available by mail and must be requested by March 1.

Dream Deferred Essay Contest on Civil Rights in the Mideast

Hands Across the Mideast Support Alliance
263 Huntington Avenue, #315
Boston MA 02115
Phone: 617-266-0080
Email: info@hamsaweb.org
http://www.hamsaweb.org
Purpose: To award American and Middle Eastern youth who have written outstanding essays on civil rights in the Middle East.
Eligibility: Entrants must be age 25 or younger at the time of the entry deadline. They must be living in an Arab League nation, the U.S., Afghanistan or Iran. They must write an essay of 600 to 1,500 words responding to one of the civil rights topics presented in the official entry rules. Selection is based on the strength and relevance of the essay response.
Amount: $500-$2,000.
Number of Awards: 10.
Deadline: May 27.
How to Apply: Applications are available online. The completion of an online essay submission form is required.

DuPont Challenge Science Essay Award

c/o General Learning Communications
900 Skokie Boulevard, Suite 200
Northbrook IL 60062
Phone: 847-205-3000
http://thechallenge.dupont.com
Purpose: To promote interest in scientific studies.
Eligibility: Applicants must be full-time students between grades 7 and 12 in a U.S. or Canadian school and write a 700- to 1,000-word essay about a scientific or technological development that interests them.
Amount: $200-$5,000.
Number of Awards: Varies.
Deadline: January 31.
How to Apply: Applications are available online.

Education Exchange College Grant Program

ACCEL/Exchange Network
250 Johnson Road
Morris Plains NJ 07950
Phone: 800-519-8883
Email: eftbasupport@fiserv.com
http://www.accelexchange.com
Purpose: To assist hardworking and talented students in the pursuit of higher education.
Eligibility: Applicants must be graduating seniors and U.S. citizens. Selection criteria include scholastic achievement, extracurricular activities, character, leadership, essay and financial need. Judging is done from March to May of each year, and recipients are announced in June.
Amount: Varies.
Number of Awards: Varies.
Deadline: Varies.
How to Apply: Applications are available from ACCEL/Exchange member institutions.

Educational Advancement Foundation Merit Scholarship

Alpha Kappa Alpha Educational Advancement Foundation Inc.
5656 S. Stony Island Avenue
Chicago IL 60637
Phone: 800-653-6528
Email: akaeaf@akaeaf.net
http://www.akaeaf.org
Purpose: To support academically talented students.

Eligibility: Applicants must be full-time college students at the sophomore level or higher, including graduate students, at an accredited school. They must have a GPA of at least 3.0 and demonstrate community involvement and service. The program is open to students without regard to sex, race, creed, color, ethnicity, religion, sexual orientation or disability. Students do NOT need to be members of Alpha Kappa Alpha. The application deadline is April 15 for undergraduates and August 15 for graduates.
Amount: Varies.
Number of Awards: Varies.
Deadline: April 15 (undergraduates) and August 15 (graduates).
How to Apply: Applications are available online.

Family Travel Forum Teen Travel Writing Scholarship

Family Travel Forum
891 Amsterdam Avenue
New York NY 10025
Phone: 212-665-6124
Email: editorial@travelbigo.com
http://www.travelbigo.com
Purpose: To aid college-bound students who have written the best travel essays.
Eligibility: Applicants must be members of the TravelBIGO.com online community and be between the ages of 13 and 18. They must be in grades 8 through 12 and must be attending a U.S. or Canadian high school, U.S. or Canadian junior high school, U.S. home school or an American school located outside of the U.S. They must submit an essay about a significant travel experience that occurred within the past three years and that happened when the applicant was between the ages of 12 and 18. Selection is based on originality, quality of storytelling and grammar.
Amount: $200-$1,000.
Number of Awards: 3.
Deadline: August 1.
How to Apply: Application instructions are available online. An essay submission form and essay are required.

Financial Connects Scholarship

Net Literacy Alliance
426 Springwood Drive
Carmel IN 46032
Email: danielkent@netliteracy.org
http://www.netliteracyalliance.org
Purpose: To encourage financial literacy amongst America's youth.
Eligibility: Applicants must be legal residents of the U.S. They must be in grades 6-12 or must be college students who are enrolled at least part-time at a U.S. two- or four-year postsecondary institution. Applicants whose video entry proposals are approved will be invited to submit a video on a topic that relates to financial literacy. Selection is based on the quality of the video.
Amount: Up to $5,000.
Number of Awards: Varies.
Deadline: June 30.
How to Apply: Application instructions are available online. Submission of applicant contact information and a video proposal, followed by the video itself if requested, is required.

FiSCA Scholarship

Financial Service Centers of America
Court Plaza South, East Wing
21 Main Street, 1st Floor, P.O. Box 647
Hackensack NJ 07602
Phone: 201-487-0412
Fax: 201-487-3954
Email: info@fisca.org
http://www.fisca.org
Purpose: To help collegebound high school seniors from areas served by FiSCA centers.
Eligibility: Applicants must be high school seniors. Selection is based on leadership, academic achievement and

financial need. There are more than 7,000 locations nationwide.
Amount: At least $2,000.
Number of Awards: At least 2.
Deadline: April 8.
How to Apply: Applications are available online.

Fountainhead Essay Contest

Department W
P.O. Box 57044
Irvine CA 92619-7044
Phone: 949-222-6550
Email: essay@aynrand.org
http://www.aynrand.org
Purpose: To honor high school students who distinguish themselves in their understanding of Ayn Rand's novel *The Fountainhead.*
Eligibility: Applicants must be high school juniors or seniors who submit a 800-1,600 word essay which will be judged on both style and content with an emphasis on writing that is clear, articulate and logically organized. Winning essays must demonstrate an outstanding grasp of the philosophic and psychological meaning of *The Fountainhead.*
Amount: $50-$10,000.
Number of Awards: 236.
Deadline: April 26.
How to Apply: Application request information is available online.

Frank Newman Leadership Award

Campus Compact
45 Temple Place
Boston MA 02111
Phone: 617-357-1881
Email: campus@compact.org
http://www.compact.org
Purpose: To provide scholarships and opportunities for civic mentoring to students with financial need.
Eligibility: Emphasis is on students who have demonstrated leadership abilities and significant interest in civic responsibility. Students must attend one of the 1,000 Campus Compact member institutions and be nominated by the Campus Compact member president.
Amount: Varies.
Number of Awards: Varies.
Deadline: February 28.
How to Apply: Nominations must be made by the Campus Compact member president.

From Failure to Promise Essay Contest

P.O. Box 352
Olympia Fields IL 60461
Phone: 708-252-4380
Email: drcmoorer@gmail.com
http://www.fromfailuretopromise.com
Purpose: To aid college students.
Eligibility: Applicants must be high school seniors, undergraduate students or graduate students who are enrolled at or planning to enroll at an accredited U.S. college or university. They must have a GPA of 3.0 or higher. They must submit an essay discussing how the book "From Failure to Promise: An Uncommon Path to Professoriate" has impacted their pursuit of success. Selection is based on essay originality, presentation and quality of research.
Amount: $500-$1,500.
Number of Awards: 3.
Deadline: July 31.
How to Apply: Applications are available online. An application form, transcript and essay are required.

Fulbright Grants

U.S. Department of State
Office of Academic Exchange Programs, Bureau of Educational and Cultural Affairs
U.S. Department of State, SA-44
301 4th Street SW, Room 234
Washington DC 20547
Phone: 202-619-4360

Email: academic@state.gov
http://fulbright.state.gov
Purpose: To increase the understanding between the people of the United States and the people of other countries.
Eligibility: Applicants must be graduate students, scholars or professionals. Funds are generally used to support students in university teaching, advanced research, graduate study or teaching in elementary and secondary schools.
Amount: Varies.
Number of Awards: Varies.
Deadline: October 17.
How to Apply: Applications are available online.

GE-Reagan Foundation Scholarship Program

Ronald Reagan Presidential Foundation
40 Presidential Drive, Suite 200
Simi Valley CA 93065
Phone: 800-310-4053
Email: info@applyists.com
https://aim.applyists.net/RonaldReagan
Purpose: To reward students who demonstrate leadership, drive, integrity and citizenship.
Eligibility: Applicants must be high school seniors and pursue a bachelor's degree at an accredited U.S. college or university the following fall. Students must demonstrate strong academic performance (3.0 or greater GPA or equivalent), demonstrate financial need and be a U.S. citizen. Funds may be used for student tuition and room and board.
Amount: $10,000.
Number of Awards: Up to 20. Scholarship may be renewable.
Deadline: February 17.
How to Apply: Applications are available online. Students must be nominated by an eligible community leader, such as a high school principal, elected official or executive director of a nonprofit organization and be recommended by an authority figure, such as a student activity advisor, community service coordinator, coach, employer, teacher, counselor or religious leader.

Gen and Kelly Tanabe Student Scholarship

2713 Newlands Avenue
Belmont CA 94002
Phone: 650-618-2221
Email: tanabe@gmail.com
http://www.genkellyscholarship.com
Purpose: To assist high school, college and graduate school students with educational expenses.
Eligibility: Applicants must be 9th-12th grade high school students, college students or graduate school students who are legal U.S. residents. Students may study any major and attend any college in the U.S.
Amount: $1,000.
Deadline: July 31.
How to Apply: Applications are available online.

Get Up Get Active Scholarship

Get Up Get Active
Email: scholarship@getupgetactive.org
http://www.getupgetactive.org/scholarship
Purpose: To educate youth about the importance of getting active.
Eligibility: Applicants must be current high school students or current or future college or graduate students and be U.S. citizens or permanent residents.
Amount: $1,500.
Number of Awards: 2.
Deadline: October 31.
How to Apply: Applications are available online. An application form and short essay are required.

Global Citizen Awards

EF Educational Tours
EF Center Boston

One Education Street
Cambridge MA 02141
Phone: 617-619-1300
Email: marisa.talbot@ef.com
http://www.eftours.com
Purpose: To help students reflect on their place in the world through writing and then have a chance to experience it first-hand.
Eligibility: Applicants must be college-bound high school sophomores and juniors in the U.S. or Canada nominated by their schools and must write an essay on a topic related to global citizenship. The award involves a paid educational trip to Europe.
Amount: Educational tour expenses.
Number of Awards: Varies.
Deadline: February 1.
How to Apply: Applications are available online.

Go! Study Abroad Scholarship

2680 Bancroft Way
Berkeley CA 94704
Phone: 415-796-6456
Email: scholarship@gooverseas.com
http://www.gooverseas.com/study-abroad/
Purpose: To aid students who have been accepted into a study abroad program.
Eligibility: Applicants must be current college or graduate students who have been accepted into a study abroad program for the coming academic year. Selection is based on application creativity and display of analytical thinking.
Amount: $1,000.
Number of Awards: Varies.
Deadline: September 15.
How to Apply: Applications are available online. An application form and essay are required.

Hire a Licensed Contractor Public Service Announcement Video Contest

National Association of State Contractors Licensing Agencies
Attention: PSA Contest
23309 North 17th Drive
Unit 1, Building 110
Phoenix AZ 85027
Phone: 623-587-9354
Email: info@nascla.org
http://www.nascla.org
Purpose: To raise awareness of the importance of hiring a licensed contractor.
Eligibility: Applicants must be current college students and may enter the contest individually or in groups. They must submit a brief videotaped public service announcement on the topic of hiring a licensed contractor. Selection is based on video production quality, originality and message effectiveness.
Amount: $750-$3,000.
Number of Awards: 3.
Deadline: January 30.
How to Apply: Entry instructions are available online.

Holocaust Remembrance Project Essay Contest

Holland and Knight Charitable Foundation
P.O. Box 2877
Tampa FL 33601
Phone: 866-HK-CARES
Email: holocaust@hklaw.com
http://holocaust.hklaw.com
Purpose: To reward high school students who write essays about the Holocaust.
Eligibility: Applicants must be age 19 and under who are currently enrolled as high school students in grades 9 to 12 (including home-schooled students), high school seniors or students who are enrolled in a high school equivalency program and be residents of either

the United States or Mexico or United States citizens living abroad. Applicants should submit essays about the Holocaust and entry forms. Every essay must include works cited, a reference page or a bibliography. First place winners will receive free trips to Washington, DC.

Amount: Up to $5,000.

Number of Awards: Varies.

Deadline: April 15.

How to Apply: Essays may be submitted online.

Intel Science Talent Search

Intel Corporation and Science Service
Society for Science and the Public
1719 North Street NW
Washington DC 20036
Phone: 202-785-2255
Email: sciedu@sciserv.org
http://www.societyforscience.org/sts/

Purpose: To recognize excellence in science among the nation's youth and encourage the exploration of science.

Eligibility: Applicants must be high school seniors in the U.S., Puerto Rico, Guam, Virgin Islands, American Samoa, Wake or Midway Islands or the Marianas. U.S. citizens attending foreign schools are also eligible. Applicants must complete college entrance exams and complete individual research projects and provide a report on the research.

Amount: $7,500-$100,000.

Number of Awards: 40.

Deadline: Varies.

How to Apply: Applications are available by request.

Jeannette Rankin Foundation Award

1 Huntington Road, Suite 701
Athens GA 30606
Phone: 706-208-1211
Email: info@rankinfoundation.org
http://www.rankinfoundation.org

Purpose: To support the education of low-income women 35 years or older.

Eligibility: Applicants must be women 35 years of age or older, plan to obtain an undergraduate or vocational education and meet maximum household income guidelines.

Amount: Varies.

Number of Awards: Varies. Scholarship may be renewable.

Deadline: March 1.

How to Apply: Applications are available online or by sending a self-addressed and stamped envelope to the foundation.

Joe Foss Institute Essay Scholarship Program

14415 North 73rd Street
Suite 109
Scottsdale AZ 85260
Phone: 480-348-0316
Email: scholarship@joefoss.com
http://www.joefoss.com

Purpose: To encourage students to befriend a military veteran.

Eligibility: Applicants must be U.S. citizens and be high school students or recent high school graduates. They must befriend a veteran and submit a 1,500-word essay on the experience. Previous grand prize winners are ineligible. Selection is based on essay creativity, clarity and grammar.

Amount: $250-$5,000.

Number of Awards: 3.

Deadline: October 9.

How to Apply: Entry instructions are available online. A cover sheet and essay are required.

John F. Kennedy Profile in Courage Essay Contest

John F. Kennedy Library Foundation
Columbia Point
Boston MA 02125
Phone: 617-514-1649
Email: profiles@nara.gov
http://www.jfkcontest.org

Purpose: To encourage students to research and write about politics and John F. Kennedy.
Eligibility: Applicants must be in grades 9 through 12 in public or private schools or be home-schooled and write an essay about the political courage of a U.S. elected official who served during or after 1956. Essays must have source citations. Applicants must register online before sending essays and have a nominating teacher review the essay. The winner and teacher will be invited to the Kennedy Library to accept the award, and the winner's teacher will receive a grant. Essays are judged on content (55 percent) and presentation (45 percent).
Amount: $500-$10,000.
Number of Awards: Up to 7.
Deadline: January 7.
How to Apply: Applications are available online. A registration form and essay are required.

KFC Colonel's Scholars Program

P.O. Box 725489
Atlanta GA 31139
Phone: 866-532-7240
Email: kfcscholars@act.org
http://www.kfcscholars.org
Purpose: To assist students with financial need in obtaining a college education.
Eligibility: Applicants must be high school seniors who are enrolling in a public college or university within their state of residence and pursuing a bachelor's degree. They must also have a GPA of 2.75 or higher and demonstrate financial need. The award is up to $5,000 per year and renewable for up to four years. To renew the scholarship, recipients must maintain a 2.75 minimum GPA, take a minimum of 12 credit hours per semester and during the second year of funding work an average of 10 hours per week.
Amount: Up to $5,000.
Number of Awards: At least 75. Scholarship may be renewable.
Deadline: February 8.
How to Apply: Applications are available online. An online application is required.

Kohl's Kids Who Care Scholarship

Kohls Corporation
N56 W17000 Ridgewood Drive
Menomonee Falls WI 53051
Phone: 262-703-7000
Fax: 262-703-7115
Email: community.relations@kohls.com
http://www.kohlscorporation.com/CommunityRelations/scholarship/index.asp
Purpose: To recognize young people who volunteer in their communities.
Eligibility: Applicants must be nominated by parents, educators or community members. There are two categories: one for kids ages 6-12 and another for ages 13-18. Nominees must be legal U.S. residents who have not graduated from high school.
Amount: Up to $10,000.
Number of Awards: At least 2,100.
Deadline: March 15.
How to Apply: Applications are available online and at Kohl's stores.

Kymanox's James J. Davis Memorial Scholarship for Students Studying Abroad

Kymanox
Attn.: Scholarship Administrator
2220 Sedwick Road, Suite 201
Durham NC 27713
Phone: 847-433-2200
http://www.kymanox.com/scholarship
Purpose: To assist students who are interested in studying abroad.
Eligibility: Applicants must be U.S. citizens or permanent residents planning to attend an accredited program abroad for at least eight weeks. Strong

preference is given to students with financial need, and preference is given to applicants studying in a non-English speaking country and to applicants who are majoring in engineering, math or science.

Amount: $1,000.

Number of Awards: 1.

Deadline: March 2.

How to Apply: Applications are available online. An application form, an acceptance letter to a study abroad program, a financial needs verification form and an essay are required.

Leaders and Achievers Scholarship Program

Comcast
1500 Market Street
Philadelphia PA 19102
Phone: 800-266-2278
http://www.comcast.com

Purpose: To provide one-time scholarship awards of $1,000 each to graduating high school seniors. Emphasis is on students who take leadership roles in school and community service and improvement.

Eligibility: Students must be high school seniors with a minimum 2.8 GPA, be nominated by their high school principal or guidance counselor and attend school in a Comcast community. See the website for a list of eligible communities by state. Comcast employees, their families or other Comcast affiliates are not eligible to apply.

Amount: $1,000.

Number of Awards: Varies.

Deadline: Unknown.

How to Apply: Applications are available from the nominating principal or counselor.

Letters About Literature Contest

P.O. Box 5308
Woodbridge VA 22194
Email: programdirector@lettersaboutliterature.org
http://www.lettersaboutliterature.org

Purpose: To encourage young people to read.

Eligibility: Applicants must be U.S. legal residents. They must be students in grades 4 through 12 and must be at least nine years old by the September 1 that precedes the contest deadline. They must submit a personal letter to an author about how that author's book or work has impacted them. Selection is based on letter originality, grammar and organization.

Amount: $100-$500.

Number of Awards: 18.

Deadline: January 6.

How to Apply: Entry instructions are available online. An entry coupon and letter are required.

Lions International Peace Poster Contest

Lions Club International
300 W. 22nd Street
Oak Brook IL 60523-8842
Phone: 630-571-5466
Email: pr@lionsclubs.org
http://www.lionsclubs.org

Purpose: To award creative youngsters with cash prizes for outstanding poster designs.

Eligibility: Students must be 11, 12 or 13 years old as of the deadline and must be sponsored by their local Lions club. Entries will be judged at the local, district, multiple district and international levels. Posters will be evaluated on originality, artistic merit and expression of the assigned theme.

Amount: $500-$5,000.

Number of Awards: 24.

Deadline: November 15.

How to Apply: Applications are available from your local Lion's Club.

LiveCitizen Scholarship Competition

LiveCitizen
633 West Fifth Street
Suite 6800
Los Angeles CA 90071
http://www.livecitizen.com

Purpose: To recognize outstanding video essayists.

Eligibility: Applicants must be U.S. legal residents who are age 16 or older. They must be enrolled in or planning to enroll in an accredited post-secondary institution. They must submit a brief video and accompanying essay on a topic that has been determined by the sponsor. Selection is based on the overall strength of the entry and on votes from online users.

Amount: $2,500.

Number of Awards: 3.

Deadline: June 7.

How to Apply: Entry instructions are available online. An entry form, video and essay are required.

LULAC General Awards

League of United Latin American Citizens
2000 L Street NW, Suite 610
Washington DC 20036
Phone: 202-835-9646
Email: scholarships@lnesc.org
http://www.lnesc.org

Purpose: To provide assistance to students who are seeking or plan to seek degrees.

Eligibility: Applicants do not need to be Hispanic or Latino. Students must have applied to or be enrolled in a two- or four-year college or graduate school and be U.S. citizens or legal residents. Grades and academic achievement may be considered, but emphasis is placed on motivation, sincerity and integrity as demonstrated by the interview and essay.

Amount: $250-$1,000.

Number of Awards: Varies.

Deadline: March 31.

How to Apply: Applications are available online.

Marshall Scholar

Marshall Aid Commemoration Commission
Email: info@marshallscholarship.org
http://www.marshallscholarship.org

Purpose: Established in 1953 and financed by the British government, the scholarships are designed to bring academically distinguished Americans to study in the United Kingdom to increase understanding and appreciation of the British society and academic values.

Eligibility: Applicants must be U.S. citizens who expect to earn a degree from an accredited four-year college or university in the U.S. with a minimum 3.7 GPA. Students may apply in one of eight regions in the U.S.

Amount: Varies.

Number of Awards: 40.

Deadline: October.

How to Apply: Contact your regional center at the address listed on the website.

Mediacom World Class Scholarship Program

Mediacom
3737 Westown Parkway, Suite A
West Des Moines IA 50266
Email: scholarship@mediacomcc.com
http://www.mediacomworldclass.com

Purpose: To aid students in Mediacom service areas.

Eligibility: Applicants must be graduating high school seniors. Those who have earned college credits may apply as long as they have not yet graduated high school. Applicants may not be children of Mediacom employees and must live in areas serviced by Mediacom.

Amount: $1,000.

Number of Awards: Varies.

Deadline: February 15.

How to Apply: Applications are available online. An application form, essay, transcript and two reference forms are required.

Mensa Education and Research Foundation Scholarship Program

1229 Corporate Drive West
Arlington TX 76006
Phone: 817-607-5577
Email: info@mensafoundation.org
http://www.mensafoundation.org

Purpose: To support students seeking higher education.

Eligibility: Applicants do not need to be members of Mensa but must be residents of a participating American Mensa Local Group's area. They must be enrolled in a degree program at an accredited U.S. college or university in the academic year after application. They must write an essay explaining career, academic or vocational goals.

Amount: Varies.

Number of Awards: Varies.

Deadline: January 15.

How to Apply: Applications are available online in September. Please do NOT write to the organization to request an application.

Most Valuable Student Scholarships

Elks National Foundation Headquarters
2750 North Lakeview Avenue
Chicago IL 60614
Phone: 773-755-4732
Email: scholarship@elks.org
http://www.elks.org

Purpose: To support high school seniors who have demonstrated scholarship, leadership and financial need.

Eligibility: Applicants must be graduating high school seniors who are U.S. citizens and who plan to pursue a four-year degree on a full-time basis at a U.S. college or university. Male and female students compete separately.

Amount: $1,000-$15,000.

Number of Awards: 500.

Scholarship may be renewable.

Deadline: Unknown.

How to Apply: Contact the scholarship chairman of your local Lodge or the Elks association of your state.

Nancy Reagan Pathfinder Scholarships

National Federation of Republican Women
124 N. Alfred Street
Alexandria VA 22314
Phone: 703-548-9688
Email: mail@nfrw.org
http://www.nfrw.org/programs/scholarships.htm

Purpose: To honor former First Lady Nancy Reagan.

Eligibility: Applicants must be college sophomores, juniors, seniors or master's degree students. Two one-page essays and three letters of recommendation are required. Previous winners may not reapply.

Amount: $2,500.

Number of Awards: 3.

Deadline: June 1.

How to Apply: Applications are available online. An application form, three letters of recommendation, transcript, two essays and State Federation President Certification are required.

National College Match Program

QuestBridge
120 Hawthorne Avenue
Suite 103
Palo Alto CA 94301
Phone: 888-275-2054
Email: questions@questbridge.org
http://www.questbridge.org

Purpose: To connect outstanding low-income high school seniors with admission and full four-year

scholarships to some of the nation's most selective colleges.
Eligibility: Applicants must have demonstrated academic excellence in the face of economic obstacles. Students of all races and ethnicities are encouraged to apply. Many past award recipients have been among the first generation in their families to attend college.
Amount: Varies.
Number of Awards: Varies.
Scholarship may be renewable.
Deadline: September 30.
How to Apply: Applications are available on the QuestBridge website in August of each year. An application form, two teacher recommendations, one counselor recommendation (Secondary School Report), a transcript and SAT and/or ACT score reports are required.

National D-Day Museum Online Essay Contest

945 Magazine Street
New Orleans LA 70130
Phone: 504-527-6012
Email: info@nationalww2museum.org
http://www.ddaymuseum.org
Purpose: To increase awareness of World War II by giving students the opportunity to compete in an essay contest.
Eligibility: Applicants must be high school students in the United States, its territories or its military bases. They must prepare an essay of up to 1,000 words based on a topic specified by the sponsor and related to World War II. Only the first 500 valid essays will be accepted.
Amount: $250-$1,000.
Number of Awards: 6.
Deadline: March 30.
How to Apply: Applications are available online. An essay and contact information are required.

National Merit Scholarship Program and National Achievement Scholarship Program

National Merit Scholarship Corporation
1560 Sherman Avenue, Suite 200
Evanston IL 60201-4897
Phone: 847-866-5100
http://www.nationalmerit.org
Purpose: To provide scholarships through a merit-based academic competition.
Eligibility: Applicants must be enrolled full-time in high school, progressing normally toward completion and planning to enter college no later than the fall following completion of high school, be U.S. citizens or permanent legal residents in the process of becoming U.S. citizens and take the PSAT/NMSQT no later than the 11th grade. Participation in the program is based on performance on the exam.
Amount: $2,500.
Number of Awards: Varies.
Scholarship may be renewable.
Deadline: Varies.
How to Apply: Application is made by taking the PSAT/NMSQT test.

National Oratorical Contest

American Legion
Attn.: Americanism and Children and Youth Division
P.O. Box 1055
Indianapolis IN 46206
Phone: 317-630-1249
Email: acy@legion.org
http://www.legion.org
Purpose: To reward students for their knowledge of government and oral presentation skills.
Eligibility: Applicants must be high school students under the age of 20 who are U.S. citizens or legal residents. Students first give an oration within their state and winners compete at the national level. The oration must

be related to the Constitution of the United States focusing on the duties and obligations citizens have to the government. It must be in English and be between eight and ten minutes. There is also an assigned topic which is posted on the website, and it should be between three and five minutes.
Amount: Up to $18,000.
Number of Awards: Varies.
Deadline: Local American Legion department must select winners by March 12.
How to Apply: Applications are available from your local American Legion post or state headquarters. Deadlines for local competitions are set by the local Posts.

Natural Disaster PSA Video Contest

National Association of State Contractors Licensing Agencies
Attention: PSA Contest
23309 North 17th Drive
Unit 1, Building 110
Phoenix AZ 85027
Phone: 623-587-9354
Email: info@nascla.org
http://www.nascla.org
Purpose: To encourage people to hire a licensed contractor after a natural disaster.
Eligibility: Applicants must be current college students. They can enter the contest individually or in groups. They must submit a brief, original videotaped public service announcement on the topic of hiring a licensed contractor after a natural disaster. Selection is based on video production quality, creativity and message effectiveness.
Amount: $750-$3,000.
Number of Awards: 3.
Deadline: January 30.
How to Apply: Entry instructions are available online. An entry form, participant release form(s) and two copies of the entry video in DVD format are required.

Odenza Marketing Scholarship

Odenza Vacations
4664 Lougheed Highway
Suite 230
Burnaby BC V5C5T5
Phone: 877-297-2661
http://www.odenzascholarships.com
Purpose: To aid current and future college students who are between the ages of 16 and 25.
Eligibility: Applicants must be U.S. or Canadian citizens who have at least one full year of college study remaining. They must have a GPA of 2.5 or higher. Selection is based on the overall strength of the essays submitted.
Amount: $500.
Number of Awards: Varies.
Deadline: March 30.
How to Apply: Applications are available online. An application form and two essays are required.

Off to College Scholarship Sweepstakes

SunTrust
P.O. Box 27172
Richmond VA 23261-7172
Phone: 800-786-8787
http://www.suntrusteducation.com
Purpose: To assist a student for the first year of expenses at any accredited college.
Eligibility: Applicants must be high school seniors who are at least 13 years old and plan to attend a college accredited by the U.S. Department of Education the following fall. U.S. residency is required. Financial need and academic achievement are not considered. Note that this is a sweepstakes drawing every two weeks.
Amount: $1,000.
Number of Awards: 15.
Deadline: May 13.
How to Apply: Applications are available online. Mail-in entries are also accepted. Contact information is required.

OP Loftbed Scholarship Award

P.O. Box 573
Thomasville NC 27361-0573
Phone: 866-567-5638
Email: info@oploftbed.com
http://www.oploftbed.com
Purpose: To reward students who excel at creative writing.
Eligibility: Applicants must be U.S. citizens who plan to attend an accredited college or university in the upcoming school year. They must have a mailing address in the United States. Essay topics and requirements vary from year to year.
Amount: $500.
Number of Awards: Varies.
Deadline: July 31.
How to Apply: Applications are available online.

Optimist International Essay Contest

4494 Lindell Boulevard
St. Louis MO 63108
Phone: 314-371-6000
Email: programs@optimist.org
http://www.optimist.org
Purpose: To reward students based on their essay-writing skills.
Eligibility: Applicants must be under 18 years of age as of December 31 of the current school year and application must be made through a local Optimist Club. The essay topic changes each year. Applicants compete at the club, district and international level. District winners receive a $2,500 scholarship. Scoring is based on organization, vocabulary and style, grammar and punctuation, neatness and adherence to the contest rules. The club-level contests are held in early February but vary by club. The deadline for clubs to submit their winning essay to the district competition is February 28.
Amount: $2,500.
Number of Awards: Varies.
Deadline: Early February.
How to Apply: Contact your local Optimist Club.

Patriot's Pen Youth Essay Contest

Veterans of Foreign Wars
406 W. 34th Street
Kansas City MO 64111
Phone: 816-968-1117
Email: kharmer@vfw.org
http://www.vfw.org
Purpose: To give students in grades 6 through 8 an opportunity to write essays that express their views on democracy.
Eligibility: Applicants must be enrolled as a 6th, 7th or 8th grader in a public, private or parochial school in the U.S., its territories or possessions. Home-schooled students and dependents of U.S. military or civilian personnel in overseas schools may also apply. Foreign exchange students and former applicants who placed in the national finals are ineligible. Students must submit essays based on an annual theme to their local VFW posts. If an essay is picked to advance, the entry is judged at the District (regional) level, then the Department (state) level and finally at the National level. Essays are judged 30 percent on knowledge of the theme, 35 percent on development of the theme and 35 percent on clarity.
Amount: Up to $10,000.
Number of Awards: Varies.
Deadline: November 1.
How to Apply: Applications are available online or by contacting the local VFW office.

Philanthro Scholarship

Philanthro Productions Inc.
3701 Overland Avenue, Suite 133
Los Angeles CA 90034
Email: scholarship@philanthroproductions.org
http://www.philanthroproductions.org
Purpose: To raise awareness of philanthropic responsibility.

Eligibility: Applicants must be college students who have demonstrated academic achievement and community involvement. Selection is based on leadership ability, personal qualities and professional potential.
Amount: Varies.
Number of Awards: Varies.
Deadline: Varies.
How to Apply: Applications are available online. An application form and supporting materials are required.

Phoenix Scholarship Program

159 Concord Avenue, Suite 1C
Cambridge MA 02138
Email: phoenixawards@gmail.com
Purpose: To provide financial assistance to deserving high school seniors who plan to seek higher education.
Eligibility: Applicants must be U.S. high school seniors or they must have graduated within 13 months prior to the application deadline. They must have a 2.75 or higher GPA and have already taken the SAT or ACT. They must be in good standing with their high school, possess good moral character and plan to enroll in an accredited college or university upon graduation.
Amount: Varies.
Number of Awards: Up to 4.
Deadline: April 30.
How to Apply: Applications are available via email.

Principal's Leadership Award

Herff Jones
c/o National Association of Secondary School Principals
1904 Association Drive
Reston VA 20191
Phone: 800-253-7746
Email: carrollw@principals.org.
http://www.principals.org/awards/
Purpose: To recognize students for their leadership.
Eligibility: Applicants must be seniors and nominated by their high school principal. Each principal can nominate one student leader from the senior class. Application packets are mailed each fall to every secondary school.
Amount: $1,000-$12,000.
Number of Awards: 100.
Deadline: December 2.
How to Apply: Nomination forms are available online.

Proof-Reading.com Scholarship

Proof-Reading Inc.
12 Geary Street, Suite 806
San Francisco CA 94108
Phone: 866-433-4867
Email: support@proof-reading.com
http://www.proof-reading.com
Purpose: To recognize outstanding student essayists.
Eligibility: Applicants must be U.S. legal residents who are full-time students at an accredited four-year U.S. college or university. They must have a GPA of 3.5 or higher and must submit an essay on a topic that has been determined by the sponsor. Selection is based on the overall strength of the application and essay.
Amount: $1,500.
Number of Awards: 1.
Deadline: June 1.
How to Apply: Applications are available online. An application form and essay are required.

Prudential Spirit of Community Awards

Prudential Financial Inc.
751 Broad Street, 16th Floor
Newark NJ 07102
Phone: 877-525-8491
Email: spirit@prudential.com
http://spirit.prudential.com
Purpose: To recognize students for their self-initiated community service.
Eligibility: Applicants must be a student in grades 5-12 and a legal resident one of the 50 states of the U.S.

or District of Columbia and engaged in a volunteer activity that occurred at least in part after the year prior to date of application.

Amount: $1,000-$5,000.

Number of Awards: 102.

Deadline: November 1.

How to Apply: Applications are available online.

Religious Liberty Essay Scholarship Contest

Baptist Joint Committee for Religious Liberty
Essay Contest
200 Maryland Avenue, NE
Washington DC 20002
Phone: 202-544-4226
Email: ccrowe@bjconline.org
http://www.bjconline.org

Purpose: To reward students who have written outstanding essays on religious liberty.

Eligibility: Applicants must be high school juniors or seniors. They must submit an 800- to 1,200-word essay on a sponsor-determined topic relating to religious freedom. Selection is based on the overall strength of the essay.

Amount: $100-$1,000.

Number of Awards: 3.

Deadline: March 15.

How to Apply: Application instructions are available online. A registration form and an essay are required.

Return 2 College Scholarship

R2C Scholarship Program
http://www.return2college.com/awardprogram.cfm

Purpose: To provide financial assistance for college and adult students with college or graduate school expenses.

Eligibility: Applicants must be college or adult students currently attending or planning to attend a two-year or four-year college or graduate school within the next 12 months. Students must be 17 years or older and U.S. citizens or permanent residents. The award may be used for full- or part-time study at either on-campus or online schools.

Amount: $1,500.

Number of Awards: Varies.

Deadline: August 31.

How to Apply: Applications are available online.

Rhodes Scholar

Rhodes Scholarship Trust
Attn.: Elliot F. Gerson
8229 Boone Boulevard, Suite 240
Vienna VA 22182
Email: amsec@rhodesscholar.org
http://www.rhodesscholar.org

Purpose: To recognize qualities of young people that will contribute to the "world's fight."

Eligibility: Applicants must be U.S. citizens between the ages of 18 and 24 and have a bachelor's degree at the time of the award. The awards provides for two to three years of study at the University of Oxford including educational costs and other expenses. Selection is extremely competitive and is based on literary and scholastic achievements, athletic achievement and character.

Amount: Full Tuition plus stipend.

Number of Awards: 32.

Deadline: October 5.

How to Apply: Applications are available online.

Ronald McDonald House Charities U.S. Scholarships

Ronald McDonald House Charities
1321 Murfreesboro Road, Suite 800
Nashville TN 37217
Phone: 855-670-ISTS
Email: contactus@applyists.com
http://www.rmhc.org

Purpose: To help high school seniors attend college.

Eligibility: Applicants must be high school seniors less than 21 years of age who have a minimum 2.7 GPA. They must be U.S. residents who live in a participating Ronald McDonald House Charities chapter's geographic area. A list of chapters is on the website. There are four scholarships available: RMHC®/Scholars®, RMHC®/Asia®, RMHC®/African-American Future Achievers® and RMHC®/HACER®.
Amount: $1,000.
Number of Awards: Varies.
Deadline: January 27.
How to Apply: Applications are available online. An application form, transcript, personal statement, letter of recommendation and parents' tax forms are required.

Ronald Reagan College Leaders Scholarship Program

Phillips Foundation
7811 Montrose Road, Suite 100
Potomac MD 20854
Phone: 301-340-7788
Email: jhollingsworth@phillips.com
http://www.thephillipsfoundation.org
Purpose: To recognize students who demonstrate leadership on behalf of freedom, American values and constitutional principles.
Eligibility: Applicants must be enrolled full-time at any accredited, four-year degree-granting institution in the U.S. or its territories. Applicants may apply for a Ronald Reagan College Leaders Scholarship Program grant during their sophomore or junior year in high school. Selection is based on merit and financial need.
Amount: Up to $10,000.
Number of Awards: Varies. Scholarship may be renewable.
Deadline: January 17.
How to Apply: Applications are available online.

Rotary International Ambassadorial Scholarship Program

One Rotary Center
1560 Sherman Avenue
Evanston IL 60201
Phone: 847-866-3000
Email: scholarshipinquiries@rotaryintl.org
http://www.rotary.org
Purpose: To further international understanding and friendly relations among people of different countries.
Eligibility: Applicants must be citizens of a country in which there are Rotary clubs and have completed at least two years of college-level coursework or equivalent professional experience before starting their scholarship studies. Initial applications are made through local clubs. Students must be proficient in the language of the proposed host country.
Amount: Varies.
Number of Awards: Varies.
Deadline: February 1.
How to Apply: Applications are available through your local Rotary club or online.

Ruth Stanton Community Grant

Action Volunteering
Ruth Stanton Community Grant
P.O. Box 1013
Calimesa CA 92320
Email: painter5@ipsemail.com
http://www.actionvolunteering.com
Purpose: To aid those who are interested in furthering their community service activities.
Eligibility: Applicants must be active in community service work. Selection is based on the overall strength of the application.
Amount: $500.
Number of Awards: Varies.
Deadline: May 31.

How to Apply: Application instructions are available online. An essay and letter of recommendation are required.

Samuel Huntington Public Service Award

National Grid
25 Research Drive
Westborough MA 01582
Phone: 508-389-2000
http://www.nationalgridus.com

Purpose: To assist students who wish to perform one year of humanitarian service immediately upon graduation.

Eligibility: Applicants must be graduating college seniors, and must intend to perform one year of public service in the U.S. or abroad. The service may be individual work or through charitable, religious, educational, governmental or other public service organizations.

Amount: $10,000.

Number of Awards: Varies.

Deadline: January 18.

How to Apply: Applications are available online.

SanDisk Foundation Scholarship Program

c/o International Scholarship and Tuition Services Inc. (ISTS)
1321 Murfreesboro Road, Suite 800
Nashville TN 37217
Phone: 855-670-ISTS
Email: contactus@applyists.com
http://www.sandisk.com/about-sandisk/corporate-social-responsibility/community-engagement/sandisk-scholars-fund

Purpose: To assist students who have demonstrated leadership or entrepreneurial interests.

Eligibility: Applicants must be a high school senior or college freshman, sophomore or junior and must attend or plan to attend a full-time undergraduate program. Students must also demonstrate financial need. Up to 27 $2,500 renewable awards will given to students from the general public, up to three $2,500 renewable awards will be given to dependents of SanDisk employees and two students will be chosen as SanDisk Scholars and awarded full tuition scholarships for up to four years.

Amount: $2,500 to full tuition.

Number of Awards: Up to 30. Scholarship may be renewable.

Deadline: March 15.

How to Apply: Applications are available online. An application form and an essay are required.

Scholarship Drawing for $1,000

Edsouth
eCampusTours
P.O. Box 36014
Knoxville TN 37930
Phone: 865-342-0670
Email: info@ecampustours.com
http://www.ecampustours.com

Purpose: To assist students in paying for college.

Eligibility: Eligible students include U.S. citizens, U.S. nationals and permanent residents or students enrolled in a U.S. institution of higher education. Winners must be enrolled in an eligible institution of higher education, as stipulated in the eligibility requirements, within one year of winning the award. Scholarship awards will be paid directly to the college.

Amount: $1,000.

Number of Awards: 2.

Deadline: March 31.

How to Apply: Applications are available online or by mail. Registration with eCampusTours is required.

Scholarships for Student Leaders

National Association for Campus Activities
13 Harbison Way
Columbia SC 29212
Phone: 803-732-6222
Email: info@naca.org
http://www.naca.org
Purpose: The NACA foundation is committed to developing professionals in the field of campus activities.
Eligibility: Applicants must be current undergraduate students who hold a significant campus leadership position, have made significant contributions to their campus communities and have demonstrated leadership skills and abilities.
Amount: Varies.
Number of Awards: Varies.
Deadline: November 1.
How to Apply: Applications are available online.

Shepherd Scholarship

Ancient and Accepted Scottish Rite of Freemansonry Southern Jurisdiction
1733 16th Street NW
Washington DC 20009-3103
Phone: 202-232-3579
http://www.srmason-sj.org
Purpose: To provide financial assistance to students pursuing degrees in fields associated with service to country.
Eligibility: Applicants must have accepted enrollment in a U.S. institution of higher learning. No Masonic affiliation is required. Up to four letters of recommendation will be considered. Selection is based on "dedication, ambition, academic preparation, financial need and promise of outstanding performance at the advanced level."
Amount: $1,500.
Number of Awards: Varies. Scholarship may be renewable.
Deadline: April 1.
How to Apply: Applications are available online.

Siemens Competition in Math, Science and Technology

Siemens Foundation
170 Wood Avenue South
Iselin NJ 08330
Phone: 877-822-5233
Email: foundation.us@siemens.com
http://www.siemens-foundation.org
Purpose: To provide high school students with an opportunity to meet other students interested in math, science and technology and to provide monetary assistance with college expenses.
Eligibility: Students must submit research reports either individually or in teams of two or three members. Individual applicants must be high school seniors. Team project applicants must be high school students but do not need to be seniors. Projects may be scientific research, technological inventions or mathematical theories.
Amount: $1,000-$100,000.
Number of Awards: Varies.
Deadline: October 3.
How to Apply: Applications are available online.

Simon Youth Foundation Community Scholarship

225 W. Washington Street
Indianapolis IN 46204
Phone: 800-509-3676
Email: syf@simon.com
http://www.syf.org
Purpose: To assist promising students who live in communities with Simon properties.
Eligibility: Applicants must be high school seniors who plan to attend an accredited two- or four-year college, university or technical/vocational school full-time. Scholarships are awarded

without regard to race, color, creed, religion, gender, disability or national origin, and recipients are selected on the basis of financial need, academic record, potential to succeed, participation in school and community activities, honors, work experience, a statement of career and educational goals and an outside appraisal. Awards are given at every Simon mall in the U.S.
Amount: $1,400-$28,000.
Number of Awards: Varies.
Deadline: Unknown.
How to Apply: Applications are available online.

STA Travel Inc. Scholarship

Foundation for Global Scholars
12050 North Pecos Street, Suite 320
Westminster CO 80234
Phone: 303-502-7256
Email: kbrockwell@foundationforglobalscholars.org
http://www.foundationforglobalscholars.org
Purpose: To aid students who wish to study abroad.
Eligibility: Applicants must be U.S. or Canadian citizens who are attending a North American college or university. They must be able to apply transfer credits from studying abroad to their current degree program. They must be majoring in foreign language, international business, travel, tourism, international studies, film, art or photography.
Amount: Varies.
Number of Awards: Varies.
Deadline: Varies.
How to Apply: Applications are available online.

Stephen J. Brady STOP Hunger Scholarship

Sodexo Foundation
9801 Washingtonian Boulevard
Gaithersburg MD 20878
Phone: 800-763-3946
Email: stophunger@sodexofoundation.org
http://www.sodexofoundation.org
Purpose: To aid students who have been active in the movement to eradicate hunger.
Eligibility: Applicants must be U.S. citizens or permanent residents. They must be students in kindergarten through graduate school who are enrolled at an accredited U.S. institution. They must have been active in at least one unpaid volunteer effort to end hunger during the past 12 months.
Amount: $5,000.
Number of Awards: Varies.
Deadline: December 5.
How to Apply: Applications are available online.

Stuck at Prom Scholarship

Henkel Consumer Adhesives
32150 Just Imagine Drive
Avon OH 44011-1355
http://www.stuckatprom.com
Purpose: To reward students for their creativity with duct tape.
Eligibility: Applicants must attend a high school prom as a couple in the spring wearing the most original attire that they make from duct tape. Both members of the couple do not have to attend the same school.
Amount: $500-$3,000.
Number of Awards: Varies.
Deadline: June 13.
How to Apply: Applications are available online.

Student Activist Awards

Freedom from Religion Foundation
P.O. Box 750
Madison WI 53701
Phone: 608-256-5800
Email: info@ffrf.org
http://www.ffrf.org
Purpose: To assist high school and college student activists.

Eligibility: Selection is based on activism for free thought or separation of church and state.
Amount: $1,000.
Number of Awards: Varies.
Deadline: Varies.
How to Apply: Contact the organization for more information.

Study Abroad Grants

Honor Society of Phi Kappa Phi
7576 Goodwood Boulevard
Baton Rouge LA 70806
Phone: 800-804-9880
Fax: 225-388-4900
Email: awards@phikappaphi.org
http://www.phikappaphi.org
Purpose: To provide scholarships for undergraduate students who will study abroad.
Eligibility: Applicants do not have to be members of Phi Kappa Phi but must attend an institution with a Phi Kappa Phi chapter, have between 30 and 90 credit hours and have at least two semesters remaining at their home institution upon return. Students must have been accepted into a study abroad program that demonstrates their academic preparation, career choice and the welfare of others. They must have a GPA of 3.5 or higher.
Amount: $1,000.
Number of Awards: 50.
Deadline: April 1.
How to Apply: Applications are available online.

SuperCollege Scholarship

SuperCollege.com
2713 Newlands Avenue
Belmont CA 94002
Email: supercollege@supercollege.com
http://www.supercollege.com/scholarship/
Purpose: SuperCollege donates a percentage of the proceeds from the sales of its books to award scholarships to high school, college, graduate and adult students.
Eligibility: Applicants must be high school students, college undergraduates, graduate students or adult students residing in the U.S. and attending or planning to attend any accredited college or university within the next 12 months. The scholarship may be used to pay for tuition, books, room and board, computers or any education-related expenses.
Amount: $1,500.
Number of Awards: 1.
Deadline: July 31.
How to Apply: Applications are available online.

Talbots Scholarship Foundation

Scholarship Management Services, Scholarship America
One Scholarship Way
P.O. Box 297
Saint Peter MN 56082
Phone: 507-931-1682
http://www.talbots.com/scholarship
Purpose: To provide scholarships for women returning to college.
Eligibility: Applicants must be female U.S. residents who have earned their high school diploma or GED at least 10 years ago and who are now enrolled or planning to attend undergraduate study at a two- or four-year college or university or vocational-technical school. The deadline is January 2 or when the first 1,000 applications are received, whichever is earlier.
Amount: $15,000-$30,000.
Number of Awards: 11.
Deadline: January.
How to Apply: Applications are available online.

The Lowe's Scholarship

Lowe's Company
1000 Lowe's Boulevard
Mooresville NC 28117

Phone: 800-44-LOWES
http://www.lowes.com
Purpose: To help young people in the communities where Lowe's does business.
Eligibility: Applicants must be high school seniors who will enroll in an accredited two- or four-year college or university in the United States. Leadership ability, community involvement and academic achievement are considered when making the selection for the scholarships.
Amount: $2,500.
Number of Awards: 140.
Scholarship may be renewable.
Deadline: February 28.
How to Apply: Applications are available online.

Truman Scholar

Truman Scholarship Foundation
712 Jackson Place NW
Washington DC 20006
Phone: 202-395-4831
Email: office@truman.gov
http://www.truman.gov
Purpose: To provide college junior leaders who plan to pursue careers in government, non-profits, education or other public service with financial support for graduate study and leadership training.
Eligibility: Applicants must be juniors, attending an accredited U.S. college or university and be nominated by the institution. Students may not apply directly. Applicants must be U.S. citizens or U.S. nationals, complete an application and write a policy recommendation.
Amount: Up to $30,000.
Number of Awards: Up to 68.
Deadline: February 7.
How to Apply: See your school's Truman Faculty Representative or contact the foundation.

U.S. Bank Scholarship Program

U.S. Bancorp Center
800 Nicollet Mall
Minneapolis MN 55402
Phone: 800-242-1200
http://www.usbank.com/student-lending/scholarship.html
Purpose: To support graduating high school seniors.
Eligibility: Applicants must be high school seniors who plan to attend or current college freshmen, sophomores or juniors attending full-time an accredited two- or four-year college and be U.S. citizens or permanent residents. Recipients are selected through a random drawing.
Amount: $1,000.
Number of Awards: 40.
Deadline: March 31.
How to Apply: Applications are only available online.

U.S. JCI Senate Scholarship Grants

106 Wedgewood Drive
Carrollton GA 30117
Email: tom@smipc.net
http://www.usjcisenate.org
Purpose: To support high school students who wish to further their education.
Eligibility: Applicants must be high school seniors and U.S. citizens who are graduating from a U.S. accredited high school or state approved home school or GED program. Winners must attend college full-time to receive funds.
Amount: $1,000.
Number of Awards: Varies.
Deadline: Varies.
How to Apply: Applications are available from your school's guidance office.

Undergraduate Transfer Scholarship

Jack Kent Cooke Foundation
44325 Woodridge Parkway
Lansdowne VA 20176
Phone: 800-498-6478
Email: jkc-u@act.org
http://www.jkcf.org
Purpose: To help community college students attend four-year universities.
Eligibility: Applicants must be students or recent alumni from accredited U.S. community colleges or two-year institutions who plan to pursue bachelor's degrees at four-year institutions. Applicants may not apply directly to the foundation but must be nominated by the Jack Kent Cooke Foundation faculty representatives at their institutions. The award is based on academic merit and unmet financial need. A GPA of 3.5 or higher is required.
Amount: Up to $30,000.
Number of Awards: About 50. Scholarship may be renewable.
Deadline: December 6.
How to Apply: Applications are available online.

Violet Richardson Award

Soroptimist International of the Americas
1709 Spruce Street
Philadelphia PA 19103
Phone: 215-893-9000
Email: siahq@soroptimist.org
http://www.soroptimist.org
Purpose: To recognize young women who contribute to the community through volunteer efforts.
Eligibility: Applicants must be young women between the ages of 14 and 17 who make outstanding contributions to volunteer efforts. Efforts that benefit women or girls are of particular interest.
Amount: Up to $2,500.
Number of Awards: Varies.
Deadline: December 1.
How to Apply: Contact your local Soroptimist club.

Visine Students with Vision Scholarship Program

Johnson & Johnson Healthcare Products
c/o International Scholarship and Tuition Services Inc.
1321 Murfreesboro Road, Suite 800
Nashville TN 37217
Phone: 855-670-ISTS
Email: contactus@applyists.com
http://www.visine.com/scholarship
Purpose: To assist students who have a vision and communicate their vision through an essay or video.
Eligibility: Applicants must be a high school senior or a college freshman, sophomore or junior and demonstrate involvement in school activities and/or community service. Students must also demonstrate financial need and have a clear vision and be able to communicate their vision through a well-written essay or video presentation. A minimum GPA of 2.8 is required.
Amount: $5,000.
Number of Awards: Up to 10.
Deadline: April 16.
How to Apply: Applications are available online.

Voice of Democracy Audio Essay Contests

Veterans of Foreign Wars
406 W. 34th Street
Kansas City MO 64111
Phone: 816-968-1117
Email: kharmer@vfw.org
http://www.vfw.org
Purpose: To encourage patriotism with students creating audio essays expressing their opinion on a patriotic theme.
Eligibility: Applicants must submit a three- to five-minute audio essay on tape or CD focused on a yearly theme. Students must be in the 9th to 12th

grade in a public, private or parochial high school, home study program or overseas U.S. military school. Foreign exchange students are not eligible for the contest, and students who are age 20 or older also may not enter. Previous first place winners on the state level are ineligible.
Amount: Up to $30,000.
Number of Awards: Varies.
Deadline: November 1.
How to Apply: Applications are available online but must be submitted to a local VFW post.

Watson Travel Fellowship

Thomas J. Watson Fellowship
11 Park Place
Suite 1503
New York NY 10007
Phone: 212-245-8859
Fax: 212-245-8860
Email: tjw@watsonfellowship.org
http://www.watsonfellowship.org
Purpose: To award one-year grants for independent study and travel outside the U.S. to graduating college seniors.
Eligibility: Only graduating seniors from the participating colleges are eligible to apply. A list of these colleges is available online. Applicants must first be nominated by their college or university. An interview with a representative will follow. Recipients must graduate before the fellowship can begin.
Amount: Up to $35,000.
Number of Awards: Varies.
Deadline: November 9.
How to Apply: Interested students should contact their local Watson liaison to begin the application process.

Wendy's High School Heisman Award

Wendy's Restaurants
Phone: 800-205-6367
Email: wendys@act.org
http://www.wendyshighschoolheisman.com
Purpose: To recognize high school students who excel in academics, athletics and student leadership.
Eligibility: Applicants must be entering their high school senior year and participate in one of 27 officially sanctioned sports. Eligible students have a minimum 3.0 GPA. Selection is based on academic achievement, community service and athletic accomplishments.
Amount: Varies.
Number of Awards: Varies.
Deadline: October 2.
How to Apply: Application forms are available online.

Where's FRANKIE the Diploma Frame? Scholarship Photo Contest

Church Hill Classics
594 Pepper Street
Monroe CT 06468
Phone: 800-477-9005
Fax: 203-268-2468
Email: info@diplomaframe.com
http://www.framemyfuture.com
Purpose: To help college students pay for textbooks.
Eligibility: Applicants must be U.S. legal residents, be age 18 or older and be full-time college students or family members of full-time college students. They must submit a photo depicting the sponsor's mascot participating in a summer activity. Selection is based on the creativity and popularity of the photo.
Amount: $500.
Number of Awards: 3.
Deadline: August 13.
How to Apply: Entry forms are available online.

William E. Simon Fellowship for Noble Purpose

Intercollegiate Studies Institute (ISI)

3901 Centerville Road
Wilmington DE 19807
Phone: 800-526-7022
Fax: 302-652-1760
Email: simon@isi.org
http://www.isi.org

Purpose: To aid graduating college seniors who are committed to strengthening civil society in some way.

Eligibility: Applicants must be graduating college seniors who have self-directed plans to improve society in some noble way. Selection is based on academic achievement, recommendations, extracurricular activities and stated project goals.

Amount: $40,000.

Number of Awards: 3.

Deadline: January 17.

How to Apply: Applications are available online.

Win Free Tuition Giveaway

Next Step Magazine
86 W. Main Street
Victor NY 14565
Phone: 800-771-3117
Email: webcopy@nextstepmag.com
http://www.nextstepmagazine.com/winfreetuition

Purpose: To support higher education.

Eligibility: Entrants must be legal residents of the U.S. and Canada (except for Puerto Rico or Quebec) who are age 14 or older. They must also be planning to enroll or currently enrolled in college by September 30 three years after the application date. This is an annual sweepstakes drawing for one year's tuition up to $10,000 and 11 monthly drawings for a $1,000 scholarship.

Amount: $1,000-$10,000.

Number of Awards: 12.

Deadline: June 30.

How to Apply: Applications are available online.

Women's Opportunity Awards Program

Soroptimist International of the Americas
1709 Spruce Street
Philadelphia PA 19103
Phone: 215-893-9000
Fax: 215-893-5200
Email: siahq@soroptimist.org
http://www.soroptimist.org

Purpose: To assist women entering or re-entering the workforce with educational and skills training support.

Eligibility: Applicants must be attending or been accepted by a vocational/skills training program or an undergraduate degree program. Applicants must be the women heads of household who provide the primary source of financial support for their families and demonstrate financial need. Applicants must submit their application to the appropriate regional office.

Amount: Up to $10,000.

Number of Awards: Varies.

Deadline: December 1.

How to Apply: Applications are available online.

Writing Awards

Scholastic
557 Broadway
New York NY 10012
Phone: 212-343-6100
Email: a&wgeneralinfo@scholastic.com
http://www.artandwriting.org

Purpose: To reward creative young writers.

Eligibility: Applicants must be in grades 7 through 12 in U.S. or Canadian schools and must submit writing pieces or portfolios in one of the following categories: dramatic script, general writing portfolio, humor, journalism, nonfiction portfolio, novel, personal essay/memoir, poetry, science fiction/fantasy, short story and short short story.

Amount: Varies.

Number of Awards: Varies.
Deadline: Varies.
How to Apply: Applications are available online.

Young Naturalist Awards

American Museum of Natural History
Central Park West at 79th Street
New York NY 10024
Phone: 212-769-5100
Email: yna@amnh.org
http://www.amnh.org

Purpose: The Young Naturalist Awards is an inquiry-based research competition that challenges students in grades 7 to 12 to complete an investigation of the natural world. It encourages students to explore an area of science that interests them, typically in life science, Earth science, ecology or astronomy.
Eligibility: Applicants must be students in grade 7 through 12 who are currently enrolled in a public, private, parochial or home school in the United States, Canada, U.S. territories or in a U.S.-sponsored school abroad. Each essay is judged primarily on its scientific merits.
Amount: Up to $2,500.
Number of Awards: 12.
Deadline: March 1.
How to Apply: Applications are available online.

Young People for Fellowship

2000 M Street, NW
Suite 400
Washington DC 20036
Phone: 202-467-4999
Email: zdryden@fpaw.org
http://www.youngpeoplefor.org

Purpose: To encourage and cultivate young progressive leaders.
Eligibility: Applicants must be undergraduate students and be interested in promoting social change on their campuses and in their communities. Selection is based on the overall strength of the application.
Amount: Varies.
Number of Awards: 150.
Deadline: January 31.
How to Apply: Applications are available online. An application form is required.

Young Scholars Program

Jack Kent Cooke Foundation Young Scholars Program
301 ACT Drive
P.O. Box 4030
Iowa City IA 52243
Phone: 800-498-6478
Fax: 703-723-8030
Email: jkc@jackkentcookefoundation.org
http://www.jkcf.org

Purpose: To help high-achieving students with financial need and provide them with educational opportunities throughout high school.
Eligibility: Applicants must have financial need, be in the 7th grade and plan to attend high school in the United States. Academic achievement and intelligence are important, and students must display strong academic records, academic awards and honors and submit a strong letter of recommendation. A GPA of 3.5 is usually required, but exceptions are made for students with unique talents or learning differences. The award is also based on students' will to succeed, leadership and public service, critical thinking ability and participation in the arts and humanities. During two summers, recipients must participate in a Young Scholars Week and Young Scholars Reunion in Washington, DC.
Amount: Varies.
Number of Awards: 60.
Scholarship may be renewable.
Deadline: April 16.
How to Apply: Applications are available online and at regional talent centers.

A

B

C

D

E

J

K

L

M

About the Authors

Harvard graduates Gen and Kelly Tanabe are the founders of SuperCollege and the award-winning authors of twelve books including *The Ultimate Scholarship Book, Get Free Cash for College, The Ultimate Guide to America's Best Colleges, Get into Any College* and *Accepted! 50 Successful College Admission Essays.*

Together, Gen and Kelly were accepted to every school to which they applied, all of the Ivy League colleges, and won over $100,000 in merit-based scholarships. They were able to leave Harvard debt-free and their parents guilt-free.

Gen and Kelly give workshops at high schools across the country and write the nationally syndicated "Ask the SuperCollege.com Experts" column. They have made dozens of appearances on television and radio and have served as expert sources for respected publications including *U.S. News & World Report, USA Today, The New York Times,* the *New York Daily News,* the *Chronicle of Higher Education, Money, Woman's Day* and *CosmoGIRL.*

Gen grew up in Waialua, Hawaii, and was the first student from Waialua High School to attend Harvard. He graduated magna cum laude from Harvard with a degree in both History and East Asian Studies.

Kelly attended Whitney High School, a nationally ranked public high school in her hometown of Cerritos, California. She graduated magna cum laude from Harvard with a degree in Sociology.

Gen, Kelly, their sons Zane and Kane and their dog Sushi live in Belmont, California.